THE
OPTICAL PAPERS OF
ISAAC NEWTON

THE
OPTICAL PAPERS OF
ISAAC NEWTON

VOLUME I
THE OPTICAL LECTURES
1670–1672

EDITED BY
ALAN E. SHAPIRO

The right of the
University of Cambridge
to print and sell
all manner of books
was granted by
Henry VIII in 1534.
The University has printed
and published continuously
since 1584.

CAMBRIDGE UNIVERSITY PRESS

CAMBRIDGE
LONDON   NEW YORK   NEW ROCHELLE
MELBOURNE   SYDNEY

CAMBRIDGE UNIVERSITY PRESS
Cambridge, New York, Melbourne, Madrid, Cape Town, Singapore,
São Paulo, Delhi, Dubai, Tokyo, Mexico City

Cambridge University Press
The Edinburgh Building, Cambridge CB2 8RU, UK

Published in the United States of America by Cambridge University Press, New York

www.cambridge.org
Information on this title: www.cambridge.org/9780521155090

First published 1984
First paperback edition 2010

A catalogue record for this publication is available from the British Library

Library of Congress Cataloguing in Publication Data

Newton, Isaac, Sir, 1642-1727.
The optical papers of Isaac Newton.
Bibliography: v. 1, p.
Includes index.

Contents: v. 1. The optical lectures, 1670-1672.
1. Optics- Early works to 1800. 2. Optics-History-
Sources. Shapiro, Alan E. (Alan Elihu), 1942-
II. Title.
QC353.N555 1983 535 82-14751

ISBN 978-0-521-25248-5 Hardback
ISBN 978-0-521-15509-0 Paperback

TO MY MOTHER AND FATHER

Contents

List of Plates

Preface

Newton's contributions to the science of optics—his discovery of the unequal refractions of rays of different color, his theory of color, and his investigations of "Newton's rings," to mention only a few of the most noteworthy—place him among the premier contributors to that science. Yet, as fundamental, even revolutionary, as his optical investigations were, he did not succeed in creating an entire, new science based on novel principles, as was true of his mechanics and mathematics. Consequently, the status of his optical work has suffered greater vicissitudes, depending in particular upon the prevailing physical theory of light, and has varied from virtual dominance in the eighteenth century to relative neglect in the mid-nineteenth century. In our own century interest in and esteem for Newton's optics, especially the readily accessible *Opticks,* is once again at a high level, and it now appears to be sustained by a genuine historical appreciation rather than by a concern for its contemporary scientific relevance. Today we recognize that his work on optics offers unique rewards in its exciting, innovative conjunction of physical theory, experimental investigation, and mathematics, and in the revealing glimpse that it provides of a crucial period in the evolution of experimental science. Perhaps because of its less exalted status in the trinity of Newtonian sciences and, no doubt, also because of the usual train of historical accidents, his optical papers have eluded the initial wave of multivolume Newton editions, the *Correspondence, Mathematical Papers,* and *Principia.* With this, the first of a three-volume edition of *The Optical Papers of Isaac Newton,* Newton's optical work at last takes its rightful place in modern historical scholarship.

The papers published in this edition are Newton's two treatises, the *Optical Lectures* and *Opticks;* his optical correspondence; and his notes, essays, and calculations. Most of the papers are in the Portsmouth Collection, which was deposited in the Cambridge University Library in 1888; and most of the rest, namely, the correspondence, are at the Royal Society. All of the papers in this edition are published from the original manuscript or the closest extant copy, but the edition could with some justification also be called one of his works and papers, since most of the principal texts have already been published in some form or other. I have, however, chosen to publish the texts as Newton actually composed them, not simply for the sake of textual purity (as admirable a goal as that may be), but rather so that the evolution and formulation of Newton's ideas may be followed through early versions, drafts, and revisions in the papers. The principal aim of this edition is to

document the development and expression of Newton's optical thought, and in this respect it will differ fundamentally from the eighteenth-century editions of his optical works, the primary aim of which was to present Newton's definitive scientific position. This was perfectly justified for these editions, which consisted solely of his published works (the posthumous edition of the *Optical Lectures,* the letters published in the *Philosophical Transactions,* and the *Opticks*), for Newton's optical work was then at the high point of its influence and the major scientific—not yet historical—source for that science. But a different sort of edition, a truly historical one, is now required. By gathering together Newton's papers—both published and unpublished, in order to fill in the gaps in the historical record—by ordering them, providing technical and historical commentary, and when necessary a translation, I believe that our perception of Newton's optics will inevitably be altered and our understanding enhanced.

The present volume contains the *Optical Lectures* (1670–2) in its two successive versions together with translations. I had intended to include in this volume the few earlier, surviving optical essays from the 1660s, but because of the volume's size (at least as initially planned in the large format of the companion Newton editions), I reluctantly decided to disturb their proper chronological sequence and postpone their publication to the next volume. Nonetheless, it is appropriate that the *Optical Lectures* stands by itself, for it is Newton's first physical treatise and it has been woefully ignored. This neglect is somewhat surprising, since it is the carefully conceived work of a mature—if even still young—scientist synthesizing investigations carried out over a period of five or six years, in which he also attempts to lay a new, more secure foundation for the physical sciences. It has certainly not received the attention usually accorded to the early work of a great scientist, such as Galileo's *De motu* and Descartes's *Le Monde.* This edition of the *Optical Lectures* with its translation and commentary is intended to make this major work accessible to a broader audience.

The remaining two volumes will contain the manuscript of the *Opticks* together with its drafts; Newton's various notes, essays, and calculations; and his optical correspondence. I have decided to re-edit the optical correspondence despite its inclusion in the *Correspondence* for a number of reasons: Although they have the form of letters, they are actually scientific papers, nearly all of which were published in the *Philosophical Transactions* or read to the Royal Society, and, as has been long recognized, they form an essential part of his optical writings; important drafts of many of these letters remain unpublished; and with more detailed annotations they will be more fully integrated within the corpus of his optical works and will assume their proper significance. Barring more than the usual disturbances in my life, I hope to be able to complete the edition in about a decade.

This volume, indeed the entire three-volume edition, owes its very existence to D. T. Whiteside, who, as it were, coaxed me into it. Long ago, when I was in Cambridge still working on Newton's antagonists, the early wave theorists, he suggested that I look at Newton's *Optical Lectures* in the Manu-

scripts Room of the Cambridge University Library and assured me that I would find it interesting—perhaps so interesting that I would undertake an edition of it. When I returned to Cambridge three years later to work on this edition of the *Lectures,* Whiteside then suggested that as long as I was doing the *Lectures,* I might as well do a complete edition of the optical papers. Having already had a brief taste of editing, I was not now so eager to commit myself, knowing that the task ahead would take me at least to the age of fifty—a terrifying thought to someone thirty-two years old. I have since often recalled his enigmatic counsel to me that if I really knew what was involved in doing an edition I would not undertake it, but that after having done it I would not regret it. I now understand this: The years of tedium and frustration have been more than compensated by the satisfaction and exhilaration of expounding Newton's thought and attaining a deeper appreciation of it. Whiteside's role did not end with the conception of the edition, for in the following years he has provided advice on all variety of matters and, when necessary, encouragement; and we have in long letters discussed the fine points of Newton's optics. To be sure, I have not always accepted his advice or views—nor those of the many others who have generously helped me—and I alone remain responsible for the edition and its inevitable errors.

I. B. Cohen, J. A. Lohne, and Richard S. Westfall provided early but crucial encouragement and guidance and, just as important, the published legacy of their contributions to Newtonian scholarship. In preparing this volume, I have always attempted to cite earlier relevant scholarship, but I have also drawn upon it in ways that cannot be fully documented. Specifically, I frequently turned to the work of my direct predecessors for aid in transcribing, translating, and understanding particular points of the *Lectures,* beginning with the anonymous translator of 1728 and the anonymous editor of the *editio princeps* of 1729, and then proceeding through Samuel Horsley, S. I. Vavilov, and Whiteside. At various points I received valuable assistance from Marshall Clagett, W. Louis Fowlks, Albert Van Helden, David C. Lindberg, Ernan McMullin, G. A. J. Rogers, Donald Ross, and A. I. Sabra. Morton Hamermesh was exceptionally generous to me in translating the notes from Vavilov's Russian translation of the *Lectures,* and Susi Jeans in letting me see the manuscript of her book on Newton's music theory and helping me with problems in seventeenth-century music theory. I have, of course, used numerous libraries and been assisted by still more numerous librarians; I fully appreciate their efforts and recognize that without them historical scholarship of this sort would be impossible. Above all I must acknowledge the Syndics of the Cambridge University Library for their permission to publish the texts of Newton's manuscripts that are in their custody, and for readily making their facilities and services available to me. I have spent many months in their Manuscripts Room, and A. G. Purvis and his colleagues have helped to make my visits there fruitful ones. I further acknowledge the permission of the University Library, Keele, England, to reproduce a Newton manuscript in their collection. The staff of my own University Libraries has over the years also provided invaluable service.

Without the financial assistance provided by the National Science Foundation and the University of Minnesota, I would not have been able to undertake this edition. At various times my university, especially the Graduate School, provided crucial support—for example, to begin this edition and later to employ Madeleine Henry as a research assistant during 1976–7. She served with diligence and rescued me from not a few blunders in the transcription and translation. I was ably assisted in seeing this volume through the press by Frederick Fellows. Maurine Bielawski faithfully typed and retyped the entire manuscript, but to me equally as important as her secretarial skills was her respect for the task at hand. She made a difficult undertaking somewhat less difficult, and I am grateful for that.

A.E.S.

Minneapolis, Minnesota
October 1982

Editorial Note

This edition of Newton's *Optical Lectures* was complicated by the existence of two significantly different versions: an untitled first draft (now University Library, Cambridge, Add. MS 4002), which we will call his *Lectiones opticae,* and a major revision, "Opticae, Pars 1ª/2ᵈᵃ" (ULC, MS Dd.9.67), which we will call his *Optica.* Where there is no necessity to distinguish between the two, they will be referred to collectively as the *Optical Lectures,* although they are usually known by the Latin title, *Lectiones opticae,* under which the revision was published shortly after Newton's death. The *Optica* includes the greater part of the *Lectiones,* but we nonetheless decided to publish the entirety of both versions, with a translation, in their proper historical sequence. Combining the two, as we had initially intended, was ultimately rejected. It corrupted and muddled the two texts, was very confusing to follow because of their different arrangements, and necessarily consigned one to a subsidiary status, whereas both must be granted equal historical significance. On the other hand, the existence of two different versions of the same text provides us with the exciting opportunity of following the development of Newton's optical ideas in one short, fruitful period.

In accordance with the principles of modern scholarship, the published texts faithfully follow the manuscripts, in particular, in reproducing spelling, punctuation, and capitalization. Other than for the expansion of contractions (such as the enclitic *que*) and ligatures and the addition of parentheses to Newton's superscript alphabetical notes, all editorial emendations are enclosed within square brackets. Since Newton's *Optical Lectures* are, above all, scientific texts, we decided it would be pointless to clutter them by noting every change in the manuscript or every variant between the two. Accordingly, word order, tense, mood, spelling, and the like have for the most part been ignored in the textual notes. Even by thus limiting the notes to changes and variants that affect the meaning, the reader will more likely than not find that we have erred on the side of comprehensiveness. We have included facsimiles of corresponding pages from each manuscript, as well as of the sole surviving draft sheet, so that our editorial conventions and the accuracy of our transcription can be assessed directly. Variants between the two versions are indicated in the *Optica,* except for particularly important ones that are also noted in the *Lectiones.* In addition, since the 1729 *editio princeps* of the *Optica* served as the text for all subsequent editions and translations, we have also noted its more significant variants from Newton's manuscript, although the two have not been scrupulously collated. In general we have

attempted to preserve the layout of the manuscripts in the transformation to the printed page, but a few changes have been introduced for clarity. When breaking up an equation could be ambiguous or confusing, it is displayed, although Newton's notation and arrangement is otherwise maintained; and the figures for the *Optica,* which are gathered together at the end of each part in the manuscript, are inserted in the appropriate place in the text, just as Newton did in the *Lectiones opticae.* The figures, strictly considered, are part of the text and belong with the Latin, but for the sake of better balance of facing pages they have in some cases been placed in the translation. The postils and article numbers, unless in square brackets, are Newton's. Divisions in pagination of the manuscripts are indicated in the text by a virgule (/), and the page numbers are printed alongside in the inside margin. The subscript 2 preceding the page numbers in Part II of the *Optica* indicates, as is customary, the second sequence of page numbers in the manuscript.

The format adopted for the textual notes is, we trust, clear and simple, but a few conventions require elaboration. The lemma, that is, the word or phrase that is the subject of such a note and is specified at the beginning of the note within a single bracket (]), is omitted when it consists of one word. The ellipsis (. . .) is used in these notes in two distinct ways: If it is repeated after the lemma, then the word(s) represented by the ellipsis is not at all affected by the reading following the lemma; but if it is not repeated after the lemma, then the entirety of the lemma, including the word(s) represented by the ellipsis, is affected by the subsequent reading. The note "originally continued" supplies a deletion that initially followed immediately in the text; whereas "originally" furnishes the original reading of the text before any variety of changes may have been made, such as additions, deletions, and the writing over of words. In the textual notes the *Lectiones opticae* is indicated by I and the *Optica* by II. For convenience the numbering of the footnotes begins anew with each lecture, even though Newton's division into lectures is somewhat arbitrary; and these numbers run consecutively only in the Latin text, since some untranslated notes to the Latin are not inserted in the English translation.

By extending Newton's own sequence of numbered articles in the *Lectiones* and Part I of the *Optica* to the entirety of both works, we have developed a reference system whereby any article in either version can be readily and unambiguously distinguished and related to the other version. In this cross-reference system we prefix the Roman numerals I or II to the article numbers in Parts I and II of the *Optica,* so that §3 in Parts I and II are designated §I, 3 and §II, 3, respectively. The absence of a Roman numeral in a cross reference indicates that the article is in the *Lectiones.* This notation is also applied to lecture and figure numbers, so that, for example, Lecture 4 refers to the *Lectiones* and can be distinguished at a glance from Lecture I, 4 in Part I of the *Optica* and Lecture II, 4 in Part II. To complete this cross-reference system we use the following symbols:

    =   The corresponding articles are essentially equivalent, and any differences between them are less than one sentence.

≈   The corresponding articles are roughly equivalent, but in revision a sentence or more was rewritten, deleted, or added.

→   The corresponding articles treat the same topic, but in revision were completely rewritten.

Thus the references 62 = II, 53 and II, 53 = 62 signify that §62 in the *Lectiones* is virtually identical to §53 in Part II of the *Optica*; this notation is further abbreviated by the omission of the Roman numerals in the postils of the *Optica* where the Part is readily apparent from the running head, becoming in this case 53 = 62. To supplement this system for major revisions, a single rule in the margin of the *Lectiones* marks rewritten or deleted passages, and a double bar in the *Optica* marks rewritten or added passages. If an article number is not followed by another corresponding one, in the *Lectiones* it indicates that it was omitted in revision, whereas in the *Optica* it is a new article. All changes of a sentence or more and their significance are discussed in the Synopsis.

To make Newton's *Optical Lectures* accessible to the modern reader we have provided a translation and commentary. The translation is aimed at a rendering between the literal and modern. We hope that those who want simply to grasp the main line of Newton's thought without necessarily having to rely upon the Latin will feel comfortable with this translation without forgetting that his *Lectures* was written in another language over three centuries ago; we hope, too, that scholars will find it a ready guide to Newton's clear Latin prose. The technical commentary, including the Synopsis, is intended as an aid—and not a substitute—to following Newton's work. Utilizing as much as possible Newton's own sources, the historical notes and the Introduction aim both to place his ideas within the context of his own time and also to trace the investigations and ideas presented in the *Lectures* through the course of his writings on optics, which span more than half a century. Thus this volume may also serve as a guide to all of Newton's relevant optical writings. In later volumes, which will contain works and papers published in his lifetime and having far greater historical influence, we will be more concerned with the reception of his work and his interactions with the contemporary scientific community.

J. A. Lohne, in his pioneering paper on the editing of scientific diagrams, admonishes Newton's editors to bestow as much attention on preparing the figures for publication as they customarily give to the text.[1] He stresses the necessity of paying constant attention to the text while preparing the drawings and of "minutely collating" them with the originals. These sound and apparently objective procedures have been followed here, but they disguise the silent editorial judgment that must be continually exercised in preparing a proper copy of the figures, for the aim is to reproduce the figures not exactly but intelligently and clearly. Just as no attempt is made to reproduce in the printed text the character of the handwriting (such as the spacing of the lines

---

(1) Johannes A. Lohne, "The increasing corruption of Newton's diagrams," *History of Science*, 6 (1967):69–89.

or the size and form of the letters), so in modern drawings straight lines must be drawn straight, parallel lines parallel; small angles must be perceptible; and other features must be drawn according to Newton's intentions as determined from both the text and the figure. The resulting figures will often look different from Newton's, but the difference will not be as great as that between the printed text and the manuscript. For this reason and to save unnecessary labor and expense, we have used the same figure for both versions when the differences between the two were judged to result simply from the vagaries of drawing and copying and were thus inconsequential; otherwise two separate figures were drawn. The figures in the *editio princeps* and all subsequent editions of the *Optical Lectures* are unreliable—sometimes even caricatures of the originals—and are definitely not up to the standards exercised in preparing the text; accordingly, we will for the most part pass silently over their inadequacies in the notes.

Abbreviated References

Add.   Additional Manuscript, University Library,
                            Cambridge.

Correspondence   *The Correspondence of Isaac Newton*, ed. H. W.
                            Turnbull et al., 7 vols. (Cambridge, 1959–77).

Editio princeps   *Isaaci Newtoni . . . lectiones opticae, annis
                            MDCLXIX, MDCLXX & MDCLXXI. In scholis
                            publicis habitae: et nunc primum ex MSS. in lucem
                            editae* (London, 1729).

Lectiones XVIII   Isaac Barrow, *Lectiones XVIII, Cantabrigiae in
                            scholis publicis habitae; in quibus opticorum
                            phaenomenωn genuinae rationes investigantur, ac
                            exponuntur* (London, 1669), in *The Mathematical
                            Works of Isaac Barrow, D.D.*, ed. William Whew-
                            ell, vol. 2, (Cambridge, 1860). All page references
                            are to this edition.

Mathematical Papers   *The Mathematical Papers of Isaac Newton*, ed. D. T.
                            Whiteside, 8 vols. (Cambridge, 1967–81).

Opticks   Isaac Newton, *Opticks: Or, a Treatise of the Reflex-
                            ions, Refractions, Inflexions and Colours of Light*
                            (London, 1704).

Opticks (1721)   Isaac Newton, *Opticks*, 3rd ed. (London, 1721).

Phil. Trans.   Royal Society of London, *Philosophical Transactions*.

I   Add. 4002, here called the *Lectiones opticae*.

II   ULC Dd.9.67, here called the *Optica*.

Introduction

In the autumn of 1669, when Isaac Newton, not yet twenty-seven years old, was appointed Lucasian Professor, he chose to deliver his inaugural series of lectures (1670–2) on refractive optics, expounding his new theory of light and color with its radical claim that sunlight is a mixture of rays of different colors, each refracted differently. In these lectures Newton was attempting to codify a wealth of investigations that he had pursued during the preceding half decade. Since the resulting treatise, the *Optical Lectures,* was his first and most comprehensive account of his theory of color, he naturally drew upon it in his later writings. It served as the immediate source for his "New theory about light and colors" (1672) in the *Philosophical Transactions,* his first public statement of his theory outside the Cambridge lecture halls. And twenty years later it remained the foundation for the "definitive" statement of his theory in Book I of the *Opticks.* By reason of its central position in his early writings on optics, the *Optical Lectures* provides a natural focus for interpreting his initial researches on color and refraction. From these notes and essays on optics, together with his few autobiographical statements, it is possible to sketch the course of Newton's early optical research from the latter part of 1664 through the early 1670s. Because the surviving documentary evidence is rather scant, the following sketch must perforce be done in broad strokes.[1]

---

(1) Other modern accounts of the early development of Newton's theory of color, by no means agreeing with one another or with that presented here, are: A. Rupert Hall, "Sir Isaac Newton's note-book, 1661–65," *Cambridge Historical Journal,* 9 (1948):239–50 and "Further optical experiments of Isaac Newton," *Annals of Science,* 11 (1955):27–43; Richard S. Westfall, "The development of Newton's theory of color," *Isis,* 53 (1962):339–58; Johannes A. Lohne, "Isaac Newton: the rise of a scientist 1661–1671," *Notes and Records of the Royal Society of London,* 20 (1965):125–39; Johannes A. Lohne and Bernhard Sticker, *Newtons Theorie der Prismenfarben. Mit Übersetzung und Erläuterung der Abhandlung von 1672* (Munich, 1969); Maurizio Mamiani, *Isaac Newton filosofo della natura. Le lezioni giovanili di ottica e la genesi del metodo newtoniano* (Florence, 1976); and John Hendry, "Newton's theory of colour," *Centaurus,* 23 (1980):230–51. The introductions to the first three volumes of Newton's *Mathematical Papers* contain much valuable documentation on Newton's activities in the period from 1664 to 1673. Westfall's presentation of Newton's early researches in *Never at Rest: A Biography of Isaac Newton* (Cambridge, 1980) is the closest to our own, particularly in dividing Newton's discovery of his theory of color into two distinct phases, but his work appeared too late for us to use in the detailed preparation of this volume.

1. Newton's Introduction to Contemporary Optics and His Optical Investigations until 1666

Even when young, Newton seems to have been fascinated by color. While in his teens he took extensive extracts on the making of paints of various colors from John Bate's popular manual, *The Mysteryes of Nature, and Art*.[2] There is, however, no theoretical underpinning to Bate's treatment of color, and Newton's notes likewise remain just a compilation of recipes. When he entered Cambridge in June 1661, Newton encountered a scholastic curriculum, albeit a somewhat moribund one. A surviving undergraduate notebook contains, among other entries, his notes from Johannes Magirus's *Physiologiae peripateticae libri sex*, an introduction to Aristotelian natural philosophy which, at least insofar as color theory is concerned, is rather superficial. Newton never seems to have seriously pursued Aristotelian color theory, and it is perhaps with symbolic significance that his notes on Magirus break off precisely at Magirus's account of apparent colors.[3] Subsequently—in §25 = II, 2 of his *Lectures*—he was vigorously to attack, even mock, Aristotelian color theory both as he encountered it in Magirus's introductory text and as it was filtered throught the anti-Aristotelian critiques by the mechanical philosophers, Walter Charleton, Robert Boyle, and René Descartes, whose works he also studied in his undergraduate years. Despite widespread reaction against Aristotelian concepts of light and color, after two millennia they remained a significant element in seventeenth-century color theory, and so it is necessary here to present an epitome of them.

In his *De sensu* Aristotle explained that light arises when fire is present in a transparent medium such as air or water (darkness occurs when fire is ab-

---

(2) John Bate, *The Mysteryes of Nature, and Art: Contained in Foure Severall Tretises, the First of Water Workes, the Second of Fyer Workes, the Third of Drawing, Colouring, Painting, and Engraving, the Fourth of Divers Experiments, as wel Serviceable as Delightful: Partly Collected, and Partly of the Authors Peculiar Practice, and Invention* (London, 1634). Newton's notes "Of Drawing" are in a notebook he bought in 1659 (now in the Pierpont Morgan Library in New York) and extracts from it were published by D. E. Smith, "Two unpublished documents of Sir Isaac Newton," in *Isaac Newton, 1642–1727*, A Memorial Volume Edited for the Mathematical Association by W. J. Greenstreet (London, 1927), pp. 16–34. A typical note is that on preparing "A light greene": "Temper Verdigrease, & white lead 2 Verdigrease as much yellow berrys & a little white" (ibid., p. 20). E. N. Andrade first recognized that these notes are taken from Bate and argued that they are from the third edition of 1654 in "Newton's early notebook," *Nature*, 135 (1935):360. G. L. Huxley discusses the role of this book, which Newton probably obtained shortly after its publication, in directing his boyhood interests, in "Two Newtonian Studies. I. Newton's boyhood interests," *Harvard Library Bulletin*, 13 (1959):348–54.

(3) Newton's notes (Add. 3996, ff. 16ʳ–26ʳ) from Johannes Magirus, *Physiologiae peripateticae libri sex cum commentariis* (Cambridge, 1642) are most likely taken from this edition. His summary of Magirus's explanation of the rainbow (Bk. IV, Ch. V, §§14, 19, 22–5, pp. 161–2) fairly reflects its unsophisticated level: "The rainbow is a multicolored arc appearing by the reflection of the rays of the opposite sun in an opaque and concave dewy cloud. And it is either solar or lunar. Its principal colors are red at the top, green in the middle, and blue at the bottom. [Iris est arcus multicolor in nube roridâ opaca & concava ex radiorum solis oppositi reflectione apparens. Estque vel solaris vel lunaris. Ejus colores pr[a]ecipui sunt Puniceus supremus, viridis medius, Ceruleus infimus.]" (Add. 3996, f. 26ᵛ.)

sent), and that color is visible only in light. The colors of bodies originate from the fire and the transparent contained within them:

> ... the transparent in so far as it is found in bodies (and it exists in all in varying degrees) causes them to be endowed with colour. But since it is in a bounding surface that colour is found, it is in the surface of this—the transparent—that colour exists. *Colour then is the extremity of the transparent in a determinately bounded body*; and it is found in all bodies alike, both in transparent substances themselves, such as water and anything similar to it, and in those which appear to have a surface colour of their own. Consequently, that, which in air causes light [namely, fire], may be present in the transparent medium or it may not, *i.e.* may be awanting.
>
> Thus, just as we can explain light and darkness respectively by the presence or absence of this cause [fire] in the air, so in the case of solid bodies we can account for the existence of black and white colour.[4]

Color is conceived to be like light, except that it is contained or bounded within a body and requires external light to be perceived; it was still commonly held in the seventeenth century, for example, by Johannes Kepler, Isaac Vossius, and Marcus Marci, that there is light within colored bodies.[5] After explaining the origin of white and black, Aristotle derives his five chromatic colors from various mixtures of white and black. Magirus explains that "White and black are the extreme colors ... The intermediate colors are produced from a mixture of the extremes, some of these inclining more to white and others to black ... yellow is generated whenever much white is mixed with much less black, and hence brightness with some obscurity emerges."[6] Magirus continues with the generation of the remaining colors and explains some in terms of white and black and others, equivalently, in terms of light and transparent or opaque. Still another formulation that utilizes the four elements—water and air (transparent), fire (luminous), and earth (opaque)—makes it particularly clear that in the Scholastic view colors are essential qualities of bodies, since all bodies are composed of the four elements.

To account for the colors of the rainbow, in his *Meteorologica* Aristotle adopted a fundamentally different sort of explanation than that for colored bodies. He taught that the rainbow's colors arise from a weakening and darkening of light (or, for Aristotle himself, of the visual ray) by distant, dark, black clouds; the upper band, for example, is red because it is the largest and reflections from it are least weakened. Elaborating these two different modes of explanation into a fundamental dichotomy, medieval Aris-

(4) Aristotle, *De sensu and De memoria,* ed. and trans. G. R. T. Ross (Cambridge, 1906), 439 b 9–19, p. 57; Newton quotes the passage in italics in §25 = II, 2.

(5) On the history of color theory in the sixteenth and seventeenth centuries see J. MacLean, "Geschiedenis van de Kleurentheorie in de zestiende Eeuw.," *Scientiarum Historia,* 9 (1967):23–39; "Kleurentheorie in de Periode 1600–1635," ibid., pp. 126–47; "De Kleurentheorie van de Aristotelianen en de Opvattingen van de la Chambre, Duhamel en Vossius in de Periode 1640–1670," *Scientiarum Historia,* 10 (1968):208–25; "De Kleurenleer van de Aanhangers der Corpusculairtheorie," *Scientiarum Historia,* 12 (1970):1–22.

(6) Magirus, *Physiologiae peripateticae,* Bk. VI, Ch. VII, p. 322.

totelian commentators introduced the distinction between apparent or emphatical colors, which are transient colors appearing without a colored body, as in the rainbow or a prism, and real or true colors, which are intrinsic qualities of colored bodies. Apparent colors in this Scholastic view are some modification of incident light, as in the rainbow, whereas real colors are forms or inherent qualities of bodies that are distinct from incident light, which is simply the vehicle for bearing color to the eye.[7]

The mechanical philosophers of the seventeenth century rebelled against these distinctions. Although they were a diverse group, they were in agreement in accepting the two tenets that all natural bodies consist solely of particles of matter of various sizes, shapes, and motions and that all knowledge of bodies derives from sensation that is simply motion propagated through the nerves to the brain. Bodies, they held, do not possess substantial forms or essential qualities but merely a disposition to excite a particular motion or sensation. The most systematic and influential exponent of this view was Descartes, whose works Newton studied with the utmost care and who influenced him to a degree that can scarcely be exaggerated. But it was Boyle who, in his *Experiments and Considerations Touching Colours,* most thoroughly applied the program of the mechanical philosophy to the problem of color. The mechanical philosophers rejected the Scholastic distinction between real and apparent colors, since they are each sensations of color produced by light falling on the eye, and the source of a sensation is irrelevant to the fundamental nature of that sensation. As Boyle observed, "since we judge other Sensible Qualities to be True ones, because they are the proper Objects of some or other of our Senses, I see not why Emphatical Colours, being the proper and peculiar Objects of the Organ of Sight, and capable to Affect it as Truly and as Powerfully as other Colours, should be reputed but Imaginary ones."[8] The elimination of this distinction allowed Newton to found his theory of color almost exclusively on experiments with prisms. Likewise, the mechanical philosophers rejected the distinction between light and color, since all light is perceived as colored, that is, "Light it self produces the sensation of a Colour . . . as it produces such a determinate kind of local motion in some part of the brain."[9] They were not, however, as rebellious in explaining the origin of colors, assuming them to be some modification or alteration of white light. This assumption was a natural extension of the Aristotelian theory of apparent colors to all colors; and indeed by everyday experience it does seem that when sunlight falls upon a body it is so modified that it causes particular sensations of color. There was, however, no agreement as to the nature of this modification. Boyle held

---

(7) Robert Boyle in *Experiments and Considerations Touching Colours . . . the Beginning of an Experimental History of Colours* (London, 1664) succinctly stated the Scholastic position, "the Peripatetick Schools, though they dispute amongst themselves divers particulars concerning Colours, yet in this they seem Unanimously enough to Agree, that Colours are Inherent and Real Qualities, which the Light doth but Disclose, and not concurr to Produce" (Pt. I, Ch. V, p. 84).

(8) Ibid., Pt. I, Ch. IV, p. 76.   (9) Ibid., Pt. I, Ch. II, p. 11.

... Colour to be a Modification of Light ... yet I propose it but in a General Sense, teaching only that the Beams of Light, Modify'd by the Bodies whence they are sent (Reflected or Refracted) to the Eye, produce there that Kind of Sensation, Men commonly call Colour; But whether I think this Modification of the Light to be perform'd by Mixing it with Shades, or by Varying the Proportion of the Progress and Rotation of the *Cartesian Globuli Caelestes,* or by some other way which I am not now to mention, I pretend not here to Declare.[10]

Newton himself would adopt modification theories of color in his preparatory years of 1664–5, but afterward he was to wage a constant battle against "the hypotheses of the philosophers."

In early 1664, his last year as an undergraduate, Newton set out on a course of reading in the new philosophy. Approximately two-thirds of the way through the same notebook in which he had earlier entered his notes out of Magirus's text and other required undergraduate books, he started a new set of notes, entitled "Qu[a]estiones Philosoph[i]cae." These were divided into forty-five topics, such as "Of Attomes," "Of Gravity & Levity," and "Of light," and under each he entered his own thoughts and notes from such works as Thomas Hobbes's *De corpore,* Charleton's *Physiologia,* Descartes's *Principia philosophiae* and *Meteores,* and Boyle's *Touching Colours.*[11] The entries in the "Quaestiones" were probably made in 1664 and 1665. Perhaps the most important entry, almost of essay length, is "Of Colours," in which Newton takes some significant steps toward his theory of color. "Of Colours" was written largely under the influence of Boyle's *Touching Colours,* and well over half of it consists of numbered entries (§§10–51) extracted directly from that work. This helps to date "Of Colours" to sometime between early 1664, the publication date of *Touching Colours,* and early 1665, the date of Robert Hooke's *Micrographia,* which Newton does not yet seem to have read. The principal aim of Boyle's *Touching Colours,* which reflected his chemical interests, was to establish that the colors of natural bodies arise from a modification of light and, in particular, to explain those colors and their alteration by chemical and mechanical means in terms of the arrangements of the bodies' parts. He

---

(10) Ibid., Pt. I, Ch. V, p. 90. In his own readings Newton was to encounter both examples of modifications suggested by Boyle. Walter Charleton, *Physiologia Epicuro-Gassendo-Charltoniana: Or a Fabrick of Science Natural, upon the Hypothesis of Atoms, Founded by Epicurus, Repaired by Petrus Gassendus, Augmented by Walter Charleton* (London, 1654) was a comprehensive introduction to atomism with four large chapters devoted to vision, color, and light. Providing a good, if neither original nor profound, survey of contemporary optics, it taught that the intermediate (chromatic) colors "are but the off-spring of the Extreme, arising from the intermission of Light and shadow, in various proportions" (pp. 191–2). Descartes's explanation of color by rotating light globules is presented below.

(11) Richard S. Westfall describes the "Quaestiones" (Add. 3996, ff. 88ʳ–135ʳ) and gives excerpts from it in "The foundations of Newton's philosophy of nature," *British Journal for the History of Science,* 1 (1962):171–82, esp. note 5 for the books that Newton cited; A. Rupert Hall also has excerpts in his "Newton's note-book;" and J. E. McGuire and Martin Tamny have prepared a fully edited version of it.

included a few prismatic experiments, but they were introduced as much to study color mixing as to study the origin of prismatic colors. A second aim of his book was to provide an extensive series of experiments as an aid to establishing "a sound and comprehensitve Hypothesis" of color,[12] and in this he found an ideal student in Newton.

Newton's "Of Colours," like Boyle's *Touching Colours,* is devoted almost exclusively to the colors of bodies.[13] Yet Newton was by no means simply following Boyle, since he had already set out his own path, discovering that rays of different colors are unequally refracted, while also entertaining the notion that white light is compound. He discovered unequal refrangibility by viewing colors directly through a prism, an experiment with no counterpart in Boyle or any of Newton's other early readings. In his first experiment (§2) he observed the boundary of a bicolored card, initially with one half white and one half black; and in his second one (§3) he observed a bicolored thread, with one half red and one half blue. He interpreted the boundary colors that are seen through the prism in the first experiment as arising from the rays' separation by their unequal refraction in the prism, "soe y^t ye slowly moved rays being seperated from ye swift ones by refraction, there ariset 2 kinds of colours viz: from ye slow ones blew, sky colour, & purples. from ye Swift ones red, yellow   & from them wch are neither moved very swift nor slow ariseth greene but from ye slow & swiftly moved rays mingled ariseth white grey & black."[14] Similarly, he explained the divided image of the thread viewed through the prism "by reason of unequall refractions in ye 2 differing colours."[15]

Having established the unequal refrangibility of different colors to his own satisfaction, Newton tried out a series of hypotheses to explain the colors of natural bodies. First (§4), he considered that "rednes yellownes &c are made in bodys by stoping ye Slowly moved rays w^{th}out much hindering of ye motion of ye swifter rays. & blew greene & purple by diminishing ye motion of ye swifter rays & not of ye slower."[16] This explanation invokes a modification theory, for the rays' motion is diminished by the bodies, thus causing their color and degree of refrangibility to be continually mutable as they successively encounter different bodies. In the following articles Newton suggests a number of different modification theories, which, however, are all based on the collisions of light corpuscles with the particles of bodies. In the last of these models (§8) he attempts to derive a quantitative measure for color using the laws of impact, thereby beginning his quest for a mathematical theory of color that would be a goal of all of his work on optics, especially the *Optical Lectures.* The final forty-two articles of the essay are extracted from Boyle's *Touching Colours* and deal exclusively with colored bodies, particularly the color changes in chemical reactions, as in the following typical example: "29. A just quantity of Oyle of Tartar

---

(12) Boyle, *Touching Colours,* Pt. I, Ch. V, p. 89.
(13) The only exception is Newton's observation of phosphenes, but this experiment too derives from Boyle and Descartes; see Lect. 3, note (14), this volume.
(14) Add. 3996, f. 122ʳ; the gap is Newton's.   (15) Ibid.   (16) Ibid.

poured into a strong solution of french verdigrease turnes it from greene to blew; a Lixivium of pot ashes turnes it to a lighter blew, & spirit of Urin, or Harts-horne make other blews."[17]

"Of Colours" represents Newton's thought at a unique transitional stage. He had already abandoned the strict modification theories of Boyle and Descartes and boldly conceived white light to be compound; but he still considered that colors could arise from some modification (an alteration of its motion) suffered by light in falling upon a body. Likewise, he recognized that some colors are more refrangible than others, but not that the sun's direct light consists of unequally refrangible rays. Indeed, Newton only implicitly takes into account the sun's direct light, since all his own experiments and hypotheses on color treat light reflected from bodies and never the sun's direct light. "Of Colours" is devoted to the colors of bodies and not to prismatic colors. When Boyle described a number of experiments with light projected through a prism—"the usefullest Instrument Men have yet imploy'd about the Contemplation of Colours"[18]—Newton was insufficiently interested in them to take any notes other than to incorporate some of the results into an entry on color mixing (§12). "Of Colours" does not contain a single experiment with light projected through a prism; but once Newton even considered such an experiment, as he did early in 1666, and posed the question of what would happen when sunlight passed directly through a prism, he arrived straightaway at his theory of color.

In his preparatory years of 1664–5 Newton's interest in optics was by no means limited to color, as he was also immersed in the study of geometrical optics, which began with his perusal of Descartes's *Geometrie* and *Dioptrique*.[19] Here he learned of the sine law of refraction, the foundation of all of his optical investigations, no later than September 1664, which is the date entered in his "Waste Book" for calculations on reflection and refraction, including the properties of Cartesian ovals.[20] Descartes's refraction model, which was based solely on mechanical parameters—the velocities of the light corpuscles and "impulses" normal to the refracting surface—made a

---

(17) Ibid. f. 133ᵛ; compare Boyle, *Touching Colours*, Pt. III, Expt. XXII, pp. 251–4.

(18) Ibid., Pt. III, Expt. XV, p. 227; for other prismatic experiments see Expts. IV–VI, XIV; and also Lect. 4, note (31).

(19) Descartes's *La Dioptrique*, *La Geometrie*, and *Les Meteores* were appended to his *Discours de la methode pour bien conduire sa raison, & chercher la verité dans les sciences* and published in 1637 as exemplars of that method; but since Newton could read French only with difficulty, he read them in Latin translation. An authorized Latin rendition of the scientific works, often differing significantly from the French, was published in Amsterdam in 1644, *Specimina philosophiae: seu dissertatio de methodo rectè regendae rationis, & veritatis in scientiis investigandae: dioptrice, et meteora. Ex gallico translata, & ab auctore perlecta, variisque in locis emendata*; this went through a number of editions, and Newton read that published at Paris in 1656. The reprints of the 1644 Latin translation, along with the original French, will be cited in Vol. 6 of *Oeuvres de Descartes*, ed. Charles Adam and Paul Tannery, 13 vols. (Paris, 1897–1913); hereafter *Dioptrice* and *Meteora*. Newton read the *Geometrie* in Frans van Schooten's much amplified second Latin edition in two volumes (Amsterdam, 1659–61).

(20) *Mathematical Papers*, 1:551–8. Newton returns to Cartesian ovals in the *Optica*, Pt. I, Prop. 34.

profound and enduring impression on Newton, both as a persuasive example
of how all nature might be reduced to similar mechanical principles and also
in its own right as a mathematical model of a general optical law, which
Newton quickly incorporated into his own investigations of refraction and
dispersion. In his essay "Of Refractions" (late 1665 or early 1666) Newton
developed Descartes's investigations of the refracting surfaces formed by the
revolution of conics about their principal axis and improved upon his lens-
grinding machine from the *Dioptrique*.[21] In addition, he carefully studied
Descartes's account of the primary and secondary rainbows in the *Meteores*.
In the *Optical Lectures* he gives Descartes full credit for his determination of
their radii and generalizes his solution to higher-order bows, but he sets up
his account of their colors as the archetypal modification theory.[22] Adopting
a traditional comparison, Descartes explained the rainbow's colors by analogy to the colors cast by a prism. From a single experiment with prismatic
spectra he deduced that for colors to be generated "at least one [refraction] is
required, and indeed one such that its effect is not destroyed by another
contrary one, for experience teaches that if the [refracting] surfaces . . . were
parallel, the rays, being straightened just as much in the one as they were
bent in the other, would depict no colors . . . And moreover I observed that
shadow or the limitation of light is also required."[23] To explain the generation of spectral colors, he assumed that the little balls or globules that compose the aether rotate uniformly when transmitting a direct beam of sunlight,
but when the beam is refracted their rate of rotation changes, because at the
edges of the beam, at the boundary of light and shadow, they encounter balls
moving more swiftly or more slowly than they do. He then ascribed different
colors to different rotations: "those which endeavor to rotate much more
vigorously make the color red, and those only a little more vigorously make
yellow. And on the contrary . . . green appears where they rotate not much
slower than usual, and blue where they rotate very much slower."[24] In a like
way, in Discourse I of the *Dioptrique* he explained the colors of natural
bodies by the various rotations that the little balls acquire upon reflection
from the variously textured surfaces of bodies. Newton frequently set forth
Descartes's account of color, but only to refute it. Nonetheless, whether
following or refuting Descartes he was profoundly influenced by him, and
Cartesian stigmata can be recognized throughout Newton's writings.

   Hooke's *Micrographia* was published in January 1665, and sometime during that year Newton read it attentively, taking seven closely written pages of
notes.[25] In his two long observations devoted to color, Hooke presented the

   (21) "Of Refractions" (Add. 4000, ff. 26ʳ–33ʳ) is discussed by Hall in "Further experiments,"
pp. 36–43; while most of the text is printed in *Mathematical Papers*, 1:559–76.
   (22) See the *Optica*, §§I, 144–51 and II, 136–9.
   (23) Descartes, *Meteora*, Ch. VIII, §5, *Oeuvres*, 6:702, 330–1.
   (24) Ibid., §7, pp. 704, 333–4.
   (25) Robert Hooke, *Micrographia: Or Some Physiological Descriptions of Minute Bodies
made by Magnifying Glasses. With Observations and Inquiries Thereupon* (London, 1665).
Newton's notes "Out of Mʳ Hooks Micrographia" (Add. 3958, ff. 1ʳ–4ʳ) are in Geoffrey Keynes,

most comprehensive treatment of color yet given by a mechanical philosopher. According to his wave, or vibratory, theory of light the front of a pulse (a physical ray with finite width), which is normal to its sides before refraction, becomes oblique to them afterward. By means of these oblique pulses, which are simply modifications of the incident white light, he explained the generation of his two fundamental colors: "Blue is an impression on the Retina of an oblique and confus'd pulse of light, whose weakest part precedes, and whose strongest follows. And . . . Red . . . [one] whose strongest part precedes, and whose weakest follows."[26] He applied this theory to a broad range of phenomena—prismatic colors, the colors of thin films, and the colors of natural bodies—and imaginatively supported it with numerous experiments and observations. Although Newton's notes show that he was frequently critical of the *Micrographia,* he also drew much from it, particularly from Hooke's explanation of the colors of thin films by means of vibrations in the aether, his descriptions of a lens-grinding machine and a refractometer, his numerous observations on the colors of natural bodies, and his account of "inflection," or the curved path of rays in inhomogeneous media such as the atmosphere.[27]

By the latter part of 1665 Newton had acquired a firm command of contemporary optics through his studies of the works of Boyle, Descartes, and Hooke; and he had even advanced it somewhat through his own investigations. There is no sign that Newton's serious reading went beyond these three or that he ever gave more than a cursory glance to contemporary works not in the new philosophical tradition, such as della Porta's *De refractione,* Marci's *Thaumantias,* Kircher's *Ars magna lucis et umbrae,* and Vossius's *De lucis natura,* or to the classical treatises of Alhazen and Witelo, or even to the more mathematical treatises of Kepler and Cavalieri. John Conduitt's story that Newton had read Kepler's "Opticks" as a freshman is entirely unsupported and scarcely reliable.[28] Newton, to be sure, knew the elements of Kepler's theory of vision, if not his more technical accounts of refraction and lenses, but the evidence indicates that he learned it indirectly, from Descartes's *Dioptrique,* which fully incorporated Kepler's theory, and also possi-

---

A Bibliography of Dr. Robert Hooke (Oxford, 1960), pp. 97–108, and in *Unpublished Scientific Papers of Isaac Newton. A Selection from the Portsmouth Collection in the University Library, Cambridge,* ed. A. Rupert Hall and Marie Boas Hall (Cambridge, 1962), pp. 400–13.

(26) Hooke, *Micrographia,* p. 64. On Hooke's theory of color see Alan E. Shapiro, "Newton's definition of a light ray and the diffusion theories of chromatic dispersion," *Isis,* 66 (1975):194–210, esp. 197–9.

(27) In his letter to Henry Oldenburg on 21 December 1675, defending himself against Hooke's charges of plagiarism, Newton presented his own evaluation of his indebtedness to Hooke; *Correspondence,* 1:404–6; see also the drafts of this letter, Add. 3970, ff. 531, 532. Oldenburg was Secretary of the Royal Society and editor of the *Philosophical Transactions.*

(28) Only excerpts from Conduitt's memoirs of Newton have ever been published. They were the basis for Fontenelle's *éloge* of Newton; reprinted in *Isaac Newton's Papers & Letters on Natural Philosophy and Related Documents,* ed. I. Bernard Cohen (Cambridge, Mass., 1958), pp. 444–74, esp. 446. Extracts from Conduitt's manuscript draft are in *Mathematical Papers,* 1:15–19, esp. 15.

bly from Christoph Scheiner's *Oculus*.[29] Scheiner described the anatomy and physiology of the eye, utilized the camera obscura (which forms an essential component of Newton's later optical experiments), and gave a thorough argument supported by numerous experiments for Kepler's theory of the retinal image, though without giving proper credit to Kepler.

2. Newton's Optical Investigations, 1666–1670

Newton began his "New theory about light and colors" with a "historicall narration," his only autobiographical account of his discovery:

> ... in ye beginning of ye year 1666 (at wch time I applyed my selfe to ye grinding of Optick glasses of other figures then Sphericall) I procured me a triangular glasse Prisme to try therewth ye celebrated phaenomena of colours. And in order thereto having darkned my chamber & made a small hole in my window-shuts to let in a convenient quantity of ye Sun's light, I placed my Prism at its entrance, that it might be thereby refracted to ye opposite wall ... applying my selfe to consider ... [the colours] more circumspectly, I became suprized to see them in an oblong form wch according to ye received lawes of refraction, I expected should have been circular ... Comparing ye length of this Coloured Spectrum wth its breadth I found it about five times greater, a disproportion soe extravagant that it excited me to a more then ordinary curiosity of examining from whence it might proceed ... [30]

Newton would have us believe here and in what follows that he discovered his theory of color by his experiment on the elongated spectrum while applying himself to the grinding of nonspherical lenses. Although his account is undoubtedly in part an embellished historical reconstruction, making the discovery appear like a "Baconian induction from experiments,"[31] in many of its essentials it does agree with the surviving manuscripts. His essay "Of Refractions," the greatest part of which treats the grinding of hyperbolic and parabolic lenses and which was composed before he discovered his theory, evidently represents his endeavor to grind nonspherical lenses around the beginning of 1666. Although it appears that this was nothing but a theoretical study and that Newton had not yet actually ground any lenses, his suggestion that he discovered his theory while engaged in lens grinding is an entirely plausible one: During the course of his study of grinding nonspherical lenses,

---

(29) Christoph Scheiner, *Oculus hoc est: fundamentum opticum* ... [1619] (London, 1652). Newton owned this edition but it is not known when he acquired it; see John Harrison, *The Library of Isaac Newton* (Cambridge, 1978), no. 1459, which also has an informative introduction on Newton's use and acquisition of books. In general, it cannot be determined when Newton acquired a particular book.

(30) Newton to Oldenburg, 6 February 1671/2, Add. 3970, ff. 460^r-6^r, esp. 460^r. The original letter has not been found, and this is the copy retained by Newton in the hand of his college roommate, John Wickins. The text in *Correspondence*, 1:92–102, is a conflation of this copy and the published version, "A letter of Mr. Isaac Newton ... containing his new theory about light and colors," *Phil. Trans.*, 6 (19 February 1671/2):3075–87.

(31) Lohne, "The rise of a scientist," p. 138.

that is, in his attempt to eliminate spherical aberration, his attention turned to the other serious distortion—chromatic aberration—and only then did the significance of his earlier discovery of the unequal refrangibility of different colors dawn upon him.[32] If—he perhaps reasoned—the sun's light directly as it comes from the sun, before it is reflected from any colored body, consists of rays of different color and degrees of refrangibility, then each color would be bent a different amount and brought to a different focus. To test this conjecture, he for the first time passed a sunbeam directly through a prism to see if the spectrum was elongated rather than circular. Prismatic spectra, the "celebrated phaenomena of colours," had been described in one form or another in virtually every work on optics of the preceding century, and Newton certainly encountered them in his readings of Hobbes, Charleton, Descartes, Boyle, and Hooke, but his own experimental arrangement of passing a narrow beam through a prism at minimum deviation was an original— and by no means obvious—one chosen specifically to test quantitatively for unequal refrangibility.

Newton, then, discovered his theory in two distinct phases: First, in the "Quaestiones" he discovered unequal refrangibility, just one component, albeit a crucial one, of his theory of color;[33] and then in the beginning of 1666

(32) It would seem from Newton's account book (now in the Fitzwilliam Museum in Cambridge) that he did not buy his equipment to grind lenses until the spring of 1667, when he recorded purchases of a "Lath & Table ——— 0-15-0," "Iron worke for it ——— 0-9-0," and "Drills, Gravers, a Hone & Hammer, & a Mandril ——— 0.5.0." However, following the main portion of "Of Refractions" on the grinding of nonspherical lenses, Newton added two briefer sections on spherical lenses; a first "To Grinde Sphaericall optick Glasses" (Add. 4000, f. 31; excerpted in Hall, "Further experiments," p. 42), which seems to reflect Hooke's design for a lens-grinding machine in the *Micrographia* (sigs. e[2–3]); and a method how to test "If ye Glasses of a Telescopee bee not truely ground" (Add. 4000, ff. 32^v–3^r; excerpted in Hall, "Further experiments," pp. 42–3). Newton may have tested his method of detecting deviations from a perfect spherical shape with a ready-made lens, which would have called his attention to chromatic aberration just as well as grinding nonspherical lenses.

(33) The unequal refrangibility of rays of different colors was experimentally discovered some sixty years earlier by Thomas Harriot, whose work remained unpublished and virtually unknown until recent times. No comprehensive account of it has yet been published, but see Johannes A. Lohne, "Thomas Harriott (1560–1621), the Tycho Brahe of optics. Preliminary notice," *Centaurus*, 6 (1959): 113–21, esp. 119–20, and "Newton's 'proof' of the sine law and his mathematical principles of colors," *Archive for History of Exact Sciences*, 1 (1961):389–405, esp. 394. The prior discovery of unequal refrangibility has also been attributed to Marcus Marci. Marci, however, did not recognize that at the same angle of incidence rays of different color are refracted differently, but rather he attributed the unequal refraction of rays falling upon a prism solely to the half-degree difference in their angle of incidence arising from the sun's finite size: "It follows that all that variety of color which is caused by refraction in a triangular prism is contained within half of one degree," (*Thaumantias: Liber de arcu coelesti deque colorum apparentium natura, ortu, et causis* ... (Prague, 1648), Theorem XV, Corollary, p. 98). To determine the "dispersion," Marci calculated the different angles of refraction resulting from angles of incidence that differed by half a degree; for instance, in Theorem XXXII, pp. 107–8. Since Marci neither used the sine law of refraction (but an interpolation scheme based on Kircher's refraction table for water) nor recognized that the index of refraction for glass and water differed, his theory could have had but little empirical value. See Carl B. Boyer, *The Rainbow: From Myth to Mathematics* (New York/London, 1959), pp. 220–1.

his "theory of colour." An incidental remark made by Newton in the *Optical Lectures* corroborates that he came upon his theory in two phases. In §2 = I, 2 he sets forth his discovery that "the refrangibility of all rays is not the same," and to prove it, in the next article he introduces his experiment on the elongated spectrum by noting that it "first presented me the opportunity to think out the rest," that is, the rest of the theory beyond the existence of unequal refrangibility, principally that different colors and degrees of refrangibility are innate to the sun's light, and that there is a unique, unchangeable correspondence between them.

After describing in the "New theory" the experiments and reasoning that he would have us believe led him in such a straightforward way to conclude that "light consists of rayes differently refrangible," he relates:

> When I understood this, I left of[f] my aforesaid glass-workes, for I saw that yᵉ perfection of Telescopes was hitherto limited, not so much for want of glasses truly figured according to yᵉ prescriptions of Optick Authors ... as because that light it selfe is a heterogeneous mixture of differently refrangible rayes. So that were [a] Glasse so exactly figured as to collect any one sort of rayes into one point, it could not collect those also into the same point which having yᵉ same incidence upon yᵉ same medium are apt to suffer a different refraction ... This made me take reflections into consideration; & finding them regular, so that yᵉ angle of reflection of all sorts of rayes was equall to their angle of incidence I understood that by their mediation Optick instruments might be brought to any degree of perfection imaginable ... Amidst these thoughts I was forced from Cambridge by the intervening Plague & it was more then two years before I proceeded further ... [34]

Newton left Cambridge in June 1666 so that according to this narrative he constructed his first reflecting telescope in the second half of 1668. This chronology is supported both by his account book, which records purchases in the summer of 1668 of two furnaces and putty—for casting and polishing his mirrors—and also by his letter to "a friend" half a year later, containing his earliest extant description of his telescope.[35]

Devising and performing the experiments to test his new theory must have taken a number of weeks at the least, but there is no direct information on this, since Newton breaks off his "historicall narration" in the "New theory" after he demonstrates that the sun's light consists of unequally refrangible rays. Sometime between the beginning of 1666 and 1669, but most probably closer to the former than the latter, Newton wrote up the experiments from

(34) Add. 3970, ff. 461ᵛ–2ʳ = *Correspondence*, 1:95–6 = *Phil. Trans.*, 6 (1671/2): 3079–80. Newton made a second, improved telescope in the autumn of 1671. In December he sent it to the Royal Society, where it was so well received that they elected him a Fellow, and, in turn, he communicated his "New theory" to them. A description of this second reflector was drawn up by Oldenburg, revised by Newton, and then published as "An accompt of a new catadioptrical telescope invented by Mr. Newton," *Phil. Trans.*, 7 (1672):4004–7; see *Correspondence*, 1:73–80; and also A. A. Mills and P. J. Turvey, "Newton's telescope: an examination of the reflecting telescope attributed to Sir Isaac Newton in the possession of the Royal Society," *Notes and Records of the Royal Society of London*, 33 (1979):133–55.

(35) Newton to "a friend," 23 February, 1668/9, *Correspondence*, 1:3–4.

his "age of invention" in an essay again entitled "Of Colours."[36] There is no statement of the theory and little theoretical interpretation, but cautiously reading backward from the later accounts, especially the *Optical Lectures,* it is clear that he already had the main features of his theory, since the essay contains many of the fundamental experiments of the *Optical Lectures.* After opening with the principal phenomena of his theory of the colors of natural bodies, he repeats the experiments of the "Quaestiones" of viewing bicolored cards and strings through a prism, and then he records measurements of the elongated spectrum.[37] He follows this with an experiment in which differently colored bodies are illuminated with pure spectral colors in a dark room and seen to possess only the color of the illuminating light, thereby refuting modification theories of the colors of natural bodies such as he himself held in the "Quaestiones."[38] The relation of particular experiments in "Of Colours" to those in the *Lectures* is indicated in the notes, but we can observe that besides his initial investigations of "Newton's rings,"[39] this essay includes such fundamental experiments as the composition of white light from spectral colors, and what would become the *experimentum crucis* but showing, as in the *Optical Lectures,* that differently colored rays are refracted unequally.

The slim essay "Of Colours" is the only record of Newton's elaboration of his theory of color between 1666 and his inaugural Lucasian lecture in January 1670, but his research on geometrical optics in this period is better documented. He added some notes to "Of Refractions" after the discovery of his theory of color, and one set shows that he was attempting to design a compound achromatic lens, contrary to the implication of the *Opticks,* years later, that such a lens was impossible.[40] Also appended to "Of Refractions" was a table of refractions and dispersion for glass, "christall," and water that was constructed according to the refraction model later expounded in the *Optical Lectures,* §§115–17 = I, 42–4.[41] In still another set of appended calculations, Newton determined the spherical aberration of a plano-convex lens (returning to it in Proposition 31 of Part I of the *Optica*);[42] and according to the historical narration of the "New theory" he had also determined

(36) Add. 3975, ff. 2^r–11^v; excerpted in Hall, "Further experiments," pp. 27–36.

(37) Add. 3975, §§1–7, f. 2.   (38) Ibid., §10, f. 3^r.

(39) Newton's essay "Of ye coloured circles twixt two contiguous glasses" (Add. 3970, ff. 350^r–3^v), which Westfall dates as "closer to 1672 than to 1666," represents a more detailed investigation of "Newton's rings," but not until the spring of 1672 did he prepare a full account of them; see Richard S. Westfall, "Isaac Newton's coloured circles twixt two contiguous glasses," *Archive for History of Exact Sciences,* 2 (1965):181–96.

(40) The last four pages of "Of Refractions" were torn out of Newton's notebook (Add. 4000) about the time of his death and are now in private possession, but the mathematical portions are published in *Mathematical Papers,* 1:572–6; for the compound lens see pp. 575–6, and also Zev Bechler, "A less agreeable matter: the disagreeable case of Newton and archromatic dispersion," *British Journal for the History of Science,* 8 (1975):101–26.

(41) This table forms the final page of the essay as it now exists in the original notebook (Add. 4000, f. 33^v); it has been printed from John Collins's 1669 copy (now in Shirburn Castle) in *Correspondence,* 1:103.

(42) *Mathematical Papers,* 1:572–4; see also Lect. I, 13, note (13).

the chromatic aberration in the beginning of 1666.[43] At this same time, when he was pondering the design of a reflecting telescope, he came across James Gregory's *Optica promota,* which set forth the "Gregorian" design and more generally treated the imaging properties of mirrors and lenses, particularly those of conic transverse section.[44] Newton, however, seems to have acquired much of his knowledge of contemporary telescopic design and use from the early numbers of the *Philosophical Transactions,* which began publication in March 1665 in the midst of an exciting period for telescopic astronomy. Newton's account book shows that he purchased the *Philosophical Transactions* in 1667, and probably shortly thereafter he took thirteen pages of notes, "Out of Philosophicall Transactions," which summarized the contents of the first twenty-four numbers.[45]

It is uncertain when Newton's earliest contacts with Isaac Barrow, the first Lucasian Professor, occurred,[46] but in 1667 and 1668 he must have attended Barrow's lectures on optics. These began with the basic elements of geometrical optics but moved well beyond to treat such problems as the focal properties of lenses, the rainbow, the determination of the caustic condition for a spherical surface, and the location of the image of a point seen across a plane refracting surface. Newton by 1667 was already sufficiently well versed in geometrical optics to appreciate Barrow's mathematically sophisticated investigations and to pursue them actively in the ensuing years. By early 1669 Barrow had sufficient confidence in Newton's abilities and judgment to trust him to proofread his optical lectures, which appeared later that year. Barrow was reluctant to publish his lectures but yielded to John Collins's importunities and entrusted the care of his "offspring" to friends: "our colleague Mr. Isaac Newton, a man of outstanding genius and remarkable knowledge, reviewed the copy, noting a number of things to be corrected and also added some things from his own stock, which you will see annexed here and there to our own things with praise. The other . . . Mr. John Collins took care of

(43) See Lect. I, 15, note (9).

(44) On 4 May 1672 Newton wrote Oldenburg that "when I first applied my selfe to try the effects of reflexions, Mr Gregory's *Optica Promota* (printed in ye yeare 1663) being faln into my hands," its design was duly considered—and rejected; see *Correspondence,* 1:153. For a synopsis of James Gregory, *Optica promota, seu abdita radiorum reflexorum & refractorum mysteria, geometrice enucleata . . .* (London, 1663), see *James Gregory Tercentenary Memorial Volume. Containing his Correspondence with John Collins and his Hitherto Unpublished Mathematical Manuscripts, together with Addresses and Essays Communicated to the Royal Society of Edinburgh, July 4, 1938,* ed. Herbert Westren Turnbull (London, 1939), pp. 454–9.

(45) Newton's account book records the purchase of "Philosophical Intelligences——— 0.9.6," which is presumably the *Philosophical Transactions;* his notes on its early numbers are now Add. 3958, ff. 9^r–15^r.

(46) Barrow's early influence on Newton traditionally seems to have been placed at too early a date and its significance has been overestimated. He was not Newton's undergraduate tutor, but Newton might have attended his lectures in 1665; see *Mathematical Papers,* 1:10, note 26; 3:xiii–xv. A note written by Newton in 1718, where "he" seems to refer to Barrow, says, "Upon account of my progress in these matters he procured for me a fellowship in Trinity College in the year 1667 & the Mathematick Professorship two years later" (I. Bernard Cohen, *Introduction to Newton's 'Principia'* (Cambridge, Mass., 1971), p. 306, note 14).

the publication at very great trouble to himself."[47] Newton did indeed suggest alternative constructions for two particular problems, one for locating the image point of a lens and the other for the caustic point.[48] At the conclusion of Lecture XII, containing his derivation of the primary radius of the rainbow, Barrow offered some "guesses" about color which, he held, derived from the rarity and density of light together with an intermixture of shadow.[49] Even if Newton had fully explained his theory of color to Barrow and Barrow had fully understood it by early 1669—and this seems unlikely—the publication of Barrow's "guesses" would not have subjected him to any public ridicule as has been frequently suggested, for they were well within the mainstream of contemporary ideas.[50] After all, Newton's ideas were the ones that were so radical and subject to ridicule.

When Barrow's *Lectiones XVIII* appeared in 1669, it was, despite its somewhat tedious presentation, the most advanced treatise on geometrical optics then available. A large part of Newton's investigations of the refractions of monochromatic rays in his own *Optical Lectures* is an extension of Barrow's earlier work. "As for the primary elements of optics," the anonymous editor of the *editio princeps* observed in 1729, "our author everywhere follows Barrow's optical lectures, and Newton pursues further what he wrote about light as a whole and applies it to the different refrangibility of rays, a matter unknown to Barrow but fully approved by him once it was explained to him by our author."[51] Barrow's influence is particularly evident in Part I of the *Optica* in the extended treatment in Section 4 on refraction at curved surfaces, and in the revision of Lectures 8 and 9 in Section 3 where, apparently in an attempt to condense his treatment, Newton abandoned many of his own solutions in favor of Barrow's.

(47) *Lectiones XVIII,* "Epistola ad Lectorem," p. 6.

(48) Newton's two constructions are to be found in Lecture XIII, §XXVI and at the conclusion of Lecture XIV of Barrow, *Lectiones XVIII*; see Lect. I, 13, note (6) and Lect. I, 14, note (3), this volume.

(49) For instance: "*Red* is that which pours out all around condensed and more than usually compressed light, but broken and interrupted by shadowy interstices . . . *Blue* is that which emits a rare light or a light excited with a rather sluggish force, such appear to be bodies which consist of white and black particles arranged alternately . . . " (*Lectiones XVIII,* Lect. XII, §XVII, p. 108).

(50) Thus Barrow's ideas on color are treated seriously, with no hint of ridicule, in the anonymous review of his *Lectiones XVIII* in the *Journal des sçavans* (18 November 1675):268–71. Compare Frank E. Manuel, *A Portrait of Isaac Newton* (Cambridge, Mass., 1968), p. 96; and for a general assessment of this question see I. Bernard Cohen, *Franklin and Newton. An Inquiry into Speculative Newtonian Experimental Science and Franklin's Work in Electricity as an Example Thereof.* Memoirs of the American Philosophical Society, 43 (Philadelphia, 1956), pp. 49–53.

(51) *Isaaci Newtoni, eq. aur. in academiâ Cantabrigiensi matheseos olim professoris Lucasiani lectiones opticae, annis MDCLXIX, MDCLXX & MDCLXXI. In scholis publicis habitae: et nunc primum ex MSS. in lucem editae* (London, 1729), p. viii; hereafter *editio princeps.* The source for Barrow's approval is a letter from Collins to Francis Vernon and Richard Towneley (?) on 26 December 1671: "Dr Barrow tells me that Mr Newton . . . will send up 20 Optick Lectures, which Dr Barrow reckons one of the greatest performances of Ingenuity this age hath affoarded" (*Mathematical Papers,* 3:23); the letter had been published in Latin translation in the *Commercium epistolicum* (London, 1712).

3. The Lucasian Optical Lectures, 1670–1672

Upon his appointment as Barrow's successor to the Lucasian chair in the late autumn of 1669, Newton was confronted with developing a series of lectures to begin the following January. In a natural extension of Barrow's prior series of optical lectures he took the opportunity to make the first formal presentation of his new mathematical science of color. The Lucasian Professor was required to give one lecture for about one hour each week during the term and to submit annually not fewer than ten of those lectures to the Vice-chancellor for deposit in the University Library for public use. Newton complied with this regulation somewhat tardily in October 1674, when he delivered to the Vice-chancellor his *Optica* divided into two parts with a total of thirty-one lectures.[52] According to the marginal notations, the first lecture of Part I was delivered in January 1670, at the beginning of the Lent term, and Lecture 9 of Part I and Lectures 4 and 14 of Part II opened the Michaelmas terms (beginning in October) of 1670, 1671, and 1672. Newton himself, however, retained possession of another, much shorter version of his lectures, the *Lectiones opticae*, which on his death passed with the rest of his papers to his niece, Catherine Conduitt, and thence into the possession of the Portsmouth family. There it lay ignored until 1888 when, together with all of Newton's other scientific papers, it was given to Cambridge University.[53] For nearly another century this manuscript languished unstudied in the Cambridge University Library until Richard S. Westfall in 1963 publicly called attention to it.[54] This shorter version of the *Optical Lectures* is divided into eighteen lectures, with Lecture 1 ostensibly having been delivered in January 1670 and Lecture 9 in July 1670; it is also the earlier version, since virtually all the changes made in it are incorporated into the *Optica*. And not only do the dates of the two versions directly conflict, for according to the *Optica* its ninth lecture was delivered in October 1670, but also their contents differ, for Lecture 9 of the *Lectiones* treats the determination of the index of refraction of monochromatic rays (a topic that according to the *Optica* was treated much earlier, in the first term of the lectures), whereas the ninth lecture of the *Optica* treats the refraction of polychromatic rays.

---

(52) On 21 October 1674 Robert Peachey, the university librarian, recorded on the flyleaf of the volume (ULC MS Dd.9.67) containing the *Optica*: "These Lectures of Mr Isaac Newton ye Mathematick Profissor were delivered by him into ye hands of Dr Spencer ye Vice-chancellor of the University & by Mr Vice-chancellor delivered unto mee for to place in ye University Library according to ye order of ye Founder of that Lecture." For Newton's duties as Lucasian Professor and its statutes see *Mathematical Papers*, 3:xviii–xxvii.

(53) This manuscript (Add. 4002) was clearly described as "An early copy (MS.) of the Lectiones Opticae, Jan. 1669" in *A Catalogue of the Portsmouth Collection of Books and Papers Written by or Belonging to Sir Isaac Newton, the Scientific Portion of Which Has Been Presented by the Earl of Portsmouth to the University of Cambridge*, prepared by H. R. Luard, G. G. Stokes, J. C. Adams, and G. D. Liveing (Cambridge, 1888), Sect. VII, no. 16, p. 48.

(54) Richard S. Westfall, "Newton's reply to Hooke and the theory of colors," *Isis*, 54 (1963):82–96.

The most striking difference between the *Lectiones opticae* and the *Optica* is the interchange of the parts on refraction and color, thereby almost completely inverting the order of the lectures. Both versions begin with a demonstration that the sun's light consists of unequally refrangible rays (the first two lectures of the *Lectiones* and the first three of the *Optica*), but in the *Lectiones* Newton follows this with the "dissertation on colors," and then in Lecture 9 with the "dissertation on the measure of refractions" and (in Lecture 12) "propositions flowing from there." In the *Optica* Newton inverted the order of these now greatly expanded "dissertations" and then divided the work into two parts, with Part I dealing exclusively with refraction and Part II with color. Writing to Henry Oldenburg on 6 July 1672, Newton laid down the injunction that his theory should be examined by "a due Method; the Laws of Refraction being throughly inquired into & determined before the nature of colours be taken into consideration."[55] He advocated this order, which he had already adopted in the *Optica,* apparently because he had based his theory on the decomposition of light into its component colors by their unequal refractions; and an adequate prior mathematical treatment of their refractions allowed the possibility of developing an exact, complete mathematical theory of color founded upon refraction. Nonetheless, in itself the interchange of the two parts had no significant effect on either their content or structure or on their relation to one another.

The most significant difference between the *Lectiones* and the *Optica,* in fact, lies in all the new material Newton added to the *Optica,* thus making it about forty-five percent larger. These additions are divisible into three broad categories, those that present material promised in the *Lectiones* but never actually treated, such as Part II, Lectures 9 (on the colors of natural bodies) and 16 (on the rainbow); those that introduce new subjects, such as Part II, Lectures 3 (on the immutability of color) and 11 (on the musical division of the spectrum); and those that elaborate on the *Lectiones* with new experiments or arguments, such as §§9 and 10 in Part II (on the intersection of inclined spectra). Besides the reordering and the additions, in those articles from the *Lectiones* incorporated into the *Optica* there are a large number of mostly less significant additions, deletions, transpositions, and other manner of revisions. Put more positively, about eighty-five percent of the *Lectiones* was incorporated into the *Optica* with minimal change. The major exceptions are Lectures 9 and 10 of the *Lectiones* (on refraction at plane surfaces), which were completely redone as Lectures 8 and 9 of Part I of the *Optica.*

With two such different versions of Newton's lectures, which—if either—is the truer record of his actual lectures and their dates? It would be naive to believe that either version is a verbatim record of his lectures, for it must be recognized that there is a difference between lectures prepared for students and polished revisions intended for the public, as the statutes of the Lucasian Professor required. Since no other account of his lectures, such as his own or a student's notes, has been found, we must try to establish some guidelines to

(55) *Correspondence,* 1:209.

determine their date, contents, and order of delivery. We must, in particular, not automatically assume that the earlier *Lectiones* is in general a more accurate record. Except for that of the inaugural lecture, January 1670, the dates in neither version can be trusted. In particular, the dates in the *Lectiones* were apparently not initially present in the manuscript but were added later in the same dark ink as the other revisions made in the fall of 1671. Moreover, exhibiting the academic's traditional disregard for the minutiae of university regulations, Newton seems to have divided the text of the *Optica* into lectures and assigned them dates somewhat arbitrarily to satisfy the regulation that at least ten lectures per year be deposited. In the one case where we do have an independent, contemporary check on the date of one of Newton's lectures—John Flamsteed's dated copy of Newton's autograph elementary algebra notes, which were given to him at one of Newton's lectures in "Midsummer 1674"—the topic in question straddles two of the "Octob. 1674" series of lectures in the copy deposited by him ten years later.[56] It therefore seems fruitless to attempt to date the occasions when individual lectures in either version of the *Optical Lectures* were—or, it may be, were not—delivered.

Dating the composition of the two versions of the *Optical Lectures*, however, provides further understanding of their relation. On 30 April 1672 John Collins wrote to Newton: "A little before Christmas the Reverend Doctor Barrow informed me you were buisy in enlarging your generall method of Infinite Series's or quadratures, and in preparing 20 Dioptrick Lectures for the Presse . . ."[57] Since the "20 Dioptrick Lectures" corresponds reasonably well to the eighteen lectures of the *Lectiones opticae*, it seems that Newton was then hard at work in revising them for press as his *Optica*. This agrees with Newton's subsequent recollections in his letter to Oldenburg for Leibniz of 24 October 1676 (the "epistola posterior") that "five years ago when, being urged by friends, I formed a plan to publish a treatise on the refraction of light and on colors [*tractatum de refractione lucis et coloribus*] which I then had ready, I began to think again about these series, and I composed a treatise on them also so that I might publish both together."[58] On a number of other occasions Newton refers simply to a treatise on colors written in 1671 without distinguishing between the *Lectiones* and *Optica*. For instance, on 18 August 1676 he explained to Oldenburg that

> . . . before I wrote my first letter to you about colours [i.e., the "New theory"] I had taken much pains in trying experiments about them &

---

(56) The text of Newton's notes is in *Mathematical Papers*, 5:32–3. Also discussed there on pp. 4–6 is the general problem of dating his lectures on algebra; see in addition Cohen, "Supplement III, Newton's Professorial Lectures," *Introduction*, pp. 302–9. Mamiani, *Newton*, pp. 15–26, also attempts to establish a chronology of Newton's optical lectures.

(57) *Correspondence*, 1:146. The earliest reference to Newton's plan to publish his lectures is in Collins's letter of 26 December 1671, cited in note (51); see also Collins's letter to James Gregory of 23 February 1671/2 in Turnbull, *Gregory Tercentenary Volume*, p. 218. The "generall method of Infinite Series's or quadratures" is Newton's tract "De methodis serierum et fluxionum," which he composed in the winter of 1670–1 and "enlarged" during the following winter; see *Mathematical Papers*, 3:3–9.   (58) *Correspondence*, 2:114.

written a Tractate on that subject wherein I had set down at large ye principall of ye experiments I had tryed; amongst which there happened to be the principal of those experiments wch Mr Lucas has now sent me. And as for ye Experiments set down in my first letter to you, they were only such as I thought convenient to select out of that Tractate.[59]

This "Tractate" must refer to the *Optica,* since many of the experiments contained in the "New theory" and invoked by Lucas, such as those on the colors of natural bodies and looking directly through prisms, are only in that version and not in the *Lectiones.*[60]

We can thus conclude fairly confidently that the first version of the *Optical Lectures,* the *Lectiones,* was completed by about October 1671, when revision of its text for publication had already begun; and that this revision, the *Optica,* was essentially in its finished form by early February 1672, when the "New theory" was communicated to the Royal Society. There are some hints, however, that the copy of the *Optica* that was deposited in the University Library in October 1674 was not written out until some time after early 1672.[61] Over what precise period of time before October 1671 the *Lectiones* was composed is far less certain. It would appear that it was not a very considerable period, if we attach weight to the fact that in later years Newton always spoke of having written but a single treatise on colors and refraction, as if he considered the two versions of the *Lectures* to be two successive attempts at one treatise. Newton evidently had become dissatisfied with his first draft, the *Lectiones,* and set it aside to start anew, for it ends abruptly without any conclusion, and we know that he had decided for methodological reasons to invert the sequence of its two principal parts. Moreover, in the course of the *Lectiones* he announces many topics he intends to treat but never does: in §1 "several things" about the theory and practice of telescopes and microscopes; in §2 the colors of the rainbow; in §28 the colors of natural bodies; in §77 the colors seen looking directly through a prism; and in §124 "some things" on the structure of the eye and vision.[62] It is not uncommon for authors to make unfulfilled promises, but the omission of any treatment of the colors of natural bodies is particularly embarrassing, since it leaves one of his four propositions on color completely unproved. All of these promised topics are taken up—to a greater or lesser extent—in the *Optica.*

---

(59) Ibid., p. 79; in response to Anthony Lucas's letter to Oldenburg of 27 May 1676, ibid., pp. 8–12. Some forty years later Newton wrote that, "In 1666 I came upon the theory of colours, and in 1671 I was preparing a tract concerning this subject [*Tractatum de hac re*], and a second about the method of series & of fluxions intending to publish them. But certain disputes which arose soon after discouraged me from this design till the year 1704" (Cohen, *Introduction,* p. 292).

(60) By a similar argument it is also reasonably certain that Newton is not referring to "Of Colours," as is suggested by Turnbull in *Correspondence,* 2:81, note 5.

(61) See Lect. I, 3, note (8) and Lect. II, 10, note (32).

(62) Another, though certainly less compelling, reason for believing that the *Lectiones* was finished shortly before Newton undertook the revision is that the revisions in the *Lectiones* are made in an ink virtually indistinguishable from that used at the conclusion of the manuscript but noticeably darker than that of the earlier portions.

There they account for a large part of its new material and make it a more comprehensive record of the topics Newton initially intended to cover in his lectures, and probably also of those he actually delivered. The appearance in the *Optica* of a topic not treated in the earlier *Lectiones* does not necessarily imply that it was one to which he had only newly come; for he had started his investigations of many of these subjects even before he began to deliver his lectures in January 1670. For instance, in Part II the colors of natural bodies (Lecture 9), and the colors seen looking directly through a prism and those produced by parallel surfaces (in Lectures 12 and 13) were all touched upon in the second essay "Of Colours"; and in Part I much of the new Section 4 on refraction at curved surfaces is to be found in nascent form in the essay "Of Refractions." On the other hand, Lecture 3 in Part II of the *Optica,* on the immutability of color, represents a conceptual advance beyond the *Lectiones* and most likely was not presented in his earlier lectures. It should, however, always be borne in mind that it is possible that Newton may have repeated some of the series of lectures in other terms and presented yet other ones, for example, on his reflecting telescope, which he chose not to include in the deposited *Lectures,* for there is no record of the lectures he was required to deliver in the other terms of the academic year.

It would, accordingly, be futile for us to attempt anything more in establishing either version of the *Optical Lectures* as an accurate historical record for, as we have seen, the dates in both are unreliable and their contents can serve only as a general outline of his lectures. Our only safe ground is to consider them, as Newton himself did, to be successive drafts of a "treatise on the refraction of light and on colors," the last of which he chose to be the permanent, polished record of his Lucasian lectures for 1670–2.

4. The Fate of the *Optical Lectures* and Its Relation to Newton's Later Optical Writings

Whatever uncertainty exists about the verbatim faithfulness and the dating of the *Optical Lectures,* it is certain that in the winter of 1671–2 Newton intended to publish it, together with his mathematical treatise "De methodis serierum et fluxionum," and that he was then hard at work preparing them for publication. Yet just a few months later he wrote Collins that "Your kindnesse to me also in profering to promote the edition of my Lectures wch Dr Barrow told you of, I reccon amongst the greatest . . . But I have now determined otherwise of them; finding already by that little use I have made of the Presse, that I shall not enjoy my former serene liberty till I have done with it . . ."[63] Newton had in the interim firmly resolved to preserve his solitude and equanimity, which he felt were so disturbed by the controversies following upon the "New theory," and he never relented in his decision to

---

(63) Newton to Collins, 25 May 1672, *Correspondence,* 1:161; in response to Collins's letter of 30 April 1672, cited at note (57).

suppress his *Optical Lectures*. By the time he recovered his equanimity, in late 1675, and once again judged the world to be receptive to his theories, much of the *Optica* was already outdated. In September 1672 Newton had decided to recast his theory in a more formal structure "in imitation of the Method by wch Mathematicians are wont to prove their doctrines."[64] The next year, in outlining his restructured theory for Christiaan Huygens,[65] he recognized that it needed a more rigorous proof, especially the demonstrations of the innateness and immutability of color; and by then he no longer had complete confidence in his dispersion law, upon which so much of his mathematical theory of color, that is, Part I, was founded.[66] Instead, Newton was planning a work very much like the later *Opticks,* to consist of a "discourse about ye colours of ye Prism"[67] (corresponding to Book I) together with the papers on the colors of thin films that he had sent to the Royal Society in December 1675 (corresponding to Book II). In this newly projected work, then, the sections of the *Optica* on color were to be extensively rewritten and its mathematical part omitted. There is no evidence that Newton wrote such a discourse during this period, but when in the early 1690s he eventually composed the *Opticks,* he in essence followed the plan he had proposed in the mid-1670s, though he added a brief third book on diffraction and a number of appended queries. When, after still another postponement, the *Opticks* was finally published in 1704, Newton felt it necessary in the advertisement to warn that, "If any other Papers writ on this Subject are got out of my Hands they are imperfect, and were perhaps written before I had tried all the Experiments here set down, and fully satisfied my self about the Laws of Refractions and Composition of Colours. I have here Published what I think proper to come abroad . . ."[68] He is here *inter alia* surely referring to his *Optica,* deposited thirty years earlier in the Cambridge University Library; "From whence," its anonymous eighteenth-century translator relates, "many Copies have been taken, and handed about by the Curious in these Matters."[69] During his lifetime Newton's disavowal was respected by eager

(64) Newton to Oldenburg, 21 September 1672, ibid., p. 237.

(65) See Newton's letter to Oldenburg for Huygens, 23 June 1673, ibid., pp. 292–4.

(66) See Alan E. Shapiro, "The evolving structure of Newton's theory of white light and color," *Isis,* 71 (1980):211–35, and "Newton's 'achromatic' dispersion law: theoretical background and experimental evidence," *Archive for History of Exact Sciences,* 21 (1979):91–128.

(67) Newton to Oldenburg, 11 May 1676, *Correspondence,* 2:6. Newton first broached this new discourse in his letter to Oldenburg of 25 January 1675/6, ibid., 1:414. For a thorough account of Newton's plans for publication in the 1670s see A. Rupert Hall, "Newton's first book (I)," *Archives internationales d'histoire des sciences,* 13 (1960):39–54; and also Westfall, "Newton's reply to Hooke," and "Newton defends his first publication: the Newton-Lucas correspondence," *Isis,* 57 (1966):299–314.

(68) *Opticks,* p. [iii]. All references to the Opticks will be to the first edition of 1704, unless otherwise indicated, as well as to the manuscript of the *Opticks* (Add. 3970, ff. 17–78, 91–233, 359), which will be published in the third volume of this edition. If a passage is in the manuscript copy—and the advertisement is not—quotations will be taken from that copy.

(69) Isaac Newton, *Optical Lectures Read in the Publick Schools of the University of Cambridge, Anno Domini, 1669. By the late Sir Isaac Newton, then Lucasian Professor of the Mathematicks. Never before Printed. Translated into English out of the Original Latin* (Lon-

members of the Newtonian circle, but an English translation of Part I appeared in 1728, the year after his death, followed in the next year by the *editio princeps* of the complete Latin text of the *Optica*.[70]

These editions of the *Optical Lectures* were published partly as an act of homage—"it is superfluous to recommend this treatise to readers, for why

don, 1728), p. iv. The following manuscript copies of the *Optica* have been identified: Roger Cotes's copy "Descripsi ex Autographo anno 1701/2" (Trinity College Library, Cambridge, MS R.16.39), which also contains copies of Newton's lectures on arithmetic and algebra, and on the system of the world; a copy given by Newton to David Gregory, probably in 1701, and now in private possession according to Whiteside in *Mathematical Papers,* 3:476; an incomplete copy of Gregory's copy made by a Mr. Meldrum and another person (British Library, MS Sloane 3208, ff. 3–45, 68–95; see f. 68ʳ for the attribution to Meldrum and its derivation from Gregory's copy); and a copy whose original owner is unknown (University of Keele Library, MS Turner D2.16). Upon his death in 1716, Cotes's papers, including his copy of the *Optica,* passed on to his younger cousin, Robert Smith. Smith, who died in 1768, bequeathed these papers to Edward Howkins who, in turn, left them in 1779 to Trinity College. After this volume was completed, D. T. Whiteside informed us (21 June 1982) that there is in the Fellows' Library of Clare College, Cambridge, a copy of the *Optica* that belonged to Charles Morgan (Master of Clare College from 1726 to his death in 1736) as well as papers by Cotes and Smith that are related to the *Optica*. In a later volume we will include an account of our examination of these manuscripts. William Whiston may have taken a copy for use in his own lectures; see note (74).

(70) Both editions were rushed into print after Newton's death on 20 March 1727; the preface of the *Optical Lectures Read in the Publick Schools* is dated 29 June 1727, but the publication of the *editio princeps* was delayed for corrections after it was already printed. Whiteside has plausibly suggested that the translator was "Henry Pemberton, though William Jones (or even James Wilson?) is an outside possibility" (*Mathematical Papers,* 3:445, note 1), and they may also be possibilities for the editor of the Latin text. Whoever the editor was, he had his difficulties in establishing an accurate text: "Newton once gave to [David] Gregory, Savilian Professor of Astronomy, a copy of these lectures, from which was made that copy whence our edition is printed. On comparing it with Gregory's copy we found it to have been transcribed with the greatest faithfulness and care, and indeed we had no doubt but that Gregory's copy was perfect, since he had received it from Newton himself. Yet, after our edition had gone to press, we heard that that copy preserved in the Cambridge University Archives was more perfect, and having obtained a copy of it and comparing them we found that to be true. We have therefore noted the differences, together with typographical errors, at the end of the book" (*Editio princeps,* pp. ix–x). This text, with its five pages of "addenda & corrigenda," has been the basis of all later editions (see note (78)) and indeed it is quite serviceable for the purpose, which is more than can be said for the English translation. Assuming that (as it claims to be) the British Library copy is a reasonably accurate copy of Gregory's manuscript, if that copy is compared with the *editio princeps* alone without its appended corrections, it becomes evident that its text is not a faithful printing of Gregory's (imperfect) manuscript and that the editor has silently attempted to repair its obvious deficiencies, for his editorial emendations are not to be found in the British Library copy. Moreover, by comparing the translation with the text alone of the *editio princeps,* it is also evident that the translator used the same edited Gregory manuscript, which suggests that the two men cooperated. This suggestion is reinforced by the fact that Part I of the *editio princeps* used the plates from the translation, although some corrections were made (compare, for instance, Fig. 15 where the quadrant was reversed!). It should also be emphasized that the "addenda & corrigenda" were not prepared from the Cambridge manuscript itself but from a copy of it. In the preface the editor quotes a note at the conclusion of Pt. I, Prop. XVI, Case 1, "written by someone unknown in the margin of the Cambridge codex" (*Editio princeps,* p. xi), but there is no such note in the deposited manuscript or any of the publicly accessible copies that we have examined, so that the *editio princeps* was prepared from a still unlocated copy.

should we praise a work whose author is the great Newton"[71]—but also as an act of scholarship, particularly in an attempt to bring the mathematical part to light. The unidentified person who translated Part I into English did so because "it contains but little in common, with what has been already printed [by Newton] . . . and such as will even at present appear entirely new," while he judged that Part II on "the *Doctrine of Colours* . . . was left imperfect; but has been since published in the *Opticks* by Sir *Isaac* himself with great Improvements."[72] The editor of the Latin edition likewise emphasized the significance of the geometrical demonstrations in Part I, because in the *Opticks* Newton "seems to have been as careful as possible not to mix geometrical demonstrations with philosophical arguments, and where it was necessary to set forth a mathematical proposition, its demonstration scarcely ever occurs."[73] Indeed, the only excerpts from the *Optical Lectures* to be published in Newton's lifetime were three propositions and a lemma that William Whiston, Newton's successor as Lucasian Professor, included in his own Lucasian lectures as an aid to students, "since it hath seem'd good to that great Man, to propose certain Propositions in that Book [the *Opticks*] without their Demonstrations."[74] The editor of the *editio princeps,* however, also perceptively recognized that with respect to color "many things are found in each with the same meaning, but are explained in a different manner."[75]

It is not our aim here to analyze in detail the relation between the *Optical Lectures* and *Opticks.* We should note, however, that though the essential elements of the theory of color and its supporting experiments are substantially the same in the two works, the structure of the theory itself and its demonstration differ in many fundamental ways. One of the more striking differences is the emphasis given to demonstrating unequal refrangibility in the *Opticks,* which devotes the whole of Part I of its first book to this topic, in contrast to the relatively brief treatment of it in the beginning of the *Optical Lectures.* This undoubtedly reflects Newton's firm conviction, foreshadowed in his interchange of the two parts of the *Optical Lectures* in revision, that "the Laws of Refraction be . . . throughly inquired into & determined before the nature of colours be taken into consideration." As a consequence of the difference in basic structure, many of the experiments that in the *Lectures* were used to establish the unequal refrangibility of different colors are made to serve a different purpose in the *Opticks,* namely, to

(71) *Editio princeps,* p. v.

(72) Newton, *Optical Lectures Read in the Publick Schools,* p. iv.

(73) *Editio princeps,* p. vii.

(74) William Whiston, *Sir Isaac Newton's Mathematick Philosophy More Easily Demonstrated . . . Being Forty Lectures Read in the Publick Schools at Cambridge* (London, 1716), p. 267, which is a translation of his *Praelectiones physico-mathematicae Cantabrigiae in scholis publicis habitae. Quibus philosophia illustrissimi Newtoni mathematica explicatius traditur, & facilius demonstratur* (Cambridge, 1710), p. 227. Whiston inserted Lemma 10, and Props. 32 (in part), 35, and 36 in lectures he delivered on 8 April and 2 June 1706. In 1676 William Briggs had recounted two of Newton's explanations in the *Optica;* see Lect. II, 15, notes (8), (9).

(75) *Editio princeps,* p. vi.

establish the existence of unequal refrangibility (for example, §32 ≈ II, 7). Moreover, in accordance with his earlier decision to formulate the theory "in imitation of the Mathematicians," the presentation in the *Opticks* is more formal, following the outline drawn up for Huygens in June 1673 more closely than the *Lectures* or the "New theory."[76] During the intervening years important elements of his theory of color changed. For instance, he clarified the concepts of simple and compound colors, restricted the theory to sunlight and not all white light, and attempted to improve his demonstrations that colors are both innate to the sun's direct light and immutable.[77]

After the publication of the Latin *editio princeps* in 1729, the *Optical Lectures* was published six more times in the eighteenth century: in the second volume of Giovanni di Castiglione's edition of Newton's *Opuscula* in 1744, in the third volume of Samuel Horsley's edition of the *Opera omnia* in 1782, and four times in a combined edition that also contained Latin translations of the *Opticks* and of his optical papers and letters from the *Philosophical Transactions* in 1747, 1749, 1765, and 1773.[78] Even a cursory examination of the literature on optics in the century after 1729 shows that the *Optical Lectures* played a subsidiary or supplementary role to the *Opticks*, which by far dominated that literature. This, of course, is to be expected, for the *Opticks* treated a broader range of optical phenomena and was judged to have superseded the *Optical Lectures*, particularly since in its advertisement Newton disavowed all his previous writings on optics. The *Opticks* also contained a bare minimum of geometrical optics.

During the more than half century that passed before Newton's *Optical Lectures* was finally published, many of its most notable mathematical achievements were to be discovered and made public by others: for example, the investigation of spherical aberration by Huygens, the determination of the general diacaustic condition by Jakob Bernoulli and Guillaume de L'Hospital, and the computation of the radius of a rainbow of any order by Edmond Halley.[79] Consequently, when it was published, the *Lectures* seemed far less original and interesting than it would have been fifty years earlier. And just nine years after the publication of the *editio princeps*, Robert Smith brought out his *Compleat System of Opticks*, a compendious treatise that assimilated

---

(76) See Mamiani, *Newton*, pp. 191–212.   (77) See Shapiro, "Evolving structure."

(78) The 1729 *editio princeps* is reprinted in *Isaaci Newtoni, equitis aurati, opuscula mathematica, philosophica et philologica*, ed. Giovanni di Castiglione, 3 vols. (Lausanne/Geneva, 1744), 2:73–275; and again in *Isaaci Newtoni opera quae exstant omnia*, ed. Samuel Horsley, 5 vols. (London, 1779–85), 3(1782):249–437. Horsley, it should be noted, altered the correct article numbers of the *editio princeps*. For the combined editions see Peter Wallis and Ruth Wallis, *Newton and Newtoniana 1672–1975. A Bibliography* (Folkestone, Kent, 1977), nos. 183, 184, 184.5, 185; we have seen only 184 (Padua, 1749) and 185 (Padua, 1773), which bear the same title, *Isaaci Newtoni optices libri tres: accedunt ejusdem lectiones opticae, et opuscula omnia ad lucem & colores pertinentia sumpta ex transactionibus philosophicis*. The only complete translation of the *Optica* is in Russian by S. I. Vavilov and is based on Horsley's edition, *Lekcii po Optike* (Moscow/Leningrad, 1946).

(79) See Lect. I, 13, note (13) for Huygens; Lect. I, 14, note (2) for Bernoulli and L'Hospital; and Lect. I, 15, note (5) for Halley.

and advanced much of the earlier mathematical literature on optics, and thus made it virtually superfluous.[80] Though the appearance of the *Compleat System* made it even less likely that anyone would turn to Newton's *Optical Lectures* for enlightenment, Smith did incorporate a number of Newton's results that otherwise might have been forgotten. Thomas Young, for instance, was led back to Newton's pioneering investigation of astigmatism through Smith.[81] In the century following its publication the *Optical Lectures* was cited often enough, usually on subjects not at all or only superficially treated in the *Opticks,* such as the colors seen when looking through a prism and the eye's chromatic aberration.[82] Thereafter, particularly with the complete rejection of Newton's emission theory of light by the mid-nineteenth century, his *Optical Lectures* was all but forgotten and consigned to the bookshelves with other "dead" scientific literature. Today we can appreciate it as an invaluable document of Newton's investigations of optics that reveals his ideas in the midst of his most productive period of research. In the inevitable comparison with the *Opticks,* which recounts research for the most part carried out twenty to thirty years earlier and since refined—sometimes overrefined—the *Lectures* must be judged neither as carefully developed nor as polished. But whatever polish it may lack is more than compensated for by its vitality, as Newton boldly attempts in the following pages to create a new, mathematical science of color.

(80) Robert Smith, *A Compleat System of Opticks in Four Books, viz. A Popular, a Mathematical, a Mechanical, and a Philosophical Treatise. To Which Are Added Remarks upon the Whole,* 2 vols. (Cambridge, 1738). The many translations of this work, particularly Abraham Gotthelf Kästner's German in 1755, but also the Dutch in 1753, and the two French in 1767, were in part responsible for its broad influence.

(81) See Lect. I, 14, note (6).

(82) For example, see Lect. II, 12, note (10), and Lect. II, 14, note (14), this volume. Readers of the posthumous "fourth edition" of the *Opticks* (London, 1730) or the edition that Horsley included in Newton's *Opera omnia,* Vol. IV (London, 1782) would also have been directed to the *Optical Lectures* by the many references to it inserted in these editions.

Synopsis of the
Lectiones opticae and *Optica*
and Their Major Differences

This synopsis of the *Optical Lectures* in modern terminology is intended as a preliminary guide, since Newton's aims and strategy in each lecture, or group of lectures, are not always immediately apparent. It also describes and discusses in one location the significance of all the major differences (that is, a sentence or more) between the two versions. The order of presentation follows that of the *Lectiones,* with a description of material added in the *Optica* inserted at the corresponding location.

Lectures 1, 2 ≈ Lectures I, 1–3 (§§1–23 ≈ I, 1–24)

After briefly paying tribute to Barrow and deriding efforts to improve refracting telescopes by the use of nonspherical lenses, Newton devotes the first two lectures to laying the foundation for the whole of his *Lectures*: a demonstration that direct sunlight consists of rays that differ in their degree of refrangibility.[1] Virtually the entire burden of his demonstration is borne by an analysis of the elongated spectrum formed by passing a narrow beam of sunlight through a prism (§3 = I, 3). Newton's major insight, and the key to his demonstration, was to recognize that when a prism is placed symmetrically with respect to the incident and emergent beams, or at minimum deviation, the sun's image would be circular rather than elongated if all rays were refracted equally (§4 = I, 4). An exact solution for the shape of the sun's image with monochromatic rays is exceedingly difficult, involving a finite source and aperture and rays incident out of the principal plane; but he is able to demonstrate that under particular conditions, such as with a point aperture, the image is nearly circular (§§5–7 = I, 5–7). This was sufficient for his purpose, for he had found the spectrum's length to be five times its breadth (§§8, 9 = I, 8, 9), thus making small deviations from the assumed conditions inconsequential.

Newton begins Lecture 2 by describing the shape of the spectrum to be an

---

(1) §2 ≈ I, 2: Newton advanced a broader concept of light ray that applies to a pencil of rays and not simply to a single line; in the *Lectiones opticae* this new concept was not introduced until §21 = I, 22.

oblong bounded by straight edges and semicircular ends and, he argues, formed by innumerable overlapping circular images of the sun, each consisting of rays of a different refrangibility (§§11–12 ≈ I, 11–12).[2] To justify this interpretation he shows (§§13–16 ≈ I, 13–16)[3] how the straight edges of the spectrum can be made more distinct by using a lens, while also explaining why the semicircular ends always remain confused.[4] The thrust of the remainder of the lecture describes how to decrease the effective size of the source, and thus the circular images, and to approach the ideal spectrum—a straight line with no breadth—formed by a point source. By this mode of demonstration, culminating in the observation of Venus's spectrum (§§19–21 = I, 20–2), he makes the actually observed shape of the sun's spectrum inessential to his proof that its elongation is caused by unequal refrangibility. Newton ends this portion of the *Lectures* with arguments and experiments—significantly expanded in the *Optica* (§22 ≈ I, 23–4)[5]—against explanations of the elongation that attribute it to anything but unequal refrangibility. In the last article he adduces additional experiments in favor of his theory, which include an observation of the spectrum of a star that is evidently a prediction and not an actual observation (§23 ≈ I, 24).[6]

Lecture 3 ≈ Lectures II, 1, 2 (§§24–39 ≈II, 1–18)

At this point, which is the beginning of the "Dissertation on colors" of the *Lectiones opticae* but Part II of the *Optica*, Newton instituted the most

---

(2) §11 ≈ I, 11: When Newton changed the division into lectures in the *Optica*, he deleted the opening paragraph of Lecture 2 summarizing the preceding lecture, and he retained only the last sentence, with some revisions, for the opening of §I, 11.

(3) §14 ≈ I, 14: Newton eliminated his observation in the second deleted sentence that the length of the central portion of the spectrum is about three times its breadth, whereas its total length is more than five times greater when the fainter ends are included; the first deleted sentence was incorporated into the revision of §I, 14.

(4) §I, 17, New: An observation of the shadows of clouds cast through the entire length of the spectrum was added as additional evidence that the spectrum is formed by overlapping circular images of the sun.

(5) §22 ≈ I, 23–4: Newton added an extensive series of arguments and experiments, including his crossed-prism experiment, to reject such causes for the elongation of the spectrum as the varying thickness of the prism and the size of the hole, as well as alternative explanations to unequal refrangibility such as a diffusion or splitting of the rays. Also, he transferred the concluding paragraph of §22 without any change to the beginning of §I, 24.

(6) §23 ≈ I, 24: The concluding sentences of §23 were rewritten in §I, 24 to reflect the interchange of the two parts of the *Lectures* that occurs here. In evaluating the experiments in the *Optical Lectures* one should always bear in mind Newton's caveat in the *Opticks*, for it seems to apply equally to the *Lectures*: "in the description of these Experiments I have set down such circumstances by wch either [1] the phaenomenon might be rendred more conspicuous or [2] a Novice might more easily try them or [3] by which I did try them only . . . Concerning all wch this one admonition may suffice" (*Opticks*, Bk. I, Pt. I, Prop. I, Scholium, p. 17 = Add. 3970, f. 30^r). Consequently, it cannot be taken for granted that Newton performed all of his experiments as described, for his account may fall into one of his first two categories.

striking change of the revision by interchanging the order of the *Optical Lectures* so that the quantitative and mathematical theory of refractions precedes the theory of color, thus allowing him to develop his theory of color as a mathematical theory. He begins the "Dissertation on colors" in §24 ≈ II, 1[7] by reiterating his inaugural remarks on the defects of contemporary telescopes and the impediment presented by chromatic aberration, and in prelude to his own theory he vigorously attacks both Aristotelian and more recent modification theories of color (§§25–6 = II, 2–3). He then presents his theory in four propositions, which, however, he fundamentally reformulated in the *Optica* in five propositions: He added two new ones, revised two, omitted one, and left only one unchanged (§28 ≈ II, 5). The first proposition, that to differently refrangible rays there correspond different colors, is the same in each version, though the statement in the *Optica* is more complete. The converse of this, that rays differing in color differ in refrangibility— Proposition 2 of the *Lectiones opticae*—is formally omitted in the *Optica* but is nonetheless proved exactly as in the *Lectiones*. The new Proposition 2 of the *Optica*, which establishes the immutability of color, represents the most fundamental addition to the theory, since it is the basis for his proof in the following proposition that colors are innate to sunlight. Proposition 3, on the compound nature of white and other colors, is the same in each version, although it is reformulated in the *Optica*. Proposition 4 in the *Optica*, that spectral colors can be compounded of their neighboring colors, is new and contains a definition of "primitive" or monochromatic colors, which were left undefined in the first version. Finally, the last proposition, on the colors of natural bodies, is essentially identical in the two versions. Newton avows (§29 ≈ II, 6)[8] that he will treat these propositions "by experiments or demonstratively," not "hypothetically," and makes a powerful plea for mathematical natural science, while offering his new, mathematical theory of color as an example of the value of mathematics in natural philosophy. Thus, at the beginning of his career he had already clearly formulated a program for the reform of natural science that would come to full fruition in his *Philosophiae naturalis principia mathematica*, that is, *The Mathematical Principles of Natural Philosophy*.

Newton rapidly disposes of Proposition 1 of the *Lectiones* in §30, since he had already established it in the opening lectures; and in §31 he turns to its converse, Proposition 2, that different colors are unequally refracted.[9] To demonstrate this he introduces his crossed-prism experiments (§§32–6 ≈ II,

---

(7) §24 ≈ II, 1: In a new sentence Newton elaborates upon a vague comment, by observing that if unequal refrangibility were known, the defects of telescopes would not be attributed to the spherical shape of lenses.

(8) §29 ≈ II, 6: In a straightforward transposition the first sentence of §30 was in the *Optica* made the last sentence of §II, 6, which otherwise is the same as §29.

(9) §§30, 31 → II, 7; and §32 ≈ II, 7, 8: Apparently intent on simplifying the formulation of his theory by combining Props. 1 and 2 of the *Lectiones* into one proposition in §II, 7, Newton combined and rewrote §§30, 31, and parts of §32. But since Prop. 1 does not entail its converse, he was compelled to reintroduce the former Prop. 2 in the addition to §II, 7, which thereafter proceeds through §II, 8, just as §32.

8–14),[10] where spectra cast on a second, transverse prism become inclined to their initial orientation because the blue end is always refracted more than the red. Initially he places the second prism transverse to the first one to minimize the unequal incidence arising from the refraction of the first prism; but by passing the refracted rays through two holes far apart so that they fall on the second prism at very nearly the same angle of incidence, he eliminates the requirement for any particular orientation of the second prism and arrives at an experimental arrangement virtually identical to the *experimentum crucis* of the "New theory" (§§37–8 = II, 15–16).[11]

Lecture II, 3 (§§II, 19–26)

In the *Optica* Newton here inserts his new Proposition 2, on the immutability of monochromatic colors. He establishes this by first separating the spectral colors from one another and then demonstrating that the more completely they are separated the smaller are their changes after additional refractions. He first separates the colors with two parallel prisms and observes some color change, because the adjacent colors are still intermingled, but when he adds two more prisms, he is unable to detect any further sensible change (§§II, 19–20). He then shows that by using a lens and narrower holes the resolution of the spectral colors is so improved that with just two prisms only various gradations of the same color appear (§§II, 21–2); and he argues that in the limit of indefinitely small holes, there would be no color change whatsoever. To study the nature of the monochromatic colors in greater detail and better resolve them (§§II, 23–5), Newton replaces the circular aperture with a triangular one. Yet he finds that the colors cast through the vertex are sensibly the same as those cast through the base where they overlap, and he identifies twelve of the "more prominent" gradations of monochromatic color observed in this way. He completes the proposition by noting that the color of light cannot be changed by reflection from colored bodies.

Lectures 4, 5 ≈ Lectures II, 4–7 (§§40–63 ≈ II, 27–54)

In these lectures Newton carries out the first part of his demonstration of Proposition 3, that white, in particular sunlight, is composed of rays of every

(10) §33 ≈ II, 9; and §§II, 10, 11, New: In the sentence added at the end of §II, 9, Newton correctly notes that the inclined spectra cast by varying the distance between two prisms intersect at a common point, but in the new §II, 10, he erroneously argues that that point is the place of the sun's direct image. In §II, 11, he returns to a point already made (§I, 23 ≈ 22), that the crossed-prism experiments establish that the refractions occur according to a definite law and not by a diffusion or expansion, and he adds that the refractions are the same both before white light is separated into colors and afterward.

(11) §II, 17, New: In the *Optica* Newton added a new experiment that uses a mirror rather than two holes far apart to ensure that all rays fall on the second prism at the same angle of incidence.

color,[12] by showing five different ways to make white from a mixture of spectral colors: (i) colors from three prisms are cast onto a screen where they are mixed (§§41–2 ≈ II, 28–9); (ii) one face of a prism is covered with an opaque paper with six slits, each functioning as one of the prisms in the preceding experiment, and then the colors from the various slits mix on a screen (§§43–4 = II, 30–1); (iii) light scattered from a screen on which a spectrum has been projected is received on a second screen where the scattered rays mix (§§52–3 ≈ II, 41–2);[13] (iv) the colors dispersed by a prism are transmitted through a lens and brought together at its focus (§§54–8 = II, 43–7); and (v) in a variant of the preceding way a mirror is substituted for the lens (§59 = II, 48).[14] He also illustrates the compound nature of white by a mixture of colored powders (§51 = II, 39), and in the *Optica* (§II, 40) by a froth of soap bubbles.[15]

Newton devotes considerable effort to countering the possible objection (drawn from Descartes's and Hooke's theories) that colors are destroyed in the generation of white and are not simply mixed (§45 = II, 32). He presents numerous arguments and demonstrations opposed to this view, especially in §§46–51 ≈ II, 33–9, which analyze the second method of mixing white, and two particularly significant ones emerge. First, according to the principle of superposition, rays of different color cannot act upon and thus destroy one another when they cross, for otherwise all visible colors would blend together and distinct vision would be impossible (§§48 = II, 35 and 53 ≈ II, 42). The second fundamental argument to emerge is essentially a geometrical one, namely, that whether white or colors appear depends on the inclination and coincidence of the rays. Thus if a hole is made in the screen where in the second method the colors from the slits meet to produce white, he shows that upon emerging from that hole the light rapidly separates into colors, which could only have been mixed and not destroyed in compounding the white

---

(12) §§40 → II, 27, 28; and 41 ≈ II, 28: The differences in the introductions to Prop. 3 (§40 → II, 27) are not substantive, and a new transition was partly necessitated by the addition in revision of Lecture II, 3. If we are to take Newton literally, however, two minor historical points are revealed: He originally intended to explain the origin of prismatic colors before demonstrating Prop. 3 (§40); and the various ways of compounding white are presented in their order of conception (§II, 27). In the *Optica* the final sentence of §40 became the opening one of the next paragraph, §II, 28. Also in the *Optica* this lecture is dated October 1671, whereas according to the *Lectiones* it was delivered between January and July 1670.

(13) §§52–3 ≈ II, 41–2: Besides eliminating the introductory sentence from the first paragraph of §52, in §II, 41, Newton added a summary of the ways in which he had mixed white. In §II, 42, he added an observation that if any color is removed from a mixture compounding white, the white vanishes.

(14) §II, 49, New: Newton suggests that instead of a lens or mirror, two prisms may be used to focus the prismatic colors.

(15) §§89, 91 ≈ 90, 92: Newton adds two more ways to make white, which he did not remember earlier. He probably related so many "ways of producing *whitenesse by mixtures*" because, as he explained to Hooke on 11 June 1672, "I perswade my selfe that this assertion above the rest appears *Paradoxicall*, & is with most difficulty admitted" (*Correspondence*, 1:183 = *Phil. Trans.*, 7 (1672):5099).

($50 = II, 37).[16] In the last articles of this pair of lectures ($$60–3 ≈ II, 51–4),[17] Newton argues that the white light evident immediately upon emergence from a prism is compound even though the colors are not yet apparent. Perhaps his most ingenious demonstration that colors are not destroyed in generating white is one which he added in the *Optica* ($II, 50): When a toothed wheel inserted in a light beam between a prism and a lens is turned slowly a succession of different colors is perceived at the focus, but when it is turned rapidly white is seen there, even though all colors are never simultaneously present.

Lecture 6 ≈ Lectures II, 7, 8 ($$64–76 ≈ II, 55–70)

Newton now applies himself to the second and more difficult part of his demonstration of Proposition 3, namely, to show that the sun's direct light is compounded of colors even before they are apparent ($64 ≈ II, 55).[18] He bases his demonstration on the phenomenon of total reflection, for, as he discovered, the critical angle of reflection varies for each color. In the first and simplest experiment ($65 = II, 56), a beam of sunlight is partially reflected and partially refracted at the base of a prism. As the prism is rotated the colors are totally reflected in sequence, and the reflected and transmitted beams change color until, when the red rays are at last totally reflected and the transmitted beam vanishes, the reflected beam is restored to white. Newton argues ($$65–6 = II, 56–7 and $70 = II, 61), implicitly appealing to an emission theory of light, that this reveals that the colors are in the rays as they arrive from the sun, since they preserve and exhibit the same color whether they are reflected or refracted. Furthermore, this shows that reflected light is compound, since white is restored when the last color, red, is totally reflected. To make this interpretation still more certain he introduces three variants of this basic experiment ($$67–9 = II, 58–60), one of which is an exact analog of the *experimentum crucis,* but with total reflection replacing the second refraction.

In a variation on the initial experiment ($$71–2 = II, 62–3), which Newton considers "visually more pleasing while being equally scientific," he forms a parallelepiped from two triangular prisms to establish that light transmitted through parallel surfaces is compound. After briefly demonstrat-

(16) $II, 38, New: Extending a technique of the preceding article ($II, 37 = 50) that depends on the inclination of the rays to one another, Newton inserts a mirror so obliquely to the rays at the point where they compound white that the reflected light becomes, with a very slight change of the mirror's inclination, either red or blue.

(17) $60 ≈ II, 51: The revision of the opening sentence is strictly stylistic.

(18) $64 ≈ II, 55: In the revised introduction to the demonstration that colors are innate to the sun's direct light, Newton suggests that since the different degrees of refrangibility are innate to the sun's direct light and the same color always corresponds to the same degree of refrangibility, it is only to be expected that colors too are innate, as they both probably depend on some common cause.

ing that light transmitted through two contiguous parallel plates is also compound ($\S\S73-5 \approx$ II, 64–5),[19] he draws a rather modest conclusion. He claims that these experiments prove, at the least, the compound nature of reflected light, light transmitted through parallel surfaces, and, he adds in the *Optica*, refracted light. But, he urges, no reasonable person doubts that such light is identical to the sun's direct light, and it too must be compound. Now he abruptly introduces a new, more certain argument, founded on the principle of color immutability and unrelated to everything that preceded in the lecture: Since colors are absolutely immutable by reflection or refraction— and this is not proved in the *Lectiones opticae*—and a beam of sunlight exhibits colors after refraction, then it necessarily follows that those colors are innate to the light just as it comes from the sun, though they are not yet apparent. This conclusion ($\S\S75-6$) was completely rewritten and clarified in the *Optica* ($\S\S$II, 66–7). Newton now stresses the practical significance of his demonstration that all reflected and refracted light is compound, since light used in experiments is always of this sort. His most important advance, however, was to recognize that color immutability must be experimentally established for the "primary" refraction before colors appear, whereas in Proposition 2 it had been demonstrated only afterward.

The exposition of the theory of color in the *Lectiones* terminates at this point with the final proposition on the colors of natural bodies left unproved. In the *Optica* Newton concludes the proof of Proposition 3 by briefly explaining why the sun's light is yellowish rather than white ($\S$II, 68), and then by showing that black is compounded from all colors, gray from white and black, and all other compound colors from the painters' primaries, red, yellow, and blue ($\S$II, 69). Despite the need for some restrictions and the brevity of its demonstration, Proposition 4, that spectral colors can be compounded from their neighboring colors, is an important contribution to the theory of compound colors and displays Newton's keen experimental skill ($\S$II, 70).

Lecture II, 9 ($\S\S$II, 71–6)[20]

Newton now turns to his fifth and final proposition, that natural bodies derive their color from the sort of rays they reflect most. By the principle of color immutability, the color of a ray cannot be changed by reflection, so that bodies can appear of only the color of the rays illuminating them ($\S$II, 71). To explain why all bodies are not therefore the same color in daylight, as this principle alone would demand, he adds that bodies reflect more rays of their own daylight color than others ($\S$II, 72). After demonstrating this by illuminating various bodies with monochromatic light, he moves beyond this

(19) $\S\S74-5 \approx$ II, 65: In revision $\S74$ and the first two sentences of $\S75$ were combined with no fundamental change into one, more coherent paragraph in $\S$II, 65, and a concluding sentence directed against modification theories was added.

(20) For $\S\S$II, 77–80 $\approx$ 93–6, which complete this lecture, see note (27).

phenomenological account and attributes two distinct powers to bodies: to reflect rays and to transmit them (§II, 73). These rays are complementary, for the rays that are not reflected pass through the body, and he illustrates this with the colors of such substances as gold leaf, which reflects yellow light and transmits blue. Newton did recognize that most bodies are not of this sort but are the same color all around, and to explain this he introduces a third power—and a new concept in optics—selective absorption (§II, 74). He completes the proposition by explaining several previously improperly or unexplained phenomena, for instance, that passing light through superposed glass plates of different colors is not true, additive mixing but "subtractive" color mixing (§§75–6).

Lectures 7, 8 ≈ Lectures II, 9, 10 (§§77–96 ≈ II, 77–92)

This concluding pair of lectures on color is the most heavily revised of the first part of the *Lectures*. None of the many changes is in itself of great significance, yet together these changes result in a clearer, more polished, and better-organized presentation. Lecture 7 presents no new principles but applies those already introduced to explain the spectra cast by prisms onto a screen. The first sequence of articles is an elementary account of the formation of the spectrum and in all probability was omitted from the *Optica* because it was so elementary and largely repeated earlier material (§§78–81).[21] The next sequence is a more interesting application of his theory and treats the boundary colors produced when a beam of light is terminated on only one side (§§82–3 ≈ II, 82–3).[22] Passing light through a broad slit, which is gradually narrowed until a complete spectrum is formed, Newton explicates the transition from the polychromatic boundary colors to the monochromatic spectral colors, while emphasizing that the spectral green that emerges is a monochromatic color and not, as was then generally believed, compounded of the adjoining yellow and blue (§84 ≈ II, 84).[23] He then considers other terminations of the incident beam, until he arrives at the canonical case of a small circular hole (§§85–8 ≈ II, 85–8).[24]

Lecture 8 contains a miscellany of topics. It begins with a new way to compound white from two adjacent prisms that serves to counter the view that colors derive from a mixture of light and shadow (§§89–90 ≈ II, 90–

---

(21) §77 → II, 81: Newton completely rewrote and improved this introductory article, but the changes are essentially stylistic. In the *Lectiones* he states that he will explain the formation of prismatic colors because of their renown and for the sake of completeness; whereas in the *Optica* he more candidly admits that he will treat them for the sake of "the careless and those burdened by prejudice" who might otherwise have difficulty with them.

(22) §82 ≈ II, 82: The new introduction to this article better motivates the study of boundary colors, but there are no substantive changes.

(23) §84 ≈ II, 84: Newton added a more precise specification of the relation of green to the other spectral colors.

(24) §87 ≈ II, 87: The changes are again just stylistic.

1);[25] this is followed by a somewhat awkward experiment—never again invoked—to produce white by reflecting complementary boundary colors so that they coincide (§91 = II, 92). In a brief parting shot against "the hypotheses of the philosophers," namely, Descartes's and Hooke's theories, Newton shows that by intercepting any part of the beam he can make any color appear on either side, independent of the boundary of light and shadow (§92 ≈ II, 89).[26] He concludes by describing an eight-foot-long instrument, made from two prisms and an intervening lens, to disperse and then recombine the sun's light (§§93–6 ≈ II, 77–80).[27] By means of this instrument all experiments performed earlier with the sun's direct light may be repeated with this reconstituted—and demonstrably compound—light, but in the *Optica* (§II, 80 ≈ 96) he concedes that the instrument is very troublesome and its effects are neither distinct nor evident.

Lecture II, 11 (§§II, 93–101)

Newton commences his musical division of the spectrum by carefully measuring both the position of the boundaries and the place of greatest perfection of each of the five prinicipal colors (§§II, 94–6). Judging the five colors not to be "elegantly proportioned to one another" and desiring to establish "a more refined symmetry," he adds two new colors: orange between red and yellow, and indigo between blue and violet (§II, 97). He then "discovered" that the spaces occupied by these seven colors appeared to be divided just as a string sounding the individual degrees of the octave (§II, 98). He freely admits that the colors' boundaries correspond equally well to a string divided geometrically (that is, an equal-tempered scale, which, however, Newton did not consider to be "musical") and that he could not detect the difference between the two divisions. Nonetheless, he prefers the musical division because of the analogy between the harmonies of colors and sounds (§II, 99). Finally, he explains that the musical division can serve, at least approximately, as a partial dispersion law; that is, in all substances the dispersions,

---

(25) §89 ≈ II, 90: In revision Newton deleted the opening remark that he had not remembered this new method of compounding white earlier and so could not then satisfactorily reject the view that colors arise from a mixture of light and shadow.

(26) §§92–3 ≈ II, 89: Shifting this article to immediately follow §II, 88 = 88 improves the structure of the *Lectures*, since its argument depends directly on the experiment presented in the preceding articles. Because of this transposition the initial sentence of §93 was moved to the conclusion §II, 89. The textual changes that were made only in part clarify and improve his argument against "the philosophers."

(27) §§93–6 ≈ II, 77–80: The organization of the *Lectures* was also improved by this transposition, for the description of this instrument should immediately follow the conclusion of the formal theory, since it is intended as a means of repeating with reconstituted sunlight all earlier experiments with direct light. With the transposition Newton rewrote the introductory remarks to this experiment in §II, 77 ≈ 93 without any substantive changes; but in §II, 79 ≈ 95 he added an experiment on the color of cinnabar in lights of various colors, and in §II, 80 ≈ 96 he acknowledged the instrument's limitations.

dn, of the seven colors have the same proportion to one another as the division of the spectrum (§II, 100).

Lecture II, 12 (§§II, 102–17)

Continuing to apply his theory to a variety of phenomena, Newton now moves from projected images to the more difficult case of images viewed directly through a prism. He begins with the only explanation he would ever publicly present of the boundary colors seen through a prism on the edges of objects (§§II, 103–6), perhaps the phenomenon most commonly invoked by his eighteenth-century critics. In rapid succession he explains a variety of images seen through a prism—such as the separated or broken image of a thread painted blue on one half and red on the other—and the complex displacements that images of projected spectra can be made to undergo depending on the orientation of the viewing prism (§§II, 107–14). Finally, he describes the striking phenomenon of the bluish bow seen at the base of a prism when daylight is reflected internally from its base (§§II, 116–17).

Lecture II, 13 (§§ II, 118–26)

Newton rejects the belief of "the philosophers"—again, Descartes and Hooke—that no colors are produced when light passes through parallel refracting surfaces by showing that colors are, in fact, produced by parallel surfaces, even if they are very slight and easily escape notice (§§II, 118–20). To observe such small effects, he suggests using glass vessels having parallel sides and filled with about one foot of water. After describing the colors seen when objects are viewed through a parallelepiped (§§II, 121–2), he then summarizes the differences between colors generated by prisms and parallelepipeds (§II, 123). He concludes this short lecture by explaining various phenomena observed with prisms (§§II, 124–6), for instance, that light reflected internally from the base of a large isosceles prism generates colors no differently than light transmitted through a parallelepiped.

Lectures II, 14–16 (§§II, 127–40)

In the concluding section of the *Optica* Newton considers the colors generated by refractions at curved surfaces, namely, lenses, the eye, and raindrops, or the rainbow. He first describes the chromatic aberration of a plano-convex lens and gives a simple physical derivation and numerical estimate of its magnitude (§§II, 127–8). Observing that the eye is a lens of sorts, which should likewise suffer from chromatic aberration, he presents a simple experimental demonstration of its existence (§§II, 129–32). In the last article of Lecture 14 and in all of Lecture 15 (§§II, 133–5) Newton indulges in the sort

of speculative or hypothetical natural philosophy that he frequently and vigorously decried yet could not always resist. Exhibiting a firm command of Cartesian natural philosophy, he explains the cause of the colored circles or coronas that Descartes saw around a candle after he had pressed his eye shut for a long time. While Newton recognizes that an infinity of causes may be devised to explain these colored circles, he ascribes them to refractions in wrinkles impressed on the cornea and, invoking the principles of hydrostatics, rejects Descartes's own suggestion that they are impressed on the crystalline lens. He concludes the *Optica* in Lecture 16 with a far more notable achievement, an explanation of the dimensions and colors of the rainbow based on the mathematical results derived in Part I.

Lectures 9, 10 ≈ Lectures I, 4–6 (§§97–114 ≈ I, 25–41)

We return to the *Lectiones opticae*. Newton begins his "dissertation on the measure of refractions,"[28] which constitutes the next three lectures, with an explanation of Descartes's sine law of refraction, which he extends—without experimental demonstration—to rays of each color (§§98–100 ≈ I, 25–7).[29] He then proposes two new methods to measure the index of refraction. First, for fluids he presents his "beam and bucket" refractometer, consisting of a glass-bottomed bucket on a very long wooden beam, but he finds this instrument "troublesome" and gives no measurements made with it (§101 = I, 28). Next, in two lemmas he derives the equations for his preferred method to measure the index of refraction, that of minimum deviation in prisms, one of his most important contributions to quantitative experimental optics (§102 = I, 29). After relating a measurement for the mean-refrangible or green rays made with a prism and a one-foot quadrant (§103 ≈ I, 30),[30] he carefully explains the principal advantage of this method: Small errors in the position of the prism are inconsequential, since at minimum deviation the errors too are at a minimum (§104 = I, 31). He closes Lecture 9 by deriving the now common rule for determining the relative index of refraction of any two media, such as glass and water, when their indices with respect to a third medium, such as air, are known (§§105–6 = I, 32–3).

Newton opens Lecture 10 by extending the method of minimum deviation to fluids with the use of hollow prisms with glass sides, and he illustrates this method by a measurement of the mean index of refraction of water (§§107–8 = I, 34–5). To establish the validity of this method he demonstrates that

(28) §97, Omitted: The introductory paragraph to "the dissertation on the measure of refractions" became unnecessary when in revision the two parts of the *Lectures* were interchanged.

(29) §98 ≈ I, 25: The only substantive, albeit minor, changes in this rewrite are that in the *Optica* Newton added that "men of old" considered not only the angles of incidence and refraction but also the deviation, to be approximately in a constant ratio, while he deleted the statement that for glass they took that ratio to be three to two.

(30) §103 ≈ I, 30: In the *Optica* Newton added a measurement for the index of refraction for the rays between green and blue.

the glass plates do not alter the total refraction of the water alone (§109 = I, 36). He then advances to the next phase of his investigation of refraction: to determine the indices of refraction of the extreme rays, or the chromatic dispersion (§§110–11 ≈ I, 37–8).[31] Rather than readjusting the prism and separately determining the index of refraction of the extreme rays at minimum deviation, he simplifies the experiment and calculations and finds these values simultaneously: When the prism is placed at minimum deviation for the mean refrangible rays, he measures the length of the spectrum and thereby determines the angular dispersion. He presents a sample measurement and calculation for the dispersion of glass. The lecture ends with a demonstration of the simple rule that, when two media are traversed in one direction, the index of refraction is the reciprocal of that when they are traversed in the opposite direction (§§112–14 = I, 39–41).

Lecture 11 ≈ Lecture I, 7 (§§115–23 ≈ I, 42–50)

Newton concludes his "dissertation on the measures of refractions" in Lecture 11 by setting forth a dispersion law (§115 = I, 42), which serves as the foundation for much of the rest of the *Lectures*. He freely admits that it is a purely theoretical construct that he has not yet experimentally tested (§116 → I, 43).[32] Though he presents his dispersion law solely in mathematical terms without any mechanical interpretation, it is evidently a modification of Descartes's projectile model for a single sort of ray extended to apply to polychromatic rays. It represents the very ideal of a rational optics, for the indices of refraction of rays of every color in any medium can be determined with only a single measurement, as Newton illustrates with water (§117 = I, 44). The only experiments that need to be performed, but once, are a measurement of the dispersion in any one medium to determine the parameters of the model (he uses his earlier measurement for glass) and then the mean index of refraction of the medium whose dispersion is required. He invokes this dispersion law only at grazing incidence to determine indices of refraction; at all other angles of incidence he uses an alternative construction that incorporates it (§§118–21 = I, 45–8). He then shows that according to this construction after any number of refractions in successive media separated by plane parallel surfaces, a homocentric pencil of polychromatic rays will always remain homocentric—a rather convenient, if fanciful, physical property (§§122–3 ≈ I, 49–50).[33]

(31) §110 ≈ I, 37: In this revision Newton proposes two ways to test the sine law of refraction for rays of each color, while simultaneously determining their indices of refraction, but rejects these direct methods as unnecessary.

(32) §116 → I, 43: Though it was rewritten, this confession is essentially the same in both versions. In the *Lectiones,* however, he proclaims that he will derive the dispersion law from its proper foundations (but he does not fulfill that promise); whereas in the *Optica* he avows that he will correct the law if he finds it to be false.

(33) §123 ≈ I, 50: Newton deleted the two concluding sentences to the "discourse on the measure of refractions" in which he explains that the prolixity of his treatment of the laws of refraction was necessary because he had to treat the entire subject anew.

Lectures 12–17 ≈ Lectures I, 8–11 (§§124–77 ≈ I, 51–102)

In these lectures on refraction at a single plane surface Newton attempts to
uncover the physical implications of the laws of refraction, the sine law and
his dispersion law, by a thorough mathematical analysis. Since that dispersion law was so tenuously founded and is the starting point for much of his
analysis, these lectures are now as notable for their mathematical analyses as
for their contributions to optics.

Lecture 12[34] is ostensibly devoted to the refraction of monochromatic
rays, but the first two propositions are quite minor (§§125–7),[35] and the
lecture is actually devoted to the single problem in Proposition 3[36] of determining the position of the image of a luminous point viewed obliquely across
a plane refracting surface (§§128–30 ≈ I, 61–2).[37] Newton's recognition
here that there are two image points effectively begins the study of astigmatism; and his derivation of the primary image point was more direct, albeit
lengthier, than the earlier solution by Barrow. Newton sends his audience to
Barrow's *Lectiones XVIII* for a more detailed investigation of the refraction
of monochromatic rays as he turns to polychromatic radiation (§131 = I,
63). Lecture 13 begins with a number of essentially barren propositions,
concerned largely with the rays' relative order and inclination, (§§132–
40),[38] although there is an interesting Scholium on the difficulty of defining
the extremities of the spectrum (§136—omitted in the *Optica*). Finally, however, it rallies to its main problem, a natural extension of Proposition 3: to
determine the shape of the extended image of a point source due to the
varying index of refraction when the point is viewed across a plane surface
(§142–4 ≈ I, 73–5).[39] He elegantly demonstrates that the images of the
point lie along a Dioclean cissoid. As a preliminary, Newton presents his own

(34) §124 → I, 51: In the new introductory paragraph in the *Optica*, Newton is no longer
apologetic about treating the mathematics of refraction.

(35) §§125–6 → I, 54: In the *Optica* Newton combined and generalized two propositions on
the relation between the angles of incidence and refraction by including the deviation, but he
now omitted their proof.

(36) Propositions transferred from the *Lectiones opticae* to the *Optica* do not have the same
number in the *Optica* because of the addition and deletion of other propositions and also
because of the adoption in the *Optica* of a different, continuous system of numbering. The
content and number of all the lemmas is identical in the two versions.

(37) §§129–30 ≈ I, 61–2: Newton vastly shortened his treatment of this problem by eliminating both a long derivation of the location of the primary image point (§130 → I, 61) and also a
somewhat confused physical justification for taking that point as the place of the image (§129 ≈
I, 62)—most likely because Barrow had already treated both.

(38) §§134 → I, 71, and 138 → I, 64: There is no fundamental difference in the content of these
propositions, although they are completely rewritten: to describe parallel refracted polychromatic rays (§134), and to show that polychromatic rays refracted from one point to another
point lie in the same plane (§138).

(39) §§142–3 ≈ I, 73–4: Newton recognized in revision that he could eliminate a long
justification for an approximation in §142 by assuming the angular dispersion to be infinitesimally small. The discussion in §143 of the apparent magnitude of the chromatic aberration of a
point source was clarified in the *Optica*, partly by invoking a proposition (§I, 69 = 162) that
now came earlier in the *Lectures*; see note (43).

solution to the popular "anaclastic problem," namely, to describe the path of a ray passing from a given point in one medium to a given point in a second medium, or to find the point of refraction (§141 → I, 58).[40]

In the *Optica* Newton thoroughly revised these lectures (Lecture 8 and part of Lecture 9, §§I, 51–75)—in fact, they are the most extensively revised part of the *Lectures*—undoubtedly intending to make them shorter and more incisive. He was successful in this, but he paid a rather dear price, for he now presented his propositions on the refraction of monochromatic rays (§§I, 52–62) as a codification of Barrow's earlier investigation (§I, 52) and eliminated much of the most exciting and innovative material, namely, his solutions to the place of the primary image and the anaclastic problem. The new propositions that he added are not particularly significant and deal mostly with drawing refracted rays under specified conditions.

In the next two pairs of Lectures, 14, 15 and 16, 17, Newton continues his attempt to create a rational science of color by investigating the variation of angular dispersion as the index of refraction and hence the chromatic dispersion of the refracting media vary.[41] After setting out six lemmas (§§146–58 = I, 66–7, 77, 80–9)[42] he first proves that the angular dispersion, dr, continually increases as the angle of incidence increases (§§160–2 ≈ I, 68–9);[43] and as this proposition alone of the eight in this lecture is independent of his, or any, dispersion law, it alone remains universally valid. Proposition 2 establishes that the chromatic dispersion, dn, increases at a faster rate than the index of refraction, n (§§163–4 = I, 78–9).[44] In the next four propositions he demonstrates that the greater the difference in the optical density of two media the greater the angular dispersion (§§165–9 = I, 90–4), except when the refraction is made into a dense medium which continually gets denser. For this last case, Proposition 6, Newton neatly establishes that the dispersion must attain a maximum, and then he determines the value of the index of refraction where that maximum occurs (§§170–5 = I, 95–100). In this proposition Newton at last proposes an experimental test of his theory, only to reject it as insufficiently sensitive (§172 = I, 97). These lectures end with two propositions that qualitatively

---

(40) §141 → I, 58: In the *Optica* Newton abandoned his solution to the anaclastic problem in favor of Barrow's.

(41) §145 → I, 76: Newton shortened the introduction to these lectures by virtually eliminating his summary of the preceding lecture; and, because of the shift of the first proposition (see note (43)) he was able to announce a more coherent theme.

(42) §159, Omitted: The need for a transition from the lemmas to the propositions was eliminated, since in the *Optica* two propositions were interspersed among the lemmas.

(43) §161, Omitted: Newton undoubtedly transferred Prop. 11 (§§160, 162 = I, 68–9), together with its requisite Lemmas 1, 2 (§§146–7 = I, 66–7), from this sequence of propositions on the variation of angular dispersion to give the remaining ones a more unified theme, since it was the only one in which that variation did not arise from a variation of the index of refraction. At the same time he deleted the long, somewhat rambling Scholium (§161) in which he qualitatively discussed the variation of angular dispersion for convergent rays under various conditions.

(44) Newton probably shifted Prop. 2 to follow Lemma 3 in order to relate the lemmas more immediately to optics.

describe the variation of the inclination of convergent rays as the media's index of refraction varies (§§176–7 = I, 101–2).

Lecture 18 = Lecture I, 12 (§§178–90 = I, 103–15)

This brief lecture, carried over unchanged into the *Optica*, treats refraction in prisms; and since Newton frequently simply invokes the analogous propositions for single planes, many of his demonstrations are quite perfunctory, as is true of Proposition 1, to locate the image viewed through a prism (§§179–80 = I, 104–5). Though Newton does not himself present this easily derived result, he does relate the simple approximation that near minimum deviation the image is located as far behind the prism as the object. Similarly, in Propositions 4 and 5 he demonstrates by appealing to earlier propositions that the greater a prism's angle or its index of refraction the greater is the angular dispersion (§§183–4 = I, 108–9). With the aid of two lemmas he proves in Propositions 6 and 7 that the minimum of deviation and (erroneously) the minimum of dispersion occur when the rays are refracted equally upon entering and leaving the prism (§§185–9 = I, 110–14). Newton arrived at all his results on refraction in prisms by cleverly applying symmetry considerations, rather than by deriving the general equation for a ray passing through a prism at any angle of incidence, and this finally led him astray in his proof for the minimum of dispersion. The *Lectiones opticae* ends abruptly—apparently abandoned to start anew with the *Optica*—with an estimate of the chromatic aberration of the image of a point seen through a prism without treating many previously promised subjects (§190 = I, 115).

Lectures I, 13–15 (§§I, 116–54)

Section 4, on refraction at curved surfaces, the conclusion of the mathematical part of the *Optica*, is its highpoint, an intimate blend of mathematics and physics consistently yielding novel, significant results. Newton's success undoubtedly derives from his nearly exclusive concern with monochromatic rays, for only the last of these nine propositions deals with polychromatic radiation. He effectively begins this section in Proposition 29 by determining the image point (in a form equivalent to the Gaussian formula) for paraxial rays incident upon a single spherical surface; and then in Proposition 30 he extends this result to any curved surface by substituting the center of curvature (determined in Lemma 9) in the immediate neighborhood of the incident rays for the center of the spherical surface. In Proposition 31 Newton applies many of his newly wrought mathematical methods, such as series expansions and the determination of extrema, to find the longitudinal spherical aberration for rays incident on the plane face of a plano-convex lens, and then the circle of least confusion. Because of its algebraic formulation, this proposition is particularly accessible to the modern reader and provides a fine example of

Newton's application of mathematics to physics. In the next proposition he elegantly derives the location of the primary image point, or caustic locus, for rays obliquely incident upon a spherical surface while also noting the existence and location of the secondary image point; Proposition 33 extends this result to any curved refracting surface. In Proposition 34 he presents his own solution to a problem posed and solved by Descartes: to find the aplanatic surface (a Cartesian oval) that refracts rays perfectly from a given point to a given point. Pursuing the Cartesian theme, in Propositions 35 and 36 he derives the radii of the primary and secondary rainbows, and then moving beyond all his contemporaries he generalizes his solution to bows of any order. And to conclude, Newton in Proposition 37 calculates the chromatic aberration to show that it is much more enormous—some 1500 times greater—than spherical aberration, and once again stresses the significance of his discovery of unequal refrangibility for practical optics.

Concordance of Article Numbers

From the *Lectiones opticae* to the *Optica*

Lecture 1
1 = I, 1
2 ≈ I, 2
3 = I, 3
4 = I, 4
5 = I, 5
6 = I, 6
7 = I, 7
8 = I, 8
9 = I, 9
10 = I, 10

Lecture 2
11 ≈ I, 11
12 = I, 12
13 = I, 13
14 ≈ I, 14
15 = I, 15
16 = I, 16
17 = I, 18
18 = I, 19
19 = I, 20
20 = I, 21
21 = I, 22
22 ≈ I, 23, 24
23 ≈ I, 24

Lecture 3
24 ≈ II, 1
25 = II, 2
26 = II, 3
27 = II, 4
28 ≈ II, 5
29 ≈ II, 6
30 → II, 6, 7
31 → II, 7
32 ≈ II, 7, 8
33 ≈ II, 9
34 = II, 12
35 = II, 13

36 = II, 14
37 = II, 15
38 = II, 16
39 = II, 18

Lecture 4
40 → II, 27, 28
41 ≈ II, 28
42 = II, 29
43 = II, 30
44 = II, 31
45 = II, 32
46 = II, 33
47 = II, 34
48 = II, 35
49 = II, 36
50 = II, 37
51 = II, 39
52 ≈ II, 41
53 ≈ II, 42

Lecture 5
54 = II, 43
55 = II, 44
56 = II, 45
57 = II, 46
58 = II, 47
59 = II, 48
60 ≈ II, 51
61 = II, 52
62 = II, 53
63 = II, 54

Lecture 6
64 ≈ II, 55
65 = II, 56
66 = II, 57
67 = II, 58
68 = II, 59
69 = II, 60

70 = II, 61
71 = II, 62
72 = II, 63
73 = II, 64
74 ≈ II, 65
75 ≈ II, 65, 66
76 → II, 67

Lecture 7
77 → II, 81
78
79
80
81
82 ≈ II, 82
83 = II, 83
84 ≈ II, 84
85 = II, 85
86 = II, 86
87 ≈ II, 87
88 = II, 88

Lecture 8
89 ≈ II, 90
90 = II, 91
91 = II, 92
92 ≈ II, 89
93 ≈ II, 89, 77
94 = II, 78
95 ≈ II, 79
96 ≈ II, 80

Lecture 9
97
98 ≈ I, 25
99 = I, 26
100 = I, 27
101 = I, 28
102 = I, 29
103 ≈ I, 30

104 = I, 31
105 = I, 32
106 = I, 33

Lecture 10
107 = I, 34
108 = I, 35
109 = I, 36
110 ≈ I, 37
111 = I, 38
112 = I, 39
113 = I, 40
114 = I, 41

Lecture 11
115 = I, 42
116 → I, 43
117 = I, 44
118 = I, 45
119 = I, 46
120 = I, 47
121 = I, 48
122 = I, 49
123 ≈ I, 50

Lecture 12
124 → I, 51
125 → I, 54
126 → I, 54
127
128
129 ≈ I, 62
130 → I, 61

Lecture 13
131 = I, 63
132
133
134 → I, 71
135
136

137
138 → I, 64
139
140
141 → I, 58
142 ≈ I, 73, 74
143 ≈ I, 74
144 = I, 75

Lectures 14, 15
145 → I, 76
146 = I, 66
147 = I, 67
148 = I, 77
149 = I, 80

150 = I, 81
151 = I, 82
152 = I, 83
153 = I, 84
154 = I, 85
155 = I, 86
156 = I, 87
157 = I, 88
158 = I, 89
159
160 = I, 68
161
162 = I, 69
163 = I, 78
164 = I, 79

165 = I, 90
166 = I, 91

Lectures 16, 17
167 = I, 92
168 = I, 93
169 = I, 94
170 = I, 95
171 = I, 96
172 = I, 97
173 = I, 98
174 = I, 99
175 = I, 100
176 = I, 101
177 = I, 102

Lecture 18
178 = I, 103
179 = I, 104
180 = I, 105
181 = I, 106
182 = I, 107
183 = I, 108
184 = I, 109
185 = I, 110
186 = I, 111
187 = I, 112
188 = I, 113
189 = I, 114
190 = I, 115

From the *Optica,* Part I to the *Lectiones opticae*

Lecture 1
1 = 1
2 ≈ 2
3 = 3
4 = 4
5 = 5
6 = 6
7 = 7
8 = 8

Lecture 2
9 = 9
10 = 10
11 ≈ 11
12 = 12
13 = 13
14 ≈ 14
15 = 15
16 = 16
17
18 = 17
19 = 18

Lecture 3
20 = 19
21 = 20
22 = 21
23 ≈ 22
24 ≈ 22, 23

Lecture 4
25 ≈ 98

26 = 99
27 = 100
28 = 101

Lecture 5
29 = 102
30 ≈ 103
31 = 104
32 = 105
33 = 106
34 = 107
35 = 108

Lecture 6
36 = 109
37 ≈ 110
38 = 111
39 = 112
40 = 113
41 = 114

Lecture 7
42 = 115
43 → 116
44 = 117
45 = 118
46 = 119
47 = 120
48 = 121
49 = 122
50 ≈ 123

Lecture 8
51 → 124
52
53
54 → 125, 126
55
56
57
58 → 141
59
60
61 → 130
62 ≈ 129
63 = 131
64 → 138
65
66 = 146
67 = 147
68 = 160
69 = 162

Lecture 9
70
71 → 134
72
73 → 142
74 ≈ 142, 143
75 = 144
76 → 145
77 = 148
78 = 163

79 = 164
80 = 149
81 = 150
82 = 151

Lecture 10
83 = 152
84 = 153
85 = 154
86 = 155
87 = 156
88 = 157
89 = 158
90 = 165
91 = 166
92 = 167
93 = 168
94 = 169

Lecture 11
95 = 170
96 = 171
97 = 172
98 = 173
99 = 174
100 = 175
101 = 176
102 = 177

Lecture 12
103 = 178
104 = 179

105 = 180	113 = 188	*Lecture 14*	*Lecture 15*
106 = 181	114 = 189	135	148
107 = 182	115 = 190	...	...
108 = 183		147	154
109 = 184	*Lecture 13*		
110 = 185	116		
111 = 186	...		
112 = 187	134		

From the *Optica,* Part II to the *Lectiones opticae*

Lecture 1	29 = 42	58 = 67	87 ≈ 87
1 ≈ 24	30 = 43	59 = 68	88 = 88
2 = 25	31 = 44	60 = 69	89 ≈ 92, 93
3 = 26	32 = 45	61 = 70	90 ≈ 89
4 = 27	33 = 46		91 = 90
5 ≈ 28	34 = 47	*Lecture 8*	92 = 91
6 ≈ 29, 30	35 = 48	62 = 71	
	36 = 49	63 = 72	*Lecture 11*
Lecture 2	37 = 50	64 = 73	93
7 → 30, 31, 32		65 ≈ 74, 75	...
8 ≈ 32	*Lecture 5*	66 → 75	101
9 ≈ 33	38	67 → 76	
10	39 = 51	68	*Lecture 12*
11	40	69	102
12 = 34	41 ≈ 52	70	...
13 = 35	42 ≈ 53		117
14 = 36	43 = 54	*Lecture 9*	
15 = 37	44 = 55	71	*Lecture 13*
16 = 38		72	118
17	*Lecture 6*	73	...
18 = 39	45 = 56	74	126
	46 = 57	75	
Lecture 3	47 = 58	76	*Lecture 14*
19	48 = 59	77 ≈ 93	127
20	49	78 = 94	...
21	50	79 ≈ 95	133
22		80 ≈ 96	
23	*Lecture 7*		*Lecture 15*
24	51 ≈ 60	*Lecture 10*	134
25	52 = 61	81 → 77	135
26	53 = 62	82 ≈ 82	
	54 = 63	83 = 83	*Lecture 16*
Lecture 4	55 ≈ 64	84 ≈ 84	136
27 → 40	56 = 65	85 = 85	...
28 ≈ 40, 41	57 = 66	86 = 86	140

Optical Lectures

[Lectiones opticae][1]

Jan. 1669[2]
Lect 1
1 [= I, 1]. Incepti ratio.

Inventio Telescopiorum nupera plerosque Geometras ita exercuit, ut nihil in Optica non tritum, nullum inventioni praeterea locum alijs reliquisse videantur. Et insuper cùm dissertationes quas hic non ita pridem audivistis, tantâ rerum Opticarum varietate, novorum copiâ, et accuratissimis eorundem demonstrationibus fuerint compositae;[3] frustranei fortè videantur conatus et labor inutilis, si ego scientiam hanc iterum tractandam suscepero. Verùm cùm Geometras in quadam lucis proprietate, quae ad Refractiones spectat hucusque hallucinatos videam, dum demonstrationibus suis Hypothesin quandam Physicam haud benè stabilitam tacitè supponunt: non ingratum me facturum judico, si principia Scientiae hujus examini severiori subjiciam, et quae ego de ijs simul excogitavi, et experientia multiplici habeo comperta, subnectam ijs, quae Reverendus meus Antecessor hic loci postrema dixit.

Imaginantur sibi Dioptrices studiosi, quòd Perspicilla[4] ad quemlibet perfectionis gradum perduci possent, si modò vitris dum perpoliuntur, geometricam, quam vellent, figuram communicare concederetur. Et in eum finem instrumenta varia fuerunt excogitata, quibus vitra in figuras Hyperbolicas vel etiam Parabolicas contererentur; sed exacta figurarum istarum fabricatio nemini hucúsque successit.[5] Scilicet aratur littus; et nè labores suos in negotio

---

(1) Add. 4002, pp. 1–128; the remaining two thirds of this bound volume are blank. Newton numbered the figures only through the fifty-fourth, and we have extended his numbering to the remaining twenty-eight figures; and when he omitted figure numbers in textual references, we have inserted them in square brackets. A high-quality facsimile of this untitled autograph—called here the *Lectiones opticae*—was published by the Cambridge University Library in 1973 as *The Unpublished First Version of Isaac Newton's Cambridge Lectures on Optics 1670–1672*. The figures on page 129 of the facsimile do not belong with the text but are loose inserted slips that in the 1960s were attached to that page for preservation. We discuss the background to this manuscript in the Introduction, §3.   (2) That is, 1669/70 or New Style 1670.

(3) Barrow delivered his optical lectures in 1667 and 1668; see the Introduction, note (46).

(4) Newton used both "perspicillum" and "telescopium" for "telescope" in accordance with contemporary usage; for example, the Latin translation of Descartes's *Dioptrique* used both of these terms as well as "specillum."

(5) Descartes demonstrated in his *Dioptrique*, Discourse VIII, that lenses ground to conoids of revolution focused light perfectly without the aberration accompanying spherical lenses. He described a series of telescopes with hyperbolic lenses in the next discourse, and in Discourse X he described a machine to grind these hyperbolic lenses and urged artisans not to be discouraged by the difficulty of the task. Despite Descartes's own lack of success and the difficulties encountered by others, attempts to grind conoids of revolution continued in the ensuing decades. Hooke, for instance, in the preface to his *Micrographia* (sig. e[3]) observed that "if Glasses could be made of those kind of Figures, or some other, such as the most incomparable Des Cartes has invented, and demonstrated in his Philosophical and Mathematical Works, we might hope for a much greater perfection of Opticks then can be rationally expected from spherical ones." New-

[Optical Lectures][1]

The recent invention of telescopes has so occupied most geometers that they seem to have left to others nothing in optics untouched nor any room for further discovery. Moreover, since the lectures that you heard here not so long ago brought together such a great variety of optical topics and a vast quantity of discoveries with their very accurate demonstrations,[3] it might perhaps seem a vain endeavor and futile effort for me to undertake to treat this science again. But since I observe that geometers have hitherto erred with respect to a certain property of light pertaining to its refractions, while they implicitly assume in their demonstrations a certain not well established physical hypothesis, I judge it will not be unappreciated if I subject the principles of this science to a rather strict examination, adding what I have conceived concerning them and confirmed by numerous experiments to what my reverend predecessor last delivered in this place.

Those knowledgeable in dioptrics imagine that telescopes[4] may be brought to any degree of perfection, provided that it were possible to impart any desired geometrical figure to the lenses while they are being finely polished. For this purpose various instruments have been devised for grinding lenses into hyperbolic or even parabolic figures, but no one has yet succeeded in exactly forming those shapes.[5] It is, to be sure, a futile endeavor, and lest

---

ton was sufficiently impressed by Descartes's advocacy of hyperbolic and parabolic lenses that in the winter of 1665–6 he undertook his essay "Of Refractions," which described various methods for grinding such lenses; see the Introduction, note (21).

Newton was aware of other attempts to grind nonspherical lenses, for in his notes, "Out of Philosophicall Transactions," he succinctly entered: "Hevelius promiseth to communicate his way of grinding glasses of conick sections in a dish of a sphaere. Hugens promiseth to attempt it & Du Son is imployed in it" (Add. 3958, f. 10ʳ). This is an accurate summary of Oldenburg's report, "Of Monsieur Hevelius's promise of imparting to the world his invention of making optick glasses; and of the hopes given by Monsieur Hugens of Zulichem, to perform something of the like nature; as also of the expectations, conceived of some ingenious persons in England, to improve telescopes," *Phil. Trans.*, 1 (1665):98–9; see also Oldenburg's report in the next number, also noted by Newton, "Of Monsieur de Sons progress in working parabolar glasses," ibid., pp. 119–20. With his keen interest in this problem, Newton probably also took note of related projects described in later numbers of the *Philosophical Transactions* such as that by Francis Smethwick, "Of the invention of grinding optick and burning-glasses, of a figure not-spherical," *Phil. Trans.* 3 (1667/8):631–2. Smethwick made an eyepiece with three nonspherical plano-convex lenses for his telescope, which the members of the Royal Society judged to represent objects "free from Colours." Newton, however, never took note of the development of the compound achromatic eyepiece. Modern scholarship supports Newton's claim that no one had succeeded in grinding conical lenses; see Albert van Helden, "The telescope in the seventeenth century," *Isis*, 65 (1974):38–58, esp. 45.

desperato diutiùs insumant, ijs audeo spondere, quòd licèt omnia fierent perquam felicitèr, nihil minùs tamen quàm votis suis responderent: etenim vitra
licèt efformentur secundum figuras in istum finem optimas, quae possunt excogitari, tamen non duplo plus praestabunt quàm sphaerica, aequali politurâ
perfecta. Haec autem non ideò loquor, quasi peccatum esse a scriptoribus
Optices contenderem; illi enim omnia pro intentione demonstrationum suarum accuratè quidem et verissimè dixerunt, sed aliquid tamen idque maximi
momenti reliquerunt posteris inveniendum. Scilicet in refractionibus irregu/la- 2/
ritatem quandam reperio, quae omnia perturbat, et non solum efficit, ut figurae Conicarum[6] Sphaericas non multùm superent, sed etiam ut sphaericae
multò minus praestent quàm praestarent, si dicta refractio esset uniformis.

Itaque in Dioptricâ pedem fixi non ut pertractarem de integro, sed tantùm,
ut hanc de natura lucis proprietatem rimarer primò, deinde ut ostenderem
quantum ex hac proprietate perfectio Dioptrices impeditur, & quo pacto
incommodum istud, quatenus natura rei sinit, devitetur. Ubi nonnulla proferam quae ad Telescopiorum, juxta et Microscopiorum, tum Theoriam tum
Praxin spectant; ostendens quod Optices summa perfectio (praeter opinionem
receptam) ex Dioptrica et Catoptrica mixtis petenda est.[7] Ac interea discrimen colorum et eorum genesin a Prismatibus, et corporibus etiam coloratis
fusè explicabo.[8]

2 [≈ I, 2]. Quòd omnium radiorum non est eadem refrangibilitas.

De luce itaque compertum habeo, quòd radij ejus quoad quantitatem refractionis ab invicem differunt: Ex ijs qui omnes habent eundem angulum
incidentiae, alij angulum refractionis aliquantò majorem alijs habebunt. Plenioris illustrationis gratiâ [fig 1], sit *EFG* superficies quaelibet refringens puta

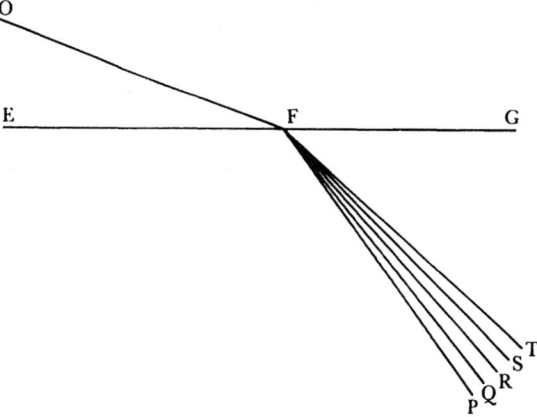

Figure 1

vitrea; et ducatur quaevis *OF* huic occurrens in *F*, et cum ea efficiens angulum *OFE* acutum: Concipe etiam radios solares per istam lineam *OF* sibi
continuò successivos fluere, ita ut alij post alios in punctum *F* impingant,
ibidemque in medium densius refringantur.[9] Jam ex opinione receptâ hi radij
eandem habentes incidentiam, eandem quoque omnes refractionem habere

they devote their efforts any longer to a hopeless occupation, I venture to promise them that even if everything were to turn out thoroughly successfully, it still would fall far short of their expectations. For even if lenses were fashioned according to the best shapes that can be conceived for that purpose, they will nonetheless perform no more than twice as well as spherical ones polished equally perfectly. I am not, however, pointing this out as if I maintained that those who write on optics are in error, for everything they have asserted with regard to the intent of their demonstrations is accurate and very true. However, they have left something—and of the greatest importance—to be discovered by their successors; namely, I find in refractions a certain irregularity that upsets everything, which not only causes the figures of conics[6] to be little superior to spherical ones but also makes spherical ones perform far less well than they would if this refraction were uniform.

I therefore have set foot on dioptrics not to treat it systematically anew, but only, in the first place, to examine thoroughly this property in the nature of light, and then to show how much the perfection of dioptrics is impeded by this property and how that obstacle, insofar as its nature allows, may be avoided. Here I will set forth several things concerning both the theory and practice of telescopes as well as microscopes, showing that the ultimate perfection of optics (contrary to the received opinion) must be sought in dioptrics and catoptrics combined.[7] In the meantime I will fully explain the difference of colors and their generation by prisms as well as colored bodies.[8]

Concerning light, therefore, I have discovered that its rays differ from one another with respect to the quantity of refraction: Of those rays that all have the same angle of incidence, some will have an angle of refraction somewhat larger than others. For the sake of a fuller illustration (Fig. 1), let *EFG* be any refracting surface, for example, glass, and draw any line *OF* meeting it at *F* and making with it an acute angle *OFE*; also imagine that along this line *OF* solar rays flow in continual succession to one another, so that they strike the point *F* one after another and are there refracted into the denser medium.[9] Now, according to the received view, since these rays have the same incidence, they should all also have the same refraction, suppose, in the line *FR*.

2 [≈ I, 2]. The refrangibility of all rays is not the same.

(6) II: *Conicarum Sectionum* (of conic sections).

(7) The topics Newton here promises to treat are omitted in the *Lectiones* and only partially dealt with in the *Optica*. Specifically, for the theory of telescopes and microscopes he calculates the image position and spherical and chromatic aberrations for single lenses in Pt. I, Sect. 4, but says nothing at all about practical matters; nor does he describe his "catadioptrical" or reflecting telescope.

(8) Colored bodies are only incidentally treated in the *Lectiones opticae*, but they are rather fully considered in the *Optica*, §§II, 71–6.

(9) II continues: Vel si mavis, finge parallelos radios indefinitè parùm distare ab *OF*, et incidere in puncta ipsi *F* vicinissima. (Or, if you prefer, imagine parallel rays to be an indefinitely small distance from *OF* and to fall on the points in the immediate vicinity of *F*.) On Newton's concept of light ray see §20 = I, 21 and §78.

debent, puta in lineam *FR*. At contrarium compertum habeo; scilicet quòd postquam refringuntur, divergent ab invicem; quasi quidam refringerentur in lineam *FP*, alij in lineam *FQ*, & alij in lineas *FR, FS,* & *FT*; ac alij etiam innumeri per spatia istis intermedia, ut et ultra citraque nonnulli pervagantes; prout radius quilibet ad refractionem majorem minoremve patiendam sit aptus. Invenio praeterea, quòd radij *FP* maximè refracti colores purpureos producunt et illi *FT* minimè refracti rubros, qui autem hisce intermedij pergunt *FQ, FR, FS,* ij / colores intermedios nempe caeruleos, virides et flavos generant. Et sic radij prout apti sunt ut alij alijs magis atque magis refringantur, hos ordine colores, rubrum, flavum, viridem, caeruleum, et purpureum generant, unà cum omnibus intermedijs quos in Iride liceat conspicere.[10] Unde productio colorum Prismatis et Iridis patebit facilè: sed his jam perfunctoriè[11] notatis, quae de coloribus dicenda sunt in posterum differo.[12]

3 [= I, 3]. Probatur experimento vulgari, per longitudinem imaginis coloratae.[13] Sententia nostra[14] de hac re in genere sic explicatâ, ne putetis fabulas pro veris enarratas esse, rationes et experimenta quibus isthaec innituntur, continuò proferam. Et quoniam experimentum quoddam Prismatis valde obvium mihi primò dedit occasionem excogitandi reliqua, istud primum explicabo.[15] Sit [fig 2] *F* foramen aliquod in pariete vel fenestrâ Cubiculi, per

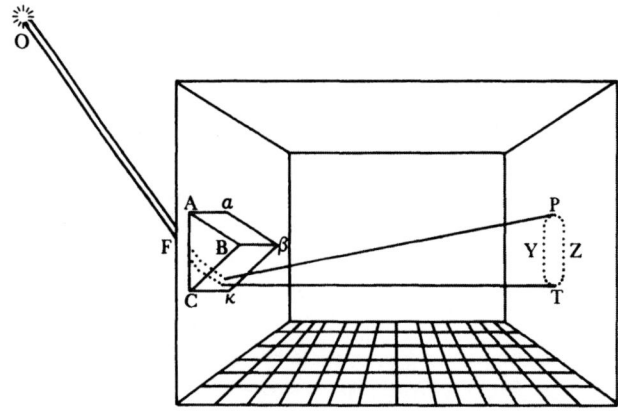

Figure 2

(10) Throughout the *Lectiones opticae* Newton divides the spectrum into five principal colors. He added orange and indigo to these when near the end of the *Optica,* in Lect. II, 11, he introduced his musical division of the spectrum; and from the "New theory" (Prop. 5) onward he used a sevenfold division alongside of the fivefold one. Of the five colors enumerated here, only the name for the most refrangible rays, purple ("purpureum"), varied until he finally opted for violet in the *Opticks.* In the early years Newton generally chose to call these rays purple, although occasionally (e.g., §28 ≈ II, 5 and §47 = II, 34) he uses the two names, "purpureum sive violaceum," interchangeably. (Once, but perhaps by mistake, in §II, 25, he even called the least refrangible or red rays purple.) He continued to equivocate in the "New theory" (adopting violet in Prop. 2, but "violet-purple" in Prop. 5) and also in his optical correspondence in the following years. It was only in the *Opticks* (Bk. I, Pt. II, Props. IV, VI) that he definitively switched to violet and explained that if red and violet are mixed, "the colour compounded shall not be any of the Prismatic colours, but a purple" (Prop. VI, Add. 3970, f. 113ʳ = *Opticks,* p. 116). Newton's indecisiveness in choosing between violet and purple was not unique because it

But I have discovered the contrary; that is, after they are refracted, they will diverge from one another, just as if some were refracted in the line *FP*, some in the line *FQ*, others in the lines *FR*, *FS*, and *FT*, and still innumerable others through the spaces between them, while some extend on both sides as well, insofar as any ray is disposed to undergo a greater or smaller refraction. Moreover, I find that the most refracted rays *FP* produce purple colors and those least refracted *FT* produce red, but those that proceed along the intermediate lines *FQ*, *FR*, and *FS* generate the intermediate colors, namely, blue, green, and yellow. Hence, insofar as the rays are so disposed that some are refracted more and more than others, they generate in order these colors, red, yellow, green, blue, and purple, together with all the intermediate ones that can be seen in the rainbow.[10] The production of the colors of the prism and rainbow will be readily evident from this, but having now perfunctorily[11] noted these things, I postpone what I have to say about colors for later.[12]

Having thus generally explained my view[14] on this subject, I will at once present the reasoning and experiments that support these things, lest you think that I have set forth fables instead of the truth. Since a certain very commonly encountered experiment with a prism first presented me the opportunity to think out the rest, I will explain that first.[15] In the wall or window of a room (Fig. 2) let *F* be some hole through which solar rays *OF*

3 [= I, 3]. A common experiment proves this by the length of the colored[13] image.

---

was then widely held that violet and purple were two different species of the same color, compounded of varying proportions of red and blue. Thus, François d'Aguilon and, following him, Athanasius Kircher considered "violaceum" to be a species of "purpureum"; see François d'Aguilon, *Opticorum libri sex philosophis juxta ac mathematicis utiles* (Antwerp, 1613), p. 40; and Athanasius Kircher, *Ars magna lucis et umbrae, in decem libros digesta* (Rome, 1646), p. 68. Descartes too once refers to the color at the extreme blue end of the spectrum as "violaceum sive purpuream" in *Meteora*, Ch. VIII, §7, *Oeuvres*, 6:704. Compare George Biernson's speculations, "Why did Newton see indigo in the spectrum?" *American Journal of Physics*, 40 (1972):526–33, esp. 530–1.

After the publication of the *Opticks*, Newton's names for the spectral colors were nearly universally adopted, but in the period preceding 1704 there was no consensus. For instance, Descartes called the colors at the two ends of the spectrum "rubrum, croceum & flavum" and "viride, caeruleum & violaceum," (*Meteora*, Ch. VIII, §5, *Oeuvres*, 6:702); Charleton described the prismatic colors as "*Vermillion, Yellow, Green,* and *Violet,* beside the inassignable variety of other *Intermediate* Colours" (*Physiologia*, p. 189); Marci designated the principal spectral colors as "puniceus, viridis, caeruleus, & purpureus" (*Thaumantias*, p. 95); Kircher chose "rubeus, puniceus, flavus, viridis, caeruleus" (*Ars magna*, Bk. I, Pt. III, Ch. IV. p. 75); and Boyle adopted "Red, Yellow, Green, Blew, and Purple," (*Touching Colours*, plate facing p. 192).

(11) Originally: in transitu (in passing).

(12) Newton returns to colors in §24 ≈ II, 1, though he treats the rainbow only in the *Optica*, Pt. I, Props. 35, 36, and Pt. II, Lect. 16.

(13) II: solaris refractae (refracted solar).

(14) Sententia nostra] Originally: Conceptionibus nostris.

(15) This experiment, "the celebrated phaenomena of colours," and its seemingly innumerable variants from the first played a fundamental role in Newton's theory of color; see the Introduction, §2. He first described it in his essay "Of colours" (§§7, 8, Add. 3975, f. 2ᵛ = Hall, "Further experiments," pp. 28–9), and it maintained its fundamental role in the "New theory" and the *Opticks*, Bk. I, Pt. I, Prop. II, Expt. III.

quod radij solares *OF* trajiciantur, reliquis ubique foraminibus diligenter obturatis, nè lux alibi ingrediatur. Ista autem obscuratio Cubiculi non est
omninò[16] necessaria, sed efficit tantùm ut Experimentum evadat aliquantò
evidentius. Deinde Prisma triangulare vitreum *AαBβCκ* ad foramen istud
applicetur, quod radios *OF* per se trajectos refringat versus *PYTZ*; quos
radios opposito pariete vel papyro aliquâ ad distantiam a Prismate satis
magnam objectâ terminatos, videbis in figuram *PYTZ* valdè oblongam efformari: cujus nempe longitudo *PT* sit quadruplex latitudinis *YZ*, et amplius. Et
hinc evinci certò videtur quòd radiorum aequalitèr incidentium alij majorem
alijs refractionem patiuntur. Nam si contrarium esset verum, praedicta Solis
imago appareret ferè orbicularis,[17] & in quadam positione Prismatis omninò
ad sensum orbicularis conspiceretur; Id quod contra omnem experientiam
est; Quocunque enim situ Prisma disposui, nunquam tamen potui efficere,
quin longitudo imaginis esset latitudinis plusquam quadrupla: angulo scilicet
Prismatis *ACB* vel *ακβ* existente graduum plus minus sexaginta.

4 [= I, 4]. Casùs in   / Quod autem quaedam datur positio Prismatis in quâ Imago Solis, ex 4/
quo radij aequè   opinione de refractionibus receptâ, appareret orbicularis, sic ostendo. Juxta
refrangibiles faciunt   foramen in Fenestra cubiculi factum, Prisma collocetur foras; vel, quod eo-
imaginem   dem recidit [fig 3], sit *EG* corpus aliquod opacum citra Prisma locatum in
orbicularem.

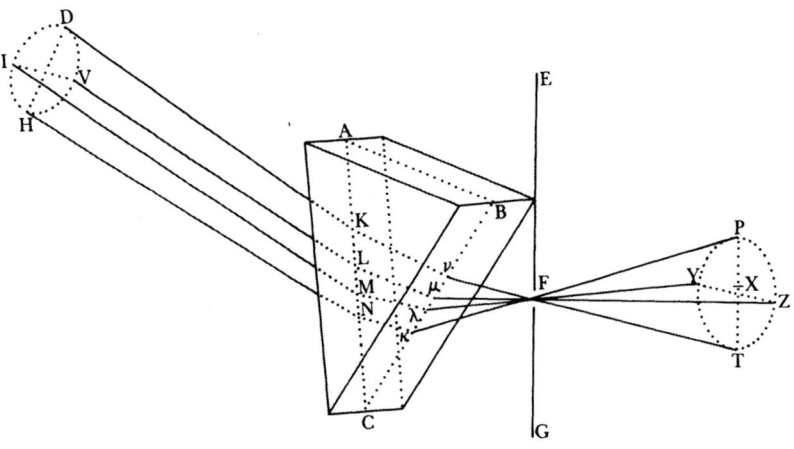

Figure 3

quo sit *F* foramen indefinitè parvum et orbiculare, per quod radij refracti in
parietem directè oppositum ad imaginem *PYTZ* ibi depingendam trajiciantur.
Et ponatur *ABC* esse planum secans *AC* & *BC* plana refringentia perpendiculariter, atque etiam transiens per foramen *F*, ut et per centrum solis *DIHV*,
quem bisecet secundum diametrum ejus *DH*, a cujus extremitatibus duo radij
DK et *HN* in eodem plano jacentes adveniant, qui postquam refringuntur *DK*
in *Kν* et *νT*, atque *HN* in *Νκ* et *κP*, utrique pergant per centrum foraminis *F*.
Et praeterea sit talis inclinatio Prismatis ad istos radios ut anguli *AKD* & *BκF*
fiant aequales.[18] Deinde sit *IV* alia solis diameter praedicto plano *ABC*
perpendicularis a cujus extremitatibus alij duo radij *VL* et *IM* adveniant, alter
IM cis planum *ABC*, qui refringatur in *Mλ* & *λY*, alter verò *VL* ultra planum

are transmitted, while other holes elsewhere have been carefully sealed off so that no light enters from any other place. This darkening of the room, however, is not absolutely[16] necessary; it only enables the experiment to turn out somewhat more clearly. Then place at that hole a triangular glass prism $A\alpha B\beta C\kappa$ that refracts the rays OF transmitted through it toward $PYTZ$. You will see these rays, terminated by the opposite wall or by some paper placed sufficiently far from the prism, formed into a very oblong figure $PYTZ$, specifically, one whose length PT is four times and more its breadth YZ. Hence, this definitely appears to establish that at equal incidence some rays undergo a greater refraction than others; for if the contrary were true, that solar image would seem almost circular, and in a certain position of the prism it would appear to the senses completely circular,[17] which is contrary to all experience. Indeed in whatever position I placed the prism, I nonetheless could never make it happen that the image's length was not more than four times its breadth, that is, with the angle of the prism ACB or $\alpha\kappa\beta$ being about 60°.

That there exists, however, a certain position of the prism in which, according to the received view of refractions, the sun's image should appear circular, I show in this way. Place the prism outside, next to the hole in the window of the room, or, what amounts to the same thing (Fig. 3), let EG be some opaque body placed on this side of the prism; and let F be an indefinitely small circular hole in it through which the refracted rays are transmitted onto the directly opposite wall depicting the image $PYTZ$ there. Assume that ABC is a plane perpendicularly cutting the refracting planes AC and BC while also passing through both the hole F and the center of the sun $DIHV$, which it bisects along its diameter DH; and from its ends let there proceed two rays DK and HN lying in the same plane, which, after they are refracted (DK into $K\nu$ and νT, and HN into $N\kappa$ and κP), both pass through the center of the hole F. Moreover, let the prism's inclination to those rays be such that the angles AKD and $B\kappa F$ become equal.[18] Next, let IV be another diameter of the sun, perpendicular to that plane ABC; and from its ends let there proceed two other rays VL and IM; letting one, IM, on this side of the plane ABC, be refracted into $M\lambda$ and λY, while letting the other, VL, on the other

4 [= I, 4]. The case in which equally refrangible rays make a circular image.

---

(16) Omitted in **II**.

(17) Where Newton here uses "orbicular," in the "New theory" he uses the English "circular"; *Correspondence*, 1:92 = *Phil. Trans.*, 6 (1671/2):3076. Later in the *Lectures*, in revising the postil to §12 = I, 12, he changed "orbiculares" to the apparently equivalent "circulares"; see also §II, 114. In the *Opticks* (Bk. I, Pt. I, Prop. II, Expt. V, p. 26 = Add. 3970, f. 37ʳ) he clarified these terms: "By a circle I understand not here a perfect Geometrical circle but any orbicular figure whose length is equal to it's breadth, & wᶜʰ, as to sense, may seem circular."

(18) The requirement that the angles AKD and $B\kappa F$ (respectively, the angle of incidence of rays from the upper part of the sun and the angle of refraction of rays from the lower part) are equal is equivalent to requiring that the prism be set at minimum deviation, for only at minimum deviation is a small increase in the angle of incidence equal (at least, to the first order) to the decrease in the angle of refraction. Lohne discusses Newton's demonstration and its relation to his observations in "Newton's 'proof' of the sine law," pp. 396–7.

istud qui refringatur in *Lμ* et *μZ*. Et praedicti quatuor radij sese omnes decussent in medio foramine *F*. Denique ponatur quòd imago lucida *PYTZ* foramen *F* directè respiciat, ita scilicet ut *FP* et *FT*, item *FY* et *FZ* aequales fiant. Dico jam quòd in illâ positione Prismatis, anguli *PFT* ac *YFZ* aequales essent, supposito quòd radij omnes aeque refringuntur qui eundem habent angulum incidentiae: Et proinde quòd imago ista, sensui saltem, debet esse orbicularis, utpote cujus diametri *PT* ac *YZ* sese decussant perpendicularitèr, et aequales istos angulos subtendunt.

5 [= I, 5].
Demonstratio istius
Casûs. Ejus pars 1ma.

Angulos autem istos *PFT* & *YFZ* aequales esse sic demonstro. Concipe radium aliquem a *P* per *κ* et *N* retrocedere, dum alius radius pergit a *D* per *K* et *v*: / Itaque cùm anguli *AKD*, & *BκF* supponuntur aequales, erunt 5/ etiam anguli per primas refractiones facti *AKv* et *BκN* aequales. Unde triangula *CKv* & *CκN* erunt similia et eorum anguli externi *κNA*, *KvB* aequales; et proinde anguli per secundas refractiones facti *ANH* & *BvF* etiam aequales. Quare cùm anguli *AKD* & *BκF*, item *ANH* et *BvF* sint aequales, eorum differentiae erunt etiam aequales, hoc est angulus *vFκ* sive *PFT* aequalis angulo quem radij *DK* et *HN* comprehendunt, sive diametro solari. Est itaque ang: *PFT* aequalis diametro solari:[19] Quare cùm praeterea demonstratum[20] fuerit quod ang: *YFZ* aequatur eidem diametro, liquebit propositum. Istud autem ut fiat, Theorema quoddam more Lemmatis praesternendum est quod cùm postea nobis forsan erit ex usu, jam facere non pigebit.[21]

6 [= I, 6]. Lemma
ad secundam partem

Sunto [fig 4] duo plana *ABCD* et *EFGH* sibimet perpendicularia quorum communis intersectio sit *KL*. Et sit *IP* radius quilibet qui in planum *ABCD* incidens ad punctum *P* ab eo refringitur in *PR*. Dico quòd sinus anguli quem radius incidens *IP* efficit cum plano perpendiculari *FH*, est ad sinum anguli

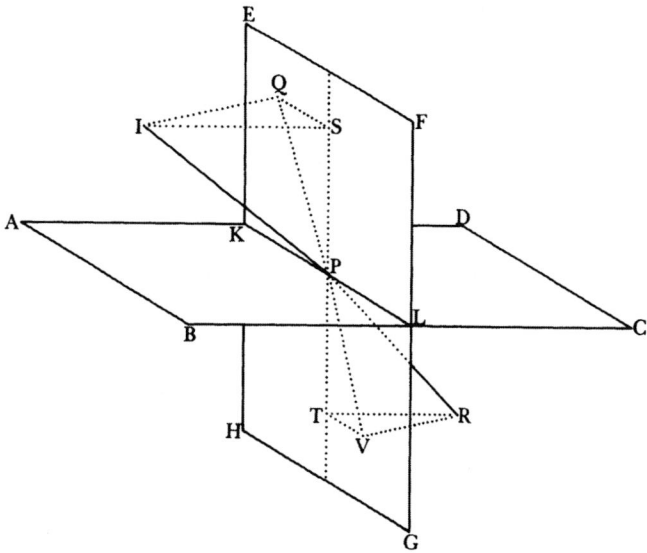

Figure 4

side of that plane, be refracted into $L\mu$ and μZ; and let the specified four rays all cross one another in the middle of the hole *F*. Finally, let it be assumed that the bright image *PYTZ* directly faces the hole *F*, specifically so that *FP* and *FT* as well as *FY* and *FZ* become equal. Now, I assert that in that position of the prism the angles *PFT* and *YFZ* are equal—supposing that all rays that have the same angle of incidence are equally refracted—and consequently that that image must be circular, at least to the senses, inasmuch as its diameters *PT* and *YZ* cross each other perpendicularly and subtend those equal angles.

That these angles *PFT* and *YFZ* are equal, however, I demonstrate as follows: Imagine some ray to proceed backward from *P* through κ to *N*, while another ray proceeds from *D* through *K* and ν. Therefore, since the angles *AKD* and *BκF* are assumed to be equal, the angles *AKν* and *BκN* made by the first refractions will also be equal. Hence the triangles *CKν* and *CκN* will be similar, and their external angles *κNA* and *KνB* will be equal; thus the angles *ANH* and *BνF* made by the second refractions will also be equal. Accordingly, since the angles *AKD* and *BκF* as well as *ANH* and *BνF* are equal, their differences will also be equal; that is, the angle *νFκ* or *PFT* will be equal to the angle contained by the rays *DK* and *HN* or to the solar diameter. The angle *PFT* is therefore equal to the solar diameter.[19] Moreover, since it will be demonstrated[20] that the angle *YFZ* is equal to the same diameter, the proposition will be evident. However, in order to do that a certain theorem, in the form of a lemma, must be premised, and since it will perhaps be useful for us afterward, it will not be troublesome to do this now.[21]

5 [= I, 5]. The demonstration of that case. Part I.

Let there be (Fig. 4) two mutually perpendicular planes *ABCD* and *EFGH* the common intersection of which is *KL*, and let *IP* be any ray that falls on the plane *ABCD* at the point *P* and is refracted by it into *PR*. I say that the sine of the angle made by the incident ray *IP* with the perpendicular plane *FH*

6 [= I, 6]. A Lemma for Part II.

---

(19) Letting *i* be the angle of incidence at the first face and *i'* the angle of refraction at the second face, and designating by *u* and *l* rays from the upper and lower parts of the sun, or from *D* and *H*, Newton has demonstrated that for rays in the principal plane (one perpendicular to the prism's refracting faces) and near minimum deviation, that is, when $i_u = i'_l$, then $i_l = i'_u$ and by subtraction $\Delta i = \Delta i'$. The rays *DKνT* and *HNκP* lie wholly in the principal plane *ABC*, whereas the rays *VLμZ* and *IMλY* make an angle of about one quarter degree (the apparent solar radius) with that plane. In his second essay "Of Colours" Newton noted, without demonstration, that "The colours should have beene in a round circle were all y^e rays alike refracted" (Add. 3975, f. 2^v = Hall, "Further experiments," p. 28). Later he twice presented abbreviated versions of this first, more essential part of his demonstration, since it is in the principal plane that the spectrum is elongated; Newton to Oldenburg for Pardies, 13 April 1672, *Correspondence*, 1:140–1 = *Phil. Trans.*, 7 (1672):4091–2; and *Opticks*, Bk, I, Pt. I, Prop. II, Expt. III.

(20) In §7 directly following.

(21) quod . . . pigebit] Deleted in II.

quem radius refractus *PR* efficit cum eodem plano, sicut sinus incidentiae ad sinum refractionis; et proinde in ratione datâ. Sumptis enim radijs *IP* et *PR* aequalibus et demissis *IQ* et *RV* ad planum *FH* perpendicularibus et praeterea ad punctum incidentiae *P* erectâ *SPT* perpendiculari ad planum refringens *BD* (quae ideò cum altero plano *FH* coincidet,) et ad istam demissis *IS* et *RT* iterum perpendicularibus: Erit *IPQ* angulus quem radius incidens *IP* efficit cum plano perpendiculari *FH*, et *RPV* angulus quem radius refractus *PR* efficit cum eodem plano: Item *IPS* angulus / incidentiae et *RPT* ang: 6/
refractionis.[22] Quare si *IP* vel *PR* supponatur radius circuli, erunt *IQ*, *RV*,

(a) 6.10[23] Elem   *IS*, et *RT* dictorum angulorum sinus. Sed *IQ* et *RV* sunt paralleli[a] propterea
(b) 28.1 Elem   quod eidem plano *FH* sunt perpendiculares. Item *IS* et *RT* sunt paralleli[b], quia jacentes in eodem plano *ISPTR* eidem rectae *ST* perpendiculàriter insistunt. Hoc est rectae *IQ, IS* quae angulum *QIS* comprehendunt sunt parallelae rectis *RV, RT* comprehendentibus angulum *VRT*. Quare isti anguli *QIS* &

(c) 10.11 Elem[24]   *VRT* sunt aequales[c]. Ductis autem *QS* & *VT* fient anguli *IQS* et *RVT* recti[d],
(d) Def:3.11 Elem   quia rectae *IQ* et *RV* plano *FH* perpendiculariter insistunt. Ergo triangula
(e) 4.6 Elem   *IQS* et *RVT* sunt similia[e]: Et *IQ . RV :: IS . RT*. Hoc est sinus angulorum quos radius incidens & refractus efficiunt cum plano aliquo *FH* ad refringens planum *BD* perpendiculari, sunt ut sinus incidentiae et refractionis; & proinde in ratione datâ. Quippe sinuum istorum rationem esse datam Cartesius edocuit, & alij deinde fuêrunt experti.[25]

Quinetiam Theorematis jam demonstrati veritas manebit salva, licèt planum *EF* plano refringenti[26] *BD* alibi perpendicularitèr insistat, quàm ad punctum refringens *P*. Exinde enim neque anguli cum radijs et plano *FH* effecti, neque ideò sinus istorum angulorum immutabuntur.

7 [= I, 7]. Pars   Hisce ita praemonstratis ad propositum jam revertor, demonstraturus scili-
secunda.   cet angulum *YFZ* (in Fig 3) diametro Solis ac proin angulo *PFT* aequari. Ex supra positis liquet quod planum *KDHNκFν* bisecat angulum radijs *IM* et

---

(22) Newton's terminology for the angle of incidence and angle of refraction (the angles made by the incident and refracted rays with the perpendicular to the refracting surface) is identical with modern terminology, and through the overwhelming influence of the *Opticks* (Bk. I, Pt. I, Defs. IV, V) he managed to redefine these terms in a way contrary to then accepted usage. In the century before the publication of the *Opticks* the angle of incidence was called either "angulus incidentiae" (the angle of incidence) or somewhat less frequently "angulus inclinationis" (the angle of inclination), with the latter apparently popularized by Kepler's prominent adoption of it in his *Dioptrice seu demonstratio eorum quae visui & visibilibus propter conspicilla non ita pridem inventa accidunt* (Augsburg, 1611), §I, p. 1 = *Gesammelte Werke,* ed. Walther von Dyck and Max Caspar, 17 vols. (Munich, 1937), 4:355. In this same period the angle that Newton called "angulus refractionis" (the angle of refraction)—and is called so now—was, insofar as we can determine, universally known as "angulus refractus" (the refracted angle). On the other hand, "angulus refractionis" was then invariably applied to the deviation, that is, the angle made by the refracted ray and the extended incident ray. See, for instance, the following works (all in Newton's library): Barrow, *Lectiones XVIII,* Lect. III, §IV, p. 38; Scheiner, *Oculus,* Bk. II, Pt. I, Ch. 3, p. 60; James Gregory, *Optica promota,* Prop. 2, pp. 8–9; and David Gregory, *Catoptricae et dioptricae sphaericae elementa* (Oxford, 1695), p. 3. Newton completely interchanged accepted usage, for he consistently calls the deviation "angulus refractus"; see for example §§153, 185, 186 = I, 84, 110, 111. He may have been influenced by Hooke's modern, English use of "angle of refraction" in his *Micrographia,* sigs. e[3], f[2].

is to the sine of the angle made by the refracted ray *PR* with the same plane as the sine of incidence is to the sine of refraction, and thus is in a given ratio.

Set the rays *IP* and *PR* equal; drop *IQ* and *RV* perpendicular to the plane *FH*; moreover, at the point of incidence *P* and perpendicular to the refracting plane *BD* erect the line *SPT* that will thus lie in the other plane *FH*; and also drop *IS* and *RT* perpendicular to that line: *IPQ* will be the angle that the incident ray *IP* makes with the perpendicular plane *FH*, and *RPV* will be the angle that the refracted ray *PR* makes with the same plane; similarly, *IPS* will be the angle of incidence and *RPT* the angle of refraction.[22] Consequently, if *IP* or *PR* is assumed to be the radius of a circle, *IQ*, *RV*, *IS*, and *RT* will be the sines of those angles. But *IQ* and *RV* are parallel,[a] since they are perpendiculars to the same plane *FH*. Likewise, *IS* and *RT* are parallel,[b] because they lie in the same plane *ISPTR* and are perpendicular to the same line *ST*. That is, the lines *IQ* and *IS* that contain the angle *QIS* are parallel to the lines *RV* and *RT* that contain the angle *VRT*. Therefore those angles *QIS* and *VRT* are equal.[c] Drawing *QS* and *VT*, the angles *IQS* and *RVT* will become right angles,[d] because the lines *IQ* and *RV* stand perpendicular to the plane *FH*. Therefore, the triangles *IQS* and *RVT* are similar,[e] and *IQ* : *RV* = *IS* : *RT*. That is, the sines of the angles that the incident ray and the refracted ray make with any plane *FH* perpendicular to the refracting plane *BD* are as the sines of incidence and refraction, and are thus in a given ratio. Indeed, Descartes has taught that the ratio of these sines is given, and others have since verified it.[25]

(a) *Elements*, XI,[23] 6

(b) *Elements*, I, 28

(c) *Elements*, XI, 10[24]

(d) *Elements*, XI, Def. 3

(e) *Elements*, VI, 4

Moreover, the truth of the theorem just demonstrated will remain valid even if the plane *EF* is perpendicular to the refracting[26] plane *BD* other than at the refracting point *P*, for neither the angles made by the rays with the plane *FH* nor consequently the sines of these angles will be changed on this account.

Thus having first shown these things, I now return to the proposition to be demonstrated, namely, that the angle *YFZ* (in Fig. 3) is equal to the sun's diameter and thus to the angle *PFT*. From what was considered above it is clear that the plane *KDHNκFν* bisects the angle formed by the rays *IM* and

7 [= I, 7]. Part II.

---

(23) Newton mistakenly cited Bk. 10 rather than 11. Newton's annotated copy of *Euclidis Elementorum libri XV. breviter demonstrati, operâ Is. Barrow* (Cambridge, 1655) is now at Trinity College, Cambridge; see Harrison, *Library of Newton*, no. 581.

(24) If two straight intersecting lines are parallel to two other intersecting lines not in the same plane, they will contain equal angles.

(25) More concisely: $IQ : RV = IP \sin \widehat{IPQ} : RP \sin \widehat{RPV}$, and

$$IS : RT = IP \sin \widehat{IPS} : RP \sin \widehat{RPT},$$

and since $IQ : RV = IS : RT$ and $IP = RP$, then $\sin \widehat{IPQ} : \sin \widehat{RPV} = \sin i : \sin r = \text{constant}$, where the angle of incidence, $i = \widehat{IPS}$, and the angle of refraction, $r = \widehat{RPT}$, are measured in the plane of refraction *ISPRV* that contains the incident and refracted rays and the normal to the refracting surface. Newton has extended Descartes's demonstration of the sine law of refraction for the plane of refraction to any plane perpendicular to the refracting surface; see Lect. 9, note (6), this volume.

(26) Originally: *refractario*.

VL utrinque jacentibus contentum. Itaque cùm iste angulus aequatur diametro Solari, angulus quem radiorum alter puta *IM* cum dicto plano facit aequabitur semidiametro Solari, cujus esto sinus α, et β sinus anguli quem radius ille refractus *Mλ* facit cum eodem plano. Jam cùm planum istud supponatur perpendiculare ad refringens planum prismatis *AC*, erit ex praecedenti Lemmate sinus α, ad sinum β sicut sinus incidentiae, ad sinum refractionis e rariori medio in medium densius. Vel e contra sicut sinus incidentiae ad sinum refractionis e medio densiori in rarius, ita erit β ad α. Quare cùm dictum planum *DHF* etiam perpendiculare sit ad alterum planum Prismatis *BC* quod radios e medio densiori in rarius / refringit; et insuper cum β 7/ supponatur sinus anguli quem radius incidens *Mλ* facit cum plano isto perpendiculari *DHF*: erit (per Lemma praecedens) α sinus anguli quem radius refractus *λF* facit cum eodem plano *DHF*. Sed α ponitur sinus semidiametri solaris; ergo ille angulus quem refractus radius *λF* facit cum plano *DHF* aequatur semidiametro solari: et ejus duplus *λFμ* sive *YFZ* toti diametro.[27] Et cùm supra fuerit ostensum, quòd angulus *PFT* sit eidem diametro aequalis, isti duo anguli *YFZ* et *PFT* erunt aequales. Q.E.D.

Jam si planum *YFZ* esset perpendiculare plano imaginis *PYTZ* aeque ac planum *PFT*, istae quatuor lineae *FP*, *FT*, *FY*, & *FZ* quae angulos aequales comprehendunt essent omnes inter se aequales, & proin subtensae *PT* & *YZ* etiam aequarentur. Sed qui rem seriò perpendet, inveniet radios collaterales *VLμFZ*, & *IMλFY* duobus reliquis *DKνFT* & *HNκFP* paulò minùs refringi; et idcircò planum *YFZ* paulò magis declinabit a radio *FP* quàm ab *FT*, secans lineam *PT* infra medium ejus punctum *X*.[28] Et sic divaricans a perpendiculari *FX* (quam concipe ductam,) erit aliquantulùm obliquum ad planum imaginis *PYTZ*, et ea de causa lineae *FY* & *FZ* erunt paulò majores quàm *FP* et *FT*, et subtensa *YZ* paulò major quàm subtensa *PT*. Sed hujus rei demonstrationem utpote longiusculam et proposito meo non omninò necessariam praetermitto: Etenim non multùm refert utrùm planum *YFZ* sit rectum ad planum imaginis *PYTZ*, vel nonnihil obliquum, hoc est, utrum *YZ* sit aequalis vel major quàm *PT*; sufficit quod nequit esse minor. Imò cùm propter ἰσοςκελέα *PFT* et *YFZ*, sit *FP . FY :: PT . YZ*, atque *FP* & *FY* sint quàm proximè aequales; tantilla erit inter *RT*[29] & *YZ* differentia ut quoad sensum pro aequalibus habeantur.

8 [= I, 8]. In isto tamen casu longitudo imaginis plusquam quadruplex est latitudinis: unde varia refrangibilitas convincitur.

Ostensus itaque casus est in quo longitudo solaris imaginis per Prisma trajectae conspiceretur aequalis ejusdem latitudini; et proinde in quo imago ista quasi orbicularis appareret, modò vera esset opinio vulgaris. Quinimò licèt positio Prismatis alia sit atque descripsi, modò radij refractionem utrinque non valde inaequalem patiantur figura tamen imaginis / eà propter 8/

---

(27) This conclusion, that the angle that the ray λF makes with the plane *DHF* equals the sun's apparent radius, is a special case of the theorem that the incident and emergent rays are equally inclined to the principal plane.

(28) Newton perhaps intuitively argued that rays inclined to the principal plane (*VLμFZ* and *IMλFY*) are refracted less than those in that plane (*DKνFT* and *HNκFP*), because in a prism only the component of the rays normal to the refracting surfaces undergoes refraction, while the parallel component passes through without a net refraction. Only in 1896 was it in fact first rigorously demonstrated that the deviation is least for rays in the principal plane; see J. Larmor,

VL lying on each side. Since, therefore, that angle is equal to the solar diameter, the angle that one of the rays, for example, *IM*, makes with that plane will be equal to the sun's radius; and let α be its sine and β the sine of the angle that the refracted ray *Mλ* makes with the same plane. Now, since that plane is assumed to be perpendicular to the prism's refracting plane *AC*, from the preceding lemma the sine α will be to the sine β as the sine of incidence to the sine of refraction from a rarer into a denser medium; or, in the opposite case from a denser into a rarer medium, the sine of incidence will be to the sine of refraction as β to α. Accordingly, since that plane *DHF* is also perpendicular to the prism's other plane *BC*, which refracts the rays from a denser into a rarer medium, and besides since β is assumed to be the sine of the angle that the incident ray *Mλ* makes with that perpendicular plane *DHF*, by the preceding lemma α will be the sine of the angle that the refracted ray *λF* makes with the same plane *DHF*. But α is assumed to be the sine of the sun's radius, and so that angle that the refracted ray *λF* makes with the plane *DHF* is equal to the sun's radius, and its double, *λFμ* or *YFZ*, is equal to the whole diameter.[27] Furthermore, since it has been shown above that the angle *PFT* is equal to the same diameter, those two angles *YFZ* and *PFT* will be equal. As was to be demonstrated.

If now the plane *YFZ*, just as the plane *PFT*, were perpendicular to the plane of the image *PYTZ*, those four lines, *FP*, *FT*, *FY*, and *FZ*, that contain equal angles would all be equal to one another, and thus the chords *PT* and *YZ* would also be equal. But whoever will carefully consider this will find that the rays on the two sides, *VLμFZ* and *IMλFY*, are refracted a little less than the other two, *DKνFT* and *HNκFP*, and that the plane *YFZ* will there-fore deviate a little more from the ray *FP* than from *FT* and intersect the line *PT* below its midpoint *X*.[28] Hence, inclining away from the perpendicular *FX* (which you should imagine to be drawn), it will be a little oblique to the plane of the image *PYTZ*, and thus the lines *FY* and *FZ* will be slightly greater than *FP* and *FT,* and the chord *YZ* slightly greater than the chord *PT*. But I am omitting the demonstration of this, as it is rather long and not completely necessary for my purpose. For it makes little difference whether the plane *YFZ* is normal to the plane of the image *PYTZ* or somewhat oblique, that is, whether *YZ* is equal to or greater than *PT*; it is sufficient that it cannot be smaller. On account of the isosceles triangles *PFT* and *YFZ*, there is *FP* : *FY* = *PT* : *YZ*; and since *FP* and *FY* are very nearly equal, the difference between *RT*[29] and *YZ* will be so small that they may be con-sidered as sensibly equal.

The case has therefore been presented in which the length of the solar image transmitted through the prism would appear equal to its breadth, and consequently one in which that image would appear nearly circular, provided that the common opinion were true. Moreover, even if the prism's position were other than I have described, as long as the rays do not undergo a particularly unequal refraction on each side, the shape of the image will

8 [= I, 8]. In this case the image's length is nevertheless more than four times its breadth: whence diverse refrangibility is proven.

"On the absolute minimum of optical deviation by a prism," *Proceedings of the Cambridge Philosophical Society,* **9** (1896):108–10.　　(29) Read (as II): *PT*; see note (34), this lecture.

vix immutabitur.[30] Nec multùm interest an corpus opacum *EG*, foramine *F* ad radios transmittendos terebratum, citra Prisma collocetur vel ultra: neque figura foraminis multùm curanda est modò sit exigua.[31] Etenim tam parvae variationes haud plus mutabunt imaginem quàm decimâ fortè vel quintâ parte diametri suae, sicut cogitanti patebit. Atque ita ut paucis tandem comprehendam omnia, liquet quod imago solis refracta utplurimùm deberet esse sensui quasi orbicularis; si modò ejusdem incidentiae in idem medium[32] refractio semper foret eadem. Sed prius repugnat experientiae, longitudine scilicet ejus latitudinem plusquam quatuor vicibus, ut dictum fuit, excedente.[33] Ergo posterius repugnat veritati; & ejusdem incidentiae refractio est varia.

9 [= I, 9]. Ejusdem rei demonstratio contractior.

Ex eodem experimento potui propositum sic breviùs indicasse. Nempe cùm ita disposuissem Prisma ut refractio radiorum tum ingredientium tum egredientium foret quasi aequalis; angulos *PFT* & *YFZ* (fig 2 vel 3) dimensus sum et inveni quidem angulum *YFZ* semissi gradûs sive diametro solis aequalem, at angulus *RFS*[34] eandem diametrum quatèr et ampliùs superavit, cui tamen aequalis esse debuisset ex parte priori demonstrationis praecedentis: Et inde planissimè liquet propositum. Verùm in eorum gratiam quae mox sequentur, oporteret demonstrasse quod illi radij quorum refrangibilitas non est dispar, efformabunt imaginem penè orbicularem. Et eâ de re mihi visum fuit demonstrationem istam etiamsi longiusculam in illustrationem hujus experimenti hic adduxisse.[35]

10 [= I, 10]. Quo pacto Prisma facilè statuatur in situ ad experienda praedicta requisito.

Verùm cùm in experiendis praedictis, eam esse positionem Prismatis supposuerim ut radij ad utramque faciem Prismatis aequaliter[36] refringantur: Conclusionis loco, dicam quâ ratione istud citò fiat, et facilè. Si Prisma teneatur in luce solari & motu lento circa suum axin convertatur, videbis colores quòs efficit, de loco in locum continuo motu translatos esse, ita quidem ut aliquando progredi, deinde verò regredi videantur. Observabis itáque / medietatem inter istos contrarios motus, quando colores modò progressi et statim regressuri, videntur sistere. Quod ubi vides siste Prisma, et in 9/

---

(30) If the prism's position differs significantly from minimum deviation, then the circular image with monochromatic rays will become as elongated as Newton found occurs with polychromatic rays. This was inadvertently demonstrated by Ignace Gaston Pardies who, while not initially recognizing Newton's requirement that the prism be placed at minimum deviation, or at an angle of incidence of about 54°, calculated that when the angle of incidence is 30°, the sun's rays, which before refraction diverged by 30′, will afterward diverge by 2°23′; see Pardies to Oldenburg, 9 April 1672, *Correspondence*, 1:132 = *Phil. Trans.*, 7 (1672):4088–9; and Newton's response, 13 April 1672, *Correspondence*, 1:140–1 = *Phil. Trans.*, 7 (1672):4091–2.

(31) In the *Optica*, §I, 23 ≈ 22, Newton adds still other factors, such as the varying thickness of the prism and the size of the hole, that do not significantly affect the image's size and so cannot cause its elongation. On Newton's use of this argument in the "New theory," see Ronald Laymon, "Newton's advertised precision and his refutation of the received laws of refraction," in *Studies in Perception,* ed. Peter K. Machamer and Robert G. Turnbull (Columbus, Ohio, 1978), pp. 231–58.

(32) in idem medium] Added.

(33) That is, the spectral length is five times its breadth, because the length is said to *exceed* its breadth by four times. Because of his choice of language, as in this postil, in §3 = I, 3, and in

nonetheless hardly be changed because of that.[30] Nor does it make much difference whether the opaque body *EG,* perforated with the hole *F* for transmitting the rays, is placed on the near or the far side of the prism; nor does the shape of the hole matter much, provided that it is small.[31] For such small alterations will scarcely change the image more than a tenth or even a fifth of its diameter, as will be clear to one who considers it. Finally, to include everything in a few words, it is manifest that generally the sun's refracted image must be sensibly nearly circular, provided that in the same medium[32] the refraction at the same incidence be always the same. But the former is contrary to experience, specifically, its length exceeded its breadth more than four times, as has been noted.[33] Therefore the latter is contrary to the truth, and the refraction at the same incidence varies.

From the same experiment I could have shown what was proposed more briefly as follows. Namely, when I had so placed the prism that the refraction of both the entering and emerging rays was almost equal, I measured the angles *PFT* and *YFZ* (Fig. 2 or 3) and found that the angle *YFZ* was indeed equal to half a degree, or the sun's diameter, but that the angle *RFS*[34] exceeded the same diameter by four times and more—which it ought to have nonetheless been equal to according to the first part of the preceding demonstration. Consequently what was proposed is most plainly evident. Yet for the sake of what will follow next, it ought to have been demonstrated that those rays whose refrangibility is not unequal will form a nearly circular image, and for this reason I considered it appropriate to have introduced that demonstration here, even if it was somewhat long, to illustrate this experiment.[35]

9 [= I, 9]. A shorter demonstration of the same thing.

Since in doing these experiments I supposed the prism's position to be such that the rays would be refracted equally at each face of the prism,[36] in place of a conclusion I will explain how that may be done quickly and easily. If the prism is held in the sun's light and turned about its axis with a slow motion, you will see the colors that it makes shifted with a continuous motion from place to place, so that in fact they appear to move now forward now backward. Therefore you will observe a mean between those opposite motions when the colors, now having moved forward and just about to move backward, appear to stand still. When you see that, stop the prism and fix it in

10 [= I, 10]. How the prism may be easily placed in the position required to do that experiment.

§14, Newton's statements on the spectral length appear to be inconsistent; but from his measurements, which are finally related in §110 ≈ I, 37, the spectral length was in fact more than five times greater than its breadth.

(34) Read (as II): *PFT.* Before settling on *P* and *T* for the red and blue ends of the spectrum, a designation consistently used in all of his optical writings from the *Lectiones opticae* through the *Opticks,* Newton evidently used *R* and *S* in a draft of the *Lectiones,* for occasionally he wrote *RS* instead of *PT,* which is used in all the diagrams—though he corrected most of these slips himself.

(35) Originally continued: Imò conclusio forsan evidentior censeatur utroque modo commonstrata. (Indeed, the conclusion may perhaps be considered more evident after being demonstrated each way.)

(36) ad . . . aequaliter] Originally: tam ingredientes tam et egredientes aequè (equally both on entering and on leaving).

eo situ fige. Dico factum. Scilicet in eo situ[37] summa refractionum utrobique factarum, sive radij emergentis ad incidentem inclinatio evadit omnium minima; quod cum accidit, refractiones utrobique sunt aequales, uti posthac demonstrabitur.[38]

Lect 2 | Ut indicarem disparitatem[1] aliquam quoad refractiones inesse luci; scilicet quòd radij ejus ex eâdem incidentiâ ijsdemque medijs refringentibus diversam refractionis quantitatem admittunt: adhibui experimentum quoddam Prismatis vulgare; quod videre licèt in Fig 2; ubi ostensum est quòd radij solares per aliquod foramen *F* et per prisma *ABC* deinceps trajecti, imaginem *PYTZ* in opposito pariete efformabunt oblongam: Quae tamen quasi orbicularis esse debuisset modò radij omnes ὁμοιοπαθῶς refringerentur. Istud autem experimentum jam repeto ut varias ejus circumstantias non minùs jucundas experienti quàm propositi nostri indicativas prosequar.

11 [≈ I, 11].
Imaginis praefatae figura describitur: quòd partim rectis, partim semicirculis terminatur.

Et primo notandum venit quòd imaginis istius figura secundum longitudinem suam lineis rectis terminata fuit, et secundum latitudinem duobus (ut ex visu potui judicare)[2] semicirculis. In fig 5 sit *PT* imago solis prismate refracta: Hanc observabam ad latera duabus lineis *AB* et *CD* sensui rectis et sibi parallelis terminari, ad extremitates autem duobus semicirculis *APC* et *BTD*. Cujus quidem eventus causa ex praemonstratis sic determinatur.[3]

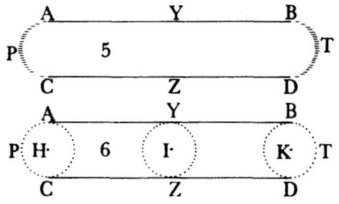

Figures 5 & 6

12 [= I, 12].
Quomodo talis evadit per orbiculares[4] imagines (quas unumquodque genus radiorum aequè refrangibilium facit) in longum dispositas.

Semicirculi illi terminales in circulos compleantur ut vides in fig 6. Et alius inscribatur circulus *YZ* istis intermedius. Jam concipe radios quosdam a Sole provenientes qui apti sunt ut incidentes aequè, etiam aequè refringantur: Illi per Prisma trajecti, ex supra demonstratis, imaginem quoad sensum (si sola posset videri)[5] circularem depingent, puta *BD*. Deinde concipe alios ejusdem solis radios / sibi etiam conformes, 10/ qui apti sunt ut prioribus paulò magis refringantur, illi itaque aliam imaginem depingent circularem puta *YZ*: Et alios etiam radios adhuc magis refrangibiles concipe, qui tertiam circularem imaginem *AC* efficiant. Denique alios

---

(37) The paragraph originally concluded: refractio radiorum est minima, hoc est summa refractionum factarum utrumque Prismatis. Sed cum ista summa est minima, refractiones istae sunt aequales uti post aliquot Lectiones facilè demonstrabitur. (the rays' refraction is least, that is, the sum of the refractions made on both sides of the prism. But when this sum is least, those refractions are equal, as will be easily demonstrated several lectures afterward.)

(38) See §186 = I, 111, where, however, it is proved that the sum of the refractions is a maximum, since the supplement of the deviation is considered there. Newton later gave slightly different instructions for setting the prism at minimum deviation: in the "New theory," *Correspondence*, 1:93–4 = *Phil. Trans.*, **6** (1671/2):3077; in his letter to Oldenburg for Line, 13 November 1675, *Correspondence*, 1:357 = *Phil. Trans.*, **10** (1675/6):501; and in the *Opticks*, Bk. I, Pt. I, Prop. II, Expt. III.

(1) Originally: difformitatem.

(2) (ut . . . judicare)] Added in revision. With this parenthetical addition Newton concedes, as he does elsewhere, that his spectrum was not bounded by perfect straight lines and semicircles. In §14 ≈ I, 14, in particular, he readily admits that its ends always appeared confused and indis-

that position. I say you are done. Namely, in that position[37] the sum of the refractions made on both sides, or the inclination of the emergent to the incident ray, proves to be least of all; and when this happens, the refractions on both sides are equal, as will be demonstrated later.[38]

In order to show that there is in light a certain inequality[1] with regard to its refractions (namely, that its rays at the same incidence and in the same refracting media admit a different quantity of refraction), I employed a certain common experiment with a prism, which may be seen in Fig. 2, where it was shown that the sun's rays, transmitted through some hole *F* and then through a prism *ABC,* will form on the opposite wall an oblong image *PYTZ* that nonetheless ought to have been nearly circular if all rays were refracted *homeopathically.* I now repeat that experiment, however, so that I may pursue its various features that are no less pleasant for the experimenter than they are informative for our purpose.

Lecture 2

In the first place, it should be noted that the shape of that image was bounded lengthwise by straight lines and in its breadth by two semicircles (insofar as I have been able to judge by sight).[2] In Fig. 5 let *PT* be the sun's image refracted by the prism. I observed this to be terminated on its sides by two lines, *AB* and *CD,* sensibly straight and parallel to each other, but at the ends by two semicircles, *APC* and *BTD.* The cause of this result is determined from what has previously been described as follows.[3]

11 [≈ I, 11]. The shape of that image, which is bounded partly by straight lines and partly by semicircles, is described.

Let those terminal semicircles be completed into circles, as you see in Fig. 6, and inscribe another circle *YZ* between them. Now imagine some rays to come from the sun that are disposed to be also equally refracted when they are equally incident; according to the above demonstration, after passing through the prism, these will depict a sensibly circular image, for example, *BD*—if it could be seen alone.[5] Then imagine other rays from the same sun, also similar to one another, that are disposed to be refracted a little more than the first; those therefore will depict another circular image, for example, *YZ.* Also imagine other still more refrangible rays that make a third circular image *AC.* Finally, conceive innumerable others more and less refrangible

12 [= I, 12]. How this arises from circular[4] images (which each kind of equally refrangible ray produces) disposed lengthwise.

---

tinct; see also, "New theory," *Correspondence,* 1:92 = *Phil. Trans.,* 6 (1671/2):3076; and *Opticks,* Bk. I, Pt. I, Prop. II, Expt. III, p. 19 = Add. 3970, f. 32ʳ. Newton's account of the shape of the spectrum, which in fact varies greatly depending on experimental conditions, has frequently been questioned; see Kuhn, "Newton's optical papers," in Newton, *Papers and Letters,* pp. 34–5; and Lohne, "Experimentum crucis," *Notes and Records of the Royal Society of London,* 23 (1968):169–99, esp. 171–3. Francis Line (or Francis Hall), who is not otherwise a generally reliable observer, and Edme Mariotte, who did not observe at minimum deviation, each saw a spectrum shaped like a teardrop, broader at the red end and narrower at the blue end; see Line to Oldenburg, 25 February 1675, in *Correspondence,* 1:335 = *Phil. Trans.,* 10 (1675):499; and Edme Mariotte, *De la nature des couleurs* [1681], in *Oeuvres de M. Mariotte . . . comprenant tous les traitez de cet auteur, tant ceux qui avoient déja paru séparément, que ceux qui n'avoient pas encore été publiés . . .* 2 vols., new ed. (The Hague, 1740) 1:208.

(3) Cujus . . . determinatur.] Originally: Id quod vix alio quàm sequenti modo explicabitur. (This will scarcely be explained by any other way than the following.)

(4) II: circulares.   (5) (si . . . videri)] Added.

innumeros cogita praedictis plus et minus refrangibiles, et illi alias etiam innumeras circulares imagines prioribus tum intermedias tum extremas efformabunt, illuminantes oblongum spatium *PYTZ* lineis[6] *AB* & *CD*, duobusque semicirculis contentum. Verùm cùm imagines illae sunt omnes ejusdem penè magnitudinis et inter lineas *AB* et *CD* in directum dispositae, istae lineae *AB* et *CD* pro rectis sibi parallelis possunt haberi et ad sensum tales videbuntur. Et sic totum spatium *PYTZ* radijs ex eâdem incidentiâ variè refractis illuminatum, partim parallelis rectis & partim semicirculis oppositis terminabitur: sicut experientiâ compertum est.[7]

13 [= I, 13]. Exinde deducitur experimentum, quo termini recti fiant distinctissimi.

Hanc autem conjecturam ut penitùs probarem, cogitabam de imagine Solis per foramen aliquod sine ullâ refractione ad distantiam magnam trajectâ, scilicet quòd malè definitur, termino existente inter lucem et tenebras minimè distincto: at si radij isti per lentem convexam transeant, cujus focus ad imaginem est, imago terminabitur distinctissimè.[8] Simili modo de radijs aequè refrangibilibus intellexi quod si per Prisma trajicerentur ad distantiam magnam, depingerent imaginem circularem malè definitam, cujus tamen terminus, mediante lente convexâ, distinctissimus evaderet. Itaque cùm vidissem terminos imaginis refractae *PYTZ* non admodum distinctos, de imaginibus *BD, YZ, AC* & reliquis circularibus oblongam istam formantibus conjiciebam quòd multò distinctiùs terminarentur per lentem convexam trajectae quàm alitèr. Et experienti res patuit: Nam rectas *AB* et *CD,* in quas imagines omnes istae circulares utrinque terminantur, vidi admodùm distinctas, quas antea confusas videram.[9]

14 [≈ I, 14]. Quare termini circulares semper apparent confusi.

Sed quod notatu valdè dignum videtur, termini circulares *APC* ac *BTD* imaginis istius semper apparuêre maximè[10] confusi, luce paulatim deficiente donec tandem in tenebras desijt. Scilicet intermedij circuli, ut *YZ*, miscentur alijs / circulis utrinque cadentibus, quibuscum, ex aliquâ sui parte, coincidunt: at extremi quidem circuli *AC* et *BD,* ex unâ tantùm parte cum alijs concurrunt, et eorum concursus continuò fit rarior, et exinde lux usque remissior dum ad extremitates *P* ac *T* deventum est. Sed et alia prodit istius rei causa, scilicet quòd radiorum maxima copia apta sit ut mediocrem refractionem patiatur; et sic in medium imaginis incidat. | Reliquorum autem radiorum numerus continuò minor est, prout eorum refrangibilitas sit alterutrinque[10] magis extrema. Et hinc in cubiculo diligentèr obscurato imaginis pars media aequali ferè luce perfusa quasi ter vicibus longior fuit quàm lata: at luce gradatim obscuriori in longitudinem plusquam quinque vicibus majorem latitudine processerit.

11/

---

(6) **II**: rectis lineis (straight lines).

(7) Newton introduces this conception of the formation of the spectrum by overlapping circular images in the crossed-prism experiment in the *Opticks,* Bk. I, Pt. I, Prop. II, Expt. V.

(8) Newton could have derived this suggestion from a number of sources: Giovanni Battista della Porta, *Magiae naturalis libri viginti* (Leyden, 1651), Bk. XVII, Ch. VI, p. 587 (Newton owned this edition of this exceedingly popular work that was first published in its twenty-book edition in 1588; see Harrison, *Library of Newton,* no. 1340); Scheiner, *Oculus,* Bk. III, Pt. I, Ch. II, p. 127; or Descartes, *Dioptrice,* Ch. V, §12, *Oeuvres,* 6:604, 126. In his essay "Of Refrac-

than the preceding, and those will also form other innumerable circular images both in between and beyond the preceding ones, illuminating the oblong space *PYTZ* contained within the lines[6]*AB* and *CD* and the two semicircles. But since those images are all nearly the same size and disposed in a straight line between the lines *AB* and *CD*, the lines *AB* and *CD* can be considered to be straight lines parallel to one another, and to the senses they will appear as such. Thus the entire space *PYTZ*, illuminated by rays diversely refracted at the same incidence, will be terminated partly by parallel straight lines and partly by opposite semicircles, just as is found by doing the experiment.[7]

So that I might fully confirm this conjecture, however, I thought about the sun's image cast through some hole to a great distance without any refraction, in particular, I thought about the fact that it is poorly defined, the border between light and darkness being very indistinct; but if those rays pass through a convex lens, whose focus is at the image, the image will be terminated very distinctly.[8] Similarly, for equally refrangible rays I perceived that if they were transmitted through a prism to a great distance, they would depict a poorly defined circular image, whose border would nevertheless turn out to be very distinct by interposing a convex lens. When, therefore, I had seen that the borders of the refracted image *PYTZ* were not very distinct, I inferred that the images *BD*, *YZ*, *AC*, and all the other circular ones forming that oblong would be terminated much more distinctly when transmitted through a convex lens than they would otherwise be. And this was evident by doing the experiment, for I saw the lines *AB* and *CD* that terminate all those circular images on both sides to be quite distinct, whereas previously I had seen them to be confused.[9]

13 [= I, 13]. From this an experiment is deduced whereby the straight edges are made very distinct.

It seems especially noteworthy, however, that the circular ends of that image, *APC* and *BTD*, have always appeared extremely[10] confused, with their light gradually decreasing until finally ending in darkness. Specifically, the intermediate circles, such as *YZ*, are mixed with other circles, which fall on each side and with which they partially coincide; but the outermost circles *AC* and *BD* in fact run into the others on only one side, and their intersection becomes continually rarer and accordingly the light continually fainter until reaching the ends *P* and *T*. This also has another cause, namely, that the greatest quantity of rays is disposed to undergo a mean refraction and thus falls in the middle of the image. | The number of remaining rays, however, is continually smaller according as their refrangibility on each side[10] is the more extreme. Hence, in a carefully darkened room, the middle portion of the image—nearly uniformly suffused with light—was about three times longer than broad, but with the gradually obscurer light it will have advanced into a length more than five times greater than its breadth.

14 [≈ I, 14]. Why the circular ends always appear confused.

tions" Newton had already noted that lenses "not onely magnify objects but render vision more distinct" (Add. 4000, f. 30ᵛ = Hall, "Further experiments," p. 41).

(9) In the account of this experiment in the *Opticks* (Bk. I, Pt. I, Prop. II, Expt. V, p. 28 = Add. 3970, f. 38ʳ) the lens is described as being placed "at that hole."

(10) Added.

15 [= I, 15].
Admonitio de figura
et situ Lentium et
Prismatum.

Caeterum ad isthaec experienda Lentes adhiberi vellem quarum foci sunt longinqui, sex fortè vel duodecim pedibus a lentibus distantes, modò tales praesto sunt: Saltem non sint minùs distantes quàm duobus. Atque etiam latera Prismatis debent esse accuratè plana. Sin latera ejus sint aliquatenus convexa tum praestat adhibere lentem cujus focus ad pedes tantùm duos tresve a se remotus est. Quibus paratis Lentem Prismati ex utravis parte vicinam colloca: ita scilicet ut radios per se trajectos directè respiciat. Deinde radij in papyrum aliquam excipiantur, quam ultrò citróque transfer, donec imaginem coloratam utrinque rectis parallelis distinctissimè terminatam videas.

16 [= I, 16]. Deíque
imagine quâdam
orbiculari.

Sed observandum est quòd cùm Prisma collocetur ultra foramen *F,* ut in Fig 3, vel ipsi quàm proximè citra; et lens magis distet ab isto foramine quàm focus lentis, quem radij in eam parallelõs incidentes efficerent, distat a lente: duplicem invenies casum in quo imago in papyrum projecta evadet distincta; alter quando radij omnes homogenei[10] qui in Lentem paralleli incidunt ita refringantur ut ad papyrum istam in eodem puncto concurrant, quod fit cùm vides imaginem coloratam, oblongam, et parallelis rectis distinctè terminatam. Alter casus est quando radij omnes homogenei[10] ab uno puncto foraminis *F* divergentes, postquam a lente refringuntur, ad unum iterum / punctum 12/ dictae papyri convergunt. Id autem accidit cùm imaginem albam, orbicularem, et undique benè definitam vides: De quo fusè dicetur alibi;[11] sufficiat hoc monitum hic dedisse, nequis haec proprijs experturus oculis, per ambiguitatem effectus incautè decipiatur, et exinde praedicta in dubium revocet.

17 [= I, 18]. Ab
imaginis figura aliud
etiam experimentum
deducitur quo fiat
multum oblongior.

Ut dictas proprietates lucis quâ potui diligentiâ perscrutarer, sequentem praeterea modum excogitavi quo illas examini subjicerem. Nempe, in fig 6, cùm magnitudo circulorum *AC, YZ, BD* dependeat a magnitudine solari: si diameter solis fieret aliquantò minor quàm nunc reverà existit, tum illi etiam circuli fierent minores, distantiâ centrorum *H, I, K* non omninò mutatâ: ut videre est in fig 7. Et sic latitudo imaginis ad ejusdem longitudinem comparata multò minor evaderet quàm antea, utrâque scilicet per eandem quantitatem diminutâ. Haec probaturus effeci ut radij Solis per duo parva foramina ab invicem longè distantia transirent antequam inciderent Prismati, quo pacto radij ab extremis partibus solis venientes excludebantur, et res perinde successit quasi diameter solis reverà fuisset diminuta. Illustrationis gratiâ (in fig 8) ϵφγ

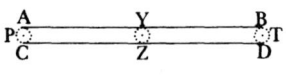

Figure 7

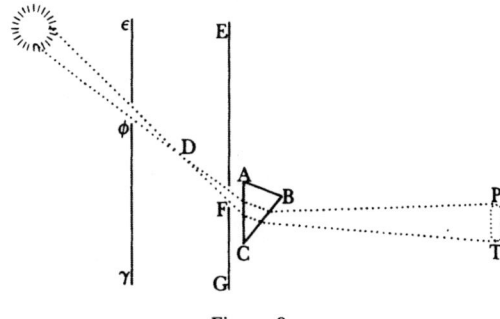

Figure 8

To do these experiments I suggest that lenses be used whose foci are far off, distant perhaps six or twelve feet from the lenses—if such lenses are readily available; in any case, they should not be distant less than two feet. In addition, the sides of the prism ought to be exactly flat; but if its sides are somewhat convex, then it is preferable to use a lens, the focus of which is only two or three feet away from it. Having procured these things, place the lens near either side of the prism such that it directly faces the rays transmitted through it. Then catch the rays on some paper while shifting it back and forth until you see a colored image terminated very distinctly on both sides with parallel straight lines.

15 [= I, 15]. Advice about the shape and position of the lenses and prisms.

It must be noted that when the prism is placed outside the hole *F* (as in Fig. 3) or else inside as close as possible to it, and the lens is farther from that hole than the lens's focus (which would be produced by parallel rays falling on it) is from the lens, you will find two cases in which the image projected onto the paper will turn out to be distinct. The first happens when all homogeneous[10] rays falling parallel on the lens are refracted such that they meet at the same point on that paper; this occurs when you see a colored oblong image terminated distinctly by parallel straight lines. The second case occurs when all homogeneous[10] rays diverging from one point of the hole *F*, after being refracted by the lens, converge again to one point on that paper; this occurs, however, when you see a white circular image well defined all around. I will discuss this case in detail elsewhere;[11] let it suffice to have given this warning here, so that anyone trying these things with his own eyes may not be unwarily deceived by the variability of the effect and then call the preceding in doubt.

16 [= I, 16]. And about a certain circular image.

In order to investigate the said properties of light as diligently as I could, I thereafter conceived of the following way whereby I might put them to a test. Namely, in Fig. 6, since the size of the circles *AC*, *YZ*, and *BD* depends on the size of the sun, if the sun's diameter were to become somewhat smaller than it now actually is, then those circles would also become smaller without at all changing the distance of their centers *H*, *I*, and *K*, as can be seen in Fig. 7. Thus the breadth of the image compared to its length would turn out to be much smaller than before, that is, both decreased by the same amount. To prove this, I made the sun's rays pass through two small holes, very far from one another, before they fell upon the prism; in this way the rays coming from the outer parts of the sun were cut out, and it succeeded just as if the sun's diameter had really been diminished. For an illustration, in Fig. 8 let

17 [= I, 18]. From the shape of the image still another experiment is deduced whereby it becomes much more oblong.

---

(11) In §§54–8 = II, 43–7.

fenestra parvo foramine ϕ penetrata, per quod radij solares cubiculum alias obscuratum ingrediantur. Deinde sit *EFG* corpus aliquod opacum perforatum ad *F,* et in medio cubiculo ita locatum ut radij iterum permeent foramen istud antequam Prisma *ABC* pone locatum attingant. Jam foraminum istorum diametro existente $\frac{1}{8}$ digiti, et eorundem distantiâ ϕF 12 pedibus, (ita scilicet ut maxima radiorum utrumque foramen permeantium inclinatio foret angulus sex minutorum ferè, hoc est quasi quinta pars diametri solaris;) atque etiam imagine [*PT*] projectâ in papyrum decem pedes a Prismate distantem, prout angustia / cubiculi tulit: inveni longitudinem imaginis esse plusquam qua- 13/ tuor[12] digitorum cum semisse et latitudinem trientis digiti, hoc est longitudinem plusquam quatuordecim[13] vicibus majorem latitudine, sicut ex praedictis oportuit evenisse.[14] Etenim cùm isti tantùm radij mittuntur intrò qui minùs quam quintâ[15] parte solaris diametri ad se invicem inclinantur, diametri *AC, YZ,* & *BD* diminutae diametro foraminis *F* debent esse quintuplo minores quàm secundum priora contingeret; ut videre est in figuris 6 & 7. Quasi a Sole essent effectae cujus diameter sit quinquies minor diametro solis nostri. Verùm si corpus opacum [ϵ]$\phi\gamma$ (fig 8) tolleretur ut radij per unum solummodò foramen *F* ad Prisma transirent, sicut in prioribus factum est: latitudo imaginis evaderet $1\frac{1}{6}^{\text{dig}}$ et longitudo plusquam 5^{dig}: angulo nempe Prismatis existente 60^{grad}, vel paulo magis eo.[16] Itaque diameter circulorum *AC, YZ,* & *BD,* qui eo quò dictum est modo imaginem constituunt, esset $1\frac{1}{6}^{\text{dig}}$. A quâ subducatur diameter foraminis nempe $\frac{1}{8}^{\text{dig}}$, & manebit $1\frac{1}{24}^{\text{dig}}$ cujus quintae parti rursus adjungatur eadem foraminis diameter sive $\frac{1}{8}^{\text{dig}}$ et prodibit $\frac{1}{3}^{\text{dig}}$, diameter circulorum *AC, YZ* & *BD* in Fig 7: quae minor est quàm diameter circulorum istorum in fig 6, quantitate $\frac{5}{6}^{\text{dig}}$. Quamobrem figura 7ma quaquaversum minor est quàm sexta, quantitate $\frac{5}{6}^{\text{dig}}$. Atque ideò longitudo ejus fit plusquam 4^{dig}, latitudo autem digiti triens. Id quod cum experientia mox[17] recensitâ quadrat. Ad eundem modum si foramina ϕ et *F* adhuc minora forent, vel si distantia ϕF foret major, imago [*PT*] oblongior evaderet. Quod idem quoque quadantenus contingeret, ex imagine [*PT*] a Prismate longiùs dissitâ. Caeterùm notandum est quod foramina ϕ et *F* ad radios directè respicientia suppono, licèt non multùm refert an situs eorum sit parùm obliquus ut in appositâ figurâ $8^{\text{mâ}}$ factum est.

18 [= I, 19]. Porrò si in hoc experimento convexam Lentem ut priùs adhibueris, cujus
Experimentum istud focus ad imaginem cadit, foramine *F* si placet dilatato vel opaco corpore *EG*
promovetur. prorsus ablato, ut radij per longinquum foramen ϕ solummodò transeant / et 14/ si foramen istud ϕ effeceris angustius quàm antea, caeteris ut priùs stantibus, imaginem valdè oblongam & pro longitudine lucidiorem videbis quàm in casu priori. Exempli gratia si diameter foraminis sit pars digiti vigesima, & si pedibus abinde duodecim Prisma cum Lente disposueris, videbis longitudi-

---

(12) Originally: trium (3[$\frac{1}{2}$]).

(13) Originally: dece[m] (ten).

(14) Newton first described this experiment in his essay "Of Colours" (§8, Add. 3975, f. 2ᵛ = Hall, "Further experiments," p. 29), where he reduced the sun's apparent diameter to less than 7 minutes and found the spectral length to be $2\frac{1}{4}$ inch and the breath $\frac{1}{8}$ inch, and so the length to be only seven and one-third times the breadth.   (15) Originally: decimâ (tenth).

$\epsilon\phi\gamma$ be a window pierced with a small hole ϕ through which solar rays may enter an otherwise darkened room. Then let *EFG* be an opaque body perforated at *F* and so placed in the middle of the room that the rays may also pass through that hole before they arrive at the prism *ABC* placed behind it. Now, when the diameter of those holes was $\frac{1}{8}$ inch and their distance ϕF was twelve feet (so that, in particular, the greatest inclination of rays passing through both holes would be an angle of nearly six minutes, that is, about a fifth part of the solar diameter); and also when the image [*PT*] was projected onto a paper ten feet away from the prism, as the narrowness of the room allowed, I found the image's length to be more than $4\frac{1}{2}$[12] inches and its breadth $\frac{1}{3}$ inch; that is, the length was more than fourteen[13] times greater than its breadth, just as it should turn out according to what was just described.[14] For since only those rays are allowed to pass inside that are inclined to one another less than a fifth[15] part of the solar diameter, the diameters *AC*, *YZ*, and *BD*, decreased by the diameter of the hole *F*, must be five times smaller than would result according to the former case, as is seen in Figs. 6 and 7. It is just as if they were produced by a sun whose diameter was five times smaller than the diameter of our sun. But if the opaque body [ϵ]$\phi\gamma$ (Fig. 8) were removed so that the rays would pass to the prism through only one hole *F*, as happened in the former experiments, then the image's breadth would turn out to be $1\frac{1}{6}$ inch and its length more than 5 inches, that is, when the prism's angle was 60° or a little more.[16] Therefore, the diameter of the circles *AC*, *YZ*, and *BD*, which form the image in the way already described, would be $1\frac{1}{6}$ inch. From this subtract the diameter of the hole, namely, $\frac{1}{8}$ inch, and $1\frac{1}{24}$ inch will remain; to its fifth part in turn add the same diameter, or $\frac{1}{8}$ inch, and this will produce $\frac{1}{3}$ inch for the diameter of the circles *AC*, *YZ*, and *BD* in Fig. 7, which is less than the diameter of those circles in Fig. 6 by a quantity of $\frac{5}{6}$ inch. Consequently, the seventh figure is less than the sixth on all sides by $\frac{5}{6}$ inch. Hence its length becomes more than 4 inches but its breadth $\frac{1}{3}$ inch, in agreement with the experiment just now[17] set forth. Likewise, if the holes ϕ and *F* were even smaller, or if the distance ϕF were greater, the image [*PT*] would become more oblong. The same thing would also occur to a certain extent when the image [*PT*] has been set farther away from the prism. It must be noted, however, that I am assuming that the holes ϕ and *F* are in a straight line with the rays, although it does not matter much whether their position is a little oblique, as is drawn in the adjoined Fig. 8.

Moreover, if in this experiment as before you use a convex lens whose focus falls at the image, with the hole *F* enlarged or, if you prefer, the opaque body *EG* completely removed so that the rays pass through only the far hole ϕ; and if you make that hole ϕ narrower than previously, with everything else remaining as before, you will see the image as very oblong and brighter in its length than in the former case. For example, if the diameter of the hole is $\frac{1}{20}$ inch, and if you place the prism with the lens twelve feet from there, you

18 [= I, 19]. That experiment is extended.

(16) There are sufficient data here to calculate the apparent solar diameter, and it turns out to be 29′51″, or about a minute smaller than the 30′56″ that results from the observations reported in §110 ≈ I, 37.   (17) Read (as II): modò (just now).

nem imaginis plusquam 80 vel 100 vicibus latitudine majorem.[18] Sed in his experiendis oportet cubiculum quaquaversus benè obturatum esse, ne lux alibi quàm per foramen ϕ ingressa perturbet imaginem & juxta circulares ejus extremitates obscuram reddat. Et praeterea si superficies Prismatis sint accuratè planae, praestat adhibere lentem quae focum ad distantiam magnam projicit, puta ad 12 aut 20[19] pedes, modò loci amplitudò sinat; quo pacto de proportionibus imaginis satiùs judicium proferas. Quod si latera Prismatis sint aliquantulùm convexa, ut ijs nonnunquam contingit quae[20] vulgò venduntur: licebit istud abque ullâ lente solum adhibere, et ejus convexitas radios vice lentis ad magnam distantiam congregabit. Quinimò si cum Prismate quolibet lentem parvam adhibeas cujus focus non sit duobus tribusvé pedibus longinquior, imaginem conspicies satis longam quidem, sed cujus latitudo tamen haud sensibilis existit: Id quod proposito nostro non minùs inservit quàm si posses de proportione longitudinis ad latitudinem ejus accuratè judicare. In istis etiam experiendis notetur praeterea quòd lens non debet ita longè post Prisma locari, quin ut possit ad omnes radios simul transmittendos extendi: nè imaginem successivè per partes tantùm observare sis coactus. Et notetur denique quòd si foramen F citra Prisma locaveris & lentem deinceps citra foramen istud, ad distantiam ab eo majorem quàm focus radiorum a foramine ϕ longinquiori manantium abest a lente; duplex erit casus in quo imago in papyrum projecta conspicietur distincta, prout radij venientes a singulis punctis foraminis F aut a singulis punctis foraminis ϕ, in totidem iterum punctis papyri colliguntur. In uno casu imago erit alba et orbicularis, uti priùs commonui[a]; in altero autem oblonga et colorata, sicut praesens experimentum exigit.[21]

(a) Numb 16

/ Jam liquet ex praefatis quod imaginis PT latitudo semper evadit eo minor 15/ quo foramen ϕ longinquum factum est angustius: ut nihil dubitandum sit quin dicta latitudo prorsus evanesceret si vice foraminis istius translucidi unum duntaxat punctum ibi lucidissimum existeret. Atque istud sic futurum esse confirmatur ex observatione non dissimili quam habui quondam de stellâ Veneris. Cubiculo nempe quaquaversus obturato, excepto foramine paulo plusquam duos digitos lato, ut tenebrosissimum efficeretur. In isto foramine vitrum objectivum Perspicilli septempedalis collocavi; latitudine ejus, ad sufficientem radiorum copiam transmittendam, duos digitos et ampliùs apertâ. Deinde ad distantiam septem pedum papyro transversè positâ, in eam vidi sideris imaginem ad instar puncti lucidi projectam. Et interposito Prismate ad distantiam pedis unius duorumvé ab istâ papyro, per quod radij trajecti aliò refringerentur: pro puncto illo lucido ad distantiam

19 [= I, 20]. Magis adhuc promovetur, per imaginem stellae Veneris.

(18) In the account of this experiment in the *Opticks* (Bk. I, Pt. I, Prop. IV, Expt. XI) Newton was able to make the image's breadth only "forty times & sometimes sixty or seventy times less then its length" (Add. 3970, f. 53ʳ = *Opticks*, p. 49). Besides the different experimental conditions reported in the *Opticks*, such as the size of the hole, the function of the experiment there differs, namely, "To separate from one another the heterogeneous rays of compound Light"; but in the *Optica*, §II, 22, Newton also invokes this experiment for that same purpose. These experiments with a narrow hole, a prism, and lens resemble a spectroscope, yet, as has frequently been noted, Newton never indicates that he saw any spectral lines. The fourth Baron Rayleigh repeated Newton's experiments and observed spectral lines even without a lens; see his excellent

will see the length of the image eighty or one hundred times greater than its breadth.[18] In these experiments it is necessary that the room be thoroughly closed off everywhere, so that light entering anywhere but through the hole ϕ does not disturb the image and make it indistinct near its circular ends. Moreover, if the prism's surfaces are exactly flat, it is preferable to use a lens that projects its focus to a great distance, for instance, to twelve or twenty[19] feet if the room's size allows it. This way you can make a better judgment of the proportions of the image. If, however, the sides of the prism are slightly convex (as sometimes happens in those[20] commonly sold), it may be used alone without any lens, and its convexity will gather the rays at a great distance like a lens. In fact, if with any prism you use a small lens whose focus is not longer than two or three feet, you will, to be sure, observe an image sufficiently long, but yet with a width that is not perceptible; this serves our purpose just as well as if you could accurately judge the proportion of the length to its breadth. It should also be noted that in doing these experiments the lens must not be placed so far behind the prism that it cannot transmit all the rays at one time, lest you be forced to observe the image only successively and piecemeal. Finally, it should be noted that if you place the hole F beyond the prism, and then you place the lens beyond that hole at a distance from the hole greater than that of the focus of the rays flowing from the farther hole ϕ from the lens, there will be two cases in which the image projected onto the paper will be seen distinctly, according as the rays coming from each point of the hole F, or from each point of the hole ϕ, are gathered again in as many points on the paper. In the one case the image will be white and circular, as I urged earlier,[a] but in the other it will be oblong and colored, just as the present experiment requires.[21]

(a) §16

Now, from the preceding it is evident that the breadth of the image PT always becomes smaller as the farther hole ϕ is made narrower, so that it cannot be doubted that the breadth would completely vanish if instead of that transparent hole there were only one extremely bright point there. That this will happen is confirmed by a not dissimilar observation I once made of the planet Venus. The room was of course closed off everywhere, except for a hole slightly more than two inches wide, in order to make it exceedingly dark. In this hole I placed the object glass of a seven-foot telescope with an aperture more than two inches wide, in order to transmit a sufficient number of rays. Then placing a paper transversely at a distance of seven feet, I saw the planet's image projected on it like a bright point. When one or two feet away from that paper a prism was inserted that refracted the transmitted rays elsewhere; instead of that bright point, distant more than a foot from there, I

19 [= I, 20]. This is extended still further by the image of the planet Venus.

analysis of this problem, "Optical topics in part connected with Charles Parsons," *Nature*, 152 (1943):676–82, esp. 679–82.

(19) aut 20] Added.   (20) ut . . . quae] Originally: uti contingit ijs qui.

(21) Originally continued: Atque de imagine ista circulari et alba postea dicam plura, sed quandem ejus quae nunc occurrit circumstantiam annotare et in transcursu jam explicatam dare placuit. (I will say more about that circular white image afterward, but it was preferable to note a feature of it that here arises and in passing to take it as already explained.)

indè plusquam pedalem, vidi lineolam licèt non valdè lucidam facilè conspicuam tamen; et cujus longitudo semissem digiti superavit; latitudo autem fuit quoad sensum nulla, saltem haud major quàm ut sentiretur. Atque idem credo de Stellis primae magnitudinis, uti de Sirio, liceat observare: praesertim si lens adhibeatur quatuor vel sex digitos lata, ut plures[22] radios transmittat.

20 [= I, 21]. Et applicatur descriptioni refractionis ad Fig 1 traditae.

Hoc experimentum quàm benè convenit cum explicatione nostrâ quam de refractione radiorum ad eundem angulum incidentium variâ, sub initio dedi, operae pretium videatur adnotare. In figurâ primâ, supposui complures radios per eandem rectam in superficiem aliquam refringentem successivè delatos esse, ibidemque alios alijs paulò magis, gradatim, refringi. Quod si fieri concipiatur, abinde sequeretur quod radij sic refracti, si corpore deinceps opaco quovis, ut papyro, interciperentur, lineolam ibi lucidam expingerent. Jam licet radij a Stellâ aliquâ venientes, non omnes in eâdem rectâ pergant; tamen, quod tantundem est, pro parallelis possunt haberi. Et quòd a lente convexâ effecti sunt convergentes antequam attingant Prisma, hoc adeò / non destruit Analogiam ut eam maximè confirmet. Etenim pro singulis radijs in eâdem rectâ pergentibus, debes tantùm concipere tot radiorum penicillos, qui omnes habent eundem axem et idem punctum concursûs: Et quòd istorum penicillorum alij magis alijs a Prismate refringuntur, ita ut eorum puncta concursus sive foci qui priùs concidêre, jam singuli cadant seorsim, lineam rectam conficientes. Ac proinde quod axes penecillorum qui, radijs puta successivis, eousque coincidebant donec attigêre Prisma, ibi per variam refractionem sint effecti divergentes, ut ad focos penicillorum in lineâ rectâ jacentes pergant.[23]

16/

21 [= I, 22]. Circumstantiâ variatâ, eidem descriptioni rursus applicatur.

Si Prisma Stellae Veneris vicinius quàm Lentem collocaveris, ut radij per illud primò trajiciantur, et a lente deinde convergentes fiant: eandem lineolam ut priùs videbis, licèt minùs conspicuam et inventu difficiliorem. Jam in hoc specimine cùm radij omnes adveniant paralleli, si aequaliter refringerentur transientes Prisma, manerent postea paralleli usque dum Lenti inciderent. Et in eâ proinde sic refringerentur ut omnes deinceps ad idem punctum pergerent. Et sic punctum lucidum[24] conspiceretur. Quare cùm vice puncti istius apparet linea, concludendum est quod omnes radij non aequaliter refringuntur.[25]

---

(22) Originally: complures (very many). Newton described this experiment in less detail in a letter to Oldenburg on 13 April 1672, responding to some queries posed by Robert Moray; *Correspondence,* 1:137 = *Phil. Trans.,* 7 (1672):4060. He returns to the spectrum of a star in §23 ≈ I, 24.

(23) Kepler, in his *Dioptrice,* §XLV, p. 17, introduced the term pencil (*penicillum*) of rays and made that concept a foundation of modern optics. Newton seems to have derived his knowledge of this concept indirectly, and indeed much of his concept of "light ray," through Gregory's *Optica promota.* Gregory began his treatise with thirty-seven definitions, of which the following are directly relevant to Newton's concept: "1. *Rays* are straight lines along which fiery corpuscles [*corpuscula ignea*] arising from luminous bodies spread outward . . . 8. *Convergent rays* are those which when extended in both directions meet in the direction of their motion, and they are called a *pencil,* and their point of intersection the *apex* of the pencil . . . 28. The *axis* of a pencil or radiant cone is that ray which is normal to the surface of a mirror or lens" (pp. 1, 3; italics

saw a small line which, though not very bright, I nonetheless easily perceived. Its length exceeded half an inch, whereas it had no sensible breadth, or at least none greater than could be perceived. I believe that the same thing might be observed for first magnitude stars, such as Sirius, especially if a lens four or six inches wide is used, so that it transmits many[22] rays.

I might usefully observe how well this experiment agrees with my explanation that I presented at the beginning concerning the different refraction of rays incident at the same angle. In the first figure I supposed that very many rays were conveyed successively along the same straight line onto some refracting surface and that there some of them were refracted gradually a little more than the others. But if this were imagined to occur, it would consequently follow that if the rays thus refracted were then intercepted by any opaque body, such as a paper, they would depict a small bright line there. Now, although the rays coming from the planet do not all proceed in the same straight line, they can nonetheless be considered parallel, which is equivalent; and because they are made to converge by the convex lens before they reach the prism, this does not destroy the analogy but rather greatly strengthens it. For instead of individual rays proceeding in the same straight line, you should imagine just as many pencils of rays all having the same axis and the same point of intersection; and because some of those pencils are refracted by the prism more than others, their points of intersection, or foci, that previously coincided, now each fall separately and form a straight line. Consequently, because the axes of the pencils coincided with the rays, which are assumed to be successive, until they reached the prism, where they were made to diverge by the varying refraction, they proceed to the foci of the pencils that lie in a straight line.[23]

20 [= I, 21]. And it is applied to the description of refraction related at Fig. 1.

If you have placed the prism rather than the lens nearer to the planet Venus, so that the rays are transmitted through that prism first and then are made to converge by the lens, you will see the same small line as before, although less visible and harder to find. Now, since in this example all rays arrive parallel, if they were equally refracted in crossing the prism, they would afterward remain parallel until they fell on the lens. Consequently, they would be so refracted in it that they would then all proceed to the same point, and thus a bright point[24] would be seen. Therefore, since a line appears instead of such a point, it must be concluded that all rays are not equally refracted.[25]

21 [= I, 22]. A changed feature is again introduced to the same description.

---

added). Newton's language reflects Gregory's quite closely, and his atomism, though only implicitly expressed, is also common to Gregory. Although Newton does not formally define light ray in his *Optical Lectures*, it is evident that he did not conceive of it as solely a mathematical line, but also physically as the locus of a stream of light corpuscles that correspond to the various colors and flow along straight lines. See §2 ≈ I, 2 and §78, and for Newton's formal definition of light ray Lect. 5, note (4).

(24) Originally continued: *non autem linea* (but not a line).

(25) Originally continued: *ad Prismata* (at prisms).

22 [≈ I, 23, 24].
Quod in adductis
experimentis
refractiones non
casu fiunt
inaequales, sed
ex inaequali
refrangibilitate.

Si jam objiciat aliquis quòd in refractionibus detur quidem irregularitas sed eam esse contingentem et non ex praevia radiorum dispositione vel ullis certis legibus ortam. Respondeo quod imago solis praefata[26] si radijs nullâ certâ lege refractis fieret oblonga, non posset in lineas rectas secundum longitudinem suam distinctè terminari: sicut ad figuram 5^{tam} ostensum est. Quinetiam non omninò deberet esse oblonga, sed parte ejus mediâ et magis splendidâ in morem orbis effingi, sensibilique termino distingui ab erraticâ luce debiliori quaquaversum dispersâ: Perinde ut Sol apparet cum nubibus penè obscuratur; Vel ut ejus imago cernitur cùm trajicitur per laminam vitream parallelis planis terminatam, et halitu vel fumo levitèr obductam, ut lux inter refringendum paululum conturbetur. Adhaec [fig 9] si duo Prismata similia *ABC* et *αβκ* juxtaponantur,[27] / secundum longitudines suas parallela, cum lateribus planis *AC* & *ακ*, item *BC* & *βκ* etiam parallelis; et si Sol transluceat utrumque in locum *Z*, ubi corpus opacum luci directè objicitur; radijs tamen ejus per orbiculare foramen *F* prius trajectis: Lux incidens in dictum *Z* apparebit distinctè orbicularis,[28] non secùs quàm si directè tenderet ab *F*, Prismatibus non omninò interpositis. Fatendum est itaque quod utriusque Prismatis conjunctim refractiones sunt regulares, et proinde etiam refractiones alterutrius. Scilicet radij illi similiter incidentes non omnes aequè refringuntur in primo Prismate *ABC*, ut neque in secundo *αβκ*: tamen cùm ea refractionis inaequalitas non contingens est sed oritur ex praevia radiorum dispositione: ideò licèt varij radij variè refringantur tamen ejusdem radij eadem erit refractionis quantitas in utroque Prismate et quantum incurvatur a priori *ABC*, tantum recurvabitur a posteriori *αβκ*. Unde radius quilibet utcunque sit refrangibilis, postquam ex utroque Prismate emerserit sibimet ipsi cùm nondum ijs inciderat fiet parallelus. Atque adeò cùm omnes ad easdem plagas tendunt ad quas liberè tenderent si Prismatibus non interciperentur; necesse est ut eandem orbicularem imaginem ad *Z* exhibeant quas illuc liberè tendentes exhiberent. Quod si imago oblonga per refractionem unici Prismatis (ut dictum est) effecta, figuram suam a radijs nullâ certâ lege divaricantibus sed forte fortuna huc illuc vage refractis acquireret: cùm refractiones binis Prismatibus geminentur, errores etiam radiorum duplo plures evaderent, ut et duplo majores. Et exinde Imago ad *Z* fieret multò oblongior, quae tamen, experientia teste, in orbem contrahitur.

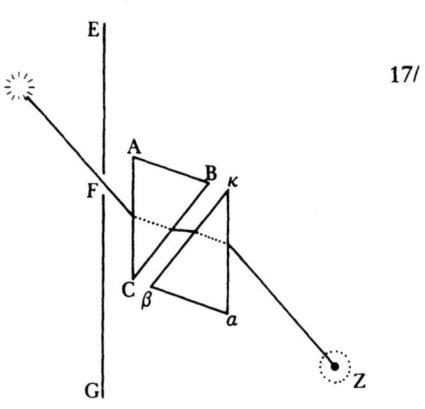

Figure 9

17/

Ex dictis opinor satis superúque constat id quod initio proposui commonstrandum: quoniam verò jucunditatem intellectui et assensum plerumque firmiorem harmonia rerum plurium[29] affert quàm unici licèt maximè scientifici

---

(26) Respondeo . . . praefata] Originally: At haec objectio penitus tolletur consideranti quod imago solis a Prismate refracta (but this objection will be completely eliminated by considering that [if] the image of the sun refracted by the prism). (27) Originally: statuantur.

If someone now objects that there is indeed an irregularity in refractions, but that it is contingent and does not arise from a previous disposition of the rays or from any definite laws, I respond that if that image of the sun[26] became oblong by rays that are refracted according to no definite law, it could not be distinctly terminated by straight lines along its length, as is shown in Fig. 5. Moreover, it ought not to be oblong at all, but in its middle and brighter part it ought to be shaped circularly with a boundary perceptibly distinguished from the scattered weaker light dispersed all around, just as the sun appears when it is almost obscured by clouds, or as its image is seen when it is transmitted through a glass plate bound by parallel planes and lightly covered with breath or smoke, so that the light is somewhat disturbed while being refracted. Furthermore (Fig. 9), if two similar prisms *ABC* and *αβκ* are placed next to one another,[27] parallel along their lengths with their plane sides *AC* and *ακ* as well as *BC* and *βκ* parallel; and if the sun shines through both to the place *Z* where an opaque body directly faces the light, with its rays first having passed through the circular hole *F*, the light falling upon *Z* will appear distinctly circular,[28] just as if it had traveled directly from *F* with no prisms at all placed in between. It must therefore be admitted that the refractions of both prisms together are regular and consequently also the refractions of each one. That is to say, those similarly incident rays are not all equally refracted in the first prism *ABC* nor in the second one *αβκ*; yet, since this inequality of refraction is not contingent but arises from a previous disposition of the rays, then although different rays are differently refracted, the quantity of refraction of the same ray will nevertheless be the same in each prism, and as much as it is bent by the first prism *ABC* it will be bent back by the second one *αβκ*. Consequently, any ray, however refrangible, after emerging from both prisms will become parallel to itself when it has not yet fallen on the prisms. Hence, since all rays tend to the same places to which they would tend freely if they were not intercepted by the prisms, it necessarily follows that they exhibit the same circular image at *Z* that they would exhibit had they freely tended there. If, however, the oblong image made by the refraction of one prism (as was said) acquired its shape from rays that are spread about not by a definite law but are refracted scatteredly here and there by mere chance, then when the refractions were repeated by two prisms, the straying of the rays would also be doubled and doubly great, and so the image at *Z* would become much more oblong; yet by the evidence of experience it is contracted into a circle.

From what I have already related I believe that I have more than sufficiently established what I initially proposed to demonstrate. Since, however, the agreement of several[29] things imparts an intellectual pleasure and a generally more assured acceptance than the evidence of a single, though

22 [≈ I, 23, 24]. The refractions in the experiments brought forward do not become unequal by chance but by unequal refrangibility.

---

(28) In two English descriptions of this experiment, Newton referred to the image as "orbicular" and "round": in the "New theory," *Correspondence*, 1:93 = *Phil. Trans.*, 6 (1671/2):3076; and in his response to Moray on 13 April 1672, *Correspondence*, 1:138 = *Phil. Trans.*, 7 (1672):4061–2. Newton later directed this experiment specifically against diffusion theories; see Lect. I, 3, note (9).   (29) Originally: *plurimorum* (very many).

argumenti testimonium: non erit abs re si in aliud experimentorum genus praecedentibus affinium experturos breviter introducam.

23 [≈ I, 24].
Perstringuntur alia experimenta prioribus affinia.

In fig 10 Sit *F* foramen valde exiguum per quod lumen Solis trajiciatur: deinde ad distantiam pro lubitu magnam statu/atur Prisma *ABC*, per quod radij transeant refracti; pro eo ut in prioribus explicui. Tum oculo pone admoto circularis foraminis *F* videbis imaginem *TP* oblongam; cujus longitudo ad latitudinem collata tanto major erit quanto foramen *F* fiat angustius. Et exinde pateat

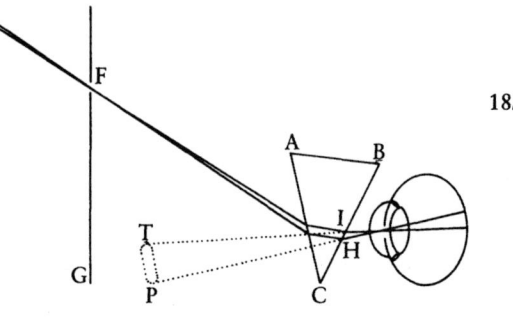

18/

Figure 10

quod radiorum alij tendentes ad oculum per *H* quasi manavissent ab *P*, sunt magis refracti quàm alij tendentes per *I*, quasi a *T* venissent. Et radijs sic in oculum non secus ingressis quàm si profluxissent ab oblongo spatio *PT*; necesse est ut spatium istud longum appareat luminosum.[30]

Sed cavendum est ne foraminis *F* tanta sit apertura ut nimiae lucis introitu laedatur oculus: imò ne tanta sit quin ut possis nudo oculo particulam solis per foramen istud quasi punctum lucidum distinctè et absque ullâ circumradiatione transpicere. Verum si lumen solis censeatur nimium, huic experiendo lumen a nubibus transmissum sufficiat; modo talis sit oculi tui dispositio ut foramen sine radijs circumcirca superfluis distinctum cernas antequam interponas Prisma: alias imaginem ejus non cernes distinctam neque debita longitudine diductam. Adhaec liceat tantundem observare si filum albens interposito Prismate aspicias, etenim filum multò latius apparebit cum in situ ad longitudinem Prismatis parallelo quàm cùm in transverso statuitur. Caeterum ut in uno comprehendam omnia, si stellam fixam primae magnitudinis mediante Prismate intuearis, ejus etiam imago conspicietur longa: At cùm radij stellarum pro parallelis habeantur, si omnes aequè refringerentur manerent etiam paralleli postquam egrediuntur Prismate, & oculum sic ingressi efficerent imaginem omnino similem stellae vel puncto lucido, & nullatenus oblongam: perinde ut fit cùm stella parallelos radios in oculum directè mittit. Videtis itaque quòd radij paralleli superficiebus planis refracti, fiunt inclinati, unde necesse est ut inaequalem refractionem patiantur. In transitu autem notetur quòd Telescopio si placeat primùm adhibito, tum ut copia lucis ad oculum transmittatur, tum ut scintillatio quâ fixae quasi coronâ / solent cingi, minuatur: et Prismate deinceps interposito, videbis albicantem lineam distinctiorem quàm prius,[31] cum latitudine vix aut ne vix quidem conspicuâ.

19/

---

(30) This experiment is succinctly described in the *Opticks,* Bk. I, Pt. I, Prop. II, Expt. IV.

(31) Originally continued: non sit major quàm ut sentiatur sive quàm apparens diameter stellae (and not greater than may be perceived, or than the apparent diameter of the star). This was then deleted in revision and changed to: eumque ferè sine aliqua sensibili latitudine (and almost without any sensible breadth); finally, it was altered to its current conclusion. Newton's

highly scientific, argument, it will not be without benefit if I briefly introduce investigators to another kind of experiment related to the preceding ones.

In Fig. 10 let *F* be a very small hole through which the sun's light is transmitted; then at a large distance chosen arbitrarily place a prism *ABC* through which the rays pass and are refracted, as I explained earlier. Then putting your eye right behind it, you will see the oblong image *TP* of the circular hole *F,* and its length compared to its breadth will be greater as the hole *F* becomes narrower. From this it is evident that those rays tending to the eye through *H,* as if they had proceeded from *P,* are refracted more than the others tending through *I,* as if they had come from *T.* Thus, since the rays entered the eye no differently than if they had flowed from the oblong space *PT,* it necessarily follows that that long space appears illuminated.[30]

23 [≈ I, 24]. Other experiments related to the previous ones are briefly considered.

You must be careful, however, that the opening of the hole *F* is not so large that your eye is hurt by the admission of too much light; in fact, it should be just large enough that with your naked eye you can see through that hole a small part of the sun like a bright point distinctly and without any surrounding radiation. If you consider the sun's light too great, light transmitted from clouds is sufficient for doing this experiment, provided that the location of your eye is such that before you insert the prism you can see the hole distinctly without any superfluous rays all around; otherwise you will see its image neither distinctly nor drawn into its proper length. Moreover, you may observe the equivalent if you look at a white thread through an interposed prism, for the thread will appear much broader when it is placed in a position parallel to the length of the prism than when it is transverse. However, to encompass everything in one, if you look through a prism at a fixed star of the first magnitude, you will also perceive its image to be long. Since the rays from stars can be considered parallel, if they were all equally refracted they would also remain parallel after they left the prism, and so entering the eye would produce an image totally like the star, or a bright point, and not at all oblong, just as what happens when the star sends parallel rays directly into the eye. You therefore see that parallel rays refracted at plane surfaces become inclined; consequently, it necessarily follows that they experience an unequal refraction. In passing, however, it may be noted that if it seems appropriate to use a telescope first, both because it transmits an abundance of light to the eye and because the twinkling that generally encircles the fixed stars like a crown is diminished, and then the prism is inserted, you will see a whitish line that is more distinct than before[31] with its breadth hardly or not even visible.

---

description of a star's spectrum here is of what one should see, and not necessarily of what he actually saw. In fact, in the concluding sentence of §19 = I, 20, he wrote only that it might be possible to observe such a spectrum, and when responding to Moray's proposed experiments on 13 April 1672, he had still not performed such an experiment: "I have sometimes designed to try how a fixt starr seen through a long Telescope would appear by interposing a Prism between the Telescope & my eye. But by the appearance of Venus viewed wth my naked eye through a Prism I presage the event," (*Correspondence,* 1:137 = *Phil. Trans.,* 7 (1672):4060).

Atque haec in praesentiâ sufficiant ad propositum nostrum stabiliendum adducta. Verùm ut innotescat quis sit harum rerum sensus plenior, et in quem finem tendunt; naturam colorum quatenus ex hisce dependent in proximo tractandam aggrediar.

Lect 3
24 [≈ II, 1].
Dissertatio de
coloribus inita.

Qui in fabricandis Telescopijs occupati sunt, de coloribus conqueruntur quibus objecta dum vitris istis mediantibus aspiciuntur tingi solent; quique eo magis augentur et apparent quo vitrum oculare[1] ex sphaeris minoribus efformatur, vel etiam quo vitrum objectivum majori latitudine radijs intrantibus patet.[2] Unde duplici incommodo implicati, impediuntur ne perspicilla ad optatum perfectionis gradum perducant: tum quod oculare vitrum ultra certos gradus parvum ad objecta magis amplianda nequeant adhibere, tum quod vitrum objectivum ultra certos limites aperire nequeant ad objecta magis lucida et perspicua reddenda. Qui gradus vel limites si non probè observentur, objecta coloribus involuta reddentur & multò minùs distincta quàm si vel minora cernerentur, ope vitri ocularis minùs convexi; vel minùs lucida, diminutâ perspicilli aperturâ. Jam cùm istae perfectiones praecipuae sint,[3] quae in Perspicillis desiderantur, nempe ut objecta magis amplient & reddant lucidiora:[4] operae pretium videtur in naturam colorum inquirere, ut investigemus tandem quid in causa sit quod ita appareant et objecta reddant indistincta.[5] Hujus enim ignorantia quamplurimos labore non exiguo sed inani tamen exercuit dum imperfectionem Telescopiorum a vitiosis vitrorum figuris ortam credentes, in istis meliori figurâ perpoliendis navârunt operam. Quod si nossent hasce colorum productiones ab alio fonte derivari, et quod in vitris quantumvis perfectis illi non secus sint apparituri; certè conatus suos / mutassent; | et laboribus istis secundum aliam methodum dispositis, Opti- 20/ cam in gradum multò perfectiorem jam promotam haberemus.

25 [= II, 2]. De
opinionibus
Philosophorum
et imprimis
Peripateticorum

Qui de coloribus hucusque disseruêre, vel id nomine tenus fecerunt ut Peripatetici, vel in eorum naturam et causas inquirere conabantur ut Epicurei et alij recentiores.[6] Quae Peripatetici de hisce tradidêre, etsi vera forent, ad nostrum tamen propositum nihil valerent: quippe dum modum quo generantur et causas unde fiunt tam varij, non omninò attingunt. Etenim illi de originibus & varijs rerum speciebus disputantes, pro causis ex quibus ipsarum existentiam et discrimen mutuantur varias quasdam formas assignarunt; verùm de particulari cujusvis formae causâ, et ratione ob quam differt ab alijs haud unquam quicquam disseruêre. Et sic ea fecerunt missa, quorum

---

(1) quique . . . oculare] Originally: Et quod isti colores eo magis augentur quo vitrum oculare vicinius (And those colors increase more as the nearer eyeglass).

(2) In the *Optica*, Pt. I, Prop. 37, Newton demonstrates that lateral chromatic aberration varies directly with both the focal length (or radius) and the aperture.

(3) Originally continued: si non solae (even if not the only).

(4) Newton's analysis accurately reflects contemporary problems of telescope design. The French astronomer Adrien Auzout rejected the accepted solution of long telescopes and advocated the same desiderata as Newton: "exhorting those, that work *Optick-Glasses,* to endeavor to make them such, that they may bear great *Apertures* and deep Eye-glasses; seeing it is not the length that gives esteem to *Telescopes* . . ." ("Considerations of Monsieur Auzout upon Mr.

solent cingi, minuatur: Et Prismate deinceps interposito, vicibus albicantem lineam ~~reddere~~ quàm prius ~~efficit~~ ~~magis quàm ut~~ ~~reddere~~ ~~quàm~~ ~~efficere~~ ~~distinctiorem~~ distinctiorem quàm prius ~~cum~~ latitudine vix aut ne vix quidem conspicuâ

Atq; hæc in præsentiâ sufficiant ad propositum nostrum stabiliendum adducta. Verùm ut innotescat quis sit horum rerum sensus plenior, et in quem finem tendunt; naturam colorum quatenus ex hisce dependent in proximo tractandam aggrediar.

Lect 3
24. Dissertatio de coloribus aggredita.

Qui in Telescopijs fabricandis occupati ~~fuerint~~ de coloribus conqueruntur quibus objecta ~~saltem binis~~ dum vitris istis mediantibus aspiciuntur, ~~apparent~~ ~~minimus~~ ex sphæris eo magis augeantur quo vitrum oculare objectivum minoribus efformatur, vel etiam quo vitrum objectivum majori latitudine radijs intrantibus patet. Vnde duplici incommodo implicati, impediuntur ne perspicilla ad optatam perfectionis gradum producant: tum quod oculare vitrum ultra certos gradus parum ad objecta magis amplianda nequeant adhibere, tum quod vitrum objectivum ultra certos limites aperire nequeant ad objecta magis ~~amplianda~~ lucida et perspicua reddenda. Qui gradus vel limites si non probe observentur, objecta coloribus involuta reddentur & multo minùs distincta quàm si vel minora cernerentur, ope vitri ocularis minùs convexi; vel minùs lucida, diminutâ perspicilli aperturâ. Jam cùm istæ perfectiones præcipuæ sint ~~reddere~~ quæ in Perspicillis desiderantur, nempe ut objecta magis amplient & reddant lucidiora: operæ pretium videtur in naturam colorum inquirere, ut investigetur tandem quid in causâ sit quod ~~ita appareant, et objecta reddant indistincta~~ ~~perturbat~~. Hujus ~~enim~~ ignorantia quamplurimos labore non exiguo sed inani tamen exercuit dum ~~istam~~ imperfectionem Telescopiorum a vitiosis vitrorum figuris ortam credentes, in istis meliori figurâ perpoliendis navârunt operam. Quod si nossent hasce colorum productiones ab alio fonte derivari, et quod in vitris quantumvis perfectis illi non secus sient apparituri; certè conatus suos

Let these things suffice for the present to establish what I have proposed. In order to reveal their fuller significance and the purpose for which they are intended I shall immediately begin to treat the nature of colors insofar as they depend on these things.

Those who are occupied with constructing telescopes complain about the colors that usually tinge objects when they are viewed through those glasses; the colors increase and are more evident as the eyeglass[1] is made from smaller spheres, or also as the object glass is opened wider to the entering rays.[2] Consequently, beset by this double obstacle they are hindered from bringing telescopes to their desired degree of perfection: both because they cannot use an eyeglass beyond certain degrees of smallness to magnify objects more; and because they cannot open the object glass beyond certain limits to make objects brighter and clearer. If these degrees or limits are not properly observed, objects will become enveloped with colors and be much less distinct than if they were seen either smaller, by using a less convex eye glass, or less bright, by decreasing the aperture of the object glass. Now, since these are the principal[3] perfections desired in telescopes, namely, that they magnify objects more and make them brighter,[4] I consider it worthwhile to investigate the nature of colors, so that we may finally discover the cause that is responsible for their appearing in this way and rendering objects indistinct.[5] Ignorance of this has indeed taxed quite a few with not a slight, but nonetheless vain, effort; while believing the imperfection of telescopes to arise from the defective shape of their glasses, they zealously directed their efforts toward finely polishing them to a better shape. | But had they known that these productions of colors derive | from another source and that they would appear no differently in glasses | however perfect, they would certainly have altered their efforts, | and by directing their labors according to a different method, we would already have advanced optics to a much more perfect state.

Those who have treated colors until now either have done it in name only, as the Peripatetics, or have endeavored to investigate their nature and causes, as the Epicureans and more recently others.[6] What the Peripatetics have taught about these things, even if true, would nevertheless be of no value for our purpose, since as yet they do not at all consider how they are generated and the causes whereby they become so diverse. Indeed, disputing about the origins and various species of things, they specify that certain different forms are the causes from which they derive their existence and difference, but hardly ever has anyone treated the particular cause of any form and how it differs from other forms. Thus they dismiss those things the explanation of

Lecture 3
24 [≈ II, 1]. The dissertation on colors begins.

25 [= II, 2]. The opinions of philosophers, and first the Peripatetics.

Hook's new instrument for grinding of optick-glasses," *Phil. Trans.*, 1 (1665):57–63, esp. 62). Newton's notes "Out of Philosophicall Transactions" (Add. 3958, f. 9ᵛ) show that he had followed this interchange with Hooke: "Auzouts rules for appertures. Reflections on Mʳ Hooks turn Lath for grinding glasses wᵗʰ Mʳ Hooks answer."

(5) quod . . . indistincta.] Originally: quae Dioptricam ita perturbat. (that so upsets dioptrics.)

(6) On Newton's knowledge of early color theory see the Introduction, §1.

explicatio videtur summum Philosophorum officium, imò quae sola mentem scientiae naturalis avidam explere possint.[7]

Attamen nè mancam tradidisse Philosophiam viderentur, effecerunt[8] ut ejusmodi disquisitiones pro maximè absurdis & ridendis habeantur, utpote quae supponunt formarum esse alias formas, et qualitates qualitatum. Itaque cùm lux definiatur esse qualitas vel forma quae dat esse lucidum, non expectandum est ut aliquid de ejus causis audiamus, vel quâ ratione ad varios colores producendos fit varia. Dicunt equidem quod plus luminis quibusdam coloribus immiscetur quam alijs: at hoc non sufficit ad eorum productionem tum quòd[9] nullus omninò color ex albedine et nigredine solummodo mixtis praeter fuscos intermedios generatur:[10] tum quòd quantitas lucis non mutat speciem coloris. Corpus enim rubrum, verbi gratiâ, semper apparet rubrum sive aspiciatur in crepusculo sive in meridie lucidissimâ. Porro autem ipsa definitio quam attribuunt coloribus adeò non pandit eorum naturam, ut eos nè nomine tenus exprimat. Ait Aristoteles χρῶμα δέ ἐστι τοῦ διαφανοῦς ἐν σώματι / ὡρισμένω πέρας.[11] Quae superficiei coloratae potiùs quàm coloris 21/ descriptio est. Illa enim dici potest extremitas perspicui in corpore terminato: at color plerumque videtur ubi nulla talis datur extremitas: ut in Iride; in Prismate; in vitris vel liquoribus perspicuis et aliquo colore leviter tinctis; in aquâ marinâ quae viridis ut-plurimum apparet, qui tamen color non in extremitate aquae, sed per totam ejus crassitiem generatur; in aere qui licèt maximè perspicuus et nullo corpore denso terminatus serenâ tamen nocte caeruleus apparet; & in flammâ, quae non minùs perspicua est, et luci pervia quàm ipse aer.[12] Sic cùm humores oculi colore aliquo tinguntur, omnia

---

(7) explere possint] Originally: quietam reddant et expleant (may put to rest and satisfy). Newton returns to a critique of Aristotelian philosophy in notes to an aborted reprint of the "New theory" from about 1677 in *Correspondence,* 1:105–6, and in Query 31 of the *Opticks* (1721), p. 377. Derek J. de Solla Price discovered this reprint, which was published by Cohen, "Versions of Isaac Newton's first published paper," *Archives internationales d'histoire des sciences,* 11 (1958):357–75. For a discussion of Newton's natural philosophy expounded in this and the following articles, see Mamiani, *Newton,* pp. 39–67.

(8) Originally: ita comparatum est (it has been so agreed).

(9) eorum . . . quòd] Originally: eas discriminandos, quia (for distinguishing them, because).

(10) For Aristotle's derivation of chromatic colors from black and white see the Introduction, note (6). Leon Battista Alberti, in an influential treatise written in 1435–6 in both a Latin and an Italian version, *De pictura* and *Della pittura* (it is not certain which is the translation), dismissed Aristotelian color theory as irrelevant to the practice of painting. He introduced four primary, chromatic colors from which all others could be derived, and he argued that white and black are not "true colors" but only moderators (*alteratores*) of the chromatic colors. Nonetheless, he did not consider gray to result from mixing white and black, since from his knowledge of actual painters' pigments, he held that whites and blacks were never pure but always partook of some chromatic color. Alberti's treatise circulated in manuscript for over a century before the Latin version was printed in 1540. See Alberti, *On Painting and On Sculpture. The Latin Texts of 'De pictura' and 'De statua,'* ed. and trans. Cecil Grayson (London, 1972), esp. pp. 44–7; and Charles Parkhurst, "Alberti's color scheme and some antecedents," in *A Conference on Color and Technique in Renaissance Painting, Italy and the North,* Temple University, September 22 and 23, 1980 (forthcoming). Parkhurst also discusses Theodoric of Freiberg's earlier, but not disseminated, views on gray. Julius Caesar Scaliger was one of the earliest to assert outright that mixing black and white yields no other color than gray (*fuscus, aut cinereum*); see *Exotericarum*

which seems the highest function of philosophy and, indeed, alone can satisfy[7] the eager mind of natural science.

But lest they seem to have handed down an imperfect philosophy, they have brought it about[8] that investigations of this kind are considered highly absurd and ridiculous, inasmuch as they assume there are other forms of forms and qualities of qualities. Therefore, since light is defined to be a quality or form that allows a luminous thing to exist, we must not expect to learn anything about its causes or how it becomes different in order to produce different colors. To be sure, they say that more light is mixed with some colors than with others; but this is insufficient for their production, both because[9] no color at all besides intermediate grays is generated from mixing white and black alone,[10] and also because the quantity of light does not change the species of the color. For a red body, for example, will always appear red whether it is viewed at twilight or at brightest midday. Moreover, the very definition that they attribute to colors so little reveals their nature that it does not even represent them in name. Aristotle says, "Color then is the extremity of the transparent in a determinately bounded body"[11]—this is a description of a colored surface rather than of color. The extremity of the transparent can of course be spoken of in a bounded body, but color frequently appears where no such extremity exists: in a rainbow; in a prism; in transparent glasses or liquids lightly tinctured with some color; in sea water, which often appears green, which color is nonetheless produced not in the extremity of the water but in its entire thickness; in air, which although exceedingly transparent and not bounded by a dense body nevertheless appears blue on a clear night; and in a flame, which is no less transparent and pervious to light than air itself.[12] Thus when the humors of the eye are

---

exercitationum liber XV. De subtilitate, ad Hieronymum Cardanum [1557] (Lyons, 1615), Exercitatio CCCXXV, §9, p. 823. As empirical investigations of color became more common in the seventeenth century, rejection of Aristotle's view became widespread. For instance, Anselm de Boodt stated that gray (*cinereum*) alone arises from a mixture of white and black, for varying the proportions of white and black results only in a brighter or darker gray; *Gemmarum et lapidum historia* (Hanau, 1609), Bk. I, Ch. XV, p. 25; Newton owned a copy of the 1636 revised edition (Harrison, *Library of Newton*, no. 245). Boyle in *Touching Colours* (Pt. I, Ch. V, p. 87) took a similar stand. Charleton, however, still deviated little from traditional Aristotelian concepts, holding that the intermediate colours "are but the off-spring of the Extreme [i.e., white and black], arising from the intermission of Light and shadow, in various proportions" (*Physiologia*, pp. 191–2). But on the first page of his first essay "Of Colours" Newton had already decisively rejected this idea: "No colour will arise out of ye mixture of pure black & white for y^n pictures drawne wth inke would be coloured[,] or printed would seeme coloured at a distance[,] & ye verges of shadows would be coloured. & lamb black and spanish whiteing would produce colours[.] whence they cannot arise from more or lesse reflection of light or shadows mixed wth light" (Add. 3996, f. 105^v).

(11) Aristotle, *De sensu*, 439 b 12–13; see the Introduction, note (4).

(12) Newton's criticism of Aristotle is not entirely fair, since he ignores Aristotle's explanation, even if inadequate, of the color or "sheen" of air and water in a passage (*De sensu*, 439 b 2–5) immediately preceding the quoted definition of color. This suggests that Newton did not directly study Aristotle's works, and it is possible that he conveniently took the Greek quotation from Magirus's *Physiologiae peripateticae* (Bk. VI, Ch. VII, p. 317), since it gives the identical quotation.

videntur eodem colore tincta, licèt extremitas perspicui sit alijs coloribus praedita. Et cùm Solem nudis oculis aspexeris modò, luminosa omnia deinceps videntur rubra, et nigra plerumque apparent caerulea; qui color erit magis conspicuus, si clausis oculis te in locum aliquem tenebrosissimum statim conferas.[13] Imò premendo oculum colores in tenebris excitare liceat;[14] quis autem vocabit illos extremitatem perspicui? Caeterùm non opus est ut has opiniones enixè refutem quae non videntur tanti,[15] neque proposito meo adversantur. Esto lux qualitas corporis lucidi, esto lumen actus perspicui,[16] et color ejus extremitas, et quicquid de istis dixerunt, esto; abinde tamen haud concipi poterit quo pacto lux refringatur, unde colores sint varij, quid in causâ sit quod in Perspicillis apparent, et quâ ratione incommodum istud devitari possit.

26 [= II, 3]. De opinionibus aliorum Philosophorum.   Ad opiniones aliorum Philosophorum quod attinet, dixêrunt colores vel ex umbrâ lucéque variè mixtis; vel ex contortione globulorum aut eorum varijs pressionibus generari; vel denique ex varijs modis quibus Medium quoddam aethereum vibratur, statuentes scilicet lucem productam esse ex impulsu vibrantis aetheris in retiformem tunicam delato. Extra oleas nimis evagarer, si has opiniones sigillatim confutandas adortus fuerim; nec opus est ut faciam cùm omnes in communi quodam errore consentant: Scilicet quod modificatio lucis, quâ singulos colores exhibet, ei non est insita ab origine suâ, sed inter reflectendum vel refringendum acquiritur. Inter radios lucis nullum contemplantur discrimen priusquam inci/dant in corpus aliquod colorificum; opi- 22/ nati tantùm quòd pro variâ dispositione corporis istius, varijs modis reflectuntur vel refringuntur et pro specie modificationis quam sic acquirunt, varia deinde colorum phantasmata spectantibus exhibent. Mixtura lucis et umbrae, gyratio globulorum, vel varia vibratio Medij non supponitur inesse radijs antecedentèr ad eorum reflexiones vel refractiones, sed per istas actiones generari creditur. Quemadmodum et Peripatetici statuunt colores a corporibus originem sumere quorum dicunt esse qualitates.[17] Attamen contrarium

27 [= II, 4]. Colorum origines et fundamenta generalia describuntur.   esse verum ex sequentibus abunde patebit. Invenio scilicet quòd modificatio lucis unde colores originem sumunt, luci connata sit, et non oritur a reflectione neque a refractione neque a qualitatibus corporum aut modis quibuslibet,[18] nec ab ijs vel destrui potest vel ullo modo mutari.[19]

---

(13) Originally: involvas (take cover). In about 1664, prompted by Boyle's observations in *Touching Colours* (Pt. I, Ch. II, pp. 16–17), and perhaps also by those of Descartes in *Dioptrice* (Ch. VI, §4, *Oeuvres*, 6:606–7, 131–2), Newton viewed the sun with his naked eye and recorded the effects, some lasting for months, in the "Quaestiones" under the entry "Immagination & Phantasia & invention" (Add. 3996, ff. 109ʳ, 125). He briefly summarized this observation in his second essay "Of Colours" (§63, Add. 3975, f. 9ᵛ) and again drew upon it for his extended description for John Locke on 30 June 1691; *Correspondence*, 3:153–4.

(14) In "Of Colours" (§9, Add. 3996, ff. 123ᵛ–4ʳ) Newton pursued Boyle's and Descartes's accounts of phosphenes and recorded his own observations; see Boyle, *Touching Colours*, Pt. I, Ch. II, p. 12; and Descartes, *Dioptrice*, Ch. VI, §3, *Oeuvres*, 6:606, 131. Newton returned to these torturous experiments in his second "Of Colours" (§58, Add. 3975, f. 9ʳ) and also briefly described phosphenes in the *Opticks*, Query 16.

(15) quae non videntur tanti] II: quae etsi verae essent tamen non sunt sufficientes (that, even if true, are nevertheless insufficient).

tinged with some color, everything appears tinged with the same color, although the extremity of the transparent is endowed with other colors. Also when you have looked at the sun with only your naked eyes, afterward all bright things seem red, and black things often appear blue, which will be more conspicuous if with your eyes shut you move[13] immediately into some very dark place. In fact, by pressing your eye you may produce colors in the dark,[14] but who will call those an extremity of the transparent? There is no need, however, for me vigorously to refute these opinions that seem of no great value[15] and are not opposed to my purpose. Let light [*lux*] be a quality of a luminous body; let light [*lumen*] be the action of the transparent and color its extremity;[16] and whatever else they have said about those things, let it be. Yet from this it can hardly be conceived how light is refracted, whence colors are different, why they appear in telescopes, and how that inconvenience can be avoided.

As for the opinions of other philosophers, they have said that colors are generated from shadow and light mixed in various ways, or from a spinning of little balls or their various pressures, or, finally, from the various ways in which a certain aetherial medium is vibrated, assuming that light is produced by an impulse of the vibrating aether carried to the retina. I would stray too far from the path were I to attempt to refute these views individually. Nor is it necessary that I do so, since they all agree in a certain common error; namely, the modification of light by which it exhibits individual colors is not innate to it from its source but is acquired by being reflected or refracted. They consider that there is no difference between the rays of light before they fall upon some color-making body and believe only that according to the varied disposition of that body they are reflected or refracted in various ways, and according to the kind of modification so acquired they thereafter exhibit to observers various sensations of color. The mixture of light and shadow, the gyration of the little balls, or the various vibrations of the medium are not assumed to be in the rays prior to their reflections or refractions, but are believed to be produced through those actions; likewise, the Peripatetics consider colors to originate in bodies, saying they are their qualities.[17] Nonetheless, it will be manifestly evident from the following that the contrary is true. Namely, I find that the modification of light whereby colors originate is connate to light and arises neither from reflection nor from refraction, nor from the qualities or any modes whatsoever of bodies,[18] and it cannot be destroyed or changed in any way by them.[19]

26 [= II, 3]. The opinions of other philosophers.

27 [= II, 4]. The origins of colors and their general principles are described.

---

(16) Newton alludes here to a traditional Scholastic distinction, where "lux" is light in a luminous body, while "lumen" is light transmitted through a transparent medium; see Magirus on "Luminis & lucis discrimen," *Physiologiae peripateticae*, Bk. VI, Ch. VII, p. 323.

(17) Quemadmodum . . . qualitates.] Added.

(18) neque a qualitatibus . . . quibuslibet] Added.

(19) Originally continued: Discrimen radijs quoad eorum refrangibilitatem inesse, antehac enarratum dedi: est et alia disparitas quatenus aliqui ad quosdam colores producendos sunt accommodati, et alij ad alios. (I fully explained earlier that there is in rays a difference with respect to their refrangibility; and there is also another inequality, since some rays are fit to produce some colors, and others other colors.)

28 [≈ II, 5]. Idque
quatuor
propositionibus
Verùm ut sententiam meam distinctiùs proferam: | Invenio primò quod radij, qui ex incidentiâ pari maximè omnium refringuntur, colores efficiunt purpureos sive violaceos; illi autem rubros, qui minimè omnium refringuntur; ac illi caeruleos, virides, et flavos, qui refringuntur mediocritèr.[20]

Secundò e contra invenio quòd radij qui purpureos colores efficiunt, ex incidentiâ pari maximè omnium refringuntur; et illi minimè omnium qui rubeos efficiunt; illi autem mediocriter qui generant caeruleos, virides, ac flavos. Hoc est invenio quòd radij pariter incidentes refractionem continuò majorem patiuntur atque adhuc majorem deinceps, prout apti sunt ad hos ordine colores rubrum, flavum, viridem, caeruleum, & violaceum generandos, unà cum omnibus eorum successivis gradibus & coloribus intermedijs.[21]

Tertiò invenio quòd ex varijs horum radiorum mixturis caeteri omnes colores producuntur: Et quod color albus fuscus et niger fit ex radijs cujusque speciei confusè mixtis.[22]

Quartò invenio quòd omnes omnium corporum colores non aliunde generantur quàm e dispositione quâdam quâ apta sunt ut alios radios reflectant et intromittant alios. Sic corpus rubrum est quod radios ad rubedinem aptos reflectit maximè, et plerosque caeteros intromittit: purpureum quod radios isti colori generando proprios reflectit, et intromittit alios: album verò quod ferè omnes reflectit, & nigrum quod / omnes intromittit, paucissimis sed 23/ omnium tamen specierum radijs repercussis.[23]

29 [≈ II, 6]. De
quibus non
hypotheticè et
probabiliter, sed ab
experimentis aut
demonstrativè
disserendum esse
promittitur.
Verùm ne videar officij limites excessisse dum naturam colorum pertrectare aggredior, qui nihil ad Mathesin attinere censeantur: non abs re erit si de ratione incepti hujus iterum commonefaciam. Nimirum tanta est inter proprietates refractionum et colorum affinitas, ut seorsim explicari nequeant. Qui alterutras rite velit cognoscere, ut alteras cognoscat necesse est. Et praeterea si de refractionibus non agerem, et earum disquisitio non esset in causâ quòd negotium de coloribus simul explicandis inceptarem: tamen generatio colorum tantam Geometriam complectitur, et eorum cognitio tantâ firmatur evidentiâ, ut vel ipsorum gratiâ possem aggredi, sic limites Mathesis nonnihil ampliaturus. Quemadmodum enim Astronomia, Geographia, Navigatio, Optica, et Mechanica pro scientijs mathematicis habentur, licèt in ijs agatur de rebus Physicis, Caelo, Terra, Navibus, luce et motu locali: Sic etiamsi colores ad Physicam pertineant, eorum tamen scientia pro Mathematicâ habenda est, quatenus ratione mathematicâ tractantur. Imò verò cùm horum accurata scientia videatur ex difficillimis esse quae Philosophus desideret; spero me quasi exemplo monstraturum quantùm Mathesis in Philosophiâ naturali valeat; et exinde ut homines Geometras ad examen Naturae strictiùs aggrediendum, & avidos scientiae naturalis ad Geometriam priùs addiscendam horter: ut nè priores suum omninò tempus in speculationibus humanae vitae nequaquam profuturis absumant, neque posteriores operam praeposterâ methodo usque navantes, a spe suâ perpetuò decidant: Verùm ut Geometris philosophantibus

---

(20) Newton treats this proposition perfunctorily in §30 → II, 6, 7.

(21) This proposition is demonstrated beginning with §31 → II, 7.

(22) The demonstration of this proposition, commencing in §40 → II, 27, includes Newton's most revolutionary conclusion, that the sun's direct light is compounded of all colors. He proves

But to present my idea more distinctly: | First, I find that those rays that at equal incidence are refracted most of all make purple or violet colors, but those that are refracted least of all produce red, and those that are refracted intermediately produce blue, green, and yellow.[20]

Second, and conversely, I find that those rays that make purple colors are at equal incidence refracted most of all, and those that make red least of all, but those that generate blue, green, and yellow are refracted intermediately. That is, I find that equally incident rays undergo a continually greater and in turn still greater refraction, according as they are disposed to generate these colors in order: red, yellow, green, blue, and violet, together with all their successive gradations and intermediate colors.[21]

Third, I find that from various mixtures of these rays all other colors are produced, and that the colors white, gray, and black are made from rays of every sort confusedly mixed.[22]

Fourth, I find that all colors of all bodies are produced in no other way but from a certain disposition whereby they are disposed to reflect some rays and to let in others. Thus a red body is one that mostly reflects rays disposed to redness and lets in most of the others; a purple body is one that reflects rays proper to producing that color and lets in the others; but a white body is one that reflects almost all rays; and a black body is one that lets in all rays but reflects very few, though ones of every kind.[23]

But lest I seem to have exceeded the bounds of my position while I undertake to treat the nature of colors, which are thought not to pertain to mathematics, it will not be useless if I again recall the reason for this pursuit. The relation between the properties of refractions and those of colors is certainly so great that they cannot be explained separately. Whoever wishes to investigate either one properly must necessarily investigate the other. Moreover, if I were not discussing refractions, my investigation of them would not then be responsible for my undertaking to explain colors; nevertheless, the generation of colors includes so much geometry, and the understanding of colors is supported by so much evidence, that for their sake I can thus attempt to extend the bounds of mathematics somewhat, just as astronomy, geography, navigation, optics, and mechanics are truly considered mathematical sciences even if they deal with physical things: the heavens, earth, seas, light, and local motion. Thus although colors may belong to physics, the science of them must nevertheless be considered mathematical, insofar as they are treated by mathematical reasoning. Indeed, since an exact science of them seems to be one of the most difficult that philosophy is in need of, I hope to show—as it were, by my example—how valuable mathematics is in natural philosophy. I therefore urge geometers to investigate nature more rigorously, and those devoted to natural science to learn geometry first. Hence the former shall not entirely spend their time in speculations of no value to human life, nor shall the latter, while working assiduously with an absurd method, perpetually fail

28 [≈ II, 5]. And that by four propositions.

29 [≈ II, 6]. It is affirmed that these propositions are to be treated not hypothetically and probably, but by experiments or demonstratively.

the proposition only for white, but adds the rest of it on gray, black, and other colors in the *Optica*, §II, 69.

(23) This proposition remains unproved and is taken up only in the *Optica*, §II, 71.

& Philosophis exercentibus Geometriam, pro conjecturis et probabilibus quae venditantur ubique, scientiam Naturae summis tandem evidentijs firmatam nanciscamur.[24]

30 [→ II, 6, 7]. De primâ propositione agitur perfunctoriè.

Itáque ad institutum redeo de coloribus secundum praecedentes quatuor propositiones explicatis disceptaturus.[25] / Et ad propositionem primam quod 24/ attinet, utpote quòd ex radijs similitèr incidentibus maximè refracti purpuram efficiant, minimè refracti ruborem, et refracti mediocriter colores mediocres,

vide Fig 2

ea pateant ex antedictis. Quippe notissimum est quòd colores Prismatis purpureus, caeruleus, viridis, flavus, & rubeus ita sese ab P versus T in ordine nominato subsequuntur, ut purpureus color ad P semper jaceat in angulo TFO radijs incidentibus OF ijsque FT ad colorem rubeum T refractis contento: atque adeò ut radij ad purpuram tendentes magis deflectant a directo cursu, sive magis refringantur quàm illi qui tendunt ad rubedinem.[26]

31 [→ II, 7]. Transitur ad secundam.

Non opus est ut hanc primam propositionem de industriâ porrò illustrandam prosequar, cùm scopus ejus et veritas in sequentibus manifestior evadet. Itaque ad secundam pergo monstraturus e contra quòd radij ex eâdem incidentiâ refractiones varias patiuntur qui varios colores producunt. scilicet quòd magis atque magis refringuntur prout colores hoc ordine rubeum, flavum, viridem, caeruleum et purpureum successivos cum omnibus eorum gradibus intermedijs generant.[27]

32 [≈ II, 7, 8]. Cui probandae adducitur experimentum.

Hoc autem ut pateat, iterum repetatur experimentum Prismatis quod in prioribus adduxeram. Nempe [fig. 11] ponatur quod[28] radij solares ad fora-

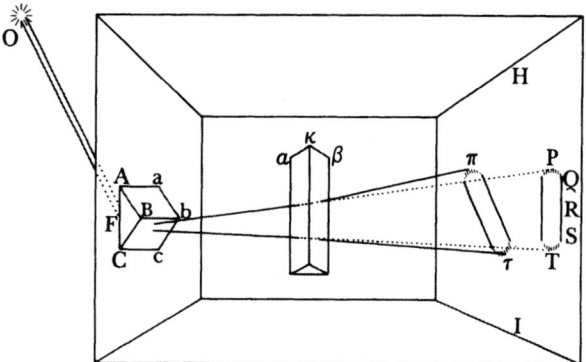

Figure 11

---

(24) Newton most likely had in mind here the "hypothetical physics" of Descartes and Hooke. Hooke, for example, in the *Micrographia* freely refers to both his own theory of color and that of Descartes as "hypotheses" (p. 59), and in his Preface (sig. *b*) he affirms that he will not offer "infallible Deductions, or certainty of *Axioms*." Newton frequently returned to the theme of the certainty of his new mathematical science of color: in the "New theory" in a passage deleted by Oldenburg from the published version (*Correspondence*, 1:96–7); in his response on 11 June 1672 to Hooke, who rejected the claim of certainty in the "New theory" (ibid., pp. 187–8); in an additional passage in an earlier draft of that letter to Hooke (Add. 3970, ff. 433ʳ–44ᵛ, esp. 443ᵛ–4ʳ; partly published in Alan E. Shapiro, "Newton's 'achromatic' dispersion law," p. 104; the draft is described by Westfall, "Newton's reply to Hooke"); and in the *Opticks*, Bk. I, Pt. II, Prop. III. On Newton's ensuing debates with Hooke, Pardies, and Huygens on the role of hypotheses, see Zev Bechler, "Newton's 1672 optical controversies: a study in the grammar of scientific dissent," *The Interactions Between Philosophy and Science*,

to reach their goal. But truly with the help of philosophical geometers and geometrical philosophers, instead of the conjectures and probabilities that are being blazoned about everywhere, we shall finally achieve a natural science supported by the greatest evidence.[24]

I now return to my original design to discuss colors treated according to the preceding four propositions.[25] As for the first proposition, namely, that of similarly incident rays, the most refracted make purple; the least refracted, red; and the intermediately refracted, the intermediate colors, these things are evident from what has already been described. For it is very well known that the prism's colors, purple, blue, green, yellow, and red, follow one another from *P* toward *T* in the order named, so that the purple color at *P* always lies inside the angle *TFO* made by the incident rays *OF* and those rays *FT* refracted to the red color *T*; and hence the rays tending to purple deviate more from a straight path or are more refracted than those tending to red.[26]

It is not worthwhile for me to continue diligently to illustrate this first proposition, since its purpose and truth will become more evident in the following. Therefore I pass to the second proposition and will demonstrate conversely that at the same incidence rays that produce different colors undergo different refractions; namely, they are more and more refracted as they produce the successive colors in this order: red, yellow, green, blue, and purple, together with all their intermediate gradations.[27]

To make this clear I again repeat the experiment with a prism that I introduced earlier. Namely (Fig. 11), assume that[28] after solar rays enter a

Margin notes:

30 [→II, 6, 7]. The first proposition is treated perfunctorily.

See Fig. 2.

31 [→II, 7]. Turning to the second one.

32 [≈ II, 7, 8]. To prove this an experiment is introduced.

---

ed. Yehuda Elkana (Atlantic Highlands, N.J., 1974), pp. 115–42; and Robert Kargon "Newton, Barrow, and the hypothetical physics," *Centaurus*, 11 (1965):46–56, who suggests that Newton may have derived his attitude toward hypotheses from Barrow. Lest one believe that Newton fully obeyed his own injunctions against "conjectures and probabilities," see Lect. II, 15.

(25) In revision this sentence was shifted to the conclusion of the preceding article, §II, 6, whereas the rest of the article was incorporated into §II, 7.

(26) This fundamental proposition appears in all later formulations of the theory: Prop. 1 in the *Optica;* Prop. 1 in the "New theory"; Prop. 2 in his letter for Huygens, 23 June 1673; and Prop. II, Bk. I, Pt. II in the *Opticks.* A proper experimental demonstration is not at all as straightforward as it is here, for the colors in Newton's unresolved spectrum are compound (§12 = I, 12), so that when a narrow portion, suppose green, is refracted a second time, the neighboring colors, such as yellow and blue will also appear. In the new Prop. 2 of the *Optica* on the immutability of color, Newton resolves this problem and shows how to separate the spectral colors and in principle to attain colors of near perfect purity. Subsequently, in the *Opticks* (Bk. I, Pt. II, Prop. II) he combined the present Prop. 1, establishing the correspondence between refrangibility and color, with that on the immutability of color into one proposition. However, in the "Fundamentum Opticae" (Add. 3970, ff. 409–10, 415–16, 394–8, 583–4, 425–6, 647–8, 407–8, 405–6, 403–4, 401–2, 399–400, 419, 422, 420–1, 411–14, 423–4, 417–18), an incomplete Latin early draft of the *Opticks* composed around 1690, Newton twice formulated and then deleted a separate proposition on the correspondence between refrangibility and color (Prop. 8, ff. 399, 422, 420) before deciding to combine it with color immutability.

(27) In the "New theory," Prop. 2, Newton asserts that the correspondence between color and refrangibility is a strict one and makes the stronger (erroneous) claim that "this Analogy 'twixt colours, and refrangibility, is very precise and strict; The Rays always either exactly agreeing in both, or proportionally disagreeing in both" (*Correspondence,* 1:97 = *Phil. Trans.,* 6 (1671/2):3081).

(28) Hoc . . . quod] II(§II, 7): Quò primam comprobem repetamus experimentum prismatis sub initio propositum, nempe (That I may prove the first, let us repeat the experiment with a prism proposed at the beginning. Namely,).

men *F* ingressi cubiculum obtenebratum,[29] a Prismate *ABC* quàm proximè foramen intus disposito refringantur, tendantes deinde versus oppositum parietem *HI* ad imaginem *PT* ibi depingendam. Et Imago illa ut vulgò notum est coloribus tingetur, | rubeo ad *T,* purpureo ad *P,* caeruleo viridique et flavo ad *Q R* & *S.* Explorandum est itáque an radij tendentes versus *P* magis refringantur quàm isti qui tendunt versus *T.* | Id quod varijs modis tentare liceat, quorum facillimum et maximè perspicuum sequentem existimo.[30] Sume aliud Prisma *αβκ* (fig [11]) et illud alicubi inter Prisma primum *ABC* et imaginem *PT* ita colloca, ut sit illi Prismati *ABC* transversum sive parallelum imagini, radiosque versus *PT* tendentes intercipiat et alioversum refringat, / puta versus *ππ.* Quo facto, imaginem *ππ* refractionibus utriusque Prismatis   25/ sic effectam, videbis ut priùs coloratam, sed in alio tamen situ dispositam: Non parallelam imagini *PT* sed secundum extremitates rubras manifestò convergentem. Jam cùm radij ad utrosque colores rubeum *T* et purpureum *P* pertinentes similiter incidant in Prisma secundum *αβκ,* si eandem praeterea refractionem paterentur, imagines *PT* ac *ππ* deberent esse parallelae.[31] Et ideò cùm non existant parallelae, sed imaginis *ππ* extremitas purpurea *π* longiùs ab alterâ imagine *PT* transferatur quàm extremitas rubea *τ*: necessariò concedendum est quòd radij ad extremitatem purpuream *P* tendentes magis refringantur quàm qui tendunt ad extremitatem rubeam *T.* Hoc est, quòd radij generantes purpuram apti sint ut magis refringantur quàm ruborem efficientes. Atque idem quoque de coloribus intermedijs eâdem ratione constabit, sicut ostendendum proposui.

33 [≈ II, 9]. Experimenti praefati circumstantia notatur.

In experiendis hisce notari poterit quòd quò viciniùs anteriori Prismati *ABC* sive quò remotiùs à pariete *HI* collocetur Prisma posterius *αβκ*: imagines *PT* ac *ππ* eo magis ab invicem distantes, etiam ad se magis inclinabuntur. Adeò ut angulum semirectum vel paulo minorem eo contineant cùm prismata collocantur ad invicem vicinissima. Cujus rei ratio facillima est consideranti quòd distantiae *Pπ* ac *Tτ* sunt in datâ quâdam ratione.[32] Sic in Fig 12 si parallelae *Pπ* ac *Tτ* sint in ratione datâ, quo majores existant eo major erit inclinatio linearum *PT* ac *ππ.*

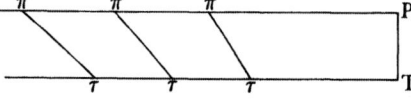

Figure 12

34 [= II, 12]. Idem instrumentis refractiones dimetientibus posse probari.[33] Tamen evidentiam experimenti jam descripti sufficere.

Sicui in potestate est instrumentum aliquod ad quantitates refractionum accuratè mensurandas paratum, nullus dubito quin istius etiam ope seorsim dimetiendo refractiones diversorum generum radiorum, facilè observabit earum differentias: licèt ego praedictis tanquam manifestissimis acquiescens, non operae pretium duxerim rem alijs modis / experiri.[34] Verùm ut cuique   26/

---

(29) Newton originally wrote "obscuratum" and then "tenebratum."

(30) This sentence concludes §II, 7, and the remainder of the article corresponds with little change to §II, 8.

(31) Both in the *Optica,* in §I, 23 ≈ 22, and in his refutation of Hooke's diffusion theory Newton considered the alternative to inclined spectra to be quadrilateral or "foursquare" rather than parallel spectra; see Newton's letter to Oldenburg for Hooke, 11 June 1672, *Correspondence,* 1:178–9 = *Phil. Trans.,* 7(1672):5092–3; and also *Opticks,* Bk. I, Pt. I, Prop. II, Expt. V.

darkened[29] room at the hole *F,* they are refracted by the prism *ABC,* which is placed inside exceedingly close to the hole, and then tend toward the opposite wall *HI,* depicting the image *PT* there. That image, as has been commonly observed, will be imbued with colors: | red at *T,* purple at *P,* and | blue, green, and yellow at *Q, R,* and *S.* It therefore must be investigated | whether the rays tending to *P* are refracted more than those tending to *T.* | This may be tested in various ways, of which I consider the following to be the easiest and clearest.[30] Take another prism *αβκ* (Fig. 11) and place that somewhere between the first prism *ABC* and the image *PT,* so that it is transverse to that prism *ABC,* or parallel to the image, and intercepts the rays tending toward *PT* and refracts them in another direction, for example, toward *ππ.* Having done this, you will see the image *ππ,* produced in this way by the refractions of both prisms, colored as before, but situated in still another position, not parallel to the image *PT* but manifestly converging along its red ends. Now, since the rays corresponding to each color, the red *T* and the purple *P,* are similarly incident upon the second prism *αβκ,* if they then experienced the same refraction, the images *PT* and *ππ* ought to be parallel.[31] Hence, since they are not parallel, but rather the purple end *π* of the image *ππ* is shifted farther from the other image *PT* than the red end *τ,* it must necessarily be concluded that the rays tending to the purple end *P* are refracted more than those that tend to the red end *T;* that is, the rays producing purple are disposed to be more refracted than those making red. For the same reason the same thing will also be evident for the intermediate colors, as I proposed to show.

In doing these experiments one will be able to observe that as the second prism *αβκ* is placed closer to the first prism *ABC* or farther from the wall *HI,* the images *PT* and *ππ* will become more distant from one another as well as more inclined to one another, until they make half a right angle or a little less when the prisms are placed as close as possible to each other. This is explained very easily if one considers that the distances *Pπ* and *Tτ* are in a certain given ratio.[32] Thus in Fig. 12, if the parallels *Pπ* and *Tτ* are in a given ratio, the greater they are, the greater will be the inclination of the lines *PT* and *ππ.*

33 [≈ II, 9]. A feature of the preceding experiment is noted.

If anyone possesses some instrument equipped for accurately measuring the quantity of refractions, I do not doubt that also with its aid in measuring the refractions of the different kinds of rays individually he will easily observe their differences; although, being content with the preceding as very clear, I considered that it was not worthwhile to test this by other ways.[34] But to

34 [= II, 12]. The same thing can be proved[33] with instruments for measuring refractions, yet the evidence of the experiment just now described is sufficient.

---

(32) The claim that the inclination of the spectra to the vertical is 45° is erroneous and derives from Newton's tacit assumption, made explicit in the *Optica* in §II, 10, that the spectra are straight lines that intersect at the position of the sun's direct image *X* in the revised Fig. II, 2; he makes the same claim in his letter to Hooke cited in note (31). Nonetheless, this error is in itself of no consequence for Newton's principal point on the different refrangibility of different colors. The ratio of the distances *Pπ/Tτ* is constant, because it is equal to the ratio of the tangents of the deviations of the blue and red rays, which is a constant for a given prism; see Lect. II, 2, note (4).

(33) Originally: comperiri (discovered).

(34) In the *Optica* (§I, 37 ≈ 110) Newton likewise rejects a direct measurement of the ratio of sines for each color individually.

magis pateat quanta sit praedictorum evidentia, quaedam quae exinde scaturiunt notatu dignissima proferre non pigebit.

35 [= II, 13]. Illud promovetur aliquantum, idque vel tribus prismatibus adhibitis,   Sit *Fφ* (fig [13]) paries vel operculum fenestrae duobus foraminibus *F* et *φ* luci pervium, ijsque digitos duos ab invicem distantibus; et intus disponantur duo Prismata *ABC DEG* in situ sibi invicem parallelo, at perpendiculari ad lineam *Fφ* per centra foraminum ductam: quae duo lucem ingressam refrin-

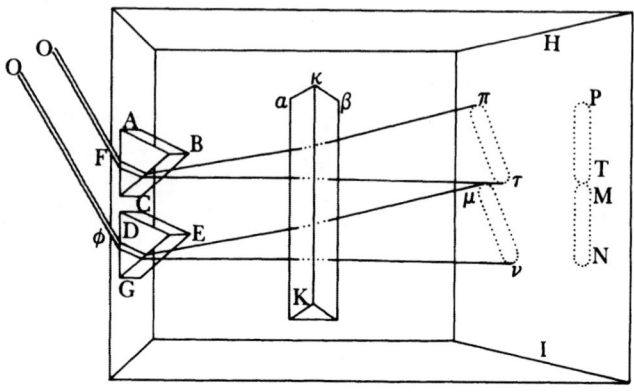

Figure 13

gant ad imagines duas *PT* et *MN* in oppositum parietem projiciendas, simili prorsus modo quo factum fuit in experimento priori. Et praeterea sint anguli Prismatum *ACB, DGE* (comprehensi planis refringentibus) aequales. Quibus ita constitutis, videbis imagines *PT* et *MN* in directum jacentes cum extremi-tatibus earum *T* et *M* contiguis. Quod si non eveniat, situs unius e Prismati-bus parùm mutandus est donec extremitates contiguas esse cernas, vel fortè nonnihil coincidentes. Purpurâ *M* et rubore *T* sic juxta positis, adhibeatur Prisma tertium *αβκ*, quod primis prismatibus et eorum imaginibus interpona-tur in situ ad lineam *Fφ* sive ad imagines dictas *PT, MN* parallelo: ita nempe ut radios utriusque Prismatis *ABC* ac *DEG* tendentes versus *PT* et *MN*[35] pariter intercipiat, eosque refringens aliò projiciat, quemadmodum ad *ππ* ac *μν*. Adeò ut quae duobus prismatibus in priori specimine facta sunt, hic videas facta tribus. His ita paratis et constitutis videbis imagines *ππ* et *μν* ab invicem disjunctas esse, quae priùs apud *PT* et *MN* fuerunt contiguae, et in directum positae: ita quidem ut purpura *μ* in extremitate imaginis *μν* magis distet ab imaginibus primis *PT* et *MN* quàm rubor *τ* in extremitate imaginis *ππ*. Id quod nullo prorsus modo potuisset accidisse, nisi[36] radij ad purpuram ge/nerandam apti aliquanto magis refringerentur ex incidentiâ pari quàm radij generantes rubedinem. Etenim cùm radij coloris utriusque pariter inci-dant in Prisma posterius *αβκ*; pariter etiam emergerent si aequaliter refringe-rentur, et exinde depingerent imagines *ππ* et *μν* prioribus *PT* et *MN* paralle-las, et in directum jacentes. Dixi radios utriusúe coloris purpurei rubeique pariter incidere in Prisma posterius *αβκ*: Quod ne moram injiciat alicui, concipiendum est quòd radij *FT* tantùm inclinantur versus extremitatem ejus *K* quantùm alteri *φM* versus extremitatem alteram *αβκ*: Et sic incident pa-

27/

make it clearer to everyone how great the evidence of the preceding is, I will not be displeased to reveal some very noteworthy things that flow from it.

Let $F\phi$ (Fig. 13) be a wall or a window shutter with two holes F and ϕ, which transmit light and are two inches from one another; inside place two prisms, *ABC* and *DEG*, in a position parallel to one another but perpendicular to the line $F\phi$, which is drawn through the center of the holes; and let these two prisms refract the admitted light, projecting two images, *PT* and *MN*, onto the opposite wall in a way wholly similar to what was done in the previous experiment. Moreover, let the prisms' angles, *ACB* and *DGE*, made by the refracting planes be equal. After this is so arranged, you will see the images *PT* and *MN* lying in a straight line with their ends *T* and *M* touching. If this does not occur, the position of one of the prisms must be changed a bit until you perceive that the ends touch or perhaps coincide somewhat. When the purple *M* and the red *T* are so juxtaposed, add a third prism, $\alpha\beta\kappa$, which is placed between the first prisms and their images in a position parallel to the line $F\phi$ or to those images *PT* and *MN*; that is, so that it may simultaneously intercept the rays tending toward *PT* and *MN*[35] from each prism, *ABC* and *DEG*, and refract and project them elsewhere, as to $\pi\tau$ and $\mu\nu$. And here you shall see done with three prisms what was done with two prisms in the previous example. Having thus prepared and arranged these things, you will see the images $\pi\tau$ and $\mu\nu$, which earlier touched at *PT* and *MN* and lay in a straight line separated from one another; so that, in fact, the purple μ at the end of the image $\mu\nu$ is farther from the first images *PT* and *MN* than the red τ at the end of the image $\pi\tau$. This could in absolutely no way have happened, unless[36] at equal incidence rays disposed to generating purple were refracted somewhat more than rays generating red. Since the rays of each color fall upon the second prism $\alpha\beta\kappa$ similarly, they would also emerge similarly if they were refracted equally, and then they would depict images, $\pi\tau$ and $\mu\nu$, parallel to the first ones, *PT* and *MN*, and lying in a straight line. I have said that the rays of each color, purple and red, fall upon the prism $\alpha\beta\kappa$ similarly. Lest this hinder anyone, it ought to be understood that the rays *FT* are inclined as much toward their end *K* as the other ones, ϕM, are inclined toward the other end $\alpha\beta\kappa$; thus they are incident similarly, or at the same angles, even if

35 [= II, 13]. The former experiment is extended somewhat, either by using three prisms.

---

(35) tendentes . . . *MN*] Added.
(36) Originally: *nisi concedatur quòd* (unless it be admitted that).

riter sive ad eosdem angulos, licèt non paralleli.[37] Siquis tamen velit efficere
ut incidant etiam paralleli, nihil aliud agendum est quàm ut alterum e prismatibus anterioribus *ABC* vel *DEG* circa suum axem paululum convertatur
donec inter *T* et *M* interiores imaginum extremitates tanta intercedat distantia quanta intersit foraminibus *F* et ϕ sive quantam isti rei sufficientem judicaverit, imaginibus ad istam distantiam in directum jacentibus. Et Prismate
$\alpha\beta\kappa$ deinceps interposito, facilè percipiet quod incidentes parallelè emergunt
inclinati, tum quòd imagines non amplius in directum jacebunt, tum quòd
purpura *M* ad majorem distantiam transferetur quàm rubedo *T*.

36 [= II, 14]. Vel   Si tria prismata non praesto sint, experimentum jam recitatum duobus
contractiùs duobus   experiri possis, idque modo magis expedito et facili. Sit *ABCDE* (fig [14])

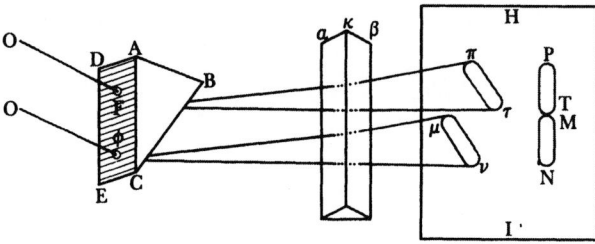

Figure 14

Prisma cujus unum latus planum *A*[*C*]*DE* papyro denigratâ tegatur duobus
parvis foraminibus *F* et ϕ luci perviâ, quorum foraminum situs esto ad longitudinem Prismatis transversus. Tum Prismate hoc ita disposito, ut radij permeantes ista foramina terminentur in oppositum quoddam planum, puta
papy/rum *HI*; transferatur ista papyrus ultra citraque donec videas imagines   28/
duas *PT* et *MN* contiguis extremitatibus in directum conjunctas, ut priùs.
Deinde altero Prismate $\alpha\beta\kappa$ interposito in situ ad alterum transverso: videbis
imagines illas *PT* et *MN* ad $\pi\tau$ et $\mu\nu$ ita translatas esse ut non amplius jaceant
in directum, rubedine τ a *PN* minùs remotâ quàm purpura μ, sicut in prioribus contingebat.

37 [= II, 15]. Idem   Est et aliud ex eodem fonte derivatum specimen haud expertu difficilius aut
aliter promovetur   minoris evidentiae. Prismate *ABC* (fig [15]) juxta foramen *F* ut prius collocato; ad distantiam convenientem (veluti duodecim pedum) statuatur aliud
Prisma $\alpha\beta\kappa$ in situ transverso respectu prioris, vel forte parallelo aut alio
quovis pro arbitrio: ita tamen ut anterius Prisma *ABC* lucem refractam et
coloratam projiciat in aliquod ex ejus planis lateribus $\alpha\delta$. Quod quidem latus
obducatur papyro denigratâ, & exiguo[38] foramine *G* per medium transfossâ,
per quod aliqui ex radijs ab anteriori prismate refractis transeant in hoc
prisma posterius: ubi cùm rursus refracti fuerint pergant ad papyrum

---

(37) That is, the rays *FT* from the upper spectrum and the rays ϕM from the lower spectrum
make equal angles with the prism's principal plane, though they fall on that plane from opposite
sides. In the following articles, Newton gradually eliminates the effect of the approximately $2\frac{1}{2}°$

not parallel.[37] Yet if anyone wants to make them also fall parallel, nothing other need be done than to rotate one of the first prisms, *ABC* or *DEG*, a little about its axis until as great a distance lies between the images' inner ends *T* and *M* as between the holes *F* and *ϕ*, or as great as is judged sufficient for the images to lie in a straight line at that distance. Inserting the prism *αβκ* again, you will easily perceive that the rays falling parallel emerge inclined, both because the images no longer lie in a straight line and because the purple *M* is shifted a greater distance than the red *T*.

If three prisms are not available, you can try the experiment just related with two, and in a more convenient and easy way. Let (Fig. 14) *ABCDE* be a prism, and cover one of its plane sides *A[C]DE* with a blackened paper having two small holes *F* and *ϕ* to transmit light; and let the holes' position be transverse to the prism's length. When this prism has been so placed that the rays passing through those holes are outlined on some opposite plane, for example, the paper *HI*, shift that paper back and forth until you see the two images *PT* and *MN* joined in a straight line with their ends touching, as before. Then inserting the second prism *αβκ* in a position transverse to the other one, you will see those images *PT* and *MN* translated to *ππ* and *μν*, so that they no longer lie in a straight line, with the red *τ* being less remote from *PN* than the purple *μ*, just as it occurred in the previous ones.

There is another example drawn from the same source that is not at all more difficult to try or less evident. As before, having placed the prism *ABC* (Fig. 15) next to the hole *F*, at an appropriate distance (for instance, twelve feet) set another prism *αβκ* in a position transverse to the first, or else perhaps parallel, or in any arbitrary way, but such that the first prism *ABC* projects the refracted colored light onto one of its plane sides *αδ*. However, cover this side with a blackened paper pierced through the middle with a small[38] hole *G* through which some of the rays refracted by the first prism may pass into this second prism where, when they have been refracted again,

36 [= II, 14]. Or more simply by two.

37 [= II, 15]. The same thing is accomplished differently.

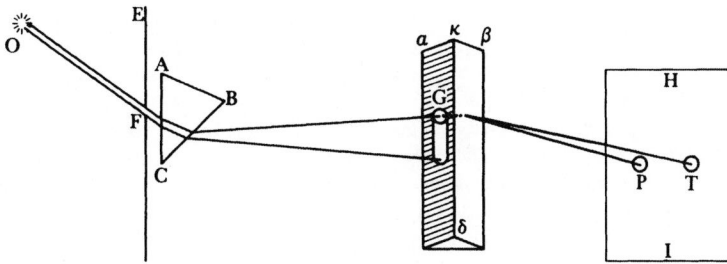

Figure 15

divergence of the rays falling on the second prism. This experiment is briefly described at the conclusion of Expt. V of Bk. I, Pt. I, Prop. II of the *Opticks*.

(38) Added.

HI abinde decem pedibus vel pluribus distantem. Quibus ita constructis et dispositis in situ illo figatur papyrus *HI* et prisma posterius αβκ. Denique prae manibus sumatur anterius Prisma *ABC* non ut moveatur a loco ejus sed motu tantùm angulari nunc huc nunc illuc paululum inclinetur, ut alios atque alios colores successivè trajiciat per foramen *G* in oppositam papyrum *HI*. Et videbis quod color quilibet diversus[38] ad locum diversum perget. Veluti cùm ea sit positio Prismatis *ABC* ut rubeum colorem projiciat in *G*, si ponatur quod ille color ab altero prismate αβκ refringatur ad *T*, tum positione Prismatis *ABC* paululum mutatâ inclinando circa axem[39] donec purpura cadat in *G*, videbis quod ille color juxta obliquiorem tramitem refringetur, puta ad *P*.[40] Et pari modo si color aliquis intermedius incidat in *G*, idem refrin/getur 29/ ad locum ipsis *P* ac *T* interjacentem. Quamobrem cùm radij cujuslibet generis pergentes a foramine *F* positione dato ad Foramen *G* positione datum, et ideò similiter incidentes in prisma posterius αβκ, refringantur ad loca diversa *P, T*, caeteraque intermedia: constat quòd inaequaliter refringuntur. Et cùm refractus *GP* observetur magis deflectere ab incidenti *FG* quàm refractus *GT*; constat quòd radij purpuram exhibentes[41] magis refringuntur quàm exhibentes[41] ruborem, caeteríque deinceps in ordine intermedio.[42]

38 [= II, 16]. Quod specimen, circumstantiâ variatâ, fit maximè scientificum.

Siqua forsan oboriatur suspicio, quod ex motu Prismatis *ABC* foraminibus *F* ac *G* interpositi, incidentia radiorum diversos colores efficientium tantùm varietur quantùm sufficiat ad efficiendam varietatem locorum *P, T*, &c: ad quos refringuntur: quamvis motus iste sit exiguus et ineptus huic effectui, tamen ut suspicio illa prorsus eximatur,[43] anterius Prisma *ABC* ad alteras partes foraminis *F* Solem versus collocandum est, ut radij incidentes in foramen *G* directè veniant a dicto foramine *F*. Eo enim pacto cùm foramina *F* ac *G* positione determinentur, positio radiorum per utrumque trajectorum determinabitur, eademque accuratè erit omnium incidentia quoscunque colores exhibentium; et tamen diversicolorum refractio non secus peragetur ad loca diversa *P, T* &c quàm modò explicui.[44]

39 [= II, 18]. Conclusio de affinitate cognitionis colorum et refractionum.

Cùm veritatem propositam sic fecerim stabilitam, ha[n]c propositionem concludam annotando[45] connexionem et affinitatem quam coloribus et refractionibus interesse dixeram: Nempe ex ostensis non solùm pateat quòd diversa colorum genera cum definitis gradibus refrangibilitatis reciprocantur:

---

(39) inclinando circa axem] Added.

(40) videbis . . . *P*.] Originally: poscatur quod ille color abinde refringetur ad *P* (let it be required that that color will be refracted from there to *P*).   (41) Originally: ostendentes.

(42) By covering the face of the second prism with a perforated paper so that the two holes *F* and *G* serve as a collimator and cause all rays to have virtually the same angle of incidence on the second prism, the orientation of the second prism, as Newton observes, has become irrelevant to the demonstration of unequal refrangibility. In the description of the *experimentum crucis* in the "New theory" (*Correspondence*, 1:94–5 = *Phil. Trans.*, 6 (1671/2):3078–9) the orientation of the second prism is left unspecified; whereas in the *Opticks* (Bk, I, Pt. I, Prop. II, Expt. VI, p. 31 = Add. 3970, f. 40ʳ) he explicitly states that its orientation is arbitrary, although his figure has the two prisms parallel. The orientation of the prism does, however, significantly affect the resolution of the colors at the second refraction, as Newton observes in §II, 19. The origin of these experiments with two refractions in two prisms may be seen in the essay "Of Colours" (§44–5, Add. 3975, f. 7ᵛ = Hall, "Further experiments," p. 35), though Newton does not fully specify the experimental arrangement. Mamiani in *Newton*, pp. 113–53, thoroughly discusses these crossed-prism experiments and traces the origin of the *experimentum crucis* to them.

they will continue to the paper *HI* ten or more feet away from it. After these have been thus constructed and set up, fix the paper *HI* and the second prism *αβκ* in that position. Finally, grasp the first prism *ABC*, not so it may be moved from its place, but only so it may be turned back and forth a bit with an angular motion to cast one and then another color successively through the hole *G* onto the facing paper *HI*. You will see that every different[38] color proceeds to a different place. For example, when the position of the prism *ABC* is such that it projects the red color to *G*, if it is assumed that that color is refracted to *T* by the second prism *αβκ*, then when the position of the prism *ABC* is changed slightly by turning it about its axis[39] until the purple falls at *G*, you will see that color refracted along a more oblique path, say to *P*.[40] Similarly, if some intermediate color falls on *G*, it will likewise be refracted to a place between *P* and *T*. Consequently, since rays of any sort whatsoever that proceed from the hole *F* in a given position to the hole *G* in a given position, and that therefore are similarly incident upon the second prism *αβκ*, are refracted to different places *P* and *T*, and all the others in between, it is certain that they are refracted unequally. Since the refracted ray *GP* is observed to be deflected more from the incident one *FG* than the refracted ray *GT*, it also is certain that rays exhibiting[41] purple are refracted more than those exhibiting[41] red, and the others in sequence are refracted in an intermediate order.[42]

If any suspicion perhaps arises that because of the motion of the prism *ABC*, placed between the holes *F* and *G*, the incidence of the rays producing the different colors is changed sufficiently to cause the diversity of the places *P*, *T*, and so forth, to which they are refracted, that motion, however, is too small and unsuitable for this effect. Nonetheless, to remove that suspicion altogether[43] the first prism *ABC* should be placed on the other side of the hole *F* toward the sun, so that the rays falling upon the hole *G* come directly from that hole *F*. Since in this way the holes *F* and *G* are fixed in position, the position of the rays transmitted through both holes will be fixed, and the incidence of all rays, whatever color they exhibit, will be exactly the same; yet the refraction of the different colors will proceed to the different positions *P*, *T*, and so forth, no differently than I have already explained.[44]

Since I have thus established the proposed truth, I shall conclude this proposition by commenting on[45] the connection and relation that I had said belong to colors and refractions. Namely, it is not only evident from what has been shown that different kinds of colors correspond to definite degrees

38 [= II, 16]. This example becomes most scientific when a feature is varied.

39 [= II, 18]. Conclusion of the examination of the relation of colors and refractions.

(43) ut . . . eximatur] Originally: ut ejus rei nulla supersit suspicio (Lest any suspicion of this matter remain).

(44) By transferring the first prism to the other side of the hole *F*, Newton has essentially arrived at the arrangement of the *experimentum crucis*, though that experiment is used to prove the existence of unequal refrangibility rather than the different refrangibility of different colors. The remaining differences in the experimental arrangement, such as setting the two prisms parallel, are minor ones, and Newton in fact makes these changes in the conceptually related experiment with unequal reflexibility in §69 = II, 60.

(45) ha[n]c . . . annotando] Originally: conclusionis loco juvabit annotare (in place of a conclusion it will help to comment on). Before settling on the final version Newton began and then deleted "claudem" (I conclude).

(a) sec 3 & 30
(b) sec 32 &c

sed et ijsdem[a] experimentis probatur dari radios diversè refrangibiles, et radios diversè refrangibiles esse diversicolores, ijsdemque[b] probatur e contra radios diversicolores esse diversè refrangibiles, et inde radios diversè refrangibiles dari. Et hinc scopus eorum quae in primis lectionibus de dispari refrangibilitate radiorum edocui, quoad causas colorum intelligendas multùm illustratur; ut pateat quod una absque alijs dilucidè tractari nequeant.

Lect. 4
40 [→ II, 27, 28].
Transitur ad
propositionem
tertiam.

/ Posteaquam ostendi radios qui producunt varios colores etiam varias 30/ refractiones pati: cogitabam de explicando modo quo colores generantur mediante Prismate. Sed istud quoniam ex praecipuis esse videatur quae de coloribus dicenda suscepi: satius esse judico me priùs ostendere veritatem tertiae propositionum quas in postremâ Lectione proposueram, quatenus albedinem concernit; eam nempe ex omnigenis coloribus posse componi; et inde lucem solis albere quòd omnes colores in eâ commisti lateant.[1] Quod cùm ostendero, genesis colorum a Prismatibus postmodum[2] satiùs et majori cum evidentiâ pandetur.

Itaque proponatur jam monstrandum quòd cùm omnes omninò colores qui virtute Prismatum generantur, debitè commiscentur sibi: color albus exinde resultabit.[3] Istud autem cùm semel deprehenderam esse verum, de varijs postea modis cogitabam quibus mistura talis perfectè fieret: ac primò rem aggressus sum cum pluribus Prismatibus ita dispositis ut colores eorum in eundem locum inciderent et sic inter se miscerentur.[4]

41 [≈ II, 28].
Modus componendi
albedinem ex
coloribus Prismatum

Sint *ABC, DEF* et *GHI* [fig 16] tria Prismata juxta se in situ parallelo ita disposita, ut alterum *DEF* sit alteris duobus *ABC* et *GHI* utrinque vicinissimis intermedium, in morem trium linearum conficientium capitalem literam graecam Ξ. Et lux per unumquodque Prisma liberè transiens excipiatur in papyrum *PT* pede uno vel duobus postpositam. Coloribus omnium prisma-

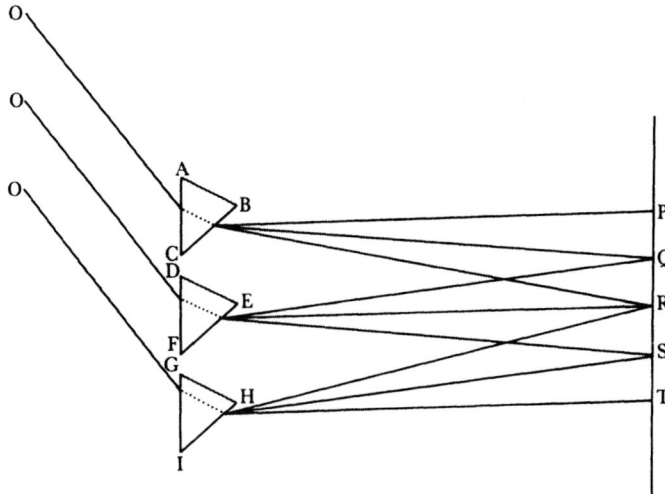

Figure 16

of refrangibility, but also it is proved by the same experiments[a] that there (a) §§3, 30
exist differently refrangible rays, and that differently refrangible rays are
differently colored; and conversely it is proved from the same experiments[b] (b) §§32, ff.
that differently colored rays are differently refrangible, and consequently
there exist differently refrangible rays. The aim of what I taught in the first
lectures about the unequal refrangibility of rays with respect to understanding the causes of colors is made much more evident, so that it is obvious that
the one cannot be clearly discussed without the other.

After I showed that the rays that produce different colors also experience Lecture 4
different refractions, I considered explaining how colors are generated by 40 [→ II, 27, 28].
means of a prism. But because that may well be one of the principal things Passing to the third proposition.
that I have attempted to describe about colors, I think it is preferable first to
demonstrate the truth of the third of the propositions that I proposed in the
last lecture insofar as it concerns white: namely, that it can be compounded
from every sort of color; and then that the sun's light is white, because all
colors mixed together are latent in it.[1] When I have demonstrated that, the
origin of colors from prisms will afterward[2] be disclosed more satisfactorily
and with greater clarity.

Therefore it is now proposed to prove that when absolutely all the colors
generated by the power of prisms are duly mixed with one another, the color
white will consequently result.[3] Moreover, once I had found that to be true,
I afterward considered various ways by which such a mixture could be made
perfectly. First, I attempted this with several prisms arranged so that their
colors fell on the same place and thus were mixed among themselves.[4]

Let *ABC, DEF,* and *GHI* (Fig. 16) be three prisms placed near each other 41 [≈ II, 28]. A way
in a parallel position so that one, *DEF,* is in between and exceedingly close to of compounding
the other two, *ABC* and *GHI,* on each side of it in the form of the three lines whiteness from
making the Greek capital letter Ξ. Let the light passing freely through each of prismatic colors.
the prisms be received on the paper *PT* placed one or two feet beyond them.

---

(1) et inde . . . lateant] Added. The strategy sketched here for establishing the nature of
whiteness, namely, first (in Lects. 4, 5) to show that white light just like the sun's light can be
compounded from all colors, and then (in Lect. 6) to show that sunlight is in fact compounded of
all colors, was adopted by Newton in all versions of his theory: in the *Optica,* Lectures II, 4–8;
in the "New theory," Props. 7, 8, though only with a partial "illustration" and no proof; in his
letter for Huygens on 23 June 1673, Props. 5, 10, (*Correspondence,* 1:293–4 = *Phil. Trans.,* 8
(1673):6091); and in the *Opticks* (Bk. I, Pt. II, Prop. V), where the two parts receive very
unequal treatment—sixteen pages for the first part but only nine lines for the second.

(2) Namely, in Lecture 7.

(3) This sentence alone, somewhat altered, was transferred from §40 to §II, 27.

(4) In the *Optica* this sentence ("ac primò . . . miscerentur") begins §II, 28.

tum sic in ipsam *PT* projectis, convertantur prismata circa proprios axes, et videbis colores istos sibi invicem accedere vel recedere. Quare convertantur donec talis sit eorum situs ut unius Prismatis *ABC* rubor, et alterius *GHI* purpura vel color indicus cum viriditate tertij *DEF* coincidant, sicut vides factum ad *R*. Et ex istis coloribus ita sibi commixtis albedinem generari cernes, colore purpureo et caeruleo juxta *P* conspecto, rubeo verò et flavo juxta *T,* et albo juxta *R* caeteros intercedente.[5]

42 [= II, 29].
Notanda quaedam
quò satiùs fiat.

/ Caeterùm in istis experiendis juvabit observare sequentia.   31/

Primò, si anguli Prismatum planis refringentibus contenti *ACB*, *DFE* et *GIH* sint inaequales; praestat ut illud Prisma cujus angulus *GIH* maximus est ponatur versus exteriorem partem anguli contenti radijs incidentibus et refractis: et istud versus interiorem cujus angulus *ACB* est minimus.

Secundò, aperturae per quas lux transmittitur per prismata debent esse magnae. Imò convenit ut transitus luci per tota Prismata pateat, obstaculo nullo adhibito. Neque opus est ut experimentum in tenebris peragatur sicut in alijs quamplurimis requiritur.

Tertiò papyrus *PT* in quam colores incidunt non nimis distare debet a Prismatibus. Sufficit distantia pedum plus minus duorum. Has autem aperturas et distantiam statuo ut colores eò meliùs commisceantur ad albedinem perfectiorem componendam.

Quartò ut colores ad *R* faciliùs etiam et satiùs comisceantur, Prisma *ABC* statuatur imprimis in situ quocunque tali ut radij tum ingredientes tum emergentes refractionem praeter propter aequalem patiantur: et in eo situ figatur. Et colores ejus ad distantiam duorum pedum excipiantur, vel ad eam potiùs ubi vides flavum ejus et caeruleum modò contiguos, albedine intermediâ tum evanescente. Postea figatur aliud Prisma *GHI* in tali situ ut purpura ejus contingat ruborem alterius *ABC*, non autem coincidat illi: et linea contactus notetur. Deinde tertium Prisma *DEF* sic fige ut ejus colorum medietas cadat in dictam lineam contactûs, quod ubi contingit facilè cognosces intercipiendo lucem ingressuram caetera prismata. Denique papyrus *PT* ultra citraque transferatur paululum donec videas albedinem perfectam in medio colorum ad *R* generari. Quam quidem albedinem ex varijs coloribus compositam esse constabit intercipiendo colores unius duorumve prismatum priusquam attingant papyrum. Nam loco albedinis eos quos non intercipis colores intueberis.

Denique[6] si velis ut colores adhuc perfectiùs misceantur, possis adhibere plura prismata modò praesto sint: tamen eventus non deerit expectationi si tria tantum adhibeas. Etenim colores cujusque Prismatis seorsim spectati non sint omninò simplices, sed viridis ejus et rubeus nonnihil miscentur in flavo: et purpureus ac viridis / in caeruleo; et sic de reliquis; quemadmodum in   32/ sequentibus ostendetur. Et inde fit quod cùm tria tantùm prismata adhibentur, non solùm tres colores rubeus viridis et indicus commisceantur in *R*, sed etiam caerulus et flavus unà cum omnibus eorum gradibus intermedijs istam albedinis compositionem ingrediantur.

---

(5) Newton first briefly described this experiment in "Of Colours" (§46, Add. 3975, f. 7ʳ), and in §89 ≈ II, 90 he sets forth a related experiment with two prisms that had slipped his mind when preparing this lecture.

When the colors of all the prisms are thus projected onto *PT,* turn the prisms around their axes, and you will see these colors approach or recede from each other. Accordingly, rotate them until they are so situated that the red of the one prism, *ABC,* and the purple or indigo color of the other, *GHI,* coincide with the green of the third, *DEF,* as you see done at *R.* From these colors thus mixed together you will see whiteness produced, with the colors purple and blue visible at *P,* but red and yellow at *T,* and white falling in between the others at *R.*[(5)]

    Moreover, in doing these experiments it will help to note the following:

    First, if the prisms' angles, *ACB, DFE,* and *GIH,* made by the refracting planes are unequal, it is better to place that prism whose angle, *GIH,* is the largest toward the exterior side of the angle made by the incident and refracted rays, and that prism whose angle, *ACB,* is the smallest toward the interior.

42 [= II, 29]. Some things to note so that it occurs more satisfactorily.

    Second, the apertures through which the light is transmitted through the prisms must be large. Indeed, it is fitting that the light's path through all the prisms be open with no obstacle being used. Nor is it necessary that the experiment be performed in darkness as is required in very many others.

    Third, the paper *PT* on which the colors fall must not be too far from the prisms; a distance of two feet more or less is satisfactory. I fixed these apertures and the distance, however, so that the colors would be better mixed together to compound a more perfect white.

    Fourth, to mix the colors at *R* more easily and also more satisfactorily, place the prism *ABC* initially in any position such that both the entering and emerging rays undergo a nearly equal refraction, and fix it in that position. Receive its colors at a distance of two feet, or rather at that distance where you see its yellow and blue just touching with the white in between then vanishing. Then fix the other prism, *GHI,* in a position such that its purple touches but does not coincide with the red of the first one, *ABC,* and note the line of contact. Next, fix the third prism, *DEF,* so that the middle of its colors falls on that line of contact, and you will readily identify where this occurs by blocking the light that enters by the other prisms. Finally, shift the paper *PT* back and forth slightly until you see a perfect white produced in the middle of the colors at *R.* Indeed, it will be established that this white is compounded of various colors by intercepting the colors of one or two prisms before they reach the paper. For, instead of white you will see those colors that you did not intercept.

    Finally,[(6)] if you want the colors still more perfectly mixed, you can use more prisms provided they are available; yet the result will not fall short of your expectations if you use only three prisms. For the colors of each prism when observed separately are not entirely simple; in fact green and red are mixed somewhat in the yellow, and purple and green in the blue, and similarly for the others, as will be shown in the following. Consequently, it turns out that when only three prisms are used, not only are three colors, red, green, and indigo, mixed together at *R,* but also blue and yellow together with all their intermediate gradations enter into that composition of white.

---

(6) Originally: *Quintò et ultimò* (Fifth and lastly).

Verùm cùm tot prismata in situ tam accurato disponere, propter motum solis et alia incommoda difficile forsan et laboriosum simul inveniatur, nisi adhibeatur Machina quaedam eâ de causâ fabricata ut ejus ope prismata desiderato situ figantur:[7] alium propterea modum profero quo ista negotio leviori, idque unico prismate periclitari poteris. Sumatur papyrus vel aliud opacum corpus attenuatum in morem laminae. Et in eo confodiantur oblongae rimae sex aut plures parallelae, quarum latitudines sint aequales distantijs aut ijs paulò majores. Deinde papyrus ista figatur alicui ex planis lateribus prismatis: Sit istud latus papyro obductum *ACED* (fig [17]), et rimae in

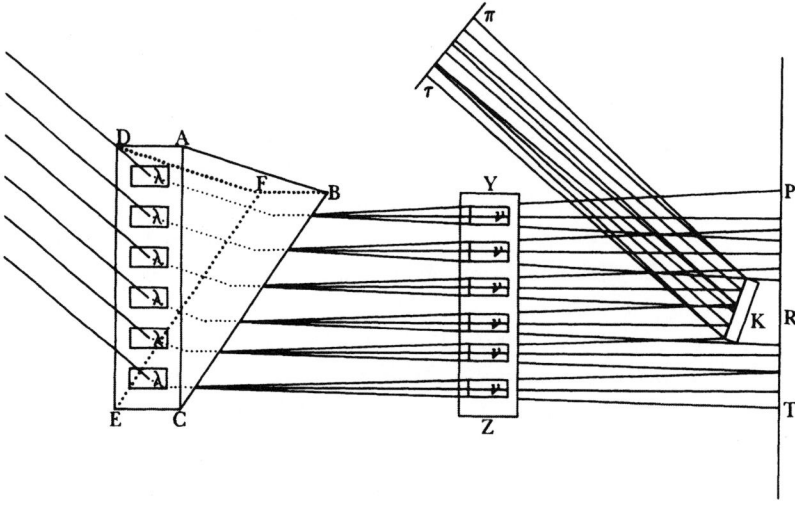

Figure 17

papyro excisae literis λ designentur; quarum situs esto parallelus ad *EC* concursum laterum refringentium prismatis sive verticem ejus. Papyrus autem debet toti isti plano *ADEC* superinduci, nequa lux alibi transmissa quàm per praedictas rimas perturbet experimentum. Tum Prisma statuatur in luce solis ut radij ejus vel per dictas rimas id ingrediantur vel postquam refracti fuerint per eas egrediantur: et in isto situ figatur. Quo facto sumatur alia papyrus *PT* quae sic teneatur a posticâ parte Prismatis ad distantiam duorum triumve digitorum ut in eam lux terminetur; et videbis tot lineas colorum quot sunt oblongae rimae λ; quarum linearum cuique tot competent colores quot solent apparere virtute Prismatum. Nempe / quaelibet rima subit officium unius e 33/ prismatibus in experimento priori adhibitis et proprios colores caeruleum rubrum caeterosque generat quasi tot essent prismata quot sunt rimae. Porro si papyrus *PT* longiùs differatur a Prismate coloratas istas lineas paulatim dilatari cernes et interjecta spatia minui donec absorbeantur a coloribus jam factis contiguis. Et si papyrus adhuc longius differatur, colores a diversis rimis effecti (rubri cum caeruleis primò deinde alij cum alijs) incipient plus plusque misceri. Et sic sese paulatim diluent, donec cùm mistura satis abso-

But it may be found both difficult and laborious to place so many prisms so accurately in position, because of the sun's motion and other inconveniences, unless some device constructed for that purpose is used, so that with its help the prisms may be fixed in the required position.[7] Accordingly, I offer another way—and with a single prism—whereby you will be able to try that with less trouble. Take a paper or other thin opaque body in the form of a sheet, and in it cut six or more parallel, oblong slits whose widths are equal to their separations or a little greater. Next attach that paper to any of the prism's plane sides. Let that side covered with the paper be *ACED* (Fig. 17) and the slits cut in the paper be designated by the letter λ; let their position be parallel to *EC,* the intersection of the prism's refracting sides, or its vertex. The paper, however, must cover that entire plane *ADEC,* so that light transmitted anywhere other than through those slits does not disturb the experiment. Then place the prism in the sun's light so that its rays either enter it through those slits or leave through them after having been refracted, and fix it in that position. After this is done, take another paper *PT* and hold it at a distance of two or three inches from the back side of the prism so that the light is terminated on it, and you will see just as many lines of colors as there are oblong slits λ. To each of these lines there correspond as many colors as usually appear by the power of prisms; namely, each slit assumes the function of one of the prisms used in the previous experiment and generates its characteristic colors, blue, red, and the others, just as if there were as many prisms as there are slits. Furthermore, if the paper *PT* is moved farther away from the prism, you will see those colored lines gradually expand and the intervening spaces diminish until they are consumed by the colors just when they become contiguous. If the paper is moved still farther away, the colors produced by the various slits will begin to be more and more mixed, first reds with blues, then other colors with still others. In this way they will gradually dilute one another until, when the mixture is sufficiently perfect, they are

43 [= II, 30].
Another way of accomplishing the same thing.

---

(7) Newton rarely mentions the difficulty of performing his experiments with a moving light source. Since the sun moves about fifteen minutes of arc in one minute of time, he had to move nimbly in order to continually readjust his apparatus. For example, in describing the closest relation to the *experimentum crucis* in the *Opticks* (Bk. I, Pt. I, Prop. II, Expt. VI, p. 31 = Add. 3970, f. 40^r), he notes that after adjusting the prisms and apertures he "returned speedily to the first Prism." When Newton projected the spectrum as far as twenty-two feet, it moved about one inch per minute and made measurements awkward. To resolve this problem he cast the spectrum onto a paper or card that could be marked, measured, and moved along with the mobile image, and—even more important—he employed an assistant or friend, as he mentions in his account of the division of the spectrum in §§II, 94–5. In 1672 he also referred to an assistant for a measurement of "Newton's rings" in "Of the colours of plated transparent substances" (Observation 14, Add. 3970, f. 521^r), repeated in Observation XIII of both the so-called "Discourse of Observations" from 1675 and Bk. II, Pt. I of the *Opticks*. Newton apparently never constructed a "machina" to deal with this problem, which in the eighteenth century was solved by the development of the heliostat. The title "Discourse of Observations," which is now widely adopted for the paper on the colors of thin films that Newton sent to the Royal Society in December 1675, was in fact first assigned by Turnbull in *Correspondence,* 1:386. .ote 1. In his correspondence with Oldenburg in 1675 and 1676 Newton referred to t]is untitled paper variously as his "discourse," "papers," and "paper of observations"; ibid., 1:358, 360, 414.

luta est, convertantur in albedinem, praeterquam in eorum extremitatibus *P* ac *T* ubi mixtura et confusio fere nulla est. Et isthaec accidunt cùm papyrus *PT* quasi ad distantiam decem vel duodecim vicibus majorem ipsâ *AC* vel *BC* latitudine planorum prisma constituentium, amoveatur. Quod si amoveatur adhuc longiùs absimilium[8] radiorum commistio fortasse perfectior evadet, sed colores purpurei et caerulei ad *P,* ac flavi rubeique ad *T* latiores fient, et interjectum spatium album minuetur, donec totum destruatur ab istis coloribus occupatum.[9]

44 [= II, 31]. In illum notae

In hisce autem experiendis cavendum est ut[10] oblonga foramina λ sint accuratè aequalia et aequalibus distantijs ab invicem dissita, nè luce magis copiosâ[11] per aliquod ingressâ quàm per caetera, colores exinde generati praevaleant caeteris et misturam perfectam conturbent: et sic vice albedinis colores apparebunt hinc illinc more fortuito sparsi. Illa verò distantia rimarum λ ut et earundem latitudo non malè statuitur fore pars digiti circiter duodecima, aut eâ fortè major si prisma satis amplum adhibeas. Quinetiam si cupias ut experimentum sit omnibus numeris absolutum, vice prismatum vitreorum vulgò venalium (quae sunt nimis gracilia) debes amplioribus uti qualia possis efficere ex laminis vitreis utrinque perpolitis et conjunctis in morem vasculi prismiformis, quod vasculum impleatur aquâ clarissimâ, et undique cemento obturetur.[12] Non multùm refert quaenam sit hujus longitudo, sufficit ut sit trium digitorum, sed refringentia latera debent esse quatuor vel / sex digitos lata aut ampliùs, ut rimae praefatae λ cum distantijs earum fiant majores et plures et magis accuratae. Sin utaris angustioribus, qualia vulgò venduntur; colores externi juxta *P* ac *T* dilatando priùs destruent interjectam albedinem quàm perficiatur per remotionem papyri *PT.* Et illa praeterea quae in totum constant ex vitro, colore aliquo ut viridi vel flavo plerumque tinguntur, et radios ita tingunt in transitu ut albedinem perfectam exhibere nequeant.

34/

45 [= II, 32]. Objectio quòd albor ex destructione non misturâ colorum generatur.

Jam verò audire videor objectionem ex receptis philosophorum opinionibus depromptam: Dicat enim aliquis quòd colores revera et propriè loquendo non miscentur sed destruuntur potiùs; idque eâ de causâ quòd umbrae vicinia, quae necessaria est ad productionem colorum, tollitur cum radij per diversas rimas trajecti commisceri incipiunt; et praeterea quòd radijs sic mixtis quorum motus inter se dissentiunt, necesse est ut isti motus destruant alterutros, quibus cessantibus color omnis perit et in albedinem convertitur. Sic Cartesianus aliquis contendat forte[13] quòd cùm globuli miscentur quorum rotationes contrariantur sibi, necesse est ut impediant sese et alternos motus destruant: Et sic alij objiciant alia.[14]

(8) Originally: fortè difformium.

(9) Newton first described this experiment, but with only four slits, in his essay "Of Colours," §47, Add. 3975, ff. 7ʳ–8ʳ. In the *Opticks* (Bk. I, Pt. II, Prop. V. Expt. XII) he replaced the perforated paper by a comb whose teeth were as wide as the intervals between them.

(10) cavendum est ut] Originally: notare vellem quòd (I wish to note that).

(11) luce magis copiosâ] Originally: pluri luce (more light).

(12) Newton frequently used hollow glass prisms filled with fluids to circumvent the imperfections—bubbles, veins, and coloration—and small size of commercially sold prisms, as well as to measure the index of refraction and dispersion of fluids; see §§66 = II, 57; 85 = II, 85; 87 ≈ II,

converted into white, except at their ends *P* and *T* where there is almost no mingling and mixing. This happens when the paper *PT* is removed to a distance about ten or twelve times greater than *AC* or *BC,* the length of the planes forming the prism. If now it is removed still farther, the mixture of unlike[8] rays will perhaps become more perfect; but the colors purple and blue at *P,* and yellow and red at *T* will become wider, and the white space in between will decrease until the entirety is destroyed when it is filled by those colors.[9]

In doing these experiments, however, one must ensure that[10] the oblong holes λ are exactly equal and placed at equal distances from each other, lest a greater quantity of light[11] enters through one hole than through the others, and the colors generated from it overwhelm the others and disturb the perfect mixture, and thus instead of whiteness colors will appear randomly scattered about. In fact, fix both the distance of the slits λ and their width, not inappropriately, at about a twelfth of an inch, or perhaps greater than that if you use a sufficiently wide prism. Moreover, if you want the experiment to be perfect in all respects, instead of the glass prisms commonly sold (which are too slender) you must use broader ones, such as those you can make from glass plates highly polished on both sides and joined together in the form of a small prism-shaped vessel; the vessel should be filled with very clear water and sealed all around with cement.[12] Its length does not matter much—three inches is sufficient—but its refracting sides must be four or six inches broad, or more, so that those slits λ together with their intervening spaces become larger, more numerous, and more accurate. But if you use narrower ones, as are commonly sold, the outer colors at *P* and *T* by their expansion will destroy the white in between before it can be perfected by moving back the paper *PT.* Those prisms, moreover, that are made wholly of glass are often tinged with some color, such as green or yellow, and they so tinge the rays in their passage that they are unable to display a perfect white.

But now I seem to hear an objection drawn from the received opinions of philosophers. Some indeed would assert that truly and properly speaking the colors are not mixed but rather are destroyed, because the adjoining shadow necessary for the production of colors is destroyed when the rays transmitted through the various slits begin to be mixed together. Besides, when the rays whose motions are opposed to one another are thus mixed, it necessarily follows that those motions destroy one another; and with the motions ceasing, all color disappears and is transformed into whiteness. Thus some Cartesian might perhaps[13] contend that when the little balls are mixed whose rotations are opposed to one another, it neccessarily follows that they impede one another and mutually destroy their motions. Similarly others would object to other things.[14]

44 [= II, 31]. Note on that experiment.

45 [= II, 32]. An objection that white is generated by the destruction and not by the mixture of colors.

87; 107 = I, 34; and II, 70; also in the *Opticks,* Bk, I, Pt. I, Expts. III, XI, XV, and Pt. II, Expts. I, VIII. In §108 = I, 35 he suggests as an alternative a hollow wood prism.

(13) Added.

(14) Newton is anticipating possible objections, derived from Descartes's and Hooke's theories of color, to his interpretation of these experiments. Like Descartes (see the Introduction,

46 [= II, 33].
Responsio
multiplex: Primò[15]
quòd illi colores non
destruuntur ex
umbrae confinio
sublato.

Sed responsio multiplex in promptu est: et imprimis inquam quòd cùm umbrae coloribus interjectae primùm evanescunt removendo papyrum *PT*, colores tamen non ideo pereunt neque minimùm immutantur donec incipiant misceri per remotiorem distantiam papyri. et albedo non producitur donec per distantiam adhuc remotiorem mistura radiorum omnis generis evadat perfecta. Unde confinium umbrae[16] non est necessarium ad colores producendos, neque albedo generatur ex isto sublato.[17]

47 [= II, 34].
Secundò, Neque
probabilitèr ex
motuum
contrarietate.

Secundo colores qui primò omnium miscentur, nimirum purpureus sive violaceus et rubeus videntur esse maximè omnium absimiles, propterea quod adversas colorum extremitates occupant. Quamobrem itáque motus eorum contrarij non destruunt sese neque color albus generatur antequam caeteri etiam colores omnes misceantur?

48 [= II, 35].
Tertiò, Quòd radij
per idem medium
confusè transientes
non agunt in se
invicem

Tertiò cuique licet observare idque nullo negotio quòd colores non omninò mutantur trajiciendo radios per / medium quantumvis luminosum. Sic colores 35/ prismatum sunt ijdem sive trajiciantur per spatium illuminatum sive tenebris involutum. Et res omnes eodem modo coloratae cernuntur, sive conspiciantur cùm lumen solis trajicitur per intermedium spatium sive cùm excluditur. Id quod secus esset si lux in lucem per idem medium transeuntem posset agere. Quinimò si radij duobus prismatibus refracti sese decussent, postquam ab invicem discreti sunt, eosdem colores efficient, quos aliàs efficerent si non omninò miscerentur. Id quod non posset evenire si radij diversis coloribus tincti sibi mutuò per eadem spacia transientibus[18] mutationem aliquam inducerent.[19]

49 [= II, 36].
Quartò Quòd albor
praefatus perit si
quilibet color e
misturâ tollatur.

Quarto, cùm in illâ distantiâ papyrum *PT* fixeris ubi colores albedinem optimè componunt: statuatur alia papyrus *YZ* ad distantiam duorum vel trium digitorum a prismate, et in eâ notentur lineae coloratae; tum exscindantur istae partes papyri in quas dictae lineae cecidêre, factis eo pacto rimis oblongis *v* parallelis et aequalibus, ut et aequè latis ac distantibus. Deinde papyrus ista *YZ* in locum suum restituatur tres digitos circiter a prismate distantem ut per rimas ejus lux colorata trajiciatur ad alteram papyrum *PT* longinquiorem. Quo facto possis observare quòd si parùm deprimas papyrum *YZ* ut purpureos colores & caeruleos superioribus[20] labris rimarum ejus impingentes intercipiat, et transmittat caeteros: albedo ad papyrum *PT* con-

---

note (23)), Hooke held that a boundary of light and shadow was necessary for the production of prismatic colors. In Hooke's wave theory, after refraction the front of a pulse becomes oblique to its sides and the leading edge becomes "dead[e]ned by the resistance of the dark or quiet *medium*," whereas the other edge becomes "stronger, having its passage already prepar'd as 'twere by the other parts preceding," thereby generating blue on the one side and red and yellow on the other; see *Micrographia*, Observation IX, p. 63. This is the case for a single ray, but Newton probably had in mind the interaction of a number of adjacent physical rays, as in a beam of sunlight, for then, according to Hooke, some of the bordering shadows are eliminated and the colors destroyed; see the Introduction, note (26).

(15) Added, as were the numbers in the next four headings.

(16) Originally: lucis et umbrae (of light and shadow).

(17) Newton returns to this argument in §92 ≈ II, 89.   (18) II: spatia transeuntibus.

(19) Alhazen (Ibn al-Haytham) first introduced the principle of superposition—that light rays can cross without affecting one another—by arguing that distinct vision would otherwise be

But a multiple response is at hand: First of all, I say that when the shadows that lie between the colors first vanish by moving back the paper *PT*, nevertheless the colors do not consequently disappear, nor are they changed in the least until they begin to be mixed on account of the more remote distance of the paper. White is not produced until, on account of a still more remote distance, the mixture of rays of every kind proves to be perfect. Consequently, the border of shadow[16] is not necessary to produce the colors nor is the whiteness generated by its elimination.[17]

46 [= II, 33]. A multiple response: First,[15] those colors are not destroyed by eliminating the border of shadow.

Second, the colors that are mixed first of all, namely, purple or violet and red, appear to be the most dissimilar of all, because they occupy opposite ends of the colors. Why therefore do their contrary motions not destroy each other and is the color white not generated before all the other colors are also mixed?

47 [= II, 34]. Second, nor probably by the contrariety of the motions.

Third, anyone can observe, and with no difficulty, that colors are not at all changed by transmitting rays through a medium however luminous. Thus the prismatic colors are the same whether they pass through an illuminated space or one enveloped in darkness. All colored things are perceived in the same way whether they are viewed when the sun's light is transmitted through the intermediate space or when it is excluded. This would not happen if light could act upon light while passing through the same medium. Indeed, if rays refracted by two prisms cross each other, after they are separated from one another they will produce the same colors that they would otherwise produce if they were not at all mixed. This could not occur if rays imbued with different colors induced some change on each other while passing through the same space.[19]

48 [= II, 35]. Third, rays passing confusedly through the same medium do not act upon one another.

Fourth, when you have fixed the paper *PT* at that distance where the colors best compose white, place another paper *YZ* two or three inches away from the prism and observe the colored lines on it. Next cut out those parts of the paper on which those lines fell, in this way making oblong slits, *v*, parallel, equal, and as wide as their separation. Then replace that paper *YZ* to its proper position about three inches from the prism so that the colored light may be transmitted through its slits to the other, more distant paper *PT*. Having done this, you can observe that if you lower the paper *YZ* a bit so that it intercepts the purple and blue colors striking the upper[20] edges of its slits and it transmits the others, the whiteness at the paper *PT* will be con-

49 [= II, 36]. Fourth, that white vanishes if any color is removed from the mixture.

---

impossible. He experimentally demonstrated it with a *camera obscura*, where he observed that a series of candles placed outside the *camera* are seen distinctly from within even though the rays from all the candles cross in the aperture; see *Opticae thesaurus*, Bk. I, §§28–9 in *Opticae thesaurus. Alhazeni arabis libri septem, nunc primùm editi . . . Item Vitellonis Thuringopoloni libri X*, ed. Friedrich Risner (Basel, 1572), p. 17. Alhazen did not prove the principle for lights of different color, but his Polish disciple, Witelo, extended the experiment by replacing the candles with colored bodies; *Perspectiva*, Bk. II, §5, ibid., p. 264. Although Newton argues here, and even more eloquently in §53 ≈ II, 42, for the principle of superposition on Alhazen's grounds— the requirement of distinct vision—it was also a fundamental principle of his color theory, following from his principle of immutability.

(20) & caeruleos superioribus] Originally (and by mistake): & rubeos inferioribus (and red . . . the lower).

vertetur in rubeum colorem, aut citrium vel flavum. Sin attollas eam ut rubei et flavi labris inferioribus intercipiantur, caeteríque perlabantur; albedo ista convertetur in purpureum indicum et caeruleum. Perinde ut fieri oporteret in mixturâ colorum: Nam unis e mixturâ sublatis alteri debent ad propriam speciem et formam restitui.

50 [= II, 37]. Quintò, quòd colores, cùm decussando segregantur iterum, ad propriam speciem redeunt.

Quintò, papyro *YZ* sublatâ, et reliquis stantibus: papyrum alteram *PT* in meditullio albedinis acu perfora ut lucis ejus albae portiuncula trajiciatur quam deinceps ex[c]ipe in aliam papyrum isti *PT* ad distantiam quatuor vel sex digitorum postpositam: et vice albedinis colores iterum apparebunt. At quomodo colores illi / de novo generari potuissent si destruerentur in pro- ductione potiùs quàm miscerentur non video.[21] Concedendum est itaque quòd tantùm miscentur: et quòd radij varijs coloribus tincti et promanantes a diversis rimis λ, λ decussant sese in dicto foramine per acum effecto, et postea divergentes ab invicem gradatim[22] segregantur et segregati proprios iterum colores depingunt:[23] quemadmodum posthac fusiùs explicabitur.[24] Ad eundem praeterea modum si speculum aliquod planum et exiguum *K* statuas in medio albedinis ad *PT* papyrum effectae, ita quidem ut aliquos ex albificantibus radijs aliorsum, veluti ad ππ reflectat;[25] lux alba sic reflexa degenerabit in colores, quos videre est ad ππ, papyrum objiciendo. Etenim radij tincti cum diversis coloribus et in albedinem ad speculum *K* commisti, inclinantur ad se invicem propterea quod adveniunt a diversis fissuris λ, λ, λ, λ, λ, λ. Atqui tantùm divergunt a speculo postquam reflectuntur quantùm anteà convergebant.[26] Divergentes itaque paulatim dissocientur ac dissociati proprios colores non secus exhibebunt quàm si nunquam fuerant commisti. Liquet ergo quòd in misturâ radiorum diversicolorum[27] suae dispositiones ad efficiendos varios colores non destruuntur; ut ut albedinem exhibeant dum commisceantur sibi.

51 [= II, 39]. Sextò, res illustratur per misturam diversicolorum pulverum. Et quòd ex pulveribus omnium colorum debitè mistis fuscus producitur.

Denique vulgò notum est quod ex pulveribus diversicoloribus inter se com- mixtis color novus[28] emergit; tamen si pulveres isti inspiciantur[29] Microsco- pijs, omnes videntur tincti cum proprijs coloribus.[30] Adeò ut ex mixturâ pulverum colores proprij non destruantur, sed permiscendo tantùm color novus elicitur. Verùm ijdem planè colores ex mixturâ colorum prismatum ac pulverum producuntur: Sic pulvis caeruleus cum flavo mixtus producit viri- ditatem, et eadem viriditas etiam producitur ex mixturâ radiorum tinctorum

36/

---

(21) non video] Added.

(22) Added.

(23) Originally: expingunt.

(24) In §62 = II, 53.

(25) Originally continued: quemadmodum factum est ad Horologia sciaterica per reflectionem construenda (just as is done in constructing a reflecting sun dial). As a boy Newton made many sundials; see Westfall, *Never at Rest*, pp. 62–3.

(26) In the added §II, 38 Newton pursues the use of an inclined mirror to exhibit colors.

(27) Originally: heterogeneorum (of heterogeneous).

(28) Originally: tertius (third).

(29) Originally: intueantur (viewed).

verted into a red, orange, or yellow color. But if you raise it so that the reds and yellows are intercepted by the lower edges, while the others pass through, that whiteness will be converted into purple, indigo, and blue. Just as ought to happen in a mixture of colors, for when one has been removed from the mixture, the others must be restored to their own species and nature.

Fifth, the paper *YZ* having been removed and everything else remaining the same, with a needle pierce the other paper *PT* in the middle of the white so that is may transmit a small portion of its white light; then receive that on another paper placed four or six inches behind *PT,* and instead of whiteness colors will appear again. But how those colors could have been produced de novo if they were destroyed in the production [of white] rather than mixed, I do not understand.[21] It must therefore be conceded that they are only mixed, and that the rays imbued with various colors and flowing from the different slits λ, λ cross each other in that hole made with the needle and diverging afterward are gradually[22] separated from one another, and once separated they again depict[23] their own colors, as will be explained more fully later.[24] Similarly, if you put some small plane mirror *K* in the middle of the whiteness produced at the paper *PT* so that it of course reflects some of the white-making rays elsewhere, such as to ππ,[25] the white light thus reflected will degenerate into colors, which can be seen at ππ by inserting a paper. For the rays imbued with different colors and mixed together into white at the mirror *K* are inclined to one another because they arrive from different openings λ, λ, λ, λ, λ, λ; yet they diverge just as much after they are reflected from the mirror as when they converged before.[26] By diverging, therefore, they are gradually separated, and once separated they will exhibit their own colors no differently than if they had never been mixed. It is therefore evident that in a mixture of diversely colored[27] rays their dispositions to produce different colors are not destroyed, howsoever they may exhibit whiteness when they are mixed together.

Finally, it has been commonly observed that when diversely colored powders are mixed together a new[28] color emerges; yet if those powders are examined[29] with microscopes, they all are seen to be colored with their own colors.[30] Consequently, their own colors are not destroyed by a mixture of the powders, but rather, by mixing, only a new color is brought forth. Clearly the same colors are produced from a mixture of the colors of prisms as well as those of powders. Thus a blue powder mixed with a yellow one produces green, and the same green is also produced from a mixture of rays

50 [= II, 37]. Fifth, when the colors are separated again by crossing, they return to their own species.

51 [= II, 39]. Sixth, the matter is illustrated by a mixture of diversely colored powders: When powders of all colors are duly mixed together gray is produced.

---

(30) Newton's source for this observation is undoubtedly Boyle's *Touching Colours,* Pt. III, Expt. XVII, pp. 238–9. In his first essay "Of Colours," he accurately summarized Boyle's account: "24. Pouder of blew bise mixed wth a greater quantity of yellow orpiment makes a greene but ye particles by a microscope are discovered to retaine theire blewnes & Yellownesse" (Add. 3996, f. 133^r). Newton also would have encountered a related observation in Hooke's *Micrographia,* Observation X, p. 78, although Hooke mixed vermillion and bice to produce purple.

cum caeruleo et flavo.[31] Et proinde non dubium est quin colores novi[32] ex
coalescentibus Prismatum coloribus,[33] non facta[32] assimilatione sed mistura
tantum, similiter[32] oriantur.[34] / Caeterùm ut nullum dubitandi locum relin- 37/
querem; effeci ut pulveres colorum principalium quos prismata generant,
rubei, flavi, viridis caerulei et purpurei in proportione certâ miscerentur: et
licèt albedo perfecta non prodibat, tamen isti colores ad sensum periêre, et
quoddam genus albedinis, fuscum et obscurum, sive mediocre inter albedinem et nigredinem perfectam producebatur. Quod nostro proposito non
minùs inservit quàm si albedo perfecta prodijsset, quandoquidem fuscus ille
ab albo perfecto tantùm differt quantitate lucis non autem specie coloris, ut
exinde pateat quòd producitur ex albo cum nigredine contemperato. Neque
expectandum est, ut mihi videtur, alium quàm fuscum colorem e tali pulverum mistura generari. Nam cùm pulveres colorati intromittant maximam
partem lucis, istam ferè solam reflectentes quae apta est ad exhibendos proprios colores, ut ostendetur postea:[35] eorum mixtura maximam quoque partem lucis intromittet. Unde pro albedine perfectâ talis color generandus est

---

(31) That mixtures of colored powders and of prismatic colors yield the same color, and in
particular that yellow and blue pigments and lights both yield green, had by 1665 become as
near as any to being a general principle of color theory. Thus Francesco Maria Grimaldi, in his
Physico-mathesis de lumine, coloribus, et iride, alijsque adnexis libri duo (Bologna, 1665), Prop.
LX, §22, p. 291, could confidently assert that "we prove that the same color results from a
mixture of two colors whether only rays of apparently colored light coincide, or two pigments
are mixed that are colored with those colors that are .exhibited by the rays individually."
Similarly, a few years later, when Newton equated the mixing of yellow and blue powders and
lights in the "New theory," Prop. 4, it was one of the few things in his paper that Pardies could
fully support; Pardies to Oldenburg, 9 April 1672, *Correspondence*, 1:133 = *Phil. Trans.*, 7
(1672):4090. Despite the nearly universal agreement, the principle is erroneous, and in general a
mixture of pigments and of lights differ, with yellow and blue lights yielding white rather than
green. Two fundamentally different physical processes are involved: A mixture of two prismatic
colors cast onto a screen or directly into the eye is truly additive, the resultant being due to the
concurrence of the two lights. In a mixture of two pigments, however, the color perceived is not
at all a mixture but the color that remains after the two substances together have absorbed, or
"subtracted," all other colors, very much like the transmission of light through two superposed,
colored glass plates or filters. In the seventeenth century there were, nonetheless, sound experimental and theoretical grounds for accepting the identity of all color mixing, however it was
achieved. Boyle, for instance, tried nine different ways to mix yellow and blue—including a
mixture of pigments—and all of them yielded green; these are recapitulated in *Touching Colours*,
Pt. III, Expt. XVII, pp. 231–6. In his essay "Of Colours" (§§11, 12, 23, 24, Add. 3996, ff. 124ᵛ,
133ʳ), Newton summarized all but one of these different ways to mix yellow and blue. Only one
of Boyle's experiments, that of superposing two spectra, was strictly additive and should have
yielded white rather than green; but with essentially uncontrolled experimental conditions, his
results are neither impossible nor surprising. Boyle himself was aware of experimental uncertainty in many of his color-mixing trials, and he noted that when he mixed spectral yellow and
blue, green only "sometimes" resulted, "for a small Errour suffices to hinder the Success"
(*Touching Colours*, Pt. III, Expt. XIV, p. 227). Support for the identity of mixing pigments and
spectral colors also came from the very sequence of the spectral colors. With orange falling
between red and yellow, and green between yellow and blue, it was quite natural to conceive that
these colors arose from the mixing of the adjacent ones and followed the same rules as the
mixing of pigments. That spectral green was a compound color was quite widely held; see

imbued with blue and yellow.[31] Consequently, it cannot be doubted that new[32] colors[33] similarly[32] arise from a coalescence of prismatic colors and are not made[32] by assimilation but only by mixture.[34] Moreover, to leave no room for doubt I caused powders of the principal colors generated by prisms, red, yellow, green, blue, and purple, to be mixed in a definite proportion. Although a perfect white did not appear, still to the senses those colors had vanished, and a certain kind of white, gray and dark, or a mean between perfect whiteness and blackness, was produced. This serves our purpose no less than if perfect whiteness had appeared; since that gray differs from a perfect white only in the quantity of light but not in the species of color, it is clear from this that it is produced from white tempered with black. Nor, it seems to me, must one expect anything other than a gray color to be generated from such a mixture of powders. Since colored powders admit the greatest part of light and reflect almost solely that part disposed to exhibit their proper color, as will be shown later,[35] a mixture of them also admits the greatest part of the light. Consequently, instead of perfect whiteness, a color ought to be generated such as is made by mixing whiteness and black-

Kircher, *Ars magna,* Bk. I, Pt. III, Ch. IV, p. 76; Boyle, *Touching Colours,* Pt. III, Expt. XIV, p. 225; and Hooke, *Micrographia,* Observation IX, p. 58. The analogy between the composition of spectral green and the mixing of yellow and blue pigments was frequently explicitly made: see Giovanni Battista della Porta, *De refractione optices parte libri novem* (Naples, 1593), Bk. IX, Prop. VII, pp. 196–7; Louis Savot, *Nova, seu verius nova-antiqua de causis colorum sententia,* (Paris, 1609), Ch. XIII, ff. 10ᵛ–11ᵛ; Grimaldi, *De lumine,* Ch. XLIII, §43, p. 360; and Mariotte, *De la nature des couleurs, Oeuvres,* 1:208. Newton, to be sure, held that spectral green was a simple color, but in Prop. 4 of the *Optica,* §II, 70, he demonstrates that it can also be compounded of the adjacent yellow and blue. Still further support came from the color-mixing rules, based on three primaries, which were gradually adopted—not least by Newton himself—in the course of the seventeenth century, for these were generally taken to be universal rules that applied to colors as qualities, regardless of their origin; see Lect. II, 8, note (14), this volume. Huygens appears to be the only one in the seventeenth century, and for a long while after, to suggest that spectral yellow and blue compound white; for Newton's interpretation of his own color-mixing experiments and his response to Huygens's suggestion, see Lect. II, 8, note (16).

Although Newton can be firmly placed within this tradition of accepting the identity of the mixing rules for lights and pigments, he rejected many of its common methods to illustrate color mixing and showed that they are, in fact, not instances of true color mixing. For example, in §II, 76 he correctly explains the transmission of light through superposed, colored glass plates to be "subtractive" mixing; see also §II, 72. The illustration of the composition of spectral colors by colored powders was a fundamental one for Newton, for he maintained that the rays of the sun's direct light—just like a heap of colored powders—do not at all act upon one another. Indeed, this idea follows directly from his principle of color immutability. About twenty years later, when he was composing the *Opticks* and considered making the principle of color immutability an axiom, he once again invoked this illustration; see Add. 3970, f. 392 = Shapiro, "Evolving structure," p. 233.   (32) Added.

(33) Originally continued: oriundi similiter sint congredientium colorum. (must arise [from a coalescence of prismatic colors] similarly to a gathering together of colors.)

(34) Et . . . oriantur.] Replaced, before the changes indicated in notes (32), (33): Adeò ut non dubium est compositos colores prismatum aliunde quàm a solâ mixtura produci. (Hence it cannot be doubted that compound prismatic colors are produced other than solely by mixture.)

(35) This is not shown in the *Lectiones opticae* but was added in the *Optica,* §§II, 71–2.

qualis efficitur ex albedine et nigredine mixtis, id est, fuscus.[36] Attamen non eo inficias quin tales fortè pulveres inveniantur, praesertim inter mineralia, qui tantum lucis reflectant ut mixti exhibeant albedinem perfectiorem quàm hactenus vidi e mixturis effectam. Caeterùm quòd pulveres coloribus tantùm quinque praecipuis tinctos miscebam non ideo cogitandum est albedinem ex quinque solis productam fuisse, sed ex omnigenis. Nam in omnium corporum coloribus alij latent principalibus commixti licèt minùs fortes ut a principali colore superati non cernantur. Sic in caeruleo pulvere latent cyaneus et indicus alijque gradus omnes usque ad viridem aut flavum fortassis ex unâ parte, et ad intensum purpureum ex alterâ: Ut ut cąeruleus eò solus appareat quòd sit caeteris longè copiosior.[37]

Experientijs hisce admonitus in mentem praeterea revocabam quòd corpuscula quae conspiciuntur in radijs solaribus huc illuc volitantia varios colores exhibent / modò quisquam ea diligenter observet in cubiculo quaquaversum   38/ luci occluso, praeter unicum foramen per quod illuminantur.[38] Et tamen cùm isti pulvisculi in acervum congregantur nullus omninò color apparet praeterquam fuscus.[39]

52 [≈ II, 41].|
Tertius modus
miscendi colores
prismatis in
albedinem.

Videtis itaque quòd possibile sit albedinem e mixtura colorum generari. | Imò quòd colores prismatum revera non destruantur ad albedinem producendam; sed commisceantur tantum, quandoquidem emergunt immutati cùm radij coeuntes decussavêre et propter divergentiam subsequentem dissociantur iterum. Adhaec cum rei dignitas postulare videatur ut nullus non moveatur lapis, praeter modos praecedentes componendi albedinem[40] lubet adhibere tertium et quartum deinde quo praedicta faciliùs experiri possis et magis fortè cum evidentiâ.

Posito quòd sol illuceat obscurato cubiculo per unicum tantùm foramen *F,* (fig [18]) cui Prisma *ABC* affigitur, ingressam lucem refringens ad *PT:* juxta colores in papyrum *PT* sic projectos teneatur alia papyrus *Z* ut illuminetur a coloratâ luce quam altera papyrus *PT* reflectit. Quo facto, papyrus *Z* sic illuminata radijs omnium colorum confusè reflexis a *PT,* apparebit

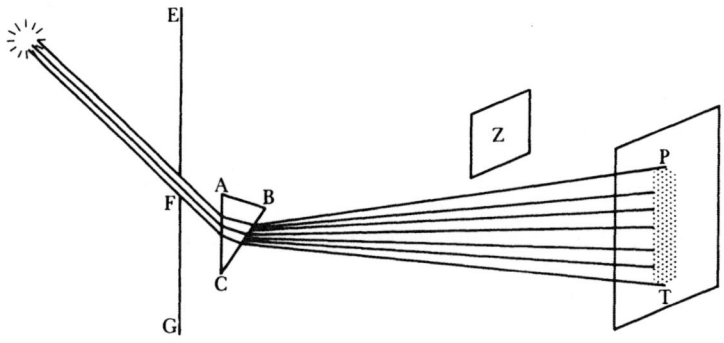

Figure 18

ness, that is, gray.[36] You may even perhaps succeed in finding such powders, especially among minerals, that reflect so much light that when mixed they exhibit a more perfect white than I have hitherto seen produced from mixtures. Moreover, because I mixed powders colored with only the five principal colors, it must not therefore be imagined that white had been produced from only five, but from all kinds. For latent in the colors of all bodies are other, though less strong, colors mixed with the principal ones, so that being overwhelmed by the principal color they are not perceived. Thus in a blue powder, cyan and indigo and all the other gradations up to green or perhaps yellow are latent on the one side, and up to an intense purple on the other side. However, blue alone is seen there, because it is far more abundant than the others.[37]

In doing these experiments, moreover, I recalled to mind the suggestion that the particles seen flitting about in the sun's rays exhibit various colors, provided that one observes them carefully in a room closed off to light everywhere except for a single hole through which they are illuminated.[38] Yet when those fine powders are gathered in a heap, no color at all appears besides gray.[39]

You will therefore see that it is possible to generate whiteness from a | mixture of colors. | Indeed, prismatic colors are in fact not destroyed in producing whiteness, rather they are only mixed, since they emerge unchanged when the rays after coming together have crossed, and by subsequently diverging they are again separated. Moreover, since the merit of the subject seems to require that no stone be left unturned, besides the preceding ways of compounding whiteness[40] it is satisfying to employ a third and then a fourth way whereby you can undertake that more easily and perhaps with greater clarity.

Assume the sun to shine into a darkened room through only a single hole *F* (Fig. 18) to which there is affixed the prism *ABC* that refracts the entering light to *PT*; near the colors thus projected upon the paper *PT* hold a second paper *Z* so that it is illuminated by the colored light that the first paper *PT* reflects. When this has been done, the paper *Z*, illuminated in this way by rays of every color reflected confusedly from *PT*, will appear white. It will be

> 52 [≈ II, 41]. A third way of mixing prismatic colors into whiteness.

(36) Since Newton's concept that white, black, and their intermediate, gray, differ only in brightness and not *in specie* was directly opposed to the contemporary view that they were distinct species of color, just like yellow, blue, and green; he elaborated this point a number of times: in his letters for Pardies on 13 April 1672, and for Hooke on 11 June 1672 (*Correspondence*, 1:141–2, 183–4 = *Phil. Trans.*, 7 (1672):4092–3, 5099–100); and in the *Opticks* (Bk. I, Pt. II, Prop. V, Expt. XV). In the *Opticks*, to compound white, or rather gray, Newton used various combinations of sometimes only two, but never more than four pigments, since each of them were themselves compounded of other colors. For Newton's view of the nature of black see §II, 69.

(37) Originally continued: Atque ita de caeteris. (Similarly for the others.)

(38) Newton is alluding to Boyle's observation in *Touching Colours* (Pt. I, Ch. III, pp. 69–70) that "Motes in yᵉ Sunne in some positions appeare of divers colors," as he had summarized it in his essay "Of Colours" (§14, Add. 3996, f. 124ᵛ).

(39) In replying to Hooke, Newton invoked this example together with still further ways to compound white, such as with soap bubbles (added in §II, 40); *Correspondence*, 1:183 = *Phil. Trans.*, 7 (1672):5099.   (40) componendi albedinem] Added.

alba. De hoc autem specimine maximè luculento et facili juvabit observare sequentia.[41]

Primò quòd auferendo papyrum *PT*, ne lucem ampliùs ad *Z* reflectat: e consequenti defectu lucis in *Z* cognoscas eam illuminari per solam lucem coloratam a *PT* reflexam.

Secundò si papyrum *Z* ipsi *PT* valdè vicinam teneas, ut una pars ejus magis illuminetur ab uno colore & alia ab alio: ipsa *Z* non apparebit alba sed ejus partes coloribus istis tingentur quibus sunt vicinissimae. Sin ipsa *Z* ad majorem a *PT* distantiam transferatur / ut omnes ejus pa[r]tes aequaliter ferè ab omnibus coloribus illuminentur; ex illâ colorum mixturâ generabitur albedo. 39/

Denique quòd albedo illa *Z* non destruendo colores sed tantùm miscendo generatur exinde pateat quod colores *PT* cernuntur beneficio radiorum non secus oculo mixtim incidentium quam papyro *Z*. Itaque si colores destruerentur potiùs quàm miscerentur ad *Z*, etiam destruerentur ad corneam tunicam vel pupillam oculi: ubi tamen certissimum est quòd miscentur tantùm, ut decussantes, postea divergant ad varias partes Retinae et sic excitent phantasmata propria.[42] Q[u]inimò si radij tincti cum diversis coloribus dum per eadem spatia confusè transeunt possent in se invicem agere et dispositiones mutare quas quilibet habent ad expingendos proprios colores: omnes omnium rerum colores conturbarentur ac se mutuò transmutarent[43] dum per aëra transmittuntur; ubique scilicet radijs aliorum corporum omnigenis coloribus tinctorum occurrentes. Et sic in coloribus visibilium nulla esset certitudo, constantia nulla.

Quartum praeterea[1] modum descripturus quo colores in albedinem misceri possint, pono quòd *ABC* (fig [19]) sit Prisma foràs ante foramen *F*

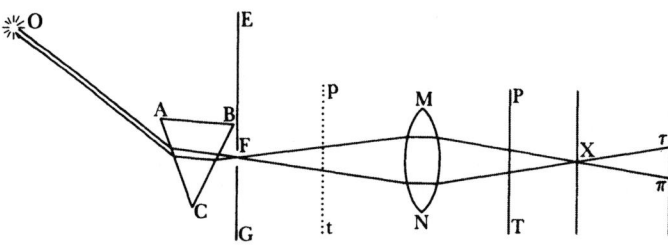

Figure 19

dispositum, quod refractam lucem in obtenebratum cubiculum transmittat versus *MN*. Tum lentem *MN* convexam sume cujus focus est[2] ad distantiam semipedis aut pedis unius duorumve (quale est objectivum vitrum Perspicilli bipedalis;)[3] et eam statue paulò plus distantem a foramine *F* quàm focus distat a se: ita scilicet ut lux colorata per eam deinceps trajiciatur, sicut videre

---

(41) This experiment was included in the *Opticks*, Bk. I, Pt. II, Prop. V, Expt. IX. In his response to Hooke, Newton described a variant using a paper painted over with five colors, although he noted it was better to use prismatic colors; *Correspondence*, 1:184–5 = *Phil. Trans.*, 7 (1672):5100–1.

useful, however, to note the following concerning this particularly excellent and easy example.[41]

First, by removing the paper *PT* so that it no longer reflects light to *Z*, you may, as a consequence of the absence of light at *Z*, perceive that it is illuminated solely by the colored light reflected from *PT*. 53 [≈ II, 42]. Notes on the same.

Second, if you hold the paper *Z* extremely close to *PT* so that one part of it is illuminated more by one color and another part by another color, then *Z* will not appear white, but its parts will be colored with those colors to which they are nearest. If now *Z* is shifted farther away from *PT* so that all its parts are illuminated almost equally by every color, white will be generated from that mixture of colors.

Finally, it is evident that the white at *Z* is generated not by destroying but only by mixing the colors, and accordingly that the colors at *PT* are perceived by means of rays falling mixedly upon the eye just as upon the paper *Z*. Therefore, if the colors were destroyed rather than mixed at *Z*, they would also be destroyed at the cornea of the eye, or rather the pupil, where nonetheless it is most certain that they are only mixed, inasmuch as upon crossing they afterward diverge to different parts of the retina and thus excite their own sensations.[42] Indeed, if rays imbued with different colors could act upon one another while passing confusedly through the same space and could change the dispositions they each have to depict their own color, then all colors of all things would be thrown into confusion, and they would mutually transmute each other[43] while they passed through the air, that is, everywhere running into rays of other bodies imbued with every sort of color. Thus there would be neither certainty nor stability in the colors of visible things.

A fourth way, moreover,[1] will be described whereby colors can be mixed into white. I assume that (Fig. 19) *ABC* is a prism placed outside in front of the hole *F*, which transmits the refracted light into a darkened room toward *MN*. Then take a convex lens *MN*, whose focus is[2] distant by half a foot, or one or two feet (such as the object glass of a two-foot telescope),[3] and place it slightly farther from the hole *F* than the focus is distant from the lens, specifically, so that the colored light is then cast through it, as is seen in the Lecture 5
54 [= II, 43]. A fourth way of accomplishing the same thing more clearly than the others.

---

(42) To illustrate the formation of the retinal image Descartes extended Witelo's experiment with a *camera obscura* (see note (19), this lecture) by placing a red, a yellow, and a blue object outside the *camera* and then setting an ox eye in its aperture to show that all the rays that crossed and were refracted in the eye cast distinct images of the differently colored objects on the retina; *Dioptrice*, Ch. V.§§1–7, *Oeuvres*, 6:600–2, 114–21.

(43) ac . . . transmutarent] Originally: atque deperirent (and they would perish).

(1) Originally: denique (finally).

(2) Originally: sibi abest.

(3) In §58 = II, 47 Newton relates that he had to use a wider lens with a longer focus. The prism is placed inside the hole, and the lens's characteristics and position vary somewhat in the accounts of this experiment in the "New theory" (*Correspondence*, 1:100–1 = *Phil. Trans.*, 6 (1671/2):3085–6), and in the *Opticks* (Bk. I, Pt. II, Prop. V, Expt. X).

est in schemate. Sit autem ejus latitudo sive apertura tanta ut omnes radios transmittat. Deinde cùm lentem in dicto situ stabilitam feceris, ponè statuatur papyrus *PT* in quam radij bis refracti terminentur. Eamque primò colloca proximè ad lentem, deinde ad majorem distantiam continuato motu transfer, et videbis / colores purpureum *P* rubeumque *T* contrahi et eousque minui 40/ dum omnes convertantur in albedinem, puta ad *X* quatuor vel sex pedes aut longiùs fortè distantem a lente, pro convexitate ejus vel positione. Deinde si papyrum adhuc longiùs transferas, colores iterum emergent sed in situ contrario, rubeo ad *τ* conspecto et purpureo ad *π*. Neque ulla inter eos ad *PT* et *πτ* differentia intercedit praeterquam quòd situs sit contrarius. Scilicet a lente *MN* effectum est ut omnes radij venientes ab aliquot punctis foraminis *F* in totidem iterum punctis congregentur ad papyrum *X*: Et sic omnes omnium specierum tum purpuram ad *P* tum rubedinem ad *T*, tum alios alibi colores efficientum convergunt ad *X* et ibi confusè miscentur ad albedinem generan-

(a) Num: 16 et 18   dam: De quâ imagine albâ et orbiculari monebam supra[a]. Postea verò cùm sese decussavêre in *X*, radij *PX* tendunt ad *π* et *TX* ad *τ*, adeo ut ijdem colores expingantur ad *P* et *π* per eosdem radios *Pπ*, et ijdem ad *T* et *τ* per eosdem *Tτ*, et sic de alijs. Unde liquet iterum quòd dispositiones, radiorum absimilium, ad diversos colores producendos non destruantur per eorum mixturam, quandoquidem eosdem expingunt cùm segregantur quos ante mixturam expingebant.

55 [= II, 44]. In   Porro si radios cujusvis coloris intercipias, interponendo corpus aliquod
eundem nota   opacum prope lentem *MN*, et caeteros facias missos: videbis non modò colores interceptos e papyris *PT* ac *πτ* tolli, sed et albedinem *X* destrui, et ejus vice colorem aliquem qualis efficitur per mixturam radiorum praeterlabentium generari. Sic si radios intercipias ostendentes rubeum ad *N*: rubedo *T* ac *τ* tolletur et albedo *X* convertetur in caeruleum. Vel si sistas tum rubeum ad *N* tum purpureum ad *M*, et intermedios flavum viridem et caeruleum praeterlapsos mittas: ex eorum misturâ viriditas producetur ad *X*. Et sic praetermittendo quos velis et sistendo alios,[4] pro arbitratu possis experiri mixturas

---

(4) This simple technique of stopping rays of any color while allowing the others to pass by and exhibit their own colors and other physical properties is one of the fundamental operations in Newton's theory of color (for other particularly clear instances see §§49 = II, 36; 87 ≈ II, 87; and §II, 50). Indeed, he incorporated it in his formal definition of light ray for Pardies on 10 June 1672, explaining that "by rays of light I understand its least or indefinitely small parts which are independent of one another; such as are all those rays which luminous bodies emit either simultaneously or successively along straight lines. For both the collateral and the successive parts of light are independent; since some parts can be intercepted [*intercipi*] without the others and be separately reflected or refracted to other places" (*Correspondence*, 1:164 = *Phil. Trans.*, **7** (1672):5014). This definition embodies the operation of intercepting or stopping (*intercipias* or *sistas*) rays of any color both "collaterally" with small opaque obstacles (as in this article) or with opaque screens with small holes (as in §69 = II, 60), and also "successively" with rotating toothed wheels (as in §II, 50). Although Newton's idea of light ray may have originated in his atomism (see Lect. 2, note (23)), the concept underlying this definition derives directly from his theory of color: the principle of the independence of rays of different color.

In composing the *Opticks* Newton elaborated his earlier definition for Pardies and made its operational significance more explicit. In the "Fundamentum Opticae" he placed the definition at the conclusion of Prop. 1, Expt. 2 (corresponding to Prop. 1, Expt. 2 in the *Opticks*, Bk. I, Pt.

figure. However, let its width or aperture be large enough to transmit all the rays. Then, when you have fixed the lens in the designated position, behind it place the paper *PT* on which the twice refracted rays are outlined. First place it very close to the lens, and then shift it with a continuous motion to a greater distance; you will see the colors purple *P* and red *T* contract and continuously diminish until they are all converted into white, suppose at *X*, at a distance of four or six feet, or perhaps more, from the lens according to its convexity or position. If you then shift the paper still farther away, the colors will emerge again but in a reversed situation, red visible at τ and purple at π. Nor is there any difference between those colors at *PT* and $\pi\tau$ except that their situation is reversed, namely, the lens *MN* causes all rays coming from the many points of the hole *F* to gather again into just as many points at the paper *X*. Thus all rays of every sort, making purple at *P* as well as red at *T* and the other colors elsewhere, converge to *X* and are there confusedly mixed to generate white; I pointed out this white and circular image above.[a] But after they have crossed one another at *X*, the rays *PX* tend to π and *TX* to τ, so that the same colors are depicted at *P* and π by the same rays, *Pπ*, and the same colors at *T* and τ by the same rays, *Tτ*, and similarly for the others. Consequently, it is again evident that the dispositions of dissimilar rays to produce diverse colors are not destroyed by their mixture, since they depict the same colors when they are separated as those they depicted before the mixture.

(a) §§16, 18

If, moreover, you intercept the rays of any color by inserting some opaque body near the lens *MN*, while allowing the others to pass by, then you will see not only the intercepted colors vanish from the papers *PT* and $\pi\tau$ but also the white at *X* destroyed and in its place some color generated such as is made by the mixture of the rays passing by. For instance, if you intercept the rays displaying red at *N*, the red at *T* and τ will vanish, and the white at *X* will be converted into blue; or if you stop both the red at *N* and the purple at *M*, and let the intermediate ones, yellow, green, and blue, pass by, from their mixture green will be produced at *X*. And thus by allowing those you choose to pass, while stopping the others,[4] you can test at will any mixtures whatso-

55 [= II, 44]. A note on this.

I) in which he shows that rays of different color are differently refrangible by reflecting light from a card painted blue on one half and red on the other. In an "annotatio" to this proposition (which became a "Scholium" in the *Opticks*), Newton observed that the experiment turns out much more clearly with pure colors from which other rays are removed by some artifice, and then he immediately introduced his first definition: "By rays of light I understand its least parts, propagated from lucid bodies both successively through the same line and simultaneously through different lines. For it is manifest that light consists of both successive and collateral parts, because at the same time light can be stopped in one place and stifled in a black body, while at another place it can proceed unimpaired; and in the same place it can at one time be stopped and stifled in a black body and afterward, by removing the body, be propagated unimpaired . . . Therefore the least light or part of light that can be stifled alone without the rest of the light, or be propagated alone, or do or suffer anything alone that the rest of the light does not do or suffer, I call a ray of light. [Per lucis verò radios intelligo minimas ejus partes tam successive per easdem lineas quam simul per diversas a corporibus lucidis propagatas. Lucem enim ex partibus tam successivis quam collateralibus constare manifestum est quia lux eodem tempore in uno loco sisti et in corpore nigro suffocari in alio illaesa pergere potest et in eodem

quaslibet et explorare qui color inde generabitur; modò pretium laboris experientiam illam judicaveris.[5]

56 [= II, 45]. Quo more radij diversicolores in albentem lentis focum convergunt. / Verum[6] cùm experimenti hujus dignitas videatur exigere ut summâ cum diligentiâ retegatur[7] et penitiùs explicetur, dum plura de coloribus simul complectitur et exhibet quàm in unico tantùm experimento solent latere: non gravabor copiosiùs ostendere quo pacto radij miscentur ad X, et nonnulla postmodum scitu non indigna exinde patefacere. Itaque concipiantur tales refractiones in Prismate fieri, ut radij incidant in varios circulos ad Lentem *MN*, qui varios gradus refractionis patiuntur; prout explicui[8] in praecedentibus.[a] Sitáque *PQRST* (fig [20]) oblonga imago composita ex istis 41/

(a) Num 12

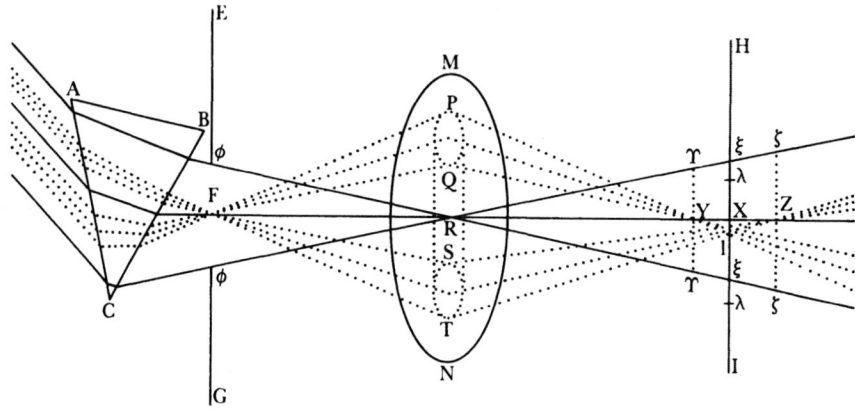

Figure 20

circulis, & in lentem projecta: quorum circulorum extremi duo sunto *PQ* purpureus et *ST* rubeus. Porrò sit $\phi F\phi$ diameter foraminis per quod lux in lentem trajicitur cujus foraminis punctum aliquod ut *F* primò consideremus a quo venientes radij dictos circulos *PQ*, *ST* totamque imaginem *PT* efformant. Et praeterea cùm radij quemlibet circulum efformantes sint conformes sibi:[9] ponatur quòd Lens sit tali figurâ praedita ut eos omnes cujusdam e circulis, puta rubei *ST* versus punctum quoddam *Z* exactè refringat. Quod fieri posse per lentem convexis Hyperbolis terminatam, ut et per lentes aliter formatas Cartesius in Dioptricâ et Geometriâ suâ edocuit.[10] Est itáque *Z* focus radiorum *FS FT* et caeterorum uniformitèr rubeorum, et recta *FZ* ducta erit axis lentis. Praeterea cùm radij *FP*, *FQ* caeteríque conficientes alterum extremum circulum *PQ* colorem purpureum ostendant & propterea magis refringantur quàm alteri tendentes ad *ST*: illi ideò convergent ad punctum quoddam aliquantò propinquius quàm *Z*, veluti ad *Y*; ut ij facilè percipient qui norunt focos lentium esse tantò propinquiores sibi, quantò major est earum / vis 42/

loco nunc sisti et in corpore nigro suffocari mox sublato corpore illaesa propagari . . . Minimam igitur lucem vel lucis partem quae sola sine reliqua luce suffocari vel sola propagari potest vel sola aliquid agere aut pati quod reliqua lux non agit aut non patitur, Radium lucis appello.]" (Add. 3970, f. 415^r.) This definition of light ray, or the least parts of light, in fact applies only to

ever and examine the color that will thereby be generated, provided that you considered that experiment worth the effort.[5]

Since the value of this experiment truly seems to require that it be disclosed[7] with the utmost diligence and explicated rather thoroughly, while it also includes and shows more about colors than usually is concealed in only a single experiment, I will not be reluctant to show more fully how the rays are mixed at X and then to reveal several things worth knowing. Accordingly, imagine such refractions to occur in the prism that the rays fall on the lens MN in different circles that experience different degrees of refraction, just as I explained[8] in the preceding.[a] Let $PQRST$ (Fig. 20) be the oblong image composed of those circles and projected onto the lens, and let its two outermost circles be PQ, purple, and ST, red. Furthermore, let $\phi F\phi$ be the diameter of the hole through which the light is cast onto the lens, and first let us consider any point of this hole, such as F, from which the rays come and form those circles PQ and ST and the whole image PT. Since then the rays forming any circle are similar to themselves,[9] let it be assumed that the lens possesses a shape such that it refracts all those rays of some one circle (for example, the red, ST) exactly toward some point Z. This can happen by means of a lens bounded by convex hyperbolas as well as by lenses shaped otherwise, as Descartes has taught in his *Dioptrics* and *Geometry*.[10] Consequently, Z is the focus of the rays FS, FT, and of the others uniformly red; and when the line FZ is drawn, it will be the axis of the lens. Moreover, since the rays FP, FQ, and the others composing the other outermost circle PQ display the color purple and are thus more refracted than the others tending to ST, they therefore converge at some point slightly closer than Z, such as at Y, as will be readily comprehended by those who know that the foci of lenses are closer to themselves the greater their refractive force is. It is therefore

56 [= II, 45]. How the diversely colored rays converge into the white focus of the lens.

(a) §12

---

monochromatic colors, which can be stopped, refracted, or acted upon alone. It does not apply to polychromatic colors, which do not act alone, for they consist of "least parts," that is, different monochromatic colors. In the manuscript of the *Opticks* (Add.3970, f. 31ʳ) Newton at first placed this definition in the same location, although he rendered it more abstract by removing the reference to black bodies, but he then decided to shift it, essentially unchanged, to the beginning of the book together with other definitions. Compare A. I. Sabra, *Theories of Light from Descartes to Newton* (London, 1967), pp. 287–90; and Shapiro, "Newton's definition of a light ray." Also see note (c) in Newton's aborted reprint of the "New theory" in *Correspondence,* 1:105, note 30.

(5) Since the colors here are barely resolved, it is not an ideal way to test color mixtures; in §II, 70 Newton describes a different method.

(6) Originally: Verum enim verò.

(7) Originally: rimetur (examined).

(8) Originally: monst[ravi] (showed).

(9) conformes sibi] II: homogenei (homogeneous).

(10) Descartes discusses the ideal curves—"Cartesian ovals"—that refract rays from a given point to a given point (conics when one of the points is at infinity) in Discourse VIII of *La Dioptrique* and in Book II of *La Geometrie*. Newton returns to this problem in Prop. 34 of the *Optica,* Pt. I.

refractiva. Liquet itáque radios in coloribus et refractionibus absimiles ad diversos focos convergere. Sed cùm eadem Lens pluribus focis haud queat adaptari, et ideò cùm *Z* supponatur focus in quem omnes radij ad circulum rubeum *ST* pertinentes exactè conveniant: radij pertinentes ad alterum circulum *PQ* purpureum, omnes in ejus focum *Y* exactè convenire nequeunt. Attamen eorum concursus juxta *Y* in axe tam proximè accuratus erit, ut quoad sensum et experientiam omnem habeatur pro accurato. Quinetiam si lens *MN* ponatur sphaericè convexa ut neuter focorum *Y* vel *Z* strictè loquendo possit esse accuratus, tamen quantum ad praesentia spectat pro accuratis habeantur. Itaque concipiendo quòd radij manantes ab *ST* convergant ad *Z* et quòd alteri manantes a *PQ* convergant ad *Y* et ibi decussantes divergant itidem: patebit quòd hi duo radiorum penicilli concurrent et miscebuntur in spatio focis *Y* et *Z* intermedio, veluti ad *l*, modò Lentis centrum *R* ponatur intermedium circulis *PQ* et *ST*. Ad eundem modum radij caeterorum generum convergent in alios focos ipsis *Y* et *Z* intermedios, ac tanto propinquiores ipsi *Y* quanto major est eorum passio refractiva. Sic focus viridiformium radiorum cadet in medio spatio veluti ad *X*; radijque caeruliformes[11] convenient citiùs inter *X* et *Y*, et flaviformes longinquius inter *X* et *Z*, ac caeteri colores intermedij in spatijs intermedijs: Eorumque penicilli sese decussabunt ultra citraque locum *l*; ita tamen ut istae decussationes sint eò densiores quantò sunt ipsi *l* viciniores, et ut spatium *Xl* sit minimum per quod omnes radij transeunt manantes ab eodem puncto *F*. Non dissimili modo radij venientes ab alio quovis puncto foraminis, ut *φ*, si sint rubriformes convergent ad *ζ*; sin purpuriformes, ad *Υ*; et ad intermedium aliquod punctum si sint intermedij generis et eorum concursus densissimus erit in loco medio, *l* veluti ad *ξλ*. Atque adeò ex radijs ab integro foramine *φF[φ]* manantibus foci maximè refrangibilium jacebunt in superficie quâdam *ΥYΥ* ad lentem proximâ, foci minimè refrangibilium jacebunt in aliâ superficie *ζZζ* a lente remotissimâ, focique mediocriter refrangibilium jacebunt in alijs intermedijs superficiebus. Et sic omnes omnium radiorum foci totum spatium *Υζζ Υ* a superficiebus istis integratum occupabunt, et in eo praecipuè penicilli decussabunt & commiscebuntur.   43/

Jam ex hâc descriptione venit observandum[12] quòd cùm papyrus *HI* teneatur in medio dicti spatij *ΥζζΥ*, ut in eam radij terminentur ubi densissimus est eorum concursus et mixtura ad albedinem generandam perfectissima: radij viridiformes tendentes ad focos in papyro sitos in eam incident intra literas *ξξ*, sed rubriformes venientes ab *ST*, ac tendentes ad focos in superficie

---

(11) viridiformium radiorum . . . radijque caeruliformes] Originally: radiorum viriditatem exhibentium . . . radijque caeruleo tincti sive (ut brevitatis gratiâ voces fingam) radij caeruliformes (rays exhibiting green . . . and the rays colored blue or (if I may invent the phrase for brevity's sake) blue-making rays). In revision Newton here deleted this brief explanation of his new terminology when he decided to advance it to §II, 19. Newton's new terms, reflecting his commitment to the mechanical philosophy in which color is taken to be nothing but a sensation excited by the rays, are here translated as "green-making," and so forth, following the definition in Bk. I, Pt. II, of the *Opticks*: "The homogeneal light & rays which appear red, or rather make Objects appear so I call rubrific or red-making, those wch make objects appear yellow green blew & violet I call yellow-making, green-making, blew-making, violet-making, and so of y^e rest" (Add. 3970, f. 96^r = *Opticks*, p. 90). In the draft of this definition in the "Fundamentum

evident that rays dissimilar in colors and refractions converge to different foci. But since the same lens is not at all able to be adapted to several foci, and since Z is assumed to be the focus at which all rays belonging to the red circle, ST, meet exactly, then rays belonging to the other, purple circle, PQ, cannot all meet exactly at their focus Y. Nonetheless, their intersection near Y on the axis will be so nearly accurate that according to sense and all experience it may be considered as accurate. Besides, if the lens MN is assumed to be spherically convex, so that neither of the foci Y or Z strictly speaking can be accurate, still, insofar as present matters are concerned, they may be considered as accurate. Consequently, by conceiving that the rays flowing from ST converge to Z, and that the others flowing from PQ converge to Y and after crossing there diverge again, it will be clear that these two pencils of rays will intersect and be mixed in the space between the foci Y and Z, such as at l, provided that the lens's center R is assumed to be between the circles PQ and ST. In the same way the other kinds of rays will converge to other foci between Y and Z, and they will converge closer to Y the greater their refractive susceptibility is. Thus, the focus of the green-making rays will fall in the middle of the space, such as at X; the blue-making rays[11] will meet sooner, between X and Y; the yellow-making rays farther off, between X and Z; and the other intermediate colors in the intermediate spaces. Their pencils will cross one another on both sides of the place l, yet in such a way that these crossings are denser the closer they are to l, and the space Xl is the smallest through which all rays flowing from the same point F pass. In a not dissimilar way, the rays coming from any other point of the hole, such as ϕ, if they are red-making, will converge to ζ, but if they are purple-making, to Υ, and to some intermediate point if they are of an intermediate kind; and their most dense intersection will be in the middle region, such as at $\xi\lambda$. Hence, of the rays flowing from the entire hole $\phi F \phi$, the foci of the most refrangible rays will be in some surface $\Upsilon Y \Upsilon$ nearest to the lens, the foci of the least refrangible ones will lie in another surface $\zeta Z \zeta$ most remote from the lens, and the foci of the mean refangible ones will lie in other intermediate surfaces. In this way all the foci of all the rays will occupy the entire space $\Upsilon \zeta \zeta \Upsilon$ made up by those surfaces, and the pencils will cross and be mixed together primarily in that space.

Now from this description it can be observed[12] that when the paper HI is held in the middle of that space $\Upsilon \zeta \zeta \Upsilon$ so that the rays are terminated on it where their intersection is densest and the mixture for generating whiteness is most perfect, the green-making rays tending to the foci situated on the paper will fall upon it between the letters $\xi\xi$, but the red-making ones coming from

57 [= II, 46]. Concerning the colors at the extremity of the focus that are scarcely visible because of their thinness.

Opticae," (Add. 3970, f. 421^r) he explained, "Lucem igitur discretam & homogeneam quae apparet rubri coloris nominabo *rubriformem* aut si mavis *rubrificam* eo quod efficit corpora omnia a quibus in sensum advenit apparere rubra[.] Sic et lucem discretam quae flava, viridis caerulea vel purpurea apparet nominabo *flavificam viridificam caerulificam & purpurificam*"; for the final words Newton first wrote "flaviformem viridiformem," and so on. See also "Of Colours," §6, Add. 3975, f. 2^v = Hall, "Further experiments," p. 28; and Newton's note (d) in his aborted reprint of the "New theory" in *Correspondence*, 1:106, note 31.

(12) Originally continued: *primò* (first).

ζZζ sitos ut dictum est, incident in papyrum intra literas λλ paulò viciniùs ad
I. Et pari modo purpuriformes incident in eundem locum λλ dum tendunt a
PQ ad focos sitos in superficie ΥΥ. Caeteri autem radij cadent in alia spatia
inter ξξ & λλ mediocria, ipsíque ξξ tantò viciniora quanto foci eorum minùs
absint a papyro. Liquet itaque quòd totum spatium ξXlλ non debet albescere,
sed pars ejus tantùm media inter literas ξ et λ interiores sita, ubi scilicet
colores omnes commiscentur: Etenim in extremitate ξ versus H radij viridiformes cadunt soli, qui proinde tingent extremitatem istam cum viriditate. Ad
alteram autem extremitatem versus I nulla miscetur viriditas sed purpura
tantum cum rubore. Qui dicta perpendet etiam facilè percipiet quòd cùm
papyrus paululum transferatur ultra citráque, colores alij praeter viriditatem
apparebunt ad extremitatem imaginis versus H, scilicet inter P et Υ purpureus
apparebit extimus, inter Υ et ξ caeruleus, et viridis ad ξ, deinde flavus inter ξ
et ζ, ac rubeus denique[13] ad ζ et postea perpetuò. Ad alteram autem imaginis
extremitatem versus I sitam rubeus erit extimus a T usque ad λ ubi commiscetur purpurae: Quae quidem mixtura dat pallidum quendam colorem nunc
ad rubeum / nunc ad caeruleum nonnihil vergentem pro variâ proportione   44/
mistorum. At ultra λ purpura semper conspicietur. Caeterùm cùm distantia
inter Υ et ζ valde parva sit et multò magis distantia inter X et l sive ξ et λ,
hoc est latitudo limbi colorati: propter summam ejus exilitatem conspectui
vix patebit, sed totum spatium ξXlλ nisi acriùs observanti apparebit album.

58 [= II, 47].   Cùm haec advertissem, experiebar deinde an responderent praeconceptis:
Dictorum colorum   et licèt malè successerat primò dum utebar angustâ lente: postea tamen cùm
observatio.   adhibui lentem eâ de causâ latiorem ut angulus XYl sive ξΥλ et inde Xl sive
ξλ hoc est latitudo dicti limbi colorati fieret major, quod optabam[14] evenit.
Adhibeatur itaque lens cujus latitudo sive apertura sit trium digitorum aut
major eo, foci autem longinquitas pro lubitu tuo pedium trium vel quatuor,
tum ea collocetur ad distantiam sex vel octo pedum a foramine φFφ, ut
colores PQRST in eam prolapsi usque ad extremitates ejus extendantur, nullis tamen praeterlabentibus. Deinde papyrus HI pone collocata transferatur
ultra citráque, et ad extremitatem imaginis versus H videbis omnes prismatum colores a purpura ad rubedinem usque gradatim successivos: sed ad
alteras imaginis partes versus I, inter purpuram ad ζ et rubedinem ad Υ
conspicuam, neque viriditas neque alius quispiam ex intermedijs coloribus
apparebit nisi fortè qui fiunt ex rubeo et purpur[e]o mixtis; Quemadmodum
ex eo cognoscas quòd cùm intercipis extremitatem purpurae, ope corporis
opaci[15] juxta lentem ad P[16] interpositi, ille limbus imaginis versus I fiet
rubeus; sin extremitas rubedinis ad T intercipiatur, limbus idem fiet purpureus. Et hinc est quòd transitus a purpura ad rubedinem ex hâc parte imaginis fit mul[t]ò celerior quàm ex alterâ versus H ubi colores omnes interveniunt.
Caeterùm cùm dictorum colorum latitudo tam exigua sit (videlicet haud
major centesimâ parte digiti,) ut nisi vitra / sint benè polita et a venis libera,   45/

---

(13) Originally: denuò (again).
(14) Originally: volebam.
(15) Added.
(16) ad P] Added.

ST and tending to the foci situated on the surface ζZζ, as has been described, will fall upon the paper between the letters λλ a little closer to *I*. In a like way, the purple-making rays will fall upon the same place λλ while they tend from *PQ* to the foci situated on the surface ϒΥϒ. The other rays, however, will fall in the other intermediate spaces between ξξ and λλ and be nearer to ξξ as their foci are less distant from the paper. It is evident, therefore, that the entire space ξ*Xl*λ ought not to become white, but only its middle part located between the inner letters ξ and λ, namely, where all the colors are mixed. For at the end ξ toward *H*, there fall only green-making rays that consequently color that end green; but at the other end toward *I* no green is mixed, but only purple together with red. Whoever carefully considers what has been described will also easily perceive that when the paper is shifted back and forth somewhat, other colors besides green will appear at the end of the image toward *H*: specifically, purple will appear outermost between *P* and ϒ, blue between ϒ and ξ, green at ξ, next yellow between ξ and ζ, and finally[13] red at ζ and ever after. At the other end of the image located toward *I*, however, red will be outermost from *T* to λ where it is mixed with purple; this mixture in fact yields a certain pale color inclining somewhat now to red and now to blue according to the different proportion of the mixtures. But beyond λ purple will always be observed. Since, however, the distance between ϒ and ζ is very small and the distance between *X* and *l* or ξ and λ (that is, the width of the colored band) is even smaller, on account of its extreme thinness the band will scarcely be evident to sight, whereas the entire space ξ*Xl*λ will appear white except to a rather sharp observer.

When I had recognized these things, I then tested whether they would agree with my preconceptions. Although the test turned out poorly at first when I employed a narrow lens, yet afterward, when I consequently used a wider lens so that the angle *XYl* or ξϒλ and hence *Xl* or ξλ (that is, the width of that colored band) would become greater, it turned out as I hoped.[14] Therefore, use a lens whose width or aperture is three inches or greater but whose focal length (as you choose) is three or four feet; then place it six or eight feet away from the hole ϕ*F*ϕ, so that the colors *PQRST* flowing into it extend up to its edges, yet with none flowing beyond. Next shift the paper *HI* placed behind it back and forth, and at the end of the image toward *H* you will see all the successive prismatic colors continuously from purple to red; but at the other parts of the image toward *I*, between the purple visible at ζ and the red at ϒ, neither green nor any of the other intermediate colors will appear, except perhaps those that arise from a mixture of red and purple. On this basis you may understand that when you intercept the extremity of the purple with the aid of an opaque[15] body inserted next to the lens at *P*,[16] that band of the image toward *I* will become red; but if the extremity of the red is intercepted at *T*, the same band will become purple. Hence it happens that the transition from purple to red on this side of the image occurs much more quickly than on the other side toward *H* where all the colors intervene. Since, however, the width of those colors is so small (specifically, not greater than one hundredth of an inch), unless the glasses are well polished and free from

58 [= II, 47].
Observation of
those colors.

et insuper experientis diligentia et curiositas solito major, fortè excidet proposito. Quamobrem in majorem evidentiam rei et experiendi copiam addo, quòd si Microscopium sumas atque ita disponas ut papyrum aliquam affixam laminae super quam objecta collocantur contemplanda, distinctè ampliat;[17] dein ita statuas ut imago lucida *ξΧΙλ* incidat in istam papyrum; colores in ejus limbo sic ampliatos videbis manifestos.[18]

59 [= II, 48].
Quintus modus albedinem componendi quarto ferè similis.

Verùm cùm mistura radiorum quoad colores absimilium non sit adeò perfecta in hoc specimine quin ut e coloribus aliqui in extremitate albedinis appareant (licèt tam exigui ut incautus fortè non advertat,) placet insuper observare quod si vice lentis refractariae speculum concavum accuratè formatum et perpolitum adhibeas, dicta mistura fiet omnibus numeris perfecta. Etenim irregularitas illa quâ refractiones ita perturbantur, in reflectionibus nulla est, sed radij quoscunque colores expingentes et utcunque refrangibiles ad eosdem tamen angulos reflectuntur in quibus incidunt. Quamobrem si *MN* (fig [21]) sit speculum Ellipticum cujus foci sint *F* et *X*: radij omnes a

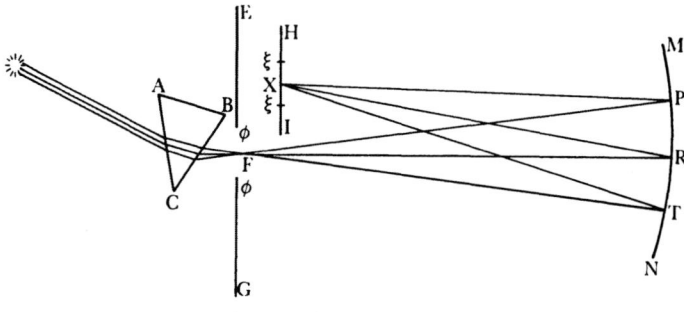

Figure 21

puncto *F* manantes cujuscunque generis sive purpuram ad *P*, sive rubedinem ad *T*, sive alios alibi colores quoscunque ad speculum exhibentes, tamen omnes accuratè convenient in eodem puncto *X*. Quinimò licet speculum *MN* non sit ex ellipticâ figurâ segmentum sed e sphaericâ: modò semidiameter sphaerae, hoc est distantia ejus a focis praedictis *F* et *X*, satis magna sit, puta trium pluriumve pedum, et distantia focorum valdè parva, puta non plusquam unius digiti: si haec inquam ponantur, radij ab *F* manantes adeò propemodum convenient in *X* ut istud *X* quoad omnem sensum pro exacto foco habeatur.[19] Et eodem modo radij manantes / ab alijs punctis ut *φ* ipsi *F* 46/ vicinis in alijs ut *ξ* ipsi *X* vicinis quàm proximè convenient. Et sic omnes omninò colores reflectentur a speculo *PT* in unumquodque punctum imaginis *ξXξ*, totamque exhibebunt albam.

60 [≈ II, 51]. Lucem egredientem prismate non secus e coloribus (licèt nondum apparentibus) componi, ac postea cum colores in idem

Porro ex his notandum est[20] quòd non solùm albedo ad Focum *X* e commisturâ radiorum omnis generis producitur, sed et ista ad foramen *φFφ* effecta cùm lux modò transierit Prisma et nondum aliqui colores apparuêre, quòd ista inquam constat ex simili mixturâ: | quandoquidem omnes radij quibuscunque coloribus affecti qui ad punctum quodvis imaginis *ξXξ* convergunt, ab alio quodam puncto foraminis *φFφ* manarunt: et sic ijdem radij ad

veins, and moreover unless the experimenter's diligence and care are unusually great, he will perhaps not succeed. Accordingly, for the greater clarity of the matter and ability to do the experiment, I add that if you take a microscope and set it so that it distinctly enlarges[17] some paper affixed to the plate on which objects to be studied are placed, and then you arrange it so that the bright image $\xi X l \lambda$ falls upon that paper, you will see the colors in its band enlarged in this way clearly.[18]

But since the mixture of rays dissimilar with respect to their colors is not perfect in this example, so that some of the colors on the extremity do not appear white (even if they are so slight that a careless person might perhaps not observe them), it is appropriate to note in addition that if instead of a refractive lens you were to use an accurately shaped and highly polished concave mirror, that mixture will become perfect in all respects. For that irregularity that so disturbs refractions does not exist in reflections; on the contrary, rays depicting any color whatsoever and however refrangible are notwithstanding reflected at the same angle at which they are incident. Accordingly, if MN (Fig. 21) is an elliptical mirror whose foci are F and X, all rays flowing from the point F—whatever kind, whether exhibiting purple at P, red at T, or any other colors at all elsewhere on the mirror—will all still meet accurately at the same point X. Indeed, even if the mirror MN is not a segment of an elliptical figure but of a spherical one, provided that the radius of the sphere (that is, its distance from the specified foci F and X) is sufficiently great, say, three or more feet, and the distance of the foci is very small, say, not more than one inch; if, I say, these things are assumed, the rays flowing from F will meet so nearly at X that that X may be sensibly considered as the exact focus.[19] In the same way, rays proceeding from other points near F, such as ϕ, will meet exceedingly closely at other points near X, such as ξ. Thus absolutely all colors will be reflected from the mirror PT to every single point of the image $\xi X \xi$, and the entire image will exhibit white.

Moreover, it ought to be noted[20] from these things that not only is the white at the focus X produced from a mixture of rays of every kind, but also that white made at the hole $\phi F \phi$ when the light has just passed through the prism and no colors have yet appeared. I say that that white consists of a similar mixture. | Since all rays—endowed with any color whatever—that converge to some point of the image $\xi X \xi$ flow from some other point of the

59 [= II, 48]. A fifth way to compound whiteness nearly similar to the fourth.

60 [≈ II, 51]. Light leaving a prism is compounded from colors, even if they are not yet apparent, no differently than afterward when the colors have been

---

(17) Originally: faciat magnam.

(18) **II**: sat manifestos. (clearly enough.)

(19) The focal property of elliptical mirrors was a classical discovery, and the study of reflection from spherical and nonspherical mirrors continued to be pursued in Newton's own day by, for instance, Barrow in his *Lectiones XVIII* and James Gregory in his *Optica promota*.

(20) notandum est] Originally: notari vellem primò (I wished it to be noted first).

spatium congregati
sunt.

61 [= II, 52].
Probatur ex eo quòd
in modo quarto et
quinto componendi
albedinem, radij non
convergunt ad idem
spatium nisi qui
divergebant ab
eodem.

62 [= II, 53]. Et
quòd divergentia
colorum a prismate
persimilis est eorum
divergentiae ab
albenti lentis foco.

utrumque spatium $\phi F\phi$ et $\xi X\xi$ miscentur, et utriusque albedinis eadem est compositio.

Atque haec clariora fient observando primò quòd rei alicujus utcunque figuratae et applicatae ad foramen $\phi F\phi$ umbra distinctè projicitur in papyrum radios excipientem ad X. Quinimò bullularum aëris in Prismate latentium (sicut vitris omnibus contingere solet) umbras videbis ad instar macularum[21] in dictam papyrum projectas. Id quod nullo pacto contingere potuisset nisi radij manantes ab aliquot punctis ipsius $\phi F\phi$ in totidem punctis rursus convenirent ad $\xi X\xi$. Et licèt non exactè conveniant in ijsdem punctis manantes ab ijsdem cùm lens refractaria vice speculi adhibeatur, ut in fig 19 & 20, & proinde colores nonnullos generent in confinio lucis et umbrae sicut fusè explicui; tamen spatium in quod conveniunt tantillum est ut pro puncto sensibili fermè habeatur.

Secundò quòd si lentem in fig [19] ita statuas ut aequidistet a focis ejus F et X[22] in medio posita, ac deinde colores excipias in papyrum PT tum ultra lentem versus X tum citra versus F alternis temporibus admotam: possis observare quòd colores eodem planè modo apparent diminuuntur, et in albedinem paulatim convertuntur dum dicta papyrus motu lento et continuo transfertur ad F, atque dum transfertur ad X. Adeò / ut divergentia colorum ab F et convergentia ad X omnino similis sit. Pari ratione si papyrus $\pi\tau$ lente[23] moveatur ad X juxta et papyrus PT[24] movetur ad F, ijdem colores in utrâque conspicientur, et eodem modo desinent in albedinem, hoc tantùm excepto quod eorum situs contrariatur propter decussationem radiorum in X. Atque adeò divergentia colorum ab utrisque F et X omninò[25] similis est. Quid itáque concludatur[26] exinde quàm quòd eodem modo commiscentur et ad F antequam divaricârunt ab invicem, et ad X ubi rursus congregantur in albedinem. Sed ut comparatio modò facta evadat illustrior ab instantiâ aliquâ, venit observandum porrò quòd cùm papyrus statuatur ipsi F contigua, & amoveatur deinde versus pt, et postea statuatur ad X et amoveatur versus $\pi\tau$: quòd, inquam, albedo ad F et X in utroque casu primò degenerabit in colores secundum extremitates ejus, dum in meditullio manet alba. Cujus rei ratio non alia est quàm quòd radij divergentes primò segregantur in confinio lucis et umbrae. Sic posito quod radij divergant a spatio $F\phi$, (fig [22])[27] alij quidem paralleli tendentes ad AB atque alij ad priores inclinati sed inter se paralleli tendentes ad CD. Prima segregatio fiet in extremitatibus juxta lineas FA et ϕD, ultimaque in medio veluti ad r. Nam lineâ pt inter $F\phi$ et r ductâ, videre est quòd parallelae juxta extremitates pq et st ab invicem segregantur, sed mixtae transeunt per intermedium spatium qs.

47/

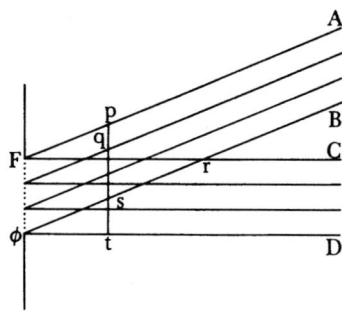

Figure 22

(21) Originally continued: geminarum et rotundarum. Newton then changed this to "geminas et rotundas" (twin and round).

(22) Newton defined the "focus" generally as "The point from wch rays diverge or to wch they converge" (Add. 3970, f. 22^r = *Opticks*, Bk. I, Pt. I, Axiom VI, p. 7). Only for parallel incident rays is his definition equivalent to the modern one of a lens's focal point; see §16 = 1, 16. In the

hole $\phi F\phi$, the same rays are thus mixed at each space, $\phi F\phi$ and $\xi X\xi$, and the composition of each white is the same.

Furthermore, these things will become clearer by observing first that the shadow of any object, however it is shaped and placed at the hole $\phi F\phi$, is distinctly projected onto the paper receiving the rays at X. Indeed, you will see the shadows of the little air bubbles that are concealed in the prism (as usually occurs in all glass) projected onto that paper in the shape of spots.[21] This could in no way occur unless the rays flowing from several points of $\phi F\phi$ came together again in just as many points at $\xi X\xi$. Although the rays flowing from the same points do not come together exactly in the same points when a refractive lens is used instead of a mirror (as in Figs. 19 and 20), and they consequently generate some colors in the boundary of light and shadow, as I have explained in detail, nevertheless, the space in which they meet is so small that it may be considered as nearly a sensible point.

Second, if you set the lens in Fig. 19 so that being placed in the middle it is equidistant from its foci F and X,[22] and then you receive the colors on the paper PT, alternately moved both on the far side of the lens toward X and on the near side toward F, then you can observe that the colors clearly appear in the same way: They are diminished and gradually converted into white, both when that paper is shifted with a slow and continuous motion to F and when it is shifted to X. Hence the colors' divergence from F and convergence to X are entirely similar. In a like manner, if the paper $\pi\tau$ is moved slowly[23] toward X, just as the paper PT[24] is moved toward F, the same colors will be perceived in each, and they will similarly end in white, with this exception only, that their position is reversed because of the crossing of the rays at X. Hence the colors' divergence from both F and X is entirely[25] similar. What therefore is to be concluded[26] from this, other than that they are mixed in the same way, both at F before they have spread out from one another and at X where they are again gathered together into white? But for the comparison just made to become clearer by some example, it appears that in addition it should be observed that when the paper is placed contiguous to F and then moved away toward pt, and when it is afterward placed at X and moved away toward $\pi\tau$, then, I say, the white at F and X will in each case initially degenerate into colors along its edges while remaining white in the middle. The reason for this is just that the diverging rays are initially separated at the boundary of light and shadow. Thus, assuming that the rays diverge from the space $F\phi$ (Fig. 22)[27]—for example, some parallel and tending to AB, and others inclined to these but parallel to one another and tending to CD—then the initial separation will occur at the edges near the lines FA and ϕD and the final one at the middle, just as at r. For when the line pt has been drawn between $F\phi$ and r, it is possible to see that the parallels near the edges pq and st are separated from one another, but that they pass through the intermediate space qs mixed together.

gathered into the same space.

61 [= II, 52]. It is proved from this that in the fourth and fifth ways of compounding whiteness only those rays converge to the same space that diverged from the same space.

62 [= II, 53]. And that the divergence of the colors from the prism is very similar to their divergence from the white focus of a lens.

present case the object F and the image X are symmetrically located on each side of the lens at a distance of twice the lens's focal length.   (23) Added.   (24) Read (as II): *pt*.

(25) Omitted in II.   (26) II: *concludendum est* (ought to be concluded).

(27) In the corresponding Fig. II, 18 in the *Optica*, the lines FC and ϕD are not perpendicular to $F\phi$ but incline downward as much as FA and ϕB incline upward.

63 [= II, 54]. Atque etiam quòd divergentes a dicto foco non secus in alium focum congregari possunt, quàm divergentes a lente.[28]

Tertiò sicut lens *MN* in fig [19] refringendo radios divergentes ab *F* facit ut convergant ad *X* et ibi conficiant albedinem: eodem modo si isti radij postquam decussavêre divergentes ab *X* iterum trajiciantur per aliam lentem $\mu\nu$ (fig [23]) priori similem et similiter positam inter focos ejus *X* et ξ, (id est, in

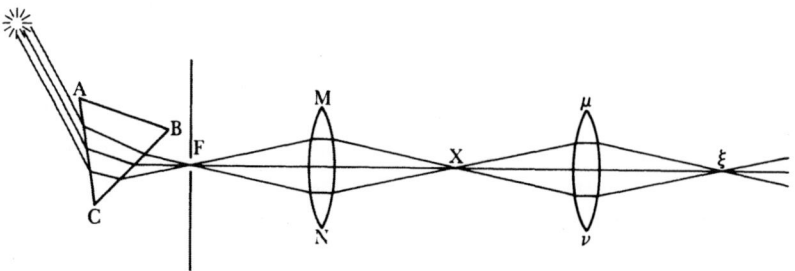

Figure 23

aequali ab utrisque distantiâ:) colores sic ad ξ secundâ vice congregati, albedinem rursus component sicut / ante composuerant ad *X*, hoc tantùm 48/ interposito discrimine, quòd apparebunt in limbo albedinis ad ξ duplo latiores quàm (e mox ostensis) appareant ad *X*; atque insuper in situ contrario. At speculis ut dictum est adhibitis quae lucem aliquoties repercutiant, isti colores erunt nulli: atque adeò penicilli *FX* et *Xξ* evadent omnino similes & similis fiet decussatio, et commistura radiorum ad *F, X,* et ξ. Concludendum est itaque quòd lux, cùm modò trajicitur per prisma, licèt albedinem exhibeat, tamen constat ex radijs heterogeneis confusè mixtis et ab invicem per divergentiam mox discessuris, qui postquam ita segregantur proprijs apparent formis, sin iterum congregantur albedinem rursus componunt, & sic praeterea in infinitum.

Lect. 6
64 [≈ II, 55]. Imò lucem e coloribus ante omnem refractionem componi

Imò verò[1] lux non solùm componitur ex omnium colorum radijs ut egreditur prismate et nondum discernitur in colores istos, sed etiam cùm nondum attigerit prisma et antecedenter ad omnem refractionem. Et inde non mirum est quòd cùm segregatur in colores virtute prismatis radios inaequaliter refringentis, et colores iterum commiscentur ope lentis aut alio quovis modo praemonstrato, quòd, inquam, rursus componant albedinem.

65 [= II, 56]. Ut ex eo pateat quòd aliqui colores reflecti possunt dum alij per prisma trajiciantur.

Verum ut hoc discrimen inesse radijs antecedentèr ad refractiones ostendam:[2] Sit *ABC* (fig [24]) prisma quod excipit radios in obscurum cubiculum per foramen *F* uno digito latum trajectos, eosque refringit[3] ad papyrum vel parietem *HI* ijs obsistentem apud *T*. Porrò autem cùm super-

---

(28) Read: prismate (prism).

(1) Imò vero] Originally: Quinetiam.

(2) Verum . . . ostendam:] In the *Optica* this article and the preceding were combined into one paragraph, and this opening phrase was incorporated into some new remarks.

(3) Originally continued: aliquà velut (some way as).

Third, just as the lens *MN* (Fig. 19) by refracting the rays diverging from *F* causes them to converge to *X* and produce white there, in the same way, if those rays, after they have crossed and diverged from *X*, are again cast through another lens *μν* (Fig. 23)—similar to the first one and similarly placed between its foci *X* and *ξ*, that is, equidistant from both—then the colors gathered in this way at *ξ* for a second time again compose white, just as they had previously composed it at *X*, with only this difference being introduced: They will appear twice as wide in the band of whiteness at *ξ* as they appear at *X* (from what was then shown), and moreover in a reversed position. But as has been described, when mirrors that reflect the light several times are used, there will be none of those colors; hence the pencils *FX* and *Xξ* will prove to be completely similar, and also the crossing and mixture of the rays at *F*, *X*, and *ξ* will become similar. It must be concluded, therefore, that when light has just passed through a prism, although it exhibits whiteness, it nonetheless consists of confusedly mixed heterogeneous rays soon about to separate from one another by diverging, which after they are thus separated, appear according to their own nature; but if they are again gathered together, they once more compound whiteness, and so on to infinity.

In fact[1] light is compounded of rays of all colors not only as it leaves the prism and is not yet separated into those colors, but also when it has not yet reached the prism and prior to any refraction. Consequently, it is not surprising that when light is separated into colors by the power of a prism to refract rays unequally, and the colors are again mixed by the aid of a lens or by any other way previously shown, that, I say, they again compound white.

But that I may demonstrate that this difference is in the rays prior to refraction,[2] let *ABC* (Fig. 24) be a prism that receives the rays cast into a dark room through a one-inch-wide hole *F* and refracts them[3] to a paper or a wall *HI* that stops them at *T*. Moreover, since the prism's surface *BC* does

63 [= II, 54]. And also that diverging from that focus they can be gathered at another focus no differently than those diverging from a lens.[28]

64 [≈ II, 55]. Light in fact is compounded of colors before any refraction.

65 [= II, 56]. As is evident because some colors can be reflected while others are transmitted through a prism.

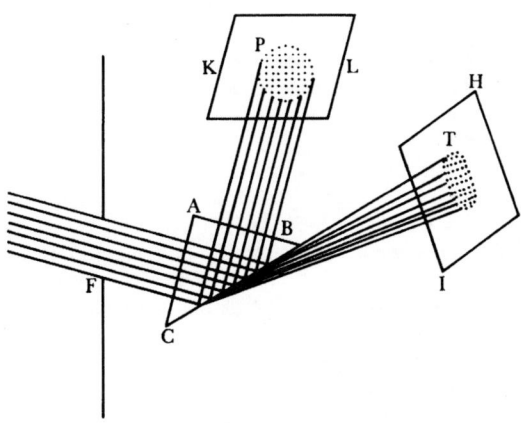

Figure 24

Hujusque rei experiendae modus adducitur.

ficies prismatis BC non omnes[4] refringat radios versus T sed et plurimos reflectat, eos apud P siste etiam cum alia papyro KL in morem albae imaginis foramini F persimili terminante. Deinde converte prisma circa axem ejus secundum ordinem literarum $ABCA$ et videbis tum amplitudinem colorum ad T, tum quantitatem lucis ad P augeri perpetuò, donec tandem cùm refractio ad planum BC fit maximè obliqua, colores ad T incipiant evanescere et reflecti ad P, / purpureus primò, deinde caeruleus viridis et flavus, ac denique[5] ruber. Cujus quidem lucis accessu imago P fiet multò lucidior quàm antea. Interea verò dum colores a T gradatim evanescunt videbis albedinem P paululum mutari & nonnihil vergere ad caeruleum, per accessum nempe purpurei et caerulei qui primò reflectuntur: at postquam caeteri etiam colores viridis flavus et ruber reflectuntur a T, albedo ad P redintegrabitur.[6] Id quod nullo modo accidisse potuisset nisi radijs prout a Sole veniunt discrimen interesse concedatur. Scilicet quòd ex ijs quidem ad efficiendos rubeum et flavum dispositi pertinaciùs et cum minori refractione penetrent superficiem BC et versus T prolabantur, dum alij ad exhibendum purpureum et caeruleum parati superficiem dictam aut penetrent languidiùs majorem refractionem patientes, aut si nequeant penetrare propter nimiam eorum obliquitatem, tum faciliùs et citiùs reflectantur ad P. Ijs primò omnium reflexis quorum potentia ad istam superficiem penetrandam sit minima, id est, purpuriformibus, et caeteris deinde suo ordine prout incidentia fit magis, obliqua, donec rubriformes ultimò reflectantur obliquitate tantâ debilitati ut non sint ampliùs potentes dictae superficiei resistentiam superare.[7] Atque haec facilè constabunt ijs qui nôrunt quòd quò major est vis refractiva superficiei cujusvis, eò citiùs et ad minorem obliquitatem radij reflectentur; et quo minor eò magis obliqui penetrabunt.[8]

66 [= II, 57].
Notanda quaedam

De hoc autem experimento juvabit observare sequentia. Primò quòd cùm praedicta variatio albedinis ad P sit admodum parva propter exuberantiam lucis albae collatae ad reflexum caeruleum: itaque cavendum est ne prismate utaris quod ex vitro conflatur tincto cum colore aliquo, ne lucem ad P reflexam ita tingat ut difficile sit dictam variationem observare. Praestat adhi-

---

(4) Originally: modò (only).

(5) Originally: denuò (after). The critical angle of reflection ι occurs when a ray refracted into a rarer medium is just parallel to the surface, or when $\sin \iota = 1/n$. Since $n_P > n_T$, where the subscripts P and T represent the extreme blue and red, then $\iota_P < \iota_T$, and the extreme blue will be reflected first and the red last.

(6) at . . . redintegrabitur] Omitted, presumably inadvertently, from II.

(7) Newton's present description of total reflection reflects his commitment to an emission or corpuscular theory of light and, in particular, Descartes's model for the reflection and refraction of light based on projectiles encountering and penetrating surfaces in *La Dioptrique*, Discourse II. On a number of later occasions Newton was more forthcoming in expounding his corpuscular theory and ideas on the physical cause of total reflection, first invoking changes in the aether density in "An Hypothesis explaining yᵉ properties of Light discoursed of in my several papers," which he sent to Oldenburg on 7 December 1675 (*Correspondence*, 1:371–6 = Thomas Birch, *The History of the Royal Society*, 4 vols. (London, 1756–7), 3:255–60); and then forces in the *Principia* (Bk. I, Prop. XCVI); and finally both forces and the aether in the *Opticks* and its successively added queries (Bk. II, Pt. III, Prop. VIII, and Queries 19, 29). While Newton's

not refract all[4] the rays toward *T* but in addition reflects very many, also stop those at *P* with another paper *KL* that terminates them in the form of a white image very similar to the hole *F*. Next rotate the prism about its axis in the order of the letters *ABCA*, and you will see both the extent of the colors at *T* and the quantity of light at *P* continually increase until finally, when the refraction at the plane *BC* becomes very oblique, the colors at *T* begin to vanish and to be reflected to *P*: first purple, then blue, green, and yellow, and last[5] red. Naturally, with the entrance of this light, the image *P* will become much brighter than before. But at the same time as the colors gradually vanish from *T*, you will see the white at *P* change slightly and incline somewhat to blue, namely, by the entrance of purple and blue, which are reflected first; but after the other colors, green, yellow, and red, are also reflected from *T*, the white at *P* will be restored.[6] This in no way could have occurred unless it be conceded that the difference is present in the rays just as they come from the sun: specifically, that some of those—disposed to produce red and yellow—more vigorously penetrate the surface *BC* with a smaller refraction and go on toward *T*, while others—fit to exhibit purple and blue—either weakly penetrate that surface and experience a greater refraction, or, if they are unable to penetrate it because of their excessive obliquity, are reflected to *P* more easily and quickly. Those are reflected first of all whose power to penetrate that surface is the least, that is, the purple-making rays, and then the others in their own order according as their incidence becomes more oblique, until the red-making rays, having been weakened by such a great obliquity that they are no longer able to overcome the resistance of that surface, are finally reflected.[7] These things will readily be evident to those who have learned that the greater the refractive force of any surface, the more quickly and at a smaller obliquity will the rays be reflected; and the smaller the refractive force, the more the oblique rays will penetrate.[8]

About this experiment, however, it will help to observe the following: First, since the aforementioned variation of white at *P* is very small, because of the abundance of white light compared to the reflected blue, you must therefore be careful that you do not use a prism that is melted from glass having some color, lest it so color the light reflected to *P* that it is difficult to

And the method of testing this is set forth.

66 [= II, 57]. Some things to be noted.

---

argument to establish that colors are in the sun's direct light may perhaps be more readily interpreted in an emission theory, it does not depend on it. The essence of his argument is the claim that the identical rays are refracted to *T*, then totally reflected to *P*, and exhibit the same color in each place.

(8) Newton first described this experiment in his essay "Of Colours" (§22, Add. 3975, f. 4ᵛ) and later alluded to it in his draft letter to Hooke in the spring of 1672 (Expt. 2, Add. 3970, ff. 439ᵛ–40ʳ); see also §95 ≈ II, 79. In the *Opticks* (Bk. I, Pt. I, Prop. II, Expt. IX), he used it to argue that the different degrees of refrangibility—rather than, as here, the colors—are innate to the sun's direct light, though the nature of the argument is the same as here. He also invoked this experiment and a related one with a compound prism (§71 = II, 62) to establish Bk. I, Pt. I, Prop. III, "The suns light consists of rays differing in reflexibility & those rays are more reflexible then others wᶜʰ are more refrangible." Newton first formally added reflexibility to color and refrangibility as a third property of light in the formulation of his theory for Huygens on 23 June 1673; *Correspondence*, 1:292–3 = *Phil. Trans.*, 8 (1673):6090.

bere prisma ex laminis vitreis tenuibus et perpolitis confectum et aquâ lympidissimâ repletum.

/ Secundò licet mutatio dicta sit parva, tamen satis est ad ostendendum 50/ quòd radij retinent eosdem colores cùm reflectuntur quos exhibent cùm trajiciuntur per superficiem *BC*; siquidem tingunt albedinem [*P*] colore suo quantum liceat tam paucis tingere. Colores itáque suos habuêre priùs, et eosdem retinent sive refringantur sive reflectantur: licèt misturis plerumque celati latent donec eruantur (non autem fiunt) virtute Prismatum.

Tertiò ex luce ad priorem speciem albedinis per reflectionem omnium colorum ab *T* restitutâ, quid aliud denotatur quàm quòd albedo ista per misturam colorum omnium reproducitur. Scilicet cùm rubor ultimò reflexus admiscetur caeteris coloribus antea reflexis, reflexorum colorum mistura tunc perfecta est ad albedinem componendam, quae superadditur albedini priùs existente in *P*.

Quartò, nequa oboriatur suspicio quòd refractiones in superficiebus *AC* et *AB* ad ingressum radiorum in Prisma et egressum factae, possint aliquid conducere ad effectus hosce producendos; juvabit observare quòd effectus ijdem producuntur, cujuscunque licèt magnitudinis statuatur angulus *ACB*; hoc est, quaecunque sit refractio superficiei *AC*; modò angulus *ABC* ponatur ejusdem magnitudinis atque angulus *ACB*: alias enim pro imagine albâ ad *P* generabuntur colores. Experimentum itaque nullatenus dependet a refractionibus superficierum *AC* et *AB*:[9] imò possis efficere quòd cùm colores partim reflectuntur ad *P* et partim trajiciuntur ad *T*, radij perpendiculariter incident in *AC* emergentáque ex *AB*, et sic neutrâ superficie refringentur; modò statuas angulum *ACB* ut et *ABC* esse 40[grad], 40′ circiter:[10] et ijdem tamen effectus producentur.

67 [= II, 58]. In majorem rei evidentiam ostenditur quosdam colores alijs facilius reflecti. Caeterùm in majorem evidentíam et explicationem modi quo praedicta fiunt, liceat experiri per lucem in colores discretam, quòd purpureus primò et caeteri deinde (quisque suo ordine) reflectuntur. Etenim (in fig. [25]) sint *ABC* et *αβκ* duo prismata / parallela quorum alterum *ABC* projicit colores in 51/

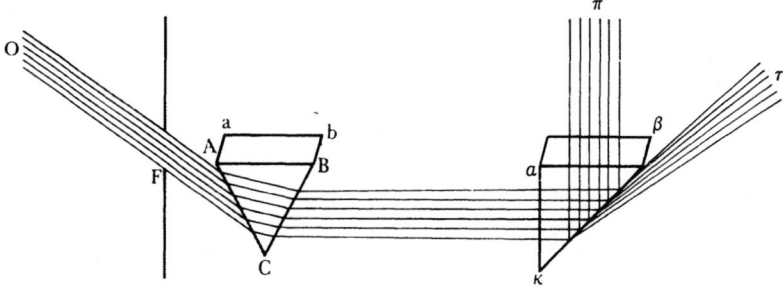

Figure 25

---

(9) Experimentum . . . *AB*:] Originally: Itaque refractio superficierum *AC* et *AB* nil interest experimento: (Therefore the refraction of the surfaces *AC* and *AB* does not at all enter into the experiment.)

observe that variation. It is better to use a prism made from thin, highly polished glass plates and filled with very clear water.

Second, although that change is small, it is nonetheless sufficient to show that the rays preserve the same colors when they are reflected as they exhibit when they are transmitted through the surface *BC*, since they color the white at [*P*] with their own color, insofar as it is possible to color with so few. Therefore they possessed their own colors previously, and they preserve the same colors whether they are refracted or reflected, although generally the colors are latent, hidden in the mixture until they are drawn out—but not made—by the power of prisms.

Third, as a consequence of the light having been restored to the former kind of white by the reflection of all colors from *T*, what else is implied other than that that white is reproduced from a mixture of all colors. Namely, when the red reflected last is mixed with the other colors already reflected, the mixture of reflected colors is then completed to compose white, which is added over and above the white previously existing at *P*.

Fourth, lest the suspicion arise that the refractions that are made in the surfaces *AC* and *AB* by the rays entering and departing from the prism can somehow combine to produce these effects, it will help to observe that the same effects are produced at whatever magnitude the angle *ACB* is set, that is, whatever the refraction of the surface *AC* is, provided that the angle *ABC* is assumed to be the same magnitude as the angle *ACB*, for, otherwise, instead of a white image at *P*, colors will be generated. The experiment, therefore, in no way depends on the refractions of the surfaces *AC* and *AB*.[9] In fact, you may arrange it so that when the colors are partly reflected to *P* and partly transmitted to *T*, the rays will fall upon *AC* and emerge from *AB* perpendicularly and will thus be refracted by neither surface, provided that you set the angle *ACB* as well as *ABC* to be approximately 40°40′;[10] and still the same effects will be produced.

However, for greater clarity and a fuller explanation of the way in which the preceding occurs, one may prove with light separated into colors that purple is reflected first and then the other colors, each in its own order. For (in Fig. 25) let *ABC* and *αβκ* be two parallel prisms, one of which, *ABC*, projects the colors onto the other, *αβκ*, twelve or more feet away. Then

67 [= II, 58]. For greater clarity of the matter, it is shown that some colors are more easily reflected than others.

(10) In §111 = I, 38 Newton determines the index of refraction for red rays passing from glass to air to be as 52400 to 80481, which makes their critical angle 40°37′, or "approximately" 40°40′; from the same measurements the critical angle for blue rays is 39°55′; also see Lect. II, 12, note (26), this volume. When a ray enters the prism *ABC* perpendicular to the face *AC* and is reflected from the base, then the angle of incidence (and reflection) at the base will be equal to the prism's base angle *ACB* or, in an isosceles prism, *ABC*. Thus Newton apparently set the prism's base angles equal to the critical angle for red rays. In the *Opticks* (Bk. I, Pt. I, Prop. II, Expt. IX) he says that the base angle should be set at 45°, but then no rays whatsoever would be refracted, since they would all be totally reflected. In his draft letter to Hooke, Newton made a related mistake when he explained that the prism's base angles should be "about 49 degrees 50 minutes" (Expt. 2, Add. 3970, f. 439ᵛ). Here he probably confused the angle of incidence, which would be about 40°10′, with its complement; see note (20), this lecture. It may be of no significance, but in Fig. 24 Newton drew the prism's base angles at about 45°, whereas in the *Optica*, Fig. II, 20, they are very nearly 40°.

alterum αβκ ad distantiam duodecim vel plurium pedum. Tum Prismate αβκ circa axem ejus secundum ordinem literarum α, β, κ, α converso donec tanta sit obliquitas radiorum in superficiem βκ incidentium ut incipiant ad π reflecti non ampliùs potentes penetrare ad τ; videbis omnes purpuriformes primò reflecti caeterosque deinde suo ordine.

68 [= II, 59]. Idem aliter ostenditur, circumstantiâ tantùm variatâ   Veruntamen quia purpuriformes radij paulò magìs refringuntur in primo prismate *ABC*, et ideò magis inclinantur ad superficiem βκ secundi prismatis αβκ quàm caeteri: poterit objici quòd eâ de causâ primò omnium reflectuntur. Quamobrem (in fig [26]) duo prismata statuantur non parallela sibi

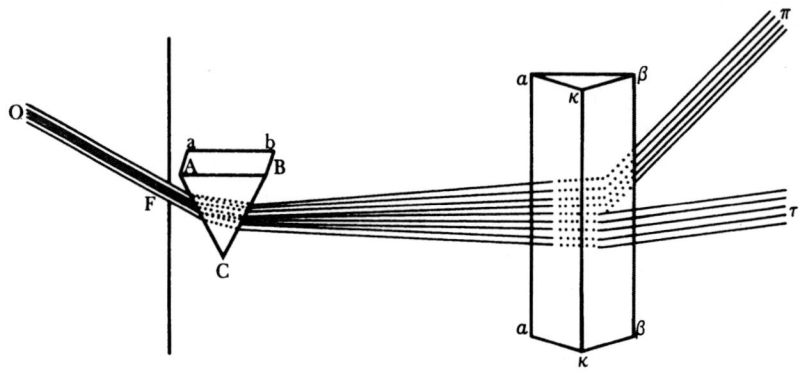

Figure 26

invicem sed in transverso[11] situ, ut omnicolores radij quasi ad eosdem angulos[12] incidant in praefatam superficiem βκ. Quo posito possis observare convertendo prisma posterius αβκ circa axem ejus secundum ordinem literarum α, β, κ, α,[13] quòd radij purpuriformes primò omnium reflectuntur, et ultimò rubriformes; coloribus ad π continuò translatis prout a τ dispareant.

69 [= II, 60]. Idem adhuc aliter.   Sunt et alij praeterea modi quibus experiri liceat quòd ex radijs similiter incidentibus quaedam genera penitus reflecti possunt dum alia partim transmittuntur. Quemadmodum si *EFG* (fig. [27]) sit operculum fenestrae ad *F* terebratum, et foras statuator Prisma *ABC* quod lucem solis[14] foramen *F* ingressuram intercipiat et refringat versus φ, ad illud φ pedibus ab *F* duodecim aut[15] longiùs postpositum statuatur opacum corpus εφγ quod lucem sistat, dempto parvo foramine φ per quod aliqua pars lucis, nempe violacea / longiùs trajiciatur ad Y. Istud autem φ non sit semisse digiti latius. Deinde 52/ prae manibus sumatur aliud Prisma αβκ et ad radios transversè positum statuatur a posticâ parte foraminis φ, circáque axem ejus convertatur donec videas lucem violaceam postquam ab ejus basi βκ obliquissimè refracta fuerit versus τ, totam a τ disparuisse modò, et ad π reflecti. Luce violaceâ tam obliquè ad π reflexâ ut ad τ statim pervasura esset modò ex angulari motu

---

(11) Originally: *perpendiculariter transverso* (perpendicularly transverse).
(12) *quasi ad eosdem angulos*] Originally: *similiter* (similarly).
(13) Read: α,κ,β,α.

turning the prism $\alpha\beta\kappa$ about its axis in the order of the letters $\alpha\beta\kappa\alpha$ until the obliquity of the rays falling upon the surface $\beta\kappa$ is so great that they begin to be reflected to π, they being no longer able to penetrate to τ, you will see all the purple-making rays reflected first, and then the others in their own order.

Nevertheless, because the purple-making rays are refracted a little more in the first prism *ABC* and are therefore more inclined than the others to the surface $\beta\kappa$ of the second prism $\alpha\beta\kappa$, it could be objected that for that reason they are reflected first of all. Consequently (in Fig. 26), set the two prisms not parallel to each other but in a transverse[11] position so that rays of all colors fall at almost the same angles[12] upon that surface $\beta\kappa$. After this has been arranged, by turning the second prism $\alpha\beta\kappa$ around its axis in the order of the letters $\alpha\beta\kappa\alpha$[13] you can observe that the purple-making rays are reflected first of all and the red-making ones last, with the colors continuously shifting to π as they disappear from τ.

68 [= II, 59]. The same thing is shown differently, with only one feature changed.

There are in addition other ways whereby one may prove that at similar incidence some kinds of rays can be totally reflected while others are partly transmitted. For instance (Fig. 27), if *EFG* is a window shutter pierced at *F*,

69 [= II, 60]. The same thing still differently.

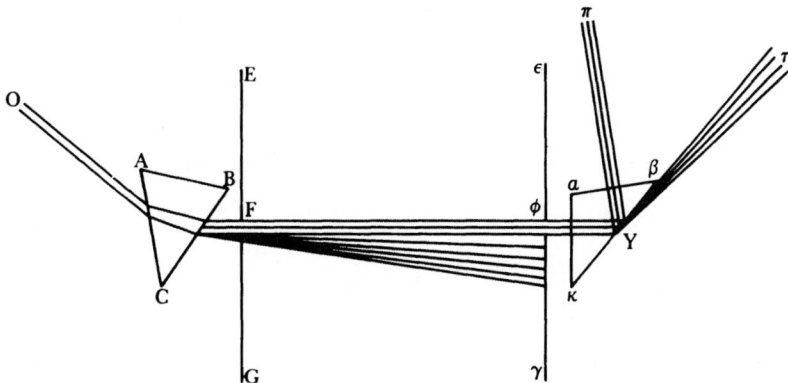

Figure 27

and outside a prism *ABC* is placed that intercepts the sun's[14] light about to enter the hole *F* and refracts it toward ϕ, then at ϕ (located twelve or[15] more feet behind *F*) place an opaque body $\epsilon\phi\gamma$ that stops the light except for the small hole ϕ through which some part of the light, namely, the violet, is transmitted farther to *Y*. Do not, however, let that hole ϕ be wider than half an inch. Then take another, readily accessible prism $\alpha\beta\kappa$ placed transverse to the rays and set it at the rear of the hole ϕ; turn it around its axis until you see that all the violet light, after it has been most obliquely refracted by its base $\beta\kappa$ toward τ, has just then vanished from τ and is reflected to π. When the violet light has been so obliquely reflected to π that it would immediately penetrate to τ, if as a result of the prism's angular motion made in the order

(14) Added.
(15) Originally continued: *viginti* (twenty).

prismatis secundum ordinem literarum α, β, κ, α[13] facto, angulus κYφ vel
minimùm augeretur: prisma istud αβκ in eo situ figatur. Tum alterum prisma
ABC motu circa axem ejus[16] nunc hac nunc illac parùm convertatur, ut
colores quos projicit in obstaculum εγ paululum attollantur, eoque pacto
omnes successivè transmittantur per foramen φ in Prisma posterius αβκ.[17]
Et videbis quod cùm flavedo transmittitur ad Y, illi radij non omnes ad π
reflectentur, sed plurimi perrumpent superficiem βκ et ad τ pertingent. Et
cum rubor ad Y transmittitur, illi radij fortiùs adhuc perrumpent ut ex copiâ
perrumpentis lucis et minori ejus refractione constet. Neque mirum videatur
quòd purpuriformes radij sint minùs potentes penetrare superficiem βκ quàm

Sec. 37   rubriformes; quandoquidem prismatibus eodem modo dispositis, antehac ostendi quòd majorem refractionem patiantur; posito scilicet angulo κYφ tanto
ut omnigeni radij possint superficiem βκ penetrare.

70 [= II, 61]. Et   Jam cùm radij qui citiùs et faciliùs reflectuntur in experimento ad Fig 24
proinde cùm e radijs   tradito (nempe purpuriformes) etiam citiùs et faciliùs reflectantur in experi-
solaribus alij alijs,   mentis duobus novissimè recitatis, cùm eadem ijsdem radijs semper eveniant;
pro specie colorum   liquet quòd hoc non fit ex contingentiâ sed ex praedispositione radiorum, et
quos postmodum   quòd antecedenter ad omnem reflectionem aut refractionem quidam ad exhi-
exhibent, faciliùs   bendos quosdam colores sunt apti et / facilius reflexibiles, alij verò alijs   53/
reflectuntur; constat coloribus et progrediendi viribus afficiuntur.[18] Neque aliud experimentis jam
lucem solis ex illis   recitatis discrimen interesse videtur, quàm quòd in primo radij omnium for-
coloribus componi.   marum, prout a sole adveniunt confusè mixti, incidant in prisma quod rubriformes transmittit et reflectit caeruliformes; in reliquis autem duobus experimentis dissimiles radij priùs discernantur ab invicem quàm incidant in dictum
prisma.

71 [= II, 62]. Alius   Adhaec lubet alium adducere modum quo dissimilitudo radiorum in luce
modus quo lux solis   solis mixtorum[19] innotescat, non multò dissimilem ei ad Fig [24] ostenso,
partim reflecti potest sed conspectui jucundiorem et aeque scientificum. In Fig [28] Sunto AαBβC
et partim refringi.   et BβDδC duo prismata ita juxta se posita et colligata, ut duo ex eorum
planis CBβ conveniant sibi et coincidant, excepto tantùm quòd nonnihil aeris
in morem tenuissimae laminae intercedat ijs: Id quòd eveniet ultrò, siquidem
haud queas prismata tam arctè constringere quin tantum intercedet aeris
quantum proposito sufficiat. Porro in majorem rei evidentiam convenit ut

Sect 90   anguli ACB et CBD sint aequales proximè, eò ut plana AαC et BβδD fiant
parallela, licèt hoc non sit omninò necessarium. His praemissis statuantur
dicta prismata juxta foramen F, ut lux ingressa per ea trajiciatur versus τ,
primò permeans superficiem AαC, deinde intermediam superficiem BβC, et

---

(16) Originally continued: angulari (angular).

(17) By placing the first prism ABC outside the hole F, as he suggested in the related experiment in §38 = II, 16, and setting the second prism parallel to the first, Newton has all but
arrived at the experimental arrangement of the *experimentum crucis*; compare Mamiani, *Newton*, p. 122.

(18) This argument is similar to that in §22 ≈ I, 23, where Newton argued that unequal
refrangibility arises from a previous disposition of the rays, though he was careful not to extend
his argument there to color as he does here. This difference arises from the fundamentally
different nature of the phenomena, for with total reflection it is always possible—as it is not with

of the letters $\alpha\beta\kappa\alpha$,[13] the angle $\kappa Y\phi$ increased even the least bit, fix that prism $\alpha\beta\kappa$ in that position. Then rotate the other prism ABC with a slight to and fro[16] motion around its axis so that the colors it projects onto the obstacle $\epsilon\gamma$ are raised a little, and in this way all the colors may be successively transmitted through the hole ϕ onto the rear prism $\alpha\beta\kappa$.[17] You will see that when the yellow is transmitted to Y not all those rays will be reflected to π, but very many of them will break through the surface $\beta\kappa$ and reach τ. When the red is transmitted to Y those rays will break through still more vigorously, as is evident from the quantity of light breaking through and its smaller refraction. Nor should it seem surprising that the purple-making rays are less able to penetrate the surface $\beta\kappa$ than the red-making ones, since when the prisms were arranged in the same way, I showed earlier that they undergo §37 a greater refraction, to be sure, with the angle $\kappa Y\phi$ having been set so large that rays of every sort could penetrate the surface $\beta\kappa$.

Now since the rays that are more quickly and easily reflected in the experiment presented at Fig. 24—namely, the purple-making ones—are also more quickly and easily reflected in the two experiments most recently recited, since the same things always occur with the same rays, it is evident that this does not occur from a contingency but from a predisposition of the rays, and that prior to any reflection or refraction some are disposed to exhibit certain colors and are more easily reflexible, whereas others are endowed with other colors and forces of proceeding.[18] Nor in the experiments just now recited does there appear to be any difference present other than that in the first one rays of every nature, just as they arrive confusedly mixed from the sun, fall upon the prism, which transmits the red-making rays and reflects the blue-making ones; whereas in the other two experiments the dissimilar rays are separated from one another before they fall upon that prism.

Moreover, it is desirable to bring forward another way whereby the dissimilarity of the rays mixed[19] in the sun's light may become known; it does not differ much from that shown at Fig. 24, but it is visually more pleasing while being equally scientific. In Fig. 28 let $A\alpha B\beta C$ and $B\beta D\delta C$ be two prisms juxtaposed and fastened together in such a way that two of their planes, $CB\beta$, meet one another and coincide, except for some air that comes between them in the form of a very thin plate. This will occur spontaneously, provided that you do not have the power to compress the prisms too closely, so that not enough air comes between them as suffices for the purpose. Moreover, for greater clarity of the experiment, it is appropriate that the angles ACB and CBD be almost equal so that the planes $A\alpha C$ and $B\beta\delta D$ may §90 consequently become parallel, although this is not altogether necessary. With these matters having been premised, set those prisms near the hole F so that the light entering through it is transmitted toward τ, by first passing through the surface $A\alpha C$, then the intermediate surface $B\beta C$, and from there proceed-

70 [= II, 61]. Consequently, since some of the solar rays are more easily reflected than others according to the species of color they exhibit afterward, it is manifest that the sun's light is compounded of those colors.

71 [= II, 62]. Another way whereby the sun's light can be partly reflected and partly refracted.

---

refraction—to identify the particular color that is totally reflected at any given angle, since the rays exhibit their color at τ before they are reflected to π.

(19) Added.

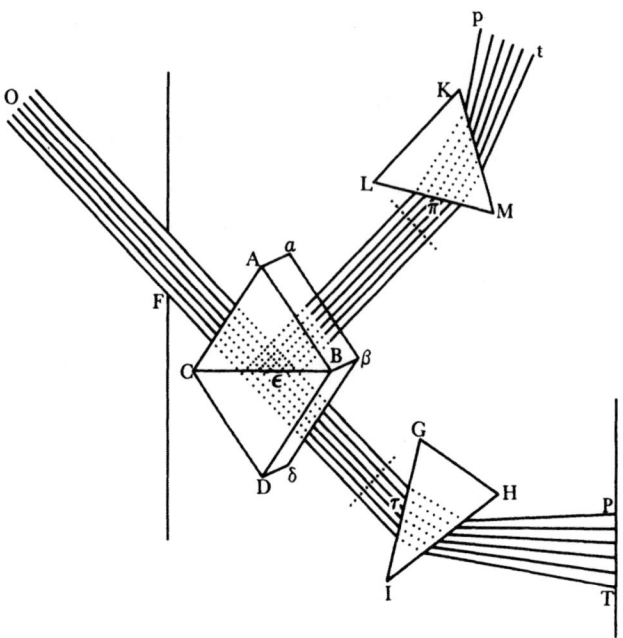

Figure 28

inde per *BβδD* prolapsa in papyrum ad τ collocatam, quam albedine tingit
tanquam si non omninò transierit prismata, sed vitrum parallelis planis *AαC*
et *BβδD* terminatum. Praeterea cùm intermedia superficies *BβC* lucem ei
incidentem non omnem transmittat àd τ sed multam reflectat, quae aliquò
exibit à prismate *ABC* per superficiem ejus *AαβB*, puta versus π: ad illud π
statuatur alia papyrus quae lucem hanc / terminet similiter albicantem. Quod 54/
ubi feceris, converte prisma quadrangulare (ex duobus triangularibus colligatis confectum) motu lento circa axem ejus secundum ordinem literarum
ABDCA: tandémque videbis quòd albedo ad π ac τ degenerabit in colores;
flavedine primò, deinde rubedine ad τ conspectâ, caeruleo autem colore ad π;
donec post intensissimam rubedinem ad τ color et lux omnis evanescat inde,
et caeruleus ad π iterum transformetur in albedinem aliquanto lucidiorem
quam antea. Utpote dum prismata circa communem axem, ut dictum est,
convertuntur, radiorum in mediam superficiem *BβC* (hoc est in laminam
aeris prismatibus interjectam) prolapsorum incidentia continuò fit obliquior,
donec tanta sit eorum obliquitas ut nequeant ampliùs penetrare dictam laminam progredíque ad τ, sed ab eâ reflectantur ad π. Quod accidet cùm angulus *FϵC* (obliquitas incidentium) sit graduum ferè quinquaginta.[20] Radij autem
purpuriformes minimè omnium potentes penetrare dictam laminam aëream,
reflectentur primò, et albedinem priùs reflexam ad π nonnihil tingent eorum
colore, dum ex radijs perlabentibus ad τ flavedo imperfecta aut potiùs color
inter flavum et viridem mediocris[21] componitur. Postea caeruleus, et viridis
deinde reflexus paulo magis tinget lucem in π cum colore caeruleo (licèt
admodum diluto propter exuberantiam albedinis commixtae,) manebítque

ing through $B\beta\delta D$ to the paper placed at τ, which it colors white, just as if it had not passed through the prisms at all but through a glass bounded by the parallel planes $A\alpha C$ and $B\beta\delta D$. Furthermore, since the intermediate surface $B\beta C$ does not transmit to τ all the light falling on it, but reflects much of it, which will pass out of the prism ABC through its surface $A\alpha\beta B$ to some place, suppose toward π; at π place another paper that similarly terminates this whitish light. When you have done this, turn the quadrangular prism (made from the two fastened triangular ones) with a slow motion about its axis in the order of the letters $ABDCA$. After a while you will see that the white at π and τ will degenerate into colors: First yellow and then red is visible at τ, but the color blue is visible at π, until finally, after the most intense red is visible at τ, all color and light vanishes from there [τ], and the blue at π is again transformed into a white somewhat brighter than before. That is to say, when the prisms are rotated about their common axis, as has been described, the incidence of the rays proceeding into the middle surface $B\beta C$ (that is, into the plate of air lying between the prisms) becomes continually more oblique, until their obliquity is so large that they are no longer able to penetrate that plate and continue to τ, but are then reflected to π. This will occur when the angle $F\epsilon C$ (the obliquity of the incident rays) is about fifty degrees.[20] The purple-making rays, however, the least able of all to penetrate that plate of air, will be reflected first, and they will slightly color the whiteness already reflected to π with their color; while an imperfect yellow, or rather a mean color between yellow and green,[21] is compounded from the rays flowing through to τ. Afterward, the blue, and then the green reflected a little more, will color the light at π with the color blue (although quite diluted because of the overabundance of white mixed with it); while the red

---

(20) The obliquity, the angle $F\epsilon C$, is the complement of the angle of incidence, which is therefore "about" forty degrees, in agreement with the value in §66 = II, 57.

(21) aut . . . mediocris] Added.

rubor in τ, qui mox per flavedinis hactenus commixtae reflectionem fiet intensior, donec ipse etiam denuò reflexus albedinem in π redintegret.[22]

72 [= II, 63].
Penitiùs hic ostenditur quinam e radijs solaribus reflectuntur et quinam transmittuntur: atque adeò hoc non casu sed praedispositione radiorum evenire

Caeterùm ut hoc specimen evadat illustratius, sumatur aliud prisma *GHI* quod a posticâ parte prismatum *ABCD* ita collocetur ut lucem $O\epsilon\tau$ per ea transmissam refringat versus *PT* et in colores permutet: violaceo in *P*, rubeo in *T* caeterisque in intermedia loca projectis. Tum prismata colligata circa communem axem (ut priùs) rotentur donec lux alba versus τ transmissa incipiat flavescere; et videbis quòd color purpureus in *P* simul evanescet. Id quòd arguit purpuriformes radios non campliùs ad prisma *GHI* pertingere, sed a superficie $CB\beta$ primò omnium ad π reflecti; et lucem $\epsilon\tau$ ideò flavescere quod purpura e misturâ tollitur quâ priùs albedinem exhibuit. Ad eundem modum si prismata *ABCD* diutiùs rotentur, / videbis reliquos colores a π ad τ[23] successivè disparere prout lux $\epsilon\tau$ plus plusque rubescit; et cùm fit ruberrima, tum solam rubedinem in τ manere. Quod manifestò convincit hanc lucem $\epsilon\tau$ non aliunde rubescere quàm quòd a radijs aliorum colorum per superficiem $CB\beta$ reflexis secernitur.

Simili ratione si cum prismate quarto *KLM* refringas radios ad π reflexos, et colores eo pacto productos et in album parietem projectos duodecim pedes aut longiùs distantem animadvertas: videbis quòd cùm lux $\epsilon\tau$ incipiat viridè flavescere, purpura in *p*, quam prisma hoc elicit e luce $\epsilon\pi$, plusquam caeteri colores augebitur, per accessum nempe purpurae quae tunc in *P* disparuit: caeterisque deinde coloribus in *pt* gradatim fiet accessus prout a *PT* disparent; donec cùm omnis color a *PT* disparuit, colores ad *pt* non campliùs augeantur. Hoc autem discrimen quo violaceus et caeruleus ad *pt* augmentum suum omne paulo citiùs obtinent quàm rubeus et flavus, tam exile est, ut nisi observator sit attentus, is aegrè advertat.

73 [= II, 64].
Tertius modus quo lux solis partim reflecti potest et partim refringi.

Ut istis denique finem imponamus, lubet alium adducere modum quo quaedam genera radiorum luci solis intermista partim transmitti possint dum alia reflectantur. Nempe si duas laminas vitreas *CB* (in fig [29]) planè perpolitas et ad invicem applicatas secundum planitiem earum connectas, easque vasi *RQ* aquae pleno immergas, extremitate superficierum juxta-positarum undique cerâ vel pice priùs obturatâ, ut aqua non interrepat et expellat aërem, qui more laminae tenuissimae, ut dictum est, interjacebit vitris. Si haec, inquam,

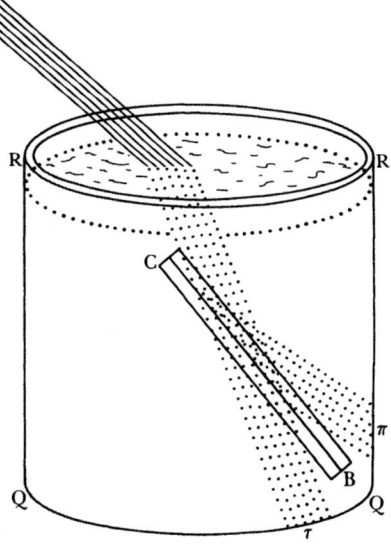

Figure 29

(22) Newton first described this experiment in his essay "Of Colours" (§24, Add. 3975, ff. 4ᵛ–5ʳ = Hall, "Further experiments," p. 31), where he noted that in this "& all such like experiments" the incident rays must be "all wholly or almost parallell" (§25, Add. 3975, f. 5ʳ);

will remain at τ and will thereafter become more intense by the reflection of the yellow previously mixed with it, until after also being reflected the red itself will restore the whiteness at π.[22]

But so that this example may become more illustrative, take another prism *GHI*, so placed behind the prisms *ABCD* that it refracts the light $O\epsilon\tau$ transmitted through it toward *PT* and changes it into colors: violet projected to *P*, red to *T*, and the other colors to the intermediate spaces. Then rotate the fastened prisms about their common axis (as before) until the white light transmitted toward τ begins to grow yellow, and you will see that the color purple at *P* will simultaneously vanish. This makes it clear that the purple-making rays no longer reach the prism *GHI* but are reflected first of all from the surface *CBβ* to π; and the light $\epsilon\tau$ therefore becomes yellow because the purple, with which it previously exhibited white, is removed from the mixture. In the same way, if the prisms *ABCD* are rotated further, you will see the remaining colors disappear successively [from *P* to *T*][23] as the light $\epsilon\tau$ becomes more and more red; and when it becomes reddest, then the red alone remains at τ. This manifestly proves that this light $\epsilon\tau$ becomes red from nothing else other than that it is separated from the rays of the other colors that are reflected by the surface *CBβ*.

72 [= II, 63]. It is more thoroughly shown here which of the solar rays are reflected and which are transmitted, and thus that this does not occur by chance but by a predisposition of the rays.

In a similar manner, if you refract the rays reflected to π with a fourth prism *KLM*, and you observe the colors produced in this way and projected onto a white wall twelve or more feet away, you will see that when the light $\epsilon\tau$ begins to become a greenish yellow, the purple at *p*, which this prism draws out from the light $\epsilon\pi$, will increase more than the other colors, namely, by the entrance of the purple that has then disappeared at *P*. And thereafter the entrance of the other colors at *pt* will occur gradually according as they disappear from *PT*, until when all color has disappeared from *PT*, the colors at *pt* no longer increase. But this difference whereby the violet and blue attain their whole increase at *pt* somewhat sooner than the red and yellow is so small that unless the observer is attentive, he will scarcely notice it.

As we finally set a limit to these things, it is appropriate to bring forth another way by which some kinds of rays intermingled in the sun's light can be partly transmitted while the others are reflected. Specifically (Fig. 29), if you join together two highly polished glass plates *CB* by connecting them to each other along their flat surfaces, and you plunge them into a vessel *RQ* that is full of water, after having first sealed the edges of the juxtaposed surfaces everywhere with wax or pitch, so that water does not creep in and drive out the air that will lie between the glasses in the form of a very thin plate (as has been described)—if, I say, these things are done—you can ar-

73 [= II, 64]. A third way whereby the sun's light can be partly reflected and partly refracted.

then in his draft reply to Hooke (Expt. 2, Add. 3970, ff. 439ᵛ–40ʳ); and once again in the *Opticks* (Bk. I, Pt. I, Prop. II, Expt. X). In "Of Colours" he had not yet added the third and fourth prisms (described in the next article), but he did describe the interference colors (ignored here) which are produced by the air film; see §§27–33, Add. 3975, ff. 5ᵛ–6ʳ (excerpted in Hall, "Further experiments," pp. 31–2), and also the *Opticks*, Bk. II, Pt. I, Observations I–III.

(23) Newton's "a π ad τ [from π to τ]" does not make sense; possible alternatives are: "a *P* ad *T*" (adopted in the translation), "a τ ad π," and "a *PT*."

fiant, possis efficere ut dictorum vitrorum *CB* talis sit situs, ut (illucente Sole)[24] aer interjectus caeruliformes radios reflectat versus π, et transmittat rubriformes versus τ;[25] atque alias omnes apparentias modò recensitas exhibeat.

74 [≈ II, 65]. Notandum Quòd colores hic fiunt a parallelis superficiebus.

Caeterùm dè hisce modis experiendi notandum venit; primò quòd colores hic producuntur a parallelis superficiebus quarum aliquae recurvant radios quantum aliae incurvant, atque adeò quae mutuos effectus destruerent, si-quos / in immutando dispositiones radiorum intrinsecas quoad eorum colores, ut opinantur Philosophi, producerent.[26]   56/

75 [≈ II, 65, 66]. Et quod lux postquam reflexa vel per parallelas superficies trajecta fuit, e coloribus componitur; quam tamen ejusdem naturae cum immediatâ solis luce credimus.

Secundò quòd lux postquam trajicitur per istas parallelas superficies, licèt alba sit, manifestò tamen constet ex dissimilibus[27] radijs, quandoquidem earum aliqua genera penitus reflecti possunt ad π dum alia ad τ partim trajiciuntur. Et eadem ratione constet albedinem reflexam similiter composi-tam esse, siquidem, ut dixi, redintegrata est, cum rubor omnium ultimus reflectitur a τ.[28] | Quis autem dubitaverit unquam quin lux a Sole directè adveniens sit ejusdem naturae cum luce reflexâ, vel per parallelas superficies trajectâ, cùm ijdem cujusque sunt effectus, et eaedem proprietates omnes. Quaelibet in idem quodvis corpus incidens, tingit cum ijsdem coloribus; quaelibet, si per prisma trajiciatur eosdem colores ostendit, et eadem in omnibus perficit. Atqui si nil aliud ostenderam quàm quòd lux reflexa vel trajecta per parallela plana (ut dictum est), albedinémque exhibens, componitur ex radijs diversorum generum, fuisset aliquid prodijsse tenus; dum causa colorum detegeretur, quos illi deinceps efficerent per prismata trajecti. At quis dubitabit quin causa colorum sit eadem sive lux rectà tendat a sole ad Prisma sive priùs reflexa ut a nubibus[29] aut trajecta per parallela plana, ut per vitream fenestram, adveniat; et ideò quòd in quoquo casu incidentes radij sunt mistura dissimilium. Verùm in summam rei certitudinem, super quàm quòd obvium est advertere, ut liqueat[30] nullam simplicis cujusvis coloris lucem, quoad colorem ejus, reflexionibus speculorum variari posse, ostendam posthac quòd istud refractionibus non potest fieri: Atque adeò (praeter rationes sparsim antedictas)[32] cùm inhaerentes dispositiones vel formae radiorum, quibus apti sunt ad exhibendum colorem aliquem destrui nequeant, vel

76 [→ II, 67]. Conclusio, quòd reflectio vel refractio non mutat radiorum dispositiones; neque adeò discrepantiam quoad colores[31] inducit.

---

(24) (illucente Sole)] Added.

(25) Unless the plates are submerged in water, which increases the critical angle at the plates' outer surface, the rays totally reflected at their interface would also be totally reflected at the outer surface and consequently would never emerge.

(26) Newton undoubtedly had Descartes and Hooke specifically in mind. Descartes had laid down as a general principle that no colors are generated by two contrary and equal refractions; see the Introduction, note (23). Although Hooke rejected this as a general principle by showing, for example, that it is invalid for spheres, he did nonetheless grant that it "does hold most true indeed, if the surfaces be plain [i.e., plane], as may be experimented with any kind of prisme where the two refracting surfaces are equally inclin'd to the reflecting" (*Micrographia*, Observation IX, p. 60); see Newton's notes on this passage in Add. 3958, f. 1ᵛ = Newton, *Unpublished Papers*, p. 403. In his draft reply to Hooke in the spring of 1672 Newton specifically identified "Mʳ Hooke himself in his Micrographia" as one who "consents to this opinion" (Add. 3970, f. 439ʳ). Barrow also adopted this principle in his *Lectiones XVIII*, Lect. II, §VII.

(27) Originally: absimilibus.

range it so that the position of those glasses *CB* is such that when illuminated by the sun,[24] the interposed air reflects the blue-making rays toward π and transmits the red-making rays toward τ.[25] And also, it may exhibit all the other appearances just now related.

Moreover, concerning these ways of experimenting it can first be noted that colors are here produced by parallel surfaces, some of which bend the rays back as much as the others bend them in, which would consequently destroy their mutual effects, if—as philosophers believe—they produced any effects by changing the intrinsic dispositions of the rays with respect to their colors.[26]

Second, after the light is transmitted through those parallel surfaces, although it is white, it nevertheless manifestly consists of dissimilar[27] rays, because some sorts of them can be totally reflected to π while others are partly transmitted to τ. By the same argument it is evident that the reflected white is similarly compounded, since, as I said, it is restored when the red is reflected from τ last of all.[28] | But who has ever doubted that light arriving directly from the sun is of the same nature as reflected light or as that transmitted through parallel surfaces, since the effects of each are the same and all their properties are the same? Any one falling upon the same body, whatever it be, colors it with the same colors; any one, if it is transmitted through a prism, shows the same colors and performs the same way in every respect. Yet, if I had showed nothing other than that light, which is reflected or transmitted through parallel planes (as has been described) and exhibits white, is compounded from rays of different kinds, it would have been somewhat of an advance if only the cause of the colors that they produce when they are transmitted successively through prisms were revealed. But who will doubt that the cause of colors is the same whether light travels directly from the sun to a prism or arrives after first having been reflected (as from clouds)[29] or transmitted through parallel planes (as through a glass window), and thus that in every case the incident rays are a mixture of dissimilar rays? But for the greatest certainty of the matter, it is in addition readily observed that clearly[30] no light of any simple color can be changed in its color by reflections from mirrors. I will show afterward that this cannot occur in refractions. Consequently, besides the reasons previously given here and there,[32] since the inherent dispositions or forms of the rays whereby they are disposed to exhibit some color can neither be destroyed nor in any way

74 [≈ II, 65]. It ought to be noted that colors are made here by parallel surfaces.

75 [≈ II, 65, 66]. And that after light has been reflected from or transmitted through parallel surfaces, it is compounded from colors; and we nonetheless believe it is of the same nature as the sun's direct light.

76 [→ II, 67]. Conclusion: Reflection or refraction does not change the rays' dispositions, and thus does not introduce a difference with respect to their colors.[31]

(28) In §92 ≈ II, 89 Newton notes that this experiment also serves to refute modification theories, and in the *Optica* he added a sentence here to that effect.

(29) priùs reflexa ut a nubibus] Originally: speculo plano priùs reflexa (after previously being reflected from a plane mirror).   (30) ut liqueat] Added.

(31) Originally continued: aut gradus refrangibilitatis (or degrees of refrangibility). Newton has for the first time invoked and recognized the essential role of the principle of immutability of color in his theory; and though he tells us in the text that he will demonstrate it afterward ("posthac"), he added it only in the *Optica* as Prop. 2, in §II, 19. This deletion shows that he initially intended to couple the principle of the immutability of color with that of degree of refrangibility, as he would later do in the "New theory," Prop. 3.

(32) Newton has a marginal note here to "Lect——"; presumably the lecture number was omitted because he had not earlier argued that colors are innate to the sun's direct light.

ullo modo virtute reflectionis aut refractionis mutari, quid aliud concludamus quàm quòd dictae dispositiones sunt insitae radijs ab eorum origine, atque / ijs, ut dicam, connatae: licèt non possunt exhibere proprios colores ante- 57/ quam heterogenei ab invicem virtute refractionum secernantur.[33]

Lect. 7
77 [→ II, 81].
Colorum vulgare
phaenomenon
explicatur,

78. Supposito quòd
plures radij per
eandem lineam ex
ordine fluunt:
Eamque vel
perpendicularem
anteriori plano
prismatis;

Hucusque fundamenta struximus, quibus apparentiae vulgares colorum prismatibus effectorum certissimè possunt explicari. Imò praeter alios complures modos eosque magis intricatos, et antehac ignotos, quos ante explicatos habuistis; horum etiam causas tam apertè passim insinuavimus et declaravimus, ut eas non opus esset jam ampliùs attingere nisi propterèa[1] ut methodum propositam retineamus; eas nempe ex principijs antè monstratis scientificè determinandi. Atque has imprimis explicandas apud me proposui noti tam propter dignitatem experimentorum, quàm celebritatem, dum sola ferè hactenus de coloribus per prismata generandis innotuêre. Ea verò sunt duplicia: vel cùm colores projiciuntur in corpus aliquod pone prisma locatum, vel cùm objecta mediante prismate cernuntur tincta coloribus.[2]

Ad enarrandum priorem casum primò [fig 30]: Sit *ABC* prisma, quod lucem solis *OF* per foramen *F* transmissam refringit et refractam projicit in papyrum *MN*. Et sit *OF* una ex rectis in quâ radij successivè adveniunt, eáque ponatur perpendicularis ad primam prismatis superficiem *AC,* incidat

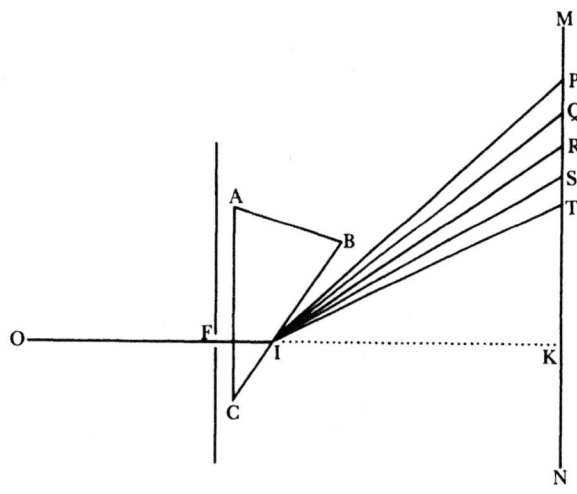

Figure 30

in secundam *CB* ad *I,* et producatur ad *K.* Et cùm omnes radij in ipsâ *OI* advenientes non patiantur eandem refractionem (per Lect 1^mam et 2^dam), ducantur *IP, IQ, IR, IS,* & *IT* pro quibusdam ex ejus refractis, et alij innumeri subintelligantur intermedij. Plures autem radios in eâdem rectâ successivos siquis aegrè concipiat: vice lineae *OF* cogitet exiguum spatium in rebus physicis[3] aequipollens lineae, in quo plures paralleli radij fluant, sed

be changed by the power of reflection or refraction, what else can we conclude but that those dispositions are innate to the rays from their origin and, as I may say, connate to them, although they cannot exhibit their own colors before the heterogeneous rays are separated from one another by the power of refractions.[33]

Thus far we have erected the foundations whereby the common appearances of colors produced by prisms can be most certainly explained. Indeed, besides several different ways—and those rather intricate and previously unknown—that you have had explained earlier, at various places we have also so clearly introduced and demonstrated their causes that it would not now be useful to treat them further except that consequently[1] we may retain the proposed method, namely, to determine them scientifically from principles previously demonstrated. I have resolved to myself to consider explaining these in the first place because of the merit and renown of the experiments, whereas until now almost these alone have been known concerning the generation of colors by prisms. These are in fact twofold: either when colors are projected onto some body located behind the prism, or when objects are perceived through an intervening prism to be tinged with colors.[2]

To explain the former case first (Fig. 30): Let *ABC* be a prism that refracts the sun's light *OF*, transmitted through the hole *F*, and projects the refracted rays onto the paper *MN*. Let *OF* be one of those straight lines along which the rays successively arrive; assume it to be perpendicular to the prism's first surface *AC* and to fall on the second one *CB* at *I*; extend it to *K*. Since all the rays arriving along *OI* do not experience the same refraction (by Lects. 1 and 2), draw *IP, IQ, IR, IS,* and *IT* for some of its refracted rays, and imagine other, innumerable, intermediate ones. If, however, someone has difficulty conceiving of several successive rays in the same straight line, let him imagine in place of the line *OF* a small space (in a physical sense[3] equivalent to the

Lecture 7
77 [→ II, 81]. The common phenomenon of colors is explained.

78. Supposing that many rays flow in succession along the same line which is either perpendicular to the first plane of the prism.

(33) Newton was not again to attempt to prove that colors are innate to the sun's direct light by invoking total reflection, the aim of this lecture, though he would invoke it for unequal refrangibility; see note (8), this lecture. Rather, he chose to pursue the certainty promised by this logically rigorous proof using the principle of color immutability. When reduced to its essentials, Newton's proof is elegant and deceptively simple: Since colors are absolutely immutable, and sunlight exhibits colors after it is refracted, then it necessarily follows that those colors are innate to sunlight before refraction, even though they are not yet apparent. This proof that colors are innate to sunlight, however, ultimately eluded Newton for want of an experimental demonstration of the principle of color immutability for the sun's direct light before it is separated into colors; see Lect. II, 3, note (18), this volume. After he finally abandoned his attempt to prove color immutability for the sun's direct light in the *Opticks*, Newton only half-heartedly attempted to demonstrate that colors are innate to sunlight with an appeal to the similarity between sunlight and white light compounded from colors (as in the *Optica*, §II, 66), and in a late addition of seven words with a last appeal to color immutability (Bk. I, Pt. II, Prop. V, p. 114 = Add. 3970, f. 111ʳ). He did, however, sketch the argument presented here later in the *Opticks*, in Bk. I, Pt. II, Prop. VII, which is more like a "General Scholium" to Book I than a mere proposition.   (1) Originally: eâ de re.

(2) Newton treats objects viewed through a prism only in the *Optica*, Lect. II, 12.

(3) in rebus physicis] Added. On Newton's concept of light ray see Lect. 2, note (23).

indefinitè parùm distantes, ut quoad sensum pro coincidentibus habeantur. Quibus praemissis constat ex Lect 4, 5, et 6 quòd isti radij non concipiendi sunt omninò similes, sed mixtura rubriformium, flaviformium, viridiformium, caeruliformium, et purpuriformium, cum omnibus eorum gradibus intermedijs. / Sed quia colores isti quinque sunt praecipui, unicum cujusque 58/ radium considerabimus. Et quia radius rubriformis est minime omnium refrangibilis (per Lect 3) sit ille *IT,* nempe cujus angulus refractus *KIT* est omnium minimus. Porrò cùm radius flaviformis paulo magis refrangibilis est, (per eandem Lect 3), erit ejus angulus refractus paulo major, puta quòd sit *KIS,* et radius *IS.* Ad eundem modum caeterorum trium radiorum, viridiformis, caeruliformis, et purpuriformis, anguli refracti sunt gradatim adhuc majores (per eandem Lect 3;[)] qui proinde radij magis atque magis divaricabunt ab *IK* puta in *IR, IQ* et *IP.* Liquet ergo quòd heterogenei radij in eâdem rectâ *OFI* advenientes commixti, segregantur per inaequales refractiones et seorsim incidunt in puncta *P, Q, R, S, T.* Et seorsim incidentes quisque pro dispositione suâ, vel formâ diversâ, diversum colorem dabit, rubeum ad *T,* flavum ad *S,* viridem ad *R,* caeruleum ad *Q* et purpureum ad *P:* Idque ex definitionibus ipsorum, quandoquidem rubriformes radios definio qui cùm soli sunt rubeum colorem efficiunt, et flaviformes qui flavum, et sic de caeteris.

79. Vel obliquam utrique.

Praeterea licèt irradiata linea *OF* non sit perpendicularis ad planum *AC* sed utcunque inclinata; tamen flaviformes, et caeteri suo ordine prout sunt magis refrangibiles magis divaricabunt ab incidentibus radijs *IK* quàm rubriformes, (id quod manifestius est quàm ut jam demonstrem). Quare tum etiam heterogenei segregabuntur per divergentiam ab invicem, et seorsim cadentes colores proprios exhibebunt.

80. Vel quòd adveniunt in pluribus lineis, ijsque vel parallelis,

Adhaec si praeter unicam irradiatam lineam *OI* plures alias ei parallelas concipiamus, pari ratione liquebit quod cujusque lineae flaviformes radij, et reliqui suo ordine magis divaricabunt ab *IK* quàm rubriformes: ita quidem ut si foramen *F* sit circulare, pro quinque punctis *P, Q, R, S, T* [fig 30 *bis*][4] tot circuli vel Ellipses πχρστ ponendae sint in quas dictarum quinque specierum radij / incident. Et sic pro innumeris alijs punctis intermedijs totam lineam *PT* constituentibus circuli vel Ellipses aliae ponendae sunt, quae component oblongam imaginem rubeam ad τ, purpuream ad π, et coloribus intermedijs ad intermedia loca tinctam.

P· ·π
Q· ·χ
R· ·ρ
S· ·σ 59/
T· ·τ

Figure 30 *bis*

81. Vel parùm inclinatis.

Quinetiam denique id ipsum fiet si radij non omnes adveniunt paralleli, sed aliquantulùm inclinati, quemadmodum ijs contingit qui manant a diversis partibus solaris disci. De quâ re satis dictum fuit in Lect: 2^{dâ}.

82 [≈ II, 82]. De varijs phaenomeni circumstantijs; Luce juxta basem prismatis terminatâ

Verùm e re futurum judico, ut experimentum hoc vulgare fusiùs prosequar, et singulas ejus circumstantias attingam. Et primò si radij non transeunt per angustum foramen *F,* sed ex unicâ tantùm parte limitantur, colores non omnes apparebunt: | Verbi gratiâ si corpus aliquod opacum *FG* (fig [31]) Soli interponatur et Prismati juxta basem ejus *AB,* quod umbram projiciat in *MP,*

---

(4) Newton neglected to number this figure.

line) in which many parallel but indefinitely close rays flow, so that they may be considered as sensibly coincident. With these premises, it is known from Lects. 4, 5, and 6 that these rays must not at all be conceived of to be similar but a mixture of red-, yellow-, green-, blue-, and purple-making rays together with all their intermediate gradations; but because those five colors are the principal ones, we will consider only a single ray of each one. Because the red-making ray is the least refrangible of all (by Lect. 3), let that be *IT*, namely, the one whose deviation *KIT* is the least of all. Moreover, since the yellow-making ray is slightly more refrangible (by the same Lect. 3), its deviation will be slightly larger; suppose it to be *KIS* and its ray to be *IS*. In the same way the deviations of the remaining three rays, green-, blue-, and purple-making, are gradually still greater (by the same Lect. 3); consequently their rays will diverge more and more from *IK*, for instance, in *IR*, *IQ*, and *IP*. It is therefore evident that the heterogeneous rays arriving mixed together along the same straight line *OFI* are segregated by the unequal refractions and fall separately on the points *P*, *Q*, *R*, *S*, and *T*. And falling separately, each one according to its own disposition or diverse nature will yield a different color: red at *T*, yellow at *S*, green at *R*, blue at *Q*, and purple at *P*. This follows from the definitions of the rays themselves, since I define red-making rays as those which, when they are alone, produce a red color, yellow-making ones as those which produce yellow, and similarly for the others.

Moreover, even if the radiant line *OF* is not perpendicular to the plane *AC* but inclined in any way, nonetheless the yellow-making rays, and the others in their proper order, according as they are more refrangible, will deviate more from the incident rays *IK* than will the red-making ones (which is clearer than I may now demonstrate). Therefore the heterogeneous rays will then also be separated by diverging from one another, and falling separately they will exhibit their own colors. 79. Or oblique to both.

In addition, if besides the single irradiated line *OI* we conceive of many others parallel to it, it will be clear for the same reason that the yellow-making rays of each line, and the others in their proper order, will diverge more from *IK* than will the red-making; so that if in fact the hole *F* were circular, instead of the five points *P*, *Q*, *R*, *S*, and *T* (Fig. 30 *bis*),[4] we must assume just as many circles or ellipses π, χ, ρ, σ, and τ in which the rays of the five specified species fall. Thus instead of those innumerable, intermediate points that form the entire line *PT*, we must assume these circles or ellipses that compose the oblong image, colored red at τ, purple at π, and intermediate colors at intermediate places. 80. Or that they arrive in many lines, either parallel.

Finally, precisely this will occur if all rays do not arrive parallel but somewhat inclined, just as happens with those that flow from different parts of the solar disk; enough was said about this case in Lecture 2. 81. Or slightly inclined.

For future purposes, however, I consider that I should pursue this common experiment in great detail and treat its individual features. In the first place, if the rays do not pass through the narrow hole *F* but are bounded on only one side, not all colors will appear. | For instance (Fig. 31), if some opaque body *FG* is placed between the sun and the prism near its base *AB* and casts a 82 [≈ II, 82]. On various features of the phenomenon: When the light is terminated near the prism's base.

colores efficiat in spatio *PT,* et lucem per-
mittat in ipsum *NT* influere:[(5)] In *PT* confinio lucis et umbrae nulli colores generabuntur praeter purpureum et caeruleum
cum varijs eorum gradibus. Et ratio est
quòd ex radijs omnium formarum qui
transeunt per extremitatem dicti corporis
opaci *FG* soli purpuriformes propter
maximam earum refractionem possunt
ad *P* usque deflecti; unde color purpureus

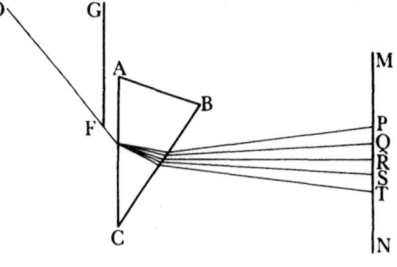

Figure 31

ibi conspicietur. Deinde caeruliformes, cùm paulò minùs refrangibiles existant, incident in totum spatium *NQ,* non potentes ulteriùs versus *M* deflecti
quàm ad *Q.* Atque ita duae radiorum species eaeque solae incident in *Q,* et
colorem ex purpureo et caeruleo compositum exhibebunt. Praeterea viridiformes minùs adhuc refringibiles, in spatio *NR* non ultra extendentur quàm
ad *R;* flaviformes autem terminabuntur in *S.* Quare tres tantùm species colorum miscebuntur ad *R,* et color ex ijs omnibus (nempe ex purpureo caeruleo
et viridi) generabitur. At cùm purpureus et viridis / commixti producant 60/

Lect 4 & 5. caeruleum, ut facilè est ex antedictis experiri; liquet colorem ad *R* non alium
vide etiam Lect[(6)] fore quàm caeruleum. Deníque cùm radij rubriformes minimè omnium refringuntur ut in spatium *NT* incidentes non magis deflectantur versus *M*
quàm ad *T,* liquet quòd in dicto spatio *NT* fiet mistura colorum omnium, et
proinde albescet: sed in ipso *S* (ubi color omnis dempto rubeo miscetur)
caeruleus ad viriditatem nonnihil vergens apparebit, sed maximè dilutus,
propterea quòd solus rubor ex albedinis compositione desit.

83 [= II, 83]. Vel   Porrò si corpus opacum $\phi\gamma$ Soli interponatur et Prismati juxta verticem
juxta verticem ejus ejus *C,* sicut videre est in schemate [32]. Inter obscuratum spatium *NT* et
lucidum *PM* cernes alios duos colores, ru-
beum in *T* et flavum in *R;* idque propter
jam dictas rationes. Quippe radij prout
apti sunt ad hos ordine colores (rubeum,
flavum, viridem, caeruleum, et violaceum)
generandos, extenduntur per spatia *MT,*
MS, MR, MQ et *MP.* Et cùm soli rubriformes extendantur usque ad *T,* caeteris
propter majorem refractionem citiùs terminatis; necesse est ut iste color in *T* sit
rubeus. Item cùm tria radiorum genera in

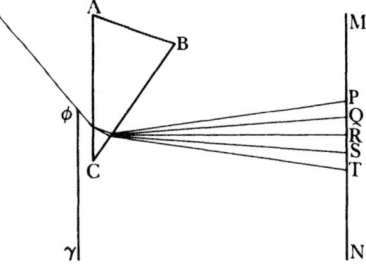

Figure 32

R incidant, color ex istis (nempe rubeo, flavo, et viridi) compositus ibidem
Lect cernetur; rubeus autem et viridis flavum constituunt,[(7)] atque adeò flavus apparebit in *R.* Praeterea cùm omnium formarum radij misceantur in *P,* et postea
perpetuò versus *M;* spatium istud *PM* apparebit album. Nec secus constat
quòd citrius in *S,* et in *Q* flavus[(8)] ad viriditatem vergens apparebit, sed adeò
dilutus tamen, et caeruleo redundans, ut nomen viriditatis non mereatur.

---

(5) Added.

shadow at *MP,* then it produces colors in the space *PT* and lets light flow[5] into *NT.* In *PT,* the boundary of light and shadow, no colors will be generated besides purple and blue together with their various gradations. The reason is that of the rays of every nature that pass by the end of that opaque body *FG,* the purple-making ones alone, because of their greatest refraction, can be deflected as far as *P,* whence the color purple will be seen there. Next, the blue-making rays, since they are somewhat less refrangible and are unable to be deflected farther toward *M* than to *Q,* will fall in the whole space *NQ.* Thus two sorts of rays, and only those, will fall at *Q,* and they will exhibit a color compounded of purple and blue. Moreover, the still less refrangible green-making rays will extend into the space *NR* no farther than *R,* while the yellow-making ones will terminate at *S.* Consequently, only three sorts of colors will be mixed at *R,* and a color will be generated from all of them, that is, from purple, blue, and green. Since, however, purple and green mixed together produce blue, as is easily proved from what was said earlier, it is evident that the color at *R* will be none other than blue. Finally, since the red-making rays are refracted least of all, so that falling in the space *NT* they are deflected no farther toward *M* than to *T,* it is clear that in that space *NT* a mixture of all colors will occur, and consequently it will become white. But at *S* (where every color except red is mixed) blue inclining somewhat to green will appear, but very diluted, since red alone is wanting from the composition of white.

Lects. 4, 5.
See also Lect.[6]

Moreover, if an opaque body *φγ* is placed between the sun and the prism near its vertex *C,* as can be seen in Fig. 32, between the darkened space *NT* and the bright one *PM* you will see two other colors, red at *T* and yellow at *R,* for the reasons already related. Specifically, according as the rays are disposed to generate these colors in order (red, yellow, green, blue, and violet), they extend through the spaces *MT, MS, MR, MQ,* and *MP.* Since only red-making rays extend as far as *T* (the others end sooner because of their greater refraction), it necessarily follows that that color at *T* is red. Likewise, since three kinds of rays fall upon *R,* the color compounded from those (that is, from red, yellow, and green) will be seen at the same place; but red and green make yellow,[7] so that yellow will appear at *R.* Moreover, since rays of every nature are mixed at *P* and thereafter continuously toward *M,* that space *PM* will appear white. No differently it is clear that orange will appear at *S* and yellow[8] inclining to green at *Q,* but yet so diluted and overflowing with blue that it does not deserve the name green.

83 [= II, 83]. Or near its vertex.

Lect.

---

(6) Lects. 4 and 5 specifically treat the composition of white from all colors. In passing, however, Newton relates—but without giving any particulars—in Lect. 4 (§51 = II, 39) that mixtures of various spectral colors can be tested by mixing colored powders, and also in Lect. 5 (§55 = II, 44) that mixtures can be tested by stopping some rays before they are brought to a focus with a prism-lens combination. That violet (Newton's "purple") and green yield blue follows from the new Prop. 4 (§II, 70), on the mixture of neighboring colors, which was added only in revision and presumably accounts for the blank lecture number. The resulting blue, however, should be rather unsaturated or pale; and in §II, 97 he explains why these two colors do not compound blue well.

(7) This yellow should be pale, just as the preceding blue.

(8) Newton in haste first wrote "caeruleus" (blue).

84 [≈ II, 84]. Vel utrinque. (Ubi viriditatis productio bella describitur.)

Tertiò si opaca duo corpora *GF* et *γφ* (fig [33]) Soli et Prismati interponantur, ut radij inter utrumque quasi per oblongam rimam prismati parallelam / transeant; atque distantia *Fφ* sit satis magna: pro utroque termino *F* et *φ* generabuntur colores, purpureus nempe ad *P* et caeruleus ad *R* per terminum *T*;[9] atque flavus ad *ρ*, ac rubeus ad *τ* per terminum *φ*, sicut modò explicatum fuit: Eritque *Tπ* spatium album utrisque coloribus interjectum. Jam si obstacula *GF* et *γφ* ad se invicem paululum admoveantur, ut intermedium spatium *Fφ* evadat angustius, isto pacto spatium album quoque *Tπ* fiet angustius, donec tandem evanescat, et colores utrinque coeant.[10] Sin spatium *Fφ* magis adhuc coarctetur,[11] viriditas in medio colorum emerget vice albedinis, quae jam evanuit. Quae quidem viriditas antea non apparuit propter commisturam radiorum heterogeneorum, quibus involuta latuit: jam verò heterogeneis istis per obstacula duo sibi propiùs admota alternè interceptis; ea paulatim detegitur, patet, et evadit perfectior; donec (cùm dictum *Fφ* satis angustum est) ab omni ferè misturâ liberatur, et eruitur, propriâque specie non minùs quàm caeteri colores elucet.[12] Et hinc in transitu colligitur, quod viriditas inter colores medietatem exactè obtinet, non magis ad rubeum vergens quàm violaceum, neque ad flavum quàm caeruleum.[13] Praeterea[14] observandum est quòd cùm praefata albedo *Tπ* per angustiam spatij *Fφ*[15] incipit evanescere, intermedij colores paulatim fiunt viciniores, flavus videlicet ad rubeum et caeruleus ad violaceum.[16] Ita ut cùm spatium *Fφ* fit valde angustum flavus ad rubeum et caeruleus ad violaceum quasi duplo vicinior evadat, quàm cùm amplitudo dicti *Fφ* permisit albedinem in medio cerni. Et ut quinque colores (viriditate jam internatâ) non occupent plus spatij quàm eorum duo priùs occupavêre. Cujus rei ratio patebit ex figuris tribus praecedentibus, contemplanti modum quo flavus ad *ρ* & caeruleus ad *R* heterogeneis radijs compositus mutatur in flavum ad *S* vel *σ* & caeruleum ad *Q* vel *χ* constantem ex solis homogeneis;[17] caeteris e misturâ[18] per angustiam spatij *Fφ* sublatis.[19]

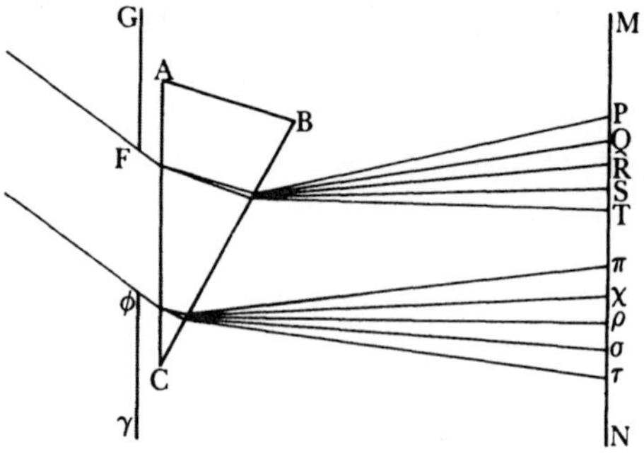

Figure 33

61/

---

(9) Read: *F*.

(10) et . . . coeant.] Originally: et utrique colores tunc facti sint contigui. (and then both colors became contiguous.)

(11) Originally continued: duo venient observanda; quorum primum est quod (two things are to be noted; the first of which is that).

(12) Newton's explanation of the emergence of a monochromatic spectral green from the boundary colors contrasts sharply with the then common view (which Hooke, for example, explained in some detail) that spectral green is compounded from the adjacent yellow and blue; see Lect. 4, note (31).

(13) In the *Optica* Newton elaborates on the intermediate character of green.

(14) Originally: Alterum (the other); see note (11), this lecture.

(15) per . . . *Fφ*] Added.

Third (Fig. 33), if two opaque bodies *GF* and *γφ* are placed between the sun and the prism so that the rays pass between both, as it were, through an oblong slit parallel to the prism, and if the distance *Fφ* is sufficiently large, then colors will be generated by each of the edges *F* and *φ*; namely, purple at *P* and blue at *R* by the edge *T*,[9] and yellow at *ρ* and red at *τ* by the edge *φ*, as was just explained; and there will be a white space *Tπ* lying between both colors. If now the obstacles *GF* and *γφ* are moved slightly toward one another so that the intermediate space *Fφ* becomes narrower, in this way the white space *Tπ* will also become narrower, until it finally vanishes, and the colors on each side coalesce.[10] But if the space *Fφ* is contracted still more,[11] green will emerge in the middle of the colors in place of the white that just vanished. This green in fact did not appear earlier, because of the mixture of heterogeneous rays in which it was enveloped and hidden. But now when these heterogeneous rays are intercepted one after another by the two obstacles brought nearer to each other, the green gradually is uncovered, stands visible, and becomes more perfect, until (when the specified *Fφ* is sufficiently narrow) it is nearly freed and extracted from the whole mixture and shines with its own species no less than the other colors.[12] It is gathered in passing from this that green occupies exactly the middle among the colors, inclining no more to red than to violet, nor to yellow than to blue.[13] Furthermore,[14] it ought to be observed that when that white *Tπ* begins to vanish because of the narrowness of the space *Fφ*,[15] the intermediate colors gradually become closer, namely, the yellow to the red, and the blue to the violet.[16] Thus, when the space *Fφ* becomes very narrow, the yellow turns out nearly two times closer to the red, as does the blue to the violet, than when the size of the specified *Fφ* allowed white to be seen in the middle. Consequently, five colors (green now springing up in between) will occupy no more space than two of them had previously occupied. The reason for this will be clear from the three preceding figures to anyone who considers the way in which the yellow at *ρ* and the blue at *R*, composed of heterogeneous rays, are changed into yellow at *S* or *σ* and blue at *Q* or *χ*, consisting solely of homogeneous rays,[17] with the others having been removed[19] from the mixture[18] by the narrowness of the space *Fφ*.

84 [≈ II, 84]. Or on both sides (where the beautiful production of green is described).

---

(16) intermedij . . . violaceum.] **II**: colores etiam paulatim contractiores apparebunt (the colors will also gradually appear more contracted.)

(17) in . . . homogeneis] **II**: in ferè uniformem flavum ad loca *S* et *σ* incidentem et in fere uniformem caeruleum ad loca *Q* et *χ* similiter incidentem (into a nearly uniform yellow falling at the places *S* and *σ*, and into a nearly uniform blue similarly falling at the places *Q* and *χ*).

(18) e misturâ] Originally: nempe (namely).

(19) **II**: magnâ ex parte sublatis (for the most part removed). In the *Opticks* (Bk. I, Pt. II, Prop. VIII) Newton combined this and the preceding two articles on the relation of boundary and spectral colors into a succinct but lucid explanation of the origin of the colors generated by prisms. The designation of the colors at the edges is somewhat different there, for he derived them from a mixture of seven principal colors (not five as here) according to his new color-mixing circle. Despite the clarity of Newton's explanation in Prop. VIII of the *Opticks*, which also contained his account of directly viewed spectral images (see Lect. II, 12), it was too

85 [= II, 85]. Vel
juxta alterutrum
triangularem
limitem

Quartò si lux terminetur obstaculo *Gγ* cujus extremitas perpendiculariter transversa est ad longitudinem Prismatis, colores omninò nulli virtute termini illius genera/buntur. Etenim ponamus parallelos radios *OF Oφ* caeterosque 62/ (in fig [34]) juxta extremitatem dictam *Gγ*[20] in Prisma *ABC* prolapsos,

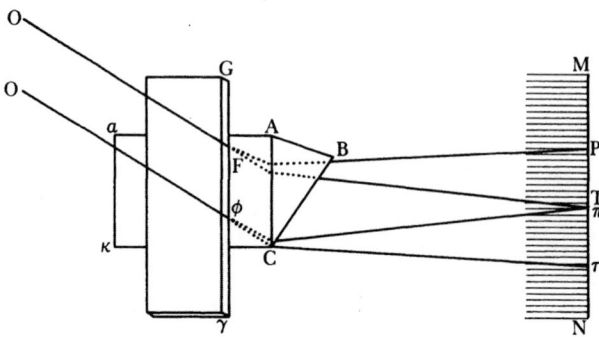

Figure 34

ibidemque refractos esse ad *PT* et *ππ*; atque *MN* esse umbram ipsius *Gγ*. Jam licèt radij purpuriformes *FP* et *φπ* magis refringantur quàm rubriformes *FT* et *φτ*, tamen istâ refractione secundum terminum umbrae factâ, ita ut ex dictis radijs nulli magis deflectant versus umbram quàm caeteri; palam est quòd ubicunque purpuriformes incidunt, rubriformes etiam incident in eundem locum: et e contra. Quod idem de radijs intermedijs pari modo concipiatur. Et sic radijs omnium specierum ubique per extremitatem umbrae commixtis, umbra benè definietur sine aliquo colore (praeter album vel fuscum ex luce et umbrâ mixtis) conspecto. Sed cavendum est, ne colores per limites prismatis *Aα* vel *Cκ* generati habeantur pro generatis a limite *Gγ*. Quamobrem de prismatibus monendum volo, quòd quae ex vitro in totum fiunt, ad examen hujus et proxime praecedentis commodè instituendum sunt nimis exigua, propterea quòd colores per extremitatem verticis et basis producti interjectum spatium album haud relinquent satis amplum in quo generatio colorum praedictis modis probetur.[21] Itaque ut prisma conficiatur ex vitris planis et benè politis, qualia ad specula conficienda adhibentur, moneo; quibus in morem cunei connexis et in vasculum dein prismiforme completis, (ut supra dictum,)[22] vasculum istud impleatur aquâ limpidissimâ et occludatur; Et sic prismata ad arbitrium ampla conficias.

86 [= II, 86]. Vel
undique. (Ubi
rursus de praefatâ
viriditate.)

Quintò, ut omnia jam uno comprehendam specimine, sit *Gγ* (fig [35])[23] corpus opacum orbiculari foramine *Fφ* unum duosve digitos lato pertusum, per quod lux in prisma trajiciatur, ubi cùm refracta fuerit, projicitur[24] deinde in papyrum / vel quodvis album corpus *MN* quasi semisse pedis a prismate 63/ postpositum, et videbis illuminatum spatium *PYTZ* rotundum ad modum foraminis *Fφ*, album in ejus medietate, et duobus semilunulis colorum terminatum, purpureo et caeruleo ad *P*, flavo autem et rubeo ad *T*: qui colores paulatim deficiunt versus *Y* et *Z* ubi nulli omninò conspiciuntur. Praeterea si papyrum ad majorem distantiam paulatim distuleris, velut ad *μν*; videbis

Fourth, if the light is terminated by an obstacle $G\gamma$ whose edge is transverse and perpendicular to the prism's length, no colors at all will be generated by virtue of that edge. For, let us assume (in Fig. 34) that the parallel rays OF, $O\phi$, and others[20] flow alongside that edge $G\gamma$ into the prism ABC and are there refracted to PT and $\pi\tau$, and that MN is the shadow of $G\gamma$. Now, although the purple-making rays FP and $\phi\pi$ are refracted more than the red-making ones FT and $\phi\tau$, nevertheless, since that refraction is made along the boundary of the shadow so that none of those rays are more deflected toward the shadow than the others, it is clear that wherever purple-making rays fall, red-making ones will also fall at the same place, and conversely. In the like way, let the same thing be understood for the intermediate rays. Thus with rays of every sort mixed together everywhere along the edge of the shadow, the shadow will be well defined without any visible color (except white or gray from a mixture of light and shadow). But you must be careful that the colors generated by the prism's edges $A\alpha$ and $C\kappa$ are not taken for those generated by the edge $G\gamma$. Consequently, I wish to advise that prisms made entirely from glass are too small for suitably carrying out an examination of this and the immediately preceding, because the colors produced by the edge of the vertex and of the base will not leave a sufficiently wide interposed white space in which the generation of colors may be tested[21] in the specified ways. I therefore recommend that a prism be made from flat, well-polished glasses such as those used to make mirrors. After joining them in the shape of a wedge and then finishing it into a small prismatic vessel (as described above),[22] fill that small vessel with very clear water and seal it. In this way you can make prisms as large as you wish.

Fifth, that I may now include everything in one example (Fig. 35),[23] let $G\gamma$ be an opaque body perforated with a one- or two-inch wide circular hole $F\phi$ through which light is cast onto a prism, where after being refracted it is then projected[24] onto a paper or any white body MN placed about half a foot behind the prism. You will see the illuminated space $PYTZ$ to be round like the hole $F\phi$, white in the middle, and bounded by two crescents of colors, with purple and blue at P, but yellow and red at T; and these colors gradually decay toward Y and Z, where none at all are visible. Moreover, if you gradually move the paper away to a greater distance, for example, to $\mu\nu$, you

85 [= II, 85]. Or near either triangular end.

86 [= II, 86]. Or on all sides (and once again about that green).

---

brief. Newton perceptively remarked in the new introduction to this lecture (§II, 81 → 77) that although an understanding of the formation of prismatic images follows directly from his theory of color, it could cause difficulty to "the careless and those burdened by prejudice." In the eighteenth century, and even in the nineteenth, the phenomenon of boundary colors was frequently invoked to refute Newton's theory of color and to show that colors are produced by the boundary of light and shadow. Johann Wolfgang von Goethe was the most prominent of Newton's critics to invoke such an argument; see Lect. II, 12, note (10). For an interesting, modern account of boundary colors see P. J. Bouma, *Physical Aspects of Colour*, 2nd Eng. ed. (New York, 1971), pp. 109–17, 173–7.

(20) Originally continued: terminatos (terminated).
(21) Originally: experiatur.　　(22) In §44 = II, 31.
(23) In the *Optica*, Fig. II, 33, Newton added the rays casting the image.
(24) Originally: trajicitur.

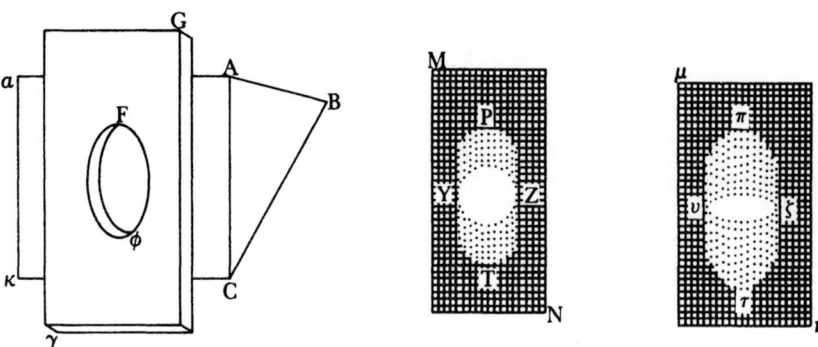

Figure 35

colores distendi et augeri, et intermediam albedinem usque comminui dum prorsus evanescat totumque spatium coloribus rubeo flavo caeruleo et purpureo tinctum appareat. Et papyrum longiùs differendo, viriditas e medio emerget et crescet tum amplitudine spatij tum perfectione speciei: Totúmque spatium coloratum distrahetur in oblongam formam. Quorum omnium rationes ex supradictis depromantur.

87 [≈ II, 87]. Vel ad distantiam aliquam a prismate.

    Adhaec si corpus opacum quo lux terminata est, collocetur a posticâ parte prismatis, colores eodem planè modo producentur ac priùs; nec quicquam refert quanta intersit prismati et corpori terminanti distantia. Verbi gratiâ [fig 36] si corpus opacum *Gγ* perforatum in *F* absit a prismate *ABC* ad distantiam pedis unius aut ampliùs; et prisma istud sit satis amplum (videlicet ex vitreis laminis, ut dictum est, confectum) | nè lux priùs discernatur in colores quàm permeet dictum foramen *F*: ista lux alba postquam transijt per ipsum *F*, non secus degenerabit[25] in colores apud *P, Q, R, S, T* quàm factum erat in praecedentibus. Scilicet ex solâ schematis contemplatione patebit modus quo radij diversorum generum inaequaliter refracti convergunt a diversis partibus prismatis ad foramen *F*, ubi (ut et hinc inde versus *G* et *γ*) componunt albedinem; sed inibi decussantes divergunt postea, diversique colores in diversa spatia *P, Q, R, S, T* tendunt. Atque haec fortè clariora fient experienti quòd cùm radij obstaculo quolibet *H* ex utravis parte prismatis intercipiantur, e coloribus *P Q R S T* aliqui tollentur. Si radios nempe vertici *C* vicinos intercipias tolles purpureum *P*, vel tolles rubeum *T*, si intercipias / eos basi *AB* vicinos, et sic de reliquis. Ita ut quoslibet pro arbitrio tuo possis tollere vel facere ut quilibet solus, appareat.[26]   64/

88 [= II, 88]. Vel alio quovis modo.

    Denique si lux ex unicâ tantùm parte pone prisma limitetur, vel si duo statuantur limites, ijque vel ad easdem vel ad oppositas partes prismatis; vel

---

(25) II: *convertetur* (it will be converted).

(26) Newton invoked this experiment in his draft reply to Hooke in 1672 to show "that the production of the Prismatic colours hath no dependance on the termination of light wth shaddow" (Expt. 4, Add. 3970, f. 440^v), a point he takes up when he returns to this experiment in §92 ≈ II, 89. In the *Opticks* (Bk. I, Pt. II, Prop. I, Expt. I), where the experiment supports the same argument, he modified the arrangement to attain a much purer spectrum by first passing

will see the colors expand and increase and the intermediate white diminish, until it vanishes completely, and the entire space appears colored with the colors, red, yellow, blue, and purple. By moving the paper farther away, green will emerge from the middle and increase in both the size of its space and the perfection of its species, and the entire colored space will be stretched out into an oblong shape. The reasons for all this may be derived from what has already been described.

Furthermore, if the opaque body that terminates the light is placed behind the prism, then colors will be produced in entirely the same way as before; nor does it at all matter how great the distance between the prism and the terminating body is. For example (Fig. 36), if the opaque body *G*γ perforated

87 [≈ II, 87]. Or at any distance from the prism.

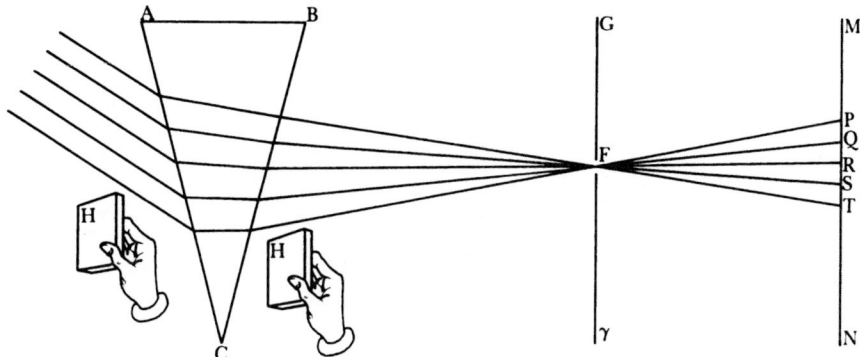

Figure 36

at *F* is at a distance of one or more feet from the prism *ABC*, and if that prism is sufficiently large (that is, one made from glass plates, as was described) | lest the light be separated into colors before it goes through that hole *F*, after that white light passes through *F*, it will degenerate[25] into colors at *P*, *Q*, *R*, *S*, and *T* just as occurred in the preceding. Namely, solely from studying the figure it will be clear how the unequally refracted rays of different kinds converge from different parts of the prism to the hole *F*, where (as well as on both sides toward *G* and γ) they compound white; but upon crossing there, they afterward diverge, and the different colors tend to different spaces *P*, *Q*, *R*, *S*, and *T*. These things will perhaps become clearer to someone doing the experiment, because when the rays are intercepted by some obstacle *H* on either side of the prism, some of the colors *P*, *Q*, *R*, *S*, and *T* will be removed. Specifically, if you intercept the rays near the vertex *C*, you will remove the purple *P*; or you will remove the red *T* if you intercept those rays near the base *AB*; and similarly for the rest. So that you can, as you wish, remove any or cause any one alone to appear.[26]

Finally, if the light is bounded on only one side behind the prism, or if two boundaries are erected either on the same or on opposite sides of the prism,

88 [= II, 88]. Or in any other way.

the incident sunlight through a distant narrow aperture and then by making the second aperture still narrower.

quocunque alio more lux terminetur; modus quo colores exinde generantur ex antedictis facilè patebit; ut jacturam temporis fecero de hâc re plura verba facturus. Quinetiam si duo vel plura prismata quocunque modo inter se disponantur, peritus Optices facilè explorabit causam.

Lect 8
89 [≈ II, 90].
E praefatis modus
deducitur albedinem
e coloribus
componendi

Sed antequam haec penitus dimitto placet insuper annotare sequentia duo: quorum primum esto modus, quo colores duorum prismatum ita commisceantur ut albedinem componant. De qua re perficienda cùm plures antehac modos ostendi, hujus non omninò memini; quippe objectioni, quòd colores non miscentur, sed ex confinio lucis sublato pereunt, et in albedinem evanescunt, in hoc minùs commode tunc satisfecisse potuissem. Sunto [fig 37] *ABC*

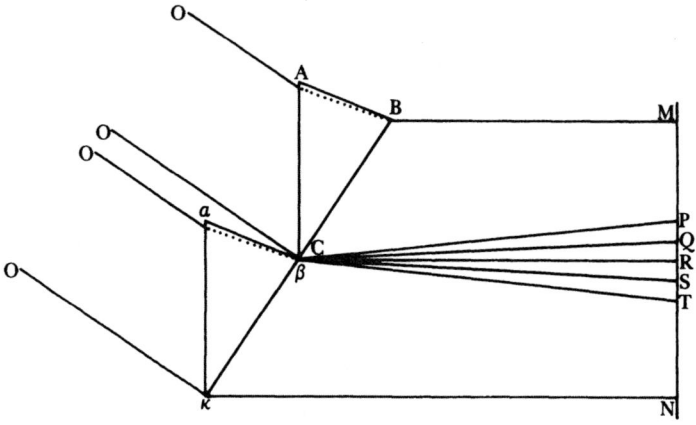

Figure 37

et *αβκ* duo prismata cum angulis verticalibus *ACB* et *ακβ* aequalibus, et ea disponantur in situ parallelo ita | ut alterius linea verticalis *C* cum *β* extremitate basis alterius conveniat, planis *BC* et *βκ* in directum jacentibus. Quo facto si sol transluceat ea in papyrum *MN*, octo vel duodecim digitos postpositam, colores quidem generabuntur ad *M* et *N* per exteriores prismatum terminos *B* et *κ*, non autem per interiores *C* et *β*, sed medium spatium *PT* totum apparebit album. Sin alterutrum prisma tollas alterius extremitas *C* vel *β* generabit colores ad *PT*; ac dein si restituas, albedo[1] etiam restituetur. Scilicet albedo ista componitur[2] e coloribus ab extremitate *C* et *β* prismatis utriusque prolapsis. Id quod facilè constet ex praefatis. Nam radij purpuriformes ab utroque / prismate refracti limitantur in eodem puncto *P*; ita ut ab 65/ uno prismate manantes incidant in *PM*; ab altero, in *PN*; et ab utroque simul in totum *MN*, non secus quàm si omnes ab unico prismate venissent. Eodem modo caeruliformes extenduntur per totum spatium *MN*: et eorum terminus communis est *Q* prout manant a diversis prismatibus. Et sic de caeteris. Quare omnigeni radij commiscentur in unaquâque parte spatij *PT*, et albedinem ideò component. Sin alterutrum prisma tollas, puta *ABC*, vel lucem ei potiùs occludas; tum radijs rubriformibus ab *MT*, flaviformibus ab *MS*, vi-

or the light is terminated in any other way whatsoever, the way in which colors are consequently generated will readily be evident from the preceding, so that I would waste time in discussing this further. Moreover, if two or more prisms are arranged among themselves in any way whatsoever, the expert optician will easily investigate the cause.

But before I completely set these things aside, it is appropriate to observe in addition the following two things, the first of which is how the colors of two prisms may be so mixed that they compound whiteness. I did not at all remember this method of accomplishing it when I previously demonstrated many ways, for in fact I could not then satisfactorily answer the objection that the colors are not mixed but, because of the elimination of the boundary of the light, are destroyed and vanish into whiteness. Let (Fig. 37) ABC and $\alpha\beta\kappa$ be two prisms with equal vertex angles ACB and $\alpha\kappa\beta$, and place them in a parallel position | so that the vertex line C of the former meets the edge of the base β of the latter with their planes BC and $\beta\kappa$ lying in a straight line. After this is done, if the sun shines through them onto the paper MN, placed eight or twelve inches behind, colors will indeed be generated at M and N by the prisms' outer edges, B and κ, but not by their inner ones, C and β, while the entire middle space PT will appear white. But if you remove either prism, the other's edge, C or β, will generate colors at PT; and then if you restore it, the white[1] will also be restored. That white is of course compounded[2] from the colors proceeding from the edges C and β of each prism. This may be readily ascertained from the preceding: For the purple-making rays refracted by each prism terminate at the same point P, so that those flowing from the one prism fall on PM, those from the other on PN, and those from both together on all of MN, just as if they had all come from a single prism. In the same way the blue-making rays extend through the entire space MN and their common terminus is Q, according as they flow from the different prisms, and similarly for the others. Consequently, rays of every kind are mixed together in every single part of the space PT, and they will thus compound white. But if you remove either prism, say ABC, or rather if you block out its light, then after the red-making rays have been removed from MT, the yellow-making from MS, the green-making from MR, the blue-

Lecture 8
89 [≈ II, 90]. From the preceding a way to compound whiteness from colors is deduced.

---

(1) Newton first mistakenly wrote "colores."
(2) Originally: *immiscetur* (mixed).

ridiformibus ab *MR*, caeruliformibus ab *MQ*, et purpuriformibus ab *MP* sublatis, manebunt rubriformes in *NT*, flaviformes in *NS*, viridiformes in *NR*, caeruliformes in *NQ*, et purpuriformes in *NP*. Adeóque purpureus apparebit in *P*, et caeruleus in *R*, ut ostendimus ante. Et simili ratione si lux occludatur alteri prismati *αβκ*, ne permeet; rubor apparebit in *T* et flavedo in *R*.[3]

Vide fig 31 et 32.

90 [= II, 91]. Utrum anguli prismatum sint aequales cognoscere. In istis autem experiendis requiritur ut anguli *ACB* et *ακβ* sint aequales. Id quod tentabis[4] si prismata secundum longitudinem eorum ita connectas ut duo ex planis dictos angulos comprehendentibus puta *BC* et *βκ* [fig 38] fiant contigua et reliqua duo *AC* et *ακ* sibi opposita. Quo facto si radij Solis ingressi foramen *F*, pergant ad eundem locum *S* cùm trajiciuntur per dicta prismata perpendiculariter ad eorum latera *AC* et *ακ*, atque cùm liberè progrediuntur, nullo interjecto obstaculo:[5] tum plana *AC* et *ακ* sunt parallela, et anguli *ACB* et *ακβ* aequales. Sin istud non eveniat, sunt inaequales: in quo casu notetur praeterea,

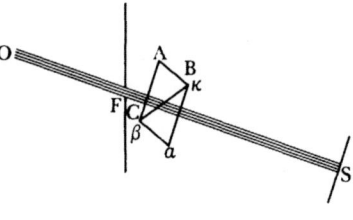

Figure 38

quòd inclinando plana *BC* et *βκ* (in fig [37]) vel ab invicem reclinando, possis albedinem in *PT* haud secus componere ac si dicti anguli fuissent aequales et plana *BC* et *βκ* in directum jacentia.

91 [= II, 92]. Alius modus commiscendi colores in albedinem, priori affinis. Quinetiam possis hoc idem cum unico tantùm prismate perficere, dummodò satis magnum sit, puta [fig 39] cujus refringentia latera *AC* et *BC* sint

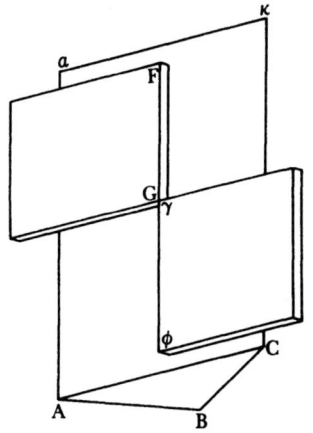

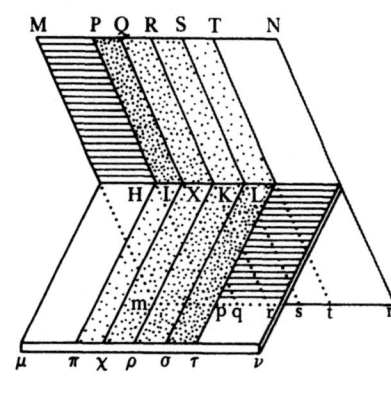

Figure 39

sex vel octo digitos lata. Etenim sint *FG* et *φγ* duo corpora opaca, plana, rectangula, et ad prismatis planum *ACκα* secundum / planitiem ejus sic applicata ut eorum angularia puncta *G* et *γ* juxta plani istius centrum se mutuò contingant[6] et latera concurrentia (quorum *FG* et *φγ* sint ad axem Prismatis parallela) ex adverso jaceant in directum.[7] Quo facto si lux refracta projiciatur in papyrum *MNX* pedes quasi duos distantem; obstaculum *FG* projiciet umbram in *MH*, purpuram efficiet in *PHIQ*, ac caeruleum colorem in *QILT*, 66/

making from *MQ*, and the purple-making from *MP*, the red-making ones will remain in *NT*, the yellow-making in *NS*, the green-making in *NR*, the blue-making in *NQ*, and the purple-making in *NP*. Hence, purple will appear at *P* and blue at *R*, as we showed before. By similar reasoning, if the light is blocked out at the other prism *αβκ* so that it does not pass through, red will appear at *T* and yellow at *R*.[3]

See Figs. 31, 32.

In doing these experiments, however, it is required that angles *ACB* and *ακβ* be equal. You will test this[4] if you join the prisms along their lengths so that two of the planes, suppose *BC* and *βκ* (Fig. 38), that contain those angles become contiguous, and the other two, *AC* and *ακ*, become opposite to one another. After this has been done, if the sun's rays entering the hole *F* continue to the same place *S* when they are transmitted through those prisms perpendicularly to their sides *AC* and *ακ*, as when they proceed freely with no interposed obstacle,[5] then the planes *AC* and *ακ* are parallel, and the angles *ACB* and *ακβ* are equal. If that does not happen, they are unequal. In this case it is to be observed, moreover, that by tilting the planes *BC* and *βκ* (Fig. 37) toward or away from one another, you may compound whiteness at *PT* just as if the specified angles were equal, and the planes *BC* and *βκ* lay in a straight line.

90 [= II, 91]. To know whether the prisms' angles are equal.

Moreover, you may also accomplish this with only a single prism, provided that it is sufficiently large, for example (Fig. 39) one whose refracting sides *AC* and *BC* are six or eight inches wide. For, let *FG* and *φγ* be two plane, rectangular, opaque bodies attached to the prism's plane *ACκα* along its surface so that their angular points *G* and *γ* touch each other near the center of that plane[6] and the concurrent sides (of which *FG* and *φγ* are parallel to the prism's axis) lie opposite one another in a straight line.[7] After this is done, if the refracted light is projected onto the paper *MNX* about two feet away, the obstacle *FG* will cast a shadow on *MH* and will make purple at *PHIQ* and a blue color at *QILT*, while it will let light pass onto *LN*. Con-

91 [= II, 92]. Another way, related to the previous one, of mixing colors into whiteness.

---

(3) In his draft letter to Hooke (Expt. 8, Add. 3970, ff. 441ʳ–2ʳ) Newton described this experiment in considerable detail and incorporated some of the material on boundary colors from the preceding articles; he also included it in the *Opticks*, Bk. I, Pt. II, Prop. V, Expt. XIII.

(4) requiritur . . . tentabis] Originally: quoniam requiritur ut anguli *ACB* et *ακβ* sint aequales, notetur quod examen aequalitatis facilimè fiet (since it is required that the angles *ACB* and *ακβ* be equal, it may be noted that a test of the equality will be most easily made).

(5) nullo interjecto obstaculo] Originally: nullâ re interjecta (with nothing interposed).

(6) juxta . . . contingant] Originally: contingant sese circa centrum dicti plani *ACκα*.

(7) (quorum . . . directum.] Originally: jaceant in directum, et eorum duo praeterea, ut *FG* et *γφ* sint parallela ad axem prismatis. (lie in a straight line and, in addition, two of them, as *FG* and *γφ*, are parallel to the prism's axis.)

et permittet lucem in *LN.* E contra verò obstaculum γφ permittet lucem in *Hm,* rubedinem efficiet in *pHIq,* ac flavedinem in *qILt,* et projiciet umbram in *Ln.* Dico jam si speculo aliquo μνX colores ex alterutrâ parte lineae *HL,* ut *HLpt,* ita reflectantur ut incidant in papyrum ad eundem locum cum coloribus *HLPT* ex alterâ parte: color omnis evanescet totumque *HLTP* apparebit album. Nam purpuriformes radij a prismate ad *PHIQ* directè tendunt, et caetera quatuor radiorum genera ad eundem locum reflectuntur a speculo, incidentes puta in *HIχπ:* Item purpuriformes & caeruliformes directè tendunt ad *QIXR* et caetera tria genera illuc reflectuntur ab *IXρχ:* Et sic de reliquis. Adeò ut omnes omnium generum radij passim per spatium *PHLT* misceantur, ibidemque componant albedinem.[8] Sed notandum est[9] quòd cùm lux reflectione semper debilitetur, radijs quamplurimis inter reflectendum amissis; exinde forsan eveniat quòd lux directa nonnihil praevalebit reflexae et color ejus dominabitur, nisi compensatio fiat ita papyrum inclinando ut directa lux paulo obliquiùs in eam incidat quàm reflexa: de quâ re facilè judicium feras ex perfectione albedinis emergentis.

92 [≈ II, 89]. Adversus philosophorum Hypotheses notae.

Alterum quod notandum venit,[10] est de modo tollendi quoslibet colores in fig 36 per interpositionem corporis *H,* quantùm nempe ista circumstantia adversatur hypothesibus Philosophorum, quae de coloribus hucusque fuerunt excogitatae. Ex illis enim positis, refracta lux ad eas / semper partes cum 67/ caeruleo et violaceo terminanda est, versus quas fit refractio; quandoquidem gyrationes globulorum ex opinione Cartesij, vel partes anteriores pulsuum Aetheris obliquè vibrantis ex aliâ quadam[11] Hypothesi, per viciniam quiescentis Medij ad eas semper partes impediuntur et hebescunt. Attamen ostensum est ad fig 36 quòd, obstaculo *H* ex utrâvis parte prismatis[12] interjecto ut radios ipsius vertici *C* vicinos intercipiat, possis violaceum et caeruleum tollere et efficere ut viridis vel flavus aut etiam ruber ad eas partes maneat extimus versus quas refractio peragitur. | Nec Hypothesis eorum tutior est, qui ponunt colores ex luce et umbrâ mistis componi; cum eadem videatur mistura quaecunque licèt color sit extremus ad easdem[13] partes *M* vel *N.*

Numb 71 & 73.

Hujusmodi etiam Hypotheses ex alijs experimentis sparsim occurrentibus everti possent, modò id meo proposito necessarium judicarem: quemadmodum ex illis ubi lucem partim reflecti posse, et partim transmitti docebam; nam lux transmissa dabat flavum vel rubeum, nec tamen ab ullo[14] quiescente Medio, vel tenebris terminabatur.[15]

---

(8) This and the preceding way of compounding white depend on the fact that the boundary colors on opposite sides are complementary.

(9) notandum est] Originally: notandum volo (I would note).

(10) Newton first wrote "volui" and then "est."

(11) aliâ quadam] II: Mʳⁱ Hookij (Mr. Hooke's). For Descartes's and Hooke's views, see Lect. 4, note (14).

(12) ex . . . prismatis] Omitted in II.

(13) For "easdem" one should perhaps read "diversas [different]," though the argument still is obscure. In revision Newton partly clarified this argument, and he was still more successful in his draft reply to Hooke; see Lect. 7, note (26), this volume. However, it was only in the *Opticks* (Bk. I, Pt. II, Prop. I, Expt. I) that he most fully and clearly explained this experiment: "any one of the colours as well as violet may become outmost in the confine of yᵉ shadow towards [M], &

versely, the obstacle $\gamma\phi$ will let light pass onto *Hm* and will make red at *pHIq* and yellow at *qILt*, while it will cast a shadow on *Ln*. I now say that if with some mirror $\mu\nu X$ the colors from either side of the line *HL*, such as *HLpt,* are so reflected that they fall on the paper at the same place as the colors *HLPT* from the other side, then all color will vanish, and the whole of *HLTP* will appear white. For the purple-making rays travel directly from the prism to *PHIQ*, and the other four kinds of rays are reflected by the mirror to the same place, falling, for instance, on *HI$\chi\pi$*. Likewise, the purple- and blue-making rays travel directly to *QIXR*, and the other three kinds are reflected there from *IX$\rho\chi$*, and similarly for the others. Hence all rays of every kind are mixed everywhere throughout the space *PHLT* and compose whiteness there.[8] It must be noted,[9] however, that since light is always weakened by reflection, with a great number of rays lost in the course of reflection, it may perhaps then turn out that the direct light will be somewhat stronger than the reflected light and its color will dominate, unless one compensates by inclining the paper so that the direct light falls on it slightly more obliquely than the reflected light. This you may easily judge from the perfection of the emergent whiteness.

There is another thing that can be noted about the way (in Fig. 36) of removing any colors by inserting a body *H,* namely, how much that circumstance is opposed to the hypotheses of the philosophers that have hitherto been devised about colors. For from those assumptions refracted light must always be terminated with blue and violet on those sides toward which the refraction occurs, since the rotations of the little balls (in Descartes's opinion) or the foremost parts of the pulses of the obliquely vibrating aether (in a certain other[11] hypothesis) are always impeded and weakened by the neighboring quiescent medium on those sides. Nevertheless, it has been shown at Fig. 36 that when the obstacle *H* is inserted on either side of the prism[12] so that it intercepts the rays near the prism's vertex *C,* you can remove the violet and blue and make green or yellow or even red remain outermost on those sides toward which the refraction is executed. | Nor is the hypothesis of those who assume colors to be compounded from light and shadow mixed more secure, since the same mixture whatever appears although a color is outermost on the same[13] sides *M* or *N.*

Hypotheses of this kind could also be overthrown from other experiments encountered here and there, if only I judged it necessary for my purpose: for instance, from those where I taught that light could be partly reflected and partly transmitted, for the transmitted light displayed yellow or red, and yet[14] it was not bounded by any quiescent medium or darkness.[15]

92 [≈ II, 89]. Notes opposed to the hypotheses of the philosophers.

§§71, 73

---

any one of them as well as red may become outmost in the confine of the shadow towards [N] & any one of them may also border upon the shadow made within the colours by the obstacle [H] intercepting some intermediate part of yᵉ light, & lastly any one of them by being left alone may border upon the shadow on either hand" (Add. 3970, f. 91ʳ = *Opticks*, p. 82). See also §§45–6 = II, 32–3.

(14) nec . . . ullo] **II:** idque in meditullio ejus ubi a nullo (especially in its middle, where. . .).

(15) In the *Optica* Newton here added a brief appeal to the principle of color immutability.

93 [≈ II, 89, 77].
Instrumentum
desc.ibitur quocum
omnia de coloribus
hactenus tradita
dilucidissimè
probentur.

Caeterùm non opus est ut Hypotheses ejusmodi refutem, quae, ex inventâ tandem veritate, suâ sponte corruent.[16] | Satiùs fecero, si proferam experimentum tandem, quo omnia qua de genesi colorum hactenus explicui, non modò probari possunt, sed etiam videri. | Quamobrem [fig 40] sit *ABCac* prisma quod radios per foramen *F* in obscuratum cubiculum transmissos refringat versus lentem *MN* ut colores quos efficit in *p, q, r, s, t,* per lentem deinde trajiciantur ad *X* et ibidem commisceantur in albedinem, sicut in praecedentibus ostendi. Deinde aliud prisma *DEGgd* priori parallelum ad locum *X*, ubi albedo redintegrata est, statuatur, quod lucem versus *Y* refringat. Hujus autem prismatis verticalis angulus *Gg* sit aequalis angulo verticali *Cc* prismatis anterioris, aut eo / forte minor, et similiter positus ut incidentes radios in parallelismum reducat[17] quos prisma anterius dispersit. His positis observabis an lux ad *Y* (pedes aliquot distans) trajecta, aeque alba maneat ac fuerit in *X*, vel sensim abeat in colores. Si penitus appareat alba, tunc prismata cum lente debitè disposuisti: sin aliqui colores ad *Y* cernantur prisma *DEG* circa suum axem eo modo parùm converti debet ut colores minuantur; et cùm penitus evanuêre et lux in totum albescit, siste prisma. Quod si nequeas hoc modo efficere quin lux inter transiendum ab *X* ad *Y* ex aliquâ sui parte transmigret in colores, lentem *MN* paulo longiùs a prismate *ABC* transfer, et loco *X* rursus invento ubi colores in albedinem accuratissimè convergunt, in eo statue prisma *DEF* ut priùs, et rursus experire an possis lucem sine coloribus ad *Y* projicere. Et cùm eo usque mutaveris positiones prismatum et lentis dum effeceris lucem ad *Y* trajectam quàm minimè possis coloratam, prismata cum lente in eo situ figantur idque vel ope trabis, ut in schemate describitur, vel tubi aut instrumenti cujusvis in eum finem fabricati.[18]

94 [= II, 78]. Ejus
usus describitur.

Cùm habeas hanc machinam e prismatibus et lente ut dictum est compositam, ope lucis per eam transmissae cuncta possis experiri quae hactenus fuerunt tradita. Haec enim lux *XY* jubari a Sole directo persimilis est, et easdem omnes apparentias exhibet, ac si a foramine *F* rectà promanasset, nullam omninò refractionem passa; Adeoque ejusdem esse constitutionis facilè credamus. Et tamen cùm in sua principia componentia, hoc est in radios diversorum generum, apud lentem *MN* discreta fuerit, facile erit modos examini subjicere quibus posthac in colores converti potest, idque tantùm sistendo hoc vel illud radiorum genus apud *MN*, ut constitutio lucis *XY* quoad ejus conversionem in colores pateat.

95 [≈ II, 79]. Et
illustratur exemplis

Quemadmodum si desideretur ut sensui planissimè pateat quòd prisma convertit lucem in colores non transmutando proprietates ejus intrinsecas, sed segregando tantum radios ad excitandum varia colorum phantasmata dispositos, ex quibus lux omnis albens constituitur: nihil aliud agendum est quam ut prisma aliquod *HIK* ita statuatur ut lucem / *XY* excipiat, et refringendo

68/

69/

---

(16) This sentence was shifted in the *Optica* to the end of the preceding article, §92 ≈ II, 89.

(17) similiter . . . reducat] Originally: ad similes partes positus, ut radios reducat (placed in similar directions so that it may restore the rays).

(18) On two later occasions, in his draft response to Hooke (Expt. 10, Add. 3970, f. 444), and in the *Opticks* (Bk. I, Pt. II, Prop. XI), Newton also concluded his presentation of his theory of color with a description of this instrument despite the serious limitations he noted in the *Optica*, §II, 80 ≈ 96.

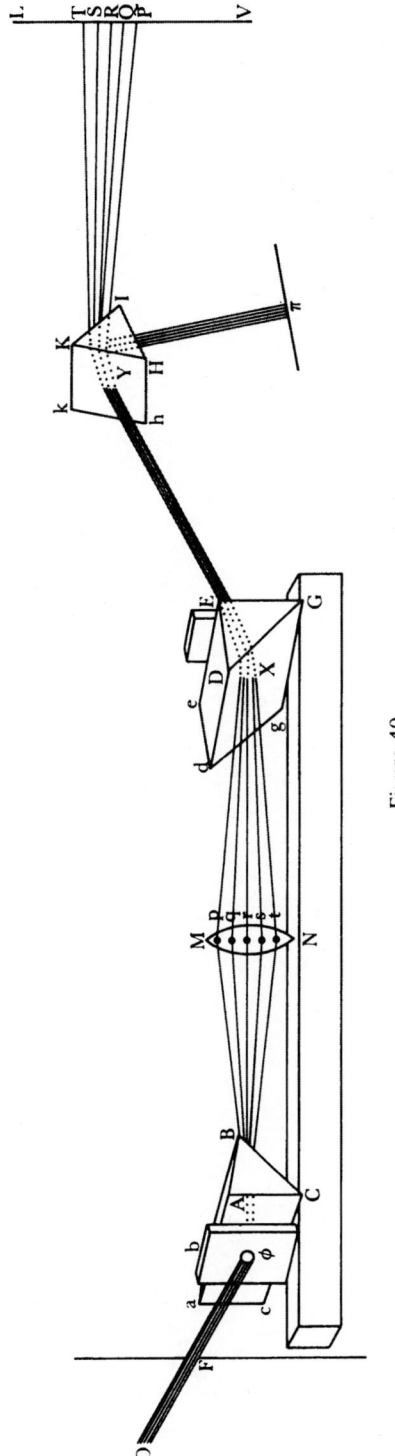

Figure 40

163

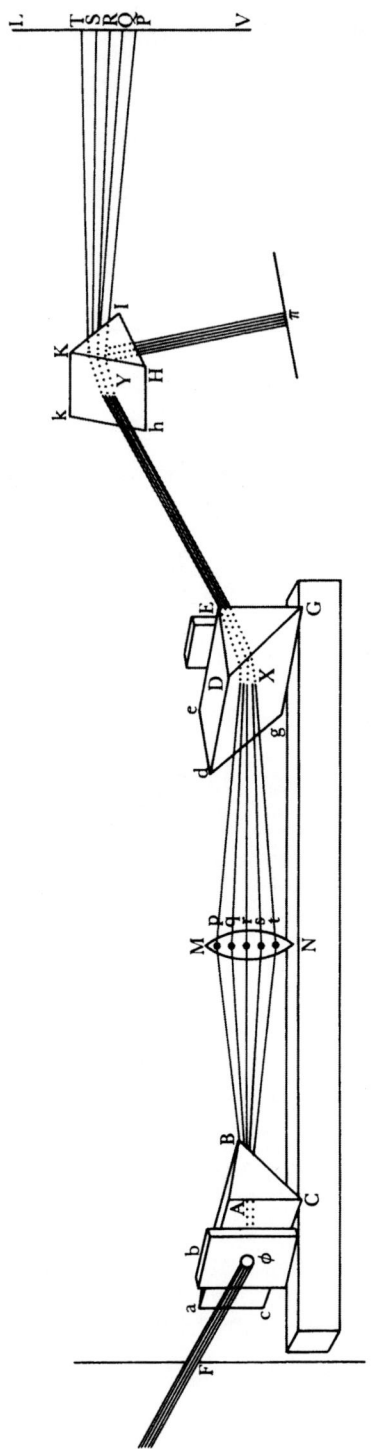

Figure 40

164

But it is unnecessary for me to refute hypotheses of this sort that will spontaneously collapse from the truth having at last been discovered.[16] | I will have done more than enough, if I finally present an experiment whereby everything that I have previously explained about the origin of colors can not only be proved but also be seen. | Wherefore (Fig. 40), let *ABCac* be a prism that refracts the rays transmitted into a darkened room through the hole *F* toward the lens *MN*, so that the colors that it makes at *p, q, r, s,* and *t* may then be transmitted through the lens to *X* and there be mixed together into white, just as I showed in the preceding. Next, parallel to the first prism at the position *X*, where the white has been reconstituted, place another prism, *DEGgd*, that refracts the light toward *Y*. However, let this prism's vertex angle *Gg* be equal to or perhaps smaller than the first prism's vertex angle *Cc* and similarly placed so that it may restore to parallelism the incident rays[17] dispersed by the first prism. After these things have been arranged, you will observe whether the light transmitted to *Y* (a few feet away) remains just as white as it was at *X* or is sensibly transformed into colors. If it appears completely white, then you have properly placed the prisms together with the lens. But if any colors are perceived at *Y*, the prism *DEG* must be turned slightly around its axis in such a way that the colors diminish. When they have completely vanished and the light becomes entirely white, stop the prism. If, however, in this way you cannot prevent the light from changing into colors in some part while passing from *X* to *Y*, shift the lens *MN* a little farther from the prism *ABC*. After again locating the place *X* where the colors converge most accurately into white, fix the prism *DEF* there as before, and once more test whether you can project the light to *Y* without colors. When you have altered the positions of the prisms and the lens to such a point that you have made the light transmitted to *Y* as little colored as you can, fix the prisms together with the lens in that position with the aid of a wooden beam (as is represented in the figure), a tube, or any other instrument constructed for that purpose.[18]

When you have constructed this device from a lens and prisms as I have described, then by means of the light transmitted through it you can test everything that has been related thus far. This light *XY* is in fact very similar to a beam of direct light from the sun and exhibits all the same phenomena as if it had proceeded directly from the hole *F* without having experienced any refraction at all. Hence we may easily believe it to be of the same constitution. Yet when it has been separated at the lens *MN* into its principal components, that is, into rays of different kinds, it will be easy to put to a test the ways by which it can afterward be converted into colors merely by stopping either this or that kind of ray at *MN*, so that the constitution of the light *XY* with respect to its conversion into colors may be manifest.

For example, if it is desired to make it very clear to sense that a prism converts light into colors, not by transmuting its intrinsic properties but only by separating the rays that are disposed to excite various sensations of colors and from which all white light is constituted, nothing other needs to be done than to place some prism *HIK* so that it receives the light *XY* and by refrac-

93 [≈ II, 89, 77]. An instrument is described whereby everything hitherto related about colors may be proved very clearly.

§54

94 [= II, 78]. Its use is described.

95 [≈ II, 79]. And it is illustrated by examples.

transmutet in colores *P, Q, R, S, T*, in papyrum aliquam *LV* procidentes. Deinde si colorem quemlibet apud lentem *MN* interposito obstaculo sistas, videbis eundem · colorem a papyro *LV* deficere. Sic purpuram *p* obstruendo,[19] disparebit purpura *P*, caeteris coloribus non omninò mutatis, (dempto fortè caeruleo, quatenus aliquid purpurae commixtum habeat.) Sic viridem *r* intercipiendo, viridis *R* evanescet: Et sic de alijs. Atque ita videre est quòd ijdem colores apud papyrum *LV* et apud lentem *MN* pertinent ad eosdem radios, ijsque non communicantur a refractione lentis[20] *HIK*, siquidem præexistebant, segregati quidem ad lentem *MN*, et congregati in luce *XY*. Ad eundem modum si cupias experimenta penitiùs rimari, quibus alia genera radiorum omninò reflecti possint, dum alia (licèt similiter incidentia) partim transmittantur:[21] Prisma *HIK* circa axem ejus converte donec altera pars colorum (violacea nempe et caerulea) postquam obliquissimè refracta fuerit versus *LV*, abinde penitus dispareat, versus *π* reflexa; parte tamen alterâ ad *LV* pervadente. Deinde si dimidium colorum rubedinem versus intercipias ad *MN*, rubor et flavus disparebunt ab *LV*, et lux ad *π* reflexa fiet admodum caerulea. Sin alterum dimidium purpuram versus intercipias, rubor apud *LV* non mutabitur sed lux in *π* (propter ablatum purpuream et caeruleum) flavescet aut rubescet. Id quod indicat purpuriformia et caeruliformia radiorum genera penitus ad *π* reflecti, dum caetera partim refringuntur ad *LV*.[22] Nec secus alia colorum phaenomena, quae prismata ab immediatâ Solis luce eliciunt, ope lucis hujus *XY* poteris experiri; et intercipiendo quodvis radiorum genus apud *MN*, eorum causas intueri.

96[≈ II, 80]. In constructionem praefati instrumenti notae quaedam.

Siquis autem velit instrumentum quale jam descripsimus ad experimenta hujusmodi instituenda conficere: lentem adhibeat latam tres digitos, et ampliùs, quae radios parallelos[23] ad focum duos pedes circiter distantem congregat: atque ita prismata distabunt octo pedibus, et conficient instrumentum satis magnum quo omnia strictiùs examini subjiciantur. Quod ad positionem lentis attinet, si prismatum anguli verticales *ACB* et *DGE* sint aequales, puto 60 vel 70[graduum], ipsa aequaliter ab utrisque distabit: sin alter angulus sit major altero, lens illi prismati vicinior collocetur cujus angulus verticalis existit major. / Et nota quòd jubar *XY* per spatium eo latius diffunditur quo lens statuitur anteriori prismati *ABC* vicinior: atque adeò siquando opus sit amplo jubare, debes tantùm efficere, ut lens sit aliquanto vicinior anteriori prismati, quàm posteriori, et adhibere prisma posterius, cujus angulus verticalis sit tanto[24] ferè minor quàm angulus verticalis anterioris. Denique si velis ut colores in lentem illam procidentes sint magis discreti et ab invicem distracti, quàm more jam descripto continget, eâ nempe de causâ ut

70/

---

(19) Originally: obsistendo (stopping).

(20) Read: prismatis.

(21) This experiment was first introduced in §65 = II, 56.

(22) In the *Optica* Newton added an experiment here on the color of cinnabar viewed in lights of different composition.

(23) radios parallelos] Originally: lucem solis (the sun's light). In his draft reply to Hooke Newton described a different lens, "4½ or 5 inches broad placed at 6 or 7 foot distance from yᵉ

tion transmutes it into the colors *P, Q, R, S,* and *T,* which fall onto some paper *LV.* Then if at the lens *MN* you stop any color by inserting an obstacle, you will see the same color depart from the paper *LV.* Thus by blocking off[19] the purple *p,* the purple *P* will disappear with the other colors not being changed at all (except perhaps blue, insofar as it may have some purple intermingled); and by intercepting the green *r,* the green *R* will vanish; and similarly for the others. Thus it is possible to see that the same colors belong to the same rays at the paper *LV* and the lens *MN* and that they are not imparted to them by the refraction of the prism[20] *HIK,* since they preexisted, having in fact been separated at the lens *MN* and gathered together into the light *XY.* In the same way, if you want to investigate thoroughly experiments in which some kinds of rays can be totally reflected while others (although similarly incident) are partly transmitted,[21] turn the prism *HIK* around its axis until one part of the colors (namely, violet and blue), after having been most obliquely refracted toward *LV,* completely disappears from there and is reflected toward π, while the other part still reaches *LV.* If then at *MN* you intercept half of the colors toward red, red and yellow will disappear from *LV,* and the light reflected to π will become completely blue. But if you intercept the other half toward purple, the red at *LV* will not be changed, but the light at π will become yellow or red because the purple and blue have been removed. This shows that the purple- and blue-making kinds of rays are totally reflected to π, while the others are partly refracted to *LV.*[22] In just the same way, with the aid of this light *XY* you will be able to test the other phenomena of colors that prisms produce from the sun's direct light and, by intercepting rays of any kind at *MN,* observe their causes.

If anyone wishes to construct an instrument such as we have just described for undertaking experiments of this kind, then he should use a lens three inches wide or larger, which gathers parallel rays[23] to a focus about two feet away. The prisms will thus be eight feet apart and will make the instrument large enough so that everything may be put to a rather strict test. As to the position of the lens, if the prisms' vertex angles, *ACB* and *DGE,* are equal, suppose 60° or 70°, the lens will be equally distant from each; but if one angle is greater than the other, place the lens nearer to that prism whose vertex angle is greater. Also note that the beam *XY* is spread through a wider space the nearer the lens is placed to the first prism *ABC.* Therefore, if you ever require a wide beam, you need only to arrange for the lens to be somewhat nearer to the first prism than to the second and to use a second prism whose vertex angle is smaller than the vertex angle of the first one in about the same proportion. Finally, if you want the colors falling on that lens to be more parted and separated from one another than will occur in the way

96 [≈ II, 80]. Some notes on the construction of that instrument.

Prism" (Add. 3970, f. 444ʳ), thus making the instrument an ungainly 12 to 14 feet. This lens seems to be the same as that described in the *Opticks,* Bk. I, Pt. II, Prop. XI.

(24) aliquanto . . . tanto] Originally: aliquoties . . . toties.

singula radiorum genera pro lubitu distinctiùs sive magis sejunctim intercipiantur; (Id quod in experimentis nonnullis necessarium duco:) Nihil aliud
agendum est, quàm ut lux per duo parva foramina *F* et *φ* ab invicem longè
(a) Sect 17   distantia priùs trajiciatur quàm incidat in prismata;[a] Vel ut alia lens non
(b) Sect 18   procul ab anteriori prismate collocetur,[b] quae apta sit ut lucem a longinquo
foramine *F* divergentem, congreget ad alteram subsequentem lentem *MN*.
His enim modis imago colorata apud *MN* fiet multùm angustior quàm antè,
(longitudine ejus vix diminutâ,) et proinde colores ejus erunt minùs commisti.

Juli 1670
Lect 9
97. Hactenus
ostensis dissertatio
de mensurâ
refractionum
subnectitur.

Enarratâ colorum genesi quos prismata in longinquum projiciunt, de ijs
jam restat dicendum quos exhibent dum transpiciuntur.[1] Quoniam verò de
mensurâ refractionum in ordine ad scientiam de opticis instrumentis visionem
perficientibus promovendam convenit me disserere, et multae sint exinde
deducendae propositiones quibus hujusmodi colorum genesis innititur et explicari debet, modò ex solis principijs antecedenter demonstratis (ut in
geometriâ solet fieri) velim determinare:[2] ideò non vobis displiceat si de
legibus refractionum nonnulla praesternam, et sic res ad mathematicam pu-
ram magis accedentes his ad physicam spectantibus interspergam.

98 [≈ I, 25]. De
mensurâ refractionis
dati generis
radiorum e quâvis
incidentiâ datâ.

Veteres quidem refractiones metiebantur per angulos quos radius incidens et
refractus cum perpendiculo refringentis plani conficiunt: quasi datam habe-
rent rationem. Ut [fig 41] si *GH* sit planum refrin-
gens cui ducitur *DCE* linea perpendicularis ad aliquod ejus punctum *C* et sit *AC* radius quilibet
incidens in ipsum *C*, & refractus in *CR*: / sup-
posuêrunt veteres quòd angulus incidentiae *ACD*
semper esset ad angulum refractionis *RCE* in
eâdem ratione; vel potiùs credidêre suppositionem
istam satis accuratam esse, modò dicti anguli sint
parvi. Sic in vitro statuebant angulum *ACD* esse
ad ang: *RCE* ut binarius numerus ad ternarium[3]
proximè: sive quòd incidentia *ACD* ferè tripla sit anguli refracti *RCF* utroque
radio comprehensi.[4] | At illa refractionum aestimatio minùs exacta deprehen-
ditur, quàm ut pro fundamento dioptrices debeat statui. Et Cartesius aliam
regulam primus excogitavit, quâ istud exactiùs determinaretur. Nempe

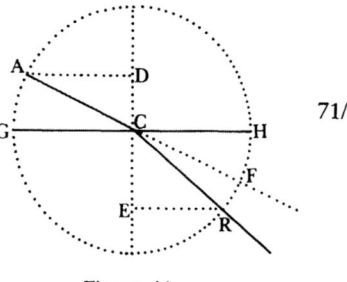

71/

Figure 41

---

(1) This topic was added in the *Optica,* Lect. II, 12.

(2) Originally: demonstrare (demonstrate).

(3) binarius . . . ad ternarium] Read: ternarius . . . ad binarium (three to two).

(4) Although Newton nowhere cites any ancient or medieval optical treatises, Claudius Ptolemy, Alhazen, and Witelo did use the angles of incidence and refraction rather than their sines,
but none of them invoked the approximation that the angle of incidence is proportional to the
angle of refraction or to the deviation. In fact, Ptolemy—whose treatise was still unpublished in
the seventeenth century, though Alhazen and Witelo followed him on this point—explicitly
formulated the inequality $i'/r' > i/r$, where $i' > i$, and r' and r are the angles of refraction
corresponding to any angles of incidence i' and i; see Albert Lejeune, *Recherches sur la catoptrique grecque d'après les sources antiques et médiévales* (Brussels, 1957), p. 157. Francesco

just now described, namely, so that the individual kinds of rays may be intercepted at will more distinctly or separately (something I consider necessary in several experiments), nothing other needs to be done than to pass the light through the two small holes *F* and *φ*, far apart from one another, before it falls on the prisms;[a] or else, not far from the first prism, to place another lens[b] that is suitable for gathering the light diverging from the farther hole *F* at the other, subsequent lens *MN*. | In these ways the colored image at *MN* will certainly become much narrower than before (with its length scarcely diminished), and consequently its colors will be less mixed together.

Since the origin of the colors that prisms project at a distance has been fully explained, it now remains to describe those that they exhibit when they are looked through.[1] But seeing that it is appropriate that I discuss in turn the measure of refractions in order to advance the science of optical instruments that perfect vision, and next many propositions are to be deduced upon which the origin of colors of this kind depends and must be explained, provided that I intended to determine[2] them solely from principles previously demonstrated (as is usual in geometry), it should not therefore displease you if I prepared some things about the laws of refractions and so interjected topics pertaining more to pure mathematics than those that concern physics.

Men of old, in fact, measured refractions by the angles that the incident and refracted ray make with the perpendicular to the refracting plane, as if they had a given ratio. For example (Fig. 41), if *GH* is the refracting plane to which the line *DCE* is drawn perpendicular at some point *C* of it, and *AC* is any ray falling at *C* and refracted into *CR*, men of old supposed that the angle of incidence *ACD* was always in the same ratio to the angle of refraction *RCE*, or rather, they believed that supposition to be sufficiently accurate provided that those angles were small. Thus in glass they considered the angle *ACD* to be to the angle *RCE* approximately as the number two to three,[3] or the angle of incidence *ACD* to be about triple the angle of deviation *RCF* comprehended by both rays.[4] | Yet to be considered as a foundation of dioptrics, that estimate of refractions is found to be less exact than it ought to be. Descartes first conceived another rule whereby that could be more exactly

(a) §17
(b) §18

July 1670
Lecture 9
97. A dissertation on the measure of refractions is added to what has been shown thus far.

98 [≈ I, 25]. The measure of refraction of rays of a given kind at any given incidence.

Maurolico, sometime before his death in 1575, apparently was the first to invoke the approximation that the angle of incidence is proportional to the deviation, except that he was unaware that it was an approximation; *Theoremata de lumine, et umbra, ad perspectivam, & radiorum incidentiam facientia. Diaphanorum partes, seu libri tres ... Problemata ad perspectivam & iridem pertinentia,* 2nd ed. (Lyons, 1613), "Diaphaneon," Bk. I, Thm. X, p. 41; this work is generally known by the title of its first edition of 1611, *Photismi de lumine.* The most likely source for Newton's knowledge of this approximation, even if secondhand, is Kepler's *Dioptrice.* Kepler assumed that in glass and crystal, up to an angle of incidence of thirty degrees, the angle of refraction is "sensibly" (*ad sensum*) proportional to the angle of incidence, and the deviation is "very nearly" (*quàm proximè*) one third of that angle; *Dioptrice,* §§VI–IIX, XII, pp. 3–4= *Gesammelte Werke,* 4: 357–8. See Lohne, "Zur Geschichte des Brechungsgesetzes," *Sudhoffs Archiv für Geschichte der Medizin und der Naturwissenschaften,* 47 (1963):152–72.

quòd dictorum angulorum sinus sunt in ratione datâ.[5] In fig 41 si centro *C* et distantiâ quâlibet *AC* circulus describatur secans radios praefatos in *A* et *R*, et ab istis punctis ad plani perpendiculum *DCE* demittantur normales *AD* et *RE*, ipsarum *AD* et *RE* proportio erit eadem perpetuò.[6] Cujus rei veritatem Author non ineleganter demonstrasset modò de causis physicis quas assumpsit nullum dubitandi locum reliquisset. Ut ut, quoniam instrumentis in istum finem accuratè constructis examinârunt aliqui, et veritati (quoad sensum) exactè convenientem adinvenêrunt:[7] non dubitabimus pro fundamento statuere; hoc solùm adhibito moderamine, quòd cùm is de quibuslibet radijs indifferenter affirmaverit, quasi omnium persimilis fuisset refractio; nos tantùm affirmamus de singulis eorum generibus, ponendo quòd radiorum aeque refrangibilium sinus refractionis sunt ut sinus incidentiae. Concipiamus aliquot genera[8] radiorum secundum lineam *AC* in fig [42] allapsa esse ad punctum *C* ibique refracta per superficiem *IH*, puta mediocriter refrangibiles radios in *CR*, minimè refrangibiles in *CT*, et maximè refrangibiles[9]

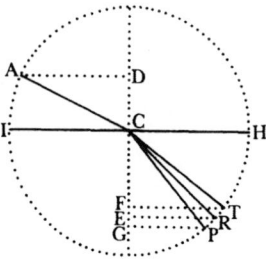

Figure 42

in *CP*; ac innumeros alios gradibus intermedijs plus minùs refringibiles per totum spatium *TCP* diffusos esse. Jam si ducatur *DCG* perpendicularis ad planum refringens *IH*, et centro *C*, distantiâ quâvis *AC* circulus ut priùs describatur secans radios dictos in *A, P, R, T*; atque ex istis punctis demittantur perpendiculares *AD, PG, RE, TF* pro sinibus angulorum *ACD, PCG, RCE, TCF*: pono quòd utcunque radij incidant, tamen semper erit / *AD* ad *PG* in   72/ eadem ratione; quâ semel cognitâ regulam habes pro refractione radiorum maximè refrangibilium[10] in eandem superficiem ad angulum quemvis incidentium mensurandâ. Et sic semper erit *AD* ad *TF* in eâdem ratione: quâ

---

(5) Newton learned the sine law of refraction through his early studies of Descartes's works; see the Introduction, note (20). The sine law, however, was first discovered in 1601 by Harriot, whose work was circulated in manuscript but was not published; see Lohne, "Thomas Harriott (1560–1621)." Sometime in the mid-1620s Wilebrod Snel rediscovered the law, but again it remained unpublished; and at about the same time Descartes and Claude Mydorge also came upon it. On the multiple discoveries of the sine law and their possible interdependence, and also for references to earlier historical studies, see Sabra, *Theories of Light,* pp. 99–103; and John Andrew Schuster, "Descartes and the scientific revolution, 1618–1634: an interpretation," Ph.D. diss. (Princeton, 1977), Ch. IV. Though Newton is quite generous to Descartes here, in the *Principia* (Bk. I, Prop. XCVI, Scholium) he attributed the law to Snel, accepting the priority claims that had been made on his behalf, most probably those made by Isaac Vossius in his *De lucis natura et proprietate* (Amsterdam, 1662), pp. 36–7.

(6) Since $A\widehat{C}D = i$ and $R\widehat{C}E = r$, and $\sin A\widehat{C}D = AD/AC$, and $\sin R\widehat{C}E = RE/RC$, therefore $\sin i/\sin r = AD/RE$, a constant, which is the sine law of refraction. In *La Dioptrique*, Discourse II, Descartes derived the law of refraction based on the model of a projectile that upon entering another medium moves with a new, constant velocity, with the component of its motion parallel to the refracting surface remaining unchanged. From this model it immediately follows that the ratio of sines is constant: if v and v' are the speeds in the upper and lower media, by hypothesis they have a constant ratio, and also by hypothesis the parallel component of the velocities is the same in the two media, or $v \sin i = v' \sin r$, so that $\sin i/\sin r = v'/v = $ constant.

determined, namely, that the sines of those angles are in a given ratio.[5] If (Fig. 41) with center *C* and any radius *AC* a circle is described cutting the specified rays in *A* and *R*, and the normals *AD* and *RE* are dropped from these points to the [refracting] plane's perpendicular *DCE*, then the ratio of *AD* and *RE* will always be the same.[6] The author would have demonstrated not inelegantly the truth of this, if only he had not left room for doubt concerning the physical causes that he assumed. Since, however, some have tested this with instruments accurately constructed for that purpose and have found it to agree exactly with the truth (at least as to sense),[7] we will not hesitate to consider it as a foundation, subject only to this qualification: Whereas he had asserted this about every ray without distinction, as if the refraction of all of them was identical, we assert this about only the individual kinds of them and assume that the sines of refraction of equally refrangible rays are proportional to the sines of incidence. Let us conceive several kinds[8] of rays to flow along the line *AC* (Fig. 42) to the point *C* and to be refracted there by the surface *IH*, namely, the mean refrangible rays into *CR*, the least refrangible into *CT*, and the most refrangible[9] into *CP*, and innumerable others more or less refrangible by intermediate gradations to be spread through the entire space *TCP*. Now, if *DCG* is drawn perpendicular to the refracting plane *IH*; and if with center *C* and any radius *AC*, a circle is described, as before, cutting the said rays in *A*, *P*, *R*, and *T*; and from those points the perpendiculars *AD*, *PG*, *RE*, and *TF* are dropped for the sines of the angles *ACD*, *PCG*, *RCE*, and *TCF*; I assume that however the rays are incident, *AD* to *PG* will nevertheless always be in the same ratio. Once this is known, you have a rule for measuring the refraction of the most refrangible[10] rays incident upon the same surface at any angle. Similarly *AD* to *TF* will always be in the same ratio, and knowing this you have a rule whereby

---

(7) Newton is most likely referring to experimental tests by Hooke and James Gregory. In the preface to the *Micrographia* Hooke described a new instrument for measuring refraction in fluids (see note (18), this lecture), and he used it to test the sine law of refraction. Although he presented no data to support his conclusion, he pronounced the sine law to be "certain." Gregory reported that he had performed many experiments to confirm the sine law. As an example of these, he published a table of his measurements of the refraction of spring water, there comparing his measured values with calculated ones assuming $n = 4/3$, but he did not indicate his experimental technique; *Optica promota*, Prop. 5, p. 13.

(8) Originally: species.

(9) mediocriter refrangibiles ... minimè refrangibiles ... maximè refrangibiles] Originally: maximè viridiformes ... intensissimè rubriformes ... intensissimè purpuriformes (the most green-making ... most intense red-making ... most intense purple-making). Whiteside has suggested that Newton replaced the colors with their corresponding degrees of refrangibility "in a fit of positivism" (*Mathematical Papers*, 3:467, note 35), but it is more likely that he made these and similar changes throughout the *Lectiones* because of the planned interchange of the two parts of the *Lectures*. Since one of Newton's aims in treating refrangibility before color in the *Optica* was to show that unequal refrangibility was a mathematical property of light that could be treated independently of color, it would have been inappropriate to identify the rays by their corresponding color, especially because that correspondence was to be demonstrated only in Part II.

(10) maximè refrangibilium] Originally: intensè purpuriformium (intense purple-making).

cognitâ, regulam habes quâcum refractio minimè refrangibilium[11] pro quavis incidentiâ determinabitur. Atque idem de ratione ipsius *AD* ad *RE,* et ad
sinum cujusvis intermedij generis concipiatur.

99 [= I, 26].
De mensurâ
refractionum
radiorum genere
differentium ex
eâdem quavis
incidentiâ.[12]

Porro autem cùm sinus *PG, RE, TF* caeterique datam habeant rationem ad
sinum *AD,* datam quoque rationem inter sese habebunt. Atque adeò si ex
unicâ observatione proportionem sinuum *PG, RE, TF* et reliquorum ad radios ex eadem incidentiâ refractos pertinentium cognoveris, regulam exinde
habebis quacum ex sinu refractionis cujusvis generis radiorum, et in istam
superficiem utcunque incidentium dato, caeterorum omnium ex eadem incidentia prolabentium sinus elicias: licèt quaenam sit eorum incidentia non
innotuerit. Quinimò si omnium *AD, TF, RE, PG* &c proportiones inter se
semel cognoscantur, habito respectu ad eadem media refringentia, regulam
habes pro caeteris omnibus exquirendis ex unicò quovis unquam dato. Itaque
quo rationes istorum sinuum investigentur,[13] convenit ut in aliquo radiorum
genere[14] proportio sinus incidentiae ad sinum refractionis primùm exquiratur: deinde ut proportiones sinuum refractionis pro radijs diversorum generum ad eundem angulum incidentium determinentur.

100 [= I, 27]. Ad
sinus incidentiae
et refractionis
conferendos
adhibetur mediocre
genus radiorum.

Ad sinus incidentiae cum sinubus refractionis conferendos commodum erit
ut medium genus eligatur, puta genus illud radiorum qui viriditatem vel
potius colorem viridi et caeruleo intermedium exhibent.[15] Credo enim illos
qui refractiones antehac mensuravêre, (sive id factum sit, ut jam dicta Hypothesis Cartesij probaretur sive alijs de causis,) credo illos, inquam, mensuram instituisse ad medietatem refractae lucis: hoc est si spatium a coloribus
occupatum spectemus, ad confinium viridis et caerulei; aut si spectemus
quantitatem lucis, ad medietatem viridis.[16] Et praeterea punctum istud pro
principali foco lentium habendum esse videtur, in quod intermedium genus
radiorum convergit: atque / etiam siquando de radijs indistinctè disserendum 73/
est, ut hactenus apud Opticae scriptores[17] consueverit, genus mediocre
commodiùs quàm extremorum aliquod pro omnibus haberi potest.

101 [= I, 28].
Modus explorandi
sinuum istorum
rationes.

Porrò cùm fortè desideretur accuratius examen dictae regulae Cartesianae,
quàm antehac instituebatur, dum varia radiorum refrangibilitas experientes
latuit: primò dicam id quo pacto non incommodè fiat. Quoniam fluidi pellucidi superficies refringentes facilè possint inclinari ad quemvis datum angulum, quod solido non est concessum, fluida in hunc finem fuerunt adhibita,
sed instrumento magis laborioso quàm opus erat, et erroribus fortè magis
obnoxio, quàm si omni apparatu privaretur, demptâ trabe cui vasculum
aquae plenum affigitur.[18] Sit itaque *HK* in fig: [43] vectis ligneus, duas

---

(11) minimè refrangibilium] Originally: rubriformium (red-making).

(12) De ... incidentiâ.] **II:** De conferendis refractionibus radiorum diversi generis. (Comparing the refractions of rays of different kinds.)

(13) Itaque ... investigentur] Originally: Quamobrem de rationibus istorum sinuum investi-
gandis quod attinet (Wherefore, as for investigating the ratios ...).

(14) Originally: specie.

(15) puta ... exhibent.] Originally: puta viridiforme, aut forte genus illud radiorum qui colorem viridi et caeruleo intermedium producunt. (for example, green-making, or perhaps that kind
of ray that produces the color ...)

the refraction of the least refrangible[11] rays will be determined for any incidence. The same thing is to be understood for the ratio of *AD* to *RE,* and for the sine of any intermediate kind.

Moreover, since the sines *PG, RE, TF,* and the others have a given ratio to the sine *AD,* they will also have a given ratio to one another. Hence, if from a single observation you know the proportion of the sines *PG, RE, TF,* and the others belonging to rays refracted at the same incidence, you will then have a rule whereby given the sine of refraction of any kind of ray, whatever its incidence on that surface, you may derive the sines of all the others proceeding at the same incidence, even if their incidence is not known. In fact, if the proportions of all of them, *AD, TF, RE, PG,* and so forth, to one another are once known with respect to the same refracting medium, you have a rule for finding all the others whenever any one is given. Therefore, to investigate the ratios[13] of these sines it is proper first to find the ratio of the sine of incidence to the sine of refraction for some kind[14] of ray, so that then the proportions of the sines of refraction for rays of different kinds incident at the same angle may be determined.

99 [= I, 26]. The measure of refractions of rays of different kinds at any identical incidence.[12]

To compare the sines of incidence with the sines of refraction it will be appropriate to choose a middle kind, for example, that kind of ray that exhibits green, or rather, the color[15] between green and blue. For I believe that those who have hitherto measured refractions (whether it was done to prove Descartes's hypothesis just described, or for other reasons), I believe, I say, that they determined their measure at the middle of the refracted light; that is, if we consider the space occupied by the colors, at the boundary of green and blue, or, if we consider the quantity of light, at the middle of the green.[16] Moreover, it appears that that point ought to be taken as the principal focus of lenses at which the intermediate kind of rays converge; and also whenever rays must be discussed indiscriminately—as has hitherto been the custom with writers[17] in optics—the middle kind, rather than any of the extreme ones, can be more properly taken for all of them.

100 [= I, 27]. To compare the sines of incidence and refraction a mean kind of ray is used.

Moreover, since a more accurate investigation of that Cartesian rule may perhaps be desired than was previously undertaken while the diverse refrangibility of rays was unknown to experimenters, I first describe how that can be conveniently done. Since the refracting surfaces of a transparent fluid can easily be inclined at any given angle (something that is impossible with a solid), fluids were used for this purpose, but with an instrument more troublesome than was necessary and perhaps more prone to errors than if it were freed from the entire apparatus, except for a beam to which a vessel full of water is attached.[18] Accordingly, let *HK* in Fig. 43 be a wooden beam, two,

101 [= I, 28]. A method for investigating the ratios of those sines.

---

(16) For a more detailed specification of the relation of green to the other colors, see §II, 84.

(17) Originally (and also II): *peritos* (those skilled).

(18) The most widely adopted, or described, method of measuring the refractions of fluids until the end of the seventeenth century derived from Ptolemy. A circular copper disk graduated along its circumference (an astrolabe was commonly used) was placed vertically in a basin of water with the water covering exactly the lower half of the disk. An object or marker on the

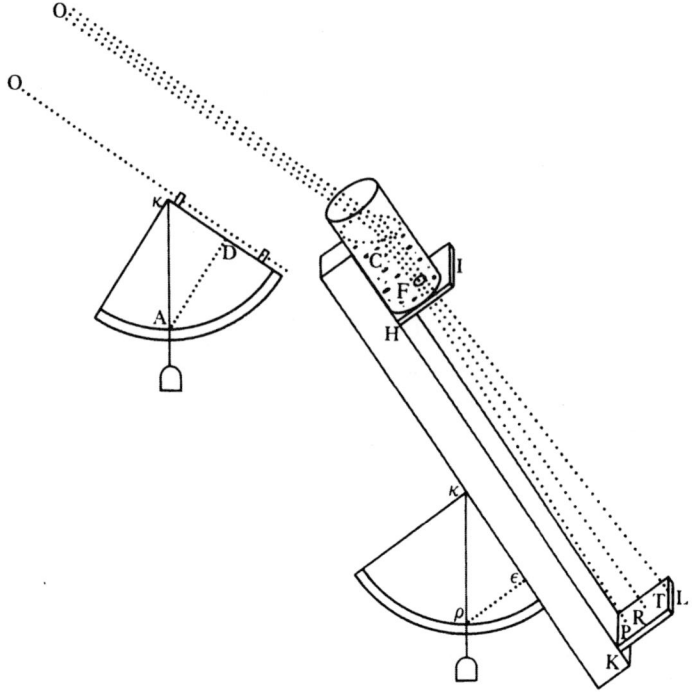

Figure 43

tresve ulnas longus aut amplius, satis crassus ne ob longitudinem ejus et
pondus minimùm inflecti queat, quadrilaterus, rectangulus, et rectus, cum
lateribus oppositis exacte parallelis. Tum lamellae duae *HI* et *KL* super unum
ejus latus ad angulos rectos erigantur, *KL* proximè ad unam extremitatem et
HI quasi quatuor digitos ab alterâ distans; quarum longitudo sit trium digitorum quatuorve, latitudo autem duorum vel trium. Deinde sumatur vasculum
aliquod cylindricum vel prismiforme *CF* duos tresve digitos latum, longum
verò quatuor, vel quinque; ejus basis super lamellam *HI* figatur cemento
aliquo duro ac tenaci et in eo situ firmetur ope trabis *HK* ultra dictam
lamellam *HI* productae. Tum trajiciatur ejus fundum in medietate, et lamella
simul, parvo foramine *F*, puta decimâ parte digiti lato: et juxta foramen istud
in alterâ lamellâ notetur punctum *R*, quod aequè distat a trabe ac dicti
foraminis centrum; ita scilicet ut linea *FR* per centrum foraminis ad *R* ducta
sit parallela longitudini trabis. Denique sumatur lamella vitrea plana polita,
et uniformiter crassa, eaque applicetur ad planitiem lamellae *HI* vasculo *CF*
obversam super foramen *F*, et cemento figatur / ita ut vasculum istud aquae 74/
(quâ repleatur) non sit pervium. Et cum normâ aliquâ fiat periculum an illa
vitrea lamella perpendiculariter insistat trabi. Quod si non contingat, corrigatur situs, donec sit exactè perpendicularis. In cujus rei gratiam convenit ut
dicta lamella vitrea sit tres vel quatuor digitos longa et lata, quò de situ ejus

three, or more yards long, sufficiently thick (so that it is not at all able to bend on account of its length and weight), quadrilateral, rectangular, and straight, with its opposite sides exactly parallel. Then erect two small plates, *HI* and *KL*, at right angles to one of its sides, with *KL* being nearly at the one end and *HI* about four inches from the other; and let their length be three or four inches, but their width two or three. Next take some small cylindrical or prismatic vessel, *CF*, two or three inches wide but four or five inches long; attach its base over the plate *HI* with some hard and gripping cement, and let the vessel be supported in that position by means of the beam *HK* extending beyond that plate *HI*. Then pierce both its base and the plate in the middle with the small hole *F*, say, a tenth of an inch wide; and opposite that hole in the other plate mark the point *R*, which is just as far from the beam as the center of that hole is, namely, so that the line *FR* drawn through the center of the hole to *R* is parallel to the length of the beam. Finally, take a flat, polished, and uniformly thick glass plate and attach it to the side of plate *HI* facing the vessel *CF* above the hole *F*, and fasten it with cement, so that the vessel is impervious to the water with which it is filled. With any square, test whether that glass plate is perpendicular to the beam, and if it is not, correct its position until it is exactly perpendicular. For this purpose it is convenient that that glass plate be three or four inches long and wide, so that its position

submerged portion of the disk was observed through a sight on the upper rim of the disk to determine the angles of incidence and refraction; see *L'Optique de Claude Ptolémée dans la version latine d'après l'arabe de l'émir Eugène de Sicile*, ed., Albert Lejeune (Louvain, 1956), Ch. V, pp. 227–30; translated in *A Source Book in Greek Science*, ed. Morris R. Cohen and I. E. Drabkin (Cambridge, Mass., 1958), pp. 274–5. Alhazen and Witelo described slightly modified versions of Ptolemy's apparatus; see Alhazen, *Opticae thesaurus*, Bk. VII, §§2, 3, pp. 231–5; and Witelo, *Perspectiva*, Bk. II, §1, Bk. X, §§4–8, pp. $_2$61–3, $_2$407–13. In his *Micrographia* (Preface, sigs. e[4]–f[2]), Hooke described a refractometer in which a fluid was placed in a small, horizontal, cylindrical vessel that had two openings for light to pass through. On a pivot at the center of a six- to seven-foot vertical beam he mounted the cylinder along its axis so that it could rotate, and to the cylindrical vessel he attached two, movable, three-foot arms with sights. Since a fluid's surface is always horizontal, he used a plumb line to define the normal to its surface, and the angles of incidence and refraction were then read directly from the movable arms. Newton's design for his instrument seems, in part, to have been inspired by Hooke's innovation of using a small movable vessel containing a fluid, whose surface is perforce always horizontal.

meliùs judicare liceat. Instrumento hoc sic fabricato, et aquâ vasi *CF* plus-quam ad medietatem ejus infusâ, illud in radijs solaribus ita statuatur, ut in superiori superficie aqueâ refracti, perpendiculariter emergant ad foramen *F*, rectàque progrediantur versus laminam *KL*; rubedine ad *T*, purpurâ ad *P*, et viridi vel confinio caerulei et viridis ad *R* incidentibus. Convenit autem ut dicta lamina *KL* dealbetur, aut albente papyro vestiatur, quò de coloribus judicium certius proferri queat. Interea verò cum Quadrante aliquo amplo, et exactè fabricato ϵκρ quaeratur inclinatio trabis *HK* ad horizontem, et habebis angulum refractionis ϵκρ, et ejus sinum ϵρ.[19] Tum solis altitudo statim in-quiratur, ejusque complementum ad 90^{grad} [*AκD*] erit angulus incidentiae, et *AD* sinus.[20] Quibus sinubus ad invicem collatis, et experimento ad diversas Solis altitudines repetito, constabit an sinuum ratio semper sit eadem. Quòd si velis ut experimenta varia simul fiant, aut ad minorem incidentiam quàm sit complementum maximae altitudinis solaris:[21] vice radiorum a sole directè manantium possis adhibere reflexos.

102 [= I, 29].
Modus explorandi vim refractivam Medij cujusvis aëre circundati, praesertim verò Solidi.

Cum eandem sinuum incidentiae et refractionis rationem alicui radiorum generi utcunque in eandem quamvis superficiem incidenti perpetuò compe-tere sat exploratum fuerit,[22] proponatur exquirere rationem illam ad superfi-ciem data quaelibet Media disterminantem; idque unico experimento. Si aer sit unum ex datis Medijs, et liquor quilibet alterum, instrumentum novissimè descriptum non incommodè possit adhiberi. Sin Mediorum alterum sit soli-dum, res expeditè perficitur ad diagramma [44]. In cujus explicationem prae-mittantur sequentia duo Lemmata.

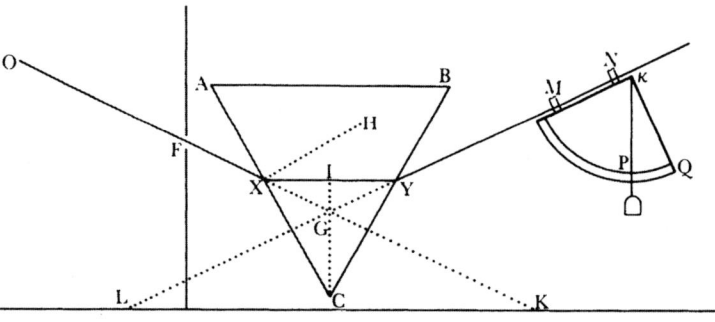

Figure 44

Lem: 1. In fig [44] Sit *ABC* prisma ex materiâ quâvis pellucidâ confectum, cujus axis sit horizonti parallelus et / perpendicularis ad radios solis, et prae-terea sit ejus positio talis ut dictos radios *OX* aeque refringat ingredientes ad *X* et egredientes ad *Y*: istud autem quo pacto debet fieri, ostensum fuit in conclusione Lect 1. Jam dico quòd angulus refractionis ad alterutram refrin-gentem superficiem, ut *AC* factae, est aequalis dimidio verticalis anguli pris-matici *ACB*. Scilicet ad punctum incidentiae *X* erigatur perpendicularis *HX* eritque *HXY* angulus refractionis ad superficiem *AC*: Porrò demittatur *CI*

75/

sect 10.

may be better judged. After thus constructing this instrument and filling the vessel *CF* more than halfway with water, place it in the sun's rays so that the rays refracted at the upper surface of the water emerge perpendicular to the hole *F* and proceed in a straight line toward the plate *KL*, with red falling at *T*, purple at *P,* and green, or else the boundary of blue and green, falling at *R.* It is convenient, moreover, to paint that plate *KL* white or to cover it with white paper, so that you may more accurately judge the colors. Meanwhile, with some large and exactly constructed quadrant ϵκρ determine the inclination of the beam *HK* to the horizon, and you will have the angle of refraction ϵκρ and its sine ϵρ.[19] Then immediately find the sun's altitude, and its complement to 90° [*AκD*] will be the angle of incidence and *AD* its sine.[20] By comparing these sines to one another and repeating the experiment at different altitudes of the sun, it will be ascertained whether the ratio of the sines is always the same. But if you wish to make various experiments at the same time, or at a smaller [angle of] incidence than the complement of the greatest solar altitude,[21] you can use reflected rays instead of rays flowing directly from the sun.

Since it has been sufficiently tested that the same ratio of the sines of incidence and refraction perpetually applies to any one kind of ray, however it is incident on any one surface,[22] it is proposed to discover that ratio at a surface separating any given media—and indeed by a single experiment. If air is one of the given media, and any fluid the other, the instrument just described can be conveniently used. But if one of the media is a solid, the matter is readily carried out according to Fig. 44. For the explanation of this, I premise the following two lemmas:

Lemma 1. In Fig. 44 let *ABC* be a prism made from any transparent material; let its axis be parallel to the horizon and perpendicular to the sun's rays; and moreover let its position be such that it equally refracts the specified rays *OX* upon entering at *X* and emerging at *Y*—and how that should be done was shown at the conclusion of Lecture 1. Now, I assert that the angle of refraction made at either refracting surface, such as at *AC*, is equal to half the prism's vertex angle *ACB*.

Specifically, at the point of incidence *X*, erect the perpendicular *HX*, and *HXY* will be the angle of refraction at the surface *AC*. Furthermore, drop *CI*

102 [= I, 29]. A way to investigate the refractive force of any medium surrounded by air, but especially of a solid.

§10

(19) The angle ϵκρ is the angle the refracted rays make with the vertical, or the perpendicular to the refracting surface, and is therefore the angle of refraction *r*.

(20) The sun's altitude is its height above the horizon, so that its complement, the angle *AκD* made with the vertical, is the angle of incidence *i*.

(21) At Cambridge the maximum solar altitude varies from approximately 61° in summer to 14° in winter (the colatitude of Cambridge plus and minus the obliquity of the ecliptic); its complement therefore varies from approximately 29° to 76°.

(22) Newton, to be sure, may have sufficiently tested this, but he has not reported the results; see Lect. I, 6, note (4).

perpendicularis in radium *XY*, et ista bisecabit angulum *YCX* propterea quod triangulum *YCX* (ob aequalitatem refractionis in *X* et *Y*) sit isosceles. Dico itaque quod angulus *HXY* et *ICX* aequantur. Nam ang: *AXY* = ang: *XIC* + *ICX* (per 32.1. Elem.)[23] sed anguli *AXH* et *XIC* sunt recti; Ergo residui *HXY*, et *ICX* aequantur. Q.E.D.

Lem 2. Adhoc si radius incidens *OX* et emergens *YN* indefinitè producantur occurrentes in *G*, et praeterea si recta quaevis *KL* horizonti parallela radijs istis interjiciatur constituens triangulum *GKL*: Et cùm refractus radius *YN* tendit sursum, si summa angulorum *LKX* et *KLY* sumatur, aut eorum differentia cùm iste *YN* tendit deorsum: Dico quòd illius summae vel differentiae dimidium unà cum angulo refractionis *HXY* aequabitur angulo incidentiae *HXG*. Nam dicta summa vel differentia aequatur angulo *NGK* (per 32 1 Elem,) hoc est angulis *GXY* + *GYX*. Et cùm triangulum *GYX* sit isosceles, dictae summae vel differentiae dimidium aequabitur angulo refracto *GXY*, qui cum angulo refractionis *YXH* constituit angulum incidentiae.[24] Q.E.D.

His praemissis, problema propositum sic perficitur. Primò mensuretur angulus verticalis prismatis *ACB*: et ejus dimidium erit angulus refractionis. Dein prismate in positione praefatâ disposito, per quod radij trajiciantur ingressi foramen *F*, ope Quadrantis *MNPQ* ampli et accurati (puta cujus pinnarum *M* et *N* distantia sit pedis unius minimùm) exploretur angulus *YLK* vel *PκQ* quem refracti radij *YMN* cum horizonte constituunt; faciendo ut mediocriter refrangibiles[25] per pinnas *M* et *N*, ad distantiam decem aut viginti / pedum a prismate trajiciantur: Et simul observetur Solis altitudo *XKL*. Qui duo anguli addantur, si refracti radij *YMN* sursum tendant, sicut in schemate describitur; alias minor subtrahatur de majori. Et summae vel differentiae dimidium unà cùm angulo refractionis priùs invento, erit angulus incidentiae, ut pateat per Lemma 2.ᵘᵐ. Denique ex angulis incidentiae et refractionis sic datis, dantur eorum sinus.[26] Q.E.F.

76/

---

(23) If any side of a triangle is extended, the exterior angle is equal to the two opposite interior angles.

(24) Since $\widehat{NGK} = \widehat{LKX} + \widehat{KLY} = \widehat{GXY} + \widehat{GYX}$ is the deviation D, this lemma states that at minimum deviation $i = \frac{1}{2}D + r = \frac{1}{2}(D + C)$, for by Lemma 1, $r = \frac{1}{2}C$, where C is the prism's refracting angle. This lemma and its proof are valid even if *KL* is not parallel to the horizon.

(25) mediocriter refrangibiles] Originally: viridiformes (green-making).

(26) With this demonstration that at minimum deviation the index of refraction n is equal to $\sin \frac{1}{2}(D + C)/\sin \frac{1}{2}C$, Newton has laid down his basic experimental technique for measuring indices of refraction. In the *Opticks* (Bk. I, Pt. I, Prop. VII, pp. 59–60 = Add. 3970, ff. 61–2), he set forth the sequence of measurements and calculations necessary to determine the index of refraction by this method, but he omitted these lemmas or any demonstrations. As it turned out, these calculations sufficed to allow a reconstruction of their theoretical foundation. In the period of intense interest in refraction and dispersion following John Dollond's construction of an achromatic lens in 1758, Alexis Claude Clairaut, in a paper read in 1761, was the first to recover, even if somewhat superficially, the basis of Newton's method; "Mémoire sur les moyens de perfectionner les lunettes d'approche, par l'usage d'objectifs composés de plusieurs matières différemment réfringentes," *Mémoires de l' académie royale des sciences*, 1756 [Paris,

perpendicular to the ray XY, and it will bisect the angle YCX, because the triangle YCX is isosceles (on account of the equality of the refractions at X and Y). I say, therefore, that angle HXY and angle ICX are equal. For $A\widehat{X}Y$ = $X\widehat{I}C$ + $I\widehat{C}X$ (*Elements,* I, 32),[23] but the angles AXH and XIC are right angles; therefore the remainders HXY and ICX are equal. As was to be demonstrated.

Lemma 2. Moreover, if the incident ray OX and the emergent ray YN, meeting in G, are extended indefinitely; if, besides, any line KL parallel to the horizon is placed between those rays and forms the triangle GKL; and if the sum of the angles LKX and KLY is taken when the refracted ray YN tends upward, or their difference when YN tends downward; I assert that half of that sum or difference together with the angle of refraction HXY will equal the angle of incidence HXG.

For the specified sum or difference equals the angle NGK (*Elements,* I, 32), that is, $G\widehat{X}Y$ + $G\widehat{Y}X$. Since the triangle GYX is isosceles, half of the specified sum or difference will equal the deviation GXY, which together with the angle of refraction YXH constitutes the angle of incidence.[24] As was to be demonstrated.

With these lemmas having been premised, the proposed problem is carried out as follows: First, measure the prism's vertex angle ACB, and half of it will be the angle of refraction. Then after the prism that transmits the rays entering the hole F has been arranged in the specified position, using a large and accurate quadrant MNPQ (for example, one whose sights M and N are at least one foot apart), determine the angle YLK or PκQ, which the refracted rays YMN make with the horizon, by causing the mean refrangible[25] rays to pass through the sights M and N ten or twenty feet away from the prism. Also, at the same time observe the sun's altitude XKL. Add these two angles if the refracted rays YMN tend upward (as is drawn in the figure); otherwise subtract the smaller from the larger. Half the sum or difference together with the angle of refraction found earlier will be the angle of incidence, as is evident by Lemma 2. Finally, from the angles of incidence and refraction having been thus given, their sines will be given.[26] As was to be done.

1762]:380–437, esp. 408–20. In 1764 Jean le Rond d'Alembert gave a complete reconstruction, going far beyond Newton's symmetrical case, and derived the equation describing the path of a ray through a prism at any angle of inclination; and then differentiating it, he showed that in the symmetrical situation the deviation is a minimum; *Opuscules mathématiques ou mémoires sur différens sujets de géométrie, de méchanique, d'optique, d'astronomie, &c.*, 8 vols. (Paris, 1761–80), Mémoire XX, 3(1764):341–413, esp. 371–9. In the following year Rudjer Josip Bošković independently, but somewhat less generally, derived Newton's method for determining the index of refraction and showed that this arrangement was also the minimum of deviation; *Dissertationes quinque ad dioptricam pertinentes* (Vienna, 1767), Dissertation I, §§142–204, pp. 72–91; a German translation of this first dissertation was published two years earlier. In 1762 Leonhard Euler had also independently derived the equation for a ray passing through a prism at any

103 [≈ I, 30].
Exemplum in
refractione
cujusdam generis
vitri.

Sic in prismate quodam vitreo dimensus sum angulum ejus maximum *ACB* et inveni esse 63^grad, 12′: Cujus dimidium *HXY* est 31^grad, 36′: ejusque sinus 5240, posito sinu 90^grad 10000. Deinde cùm altitudo Solis *OKL* observabatur esse 14^grad 4′, alter angulus *MLK* a radio *YN* ad medium viriditatis tendente conflatus[27] erat 30^grad 52′: quorum summa est 44^grad 56′ ejúsque dimidium *YXK* 22^grad 28′: quod unà cum angulo refractionis *HXY* facit 54^grad 4′ angulum incidentiae: cujus sinus est 8097. denique conferendo sinus jam inventos ut eorum proportio in minimis terminis haberetur; inveni esse ut 11 ad 17 ferè. Quare pro regulâ generali statuendum est, quòd radiorum viriditatem exhibentium[28] sinus incidentiae ex aere in vitrum quodvis aeque refractivum ac illud prisma, sit ad sinum refractionis, ut septendecim ad undecim.[29]

104 [= I, 31]. Modi
praefati
commoditas.

sect 10

Hujus autem modi commoditas in mensurandis refractionibus ex eo conjicietur, quòd instrumento nullo hic opus est, dempto Quadrante, et Prismate cujus refractio desideratur; quòd refractionem dum geminatur, facta ad *X* et *Y*, exinde certiùs metiri possis; et quòd facillimum est prisma in desiderato situ disponere, ut supra ostenditur.[30] Imò quòd parvus error a situ desiderato ferè nihili est, dum quoad sensum haud inde mutabitur angulus refractus *MGK*, ut experienti patebit. Quippe angulus iste tunc minimus est; et in quantitatibus per motum generatis cùm maximae existant vel minimae, hoc est in momento regressûs, earum motus, utplurimùm, sunt infinitè parvi. Sic verbi gratiâ in fig [45] si centro *C* describatur circulus *λLλ*, et extra eum sumatur punctum quoddam *G* ducaturque *GC* et erigatur normalis *GK*; Deinde si concipiatur, quòd punctum *λ* movetur uniformiter in illius circuli circumferentiâ, / per quod punctum recta quaedam *Gλ* circa centrum *G* rotata perpetuò transeat: manifestum est quod quo major sit angulus *CGλ* sive quo minor angulus *KGλ*, eo minor erit motus angularis ipsius *Gλ*; et cùm angulus *CGλ* sit maximus sive angulus *KGλ* minimus, hoc est in momento regressûs (rectâ *Gλ* tunc circulum in *L* tangente) motus ejus erit infinitè parvus et quoad sensum nullus, parvusque error a puncto contactûs *L* nullam sensibilem variationem in angulis istis

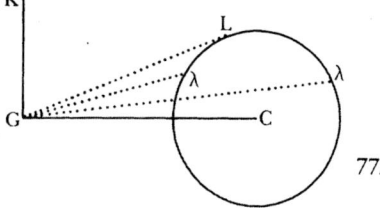

Figure 45

77/

inclination, but since he rejected the method of minimum deviation as an unsatisfactory experimental technique, he did not specifically consider that case; "Considerations sur les nouvelles lunettes d'angleterre de Mr. Dollond et sur le principe qui en est le fondement," *Mémoires de l'académie des sciences de Berlin*, (1762)[1769]:226–48 = *Opera omnia*, ser. 3, 8:102–21, esp. §§10–12, p. 106–7. Although Robert Smith in his *Compleat System of Opticks* ("General Remarks," §§391–8, 2:66–7) reproduced Newton's lemmas directly from the *Lectures*, together with §104 = 1, 31, the recovery of Newton's method of minimum deviation seems rather to have followed straight from the clues in the *Opticks* (except perhaps for d'Alembert who cites Smith's work), whence it passed rapidly into the general optical literature.

(27) a . . . conflatus] Originally: a radio viridiformi *YN* conflatus (made by the green-making ray *YN*).

(28) viriditatem exhibentium] Originally: viridiformium (green-making).

In this way I measured in a certain glass prism its largest angle *ACB* and found it to be 63°12′, half of which, *HXY*, is 31°36′, and its sine is 5240, assuming the sine of 90° to be 10000. Then since the sun's altitude *OKL* was observed to be 14°4′, and the other angle, *MLK*, made by the ray *YN* tending to the middle of the green,[27] was 30°52′, their sum is 44°56′, and its half, *YXK*, is 22°28′, which together with the angle of refraction *HXY* makes the angle of incidence 54°4′, whose sine is 8097. Finally, comparing the sines just found so that their proportion may be in the least terms, I found it to be about 11 to 17. Accordingly, it must be considered as a general rule that for rays exhibiting green,[28] the sine of incidence from air into any glass equally as refractive as that prism is to the sine of refraction as 17 to 11.[29]

103 [≈ I, 30]. An example of the refraction of a certain kind of glass.

The advantage of this method of measuring refractions will be inferred from the following: No instrument is needed here except a quadrant and the prism whose refraction is sought; you can measure the refraction more precisely, because it is doubled at *X* and *Y*; and the prism is very easily placed in the required position, as was shown above.[30] Indeed, a small departure from the required position is almost inconsequential, insofar as the deviation *MGK* will not be perceptibly changed from this, as will be clear to anyone trying it. Namely, that angle is then a minimum, and in quantities generated by motion, when they are either a maximum or minimum—that is, in the moment of regression—their motions are generally infinitely small. Thus for example (Fig. 45), if the circle *λLλ* is described with the center *C*, some point *G* is chosen outside of it, *GC* is drawn, and the normal *GK* is erected; then if one imagines the point *λ* to move uniformly in the circumference of that circle, and a certain straight line *Gλ* turning about its center *G* always to pass through that point, it is evident that the greater the angle *CGλ*, or the smaller the angle *KGλ*, the smaller will be the angular motion of *Gλ*. When the angle *CGλ* is a maximum, or the angle *KGλ* a minimum, that is, in the moment of regression (the line *Gλ* then being tangent to the circle at *L*), its motion will be infinitely small and imperceptible; and a small departure from the point of contact *L* will produce no perceptible variation in those angles *KGL* and

104 [= I, 31]. The advantage of the preceding method.

§10

(29) Newton arrived at the approximation 17/11 by the contemporary technique of converting to a unit continued fraction,

$$\frac{8097}{5240} = 1 + \frac{1}{5240/2857} = 1 + \frac{1}{1 + 2857/2383} \cdots = 1 + \frac{1}{1+} \Big/ \frac{1}{1+} \Big/ \frac{1}{5+} \Big/ \frac{1}{36+} \Big/ \frac{1}{2+} \Big/ \frac{1}{6}.$$

The large term 36 indicates that to a very close approximation

$$\frac{8097}{5240} \approx 1 + \frac{1}{1+} \Big/ \frac{1}{1+} \Big/ \frac{1}{5} = \frac{17}{11}.$$

The approximation 31/20 reported in the "New theory" is based on these measurements but is not as close an approximation to the measured value (8097/5240 = 1.5452) as 17/11, although it is closer to the traditional 3/2 for glass; *Correspondence*, 1:93 = *Phil. Trans.*, 6 (1671/2):3077.

(30) ut supra ostenditur.] Originally: ut constat ex conclusione Lectionis primae. (as is evident from the conclusion of the first Lecture.)

KGL et *CGL* producet. Et ad eundem fere[31] modum parva convolutio prismatis haud omnino mutabit angulum *MGK*, cùm iste sit minimus sive
complementum[32] ejus maximum. Quòd si prisma disponeretur in quovis alio
situ quàm hic describitur, (puta cùm radij perpendiculariter ingressi, ad
egressum duntaxat refringuntur,)[33] minimus error ab isto desiderato situ
multùm mutaret angulum refractum, et sic experientia foret incertitudini, et
erroribus multò magis obnoxia.[34]

105 [= I, 32].
Regula de
investigandâ
refractione
Mediorum sibi ipsis
contiguorum
quorum aeri
contiguorum
refractiones
cognoscuntur.

In majorem hujus rei copiam, quia dantur aliqui casus ubi refractiones per
modos jam descriptos haud possint mensurari, (ut cùm refractio fit ex vitro
in chrystallum, ex aquâ in vitrum, vel ex uno liquore in alium;) Et nequa
omninò sit refringens superficies cujus refractio nequit investigari, problema
sequens lubet proponere.

Datis refractionibus quas duo Media alicui tertio contigua conficiunt, illorum sibi ipsis contiguorum refractionem invenire.

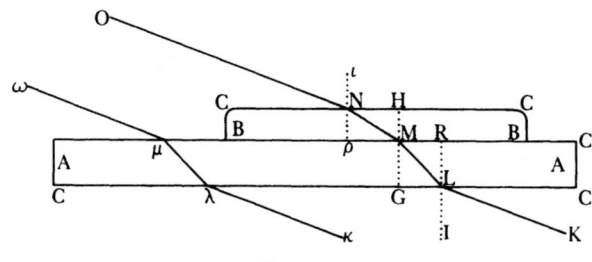

Figure 46

In fig [46] Sunto duo media proposita *A* et *B*, quorum superficiei disterminantis refractio quaeritur, et sit *C* Medium tertium cujus superficiei ipsis *A* et
B contiguae refractiones dantur sitque sinus incidentiae ad sinum refractionis
ex Medio *C* in Medium *A* sicut *I* ad *R*; et sinus incidentiae ad sinum refractionis ex eodem Medio *C* in alterum Medium *B* sicut ι ad ρ. Dico quod est[35]
$I \times \rho : R \times \iota ::$ sinus incidentiae ad sinum refractionis ex Medio *B* in Medium *A*.[36]

Verbi gratiâ proponatur investigatio refractionis ex aquâ in vitrum, datâ
refractione ex aëre in utrumque. Sitque sinus incidentiae ex aëre in vitrum ad
sinum refractionis ut 17 ad 11 et sinus incidentiae ex aere in aquam ad sinum
refractionis ut 4 ad 3. Quare sinus istos multi/plicando reciprocè, erit ut 17 × 78/

---

(31) Added.   (32) Read: supplementum (supplement).

(33) This is probably a criticism of Descartes who determined the index of refraction of glass
by letting light rays fall perpendicularly on one face of a right-angled prism and measuring the
refraction at the second face; *Dioptrice*, Ch. X, §§1–2, *Oeuvres*, 6:642–4, 211–14.

(34) Newton's assessment of the advantages and accuracy of his method of minimum deviation is borne out by an analysis of his extensive measurements of indices of refraction in his table
of refractive powers in the *Opticks*, Bk. II, Pt. III, Prop. X, p. ₂73 = Add. 3970, f. 192ʳ. What
could perhaps be considered the crudity of his instruments is more than compensated for by his
choice of minimum deviation where the errors too are at a minimum. Despite the inherent
difficulties in comparing older measurements of indices of refraction with modern ones, his
measurements are found to be at least as good as any others made between the early seventeenth

CGL. In almost[31] the same way, a small rotation of the prism will not at all change the angle *MGK*, since it is a minimum, or its complement[32] a maximum. But if the prism were placed in any position other than that described here (as when rays entering perpendicularly are refracted only upon leaving),[33] the least departure from that required position would greatly change the deviation, and the experiment would thus be much more liable to uncertainty and errors.[34]

For greater facility in this matter, because some cases are given where the refractions cannot be measured by the methods already described (as when the refraction is made from glass into crystal, from water into glass, or from one fluid into another), and lest there be any refracting surface whose refraction cannot be investigated, it is desirable to propose the following problem:

105 [= I, 32]. A rule for investigating the refraction of media contiguous to one another, when their refractions contiguous to air are known.

> Given the refractions that two media make when contiguous to any third one, to find their refraction when they are contiguous to each other.

In Fig. 46, let *A* and *B* be the two proposed media, and the refraction of the surface separating them is to be found; let *C* be the third medium, and let the refractions of its surface when contiguous to *A* and *B* be given; and let the sine of incidence be to the sine of refraction from the medium *C* into the medium *A* as *I* to *R*, and let the sine of incidence be to the sine of refraction from the same medium *C* into the other medium *B* as ι to ρ. I say that[35] $I \times \rho$ is to $R \times \iota$ as the sine of incidence is to the sine of refraction from the medium *B* into the medium *A*.[36]

For example, the investigation of the refraction from water into glass may be proposed, given the refraction from air into each. Let the sine of incidence from air into glass be to the sine of refraction as 17 to 11, and the sine of incidence from air into water be to the sine of refraction as 4 to 3. Consequently by multiplying those sines reciprocally, then as 17×3 to 11×4, or

---

and early nineteenth centuries; see Lohne, "Newton's table of refractive powers: origins, accuracy, and influence," *Sudhoffs Archiv für Geschichte der Medizin und Naturwissenschaften,* **61** (1977):229–47; and Shapiro, "Newton's 'achromatic' dispersion law," pp. 116–19.

(35) Originally continued: reciproci factus datorum sinuum sunt ut sinus quaesiti, hoc est (the inverse products of the given sines are as the required sines, that is).

(36) More generally, with a pair of subscripts, for example, *ab,* representing the passage of light from medium *a* to *b*; and with *a, b,* and *c* representing glass, water, and air, respectively; then this rule states that $I_{ba}/R_{ba} = (I_{ca} \times R_{cb})/(I_{cb} \times R_{ca})$, or $n_{ba} = n_{ca}/n_{cb}$, where $n = I/R$. On a separate folio that Newton had intended to include in his draft letter for Hooke in the spring of 1672, he formulated this "rule" slightly differently: "as y^c ratio of y^e given sines of incidence is to y^c ratio of y^e given sines of refraction, so are y^e desired sines of incidence & refraction to one another" (Add. 3970, f. 529^r). He has simply rearranged the terms to the equivalent rule, $I_{ba}/R_{ba} = (I_{ca}/I_{cb})/(R_{ca}/R_{cb})$. In the *Opticks* (Bk. I, Pt. II, Prop. III, Expt. VIII) he reformulated it in a still different way, $I_{ab}/R_{ab} = (I_{ac}/R_{ac})(I_{cb}/R_{cb})$, or $n_{ab} = n_{ac} \times n_{cb}$. Newton had command of this rule by the time he composed his table of refractions in 1666 or 1667 (see the Introduction, note (41)), since it is built into its very structure. In §123 ≈ I, 50, no. 2, he extends the rule to polychromatic rays. He adopted the notation I/R for the index of refraction, or ratio of sines, from Barrow (*Lectiones XVIII,* Lect. III, §II, p. 37), although in his earliest studies he used the Cartesian *d/e*; and in §112 = I, 39, he extends the Barrovian notation to rays of different color.

3 ad 11 × 4, sive ut 51 ad 44 ita sinus incidentiae ex aquâ in vitrum ad sinum refractionis.[37] Et sic cognitâ refractione ex aere in quaevis alia Media proposita, possis adipisci refractionem eorum inter se: et e contra.

106 [= I, 33]. Ejus regulae demonstratio

Caeterum demonstratio hujus non est omittenda, in quem finem praesternatur Lemma sequens. Si Media duo proposita *A* et *B* in fig [46] concipiantur esse planis parallelis terminata, contigua, et dicto Medio tertio (puta aëre) circundata et radius quilibet *ON* obliquè incidens ad *N* refringatur primò ad *M* ac deinde ad *L*, et emergens pergat ad *K*:[38] Dico quòd iste radius incidens *ON* sibi emergenti *LK* parallelus est. Cujus quidem assertionis veritas experientiâ patebit. Etenim ponatur Medium *A* esse vitrum, et Medium *B* esse aquam, Mediumque tertium circundans esse aëra: Et laminae vitreae *A* superficies *ρMR* tenuiter illinatur aquâ *B*, et statuatur parallela ad Horizontem, ut aqua consistat uniformiter crassa. Quo facto videbis quod radij per utrumque Medium *A* et *B* trajecti tendent ad easdem plagas versus quas tenderent a Sole directi.[39]

Praemisso hoc, erigantur *ιNρ*, *HMG*, et *RLI* perpendiculares ad refringentia puncta *N*, *M*, et *L*. Est ergo *ι* ad *ρ* ut sinus anguli *ONι* ad sinum anguli *MNρ*, sive anguli *NMH*. Et multiplicando rationem antecedentem per *I* fiet *I* × *ι* ad *I* × *ρ* ut sinus de *ONι* ad sinum de *NMH*. Porro est *I* ad *R*, ut sinus anguli *KLI* sive *ONι*, ad sinum anguli *MLR* sive *LMG* et multiplicando antecedentem rationem per *ι*, fiet *I* × *ι* ad *R* × *ι* ut sinus de *ONι* ad sinum de *LMG*. Jam permutando terminos utriusque proportionis fiet

$$I \times \iota . \sin ON\iota :: I \times \rho . \sin NMH.$$

Et *I* × *ι* . sin *ONι* :: *R* × *ι* . sin *LMG*. Quare ex aequalitate rationis est *I* × *ρ* . sin *NMH* :: *R* × *ι* . sin *LMG*. Et permutando

$$I \times \rho . R \times \iota :: \sin : NMH . \sin : LMG.$$

Q.E.D.

Lect 10
107 [= I, 34].
Modus dimetiendi refractiones solidorum, ad fluida accommodantur

Ex hisce sic ostensis problema non inutile proficiscitur, quo refractiones fluidorum eodem modo metiri possis ac de solidis ostensum est ad fig [44]: non adhibito instrumento *HILK*, quod in fig [43] describitur.[1] Scilicet / ex /79/ laminis vitreis in morem cunei connexis vasculum prismiforme conficiatur,

---

(37) In his draft reply to Hooke, Newton found 93/80 for the index of refraction from water into glass, since he used the somewhat cruder value 31/20 for the index of refraction of glass, but kept 4/3 for the refraction of water. In the "Fundamentum Opticae" (Prop. 9, Expt. 27, Add. 3970, f. 412ʳ) and initially in the manuscript of the *Opticks* (Bk. I, Pt. II, Prop. III, Expt. VIII, Add. 3970, ff. 99–100), he used $70\frac{1}{2}/45\frac{1}{2}$ for the index of refraction of glass at the border of blue and green and 197/147 for water and found 2961/2561 for refraction from water into glass. In the published *Opticks* (pp. 95–6), however, he again found 93/80 for refraction from glass into water, but this calculation is erroneous, since he used mean refrangible rays for glass but least refrangible ones for water.

(38) emergens pergat ad *K*] Originally: emergat in *LK* (emerges in *LK*).

(39) Huygens in 1653 invoked virtually the identical observation as the basis for his proof of this rule for relative indices of refraction in his *Dioptrica* (Pt. I, Prop. XXV, *Oeuvres complètes de Christiaan Huygens. Publiées par la société hollandaise des sciences*, 22 vols. (The Hague, 1888–1950), **13**, i:125–9), but for some reason it was omitted in the posthumous editions of

as 51 to 44, so will the sine of incidence from water into glass be to the sine of refraction.[37] Thus knowing the refraction from air into any other proposed media, you can obtain their mutual refraction, and conversely.

The demonstration of this, however, must not be omitted, and to this end the following lemma is provided beforehand. If the two proposed media *A* and *B* in Fig. 46 are conceived to be bounded by parallel planes, contiguous, and surrounded by the specified third medium (for instance, air); and if any ray *ON* incident obliquely at *N* is refracted first to *M* and then to *L* and upon emerging continues to *K*;[38] I say that that incident ray *ON* is parallel to itself emerging as *LK*. The truth of this assertion will certainly be evident from experience. For let it be assumed that the medium *A* is glass, the medium *B* is water, and the third, surrounding medium is air; and cover the surface *ρMR* of the glass plate *A* lightly with the water *B* and place it parallel to the horizon, so that the water is uniformly thick. Having done this, you will see that rays transmitted through both media *A* and *B* will tend in the same direction to which they would tend directly from the sun.[39]

With this premised, erect the perpendiculars *ιNρ*, *HMG*, and *RLI* at the refracting points *N*, *M*, and *L*. Therefore,

$$\iota : \rho = \sin O\hat{N}\iota : \sin M\hat{N}\rho \; (\text{or } N\hat{M}H);$$

and multiplying the preceding ratio by *I* it will become

$$I \times \iota : I \times \rho = \sin O\hat{N}\iota : \sin N\hat{M}H.$$

In addition, $I : R = \sin K\hat{L}I$ (or $O\hat{N}\iota$) $: \sin M\hat{L}R$ (or $L\hat{M}G$); and multiplying the preceding ratio by *ι*, it will become $I \times \iota : R \times \iota = \sin O\hat{N}\iota : \sin L\hat{M}G$. Now permuting the terms of each proportion there will result $I \times \iota : \sin O\hat{N}\iota = I \times \rho : \sin N\hat{M}H$, and $I \times \iota : \sin O\hat{N}\iota = R \times \iota : \sin L\hat{M}G$. Therefore from the equality of the ratios, $I \times \rho : \sin N\hat{M}H = R \times \iota : \sin L\hat{M}G$, and by a permutation, $I \times \rho : R \times \iota = \sin N\hat{M}H : \sin L\hat{M}G$. As was to be demonstrated.

From these things thus demonstrated a rather useful problem arises whereby you can measure the refractions of fluids in the same way as that demonstrated for solids at Fig. 44 without using the instrument *HILK* illustrated in Fig. 43.[1] Namely, make a prismatic vessel from glass plates joined

106 [= I, 33]. The demonstration of this rule.

Lecture 10
107 [= I, 34]. The method of measuring the refractions of solids is adapted to fluids.

1703 and 1728. Although Harriot arrived at Newton's rule in his private papers composed at the beginning of the seventeenth century, Emmanuel Maignan seems to have been the first to publish this obvious rule, the proof of which essentially follows from the principle of the reversibility of light rays. Maignan, however, formulated it in terms of ratios of optical densities *d*, rather than ratios of indices of refraction. He demonstrated that the refraction from medium *C* to *A* is the same as if the ray passed from medium *C* to *B* and thence to *A*, and that $d_{ca} = d_{cb} \times d_{ba}$; *Perspectiva horaria sive de horographia gnomonica tum theoretica, tum practica libri quatuor* (Rome, 1648), Bk. IV, Prop. VII, pp. 567–9; see also Lect. 10, note (13), this volume. Smith incorporated this and the preceding article, together with the account in the *Opticks*, in his *Compleat System*, "General Remarks," §§399–403, 2:67–8.

(1) That is, in §§101–2 = I, 28–9.

cujus acies sive angulus verticalis sit 80^grad circiter, vel 90.[2] Istius autem anguli quantitatem exactissimâ mensura cognitam habeto, ejusque dimidij sinum pro sinu refractionis semper statue. Quo peracto, cùm liquoris alicujus vis refractiva desideratur, vasculum cum illo liquore impleatur et in tali situ disponatur, ut acies ejus a concursu refringentium planorum constituta, sit parallela ad Horizontem et perpendicularis ad radios solares, atque ut illi radij per praefata refringentia plana trajecti refractiones ad ingressum et egressum aequales patiantur. Et ope Quadrantis, ut ostensum erat ad fig [44], exploretur angulus incidentiae, cujus sinus ad praefatum sinum refractionis erit ut sinus incidentiae ad sinum refractionis ex aere in liquorem propositum.

108 [= I, 35].
Aquae refractio prout ipse dimensus sum in specimen ejus rei adducitur

Instantiae gratiâ, ut aquae refractionem cognoscerem curavi, ut prisma ligneum conficeretur quale est *AB*κ, in fig [47] cujus ille angulus *ACB*, quem pro verticali designabam foret rectus, caeteriq́ue duo semirecti. Et effeci ut refringentia plana *A*κ et *B*κ per meditullium trajicerentur foramine *F* parallelo ad basem *A*β, per quod foramen lux itura esset; et ut tertium planum *A*β foderetur in *G* usque dum aditus ad foramen *F* transversè pertingeret. Dein sumptis duabus ex vitro lamellis, quas speculum confractum mihi subminis-

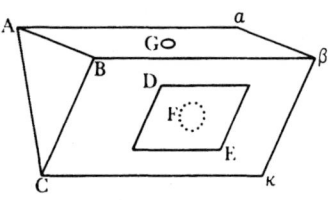

Figure 47

travit, alteram *DE* super meditullium plani *B*κ cum caemento fixi, et alteram super meditullium alterius plani *A*κ, ut meatus[3] *F* utrinque clauderetur. Tum aquam pluvialem per orificium *G* in excavatum spatium infudi, et cum operculo ex subere conciso clausi. Atque adeò aqua duabus vitreis lamellis ad angulum rectum inclinatis interjecta vices subibat aqueae prismatis habentis angulum rectum. Eas autem laminas rectum angulum exactè comprehendere ex applicatione normae cognovi, cujus ideò dimidium 45^gr pro angulo refrac-

Lem: 1.
Sect 102.

tionis habendum est. Hoc prisma dein ita statuebam ad ingressum lucis in obscurum cubiculum, ut eadem foret utrinque refractionis quantitas; et ex altitudine solis, et refractorum radiorum viridiformium inclinatione ad Horizontem, inveni angulum / refractum[4] esse 51^gr 16^min. Cujus dimidium 25^gr 38^min unà cum angulo refractionis 45^gr dabit angulum incidentiae 70^gr 38^min. 80/ Horum verò angulorum 70^gr 38^min et 45^gr sinus sunt 9434[5] et 7071 respectu sinûs 90^graduum 10000, quorum quidem numerorum ratio est paulo minor quam Cartesiana 250 ad 187 et[6] paulo major quam 4 ad 3 nempe 4,002 ad 3;[7] quae tamen a ratione $\frac{4}{3}$ tam parvâ differentiâ recedit, ut error fuerit insensibilis si posuerim esse ut 4 ad 3, idque maximè cùm aquae refractio non perpetim eadem maneat, sed a caloris vicissitudine nonnihil patiatur

---

(2) Newton described how to construct such a prism in §44 = II, 31. Though he does not state how he determined the twenty-two indices of refraction for his table of refractive powers in the *Opticks* (Bk. II, Pt. III, Prop. X), the surviving work sheets and drafts for the table (Add. 3970, ff. 304ᵛ, 362, 365–70, 380ᵛ, 472ᵛ) show that he used the method of minimum deviation with prisms having a large refracting angle; see Lohne, "Newton's table of refractive powers." For the fluids in the table he used a hollow prism with a vertex angle of 78°50′ (f. 304ᵛ), which is "about 80°" and clearly not the (exactly) 90° glass-wood prism described in the next article.

in the shape of a wedge whose edge or vertex angle is about 80° or 90°.[2] You will, however, have to know the quantity of that angle from the most exact measurement, and always set the sine of half of it for the sine of refraction. Having carried this out, when the refractive force of any fluid is required, fill the vessel with that fluid and set it in a position so that its edge, formed by the intersection of the refracting planes, is parallel to the horizon and perpendicular to the solar rays, and also so that those rays transmitted through these refracting planes undergo equal refractions upon entering and emerging. Then using a quadrant, as was shown at Fig. 44, find the angle of incidence, the sine of which will be to the specified sine of refraction as the sine of incidence is to the sine of refraction from air into the proposed fluid.

For example, to learn the refraction of water I arranged for the construction of a wooden prism such as *ABκ* (Fig. 47) whose angle *ACB* (which I designated as the vertex) was a right angle and the other two half a right angle. I caused the refracting planes *Aκ* and *Bκ* to be bored through the middle with a hole *F* parallel to the base *Aβ* for the light to pass through; and I also caused the third plane *Aβ* to be pierced at *G* until a passage extended across to the hole *F*. Then taking two glass plates, which a shattered mirror provided to me, I fastened the one plate *DE* with cement over the middle of the plane *Bκ* and the other over the middle of the other plane *Aκ* in order to seal the passage[3] *F* on both sides. Then I poured rain water through the opening *G* into the hollowed space and sealed it with a stopper cut from cork. Hence the water, placed between the two glass plates inclined at a right angle, assumed the role of a right-angled aqueous prism. But by applying a square I determined that those plates comprehended exactly a right angle, half of which, 45°, must therefore be taken for the angle of refraction. I then so placed this prism at the light's entrance into a dark room that the quantity of refraction would be the same on both sides. From the sun's altitude and the inclination to the horizon of the refracted green-making rays, I found the deviation[4] to be 51°16′, half of which, 25°38′, together with the angle of refraction, 45°, will give the angle of incidence, 70°38′. The sines of these angles, 70°38′ and 45°, are 9434[5] and 7071 with respect to sin 90° as 10000. The ratio of these numbers is in fact a little smaller than the Cartesian 250 to 187 and[6] a little more than 4 to 3, namely, 4.002 to 3.[7] Yet this deviates from the ratio 4/3 by so small a difference that the error would be insensible if I assumed it to be 4 to 3; particularly since the refraction of water does not always remain the same, but it is affected somewhat by the ·

108 [= I, 35]. The refraction of water, as I myself have measured it, is set forth as an example of this.

Lemma 1, §102

(3) Originally: *foramen* (hole).

(4) Newton first mistakenly wrote "*angulum refractionis*" (angle of refraction).

(5) Newton originally had 51°10′ for the deviation, and consequently 25°35′ for its half, 70°35′ for the angle of incidence, and 9431 for the sine.

(6) *paulo . . . et*] Added in revision with the larger value for the deviation. Descartes gave this "measured" value, without any hint as to how he determined it, in his *Meteora*, Ch. VIII, §10, *Oeuvres*, 6:706, 337.

(7) Originally: 4,001 *ad* 3. In the table of refractive powers in the *Opticks*, the index of refraction of water for yellow rays is 529/396 (= 1.3359), which necessarily gives a greater value for green rays than the present 9434/7071 (= 1.3342).

variosque densitatis gradus induat:[8] Quod idem et aeri circundanti contingit, qui a vaporibus etiam non solùm variè incrassatur, sed et arctiùs (auctâ Atmosphaerae gravitate) vel laxiùs comprimitur[.][9] Adde quòd aquarum ex diversis terrarum regionibus scaturientium aut vi solis in vapores et pluviam deinde conversarum diversae sint densitates, et internae dispositiones ad refringendum, ortae ex varijs mineralium tincturis, quas e locis subterraneis extrahunt, et exhalationibus variè crassis aut[10] copiosis, quae simul cum vaporibus in altum attolluntur.[11]

109 [= I, 36].
Praefatorum
Demonstratio.

Problematis hujus de refractionis fluidorum mensurâ sic soluti veritas constabit ex ostenso quod refractionis in hoc prismate ex aquâ et vitris composito eadem est quantitas, quae foret si vitrum tolleretur, et aqua sola maneret aëre circundata. Sit itaque *ABC* prisma in fig [48] confectum ex laminis vitreis *ACφδ*, et *BCφε* (ut dictum est,) et aquâ *δφε* repletum; et concipiatur quòd *DEF* sit aqueum prisma immediatè circundatum aere, et omninò simile aquae *δεφ* circumclusae vitro, similitérque positum. Et incidant radij paralleli *ON*, *OX* in utrumque; quorum alter *ON* refractus in *N M L*, et *K* tendit ad *H*; alter verò *OX* refractus in *X* et *Y* tendit ad *Z*. Dico jam quòd emergentes *KH* et *YZ* erunt paralleli, atque adèo quòd in utroque prismate tota refractionum quantitas erit eadem.

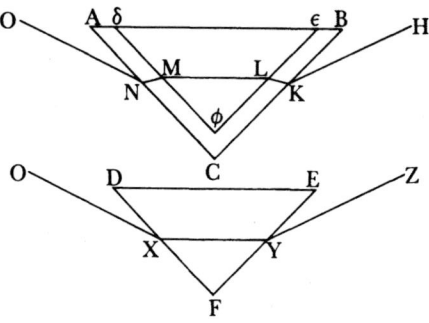

Figure 48

Etenim in fig [46] si radius *ωμ* ipsi *ON* parallelus incidat in vitream laminam *A*, emergatque in *λκ*, notum est quòd radius *λκ* erit parallelus ipsi *ωμ*, hoc est ipsis *ON* et *LK*; et cùm *λκ* et *LK* sint paralleli, erunt etiam *μλ* et / *ML* paralleli. Unde liquet propositio, quòd   81/

> Quantitas refractionis ex aëre in Medium quodvis propositum est eadem sive radij immediatè[12] ingrediantur istud Medium ex aëre (ut fit ad *ωμλ*,) sive priùs permeent aliud Medium interpositum, et parallelis planis terminatum (uti fit ad *ONML*): Et e contra. Atque idem intellige, cùm vice aeris aliud quodpiam adhibetur Medium.[13]

Quare in fig. [48] cùm paralleli radij *OX* et *ON* incidant in prismata *DFE* & *ACB* similia et similiter posita, refractionis quantitas ex aere in aquam erit eadem sive radij immediatè intrent, ut videre est ad *DEF*, sive priùs permeent lamellam vitream *AδφC*: hoc est, radius *XY* semel refractus erit parallelus ipsi *ML* bis refracto. Et ob eandem rationem cùm *XY* et *ML* sint paralleli, radij emergentes *YZ* et *KH* erunt etiam paralleli. Quare cùm radij et inci-

---

(8) Descartes (*Meteora*, Ch. VIII, §11, *Oeuvres*, 6:707, 340) had observed that the refraction of water varies with its temperature, and Hooke (*Micrographia*, Observation LVIII, p. 220) demonstrated that it varies with density.

change of heat and assumes various degrees of density.[8] The same thing happens to the ambient air, which in addition is not only variably thickened by vapors but is also compressed more closely (by the increased weight of the atmosphere) or more loosely.[9] Moreover, the density of the water flowing from different regions of the earth or transformed by the force of the sun into vapors and then into rain differs, and its internal disposition to refract also differs, because of various mineral tinctures extracted from subterranean locations and variably thick or[10] abundant exhalations that are carried aloft together with vapors.[11]

The validity of this problem of the measurement of the refraction of fluids solved in this way will be established by showing that the quantity of refraction in this prism composed of water and of glass is the same as it would be if the glass were removed and only water surrounded by air remained. Therefore (Fig. 48), let ABC be a prism made (as was described) from the glass plates $AC\phi\delta$ and $BC\phi\epsilon$ and filled with water $\delta\phi\epsilon$; and imagine that DEF is a prism of water immediately surrounded by air, entirely similar to the water $\delta\epsilon\phi$ enclosed by glass, and similarly positioned. Let there fall upon each prism the parallel rays ON and OX, of which the one ray ON, refracted at N, M, L, and K, tends to H, whereas the other OX, refracted at X and Y, tends to Z. I say now that the emergent rays KH and YZ will be parallel, and hence that the total quantity of the refractions in each prism will be the same. For in Fig. 46, if the ray $\omega\mu$ parallel to ON falls upon the glass plate A and emerges in $\lambda\kappa$, it is known that the ray $\lambda\kappa$ will be parallel to $\omega\mu$, that is, to ON and LK; and since $\lambda\kappa$ and LK are parallel, $\mu\lambda$ and ML will also be parallel. Whence the proposition is evident:

> The quantity of refraction from air into any proposed medium is the same whether the rays immediately[12] enter that medium from air (as occurs with $\omega\mu\lambda$) or first cross another interposed medium bounded by parallel planes (as occurs with $ONML$), and conversely. The same thing must be understood when in place of air any other medium is used.[13]

Accordingly, in Fig. 48, since the parallel rays OX and ON fall upon the similar and similarly positioned prisms DFE and ACB, the quantity of refraction from air into water will be the same whether the rays enter immediately, as can be seen at DEF, or first cross the glass plate $A\delta\phi C$; that is, the ray XY refracted once will be parallel to ML refracted twice. For the same reason, since XY and ML are parallel, the emergent rays YZ and KH will also be

(margin note) 109 [= I, 36]. The demonstration of the preceding.

---

(9) This observation on air is likewise drawn from Hooke (*Micrographia*, p. 221); Newton had taken extensive notes (Newton, *Unpublished Papers*, pp. 411–3) on Hooke's rich Observation LVIII.   (10) II: et (and).

(11) The distinction between vapors (smooth aqueous particles) and exhalations (irregularly shaped, nonaqueous particles) is drawn from Descartes; see *Meteora*, Ch. II, "De vaporibus & exhalationibus," or *Principia philosophorum*, Pt. IV, §LXX.   (12) Added.

(13) Atque . . . Medium.] Added. This proposition was deduced earlier by Maignan; see Lect. 9, note (39).

dentes et emergentes sint paralleli, refractio tota prismatis utriusque est eadem. Atque adeò cùm aqueum prisma aëri contiguum propter fluiditatem
aquae[14] fabricari nequeat, ejus vice liceat adhibere vitreum prisma cum aquâ
repletum.   Q.E.O.

Et sic modus generalis, quo refractiones ex aere in quaelibet proposita
Media determinentur, ostensus est; facillimus quidem et erroribus minimè
obnoxiùs, praesertim si angulus prismatis sit magnus et exactè cognitus,
Quadrans magnus et accuratus, et observatio facta longè post prisma, ubi
colores multùm dilatati faciliùs distinguuntur. Et praeterea, cùm refractiones
inter aerem et Media proposita sic experientijs determinantur; indicata est
regula,[15] quâ Mediorum eorundem sibi ipsis contiguorum refractiones eliciantur. Quod satis est in gratiam primi casûs de refractionibus dimetiendis
cùm in eodem quopiam radiorum genere proportio sinûs incidentiae, et refractionis quaeritur, ostendisse.

110 [≈ I, 37].
Radiorum diversi
generis refractiones
conferuntur,
et maxima
refrangibilitatis
differentia
investigatur[16]

Prosequendus est jam alter casus, ubi heterogeneorum radiorum refrac-
tiones sunt conferendae; | Et proportiones sinuum refractionis investigandae
sunt cùm eorum incidentia supponitur eadem. Id quòd ex ostensis quo-
dammodò praestari potest, sed convenit, ut aliquid ampliùs urgeam. | Et
quoniam de intermedijs radiorum generibus facilè esset judicium proferre, si
modò refractiones extremorum forent cognitae: satisfecero si radios maximè
omnium refrangibiles cum minimè refrangibilibus[17] comparavero. Itaque in
fig [49] sit *ABC* prisma vitreum, ita positum, ut radij tum ingredientes tum

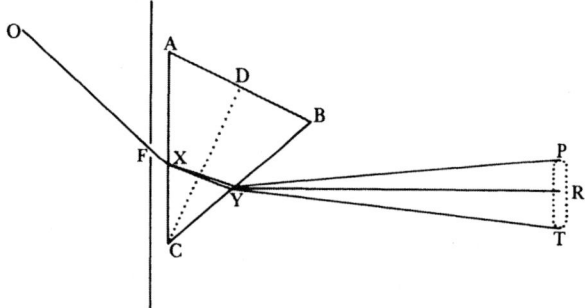

Figure 49

egredientes eandem quantitatem refractionis / ut priùs patiantur. Dies autem   82/
seligatur splendidus, et cubiculum esto valde obscurum, ut colores usque ad
ultima quae occupent spatia distinctè satis videri possint. Tunc ad distantiam
viginti pedum aut ampliùs a prismate radij excipiantur in papyrum aliquam
directè obversam, et spatij a coloribus illuminati (ut *PT*) longitudo et latitudo
mensuretur. Sic prismate adhibito cujus angulus verticalis *ACB* fuit 63^{grad}
12^{min}, et latitudine foraminis radios intromittentis existente $\frac{1}{4}^{ta}$ parte digiti: Ad
distantiam *XP* vel *XT* 22^{pedum} inveni maximam longitudinem imaginis *PT* esse
13$\frac{1}{4}^{dig}$ circiter et latitudinem 2$\frac{5}{8}^{dig}$.[18] Jam si latitudo hujus imaginis ab ejus

---

(14) propter . . . aquae] Added.

parallel. Therefore, since the incident and emergent rays are parallel, the total refraction of each prism is the same. Hence, since an aqueous prism contiguous to air cannot be constructed because of the fluidity of water,[14] one may use in its place a glass prism filled with water. As was to be shown.

Thus a general method has been presented whereby the refractions from air into any proposed medium may be determined, and indeed very easily and hardly subject to errors; especially if the angle of the prism is large and exactly known, the quadrant is large and accurate, and the observation is made far behind the prism, where the colors being greatly spread out are more easily distinguished. Moreover, when the refractions between air and the proposed media are thus determined by experiments, a rule has been disclosed[15] whereby the refractions of the same media when contiguous to one another may be derived. It suffices to have shown this for the first case of measuring refractions, when the ratio of the sine of incidence and of refraction for any one kind of ray is sought.

The other case must now be pursued, where the refractions of heterogeneous rays are to be compared, | and the proportions of the sines of refraction are to be investigated when their incidence is supposed to be the same. This can in some way be accomplished from what has already been shown, but it is appropriate that I press somewhat further ahead. | Since it would be easy to make a judgment about the intermediate kinds of rays, provided that the refractions of the extreme ones were known, it will be sufficient for me to have compared the most refrangible of all rays with the least refrangible.[17] Therefore, in Fig. 49 let *ABC* be a glass prism so placed that, as earlier, both the entering and emerging rays undergo the same quantity of refraction. Choose a bright day, however, and let the room be very dark, so that the colors can be seen sufficiently distinctly to the farthest space they occupy. Then at a distance of twenty feet or more from the prism receive the rays on a paper directly opposite them, and measure the length and breadth of the space, such as *PT*, illuminated by the colors. In this way, using a prism whose vertex angle *ACB* was 63°12′, and with the breadth of the hole admitting the rays being $\frac{1}{4}$ inch, at a distance *XP* or *XT* of twenty-two feet I found the greatest length of the image *PT* to be approximately $13\frac{1}{4}$ inches and its breadth $2\frac{5}{8}$ inches.[18] Now if the breadth of this image is subtracted from its

110 [≈ I, 37]. The refractions of different kinds of rays are compared, and the greatest difference of refrangibility is investigated.[16]

(15) In II Newton added a marginal note to §I, 32 [= 105].

(16) Originally: *primò investigatur* (first investigated).

(17) *maximè . . . refrangibilibus*] Originally: *intensè purpuriformes, et rubriformes* (the intense purple-making and red-making). See §136 where Newton distinguishes between extreme and intense colors in defining the measure of the spectral length.

(18) Newton reported these values in the "New theory" (*Correspondence*, 1:93 = *Phil. Trans.*, 6 (1671/2):3077). His earliest recorded measurement, in his second essay "Of Colours" (§7, Add. 3975, f. 2ᵛ = Hall, "Further experiments," pp. 28–9), yielded a smaller spectral length, but it was only a rough measurement. In the "Fundamentum Opticae" (Add. 3970, ff. 421ʳ, 411ʳ) and initially in the manuscript of the *Opticks* (ff. 96ʳ, 97ʳ), he recounted new measurements on this 63°12′ prism, but they scarcely differed from the values reported here. In the published version of the *Opticks*, however, he substituted a series of new measurements on different prisms, having vertex angles of $62\frac{1}{2}$° and $63\frac{1}{2}$° and significantly smaller dispersive powers; see note (31), this lecture.

longitudine subtrahatur, manebit $10\frac{5}{8}$ digiti pro longitudine quam habere debuisset si discus Solis et foraminis *F* diameter fuissent infinitè parva. Hoc est si radij advenissent omnes in eâdem rectâ *OF*. Ista itáque linea $10\frac{5\text{digitorum}}{8}$ subtendit angulum quem radij duo similiter incidentes per inaequalitatem refractionis constituunt, quorum alter maximè omnium similiter incidentium & alter minimè omnium refringitur: qui proinde angulus ex calculo reperietur 2^{grad} 18^{min}.[19] Verùm cùm angulus iste binâ refractione ad *X* et *Y* conficiatur, et praeterea cùm utraque supponatur aequalis; calculus ad hoc negotium satis accuratus ex unicâ tantùm refractione poterit institui, puta quae conficitur ad latus *BC*. Etenim si verticalis angulus *ACB* plano *DC* bisecetur, et alterum prismatis dimidium *DCB* vel *DCA* concipiatur tolli, refractio ad alterum dimidium facta (radijs *OF* obliquè incidentibus in latus *AC* et perpendiculariter emergentibus e latere *DC*, vel perpendiculariter incidentibus in latus *DC* secundum unicam quandam lineam *XY* et obliquè emergentibus e latere *BC*) refractio, inquam, sic ad alterum dimidium facta foret semissis refractionis ad integrum prisma, si modò unicum quodpiam radiorum puta mediocriter refrangibilium[20] genus spectetur[.] Quinetiam si caetera omnia radiorum genera simul spectentur, assertio illa licèt non ampliùs sit absolutè vera, tamen veritati tam proximè accedit, ut quoad sensum et calculum mechanicum[21] pro verâ habeatur. Quamobrem cùm refractiones / utrinque   83/ ad *X* et *Y* peractae computatio geometrica aegriùs institui possit, istud more ad praxin magis accommodato, ut ut mechanico, perficere non verebor, confisus id mihi vitio verti non debere, si dum computationes rebus physicis adhibeo, minutias quae operam molestè et sine fructu producerent, missas faciam. Refractionem itaque ex unicâ tantùm parte prismatis perpendam; et quoniam omnes radij, demptis viridiformibus, a dimidio *ACD* bis deberent refringi, et semel tantùm ab altero dimidio *DCB*, perpendiculariter ingressi latus planum *DC* secundum lineam *XY*: itaque in dimidio *DCB* fiat calculus, hoc est ad latus planum *BC*; supposito quòd omnibus radijs secundum eandem lineam *XY* allapsis, angulus quem maximè refrangibiles cum minimè refrangibilibus[22] postquam refringerentur a latere *BC*, constituerent, foret dimidium anguli *P,YT*, hoc est 1^{grad} 9^{min}. Jam cùm angulus incidentiae radij *XY*

Sect 103

ex praemonstratis sit 31^{gr} 36^{min} et angulus refractionis mediocris 54^{gr} 10^{min},[23] transferantur haec omnia in Schema [50] ponendo quòd *CB* sit superficies disterminans Medium vitreum versus *A*, et aereum ver-

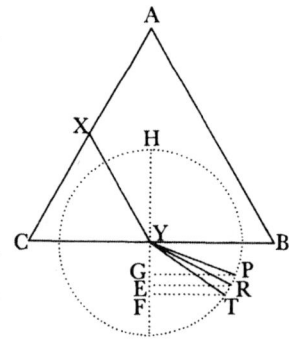

Figure 50

---

(19) That is, the angular dispersion $\Delta D = (l - b)/s = 10\frac{5}{8}$ in/22 ft = 2°18′, where $(l - b)$ is the "diminished length," or the spectral length *l* minus its breadth *b*, and *s* is the distance from the spectrum to the prism. A similar calculation for the spectral breadth gives 30′56″ for the apparent solar diameter, or "about 31′" as he reported in the "New theory"; compare Lect. 2, note (16).

(20) mediocriter refrangibilium] Originally: viridiformium (green-making).

length, $10\frac{5}{8}$ inches will remain for the length it ought to have had if the sun's disk and the diameter of the hole *F* had been infinitely small; that is, if all the rays had arrived in the same straight line *OF*. That $10\frac{5}{8}$-inch line thus subtends an angle that two similarly incident rays make because of the inequality of the refraction, one ray of which is refracted the most of all similarly incident ones and the other the least of all. This angle, consequently, will be found by calculation to be 2°18′.[19] But since that angle is made by a twofold refraction at *X* and *Y*, and moreover since both are assumed to be equal, a calculation sufficiently accurate for this purpose can be carried out from only one refraction, for instance, from that made at the side *BC*. For if the vertex angle *ACB* were bisected by the plane *DC*, and one half of the prism, either *DCB* or *DCA*, were imagined to be removed, then the refraction made at the other half (when the rays *OF* fall obliquely onto the side *AC* and emerge perpendicularly from the side *DC*, or fall perpendicularly along some one line *XY* onto the side *DC* and emerge obliquely from the side *BC*), the refraction, I repeat, thus made in the other half would be one half of the refraction in the whole prism, if only some one kind of ray, suppose the mean refrangible,[20] were considered. Moreover, if all the other kinds of rays were considered simultaneously, that assertion, although no longer absolutely true, would still so closely approach the truth that it could be taken as true with respect to sense and a mechanical[21] calculation. Consequently, since a geometric calculation of the refraction made at each side, at *X* and *Y*, can be undertaken rather difficultly, I will not be afraid to do it in a way which, however mechanical, is more suited to practice, being confident that no fault ought to be attributed to me if, when I perform computations in physical matters, I omit the minutiae that entail burdensome and fruitless work. I will accordingly consider the refraction from only a single side of the prism. And since all rays, except the green-making ones, must be refracted twice by the half *ACD*, and only once by the other half *DCB* (since they enter the plane side *DC* perpendicularly along the line *XY*), therefore let the calculation be made in the half *DCB*, that is, at the plane side *BC*, while supposing that for all rays flowing along the same line *XY*, the angle that the most refrangible rays would make with the least refrangible[22] ones after they were refracted by the side *BC* would be one half of the angle *PYT*, that is, 1°9′. Now, since the angle of incidence of the ray *XY* is 31°36′ (from what has been shown before), and the angle of the mean refraction is 54°10′,[23] transfer all these to §103 Fig. 50, while assuming that *CB* is the surface separating the glass medium

---

(21) Added. By "mechanical" Newton, following Descartes's distinction, means here an approximate solution, "taking as geometric that which is precise and exact, and mechanical that which is not" (*La Geometrie*, Bk. II, *Oeuvres*, 6:389). For Newton's mature views on the classical Greek distinction between the geometrical and mechanical, see his writings on geometry from the 1690s in *Mathematical Papers*, Vol. 7, Pt. 2.

(22) maximè . . . refrangibilibus] Originally: purpuriformes cum rub[r]iformibus (the purple-making . . . with the red-making). Newton frequently omitted an *r* from "rubriform" and henceforth it will be ignored.

(23) This measurement was added in the *Optica* in §I, 30 ≈ 103.

sus *F*, et quod angulus incidentiae *XYH* sit 31^{gr} 36^{min}: eritque angulus refractionis *RYF* 54^{grad} 10^{min}; et angulus *PYT* 1^{gr} 9^{min}, differentia nempe refractionis inter maximè refrangibiles *YP* et minimè refrangibiles *YT*.[24] Qui angulus a radio *YR* mediocriter refracto et confinium caerulei ac viridis occupante, bisecatur; Et proin ang *PYR* vel *RYT* erit 34$\frac{1}{2}$^{min} dimidium totius *PYT*. Adeoque ang: *PYE* 54^{gr} 44$\frac{1}{2}$^{min}, et ang: *TYE* 53^{gr} 35$\frac{1}{2}$^{min}. Et eorum sinus *PG* ac *TF* erunt 81656 & 80481: quorum proportione ad simpliciores numeros redactâ erit *TF* ad *PG* ut 69$\frac{1}{2}$ ad 68$\frac{1}{2}$.[25] Ad hunc modum experimenta et calculum cùm saepius instituerim, horum sinuum proportiones inter terminos 67 ad 66 et 72 ad 71[26] semper obvenerunt; sed ut plurimùm incidi in proportiones 69 ad 68, 69$\frac{1}{2}$ ad 68$\frac{1}{2}$, & 70 ad 69,[27] quarum tantilla est differentia, ut parvi intersit quaenam adhibeatur.

111 [= I, 38].
Illarum refractionum sinus ad communem sinum incidentiae conferuntur.[28]

Ratione sinuum refractionis pro extremis radiorum similiter incidentium generibus sic inventâ, eorum comparatio ad sinum incidentiae[29] simul innotescit, quippe qui paulo ante inventus est 52400.[30] Et conferendo hunc 52400 ad sinus 81656 & 80481, eorum ratio in minoribus numeris reperietur 44$\frac{1}{2}$ ad 69$\frac{1}{2}$ & 68$\frac{1}{2}$: aut / 44$\frac{1}{4}$ ad 69 & 68 fere.[31] Refractionibus nempe ex vitro 84/ in aerem peractis.

---

(24) maximè . . . *YT*.] Originally: purpuriformes *YP* et rubriformes *YT*. (the purple-making *YP* and the red-making *YT*.)

(25) *TF* . . . 68$\frac{1}{2}$] Read "*PG* ad *TF*"—a slip related to Newton having originally written "68$\frac{1}{2}$ ad 69$\frac{1}{2}$"; also, in the *Optica* he adds "circiter" (approximately). To determine these values Newton requires that the ratio *P*/*T* of the sines of refraction of the extreme rays be of the form $(x + 1)/x$. Since he found *P*/*T* = 81656/80481, this gives $x = 80481/1175 = 68.4945 \approx 68\frac{1}{2}$, and so *P*/*T* = 69$\frac{1}{2}$: 68$\frac{1}{2}$. By cleverly reducing this problem to the equivalent one of refraction in half the prism—where the angle of refraction remains unchanged, whereas the deviation and angular dispersion are simply halved—Newton was able to apply the expression he derived for the index of refraction at minimum deviation to the extreme rays, which are not quite at minimum deviation. Since the mean refrangible rays are refracted solely at the surface *CB*, by the sine-law of refraction $n = \sin i'/\sin r'$, where the angle of incidence $r' = \frac{1}{2}C$, and the angle of refraction $i' = \frac{1}{2}(D + C)$, just as in §102 = I, 29. To calculate the index of refraction for the extreme rays Newton simply replaced i' by $(i' \pm \frac{1}{4}\Delta D)$, where ΔD is the angular dispersion in the whole prism, thus giving

$$n \pm \tfrac{1}{2}\Delta n = \sin (i' \pm \tfrac{1}{4}\Delta D)/\sin r',$$

with the positive sign being taken for most refrangible rays, and the negative one for least refrangible rays. Since $\frac{1}{4}\Delta D$ is very small, this becomes $n \pm \frac{1}{2}\Delta n \approx (\sin i' \pm \frac{1}{4}\Delta D \cos i')/\sin r'$, or $\Delta n/n \approx \frac{1}{2}\Delta D \cot i'$, a now standard approximation that introduces an error orders of magnitude less than Newton's experimental error. Measuring the spectral length—rather than independently determining the index of refraction for the extreme colors by adjusting the prism for minimum deviation for each color and remeasuring the angles—minimizes the effects of random errors on the chromatic dispersion ($\Delta n = n_P - n_T$), because it is the difference of two nearly equal numbers; and a small random error in either, if they were determined independently, would cause a very large error. Newton discusses the various sources that can cause a change in the spectral length, and thus the dispersive power, in his letter to Oldenburg for Anthony Lucas on 18 August 1676; *Correspondence*, 2:76–81 = *Phil. Trans.*, 11 (1676):698–705. In the *Opticks* (Bk. I, Pt. I, Prop. VII, pp. 60–1 = Add. 3970, ff. 62–3), Newton set forth the measurements

toward *A* and the air toward *F*, and that the angle of incidence *XYH* is 31°36'. The angle of refraction *RYF* will be 54°10' and the angle *PYT* will be 1°9', namely, the difference of refraction between the most refrangible rays *YP* and the least refrangible *YT*.[24] Let this angle be bisected by the mean refracted ray *YR*, occupying the boundary of blue and green, and then the angle *PYR* or *RYT* will be $34\frac{1}{2}'$, half of the whole angle *PYT*. Hence the angle *PYE* will be $54°44\frac{1}{2}'$; the angle *TYE* will be $53°35\frac{1}{2}'$; and their sines *PG* and *TF* will be 81656 and 80481. And reducing their ratio to simpler numbers, *TF* to *PG* will be as $69\frac{1}{2}$ to $68\frac{1}{2}$.[25] When I had in this way frequently carried out the experiments and calculations, the ratios of their sines always turned out between the limits of 67 to 66 and 72 to 71;[26] but most often I came upon the ratios of 69 to 68, $69\frac{1}{2}$ to $68\frac{1}{2}$, and 70 to 69[27]—whose difference is so small that it matters little which is used.

After the ratio of the sines of refraction for the extreme kinds of rays similarly incident has thus been found, their comparison to the sine of incidence[29] becomes known at the same time, inasmuch as a little earlier it was found to be 52400.[30] Comparing this 52400 to the sines 81656 and 80481, their ratio in smaller numbers will be found to be $44\frac{1}{2}$ to $69\frac{1}{2}$ and $68\frac{1}{2}$, or nearly $44\frac{1}{4}$ to 69 and 68,[31] that is, when the refractions are made from glass into air.

111 [= I, 38]. The sines of those refractions are compared to the common sine of incidence.[28]

and calculations required to determine the chromatic dispersion according to this approximation, but without the physical justification presented here, just as he had done for the index of refraction. Once again this sketch was sufficient for Clairaut, d'Alembert, Bošković, and Euler to recover and extend Newton's method, expressing it in analytic form and deriving it by differentiating the exact ray equations which were circumvented by Newton; see Lect. 9, note (26).

(26) 67 ad 66 et 72 ad 71] Newton initially had "65 ad 66 et 70 ad 71," which he then corrected to "66 ad 65 et 71 ad 70." In revision he altered this to "66 ad 65 et $71\frac{1}{2}$ ad $70\frac{1}{2}$" before finally changing it to its current reading. By adopting Newton's assumption that the mean index of refraction $\bar{n} = R/I$ is a constant throughout this series of measurements, the values of I corresponding to his extreme measurements are found to be approximately 43 and $46\frac{1}{4}$, where R is $66\frac{1}{2}$ and $71\frac{1}{2}$. Consequently, the dispersive power, $1/\nu = 1/(R - I) = \Delta n/(\bar{n} - 1)$, ranges from $\nu \approx 23\frac{1}{2}$ to $25\frac{1}{4}$, and thereby gives Newton's own measure of his experimental error to be about $7\frac{1}{2}\%$.

(27) 70 ad 69] Originally: $68\frac{1}{2}$ ad $67\frac{1}{2}$.

(28) Originally continued: *et ad vitrum determinantur* (and determined at glass).

(29) Originally continued: *facilis est* (is easy).

(30) Namely, in §103 ≈ I, 30.

(31) Since $P/I = 81656/52400$ and P was found to be $69\frac{1}{2}$, then $I = 69\frac{1}{2}$ (52400/81656) = 44.5993 ≈ $44\frac{1}{2}$; the adjusted ratio $44\frac{1}{4}$ to 69, however, is a better approximation to the ratio of sines. In the "New theory" Newton rounded these values off still more, adopting 44 to 69 and 68; *Correspondence*, 1:95 = *Phil. Trans.*, 6 (1671/2):3079. Calculating the dispersive power, $1/\nu = \Delta n/(\bar{n} - 1)$ for this glass, we find $\nu = 24\frac{1}{4}$ to $24\frac{1}{2}$. In his table of refractions from about 1667, Newton tabulated ν, and the mean value was $24\frac{1}{4}$, as he had already adopted the ratio of sines $44\frac{1}{4}$ to 69 and 68; Add. 4000, f. 33^v = *Correspondence*, 1:103. The new measurements on this prism adopted in the "Fundamentum Opticae" and initially in the manuscript of the *Opticks* (see note (18), this lecture) yielded a nearly identical $\nu = 25$. However, the measurements on the new prisms added in the *Opticks* (Bk. I, Pt. I, Prop. VII, pp. 59–61 = Add. 3970, ff. 61–3), where the ratios of sines were 50 to 78 and 77, gave a significantly smaller dispersive power, $\nu = 27\frac{1}{2}$ for the $62\frac{1}{2}°$ prism (and, as a simple calculation shows, $\nu = 28$ for the $63\frac{1}{2}°$ prism).

112 [= I, 39].
Radiorum ex oppositis partibus refringentis[32] **superficei incidentium sinus**[33] **sunt reciprocè proportionales.**

Quòd si radij e contra ex aere in vitrum similiter incidant, proportiones sinuum nullo negotio ex jam inventis eruuntur, utpote quae sunt reciprocae. Sit I sinus[34] incidentiae e vitro in aërem, P sinus refractionis maxime refrangibilium radiorum, R mediocriter refrangibilium ac T minime refrangibilium: Dico quòd ex horum reciprocè proportionalibus si $\frac{1}{I}$ ponatur esse sinus incidentiae ex aere in vitrum erit $\frac{1}{P}$ sinus refractionis maximè refrangibilium radiorum $\frac{1}{R}$ sinus refractionis mediocriter refrangibilium, ac $\frac{1}{T}$ minime refrangibilium[.][35] Nam cùm sinus incidentiae radij maxime refrangibilis[36] e vitro in aerem sit I et sinus refractionis P, radij ejus ex aëre in vitrum per easdem lineas retroacti sinus incidentiae erit P et sinus refractionis I, siquidem jam radius est incidens qui priùs fuerit refractus.[37] Est ergo sinus incidentiae radij maxime refrangibilis[36] ex aere in vitrum utcunque incidentis ad sinum refractionis ut P ad I hoc est (applicando rationem[38] ad P) ut 1 ad $\frac{1}{P}$, hoc est (applicando ad I denuò) ut $\frac{1}{I}$ ad $\frac{1}{P}$. Et simili argumento constabit ejusmodi sinus radij mediocriter refrangibilis esse ut $\frac{1}{I}$ ad $\frac{1}{R}$ et sinus minimè refrangibilis[39] ut $\frac{1}{I}$ ad $\frac{1}{T}$. Liquet ergo quod posito $\frac{1}{I}$ communi sinu incidentiae erunt $\frac{1}{P}$, $\frac{1}{R}$, & $\frac{1}{T}$ singulorum generum respectivè sinus.[40]

113 [= I, 40].
Illustratur refractione vitri

Rem numeris illustro. Cùm $44\frac{1}{4}$ ad 69 & 68 sit ratio sinus communis incidentiae ad sinus maximè discrepantium refractionum e vitro in aerem: sinus incidentiae communis ad sinus refractionum ex aere in vitrum erit ut $\frac{1}{44\frac{1}{4}}$ ad $\frac{1}{69}$ & $\frac{1}{68}$, sive $\frac{69 \times 68}{44\frac{1}{4}}$ (= 106 ferè) ad 68 et 69. Hoc est pro radijs maximè refrangibilibus sinus incidentiae ad sinum refractionis ut 106 ad 68 & pro minimè refrangibilibus,[41] ut 106 ad 69.

114 [= I, 41].
E refractionibus extremorum generum facile est de intermedijs conjecturam facere.

Hisce sic determinatis rationes sinuum pro radijs intermedijs facilè determinantur ex cognitis colorum distantijs quas in imagine coloratâ observant[.][42] Sic radij[43] qui ad caeruleum magis[44] quàm flavum vergunt, cùm in mediam imaginem cadant intermediam rationem sinuum $44\frac{1}{4}$ ad $68\frac{1}{2}$[45] vel 106 ad $68\frac{1}{2}$[46] habebunt. Et sic de alijs.

---

(32) **II**: ejusdem refringentis (of the same refracting [surface]).

(33) incidentium sinus] Originally: secundum eandem lineam incidentium sinus refractionum (the sines of refraction [of rays] incident along the same line).

(34) **II**: communis sinus (the common sine).

(35) maximè refrangibilium ... mediocriter refrangibilium ... minime refrangibilium] Originally: purpuriformium ... viridiformium ... rubriformium (purple-making ... green-making ... red-making).

(36) maxime refrangibilis] Originally: purpuriformis (purple-making).

(37) The principle of reversibility was to be made Prop. 1 in the *Optica*, §I, 53.

(38) **II**: terminos rationis (the terms of the ratio).

(39) mediocriter refrangibilis ... minimè refrangibilis] Originally: viridiformis ... rubiformis (green-making ... red-making).

If now, conversely, the rays are similarly incident from air into glass, the proportions of the sines are derived with little difficulty from what has already been found, since they are reciprocals. Let I be the sine[34] of incidence from glass into air, P the sine of refraction of the most refrangible rays, R that of the mean refrangible ones, and T that of the least refrangible. I say that from the reciprocal proportions of these, if $\frac{1}{I}$ is assumed to be the sine of incidence from air into glass, $\frac{1}{P}$ will be the sine of refraction of the most refrangible rays, $\frac{1}{R}$ that of the mean refrangible ones, and $\frac{1}{T}$ that of the least refrangible.[35] For since the sine of incidence of a most refrangible[36] ray from glass into air is I and the sine of refraction P, then the sine of incidence of its ray going back along the same lines from air into glass will be P and the sine of refraction I, because the incident ray now was the refracted ray before.[37] Therefore the sine of incidence of a most refrangible[36] ray from air into glass, however incident, is to the sine of refraction as P to I, that is (dividing the ratio[38] by P), as 1 to $\frac{1}{P}$, that is (dividing again by I), as $\frac{1}{I}$ to $\frac{1}{P}$. By a similar argument it will be evident that the sines of this sort for a mean refrangible ray are as $\frac{1}{I}$ to $\frac{1}{R}$, and the sines of a least refrangible[39] one as $\frac{1}{I}$ to $\frac{1}{T}$. It is clear therefore that setting $\frac{1}{I}$ as the common sine of incidence, $\frac{1}{P}$, $\frac{1}{R}$, and $\frac{1}{T}$ will be the sines of the individual kinds of rays respectively.[40]

I will illustrate this with numbers. Since $44\frac{1}{4}$ to 69 and 68 is the ratio of the common sine of incidence to the sines of the most different refractions from glass into air, the common sine of incidence to the sine of refraction from air into glass will be as $1/44\frac{1}{4}$ to $1/69$ and $1/68$, or $(69 \times 68)/44\frac{1}{4}$ (≈ 106) to 68 and 69. That is, for the most refrangible rays the sine of incidence to the sine of refraction is as 106 to 68, and for the least refrangible[41] as 106 to 69.

After these have been thus determined, the ratios of the sines for the intermediate rays are easily determined from the known distances that the colors maintain in the colored image.[42] Thus the rays[43] that incline more[44] to blue than to yellow, since they fall in the middle of the image, will have the intermediate ratio of sines, $44\frac{1}{4}$ to $68\frac{1}{2}$,[45] or 106 to $68\frac{1}{2}$;[46] and similarly for the others.

112 [= I, 39]. The sines of rays incident[33] from opposite sides of a refracting surface[32] are reciprocally proportional.

113 [= I, 40]. It is illustrated by the refraction of glass.

114 [= I, 41]. From the refractions of the extreme kinds of rays, it is easy to infer the intermediate ones.

(40) This theorem extends to rays of different refrangibility a theorem formulated by Barrow for rays without distinction as to their refrangibility; *Lectiones XVIII*, Lect. II, §IX.

(41) maximè refrangibilibus ... minimè refrangibilibus] Originally: extremè purpuriformibus ... extremè rubriformibus (extreme purple-making ... extreme red-making).

(42) Newton pursues this idea in Lect. II, 11, on the musical division of the spectrum, added in the *Optica*, especially in §II, 100.

(43) Originally: viridiformes (green-making ones).

(44) II: paulo magis (slightly more).

(45) $44\frac{1}{4}$ ad $68\frac{1}{2}$] Read: $\frac{1}{44\frac{1}{4}}$ ad $\frac{1}{68\frac{1}{2}}$.

(46) II continues: circiter (approximately).

Lect 11
115 [= I, 42].
Theoremate
ostenditur ut e
refractionibus
heterogeneorum ad
vitrum vel quodvis
Medium inter se
determinatis,
possunt etiam ad
alia quaelibet Media
aeri contigua
refractiones (sine
novis experimenti
molestijs) inter se
determinari

/ Ad eundem modum quo refractiones ad vitrum determinatae sunt id 85/ ipsum posset fieri ad alia Media. Sed e re erit ut regulam jam ostendam, quâ refractionum istarum mensurae ex sinubus earum sic ad vitrum determinatis, possunt determinari ad quodlibet aliud Medium propositum, idque licèt istud sit alij Medio quàm aeri contiguum. In fig: [51] sit *AB* superficies terminans

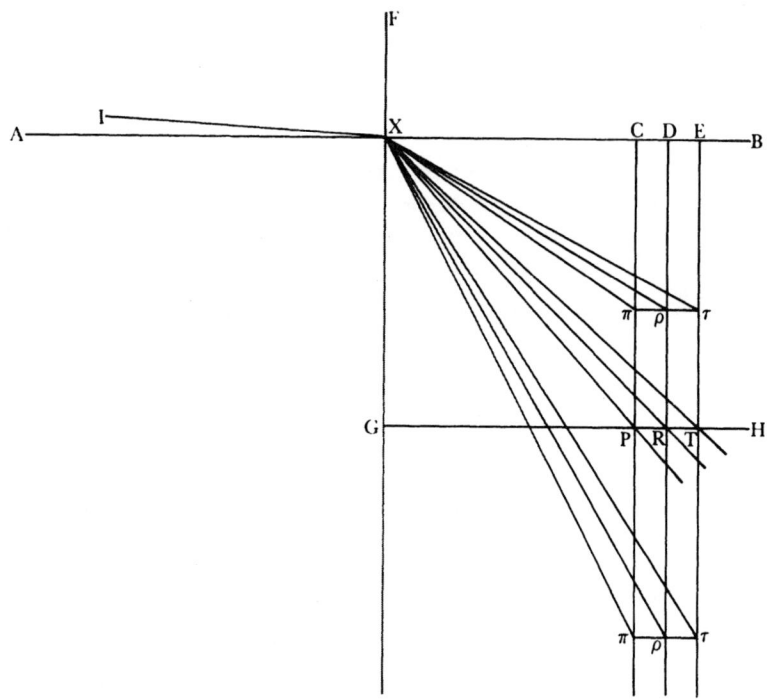

Figure 51

aerem ex parte *F*, et vitrum ex parte *G*, ad cujus aliquod punctum *X* ducatur linea *FXG* ei perpendiculariter insistens: et praeterea concipiatur rectam *IX* ad angulum *IXA* infinitè parvum duci, secundum quam omnes omnium formarum radij supponantur incidere et in *X* refringi puta mediocriter refrangibiles versus *R* maximè refrangibiles versus *P*, et minimè refrangibiles[1] versus *T*, aliosque intermedios verus intermedias plagas. Porrò ducatur linea quaevis *GH* parallela ad lineam incidentiae *IX*, hoc est perpendicularis ad *FG*. Ea verò secet radios in punctis *P R* ac *T*, a quibus demittantur *PC RD* ac *TE* perpendiculares ad refringentem superficiem *AB*. His ad vitrum sic determinatis ac descriptis, si aliud quodvis Medium in locum vitri jam concipiatur substitui, caeteris stantibus; et radij alicujus mediocriter refrangibilis[2] secundum lineam *IX* incidentis ad *X* refractus *Xρ* ducatur secans rectam *DR* in *ρ*, (Quod fieri suppono, siquidem modum quo mediocriter refrangibilium refractiones[3] ad Media quaelibet investigari possint, antehac exposui.)[4] Dein per punctum *ρ* ducatur recta *ππ* secans lineas *CP* et *ET* in *π* ac *τ* perpendiculariter, junganturque *πX* ac *τX*: Dico quòd radij maxime refrangibiles secundum dictam lineam *IX* incidentes refringentur in lineam *Xπ*, et minimè refrangibiles[5] in lineam *Xτ*, radijque cujusvis speciei quos vitrum

In the same way that the refractions were determined at glass, it could have been done at other media; but it will be useful, having determined their sines in this way at glass, if I now show a rule whereby the measures of those refractions can be determined at any other proposed medium, and even if that proposed medium is contiguous to a medium other than air. Let *AB* (Fig. 51) be a surface bounding air on the side *F* and glass on the side *G*, and at any point *X* of the surface draw the line *FXG* perpendicular to it. In addition, imagine that at the infinitely small angle *IXA* there is drawn the line *IX*, along which it is supposed that all rays of every nature are incident and refracted at *X*, suppose, mean refrangible rays toward *R*, most refrangible ones toward *P*, least refrangible[1] ones toward *T*, and other, intermediate ones toward the intermediate regions. Moreover, draw any line *GH* parallel to the line of incidence *IX*, that is, perpendicular to *FG*; but let it intersect the rays at the points *P*, *R*, and *T*, from which the perpendiculars *PC*, *RD*, and *TE* are dropped to the refracting surface *AB*. After these things have been so determined and described at glass, if any other medium is now conceived to be substituted for glass, with everything else remaining the same, and the refracted ray *Xρ* of some mean refrangible[2] one incident along the line *IX* at *X* is drawn intersecting the line *DR* in *ρ* (I assume this to be done, since I previously explained how the refractions of mean refrangible rays[3] can be found at any medium);[4] then if through the point *ρ* the line *ππ* is drawn cutting the lines *CP* and *ET* perpendicularly at *π* and *τ*, and *πX* and *τX* are joined; I say that most refrangible rays incident along that line *IX* will be refracted into the line *Xπ*, least refrangible[5] ones into the line *Xτ*, and those

Lecture 11
115 [= I, 42]. It is shown by a theorem that having determined the refractions of heterogeneous rays with respect to one another at glass or any other medium, the refractions can also be determined with respect to one another at any other medium contiguous to air without bothering anew with an experiment.

---

(1) mediocriter refrangibiles ... maximè refrangibiles ... minimè refrangibiles] Originally: viridiformes ... purpuriformes ... rubiformes (green-making ... purple-making ... red-making).

(2) mediocriter refrangibilis] Originally: viridiformis (green-making).

(3) modum ... refractiones] Originally: viridiformium refractiones ut (that the refractions of green-making rays).

(4) In §§101–2 = I, 28–9.

(5) maxime refrangibiles ... minimè refrangibiles] Originally: extremè purpuriformes ... extremè rubiformes (extreme purple-making ... extreme red-making).

refringebat ad quodlibet punctum rectae *PT,* illi ad correspondens punctum rectae *ππ* per alterum dictum Medium refringentur quod pro vitro supponitur substitui; istis punctis linearum *PT* et *ππ* habitis pro correspondentibus per quae recta quaevis ipsi *DR* parallela transit:[6] / Patet itáque modus quo 86/ refractiones quorumvis radiorum ex aëre in quodlibet Medium propositum obliquitate maximâ incidentium determinari poterunt, cognitâ unici tantùm cujusvis radiorum generis in istud Medium refractione. Et proportionibus sinuum ex obliquissima istâ refractione determinatis, eorundem radiorum refractiones dabuntur ad quamlibet aliam datam incidentiam.[7]

116 [→ I, 43]. De Theorematis illius certitudine.

Hujus autem propositionis certitudinem licèt ab experientijs nondum habeo depromptam,[8] nullus tamen dubito quin satisfaciet omnibus quas de illa licebit facere. Verùm cùm occasio de causis refractionum dicendi lata sit, veritatem ejus ex proprijs fundamentis eruere conabor, contentus interea gratis assumere.[9]

117 [= I, 44]. De proportione quarundam linearum quae computationi per hoc Theorema instituendae inserviat.

Calculum quod attinet is facilè potest institui ex hac proportionalitate quòd sinus incidentiae radij *IX* (hoc est sinus 90[graduum]) sit ad sinum refractionis (puta quae facta sit in lineam *XR,*) sicut *XR* ad *RG.* Sic ad vitrum erit *XR . RG* :: 106 . 68½, et *XP . PG* [::] 106 . 68, et *XT . TG* :: 106 . 69. Et inde deducetur quod *GP . GR . GT* :: 39 . 39½ . 40.[10] Quae proportiones

---

(6) This refraction or dispersion model assumes that at grazing incidence the projections parallel to the refracting surface of all spectra are of equal length and that the same colors always occupy equal portions of it—that is, that chromatic dispersion is a property of light and not of the refracting media. This model, which Newton used by 1666 or 1667 for his table of refractions, can also be interpreted as a natural extension of Descartes's model (see Lect. 9, note (6), this volume). The component of the velocities of the incident rays of different color, such as *XC,* *XD,* and *XE,* parallel to the refracting surface remain unchanged upon entering a new medium, while only the normal component is altered; in particular, at grazing incidence, the same normal velocity, such as *DR,* is added to rays of every color in the same medium. Although Newton here presents this model strictly as a mathematical one, in the table of refractions at the conclusion of his essay "Of Refractions," he refers to the quantities here designated as *XC* and *XE* as "the motions of the extremely heterogeneous rays . . . in Aire" (Add. 4000, f. 33[v] = *Correspondence,* 1:103). It follows from this model that at grazing incidence in all media, $\tan r_P : \tan r_R : \tan r_T = XC : XD : XE = 39 : 39\frac{1}{2} : 40$, with the constants being derived in §117 = I, 44. Alternatively, for two different media, whose indices of refraction are $n = I/R = XR/XD$ and $n' = I'/R' = X\rho/XD$,

$$\tan r'/\tan r = DR/D\rho = \sqrt{[n^2 - 1]}/\sqrt{[n'^2 - 1]},$$

since $DR = \sqrt{[XR^2 - XD^2]}$ and $D\rho = \sqrt{[X\rho^2 - XD^2]}$. Because the proportion $DR/D\rho$ is constant, independent of color, we have the refraction law $(n^2 - 1)/(n'^2 - 1) =$ constant, or $(I^2 - R^2)/(I'^2 - R'^2) =$ constant, and differentiating, we get what we call the "quadratic" dispersion law,

$$\Delta n/\Delta n' = (n - 1/n)/(n' - 1/n'), \text{ or } \Delta R/\Delta R' = R'(I^2 - R^2)/R(I'^2 - R'^2).$$

In the *Opticks* (Bk. I, Pt. II, Prop. III, Expt. VIII) Newton later proposed the refraction law $(n - 1)/(n' - 1) =$ constant, or $(R - I)/(R' - I') =$ constant, which entails what we call the "linear" or "achromatic" dispersion law,

$$\Delta n/\Delta n' = (n - 1)/(n' - 1), \text{ or } \Delta R/\Delta R' = (R - I)/(R' - I').$$

This law, Newton argued in the *Opticks,* followed from the impossibility of constructing a compound achromatic prism. However, as early as his draft letter to Hooke in the spring of 1672 (Add. 3970, f. 529[r]) he had already considered—but not yet adopted—the linear law as an

rays of any sort that glass refracted to some point of the line *PT* will be refracted to a corresponding point of the line *ππ* by that other medium that is assumed to be substituted for glass—those points of the lines *PT* and *ππ* through which any straight line parallel to *DR* passes are considered to be corresponding points.[6] It is therefore evident how the refractions of any rays incident with the greatest obliquity from air into any proposed medium can be determined by knowing the refraction of only any single kind of ray in that medium; and when the ratios of the sines have been determined from that most oblique refraction, the refractions of the same ray at any other given incidence will be given.[7]

Although I have not yet derived[8] the certainty of this proposition from experiments, nevertheless I do not doubt that it will satisfy all of them which it is possible to do with respect to it. But since this is an extended opportunity to discuss the causes of refraction, I will endeavor to elicit its truth from its proper foundations, meanwhile being content to assume it gratuitously.[9]   116 [→ I, 43]. The certainty of this theorem.

As for the calculation, it can easily be carried out from this proportionality, the sine of incidence of the ray *IX* (that is, sin 90°) is to the sine of refraction (suppose that one made in the line *XR*) as *XR* to *RG*; thus in glass *XR* : *RG* = 106 : 68½, *XP* : *PG* = 106 : 68, and *XT* : *TG* = 106 : 69. From this it will be deduced that *GP* : *GR* : *GT* = 39 : 39½ : 40.[10] Once these proportions   117 [= I, 44]. The proportion of certain lines that serve for carrying out the calculation by this theorem.

---

alternative to his quadratic dispersion law; and he clearly did not deduce it on the same grounds, for he then believed it was indeed possible to correct for chromatic dispersion; see Shapiro, "Newton's achromatic dispersion law"; and the Introduction, note (40). Newton never really abandoned the earlier quadratic dispersion law, for it is implied by the derivation of the sine law in the *Principia* (Bk. I, Prop. XCIV), and it is incorporated in the proof of the sine law in the *Opticks* (Bk. I, Pt. I, Prop. VI); see also *Mathematical Papers*, 6:422–30. Newton's inability to choose decisively between his two laws on either theoretical or experimental grounds (see note (9), this lecture) left much of the mathematical part of the *Lectures* in abeyance, and it may be one of the reasons why he never published it. For an account of Newton's refraction models and dispersion laws and their physical interpretation see Bechler, "Newton's search for a mechanistic model of colour dispersion: a suggested interpretation," *Archive for History of Exact Sciences,* 11 (1973):1–37; Hendry, "Newton's theory of colour"; and Vavilov's comments in Newton, *Lekcii po Optike,* p. 281, note 41.

(7) Newton applies this model only at grazing incidence, as he emphasizes here, to determine the indices of refraction; to determine the refractions at all other angles of incidence he introduces another construction in §118 = I, 45. Among its limitations, this model can be applied directly only for refraction into a denser medium.

(8) nondum habeo depromptam] Originally: certissimè depromptam non habeo (I have not most certainly derived).

(9) Newton never subjected his dispersion laws to systematic experimental tests. He had determined the dispersion of just two substances, "common" glass and water, whose dispersive powers differed by the same order of magnitude as the quantitative difference between his two laws and the experimental error of his measurements, so that he could neither experimentally distinguish between the two laws and perhaps confirm one of them, nor reject them. He does not treat the cause of refraction in either version of the *Lectures*.

(10) The constants of the model are readily determined from the sine law of refraction for the most and least refrangible rays: $XP/PG = I/(R - \frac{1}{2})$, and $XT/TG = I/(R + \frac{1}{2})$, where $I = 106$ and $R = 68\frac{1}{2}$, as found in §114 = I, 41. In Fig. 51 let $y = XG$, and $x = GR = XD$, and since

semel inventae possunt in eum finem asservari ut earum ope refractiones ad alia Media quàm vitra determinentur. Nam quolibet Medio proposito, sumatur $XE = 40$, $DE = \frac{1}{2}$, & $CD = \frac{1}{2}$, atque perpendicula CP, DR, & ET erigantur; Tum ex datâ sinuum refractionis radiorum mediocriter refrangibilium proportione,[11] hoc est ex datâ proportione ipsius $X\rho$ ad XD, dabitur punctum ρ et longitudo $D\rho$, cui aequales sunt $C\pi$ et $E\tau$. Punctisque π ac τ sic datis dantur rationes ipsarum $X\pi$ et XC, hoc est sinuum incidentiae et refractionis, pro radijs maxime refrangibilibus, ut et rationes ipsarum $X\tau$ et XE, hoc est sinuum incidentiae et refractionis, pro radijs minimè refrangibilibus.[12] Sic pro superficie aquam et aerem disterminante sinus isti sunt ut 68 ad 90 pro minime refrangibilibus, et ut 68 ad 91 pro maximè refrangibilibus.[13]

118 [= I, 45]. Aliud ejusdem rei peragendae Theorema

Proportionibus linearum XC, XD et XE sic inventis, mensura refractionum ex aëre in Medium quodvis propositum et ad quamlibet incidentiam factarum per aliud insuper Theorema non inelegans determinari potest. In lineâ FX (fig [52]) ad refringens planum AB perpendiculari, sumatur punctum aliquod F

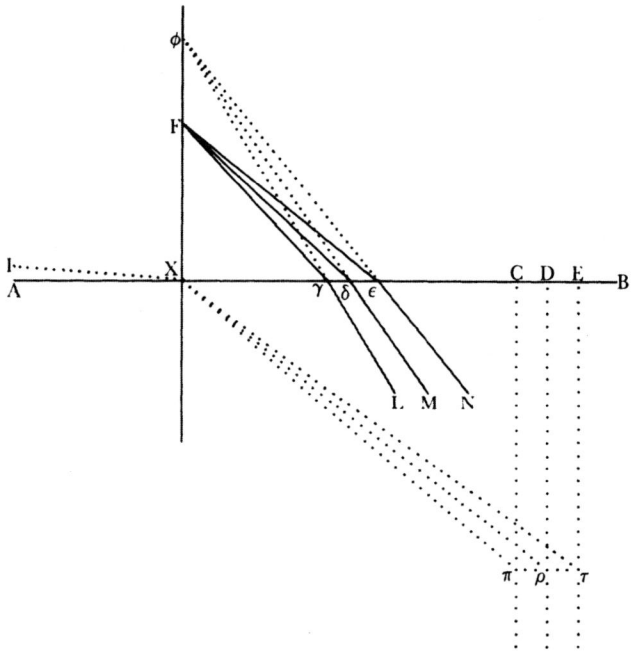

Figure 52

quod lucidum / fingatur,[14] ac ducatur quaelibet $F\delta$ secans AB in δ eaque 87/ concipiatur esse mediocriter refrangibilis[15] radius cujus refractus ex aere in Medium propositum esto δM, qui retro-ductus secet ipsam FX in ϕ. Porrò fiat $F\delta$. $FE :: X\delta$. XE[16] ($:: 39\frac{1}{2}$. 40) et $F\delta$. $F\gamma :: XD$. XC ($:: 39\frac{1}{2}$. 39.) centroque F et intervallis $F\epsilon$ et $F\gamma$ describantur circuli secantes AB in ϵ et γ jungantúrque $F\epsilon$, $F\gamma$, $\phi\epsilon$, $\phi\gamma$, et producantur $\phi\epsilon$ et $\phi\gamma$ indefinitè versus N et L. Dico jam si radius minime refrangibilis incidat secundum lineam $F\epsilon$, quòd

have been found, they can be maintained for the purpose of determining with their help the refractions in media other than glass. For in any proposed medium take $XE = 40$, $DE = \frac{1}{2}$, and $CD = \frac{1}{2}$, and erect the perpendiculars CP, DR, and ET. Then from the given ratio of the sine of refraction of the mean refrangible rays,[11] namely, from the given ratio of $X\rho$ to XD, the point ρ and the length $D\rho$, which equal $C\pi$ and $E\tau$, will be given. The points π and τ thus being given, there is given the ratio of $X\pi$ and XC, that is, of the sines of incidence and refraction for the most refrangible rays, as well as the ratio of $X\tau$ and XE, that is, of the sines of incidence and refraction for the least refrangible[12] rays. Thus for a surface separating water and air those sines are as 68 to 90 for the least refrangible rays and 68 to 91 for the most refrangible.[13]

With the proportions of the lines XC, XD, and XE having thus been found, the measure of the refractions made from air into any proposed medium and at any incidence can be determined by still another, not inelegant theorem. In the line FX (Fig. 52) perpendicular to the refracting plane AB take any point F, and imagine[14] it to be luminous. Also draw any line $F\delta$ intersecting AB in δ, and conceive it to be a mean refrangible[15] ray, whose refracted ray from air into the proposed medium is δM, which being drawn backward intersects FX in ϕ. Moreover, let

118 [= I, 45].
Another theorem for accomplishing the same thing.

$F\delta : FE = X\delta : XE$[16] $(= 39\frac{1}{2} : 40)$, and $F\delta : F\gamma = XD : XC (= 39\frac{1}{2} : 39)$;

and with center F and radii $F\epsilon$ and $F\gamma$ describe circles intersecting AB in ϵ and γ; join $F\epsilon$, $F\gamma$, $\phi\epsilon$, and $\phi\gamma$; and extend $\phi\epsilon$ and $\phi\gamma$ indefinitely toward N and L. I now say that if a least refrangible ray is incident along the line $F\epsilon$, it

---

Newton requires $CE = PT = 1$ or $PG = XC = x - \frac{1}{2}$ and $TG = XE = x + \frac{1}{2}$, then

$$\frac{(x - \frac{1}{2})^2 + y^2}{(x - \frac{1}{2})^2} = \frac{I^2}{(R - \frac{1}{2})^2} \quad \text{and} \quad \frac{(x + \frac{1}{2})^2 + y^2}{(x + \frac{1}{2})^2} = \frac{I^2}{(R + \frac{1}{2})^2}.$$

Eliminating y^2 and ignoring insignificant terms, we get the quadratic $I^2R\,x^2 - (I^2 - R^2)R^2x \approx 0$, whence (since $x \neq 0$) $x \approx R(I^2 - R^2)/I^2$. This yields $x = GR = 39.9$, and thus $GP = 39.4$, and $GT = 40.4$. These values agree perfectly with those in the table of refractions (Add. 4000, f. 33^v = *Correspondence*, 1:103). Newton probably took $GR = 39\frac{1}{2}$ here for convenience.

(11) datâ . . . proportione] Originally: datâ proportione refraction[is] viridiformium radiorum (from the given ratio of the refraction of the green-making rays).

(12) maxime refrangibilibus . . . minimè refrangibilibus] Originally: extreme pupuriformibus . . . extremè rubiformibus (extreme purple-making . . . extreme red-making).

(13) minime refrangibilibus . . . maximè refrangibilibus] Originally: rubiformibus . . . purpuriformibus (red-making . . . purple-making). In the *Optica* "proximè" (approximately) was added. Newton has simply taken the rounded-off values for the refraction from water to air from his refraction table, where he had $68\frac{1}{3}$ to $90\frac{2}{3}$ and $91\frac{1}{3}$, whereas for consistency he should have given the ratios from air to water, which are readily found to be 121.6 to 91.7 and 90.7.

(14) Originally: concipiatur (conceive).

(15) mediocriter refrangibilis] Originally: viridiformis (green-making).

(16) Read: $F\delta . F\epsilon :: XD . XE$

iste refringetur in lineam ϵN; et si maximè refrangibilis[17] incidat secundum $F\gamma$, quòd iste refringetur in ipsam γL. Et sic radij quorumlibet intermediorum generum manantes a puncto F et in puncta sibi correspondentia inter γ et ϵ incidentes, ita refringentur a Medio proposito quasi manassent omnes a puncto ϕ: Istis punctis inter C et E atque γ et ϵ habitis pro correspondentibus, quorum distantiae ab X et F respectivè, sunt in eâdem ratione cum DX ac δF.

119 [= I, 46].

Ad ejus demonstrationem duo Lemmata praelibantur.

Lemma 1.

vide fig: praecedentem

Cujus Theorematis demonstrationi praesternantur duo Lemmata sequentia.

1. Duobus punctis γ, δ in lineâ quâpiam AB sumptis, et alijs duobus ϕ et F in ejus perpendiculo FX; junctisque $\phi\delta$, $F\delta$, $\phi\gamma$, et $F\gamma$: differentiae quadratorum a duobus $\phi\delta$ et $F\delta$ concurrentibus ad δ aequabitur differentiae quadratorum ab alijs duobus $\phi\gamma$ et $F\gamma$ concurrentibus ad γ. Nam cùm $\phi\delta^q = \phi X^q + X\delta^q$, et $F\delta^q = FX^q + X\delta^q$; erit differentia $\phi\delta^q - F\delta^q = \phi X^q - FX^q$. Et ob eandem rationem est differentia $\phi\gamma^q - F\gamma^q = \phi X^q - FX^q$. Quare dictae differentiae sic aequales eidem tertio sunt aequales inter se. Q.E.D.

120 [= I, 47].

Lemma 2.

2. Si radius aliquis FG (fig [53]) incidat in superficiem AB, et refringatur versus H: Lineâ GH retro-ductâ ut secet perpendiculum FX in ϕ, dico quod $\phi[G] . FG$ :: sinus incidentiae, ad sinum refractionis. Et e contra / si $\phi G . FG$ :: sin incid . sin refract; erit $\phi[G]H$ refractus ipsius FG. Etenim sumatur $\phi K = FG$, et demittatur KL perpendicularis ad FX; quo facto, cùm ang: GFX aequetur angulo incidentiae et ang $G\phi X$ angulo refractionis, erit GX sinus incidentiae et KL sinus refractionis, habito respectu ad circulum cujus semidiameter sit FG vel ϕK. Sed est $\phi G . \phi K$ :: $GX . KL$ hoc est $\phi G . FG$ :: $GX . KL$. Q.E.D.[18]

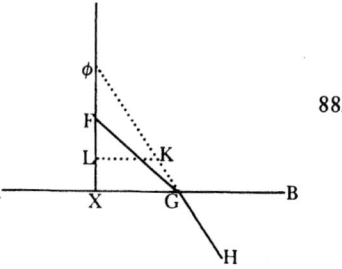

88/

Figure 53

121 [= I, 48].

Demonstratio.

His praemissis Theorema propositum sic demonstratur. In Fig 52 ducatur IX obliquissima linea secundum quam radij omnium formarum ex aëre ad punctum X incidere ponantur et in Medium propositum refringi maxime refrangibiles versus π, et minimè refrangibiles[19] versus τ, eosque lineae ad puncta D, C, et E normaliter erectae secent in punctis ρ, π, ac τ, ut explicabatur ad Fig [51]. Jam cùm istorum radiorum sinus incidentiae et refractionis statuantur[20] esse ut $X\rho$ ad XD, $X\pi$ ad XC et $X\tau$ ad XE respectivè; si praeterea demonstratum fuerit quod $\phi\delta$ ad $F\delta$, $\phi\gamma$ ad $F\gamma$ et $\phi\epsilon$ ad $F\epsilon$ respectivè sunt in eâdem ratione, (hoc est, quod $\phi\delta . F\delta$ :: $X\rho . XD$ :: sinus incidentiae, ad sinum refractionis radiorum mediocriter refrangibilium[21] et $\phi\gamma . F\gamma$ :: $X\pi . XC$ :: sinus incidentiae ad sinum refractionis radiorum maxime refrangibilium[22] &c) constabit propositum ex Lemmate secundo. Et ad

---

(17) minime refrangibilis . . . maximè refrangibilis] Originally: extremè rubiformis . . . purpuriformis (extreme red-making . . . purple-making).

(18) More directly: since sin $i = XG/FG$ and sin $r = XG/\phi G$, therefore sin i/sin $r = \phi G/FG$; Barrow had made equivalent use of this refinement in his *Lectiones XVIII*, Lect. IV, §IV, p. 47.

will be refracted into the line ϵN; and if a most refrangible[17] one is incident along $F\gamma$, it will be refracted into γL. Similarly, rays of any intermediate kind flowing from the point F and falling upon their corresponding points between γ and ϵ will be so refracted by the proposed medium as if they had all flowed from the point ϕ—those points between C and E and γ and ϵ whose distances from X and F, respectively, are in the same ratio as DX and δF, being considered as corresponding points.

To demonstrate this theorem the following two lemmas are premised.

1. Taking two points, γ and δ, in any line AB, and another two, ϕ and F, in its perpendicular FX, and joining $\phi\delta$, $F\delta$, $\phi\gamma$, and $F\gamma$, the difference of the squares of the two [lines] $\phi\delta$ and $F\delta$ meeting at δ will equal the difference of the squares of the other two, $\phi\gamma$ and $F\gamma$, meeting at γ.

For since $\phi\delta^2 = \phi X^2 + X\delta^2$, and $F\delta^2 = FX^2 + X\delta^2$, their difference will be $\phi\delta^2 - F\delta^2 = \phi X^2 - FX^2$; and for the same reason there will be the difference $\phi\gamma^2 - F\gamma^2 = \phi X^2 - FX^2$. Consequently, those differences, since they are equal to the same third one, are equal to each other. As was to be demonstrated.

2. If any ray, FG (Fig. 53), falls on the surface AB and is refracted toward H, and the line GH is drawn backward to intersect the perpendicular FX in ϕ, I say that ϕG is to FG as the sine of incidence is to the sine of refraction; and conversely, if ϕG is to FG as the sine of incidence is to the sine of refraction, then ϕGH will be the refracted ray of FG.

For take ϕK to be equal to FG, and drop KL perpendicular to FX. When this has been done, since the angle GFX is equal to the angle of incidence and the angle $G\phi X$ to the angle of refraction, GX will be the sine of incidence and KL the sine of refraction with respect to the circle whose radius is FG or ϕK. But $\phi G : \phi K = GX : KL$, that is $\phi G : FG = GX : KL$. As was to be demonstrated.[18]

With these premises, the proposed theorem is demonstrated in this way. In Fig. 52 draw IX, the most oblique line along which rays of every nature are assumed to be incident from air at the point X and to be refracted into the proposed medium—the most refrangible ones toward π and the least refrangible[19] toward τ; and let the lines erected perpendicularly at the points D, C, and E intersect them at the points ρ, π, and τ, as was explained at Fig. 51. Now since the sines of incidence and refraction of those rays are determined[20] to be as $X\rho$ to XD, $X\pi$ to XC, and $X\tau$ to XE, respectively; if moreover it were demonstrated that $\phi\delta$ to $F\delta$, $\phi\gamma$ to $F\gamma$, and $\phi\epsilon$ to $F\epsilon$, respectively, are in the same ratio (that is, $\phi\delta : F\delta = X\rho : XD$ is as the sine of incidence to the sine of refraction of mean refrangible[21] rays, and $\phi\gamma : F\gamma = X\pi : XC$ is as the sine of incidence to the sine of refraction of most refrangible[22] ones); the proposition will be established by the second lemma.

119 [= I, 46]. For its demonstration two lemmas are prefaced. Lemma 1.

See the preceding figure.

120 [= I, 47]. Lemma 2.

121 [= I, 48]. The demonstration.

---

(19) maxime refrangibiles ... minimè refrangibiles] Originally: extremè purpuriformes ... extremè rubiformes (extreme purple-making ... extreme red-making).

(20) Originally: dicebantur (were said).

(21) mediocriter refrangibilium] Originally: viridiformium (green-making).

(22) maxime refrangibilium] Originally: extremè puripuriformium (extreme purple-making).

mediocriter refrangibiles[23] quod attinet cùm $\phi\delta$ supponatur refractus ipsius $F\delta$, erit (per Lemma 2^{dum}) $\phi\delta$ ad $F\delta$ ut sinus incidentiae ad sinum refractionis, hoc est ut $X\rho$ ad $X[D]$. Sed eadem proportionalitas in caeteris radiorum generibus jam demonstranda proponitur, puta quòd sit $\phi\gamma . F\gamma :: X\pi . XC$. Scilicet est $F\gamma . F\delta :: XC . XD$, ut et $F\delta . \phi\delta :: XD . X\rho$, per Hypothesin. Quare permutando et connectendo rationes aequales est $F\gamma . XC :: F\delta . XD :: \phi\delta . X\rho$. Et quadrando $F\gamma^q . XC^q :: F\delta^q . XD^q :: \phi\delta^q . X\rho^q$, diminuendoque per terminos aequalis rationis

$$F\gamma^q . XC^q :: \phi\delta^q - F\delta^q \text{ (sive, per Lem 1, } \phi\gamma^q - F\gamma^q.) . X\rho^q - XD^q \text{ (sive } C\pi^q). \text{ }^{(24)}$$

et augendo per[25] terminos aequalis rationis $F\gamma^q . XC^q :: \phi\gamma^q . C\pi^q + XC^q$ (sive $X\pi^q$.) Denique terminorum radices extrahendo, permutandóque est $\phi\gamma . F\gamma :: X\pi . XC$. Quare $\phi\gamma$ sive γL[26] est refractus ipsius $F\gamma$ per Lemma 2^{dum}. Q.E.D. Et eodem argumento patebit quòd ϵN sit refractus radij $F\epsilon$. Deáque alijs radijs pro varijs refrangibilitatis gradibus,[27] intermedia spatia variè[28] occupantibus, idem intelligendum est.

122 [= I, 49].
Heterogeneorum refractiones a superficiebus aeri ex neutra parte contiguis Theoremate etiam determinantur

/ De refractionibus superficierum aeri contiguarum mensurandis haec satis. 89/ Quod si desideretur id ipsum ad alias superficies aeri ex neutra parte contiguas fieri, sunto [fig 54] $AB\beta H$ et $\alpha\beta\nu\mu$ duo quaelibet Media secundum planam superficiem $H\beta$ contigua, et aere circundata. Sitque AB planum ipsi $H\beta$ parallelum, et in eo sumatur punctum X, ad quod ducatur XV perpendicularis et IX obliquissima linea secundum quam (ut jam ante) radij omnium

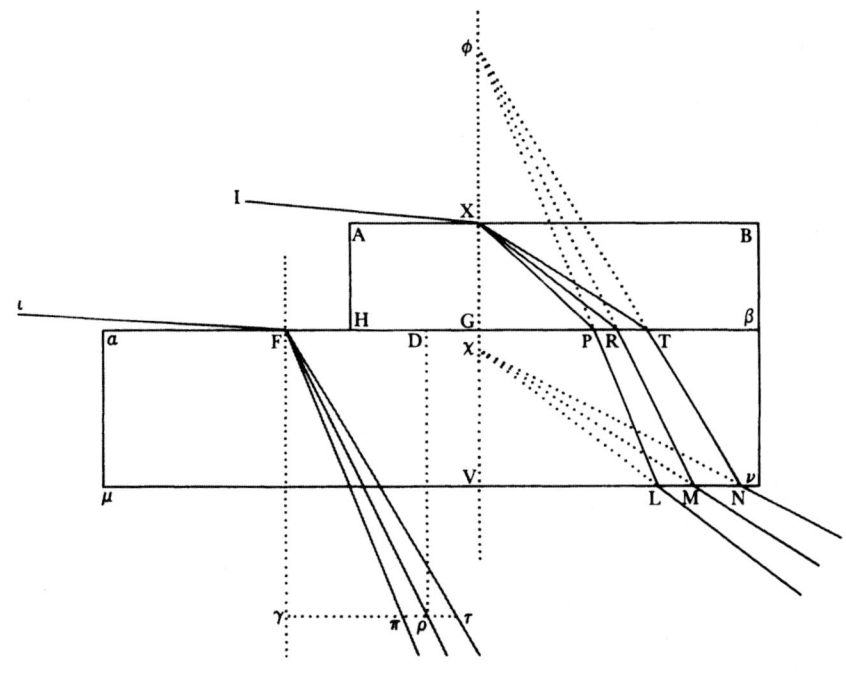

Figure 54

As for the mean refrangible[23] rays, since $\phi\delta$ is supposed to be the refracted ray of $F\delta$, then (by Lemma 2) $\phi\delta$ to $F\delta$ will be as the sine of incidence to the sine of refraction, that is, as $X\rho$ to $X[D]$. It is now proposed to demonstrate the same proportionality for the other kinds of rays, that is, that $\phi\gamma : F\gamma = X\pi : XC$. Namely, by hypothesis $F\gamma : F\delta = XC : XD$, and also $F\delta : \phi\delta = XD : X\rho$. Consequently, by permuting and equating the equal ratios, $F\gamma : XC = F\delta : XD = \phi\delta : X\rho$; and by squaring, $F\gamma^2 : XC^2 = F\delta^2 : XD^2 = \phi\delta^2 : X\rho^2$; and by diminishing the terms of the equal ratios,

$F\gamma^2 : XC^2 = \phi\delta^2 - F\delta^2$ (or, by Lemma 1, $\phi\gamma^2 - F\gamma^2) : X\rho^2 - XD^2$ (or $C\pi^2$);[24]

and by augmenting[25] the terms of the equal ratios,

$$F\gamma^2 : XC^2 = \phi\gamma^2 : C\pi^2 + XC^2 \text{ (or } X\pi^2).$$

Finally, by extracting the roots of the terms and permuting, it is $\phi\gamma : F\gamma = X\pi : XC$. By Lemma 2, therefore, $\phi\gamma$, or rather γL,[26] is the refracted ray of $F\gamma$. As was to be demonstrated. By the same argument it will be clear that ϵN is the refracted ray of $F\epsilon$. The same thing must also be understood for the other rays, which according to their various degrees of refrangibility[27] variously[28] occupy the intermediate spaces.

Let this suffice for measuring the refractions of surfaces contiguous to air. If now it is required to do this at other surfaces contiguous to air on neither side, let (Fig. 54) $AB\beta H$ and $\alpha\beta\nu\mu$ be any two media contiguous along the plane surface $H\beta$ and surrounded by air. Let the plane AB be parallel to $H\beta$; in it take a point X at which the perpendicular XV is drawn and also the most oblique line IX along which (as just before) rays of every form are

122 [= I, 49]. The refractions of heterogeneous rays by surfaces contiguous to air on neither side are also determined by a theorem.

(23) mediocriter refrangibles] Originally: viridiformes (green-making).
(24) Here and throughout the *Lectures* Newton uses the classical theory of proportions as he first learned it in Barrow's *Euclidis Elementorum*. In Bk. V, Defs. XII–XVI (pp. 92–3) five basic operations on the proportion $A : B = C : D$, are defined: XII, alternation or permutation (*alternando* or *permutando*), $A : C = B : D$; XIII, inversion (*invertendo*), $B : A = D : C$; XIV, composition (*componendo*), $A + B : B = C + D : D$; XV, division (*dividendo*), $A - B : B = C - D : D$; and XVI, conversion (*convertendo*), $A : A - B = C : C - D$. Moreover, these operations can also be combined, as here, where $F\delta^2 : XD^2 = \phi\delta^2 : X\rho^2$ becomes after permutation and division $X\rho^2 - XD^2 : XD^2 = \phi\delta^2 - F\delta^2 : F\delta^2$. In the next step of the demonstration he combines a permutation and composition.
(25) augendo per] Originally: connectendo (joining).
(26) sive γL] Added.
(27) varijs refrangibilitatis gradibus] Originally: speciebus colorum (their species of color).
(28) Added.

formarum incidant et pro gradu refrangibilitatis refringantur ad *P, R,* ac *T* aliaque intermedia locâ.[29] Horum radiorum in propositam superficiem $\alpha\beta$ sic incidentium refractiones jam quaeruntur. Atque equidem cùm viridiformium[30] refractiones ad quaslibet superficies fuerint antehac expositae, radij[31] *XR* sit *RM* refractus, et is retro-ducatur donec secet perpendiculum *XV* in ϕ. Et insuper ducantur $\phi P, \phi T$ et producantur ad *L* et *N.* Dico quòd *PL* erit refractus ipsius *XP,* ac *TN* ipsius *XT,* atque omnes aliarum formarum radij incidentes inter *P* ac *T* ita refringentur, ut postea divergant a puncto ϕ. Concipiatur enim quòd Medium $\alpha\beta\nu\mu$ longiùs versus $\alpha\mu$ producitur quàm Medium *ABβH,* ita ut ejus plani $\alpha H\beta$ pars inter *H* et α sit aeri contigua, et ad aliquod in eo punctum *F* ducatur perpendicularis *Fγ* nec non obliquissima linea ιF, secundum quam radij omnium formarum incidant, et pro gradu refrangibilitatis refringantur ad π, ρ, τ locaque intermedia[32] perinde ut effectum erat ad alterius Medij superficiem *AB.* Praeterea sumatur *FD = GR,* et ducatur *Dρ* ipsi *Fγ* parallela, ut secet radium *Fρ* in ρ, unde $\rho\gamma$ demittatur ad *Fγ* normalis, aliosque radios *Fπ* et *Fτ* secans in π ac τ. Jam cùm sit $\gamma\rho = GR$, erit etiam $\gamma\pi = GP$ et $\gamma\tau = GT$ ex ostensis ad fig [51];[33] Et insuper ex ostensis ad fig [46][34] cùm radiorum secundum *IX* et ιF lineas parallelas incidentium eadem sit refractio in Medium $\alpha\beta\nu\mu$, sive immediatè ingrediantur ex aere sicut fit ad *F,* sive priùs permeent aliud Medium ut *ABβH* parallelis planis terminatum: sequitur / quod radij alterutro modo refracti in dictum 90/ Medium $\alpha\beta\nu\mu$ sunt paralleli radijs homogeneis altero modo in idem Medium refractis; hoc est quod *Fπ* ad *PL, Fρ* ad *RM* et *Fτ* ad *TN* sunt paralleli. Quapropter si refracti radij *PL, RM,* ac *TN* retro-ducantur donec singuli occurrant perpendiculo *GX;* cum eo et basibus *GP, GR,* ac *GT* constituent triangula similia triangulis *γπF, γρF,* et *γτF,* imò et ipsis aequalia, siquidem eorum bases *γπ* et *GP, γρ* et *GR, γτ* et *GT* sibimet respectivè sunt aequales.[35] Quare cùm horum triangulorum vertices conveniant ad idem punctum *F,* illorum etiam vertices ad idem aliquod punctum ϕ convenient. Hoc est radij *PL, RM,* ac *TN* ipsorum *XP, XR* et *XT* refracti divergent omnes ab eodem puncto ϕ.   Q.E.D.

123 [≈ I, 50].
Theorema illud
notis quibusdam
promovetur

Ostenso hoc, sequentia obveniunt notanda. 1. Quòd proportio[36] sinuum incidentiae et refractionis ad superficiem *Hβ* factae, ex his facilè determinantur. Nam pro radijs maxime refrangibilibus sinus isti sunt ut ϕP ad *XP;* et pro minime refrangibilibus,[37] ut ϕT ad *XT;* &c.[38]

2. Hinc si proportiones sinuum refractionis ex aëre in duo quaelibet Media

---

(29) pro . . . locâ.] Originally: refringantur; extremitas nempe purpuriformium ad *P,* extremitas rubiformium ad *T,* et mediocritas viridiformium ad *R.* (are refracted, namely, the extremity of the purple-making ones to *P,* the extremity of the red-making to *T,* and the middle of the green-making to *R.*)

(30) II: mediocritèr refrangibilium (mean refrangible).

(31) Originally: viridiformis radij (green-making ray). See §105 = I, 32.

(32) et . . . intermedia] Originally: et refringantur extremè purpuriformes ad π, viridiformes ad ρ, et extreme rubiformes ad τ (and the extreme purple-making are refracted to π, the green-making to ρ, and the extreme red-making to τ).

(33) In §115 = I, 42.   (34) In §106 = I, 33.

incident and refracted according to their degree of refrangibility to *P*, *R*, *T*, and other, intermediate places.[29] The refractions of these rays thus incident upon the proposed surface *αβ* are now to be found. Since indeed the refractions of green-making[30] rays at any surface have been explained earlier, let *RM* be the refraction of the ray[31] *XR*, and draw it backward until it intersects the perpendicular *XV* in *φ*. Moreover, draw *φP* and *φT* and extend them to *L* and *N*. I say that *PL* will be the refracted ray of *XP*, *TN* that of *XT*, and all rays of other forms falling between *P* and *T* will be so refracted that afterward they diverge from the point *φ*. For conceive the medium *αβνμ* to be extended farther towards *αμ* than the medium *ABβH* so that the part of its plane *αHβ* between *H* and *α* is contiguous to air; and at any point *F* in it draw the perpendicular *Fγ* and also the most oblique line *ιF* along which rays of every form are incident and according to their degree of refrangibility refracted to *π*, *ρ*, *τ*, and the intermediate places,[32] just as was done at the surface *AB* of the other medium. Moreover, take *FD* equal to *GR*; draw *Dρ* parallel to *Fγ* so that it intersects the ray *Fρ* in *ρ*; and from here drop *ργ* normal to *Fγ*, intersecting the other rays, *Fπ* and *Fτ*, in *π* and *τ*. Now since *γρ* = *GR*, then also *γπ* = *GP* and *γτ* = *GT*, from what was shown at Fig. 51.[33] Furthermore, from what has been shown at Fig. 46,[34] since the refraction of the rays incident along the parallel lines *IX* and *ιF* is the same in the medium *αβνμ*, whether they enter immediately from air, as happens at *F*, or they first pass through another medium, such as *ABβH*, bounded by parallel planes: It follows that the rays refracted in the specified medium *αβνμ* in the first way are parallel to the homogeneous rays refracted in the same medium in the second way, that is, that *Fπ* is parallel to *PL*, *Fρ* to *RM*, and *Fτ* to *TN*. Therefore, if the refracted rays *PL*, *RM*, and *TN* are drawn backward until each one meets the perpendicular *GX*, they will form with it and the bases *GP*, *GR*, and *GT* triangles similar to the triangles *γπF*, *γρF*, and *γτF*, indeed, equal to them, since their bases *γπ* and *GP*, *γρ* and *GR*, and *γτ* and *GT* are respectively equal to one another.[35] Consequently, since the vertices of the latter triangles meet at the same point *F*, the vertices of the former will also meet at some one point *φ*. That is, the rays *PL*, *RM*, and *TN*, the refracted rays of *XP*, *XR*, and *XT*, will all diverge from the same point *φ*. As was to be demonstrated.

With this having been shown, the following things occur to us to be noted: 123 [≈ I, 50]. That theorem is extended by some notes.

1. The ratios of the sines of incidence and refraction made at the surface *Hβ* are easily determined from this. For instance, for the most refrangible rays those sines are as *φP* to *XP*, and for the least refrangible[37] as *φT* to *XT*, and so forth.[38]

2. Hence, if the proportions of the sines of refraction from air into any

---

(35) Since it is in general not physically true—as Newton's dispersion model demands—that *γπ* = *GP*, *γρ* = *GR*, and *γτ* = *GT*, or that the triangles are equal, the remainder of the proof does not model reality. In particular, his claim that the refracted rays diverge from the common point *φ* is mistaken.   (36) II: *proportiones*.

(37) *maxime refrangibilibus* ... *minime refrangibilibus*] Originally: *extremè purpuriformibus* ... *extremè rubiformibus* (extreme purple-making ... extreme red-making).

(38) By Lemma 2.

proposita, paribus incidentijs, dentur; proportiones sinuum refractionis ex altero Mediorum in alterum facilè dabuntur; dividendo nempe sinus posterioris Medij per correspondentes sinus anterioris. Sic cùm refractio fit ex aëre in vitrum dicti sinus sunt ut 68 . $68\frac{1}{2}$. 69; et cùm fit ex aere in aquam sunt ut 90 . $90\frac{1}{2}$. 91: Ergo cùm fit ex aquâ in vitrum erunt ut $\frac{68}{90} \cdot \frac{68\frac{1}{2}}{90\frac{1}{2}} \cdot \frac{69}{91}$, hoc est ut 281, $281\frac{1}{2}$, 282 ferè.[39]

3. Si tertium[40] aliquod Medium aëre densius postponatur Medio $\alpha\beta\nu\mu$, contingens illud in superficie $\mu\nu$, quae concipiatur plana ipsisque AB et $\alpha\beta$ parallela; et si radij divergentes a puncto ϕ (sicut modò ostensum erat) in illud incidant ad puncta L, M, et N; postquam in ijsdem refringuntur divergent rursus ab alio quodam puncto χ quod situm est in perpendiculo XG: Et sic praeterea in infinitum, quotcunque licèt Media parallelis planis ab invicem discreta sese ordine subsequantur.[41] Quod si aër immediatè succedat Medio $\alpha\beta\nu\mu$, punctum istud χ a quo emergentes radij tendunt situm erit ad V in ipsâ refringenti superficie, propterea quòd emergent paralleli ad summè obliquam lineam IX secundum quàm primùm incidebant ex aere, si modò emergere dicantur qui nunquam divaricabunt a refringenti superficie.

/ 4. Si radij ab aliquo puncto F in aëre sito divergentes, tendant ad puncta 91/ γ, δ, ϵ, pro more quem ad schema [52] explicui, et per varia deinde plana refringentia ipsíque AB parallela transeant: semper divergent omnes ab eodem aliquo puncto quod situm est in perpendiculo planorum per punctum F transeunte, non secus quàm si incidissent in planum AB advenientes in obliquissimâ lineâ IX. Et longitudines radiorum punctis refringentibus dictóque perpendiculo interceptorum sunt ut sinus incidentiae et refractionis ad singula plana quae respectant.[42] Quarum assertionum demonstrationes cùm facilè eruantur e praedictis, praetermitto, nè nimius in hâc re videar. | Et sic tandem absolvi quae de legibus refractionum dicenda esse judicabam; in quibus prolixitatem aliquam materia postulavit. Nam omnia ferè de integro hic tractanda erant, idque sedulò, cùm totius Dioptrices scientia his legibus tanquam fundamentis innitatur.

Lect 12
124 [→ I, 51].
Mensuris
refractionum sic fuse
explicatis ad
propositiones exinde
scaturientes
transitur.

Quod reliquum est in ordine ad explicandas apparentias, quas Prismata exhibent dum objectis interponuntur, id protinus aggredior. Et quibus Philosophia naturalis magis quàm Mathematica delicio est, licèt haec fortè supervacanea videantur, habito tantùm ad enarrandam colorum genesin respectu: cùm tamen ad rem opticam necessariò pertineant, quatenus refractiones radiorum refrangibilitate differentium inter se conferri debent, non potui penitùs omittere; et cùm a praesenti negotio non sint aliena, statui hic rebus

---

(39) This rule does not really depend on the preceding demonstration but rather is a direct extension to rays of different color of the rule for relative indices of refraction in §105 = I, 32, although it uses the slightly different formulation of his draft for Hooke; see Lect. 9, notes (36), (37), this volume. Newton would seem to have applied this rule, in some equivalent form, to calculate the indices of refraction from glass to water in his refraction table; Add. 4000, f. 33^v = *Correspondence*, 1:103. In his draft letter for Hooke and in the *Opticks* he carried out related calculations, but because he used the linear dispersion law to derive the dispersion of water,

two proposed media at the same incidence are given, the proportions of the sines of refraction from one of the media into the other will be easily given, namely, by dividing the sines of the second medium by the corresponding sines of the first. Thus when the refraction is made from air into glass, those sines are as 68 : 68½ : 69, and when it is made from air into water, they are as 90 : 90½ : 91. Therefore, when it is made from water into glass, they will be as 68/90 : 68½/90½ : 69/91, that is approximately as 281 : 281½ : 282.[39]

3. If any third[40] medium denser than air is placed behind the medium αβνμ in contact with it at the surface μν, conceived to be plane and parallel to *AB* and αβ; and if rays diverging from the point φ (as was just shown) fall upon it at the points *L, M,* and *N*; then after they are refracted in the same medium, they will diverge again from some other point χ located in the perpendicular *XG*. And so on in infinitum, no matter how many media separated from each other by parallel planes may follow in turn.[41] But if air immediately follows the medium αβνμ, that point χ from which the emergent rays tend will be located at *V* in the refracting surface itself, because they will emerge parallel to the most oblique line *IX* along which they were initially incident from air—if rays may be said to emerge that never diverge from the refracting surface.

4. If rays diverging from any point *F* located in air tend to the points γ, δ, and ε in the way that I explained at Fig. 52 and then pass through various refracting planes parallel to *AB*, they will all always diverge from some one point located in the planes' perpendicular that passes through the point *F*, just as if they were incident upon the plane *AB* and arrived in the most oblique line *IX*. The lengths of the rays intercepted between the refracting points and the specified perpendicular are as the sines of incidence and refraction at each plane they confront.[42] Since the demonstrations of these assertions may be easily derived from the preceding, I am omitting them, lest I seem to linger too long on this matter. | Thus I have finally completed what I judged ought to be said about the laws of refraction. The material required some prolixity in this for nearly everything had to be treated here anew and indeed diligently, since the science of all dioptrics depends upon these laws as fundamentals.

It remains, in turn, to explain the phenomena that prisms exhibit when they are set between objects, and I begin it immediately. Natural philosophy more than mathematics finds delight in these things, even if they may perhaps appear superfluous when considered solely with respect to explaining the origin of colors. Nonetheless, since they necessarily concern optical matters, insofar as the refractions of rays differing in refrangibility must be compared with one another, I could not altogether omit them; and since they are not unrelated to the business at hand, I have decided here to join more physical

Lecture 12
124 [→ I, 51].
Having so fully explained the measures of refractions, I pass to the propositions flowing from there.

made some errors in the calculation, and redetermined some parameters, the values differ somewhat.   (40) Added.
(41) For the origin of this physically invalid claim see note (35), this lecture.
(42) By Lemma 2.

magis Physicis innectere quò materiae varietas interstrata taedium relevare possit. Primò itaque refractiones contemplabor in solitariâ superficie factas, deinde refractiones geminatas, quales radijs per inclinata duo plana prismatum transeuntibus eveniunt; ac demum aliqua de constructione oculi, et ejus ad visionem dispositione levitèr attingam.[1]

125 [→ I, 54].
Obviae quaedam
conformium
radiorum
affectiones,
sequentibus
inservientes,
traduntur.

[126→ I, 54]

[127]

[128]

De radiorum semel refractorum affectionibus:
Et primò de similiter refrangibilibus agitur perfunctoriè.

Prop 1. Incidentijs aequalibus, refractiones sunt aequales. Nam sinus aequalium incidentiarum sunt aequales (per 26: 1: Elem),[2] atque adeò sinus refractionum sunt aequales (per 14: 5: Elem.)[3] Et anguli (per 7: 6: Elem).[4]
Et sic e contra[5] Refractionibus aequalibus, aequales sunt incidentiae.

Prop 2. Incidentiâ majori, major est angulus refractionis. Nam sinus incidentiae majoris est major (15 3 Elem),[6] et ergo sinus refractionis (14 5 Elem), et angulus (15 3 Elem). Similíque ratiocinio constet e contra quod angulo refractionis majori, major convenit angulus incidentiae.

Cor: Si majori incidentiâ minor sit refractio, aut contra: Id fit ob dissimilem refrangibilitatem.

/ Prop 3. Si [fig 55][7] radij homogenei *FR, Fρ*, manantes a puncto quodam   92

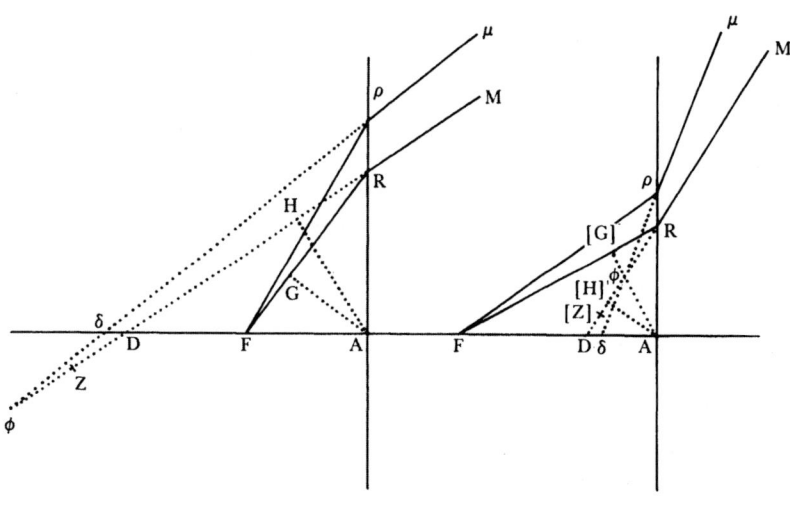

Figure 55

lucido *F*, refringantur a qualibet plana superficie *AR*, eorum refracti *RM, ρμ* ab invicem postea divergent. Nam ab *F* demittatur *FA* perpendicularis ad planum *AR*, et hinc inde producatur donec refractos radios retro-ductos secet in punctis *D*, ac δ: eruntque *RFA, ρFA* anguli incidentiae; et *RDA, ρδA* anguli refractionis. Jam cùm sit ang: *ρFA* ⊏ ang *RFA* (ex constructione,) erit etiam ang *ρδA* ⊏ *RDA* (per prop: 2 praeced:) Atque adeò ang: δρA ⊐ ang: *DRA* (per prop 32 lib 1 Elem, et ax 17)[8] hoc est δρA + DRρ ⊐ 2 rectis.

matters so that the variety of interspersed material could relieve the tedium. In the first place, therefore, I will consider refractions made in a single surface, then twofold refractions, such as happens to rays crossing through the two inclined planes of prisms, and finally I will briefly touch upon some things about the structure of the eye and its arrangement for vision.[1]

The properties of rays refracted once.
First, similarly refrangible rays are treated perfunctorily.

Proposition 1. At equal angles of incidence, the angles of refraction are equal.

For the sines of equal angles of incidence are equal (*Elements,* I, 26),[2] and thus the sines of the angles of refraction are equal (*Elements,* V, 14)[3] as are the angles (*Elements,* VI, 7).[4] Similarly and conversely,[5] at equal angles of refraction the angles of incidence are equal.

Proposition 2. At a greater angle of incidence, the angle of refraction is greater.

For the sine of a greater angle of incidence is greater (*Elements,* III, 15),[6] and therefore also the sine of refraction (*Elements,* V, 14) and the angle (*Elements,* III, 15). By a similar argument it may be established, conversely, that to a greater angle of refraction there corresponds a greater angle of incidence.

Corollary. If at a greater angle of incidence, the angle of refraction is smaller, or conversely, this occurs because of dissimilar refrangibility.

Proposition 3. If (Fig. 55)[7] homogeneous rays FR and $F\rho$ flowing from some luminous point F are refracted by any plane surface AR, their refracted rays RM and $\rho\mu$ will afterward diverge from one another.

For from F drop the perpendicular FA to the plane AR, and extend it on both sides until it cuts the refracted rays extended backward in the points D and δ; then RFA and ρFA will be the angles of incidence, and RDA and $\rho\delta A$ the angles of refraction. Now, since $\widehat{\rho FA} > \widehat{RFA}$ (by construction), there will also be $\widehat{\rho\delta A} > \widehat{RDA}$ (by Prop. 2, above). Hence $\widehat{\delta\rho A} < \widehat{DRA}$ (*Elements,* I, 32, and Axiom 17),[8] that is, $\widehat{\delta\rho A} + \widehat{DR\rho} < 2$ right angles. But the rays RD and

125 [→ I, 54].
Certain obvious properties of similar rays serving for the following are propounded.

[*126* → I, 54]

[*127*]

[*128*]

---

(1) Lect. 18, on refraction in prisms, concludes the *Lectiones,* whereas the eye and vision are touched upon in the *Optica,* Lects. II, 14–15.

(2) If two triangles have two angles and any side equal to one another, then the remaining sides and angles will also be equal to one another.

(3) Given any four magnitudes, if $A : B = C : D$ and $A = C$, then $B = D$.

(4) If two triangles have one equal angle, and the sides about the other angles proportional, then the triangles will be equiangular.

(5) Originally continued: liceat ratiocinari quod (one may reason that).

(6) The nearer a straight line in a circle is to the center, the greater it is, with the diameter being greatest.

(7) Newton omitted the lines AG and AH and the point Z from the figure on the right, and we have restored them. Also, we have omitted from the figure on the left a superfluous dotted line KR that is perpendicular to the line $AR\rho$ at the point R and extends as far as the point K on the ray $\phi\rho\mu$.

(8) According to Prop. 32, the sum of the interior angles of a triangle equals two right angles.

Sed radij *RD* et $\rho\delta$ ad eas partes concurrent ubi sunt anguli duobus rectis minores (per ax 13: 1 Elem),[9] hoc est ad partes versus punctum *F*; et proinde ad alteras partes versus *M* et μ tendentes, ab invicem divergent. Q.E.O.

[129 ≈ I, 62]

Schol: Si angulus *R$\phi\rho$*, quem refracti radij comprehendunt ponatur esse infinitè parvus, punctum istud ϕ erit limes disterminans intersectiones radiorum utrinque jacentium, quas cum radio *Rϕ* vel $\rho\phi$ efficiunt: Ita scilicet ut cùm refractio fit e Medio rariori in densius, radiorum ad partes puncti *R* adversus *A* incidentium refracti secabunt radium ϕR ad partes puncti ϕ adversus *R*, et incidentium inter *A*[10] et *R* refracti eundem ϕR secabunt inter ϕ et *R*.[11] E contra verò cùm refractio fit e densiori Medio in rarius, incidentium ad partes ipsius ϕ adversus *R*[12] refracti secabunt praefatum ϕR inter ϕ et *R*, et incidentium inter *R* et *A* refracti secabunt ad partes ultra ϕ sive ab *R* remotas.[13] Porrò cum intersectiones, quas radij utrinque cum $\phi\rho$[14] efficiunt, sint eò densiores quo sunt viciniores puncto ϕ ac in illo puncto densissimae: istud itaque ϕ pro foco radij ϕR habendum est sive pro loco imaginis illuc per refractionem translatae; habito scilicet ad eos solummodo radios respectu qui jacent in plano *FAR*, quod refringenti plano perpendiculariter insistit, transitáque / per punctum radians *F*. Nam alij refracti quorum incidentes 93/ jacent in alijs planis per puncta *F* et *R* transientibus et obliquis ad refringens planum, | radium *Rϕ* nec in puncto ϕ, nec ullibi omninò secabunt, si eos solummodò excipias quorum incidentes jacent in superficie conicâ cujus axis est *AF*, vertex *F*, et semi-angulus *AFR*; utpote qui omnes praefatum *Rϕ* in puncto *D* secabunt, quod in axe *FA* sit positum. Et hujus itaque *Rϕ* centra radiationis[15] praecipuè sunt duo, alterum ϕ a refractis jacentium in plano *FAR* effectum, et alterum a refractis jacentium in conicis superficiebus axe *DFA* angulisque *AFR, ADR* descriptis. Ad reliquos autem radios quod attinet, aliter circa *FR* quaquaversum positos, eorum refracti maximè appropinquant radio *Rϕ* alicubi inter *D* et ϕ. Adeò ut respectu oculi per cujus pupillae centrum radius *RM* transit, locus imaginis per totum spatium ϕD diffundi debeat. Vel potiùs cùm spatium ϕD sit unici tantùm puncti *F* imago, debemus unicum aliquod in eo punctum quod lucis omnis ab eo versus oculum pergentis meditullium occupet, inter puncta *D* et ϕ in mediâ circiter distantiâ interjacens, pro sensibili imagine[16] statuere. Puncti verò illius accurata determinatio, cùm omnium radiorum ab *F* versus oculi pupillam refractorum habenda sit aestimatio, problema solutu difficillimum praebebit nisi Hypothesi alicui saltem[17] verisimili, si non accuratè verae innitatur assertio. Quemadmodum cùm radij aequè multi a termino *D*, alijsque vicinis punctis, ac a termino ϕ

(9) Barrow's Axiom 13 (*Euclidis Elementorum*, pp. 7–8) is the parallel postulate of modern editions.

(10) Read: ρ.

(11) inter ϕ et *R*.] Read: ad partes ultra ϕ sive ab *R* remotas. (on the side beyond ϕ or remote from *R*.)

(12) ϕ . . . *R*] Read: *R* . . . ρ.

(13) ad . . . remotas.] Originally: eundem ϕR ad partes ejusdem ϕ ab *R* remotas. (the same ϕR on the side of ϕ remote from *R*.)

(14) Read: ϕR. For a more rigorous justification of the rays' intersection about the image point ϕ, see Barrow, *Lectiones XVIII*, Lect. V, §§XVII–XXI, pp. 60–3.

ρδ will meet on that side where the angles are smaller than two right angles (*Elements,* I, Axiom 13),[9] that is, on the side toward the point *F*; and consequently on the other side tending toward *M* and *μ*, they will diverge from one another. As was to be shown.

Scholium. If the angle *Rφρ* that the refracted rays make is assumed to be infinitely small, that point *φ* will be the limit separating the intersections of rays lying on each side made with the ray *Rφ* or *ρφ*. Namely, when the refraction is made from a rarer into a denser medium, the refracted rays of those falling on the side of the point *R* toward *A* will intersect the ray *φR* on the side of the point *φ* toward *R,* and the refracted rays of those falling between *A*[10] and *R* will intersect the same *φR* between *φ* and *R*.[11] But conversely, when the refraction is made from a denser into a rarer medium, the refracted rays of those falling on the side of *φ* toward *R*[12] will intersect the specified *φR* between *φ* and *R*, and the refracted rays of those falling between *R* and *A* will intersect it on the side beyond *φ* or remote from *R*.[13] Moreover, since the intersections that the rays on each side make with *φρ*[14] are denser the closer they are to the point *φ*, and are densest at that point, therefore that point *φ* must be considered as the focus of the ray *φR,* or rather as the place of the image shifted to there by the refraction—considering, of course, only those rays that lie in the plane *FAR,* which stands perpendicular to the refracting plane and passes through the radiating point *F*. For the other refracted rays, whose incident rays lie in other planes that pass through the points *F* and *R* and are inclined to the refracting plane, | will intersect the ray *Rφ* neither in the point *φ* nor anywhere else, except only those whose incident rays lie in the surface of the cone whose axis is *AF,* vertex is *F,* and half angle is *AFR,* inasmuch as all those will intersect the specified *Rφ* in the point *D* located on the axis *FA*. Therefore, there are principally two centers of radiation[15] of *Rφ*: the one, *φ*, made by refracted rays lying in the plane *FAR,* and the other, [*D*], made by refracted rays lying in the surfaces of the cones described with the axis *DFA* and angles *AFR* and *ADR*. As for the other rays otherwise situated all around *FR,* their refractions come nearest to the ray *Rφ* somewhere between *D* and *φ*, so that with respect to an eye through the center of whose pupil the ray *RM* passes, the image's position must be diffused through the entire space *φD*. Or rather, since the space *φD* is the image of only the single point *F,* we ought to take for the sensible image[16] some single point in it that occupies the middle of all the light proceeding from there toward the eye, and that lies approximately midway between the points *D* and *φ*. Since an estimate must be made of all rays refracted from *F* toward the pupil of the eye, the accurate determination of that point will in fact present a very difficult problem to solve, unless the claim is based on some hypothesis that is at least[17] probable, if not strictly true. For instance, since equally many rays appear to flow toward the eye from the end point *D* and other nearby points as from the end point *φ* and

[129 ≈ I, 62]

---

(15) centra radiationis] Originally: foci.
(16) Originally: foco (focus).   (17) Added.

alijsque punctis similiter sibi vicinis versus oculum videantur profluere: locús imaginis ita debet in medio istorum terminorum statui, ut angulus quem radij duo a *D* et *ϕ* ad idem quodpiam pupillae punctum convergentes includunt, a radio ab illo visionis loco ad idem pupillae punctum pergente quàm proximè semper bisecetur. Quâ Hypothesi admissâ nihil aliud agendum est, quàm ut fiat *Mϕ* + *MD* . *MD* :: *ϕD* . *DZ,* et erit *Z* locus visionis puncti *F* quaesitus; posito nempe quòd *M* sit locus oculi. Nam cùm ponatur *Mϕ* + *MD* . *MD* :: *ϕD* . *DZ,* erit divisim[18] *Mϕ* . *MD* :: *ϕZ* . *DZ.* Et proinde ductis tribus lineis a *ϕ, D,* et *Z* ad *M* vel potiùs ad punctum quodpiam huic *M* indefinitè vicinum, angulus quem externae duae continent, ab interjacente lineâ (per 3. 6. Elem)[19] / quàm proximè semper bisecabitur. | Caeterùm nè puncti *ϕ* positio gratis 94/ assumatur, placet illud insuper sequenti Methodo determinare, praesertim cùm in hac re videatur praecipuum.

[130 → I, 61]   Normalibus *AG, AH* a puncto *A* in radios dimissis, alterâ *AG* in incidentem radium *FR* et alterâ *AH* in refractum *DR,* factóque *FG* . *DH* :: *RF* . *Rϕ*; punctum *ϕ* erit locus objecti *F* post refractionem visi habito unicè ad radios in plano *FAR* jacentibus respectu. Scilicet cùm hoc punctum sit limes per interpositionem dirimens ac distinguens intersectiones radiorum utrinque positorum, nè longâ propositionum serie ad hoc demonstrandum opus sit, illorum intersectiones finitis intervallis ab *RD* distantium vix respiciam, sed radij tantùm indefinitè propinquissimi intersectionem speculando determinabo propositum, siquidem ea ipsa (ut jam dictum est) sit punctum *ϕ,* quod quaeritur. Et nè demonstratio haec, quae (dum nullis ferè[20] fundamentis praemonstratis innititur) longiuscula futura est, vos itaque taedio afficiat, lubet ut in partes aliquot sive conclusiones distinguatur.

Dico igitur imprimis quòd positis quorumlibet uniformium radiorum *FR, Fρ,* refractis *RD, ρδ* secantibus perpendiculum *FA* in *D* ac *δ*: erit

---

(18) On Newton's operations with ratios see Lect. 11, note (24).

(19) Namely, a line bisecting an angle of a triangle divides the opposite side in the same ratio as the adjacent sides, and conversely. Newton has correctly recognized that homocentric rays, or rays originating from a point, after refraction are no longer homocentric; that is, they are astigmatic and have two image points or "centers of radiation"—the one, *ϕ* (known as the primary, meridian, or tangential image point), of rays in the plane of incidence, and the other, *D* (the secondary or sagittal focal point), of rays in the specified plane surface—and that no other rays intersect the chief ray *ϕR*. The secondary image point *D* has already been determined by Lemma 2 (§120 = I, 47), that is, by the sine law, for *DR/FR* = *I/R* = *n.* The image points, as Thomas Young later recognized (see Lect. I, 14, note (6) this volume), are actually lines formed by those rays—ignored by Newton—that are in planes indefinitely close to and either parallel or perpendicular to the plane of incidence. See the historical note on astigmatism by P. Culmann, in *Geometrical Investigation of the Formation of Images in Optical Instruments,* ed. Moritz von Rohr, trans. R. Kanthack (London, 1920), pp. 201–9; and also A. G. Bennett, "Some unfamiliar British contributions to geometrical optics," in *Transactions of the International Ophthalmic Optical Congress, 1961,* British Optical Association, London, [1962], pp. 274–91.

Kepler inaugurated the modern theory of optical imagery with the introduction of the hypothesis that a virtual image is located in that place from which the rays entering the eye appear to diverge; see his *Ad Vitellionem paralipomena,* Ch. III, Prop. XVII, or *Dioptrice,* §XIX, *Gesammelte Werke,* 2:72, 4:360. (Newton later made this widely adopted principle Axiom VIII of the *Opticks.*) At the same time Kepler rejected the ancient principle of image location, namely, that the image is located at the intersection of the refracted (or reflected) ray and the *cathetus,* or

other points similarly nearby, the image's position ought to be set in the middle of those endpoints so that the angle contained by two rays converging from D and ϕ to any one point of the pupil is always bisected as closely as possible by the ray proceeding from that place of vision to the same point of the pupil. Once this hypothesis is admitted, nothing else needs to be done other than to make $M\phi + MD : MD = \phi D : DZ$, and Z will be the required place of vision of the point F, assuming, that is, that M is the eye's position. For since it is assumed that $M\phi + MD : MD = \phi D : DZ$, upon dividing[18] it will be $M\phi : MD = \phi Z : DZ$; consequently, three lines having been drawn from ϕ, D, and Z to M, or rather to any point indefinitely near M, the angle contained by the outer two will always be bisected as closely as possible by the line lying between them (*Elements*, VI, 3).[19] | Lest the position of the point ϕ be gratuitously assumed, however, it is agreeable to determine it in addition by the following method, especially since in this matter it seems important.

Dropping the normals AG and AH from the point A onto the rays, the one, AG, onto the incident ray FR and the other, AH, onto the refracted ray DR, and making $FG : DH = RF : R\phi$, the point ϕ will be the apparent place of the object F after refraction, when considering solely rays lying in the plane FAR. Namely, since this point is the limit separating and distinguishing by its insertion the intersections of the rays located on each side, in order not to need a long series of propositions to demonstrate this, I will hardly consider the intersections of those rays at a finite distance from RD; but I will achieve my purpose by examining only the intersection of the indefinitely closest ray, inasmuch as that intersection (as has already been indicated) is the required point ϕ. Lest this demonstration, which will be a little long, may consequently bore you (as it is based on almost[20] no principles previously presented), it is desirable to divide it into several parts or conclusions.

I say, therefore, in the first place that assuming the refracted rays RD and $\rho\delta$ of any uniform rays FR and $F\rho$ intersect the perpendicular FA in D and δ, then

[130 → I, 61]

---

the perpendicular from the object to the refracting (or reflecting) surface; see, for example, Alhazen, *Opticae thesaurus*, Bk. VII, §§18–19, pp. 253–6; and also Colin M. Turbayne, "Grosseteste and an ancient optical principle," *Isis*, 50 (1959):467–72. The ancient principle, however, frequently gives valid, or partly valid, results as it does here: The image point according to it is the same as Newton's secondary image point D, and also for paraxial rays (that is, rays incident near the axis, or the normal to the surface) when the primary and secondary focal points coincide. For obliquely incident rays, though, it leaves the primary image point undetermined. In Lect. IV of his *Lectiones XVIII* Barrow rigorously described the secondary image point D for a small pencil of paraxial rays, and in the following lecture he successfully tackled the more difficult problem of locating the primary image point ϕ for oblique rays, but now he rejected the existence of the secondary image point; Lect. V, §XXI, and Lect. XVI, §II, pp. 62, 136–7. Earlier, in 1653, Huygens had similarly solved the problem of locating the image for paraxial rays, although he left the problem untouched for oblique rays; *Dioptrica*, Props. IV–VII, *Oeuvres complètes*, **13**, i:19–27. Newton's unique insight in extending Barrow's solution—and thus beginning the theory of astigmatism—was to recognize that there were simultaneously two image points, or that the image of a single point is spread throughout the entire space ϕD.

(20) Added.

$$R\rho \,.\, D\delta :: \frac{FR^q}{AR + A\rho} \,.\, \frac{DR^q - FR^q}{AD + A\delta}.$$

In praecedentibus enim ostensum est (sect 120) quòd FR & RD sunt ut sinus incidentiae et refractionis,[21] et sic $F\rho$ ad $\rho\delta$ habebit eandem rationem. Quare terminos quadrando erit $FR^q \,.\, RD^q :: F\rho^q \,.\, \rho\delta^q$. et per conversam rationem $FR^q \,.\, RD^q - FR^q :: F\rho^q \,.\, \rho\delta^q - F\rho^q$. rursus per conversam rationem, subintellectâ tamen permutatione, fit

$$FR^q \,.\, RD^q - FR^q :: F\rho^q - FR^q \,.\, \rho\delta^q - F\rho^q - RD^q + FR^q.$$

Est autem $F\rho^q - FR^q$ (per sect 119) $= A\rho^q - AR^q = R\rho^q + R\rho \times 2AR = R\rho \times \overline{AR + A\rho}$. Est etiam $\rho\delta^q - F\rho^q = A\delta^q - AF^q$ & $RD^q - FR^q = AD^q - AF^q$, adeoque $\rho\delta^q - F\rho^q - RD^q + FR^q = A\delta^q - AD^q = D\delta \times \overline{AD + A\delta}$. Quare est

$$FR^q \,.\, RD^q - FR^q :: R\rho \times \overline{AR + A\rho} \,.\, D\delta \times \overline{AD + A\delta}.$$

Et applicando antecedentes ad $AR + A\rho$ et consequentes ad $AD + A\delta$ prodit

$$\frac{FR^q}{AR + A\rho} \,.\, \frac{RD^q - FR^q}{AD + A\delta} :: R\rho \,.\, D\delta.$$

Q.E.O.

Porrò si radiorum FR, $F\rho$ distantia sit indefinitè parva; Dico quod erit

$$AD \times FR^q \,.\, AR \times \overline{RD^q - FR^q} :: R\rho \,.\, D\delta.$$

Tunc enim segmenta $R\rho$, $D\delta$ pro infinitè parvis / habenda sunt, sive lineae AD, $A\delta$, ut et AR, $A\rho$ pro infinitè parùm differentibus; hoc est pro aequalibus. Evadit ergo $AR + A\rho = 2AR$, et $AD + A\delta = 2AD$. Et sic vice proportionis jam ante ostensae oritur $\frac{FR^q}{2AR} \,.\, \frac{RD^q - FR^q}{2AD} :: R\rho \,.\, D\delta$. Sive, multiplicando priorem rationem per $2AR \times AD$, est

$$AD \times FR^q \,.\, \overline{RD^q - FR^q} \times AR :: R\rho \,.\, D\delta.$$

Tertiò dico, quod est $AD^q \times FR^q \,.\, AR^q \times RD^q - AR^q \times FR^q :: R\phi \,.\, D\phi$. Nam erectâ RK ad AR normali, quae secet radium $\rho\mu$ in K: est $A\rho \,.\, A\delta :: R\rho \,.\, RK$, sive $AR \,.\, AD :: R\rho \,.\, RK$, siquidem $A\rho$ et AR, nec non $A\delta$ et AD pro infinitè parùm differentibus habentur. Et priori ratione per $AD \times FR^q$ multiplicatâ divisâque per AR orietur $AD \times FR^q \,.\, \dfrac{AD^q \times FR^q}{AR} :: R\rho \,.\, \rho K$. Quamobrem cùm supra inventum est $AD \times FR^q \,.\, AR \times RD^q - AR \times FR^q :: R\rho \,.\, D\delta$, si utriusóue permutatio subintelligatur, patebit esse

$$\frac{AD^q \times FR^q}{AR} \,.\, \rho K :: AR \times RD^q - AR \times FR^q \,.\, D\delta.$$

et multiplicando per AR permutandoque fit

$$AD^q \times FR^q \,.\, AR^q \times RD^q - AR^q \times FR^q :: \rho K \,.\, D\delta.$$

Est autem $\rho K \,.\, D\delta :: R\phi \,.\, D\phi$. Quare et

$$AD^q \times FR^q \,.\, AR^q \times RD^q - AR^q \times FR^q :: R\phi \,.\, D\phi.$$

$$R\rho : D\delta = FR^2/(AR + A\rho) : (DR^2 - FR^2)/(AD + A\delta).$$

For in the preceding it was shown (§120) that FR and RD are as the sines of incidence and refraction[21] and thus $F\rho$ to $\rho\delta$ will be in the same ratio. Consequently, upon squaring the terms, it will be $FR^2 : RD^2 = F\rho^2 : \rho\delta^2$; and by conversion of the ratio,

$$FR^2 : (RD^2 - FR^2) = F\rho^2 : (\rho\delta^2 - F\rho^2);$$

and by another conversion of the ratio (while also supplying a permutation) it becomes $FR^2 : (RD^2 - FR^2) = (F\rho^2 - FR^2) : (\rho\delta^2 - F\rho^2 - RD^2 + FR^2)$. But by §119, $F\rho^2 - FR^2 = A\rho^2 - AR^2 = R\rho^2 + R\rho \times 2AR = R\rho(AR + A\rho)$, and also $\rho\delta^2 - F\rho^2 = A\delta^2 - AF^2$, and $RD^2 - FR^2 = AD^2 - AF^2$, so that

$$\rho\delta^2 - F\rho^2 - RD^2 + FR^2 = A\delta^2 - AD^2 = D\delta(AD + A\delta).$$

Therefore, $FR^2 : (RD^2 - FR^2) = R\rho(AR + A\rho) : D\delta(AD + A\delta)$. Dividing the numerators by $AR + A\rho$ and the denominators by $AD + A\delta$, there results $FR^2/(AR + A\rho) : (RD^2 - FR^2)/(AD + A\delta) = R\rho : D\delta$. As was to be shown.

Furthermore, if the distance of the rays FR and $F\rho$ is indefinitely small, I say that $AD \times FR^2 : AR(RD^2 - FR^2) = R\rho : D\delta$. For then the segments $R\rho$ and $D\delta$ must be considered as infinitely small, or rather the lines AD and $A\delta$ as well as AR and $A\rho$, as differing by infinitely little—that is, as equal. It turns out therefore that $AR + A\rho = 2AR$, and $AD + A\delta = 2AD$. Thus, in place of the proportion shown just before there arises

$$FR^2/2AR : (RD^2 - FR^2)/2AD = R\rho : D\delta,$$

or multiplying the first ratio by $2AR \times AD$,

$$AD \times FR^2 : (RD^2 - FR^2)AR = R\rho : D\delta.$$

Third, I say that $AD^2 \times FR^2 : AR^2(RD^2 - FR^2) = R\phi : D\phi$. For when RK is erected normal to AR and intersects $\rho\mu$ in K, then $A\rho : A\delta = R\rho : RK$, or $AR : AD = R\rho : RK$, since $A\rho$ and AR and also $A\delta$ and AD are considered to differ by infinitely little. Multiplying the first ratio by $AD \times FR^2$ and dividing it by AR, there will arise $AD \times FR^2 : (AD^2 \times FR^2)/AR = R\rho : \rho K$. Consequently, since it was found above that $AD \times FR^2 : AR(RD^2 - FR^2) = R\rho : D\delta$, if a permutation is understood in both, it will be clear that

$$(AD^2 \times FR^2)/AR : \rho K = AR (RD^2 - FR^2) : D\delta;$$

and multiplying by AR and permuting, there results

$$AD^2 \times FR^2 : AR^2 (RD^2 - FR^2) = \rho K : D\delta.$$

But $\rho K : D\delta = R\phi : D\phi$, and therefore $AD^2 \times FR^2 : AR^2 (RD^2 - FR^2) = R\phi : D\phi$.

---

(21) incidentiae et refractionis] Read: refractionis et incidentiae (of refraction and incidence).

Dico denique Quòd est $FG . DH :: RF . R\phi$. Nam cùm sit

$$AD^q \times FR^q . AR^q \times RD^q - AR^q \times FR^q :: R\phi . D\phi,$$

erit divisim $AD^q \times FR^q . AD^q \times FR^q - AR^q \times RD^q + AR^q \times FR^q :: R\phi . RD$. At est $AD^q \times FR^q + AR^q \times FR^q = DR^q \times FR^q$, et $DR^q \times FR^q - AR^q \times RD^q = DR^q \times AF^q$. Quare $AD^q \times FR^q . DR^q \times AF^q :: R\phi . RD$. Ductisque extremis et medijs in se invicem fit $AD^q \times FR^q \times RD = DR^q \times AF^q \times R\phi$. et applicando ad $FR \times DR^q$ oritur $\dfrac{AD^q \times FR}{DR} = \dfrac{AF^q \times R\phi}{FR}$. Quo in proportionalitatem resoluto prodit $\dfrac{AF^q}{FR} . \dfrac{AD^q}{DR} :: FR . R\phi$. Sed (per 8. 6. El)[22] est $FR . AF :: AF . FG$; ut et $DR . AD :: AD . DH$ et proinde

$$\frac{AF^q}{FR} = FG \text{ et } \frac{AD^q}{DR} = DH,$$

atque adeò $FG . DH :: FR . R\phi$.[23]   Q.E.D.

Lect 13
[131 = I, 63]

Sed videor actum agere, et his itaque paucis[1] circa radios homogeneos in gratiam sequentium obiter[2] notatis, ut eorum penitior cognitio habeatur, Lectiones, quas Vir Reverendus Dr Barrow de ijs fusè composuit, consulendas esse moneo,[3] deque heterogeneis sive dissimiliter refrangibilibus radijs pergo actutùm disserere.

[132][4] Difformium radiorum a planâ superficie refractorum affectiones enarrantur.

Prop: 4. Si radij heterogenei manantes a puncto quodam lucido refringantur a qualibet planâ superficie, eorum refracti possunt esse paralleli vel convergentes aequè ac divergentes. Caeterùm placet, ut ad magis particularia descendam a quibus veritas hujus patebit. Et cùm a sinibus incidentiae et refractionis diversorum generum radiorum ratiocinia pendebunt; posito quòd similiter / incidentium communis sinus incidentiae brevitatis gratiâ vocetur I, sinum refractionis purpuriformium nominabo P, caeruliformium Q, viridiformium R, flaviformium S, et rubiformium T.

[133]

Prop 5. Si [fig 56] sit FT radius extremè rubiformis, ac TB subtendens

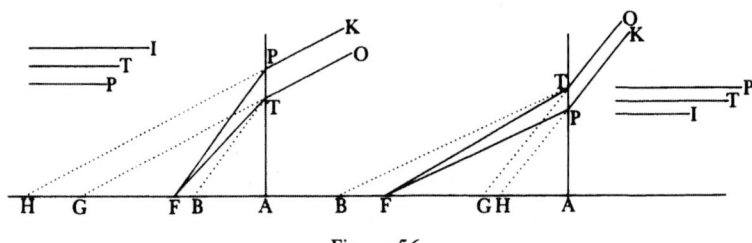

Figure 56

angulum TAF ducatur ut sit $T . P :: FT . TB$, et agatur FP huic TB parallela, quae concipiatur esse radius extremè purpuriformis: Dico quod radiorum FP et FT refracti PK ac TO erunt paralleli. Nam radijs PK TO retrorsum ductis

---

(22) The perpendicular drawn from the right angle to the hypoteneuse of a triangle forms similar triangles.

Finally, I say that $FG : DH = RF : R\phi$. For since

$$AD^2 \times FR^2 : AR^2 (RD^2 - FR^2) = R\phi : D\phi,$$

by division there will be

$$AD^2 \times FR^2 : (AD^2 \times FR^2 - AR^2 \times RD^2 + AR^2 \times FR^2) = R\phi : RD.$$

But $AD^2 \times FR^2 + AR^2 \times FR^2 = DR^2 \times FR^2$, and

$$DR^2 \times FR^2 - AR^2 \times RD^2 = DR^2 \times AF^2.$$

Consequently,

$$AD^2 \times FR^2 : DR^2 \times AF^2 = R\phi : RD.$$

When the extreme and middle terms are multiplied into one another, there results $AD^2 \times FR^2 \times RD = DR^2 \times AF^2 \times R\phi$, and on dividing by $FR \times DR^2$, there arises $(AD^2 \times FR)/DR = (AF^2 \times R\phi)/FR$. Resolving this into a proportion, there results $AF^2/FR : AD^2/DR = FR : R\phi$. But (by *Elements*, VI, 8)[22] $FR : AF = AF : FG$, and also $DR : AD = AD : DH$; consequently $AF^2/FR = FG$ and $AD^2/DR = DH$, and hence $FG : DH = FR : R\phi$.[23] As was to be demonstrated.

But I seem to be wasting your time, and therefore after these few things[1] about homogeneous rays have been noted in passing for the sake of what follows,[2] in order to acquire a more profound knowledge of them I recommend[3] that you consult the *Lectures* where Rev. Dr. Barrow has written on these things at length; and I immediately proceed to lecture on heterogeneous or unequally refrangible rays.

Lecture 13
[131 = I, 63]

Proposition 4. If heterogeneous rays flowing from some luminous point are refracted by any plane surface, their refracted rays can be parallel or convergent as well as divergent.

[132].[4] The properties of unlike rays refracted by a plane surface are described.

It is, however, my desire to penetrate to the finer details, whereby the truth of this will be evident. Since the calculations will depend on the sines of incidence and refraction of the different kinds of rays, and since the common sine of incidence of similarly incident rays is for brevity's sake called I, I will call the sine of refraction of purple-making rays P, of blue-making Q, of green-making R, of yellow-making S, and of red-making T.

Proposition 5. If (Fig. 56) FT is an extreme red-making ray; if TB, which subtends the angle TAF, is drawn so that $T : P = FT : TB$; and if FP is drawn parallel to TB and is imagined to be an extreme purple-making ray; I say that the refracted rays PK and TO of the rays FP and FT will be parallel.

[133]

---

(23) Newton's determination of the primary image point is readily put into more recognizable form: Let $i = A\hat{F}R$ and $r = A\hat{D}R$, and since $DH = (\cos r) DA = (\cos^2 r) DR$ and $FG = (\cos i) FA = (\cos^2 i) FR$, whereas $DR/FR = n$, then $R\phi/FR = DH/FG = n \cos^2 r/\cos^2 i$. This Scholium is included in Newton's *Mathematical Papers*, 3:454–61.

(1) Sed . . . paucis] II: His paucis (After these few things).

(2) in . . . obiter] Originally: in transcursu (by the way).   (3) II: hortor (I urge).

(4) In Newton's own sequence of articles this is §126.

(a) Sect 120
(b) Hypoth.

donec secent *FA* in *H* ac *G*, erit *I . T :: TG . TF*.[a] Et praeterea cùm *T . P :: TF . TB*[b], erit ex aequo *I . P :: TG . TB*. Sed est *I . P :: PH . PF*.[a]

(c) 7.6. Elem[5]

Ergo *TG . TB :: PH . PF*. Atque adeo cùm *TB* et *PF* sint parallelae[b], erunt etiam *TG* et *PH* parallelae.[c]   Q.E.O.[6]

[134 → I, 71]   Prop 6. Eodem modo [fig 57] pateat quod innumeri radij specierum inter-

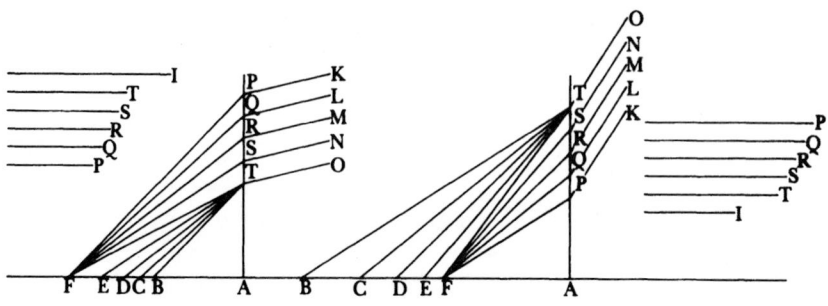

Figure 57

mediarum possunt ita duci inter *FP*, et *FT* ut eorum refracti fiant ipsis *PK* ac *TO* paralleli. Quo posito dico quod illi radij prout a rubedine ad purpuram succedunt, hoc est prout sunt magis ac magis refrangibiles, incident in punctis a *P* usque ad *T*[7] continuò successivis. Sic verbi gratiâ fiat *P . Q . R . S . T :: TB . TC . TD . TE . TF*. Et ipsis *TB, TC, TD* ac *TE* ducantur parallelae *FP, FQ, FR, FS* quae, unà cum *FT* priùs ductâ concipiantur esse radij coloribus quinque insignioribus purpureo caeruleo viridi flavo, et rubeo[8] tincti, et constabit e praecedenti propositione quod eorum refracti *PK, QL, RM, SN, TO* erunt paralleli: Sed cùm *P, Q, R, S* ac *T* ferè sint in Arithmeticâ vel potiùs in Geometricâ proportione, aut saltem sese gradatim superant longitudine, eò quòd sunt sinus refractionum ab invicem / gradatim 97/ differentium: sequitur quod *TB, TC, TD, TE, TF* sese etiam longitudine gradatim superabunt, et proinde jacebunt in ordine nominato; et ideò *FP, FQ, FR, FS*, et *FT* jacebunt in eodem ordine.

[135]   Prop 7. Radij nullius [ab *F*] extra spatium *PT* incidentis, refractus potest esse praefatis refractis parallelus: siquidem nullus datur radius magis refrangibilis quàm purpuriformis *FPK* supponitur, nec minùs refrangibilis quàm rubiformis *FTO*. Nam si talis refractus ponatur esse ipsi *PK* caeterisque parallelus; vel incidet ad partes juxta *P* et sic erit magis refrangibilis quàm *FPK* ob eandem rationem quâ *FPK* sit magis refrangibilis quàm *FQL*; vel incidet ad partes juxta *T* et sic erit minùs refrangibilis quam *FTO* ob eam rationem, qua *FTO* sit minùs refrangibilis quàm *FSN*: contra Hypothesin.

[136]   Schol: Verùm hic et in posterum notandum est quòd colorum extremitates nullibi exactè definiuntur, sed insensibili diminutione paulatim deficiunt: adeò ut haud facile sit dicere quis sit eorum terminus vel quanta sit maxima et minima radiorum refrangibilitas. Sed habitu ad experimenta et frequentes colorum apparentias respectu, convenit ut eos solummodò radios consideremus quorum numerus et vigor tantus est ut facilè sentiantur, caeteris ad extremitates summas non adnumeratis. Et sic in praecedentibus calculis,[9] ubi

For when the rays *PK* and *TO* are extended backward until they intersect *FA* in *H* and *G*, then *I* : *T* = *TG* : *TF*.[(a)] Furthermore, since *T* : *P* = (a) §120
TF : *TB*,[(b)] from the equality of the ratios there will be *I* : *P* = *TG* : *TB*. But (b) Hypothesis
I : *P* = *PH* : *PF*;[(a)] therefore, *TG* : *TB* = *PH* : *PF*. Hence, since *TB* and *PF*
are parallel,[(b)] *TG* and *PH* will also be parallel.[(c)] As was to be shown.[(6)] (c) *Elements*, VI, 7[(5)]

Proposition 6. In the same way (Fig. 57) it is clear that innumerable rays of [134 → I, 71]
intermediate kinds can be drawn between *FP* and *FT* so that their refracted
rays become parallel to *PK* and *TO*. Assuming this, I say that those rays, as
they pass from red to purple, that is, as they are more and more refrangible,
will fall continuously on the successive points from *P* to *T*.[(7)]

Thus, for example, make *P* : *Q* : *R* : *S* : *T* = *TB* : *TC* : *TD* : *TE* : *TF*; and
parallel to *TB*, *TC*, *TD*, and *TE* draw *FP*, *FQ*, *FR*, and *FS* which, together
with *FT* drawn earlier, are imagined to be colored with the five more prominent colors, purple, blue, green, yellow and red;[(8)] and from the preceding
proposition it will be manifest that their refracted rays *PK*, *QL*, *RM*, *SN*, and
TO will be parallel. Since *P*, *Q*, *R*, *S*, and *T* are nearly in arithmetic, or rather
in geometric, proportion—or at least they gradually exceed one another in
length, because they are sines of refraction gradually differing from one
another—it follows that *TB*, *TC*, *TD*, *TE*, and *TF* will also gradually exceed
one another in length and consequently lie in the order named, and thus *FP*,
FQ, *FR*, *FS*, and *FT* will lie in the same order.

Proposition 7. The refraction of no ray incident [from *F*] outside the space [135]
PT can be parallel to the specified refracted ones, since there exists no ray
more refrangible than the purple-making one, *FPK*, is supposed to be, or less
refrangible than the red-making one, *FTO*.

For if such a refracted ray is assumed to be parallel to *PK* and the others,
either it will fall on the side next to *P* and thus be more refrangible than *FPK*,
for the same reason that *FPK* is more refrangible than *FQL*, or it will fall on
the side next to *T* and thus be less refrangible than *FTO*, for the reason that
FTO is less refrangible than *FSN*, which is contrary to the hypothesis.

Scholium. In truth it ought to be noted here and in the following that the [136]
extremities of the colors are nowhere exactly defined, but they gradually
decay with an insensible decrease to such a degree that it is not easy to
establish where their end is or what the magnitude of the rays' greatest or
least refrangibility is. But with respect to experiment and the common appearances of colors, it is agreed that we will consider only those rays whose
quantity and vigor is great enough to be readily perceived, not counting the
others at the utmost extremities. Thus in the preceding calculations,[(9)] where

---

(5) If two triangles have one equal angle and the sides about the other angles are proportional,
then the triangles will be similar.

(6) The physical meaning of this proposition may perhaps be clearer from the following: By
the sine law of refraction, sin i_P/sin r_P = *I*/*P* and sin i_T/sin r_T = *I*/*T*; and since it is required that
the refracted rays be parallel, or sin r_P = sin r_T, then it must be sin i_P/sin i_T = *FT*/*TB* = *T*/*P*.

(7) *P* . . . *T*] Read: *T* . . . *P*.

(8) purpureo . . . rubeo] Newton at first inadvertently had the colors in reverse order: "praediti rubeo nempe, flavo, viridi, caeruleo et purpureo."   (9) In §§110–11 ≈ I, 37–8.

determinavimus esse $T . P :: 68 . 69$, cùm refractio fit ex vitro in aerem; ponimus T, et P esse sinus extimorum radiorum, quorum numerus est tantus, ut liquidò[10] feriant sensus; non autem radiorum qui in extremitate summâ tam pauci sunt, ut vix aut nullo modo sentiantur, et ideò non digni ut in aestimationem veniant.[11] Verùm tamen sive illi solùm radij considerentur, quorum copia tanta est ut facilè sentiantur, sive etiam alij adhuc exteriores; semper supponimus P designare sinum refractionis extremorum, qui considerantur ad extremitatem purpuream, ac T sinum eorum ad rubram: hoc est supponimus P designare sinum maximè refrangibilium ex omnibus, qui considerantur, ac T sinum eorum qui ponuntur minimè omnium refrangibiles. Atque has duas radiorum species in praecedentibus potiùs vocavimus extremè quàm intensè purpuriformes et rubiformes, eo quòd sunt omnium colorum extremitates, cùm intensissima sive perfectissima purpura et rubedo non in summis extremitatibus sed circa meditullium purpurae et rubedinis cernantur.

[137]   / Prop 8. Radij extremè purpuriformis $F\pi$ [fig 58] intra spatium PT inci- 98/

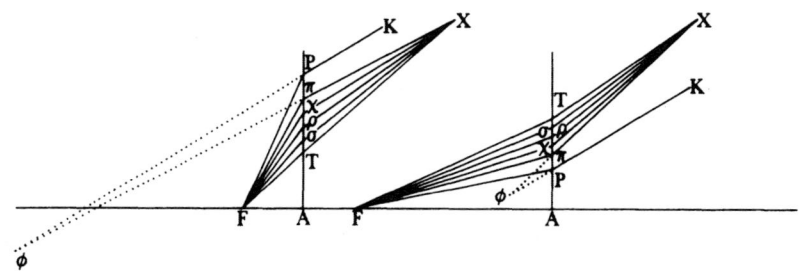

Figure 58

dentis refractus πX converget ad TX refractum radij FT extremè rubiformis et alicubi ultra refringentem superficiem secabit puta ad X. Nam cùm FP et

(a) Hypoth.
(b) Prop 3.
(c) 17.1 Elem.[12]
(d) 29.1 Elem.[13]
(e) 15.1 Elem.[14]
(f) ax 13.1 Elem.[15]

$F\pi$ sint homogenei[a], eorum refracti PK, πX divergent ab aliquo puncto velut ϕ quod citra refringentem superficiem locatum est[b]. Ergo anguli $\phi P\pi$ + $\phi\pi P$,[c] hoc est anguli $XT\pi$[d] + $X\pi T$[e] sunt minores duobus rectis. Quibus angulis ultra refringentem superficiem jacentibus, etiam πX ac TX concurrent ultra.[f]

[138 → I, 64]   Prop 9. Omnes radij ab F ad X refracti jacent in eodem plano FAX. Nam planum in quo radius incidens et ejus refractus jacent, semper perpendiculare est ad planum refringens: Sed nullum perpendiculare planum transit per puncta F et X praeter FAX. Ergo omnes radij ab F ad X refracti jacent in isto plano.

[139]   Prop: 10. Radius intermedij generis cujus refractus transit [ab F] per dictum punctum X, incidet in refringentem superficiem inter π ac T. Non enim incidet in punctum π, quia tunc eandem refractionem pateretur ac radius intensè purpuriformis, contra Hypothesin quòd sit intermedij generis. Neque

---

(10) Originally: facilè (easily).

we have determined $T : P = 68 : 69$ when the refraction is made from glass into air, we suppose that T and P are the sines of the outermost rays whose quantity is great enough to strike the senses clearly,[10] but not those of the rays at the utmost extremities that are so few as to be barely or not at all perceived and thus not appropriate to be taken into account.[11] Nevertheless, whether we consider only those rays whose quantity is so great as to be readily perceived, or even other still more exterior ones, we always suppose P to designate the sine of refraction of the outermost that are considered at the purple extremity, and T the sine of those at the red extremity; that is, we suppose P to designate the sine of the most refrangible of all those considered, and T the sine of those assumed to be the least refrangible of all. Consequently, in the preceding we have called these two species of rays extreme rather than intense purple- and red-making rays, because they are the extremities of all the colors, whereas the most intense or perfect purple and red are not perceived at the utmost extremities but near the middle of the purple and the red.

Proposition 8. The refraction πX (Fig. 58) of the extreme purple-making [137] ray $F\pi$ incident within the space PT will converge to TX, the refraction of the extreme red-making ray FT, and it will intersect it somewhere beyond the refracting surface, for example, at X.

For, since FP and $F\pi$ are homogeneous,[(a)] their refracted rays PK and πX will diverge from some point such as ϕ that is located on the near side of the refracting surface.[(b)] Therefore, the sum of the angles $\phi P\pi$ and $\phi \pi P$,[(c)] that is, of the angles $XT\pi$[(d)] and $X\pi T$,[(e)] is smaller than two right angles. Since these angles lie on the far side of the refracting surface, πX and TX will also meet on the far side.[(f)]

(a) Hypothesis
(b) Prop. 3
(c) *Elements*, I, 17[12]
(d) *Elements*, I, 29[13]
(e) *Elements*, I, 15[14]
(f) *Elements*, I, Axiom 13[15]

Proposition 9. All rays refracted from F to X lie in the same plane FAX. [138 → I, 64]

For the plane in which the incident ray and its refraction lie is always perpendicular to the refracting plane. But no perpendicular plane passes through the points F and X besides FAX. Therefore all rays refracted from F to X lie in that plane.

Proposition 10. An intermediate kind of ray whose refraction passes [from [139] F] through that point X will fall on the refracting surface between π and T.

For it will not fall at the point π, because it would then experience the same refraction as the intense purple-making ray, which is contrary to the hypothesis that it is an intermediate kind of ray. Nor will it fall beyond the

---

(11) In his letter for Lucas on 18 August 1676, Newton insisted, on the contrary, "That yᵉ utmost length of yᵉ Image from yᵉ faintest red at one end to yᵉ faintest blew at yᵉ other, must be measured" (*Correspondence*, 2:78 = *Phil. Trans.*, 11 (1676):701–2); and in the *Opticks*, he explained that he measured "the faintest and outmost" colors, but added in all editions after the first, "excepting only a little Penumbra, whose breadth scarce exceeded a quarter of an Inch" (Bk. I, Pt. I, Prop. II, Expt. III, p. 27).

(12) Any two angles in a triangle are less than two right angles.

(13) When a line intersects parallel lines, the alternate angles are equal. Newton implicitly assumes that PK is parallel to TK.

(14) When two straight lines intersect, the opposite angles are equal.

(15) See Lect. 12, note (9).

incidet extra dictum π versus P quia tunc deberet esse magis refrangibilis
(a) Cor: prop 2   quàm praefatus ille purpuriformis $F\pi$[(a)], habens nempe minorem angulum
refractionis ad majorem angulum incidentiae; Id quod adhuc magis hypothesi
contradicit. Et simili discursu patebit etiam, quòd in punctum T vel extra
illud ad partes ipsi P adversas non potest incidere. Restat ergo ut transeat
inter π ac T.

[140]   Prop 11. Heterogenei radij quorum refracti transeunt per dictum X, cadent
in spatium PT in eodem ordine in quo colores eorum sibi invicem a purpurâ
ad rubedinem succedunt. Ponatur enim quod caeruliformis radius incidat ad
χ, et cùm viriditas sit color inter / caeruleum et rubeum intermedius, eodem   99/
modo constabit quòd viridiformis radius cadet inter χ ac T, quo ostensum est
in praecedenti propositione quod radius quilibet intermediorum generum debeat inter π ac T incidere. Cadat itáque ad ρ. Et cùm flavedo sit color viridi
et rubeo interveniens, eodem quo dictum est modo constabit quòd flaviformis
radius cadet inter ρ ac T puta ad σ. Atque ita deinceps.

Si jam desideretur, ut ex datis punctis F et X, angulus $TX\pi$ determinetur:
quo id perficiam convenit ut problema sequens more Lemmatis exponatur.

[141 → I, 58]   Prob. Dato puncto planum refringens irradiante, et
alio etiam puncto, per quod refractus radius debet
transire, positio radiorum, sive refringens punctum
quaeritur. Sit [fig 59] F punctum radios ejaculans, et
R punctum ubi radius refringi debeat ut postmodum
transeat per datum X. Ab ijsdem F et X normales
FA, et $X\alpha$ ad refringens planum demittantur. Et
facto $\sqrt{II - RR}$. $I :: AF . VG$, cum latere recto VG
seorsim describatur Parabola Conica $VL\nu$, cujus vertex sit V, et axis VGH. Deinde ad axem erigatur a
vertice normalis $VK = \frac{1}{2}A\alpha$ et agatur KM ad axem
parallela, quae secet Parabolam in L, et capiatur LM
ejus longitudinis, ut sit $II - RR . RR :: \dfrac{\alpha X^q}{2VG} . LM.$
eique perpendiculariter insistens ad M ducatur MN,
ut sit $VG . VG + LM :: A\alpha . MN$, jungaturque LN

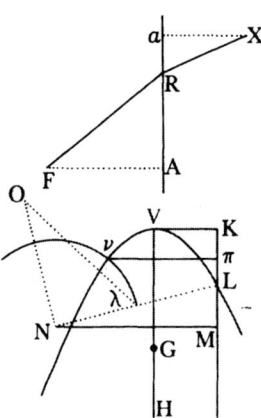

Figure 59

et ad ipsam erigatur normalis $NO = A\alpha$. et in angulo ONL inscribatur $O\lambda = NL$, centróque N et intervallo $N\lambda$ describatur circulus, secans Parabolam in
ν, unde $\nu\pi$ perpendiculariter ad LM demittatur. Denique si huic $\nu\pi$ sumatur
AR aequalis, erit R quaesitum punctum refractionis, quod radios FR, et RX
positione determinat.   Q.E.F.[(16)]

---

(16) If we set $AF = a$, $A\alpha = 2b$, and $\alpha X = c$, then the problem is to determine the point R
through which the ray must pass, or the length $AR = b + x$, such that the sine law of refraction
is satisfied:

$$I : R = \sin A\widehat{F}R : \sin \alpha\widehat{X}R = (b + x)/\sqrt{[a^2 + (b + x)^2]} : (b - x)/\sqrt{[c^2 + (b - x)^2]},$$
$$\text{or } (I^2 - R^2)x^4 + (a^2 I^2 - c^2 R^2 - 2b^2[I^2 - R^2])x^2 - 2b(a^2 I^2 + c^2 R^2)x$$
$$+ b^2 (a^2 I^2 - c^2 R^2) + b^4(I^2 - R^2) = 0.$$

point π toward P, because it would then have to be more refrangible than
that specified purple-making ray $F\pi$;[a] namely, it would have a smaller angle   (a) Prop. 2,
of refraction at a greater angle of incidence, which would contradict the   Corollary
hypothesis even more. By a similar argument it will also be obvious that it
cannot fall at the point T or beyond it in the direction opposite to P. Therefore, it remains that it passes between π and T.

Proposition 11. Heterogeneous rays whose refractions pass through the   [140]
point X will fall on the space PT in the same order in which their colors
follow one another from purple to red.

For, suppose that the blue-making ray falls at χ, and since green is an
intermediate color between blue and red, it will be evident that the green-
making ray will fall between χ and T in the same way that it was shown in
the preceding proposition that any intermediate kind of ray must fall between
π and T. Therefore let it fall at ρ. Moreover, since yellow is the color falling
between green and red, it will be evident in the same way as has been
described that the yellow-making ray will fall between ρ and T, suppose at σ;
and so on.

If now it is required to determine the angle $TX\pi$ from the given points F
and X, to accomplish this it is appropriate that I set forth the following
problem in the form of a lemma.

Problem. Given a point irradiating a refracting plane and also another   [141 → I, 58]
point through which a refracted ray must pass, the position of the rays or the
point of refraction is required.

Let (Fig. 59) F be the point emitting the rays, and R the point where a ray
must be refracted so that afterward it passes through the given point X. From
these same points F and X drop from the normals FA and $X\alpha$ to the refracting plane. Making $\sqrt{[I^2 - R^2]} : I = AF : VG$, with *latus rectum* VG separately describe the conical parabola VLv, and let its vertex be V and axis
VGH. Next, from the vertex to the axis erect the normal $VK = \frac{1}{2}A\alpha$; parallel
to the axis draw KM, which intersects the parabola in L; and take a length
LM such that $(I^2 - R^2) : R^2 = \alpha X^2/2VG : LM$. Then perpendicular to it at M
draw MN so that $VG : (VG + LM) = A\alpha : MN$; join LN, and to this erect
the normal $NO = A\alpha$; in the angle ONL inscribe $O\lambda = NL$; and with center
N and radius $N\lambda$ describe a circle intersecting the parabola in v, from which
drop $v\pi$ perpendicular to LM. Finally, if AR is taken as equal to this line $v\pi$,
R will be the required point of refraction that determines the rays FR and RX
in position. As was to be done.[16]

To solve this fourth-degree equation Newton finds the intersection of the parabola $x^2 = VGy = (aI/\sqrt{[I^2 - R^2]})y$, where V is the origin of the coordinates, and the circle $(x - PN)^2 + (y - VP)^2 = N\lambda^2$ with center N (the point P having been added to Newton's diagram). Descartes proposed this technique for solving quartics in his *La Geometrie,* Bk. III, and Newton actively pursued it from his earliest studies; see, from 1665, "Of the construction of Problems," *Mathematical Papers,* 1:492–502; and, from about 1670, "Problems for construing aequations," ibid, 2:450–517. For a detailed analysis of Newton's solution, see Whiteside's publication of this problem (ibid, 3:450–3, esp. note 5), and also Lohne, "Fermat, Newton, Leibniz und das anaklastische

Caeterùm cùm hoc idem Problema a D^re Barrow in Lect 5 rerum Opticarum eleganter solutum extet, potestis illum consulere, et ideò demonstrationem hujus constructionis / brevitatis gratiâ praetermitto.[17] Hoc autem sic   100/
praemisso, problematis priùs propositi solutio fit palam. Nempe ut angulus
TXπ e datis *F* et *X*[18] determinetur nihil aliud agendum est, quàm ut ope
problematis jam modò constructi, duo radij cum refractis suis ducantur alter
FTX extremè rubriformis, et alter *FπX* extremè purpuriformis, qui ab *F*
manantes, postquam refracti fuerint, transeant per *X*.

[142 ≈ I, 73, 74]   Schol: Verùm enim cùm anguli hujus *πXT* determinatio eò spectet, ut
noscatur quanta sit objectorum mediante refractione visorum propter inaequales absimilium radiorum refractiones confusio, perque quantum spatium
colores inde emergentes extenduntur (quemadmodum pateat concipiendo *F*
esse punctum lucidum quod oculo in *X* existente per totum angulare spatium
πXT dilatatum ac diffusum appareat:) | placet insuper modum ostendere quo
quantitas ejus uno intuitu pro quavis puncti *X* sive oculi a refringenti superficie distantiâ praeter propter innotescat, puncto *F* manente determinato.
Concipiatur itáque quòd [fig 60] a dicto *F* ad idem quodpiam refractivae

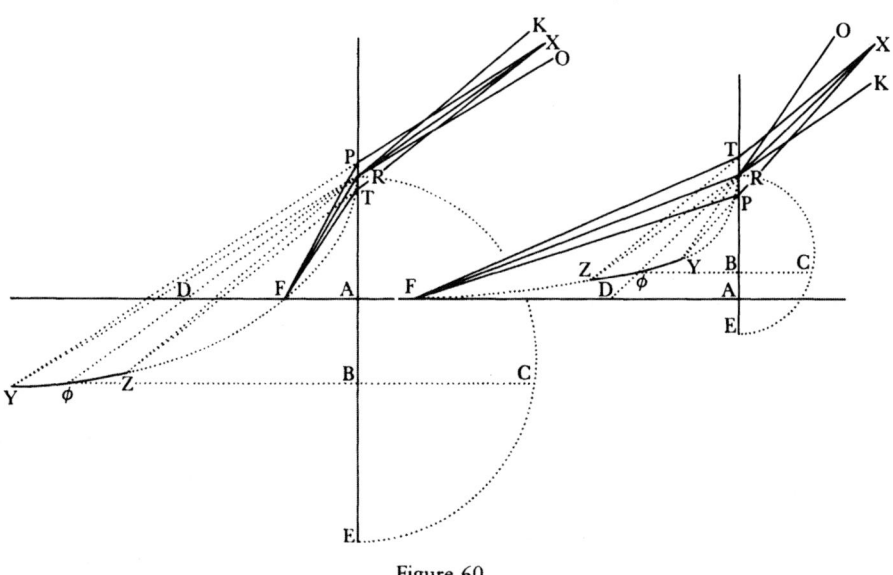

Figure 60

superficiei punctum *R*, radij omnium formarum incidant, quorum extremè
rubiformis refringatur versus *K* et purpuriformis versus *O*: et radijs hisce
retro-actis, exquirantur eorum foci, sive puncta *Y* et *Z* a quibus radij ejusdem
formae ambobus utrinque vicinissimi divergunt; per scholium ad Prop 3.
Tum ducto etiam *RX* refracto radij viridiformis similiter incidentis ab *F* ad *R*:
dico quòd ubicunque sumatur punctum *X* in lineâ novissimè ductâ *RX*,
rectae *XY XZ* abinde ad puncta *Y* et *Z* ductae comprehendent angulum *YXZ*
quàm proximè aequalem angulo quaesito / *PXT*, extra quem radij pene nulli   101/
a puncto *F* ad punctum *X* refracti divaricant. Nam cum *Y* sit focus sive

However, since this same problem is elegantly solved by Dr. Barrow in Lecture 5 of his *Optical Lectures,* you can consult that; and consequently I am passing over the demonstration of this construction for brevity's sake.[17] Having thus premised this, the solution to the problem proposed earlier becomes evident. Specifically, given F and X,[18] to determine the angle $TX\pi$ nothing other needs to be done than to draw—with the aid of the problem just now constructed—two rays together with their refractions: the one, FTX, extreme red-making, and the other, $F\pi X$, extreme purple-making, which flowing from F will pass through X after being refracted.

Scholium. But to be sure, since the determination of this angle πXT aims at [142 ≈ I, 73, 74] knowing the extent of the confusion of objects seen by an intervening refraction because of the unequal refractions of dissimilar rays, and the extent of the space through which the colors emerging from there are spread out (as may be evident by imagining F to be a luminous point which to an eye being at X appears extended and diffused through the entire angular space πXT); it is desirable in addition to show how its magnitude may become known at a glance for any point X, or rather for an eye more or less distant from the refracting surface, with the point F remaining fixed. Imagine (Fig. 60) therefore that from that point F to some one point R of the refractive surface rays of every nature fall; the extreme red-making ones are refracted toward K and the purple-making ones toward O, and having extended these rays backward, their foci are required, or the points Y and Z, from each of which the closest rays of the same nature on each side diverge (by Prop. 3, Scholium). Then having also drawn the refracted ray RX of the green-making ray similarly incident from F to R, I say that wherever the point X is taken in the line RX just drawn, the lines XY and XZ drawn from there to the points Y and Z will subtend an angle YXZ that is very nearly equal to the required angle PXT, outside of which virtually no rays refracted from the point F to the point X will spread. For since Y is the focus, or that point from which the intense

---

Problem," *Nordisk Matematisk Tidskrift,* **14** (1966):5–25. Newton later included this solution, in algebraic form, in his *Arithmetica universalis,* Problem 34 (*Mathematical Papers,* 5:242–5), a compilation made about 1683 from his earlier algebra lectures and published in 1707. (In the second edition of 1722 it is renumbered Problem XL.) Many solutions to the anaclastic problem were proposed in the seventeenth century, but of particular interest are two similar to Newton's: Harriot's solution in 1601 (see Lohne, "Dokumente zur Revalidierung von Thomas Harriot als Algebraiker," *Archive for History of Exact Sciences,* **3** (1966):185–205), and René François de Sluse's solution sent to Oldenberg on 25 July 1670 and thence to Newton (*Correspondence of Henry Oldenburg,* ed. A. Rupert Hall and Marie Boas Hall 7 (Madison, 1970):74). Lohne, however, mistakenly attributed Harriot's solution to William Lower; personal communication, Whiteside to Shapiro, 23 March 1982.

(17) See §I, 58, for Barrow's solution.

(18) e . . . X] Added.

punctum istud a quo radij intensè purpuriformes refracti divaricant qui ab *F* manantes jacent ex utraque parte radij *FRO* sibi vicinissimi; Et cùm *FPX* sit ejusmodi radius: ejus itaque refractus *PX* tendet ab eo puncto *Y*, aut saltem ab alio quodam puncto in lineâ *RY* sito, quod ipsi sit valdè propinquum; siquidem is admodum vicinus est radio *FRO* licèt non omnium vicinissimus.[19] Et simili ratione pateat quod radij *FT* extremè rubriformis refractus *TX* penè tendit a puncto *Z* quod focus est radij *RK*. Atque adeò constat quòd ang: *YXZ* quam-proximè adaequatur angulo *PXT* extra quem nulli radij punctis *F* et *X* interjecti divaricant.   Q.E.O.

[143 ≈ I, 74]   Adhaec si describatur linea quaedam curva *YϕZ* in quâ foci[20] radiorum omnigenorum jacent secundum lineam *FR* incidentium et ita refractorum in puncto *R* ut per totum angulum *KRO* divaricent: ista curva *YϕZ* non malè assimilabitur objecto lucido cujus angulus visibilis[21] ad oculum in *X* situm, sit *YXZ*, et distantia ab eodem oculo ad meditullium ejus aestimata, *ϕX*. | Sed notum est quòd visibilium apparentes magnitudines penè sunt reciprocè ut eorum distantiae: Atque adeò longitudines *ϕX* penè sunt reciprocè ut anguli *YXZ*. Instantiae gratiâ, cùm punctum *X* in ipsâ refringenti superficie ad *R* existit apparens longitudo ipsius *YZ*, erit angulus *YRZ* sive *KRO*. Est ergo *ϕX* . *ϕR* :: ang *KRO* . ang *PXT*. Quare puncto *ϕ* semel invento, (faci-

(a) Schol prop: 3.   endo nempe quod sit $\dfrac{AF^q}{FR} . \dfrac{AD^q}{DR} :: FR . R\phi$.[a]) angulus *PXT* definiens amplitudinem apparentem lucis a puncto *F* refractae facilè determinatur pro qualibet positione puncti *X* in linea *RX* ad arbitrium assumpti.

Ex his palam est, quòd quò longiùs punctum *X* distat ab *R* eò minor est angulus *KRO* et quod nullus est cum *X* cadit ad infinitam distantiam, maximus autem cùm in ipsâ refringenti superficie cadit, quemadmodum ad *R*.

[144 = I, 75]   Caeterùm curva praedicta in quâ radiorum omnis generis in puncto *R* refractorum radiationum centra locantur est Cissois vulgaris sive Dioclea, circulo accommodata cujus diameter / *RE* est ad *AR* ut *FR*q ad *AF*q.[22] Nam super   102/
diametro *RE* descripto circulo isto *RCE*, agatur quaevis recta *ϕBC* normalis ad *RE*, circuloque in *C* et curvâ in *ϕ* terminata. Et propter analoga latera similium triangulorum *RAD*, *RBϕ*, erit *AD*q . *AR* × *DR* :: *Bϕ*q . *BR* × *ϕR*. Et applicando posteriorem rationem ad *BR* fiet *AD*q . *AR* × *DR* ::   . *ϕR*. Rursusque ducendo consequentes rationum in[23] *Rϕ* et applicando ad *AR* orietur $AD^q . DR × R\phi :: \dfrac{B\phi^q}{BR} . \dfrac{R\phi^q}{AR}$. Est autem

$$\frac{AF^q}{FR} . \frac{AD^q}{DR} :: RF . R\phi$$

---

(19) In the *Optica* Newton eliminated this long justification for assuming that neighboring homogeneous rays such as *RO* and *PX* diverge from a common center of radiation *Y* by taking the angular dispersion to be indefinitely small, or the rays *FRO* and *FPX* to be indefinitely near one another.

(20) II: radiationum centra (centers of radiation).

(21) II continues: sive apparens magnitudo (or apparent magnitude).

purple-making refracted rays flowing from *F* and lying nearest to each other on each side of the ray *FRO* spread, and since *FPX* is a ray of the same kind; its refracted ray *PX* will therefore tend from that point *Y*, or at least from some other point very near it located on the line *RY*, inasmuch as it is quite near the ray *FRO*—although not the nearest of all.[19] By a similar argument it is clear that the refracted ray *TX* of the extreme red-making ray *FT* nearly tends from the point *Z*, which is the focus of the ray *RK*. Hence it is established that the angle *YXZ* is very nearly equal to the angle *PXT*, outside of which no rays lying between the points *F* and *X* spread. As was to be shown.

If, moreover, there is described some curved line *YφZ* on which lie the [143 ≈ I, 74] foci[20] of rays of every kind, incident along the line *FR* and refracted at the point *R* so that they spread through the whole angle *KRO*, that curve *YφZ* will not improperly be compared to a luminous object whose visible angle[21] to an eye located at *X* is *YXZ* and whose estimated distance from the same eye to its middle is *φX*. | It is known, however, that the apparent magnitudes of visible objects are nearly inversely as their distances, and so the lengths *φX* are nearly inversely as the angles *YXZ*. For example, when the point *X* is at *R* on the refracting surface itself, its apparent length *YZ* will be the angle *YRZ* or *KRO*. Therefore, $\phi X : \phi R = KRO : P\widehat{X}T$. Consequently, once the point *φ* is found (specifically, by making $AF^2/FR : AD^2/DR = FR : R\phi$),[a] the (a) Prop. 3, angle *PXT* defining the apparent size of the light refracted from the point *F* Scholium may be determined easily for any arbitrarily assumed position of the point *X* in the line *RX*.

From these things it is evident that the more distant the point *X* is from *R*, the smaller is the angle *KRO*; and that it is zero when *X* falls at an infinite distance, but greatest when it falls on the refracting surface itself, as at *R*.

Moreover, that curve on which are located the centers of radiation of rays [144 = I, 75] of every kind refracted at the point *R* is a common or Dioclean cissoid adapted to a circle whose diameter *RE* is to *AR* as FR^2 is to AF^2.[22] For having described that circle *RCE* on the diameter *RE*, draw normal to *RE* any line *φBC*, which is bounded by the circle at *C* and the curve at *φ*. Because of the corresponding sides of the similar triangles *RAD* and *RBφ*, there will be $AD^2 : AR \times DR = B\phi^2 : BR \times \phi R$. Dividing the latter ratio by *BR*, it will become $AD^2 : AR \times DR = B\phi^2/BR : \phi R$; and again, multiplying[23] the denominators of the ratios by *Rφ* and dividing by *AR*, there will arise $AD^2 : DR \times R\phi = B\phi^2/BR : R\phi^2/AR$. However, as before, $AF^2/FR : AD^2/DR = RF : R\phi$; and multiplying the denominators by *DR* and

---

(22) Caeterùm . . . *AF*ᑫ.] Originally (and mistakenly): Caeterùm curva praedicta in quâ foci radiorum omnis generis in puncto *R* refractorum cadant si species fortè desideretur, aio esse Cissoidem vulgarem sive Diocleam, circulo accomodatam cujus diameter *RE* est quarta continuè proportionalis ab *AF* ad *FR* seriem ordiendo. (Moreover, that curve on which the foci of rays of every kind refracted at the point *R* fall—if perhaps its species is required—I say, is a common or Dioclean cissoid adapted to a circle whose diameter *RE* is the fourth continued proportional from *AF* to *FR* in order of sequence.)

(23) ducendo . . . in] Originally: multiplicando . . . ad.

ut priùs, et consequentibus in DR atque antecedentibus in FR ductis, oritur $AF^q . AD^q :: FR^q . DR \times R\phi$ et vicissim $AF^q . FR^q :: AD^q . DR \times R\phi$. Quamobrem rationes eidem tertiae congruentes connectendo, habebitur $\dfrac{B\phi^q}{BR} . \dfrac{R\phi^q}{AR} :: AF^q . FR^q$, ducendoque antecedentes rationum in BR et consequentes in AR prodibit $B\phi^q . R\phi^q :: AF^q \times BR . FR^q \times AR$ et insuper applicando posteriorem rationem ad AF^q fiet $B\phi^q . R\phi^q :: BR . \dfrac{FR^q \times AR}{AF^q}$. Sed cum posuerim $RE . AR :: FR^q . AF^q$, erit $\dfrac{FR^q \times AR}{AF^q} = RE$, et proinde $B\phi^q . R\phi^q :: BR . RE$. ac divisim $B\phi^q . R\phi^q - B\phi^q (BR^q) :: BR . BE$. Atqui ex naturâ circuli est BC media proportionalis inter BR et BE adeoque est $BR . BE :: BR^q . BC^q$ et proinde $B\phi^q . BR^q :: BR^q . BC^q$ sive $B\phi . BR . BC \div$ Quod indicat curvam esse Cissoidem sicut ostendendum proposui.[24]

Lect 14 & 15
[145 → I, 76]

Praecedentia praelibavi ut de modo constaret quo radij difformes ab eodem puncto divergentes possint ab unicâ planâ superficie sic refringi ut ad aliud punctum puta centrum oculi fiant postea convergentes: in quâ re fundatur genesis colorum conspicuorum trans prismata, ut facilè deprehendi potest, et posthac fusiùs explicabitur. Quinetiam ut angulus quem ejusmodi convergentes radij efficiunt promptè noscatur indicavi, adèo ut cuiquam liceret doctrinam a nobis prolatam cum observationibus sensuum in ejusmodi minutijs conferre. Et in hunc finem placuit jam alia his agnata subjungere, quae obliquitas incidentium radiorum aut Mediorum densitas varia subministret. Imprimis verò Lemma unum atque alterum praesternam.

[146 = I, 66]
Lemmata quaedam[1] **ad prosequendam de difformium radiorum affectionibus doctrinam ponuntur**

Lem: 1 Quatuor lineis GB, GC, GD, GE (fig [61]) a dato puncto G ad datam lineam EB ita ductis ut sit $GB . GC :: GD . GE$: angulus BGC quem minima GB, cum alterutrâ intermediarum GC[2] constituit, major est quàm angulus DGE ab alterâ / intermediâ GD et maximâ GE constitutus. Nam centro G radio GE describatur circulus EK et radius GK ducatur con

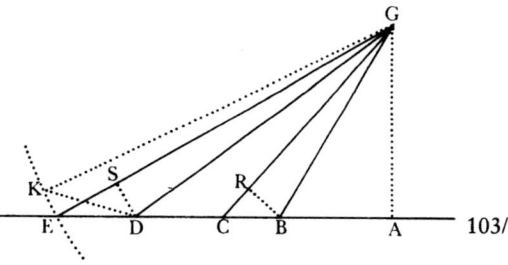

Figure 61

stituens angulum DGK aequalem angulo BGC, et puncta K D jungantur: eruntque triangula GDK, GBC similia propter aequales angulos ad G et latera circa illos proportionalia[a], nempe $GB . GC :: GD . (GE)$ GK.

(a) 6. 6 Elem & Hypoth

---

(24) This demonstration follows directly from the Scholium to Prop. 3, §130 (see Lect. 12, note (23), this volume), in which Newton found the location of the primary image point ϕ to be $\phi R/FR = n \cos^2 r/\cos^2 i = (\sin i/\cos^2 i)/(\sin r/\cos^2 r)$, where $i = \widehat{AFR}$ and $r = \widehat{ADR}$. By construction, the circle's diameter $2a = RE = FR^2 \times AR/AF^2 = FR \sin i/\cos^2 i$, and by taking R as the origin of a Cartesian coordinate system with RE as x-axis, then $\phi R = \sqrt{[x^2 + y^2]}$ and $\sin r/\cos^2 r = x\sqrt{[x^2 + y^2]}/y^2$. Upon substituting these values, the defining Cartesian equation of

the numerators by FR, there arises $AF^2 : AD^2 = FR^2 : DR \times R\phi$, and by permuting, $AF^2 : FR^2 = AD^2 : DR \times R\phi$. Consequently, by equating ratios equal to the same third ratio, there will be had $B\phi^2/BR : R\phi^2/AR = AF^2 : FR^2$; when the numerators of the ratios are multiplied by BR and the denominators by AR, there will appear $B\phi^2 : R\phi^2 = AF^2 \times BR : FR^2 \times AR$; and furthermore, dividing the latter ratio by AF^2, it will become $B\phi^2 : R\phi^2 = BR : FR^2 \times AR/AF^2$. But since I set $RE : AR = FR^2 : AF^2$, it will be

$$FR^2 \times AR/AF^2 = RE,$$

and consequently $B\phi^2 : R\phi^2 = BR : RE$, and by division,

$$B\phi^2 : R\phi^2 - B\phi^2 \text{ (or } BR^2) = BR : BE.$$

However, from the nature of a circle, BC is the mean proportional between BR and BE, and hence $BR : BE = BR^2 : BC^2$; and consequently $B\phi^2 : BR^2 = BR^2 : BC^2$, or $B\phi$, BR, and BC are in continued proportion, which indicates that the curve is a cissoid, as I proposed to show.[24]

I examined the preceding to establish how dissimilar rays diverging from the same point could be refracted from a single plane surface so that they would afterward be made to converge to another point, such as the center of the eye. The origin of colors visible across prisms is founded on this, as can be readily understood and will be explained in detail later. Moreover, I have shown how the angle that converging rays of this kind make may readily become known, so that anyone could compare in such fine points the doctrine I have revealed with observations with their senses. To this end it seemed proper now to add to these things other related ones provided by the obliquity of the incident rays or the diverse density of the media. First, however, I shall premise several lemmas.

Lectures 14, 15
[145 → I, 76]

Lemma 1. When four lines, GB, GC, GD, and GE (Fig. 61), are drawn from a given point G to a given line EB so that $GB : GC = GD : GE$, the angle BGC made by the smallest one, GB, with either of the intermediate ones, GC,[2] is greater than the angle DGE made by the other intermediate line, GD, and the largest one, GE.

[146 = I, 66]. Some lemmas[1] are proposed for pursuing the doctrine of the properties of dissimilar rays.

For with center G and radius GE describe the circle EK; then draw the radius GK making the angle DGK equal to the angle BGC; and join the points K and D. The triangles GDK and GBC will be similar because of the equal angles at G and the proportional sides around them,[a] namely, $GB : GC = GD : (GE \text{ or}) GK$. Consequently $\widehat{KDG} = \widehat{CBG}$, but $\widehat{EDG} >$

(a) *Elements*, VI, 6, & Hypothesis

the locus of the image point ϕ is found to be $x^3 = y^2 (2a - x)$, which is a Dioclean cissoid with its cusp at R. Excerpts from §§143–4 are included in Newton's *Mathematical Papers*, 3:462–5; see also Lohne, "Newton's diagrams," pp. 69–71; and on the cissoid, Thomas L. Heath, *A History of Greek Mathematics*, 2 vols. (Oxford, 1921) 1:264–6.

(1) Originally continued: *in ordine* (sequentially). This sequence of six lemmas constitutes Newton's own §127.

(2) Originally: *puta GC* (suppose, *GC*).

Quare ang KDG = ang CBG, sed ang $EDG \sqsubset$ [ang] CBG.[b] Ergo linea $KD \sqsubset ED$[c], et ang $KGD \sqsubset$ ang EGD[d]; hoc est ang $CGB \sqsubset$ ang EGD.   Q.E.D.[6]

Lemma 2. Positis istis angulis infinitè parvis, ac GA perpendiculari ad lineam EB demissâ: erit ang: EGD . ang CGB :: BA . DA. A punctis enim B ac D ad lineas GC, GE demittantur normalia BR ac DS, et erunt anguli praefati ad se invicem ut est $\dfrac{DS}{DG}$ ad $\dfrac{BR}{BG}$, ponendo nempe lineas istas BR ac DS aequipollentes esse arcubus infinitè parvis quibus anguli illi subtenduntur. Est autem BG . CG :: DG . EG ex Hypoth: ac dividendo BG . CR :: DG . ES, permutandóque BG . DG :: CR . ES. Item propter similia triangula BAG, CRB, est BA . AG :: CR . BR, et pari ratione EA vel DA . AG :: ES . DS sive AG . DA :: DS . ES. Quamobrem addendo rationes aequales est

$$BA . AG + AG . DA \ (:: BA . DA) :: CR . BR + DS . ES$$

(et permutatis terminis posteriorum rationum) :: CR . ES + DS . BR (et aequipollenti ratione pro CR . ES substitutâ)[7] :: BG . DG + DS . BR (terminisque ad invicem vicissim applicatis) :: $\dfrac{DS}{DG} . \dfrac{BR}{BG}$.

Est itáque BA . DA :: $\dfrac{DS}{DG} . \dfrac{BR}{BG}$, hoc est, ut ang EGD ad ang CGB.   Q.E.D.[8]

/Lemma 3.[9] Si a duobus punctis D, G (fig [62]) in lineâ quâpiam AD sitis, ad alia duo puncta L, N in ejus perpendiculo AN sita, ducantur quatuor rectae DN, DL, GN, GL; ratio ductarum ad punctum remotius N magis accedit ad aequalitatem quàm ratio ductarum ad vicinius punctum L. Sive est GN . $DN \sqsubset GL$. DL. Sit enim GN . DN :: GL . R, et erit GN^q . DN^q :: GL^q . R^q :: $GN^q - GL^q$. $DN^q - R^q$. Quare cùm sit $DN \sqsubset GN$, sive $DN^q \sqsubset GN^q$, erit $DN^q - R^q \sqsubset GN^q - GL^q$.

Verùm est $GN^q - GL^q = DN^q - DL^{q(a)}$ et ideo[10] $DN^q - R^q \sqsubset DN^q - DL^q$. Hoc est $DL^q \sqsubset R^q$ sive $DL \sqsubset R$. Atque adeò cum supponatur GN . DN :: GL . R, erit GN . $DN \sqsubset GL$. DL. Q.E.D.[11]

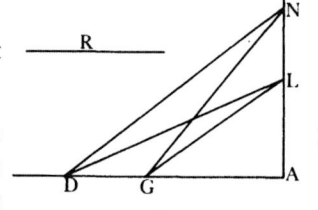

107/[ɩ

Figure 62

---

(3) If one side of a triangle is extended, the exterior angle is greater than either of the interior and opposite angles.

(4) If two straight lines are drawn through a point on a diameter of a circle, the line closer to the line through the center will be greater.

(5) If two triangles have two sides equal to one another, that angle contained by the equal sides will be greater that subtends the greater side.

(6) To illustrate the optical application of these lemmas, we can observe that by this lemma it is immediately established that as the angle of incidence increases the deviation increases, which Newton will demonstrate in Lemma 5, Cor. 1 (§153 = I, 84). If BG and DG are incident rays and CG and EG their refracted rays extended backward, where the angles ABG and ADG are the angles of incidence and the angles ACG and ADG the angles of refraction, so that the angles CGB and DGE are the deviations, then $\widehat{CGB} > \widehat{EGD}$, for by Lemma 2 (§120 = I, 47), $GB/GC = GD/GE = R/I$. Moreover, this lemma can also be applied to polychromatic rays to demonstrate, as Newton does in Prop. 1 (§160 = I, 68), that as the angles of incidence and refraction increase the angular dispersion increases. If BG and CG and DG and EG are now considered to be pairs of extreme red and blue rays refracted at two different angles of incidence, so that the

$\widehat{CBG}$.[b] Therefore $KD > ED$,[c] and $\widehat{KGD} > \widehat{EGD}$;[d] that is, $\widehat{CGB} > \widehat{EGD}$. As was to be demonstrated.[6]

Lemma 2. Assuming that those angles are infinitely small and dropping GA perpendicular to the line EB, there will be $\widehat{EGD} : \widehat{CGB} = BA : DA$.

For from the points B and D drop the normals BR and DS to the lines GC and GE, and the specified angles will be to one another as DS/DG is to BR/BG; namely, by considering those lines BR and DS to be equivalent to the infinitely small arcs that those angles subtend. By hypothesis, however, $BG : CG = DG : EG$, by dividing $BG : CR = DG : ES$; and then by permuting $BG : DG = CR : ES$. Also, because of the similar triangles BAG and CRB, there is $BA : AG = CR : BR$; and for the same reason

$$EA \text{ (or } DA) : AG = ES : DS,$$

or $AG : DA = DS : ES$. Consequently, when the equal ratios are multiplied, $BA : AG \times AG : DA (= BA : DA) = CR : BR \times DS : ES$ (and permuting the terms of the last ratio) $= CR : ES \times DS : BR$ (and substituting the equivalent ratio for $CR : ES$) $= BG : DG \times DS : BR$ (and alternately dividing the terms by one another) $= DS/DG : BR/BG$. Therefore, it is $BA : DA = DS/DG : BR/BG$, that is, as the angle EGD to the angle CGB. As was to be demonstrated.[8]

Lemma 3.[9] If from two points, D and G (Fig. 62), situated on any line, AD, four straight lines, DN, DL, GN, and GL, are drawn to two other points, L and N, situated on its perpendicular AN, the ratio of the lines drawn to the more distant point N approaches closer to an equality than the ratio of those drawn to the nearer point L, or $GN : DN > GL : DL$.

For let $GN : DN = GL : R$, and it will be

$$GN^2 : DN^2 = GL^2 : R^2 = GN^2 - GL^2 : DN^2 - R^2.$$

Consequently, since $DN > GN$, or $DN^2 > GN^2$, it will be $DN^2 - R^2 > GN^2 - GL^2$. But $GN^2 - GL^2 = DN^2 - DL^2$,[a] and thus[10] $DN^2 - R^2 > DN^2 - DL^2$, that is, $DL^2 > R^2$, or $DL > R$. Hence, since it is assumed that $GN : DN = GL : R$, it will be $GN : DN > GL : DL$. As was to be demonstrated.[11]

(b) *Elements*, I, 16[3]
(c) *Elements*, III, 7[4]
(d) *Elements* I, 25[5]
[147 = I, 67]

[148 = I, 77]

(a) §119

---

angles CGB and EGD are their angular dispersion, then $\widehat{CGB} > \widehat{EGD}$, for $GB/GC = GD/GE = P/T$. This lemma is also invoked in Prop. 3 ($\S165 = I, 90$).

(7) aequipollenti ... substitutâ] Newton first had "paribus rationibus substitutis" and then "aequalibus ... substitutis."

(8) If, as in note (6), BG and CG and DG and EG are taken to be pairs of extreme red and blue rays refracted at two different angles of incidence, where $\widehat{EGD} = \Delta r'$ and $\widehat{CGB} = \Delta r$, then by this lemma $\Delta r'/\Delta r \approx \tan r'/\tan r = BA/DA$. This lemma is used in $\S162 = I, 69$, to estimate the magnitude of the angular dispersion.

(9) Propositions 1 and 3 come next in the manuscript, but Newton wrote in the margin of p. 103: "Sequentes duae prop: numeris 1 et 3 notatae hic deleri debent et ad pag 110 & 112 transferri." (The following two propositions, designated numbers 1 and 3, are to be deleted here and transferred to pages 110 and 112.) Accordingly, they have been transferred to $\S\S160$ and 165 below, as we skip directly to p. 107 and Lemma 3.

(10) Originally continued: *multò magis* (even more so).

(11) This lemma is invoked in Props. 2, 3, $\S\S163, 165 = I, 78, 90$.

[149 = I, 80]   Lemma 4. Centro *A*, distantia quavis *AD* (in fig [63]) describatur circulus *DGγ*. Deinde centro quolibet *C* distantiâ *AC* describatur alius circulus secans rectam *AD* in *B* et circulum priùs descriptum in *G*. Tum arcus *BG* bisecetur in *F*; et *FK* demittatur ad *BD* perpendicularis: His ita constitutis, dico quòd *FK* sic perpendiculariter demissa dictam *BD* bisecabit. Junctis enim *AF, AG, BF, FG* et *FD*; in triangulis *AFG* et *AFD* anguli ad *A* sunt aequales propter aequales arcus *BF, FG* quibus

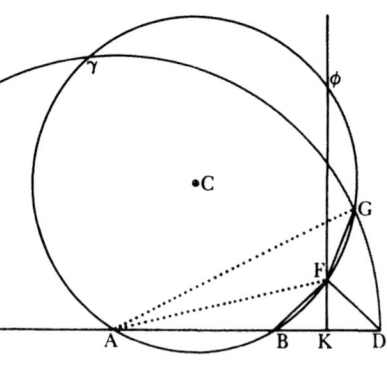

Figure 63

subtenduntur; Item latera / circa istos angulos *AD* et *AG* sunt aequalia, quippe 108/ radij ejusdem circuli. Et aliud latus *AF* habent commune. Quare etiam tertia latera *FG* et *FD* sunt aequalia. Sed est *BF = FG* propter aequalitatem arcuum quos subtendunt, adeoque *BF = FD*, et triangulum *FKB* = triang *FKD*, et inde *BK = KD*.[12]

[150 = I, 81]   Coroll: 1. Hinc recta *KF*, quae bisecat *BD* insistens ei normaliter, bisecabit etiam arcus *BG* circulorum omnium per data duo puncta *A* et *B*, transeuntium, et alicubi in *G* secantium datum circulum *DG* centro *A* intervallo *AD* descriptum. Imò et bisecabit arcus *BGγ* in altero intersectionis puncto *φ*.

[151 = I, 82]   Coroll: 2. Idem eveniet cùm *A*, et *B* coincidunt; hoc est cùm circuli *AFG* tangunt rectam *AD* in puncto *AB*. Potest etiam *B* sumi ad alteras partes ipsius *A*. In transcursu etiam notetur, quod anguli *BFK, BGD*, quos circulus *ABF* cum recta *FK* et arcu *GD* efficit, sint aequales.

[152 = I, 83]   Lemma 5. Lineis quatuor *Aβ, AB, Aγ*, et *AG* (fig [64]) circulo alicui ab eodem circumferentiae puncto ita inscriptis ut sit *Aβ . AB :: Aγ . AG*, quarum omnium *Aβ* sit minima: Dico angulum *BAG* majorem esse angulo

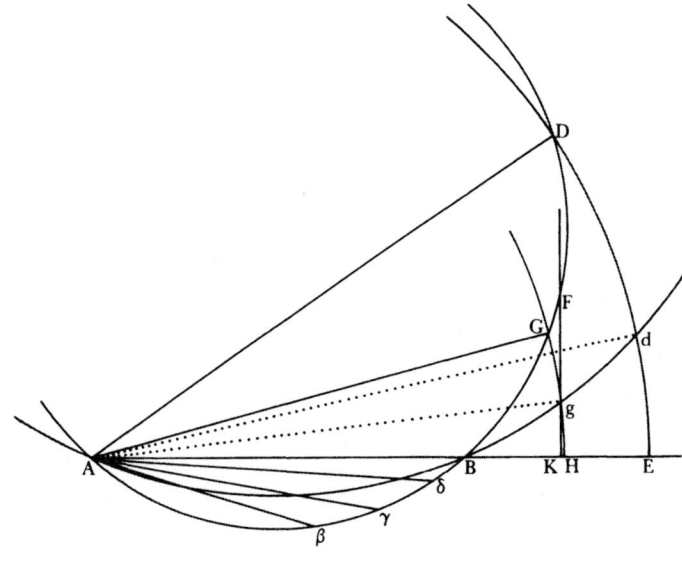

Figure 64

Lemma 4. With center *A* (Fig. 63) and any radius *AD* describe the circle [149 = I, 80]
DGγ. Next with any center *C* and radius *AC* describe another circle intersecting the line *AD* in *B* and the previously described circle in *G*. Then bisect
the arc *BG* at *F*, and drop the perpendicular *FK* to *BD*. Having so constructed these things, I say that *FK*, dropped perpendicularly in this way, will
bisect that line *BD*.

For when *AF, AG, BF, FG,* and *FD* are joined, the angles at *A* in the
triangles *AFG* and *AFD* are equal because they subtend equal arcs, *BF* and
FG. Likewise, the sides *AD* and *AG* about those angles are equal, inasmuch
as they are radii of the same circle; and they have the other side, *AF*, in
common. Consequently, the third sides, *FG* and *FD*, are also equal. But *BF* =
FG, because they subtend equal arcs, so that *BF* = *FD*; and the triangle *FKB*
is equal to the triangle *FKD*, and thus *BK* = *KD*.[12]

Corollary 1. Hence the line *KF*, which bisects *BD* and is normal to it, will [150 = I, 81]
also bisect the arcs *BG* of all circles that pass through the two given points *A*
and *B* and intersect, somewhere at *G*, the given circle *DG* described with
center *A* and radius *AD*. In fact, it will also bisect the arc *BGγ* in the other
point of intersection *φ*.

Corollary 2. It will turn out to be the same when *A* and *B* coincide, that is, [151 = I, 82]
when the circles *AFG* are tangent to the line *AD* at the point *AB*. *B* can also
be taken on the other side of *A*. In passing, it may be noted that the angles
BFK and *BGD* that the circle *ABF* makes with the line *FK* and the arc *GD*
are equal.

Lemma 5. Inscribing four lines, *Aβ, AB, Aγ,* and *AG* (Fig. 64), in any [152 = I, 83]
circle from the same point of the circumference so that *Aβ : AB = Aγ : AG*,
and letting *Aβ* be the smallest of all, I say that the angle *BAG* is greater than
the angle *βAγ*.

(12) Lemma 4 is used to prove Lemma 6. In the late 1670s Newton included this lemma in a
form similar to Corollary 1 in his "Loca plana," no. [26], *Mathematical Papers,* 4:244.

$\beta A\gamma$. Describatur enim alius circulus $ABgd$ secans priorem in punctis A et B, cujus diameter sit ad ejus ABG diametrum, sicut AB ad $A\beta$, centris utrisque ad easdem partes ipsius AB jacentibus. Dein centro A distantiâ AG describe tertium circulum GH secundo occurrentem in g. Et istud g ex constructione jacebit alicubi inter G et H, atque adeò si Ag ducatur erit angulus BAG major angulo BAg. Est autem ang: BAg = ang $\beta A\gamma$ propterea quòd AB et Ag similiter inscriptae sint circulo ABg, ac $A\beta$ et $A\gamma$ ipsi $A\beta\gamma$, habentes / nempe easdem rationes, et 109/ inter se $(A\beta \cdot A\gamma :: AB \cdot AG$ vel $Ag,)$ et ad diametros circulorum quibus inscribuntur. Cum ergo sit $BAG \sqsubset BAg = \beta A\gamma$ erit $BAG \sqsubset \beta A\gamma$.   Q.E.D.

[153 = I, 84]   Coroll 1. Hinc in eodem quovis radiorum genere, quo major est refractio, eo major erit angulus refractus. In fig [55][13] ubi est $FR \cdot RD :: F\rho \cdot \rho\delta$, erit ang $F\rho\delta \sqsubset$ ang FRD.[14]

[154 = I, 85]   Coroll. 2 Hinc etiam si sit $AG \cdot AB \sqsubset A\gamma \cdot A\beta$, multò magis erit ang $BAG \sqsubset$ ang $\beta A\gamma$. Hoc est in genere, quo majores sunt subtensae et simul quo major est inaequalitas rationis earum, eo major erit differentia angulorum quos subtendunt. Atque idem de sinibus et eorum angulis, utpote subtensarum et earum angulorum dimidijs intellige.[15]

[155 = I, 86]   Lem 6. Insuper si arcus $\gamma\delta$ ipsi $\beta\gamma$ capiatur aequalis et AD inscribatur circulo ABD quae sit ad $A\delta$ sicut AG ad $A\gamma$, caeteris stantibus: dico quòd arcus DG erit arcu GB major. Nam centro A, radio AD describe circulum DdE circulo ABg occurrentem in d et rectae AB in E. Et Ad ducatur. Jam cùm Ad, Ag et AB circulo $ABgd$ similiter inscribantur atque $A\delta$, $A\gamma$, et $A\beta$ ipsi $A\beta\gamma$, erit arcus gd = arcui Bg. Quare demissâ gK ad BE perpendiculari, et productâ donec secet arcum BD in F, ista $[g]K$ (per Lem 4) bisecabit tum rectam BE tum arcum BD. At quoniam gF ex constructione jacet extra circulum gG punctum F cadet inter G et D. Quare $DG \sqsubset DF$ sive $\sqsubset FB$, et multo magis $\sqsubset GB$.   Q.E.O.[16]

[156 = I, 87]   Coroll: 1. Hinc si arcus $\beta\delta$ non tantùm duabus sed quotcunque partibus aequalibus constet, correspondentes partes arcus BD a termino B ad terminum D sese gradatim superabunt longitudine. Adeoque si arc: $\beta\gamma$ ad arc $\gamma\delta$ habeat quamcunque rationem commensurabilem; erit arc $GD \cdot$ arc $BG \sqsubset$ arc $\gamma\delta \cdot$ arc $\beta\gamma$, siquidem numeris aequalium partium mensurantium arcus $\beta\gamma$ et $\gamma\delta$ correspondent consimiles numeri partium inaequalium constituentium arcus BG ac GD, quarum illae in GD sunt omnes parte maximâ ipsius BG majores. Quinetiam si $\beta\gamma$ ad $\gamma\delta$ habeat quamcunque rationem incommensurabilem, erit itidem $GD \cdot BG \sqsubset \gamma\delta \cdot \beta\gamma$. Nam rationum similitudines, quae quantitatibus commensurabilibus indefinitè conveniunt, eo nomine conveniunt etiam incommensurabilibus similiter affectis; quemadmodum ex Euclidea definitione similium rationum[17] ostendi posset, sed faciliùs deprehenditur / imaginando quantitates quas vocant incommensurabiles posse mensurari 110/ per partes indefinitè parvas, et sic ad naturam commensurabilium praesertim

---

(13) That is, in §128.

(14) This corollary—at last directly relating these lemmas to optics—is a late marginal addition both here and in the *Optica* and shows that Newton continued to revise the *Lectiones* even after the *Optica* had been transcribed. He invokes this corollary in Prop. 5, Case 1, §168 = I, 93.

(15) For any angle α, $\frac{1}{2}$chord $\alpha = \sin \frac{1}{2}\alpha$. This corollary is used to prove Prop. 4, §166 = I, 91.

(16) Lemma 6 is used in the proof of Lemma 8, §187 = I, 112.

For describe another circle, *ABgd,* intersecting the first in the points *A* and *B,* and let its diameter be to the diameter of *ABG* as *AB* to *Aβ,* with both centers lying on the same side of *AB.* Then with center *A* and radius *AG* describe a third circle, *GH,* meeting the second one at *g.* By construction *g* will lie somewhere between *G* and *H,* so that if *Ag* is drawn, the angle *BAG* will be greater than the angle *BAg.* The angle *BAg,* however, equals the angle *βAγ,* because *AB* and *Ag* are inscribed in the circle *ABg* similarly to *Aβ* and *Aγ* in *Aβγ,* namely, they have the same ratios both to each other (*Aβ* : *Aγ* = *AB* : *AG* or *Ag*) and to the diameters of the circles in which they are inscribed. Therefore, since $B\widehat{A}G > B\widehat{A}g = \beta\widehat{A}\gamma$, it will be $B\widehat{A}G > \beta\widehat{A}\gamma$. As was to be demonstrated.

Corollary 1. Hence, for any one kind of ray the greater the [angle of] [153 = I, 84] refraction, the greater will be the deviation. In Fig. 55[13] where *FR* : *RD* = *Fρ* : *ρδ,* it will be $F\widehat{\rho}\delta > F\widehat{R}D$.[14]

Corollary 2. Hence also, if *AG* : *AB* > *Aγ* : *Aβ,* even more so will it be [154 = I, 85] $B\widehat{A}G > \beta\widehat{A}\gamma$. That is, in general the greater the chords, as well as the greater the inequality of their ratio, the greater will be the difference of the angles they subtend. The same is to be understood for the sines and their angles, as for the halves of the chords and their angles.[15]

Lemma 6. Moreover, if the arc *γδ* is taken to be equal to *βγ,* and in the [155 = I, 86] circle *ABD* there is inscribed *AD* that is to *Aδ* as *AG* to *Aγ,* with other things remaining the same, I say that the arc *DG* will be greater than the arc *GB.*

For with center *A* and radius *AD* describe the circle *DdE* meeting the circle *ABg* in *d* and the line *AB* in *E;* and draw *Ad.* Now, since *Ad, Ag,* and *AB* are inscribed in the circle *ABgd* similarly to *Aδ, Aγ,* and *Aβ* in the circle *Aβγ,* then $\overset{\frown}{gd} = \overset{\frown}{Bg}$. Consequently, when *gK* is dropped perpendicular to *BE* and extended until it intersects the arc *BD* in *F, gK* will bisect both the line *BE* and the arc *BD* (Lemma 4). But since by construction *gF* lies outside of the circle *gG,* the point *F* will fall between *G* and *D.* Therefore $\overset{\frown}{DG} > \overset{\frown}{DF}$ or $\overset{\frown}{FB}$ and even more so $\overset{\frown}{DG} > \overset{\frown}{GB}$. As was to be shown.[16]

Corollary 1. Hence if the arc *βδ* consists not only of two but of any [156 = I, 87] number of equal parts, the corresponding parts of the arc *BD* from the end *B* to the end *D* will gradually exceed one another in length. Thus if the arc *βγ* has any commensurable ratio to the arc *γδ,* then $\overset{\frown}{GD} : \overset{\frown}{BG} > \overset{\frown}{\gamma\delta} : \overset{\frown}{\beta\gamma}$, since to the number of equal parts measuring the arcs *βγ* and *γδ* there corresponds a like number of unequal parts forming the arcs *BG* and *GD,* and those in *GD* are all greater than the greatest part of *BG.* In fact, if *βγ* has any incommensurable ratio to *γδ,* then likewise $\overset{\frown}{GD} : \overset{\frown}{BG} > \overset{\frown}{\gamma\delta} : \overset{\frown}{\beta\gamma}$. For the similarity of ratios, which applies without limitation to commensurable quantities, for that reason also applies to incommensurables similarly treated, as may be shown from the Euclidean definition of similar ratios.[17] But this is more easily understood by imagining that quantities that are called incommensurable can be measured by indefinitely small parts and so in a certain way be reduced to the nature of commensurables, particularly as to the condition of

(17) See Barrow, *Euclidis Elementorum,* Bk. V, Def. VI, p. 91, which is equivalent to Def. V in modern editions.

quoad rationum habitudines quodammodò reduci. Concipias itaque arcum βγ in aequales et indefinitè multas partes dividi, et ejusmodi tot sumi quae minus quàm unâ parte, hoc est indefinitè parùm differunt ab arcu γδ, atque adeò ipsi pro more consueto censeantur aequales. Concipe etiam *BD* in partes quales ante definivi correspondentes partibus ipsius βδ dividi; et propter tot inaequales partes majores quidem in *GD* et minores in *BG* quot sunt aequales in γδ et βγ, erit *GD . BG* ⊏ γδ . βγ.

[157 = I, 88] Coroll 2. Hinc praeterea componendo sequitur esse *BD . BG* ⊏ βδ . βγ. Nec non *GD . BD* ⊏ γδ . βδ.

[158 = I, 89] Coroll 3. Consectatur denique quòd ductis utcunque quatuor subtensis *Aβ, Aγ, Aδ, Aε* in fig [65]; et alijs quatuor *AB, AG, AD, AE* quarum singulae ad priorum singulas eandem rationem observant, (nempe *AB . Aβ :: AG . Aγ :: AD . Aδ :: AE . Aε*:) Si *AE* sit omnium maxima et *Aβ* minima, erit

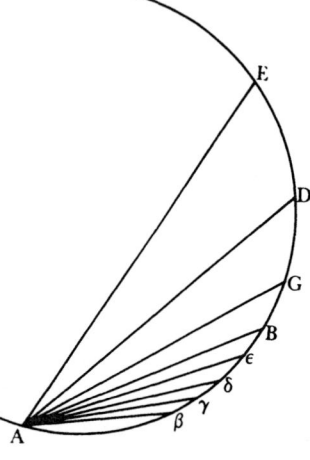

Figure 65

arcus *ED* . arc *GB* ⊏ arc εδ . arc γβ.

Nam per Coroll 1 hujus est *ED . DG* ⊏ εδ . δγ, et *DG . GB* ⊏ δγ . γβ. Et multò magis *ED . GB* ⊏ εδ . γβ. Haud secus pateat esse arc *EG* . arc *DB* ⊏ arc εγ . arc δβ. Scilicet ex coroll 2 hujus est *EG . DG* ⊏ εγ . δγ, ac *DG . DB* ⊏ δγ . δβ; et multò magis *EG . DB* ⊏ εγ . δβ.[18]

Denique quae de subtensis et earum arcubus dicta sunt, possunt etiam de sinibus et eorum arcubus aut angulis intelligi.

[159][19] Hactenus Lemmata praestravimus, ex quibus aliqua quae ad refractiones ejus-
Continuatur dem alicujus variè incidentium radiorum generis spectant, nullo negotio possent
praefatarum erui: Sed cùm apud alios demonstrata prostent, et a scopo meo videantur aliena,
affectionum mitto, déque difformium radiorum affectionibus e vestigio pergo dicere.
declaratio.

[160 = I, 68] / Prop. 1.[20] Heterogeneis radijs secundum eandem lineam incidentibus, 103/[20] quo obliquior sit eorum incidentia caeteris paribus, eo major erit differentia refractionis. In fig [66] Sit *FG* linea secundum quam duo radij incidunt,

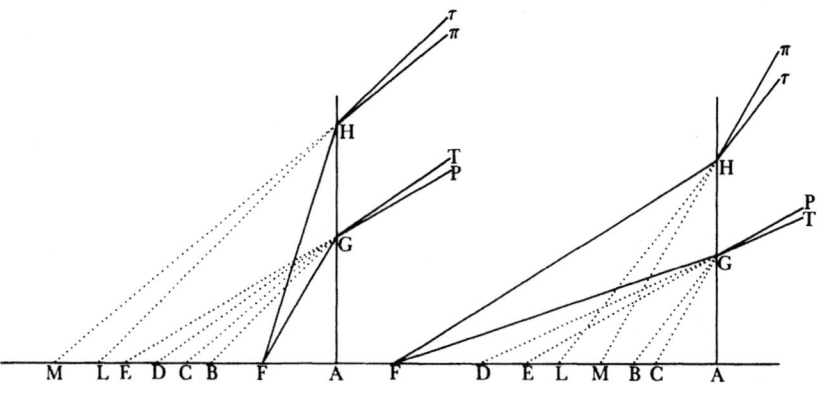

Figure 66

ratios. You may conceive, therefore, the arc $\beta\gamma$ to be divided into indefinitely many equal parts and so many of these to be taken that they differ by less than one part (that is, indefinitely little) from the arc $\gamma\delta$, and thus in the usual way they are to be considered equal. Conceive the arc BD also to be divided into parts (such as I defined before) corresponding to the parts of $\beta\delta$; and because there are as many unequal parts, indeed greater in the arc GD and smaller in the arc BG, as there are equal parts in the arcs $\gamma\delta$ and $\beta\gamma$, it will be $\overset{\frown}{GD} : \overset{\frown}{BG} > \overset{\frown}{\gamma\delta} : \overset{\frown}{\beta\gamma}$.

Corollary 2. Hence furthermore, by composition it follows that $\overset{\frown}{BD} : \overset{\frown}{BG}$ [157 = I, 88] $> \overset{\frown}{\beta\delta} : \overset{\frown}{\beta\gamma}$, and also $\overset{\frown}{GD} : \overset{\frown}{BD} > \overset{\frown}{\gamma\delta} : \overset{\frown}{\beta\delta}$.

Corollary 3. It follows, finally, that when four chords, $A\beta$, $A\gamma$, $A\delta$, and $A\epsilon$ [158 = I, 89] (Fig. 65), are arbitrarily drawn, together with another four, AB, AG, AD, and AE, each of which maintains the same ratio to each of the former ones (namely, $AB : A\beta = AG : A\gamma = AD : A\delta = AE : A\epsilon$), if AE is the greatest and $A\beta$ the smallest of all, then $\overset{\frown}{ED} : \overset{\frown}{GB} > \overset{\frown}{\epsilon\delta} : \overset{\frown}{\gamma\beta}$. For by Corollary 1 of this lemma, $\overset{\frown}{ED} : \overset{\frown}{DG} > \overset{\frown}{\epsilon\delta} : \overset{\frown}{\delta\gamma}$, and $\overset{\frown}{DG} : \overset{\frown}{GB} > \overset{\frown}{\delta\gamma} : \overset{\frown}{\gamma\beta}$, and even more so $\overset{\frown}{ED} : \overset{\frown}{GB} > \overset{\frown}{\epsilon\delta} : \overset{\frown}{\gamma\beta}$. In the same way it is clear that $\overset{\frown}{EG} : \overset{\frown}{DB} > \overset{\frown}{\epsilon\gamma} : \overset{\frown}{\delta\beta}$. Specifically, from Corollary 2 of this lemma, $\overset{\frown}{EG} : \overset{\frown}{DG} > \overset{\frown}{\epsilon\gamma} : \overset{\frown}{\delta\gamma}$, and $\overset{\frown}{DG} : \overset{\frown}{DB} > \overset{\frown}{\delta\gamma} : \overset{\frown}{\delta\beta}$, and even more so $\overset{\frown}{EG} : \overset{\frown}{DB} > \overset{\frown}{\epsilon\gamma} : \overset{\frown}{\delta\beta}$.[18]

Finally, what has been said about chords and their arcs can also be understood about sines and their arcs or angles.

We have to this point prepared lemmas from which some things concern- [159].[19] The ing the refractions of any one sort of ray variously incident could be derived exposition of those with little difficulty. But since the demonstrations are available elsewhere and properties is seem foreign to my goal, I am omitting them and immediately proceed to continued. discuss the properties of diverse rays.

Proposition 1.[20] When heterogeneous rays are incident along the same [160 = I, 68] line, with other things being equal, the more oblique their incidence, the greater will be their difference of refraction.

In Fig. 66 let FG be the line along which two rays fall. One of these, the

---

(18) Newton invokes this corollary in Prop. 5, Case 2, §169 = I, 94.

(19) This is Newton's own §128 and his last numbered article in the *Lectiones*.

(20) Following Newton's instruction in the margin ("Pete ex pag 103, 104, & 105"), Prop. 1 is here inserted from pp. 103–5; see note (9), this lecture.

quorum alter purpuriformis refringitur versus *P*, et alter rubiformis versus *T*, eritque angulus *PGT* differentia refractionis. Item sit *FH* linea obliquior quam *FG* et secundum hanc alij duo ejusmodi radij incidant quorum / purpu- 104/ riformis versus π et rubiformis versus τ refringitur, et similiter erit ang π*H*τ eorum differentia refractionis. Dico jam quòd ang π*H*τ ⊏ ang *PGT*. Demittatur enim *FA* ad refringens planum linea normalis, quae refractos radios retroactos in *D*, *E*, *L* et *M* secet. Et ad hanc a puncto *G* ducantur lineae duae *GB*, *GC* ipsis *HL* et *HM* parallelae. Jam cùm tres lineae *GF*, *GD*, *GE* (ex naturâ refractionis ante descriptâ) sint in ratione data(a), et alterae tres *HF*, *HL*, *HM* in eâdem ratione(a): proportionales erunt *HL* . *HM* :: *GD* . *GE*. Sed est *HL* . *HM* :: *GB* . *GC*, propter sim: tri: *LMH* et *BCG*. Quare *GB* . *GC* :: *GD* . *GE*. Adeóque ang *BGC* ⊏ ang *DGE*, per Lem 1. Hoc est ang *LHM* ⊏ ang *DGE*. Sive ang π*H*τ ⊏ [ang] *PGT*.   Q.E.D.(21)

(a) Sect 98 & 99 & sequ:

[161]   Schol: Propositio etiam valet de radijs heterogeneis secundum parallelas lineas incidentibus. Imò et quodammodo valebit cùm incidentia fit secundum lineas inclinatas, ita quidem ut a puncto divergentes refringantur versus aliud punctum. Quemadmodum in fig [67] si linea quaevis *VX* ducatur refringenti superficiei *AP* parallela, quae secet perpendiculum ejus *FA* in *V*; et *FPX*, *FTX*, duos dissimiliter refrangibiles radios a dato puncto *F* ad quodpiam punctum *X* in recta *VX* refractos designent: summa angulorum *PFT* et *PXT* pro differentia refractionis habenda est, quae quidem summa eo major evadet quo radij incidant obliquiores, hoc est quo punctum *X* longiùs ab *V* distet. Scilicet angulus *PXT* qui in rariori Medio existit augebitur ad certum usque terminum, et postea perpetuò diminuetur, donec *X* ad infinitam usque distantiam amoveatur. At alter angulus *PFT* in Medio densiori existens ita augebitur in perpetuum, ut utriusque simul sumpti summa etiam semper augeatur. Praeterea si punctum *X* in datâ rectâ ad superficiem *AP* normali indeterminatum existat, quo radij obliquiùs incidant, hoc est quo punctum *X* propiùs ad superficiem accedit, angulus quidem *PFT* eò major evadet ad certum solummodò terminum, et postea rursus diminuetur; sed alter angulus *PXT* ità semper in tantum augebitur, ut utriusque etiam summa, sive tota refractionis differentia augmentum simul in perpetuum adipiscatur. Et eadem plerumque observanda venient, si rectam in qua punctum *X* indeterminatum / existit obliquè positam 105/ esse respectu superficiei *AP* cogites. Aliquando tamen contrarium eveniet. Quemadmodum si positione radij *FPX* manente datâ, punctum *X* ad quod alter radiorum *FTX* vergit, concipiatur esse in lineâ *PX* indeterminatum: differentia refractionum eo minor evadet, quo dictum *X* longius amoveatur a *P*, sive quo radius *FTX* incidat obliquiùs; angulo ad *X* semper magis dimi-

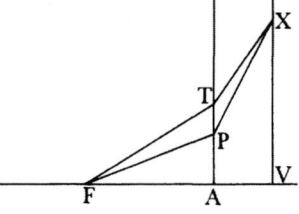

Figure 67

---

(21) For refraction to a denser medium, $dr/dn = -\sin r/n \cos r = -\sin i/n\sqrt{[n^2 - \sin^2 i]}$, and thus the difference of refraction, or the angular dispersion dr, continually increases as i increases,

purple-making, is refracted toward *P*, while the other, the red-making, is refracted toward *T*; and the angle *PGT* will be the difference of refraction. Likewise, let the line *FH* be more oblique than *FG*, and along this line let there fall two other such rays, of which the purple-making one is refracted toward *π* and the red-making toward *τ*; and similarly the angle *πHτ* will be their difference of refraction. I say now that $\pi\widehat{H}\tau > P\widehat{G}T$. For normal to the refracting plane drop the line *FA*, intersecting the refracted rays extended backward at *D*, *E*, *L*, and *M*; and to this line from the point *G* draw two lines, *GB* and *GC*, parallel to *HL* and *HM*. Now, since the three lines *GF*, *GD*, and *GE* are in a given ratio (from the nature of refraction previously described),[a] and the other three, *HF*, *HL*, and *HM*, are in the same ratio,[a] they will be proportional, *HL* : *HM* = *GD* : *GE*. But *HL* : *HM* = *GB* : *GC*, because of the similar triangles *LMH* and *BCG*, and therefore *GB* : *GC* = *GD* : *GE*. By Lemma 1 then $B\widehat{G}C > D\widehat{G}E$, that is, $L\widehat{H}M > D\widehat{G}E$, or $\pi\widehat{H}\tau > P\widehat{G}T$. As was to be demonstrated.[21]

(a) §§98, 99, ff.

Scholium. The proposition is also valid for heterogeneous rays incident [161] along parallel lines, and indeed it will be valid in a certain way when the incidence occurs along inclined lines, for example, when they diverge from a point and are refracted to another point. For instance, if in Fig. 67 one draws parallel to the refracting surface *AP* any line *VX* that intersects its perpendicular *FA* in *V*; and if *FPX* and *FTX* represent two unequally refrangible rays refracted from a given point *F* to any point *X* in the line *VX*; the sum of the angles *PFT* and *PXT* ought to be taken for the difference of refraction, and this sum will in fact become larger the more obliquely the rays fall, that is, the farther the point *X* is from *V*. Specifically, the angle *PXT* standing in the rarer medium will increase up to a certain limit, and then it will perpetually decrease until *X* is removed to an infinite distance. But the other angle, *PFT*, standing in the denser medium will perpetually increase, so that the sum of the two taken together will also always increase. Moreover, if the indeterminate point *X* stands on a given line normal to the surface *AP*, the more obliquely the rays fall (that is, the closer the point *X* approaches the surface) the greater the angle *PFT* will become, at least only up to a certain limit, and afterward it will again decrease; but the other angle, *PXT*, will always increase so much that the sum of both, or the total difference of refraction, together also perpetually achieves an increase. Generally the same things come to be observed if you imagine the line on which the indeterminate point *X* stands to be positioned obliquely with respect to the surface *AP*. Yet sometimes the contrary will occur. For example, with the given position of the ray *FPX* being maintained, if the indeterminate point *X* to which the other ray *FTX* inclines is conceived to be in the line *PX*, the difference of refraction will turn out smaller the farther that point *X* is removed from *P*, or the more obliquely the ray *FTX* falls, with the angle at *X* always decreasing

as is readily confirmed by taking the second derivative with respect to *i*. The same holds true for refraction to a rarer medium, where $dr/dn = \sin i / \sqrt{[1 - n^2 \sin^2 i]}$.

nuto, quàm alter angulus ad *F* augetur. Ad eundem modum si vice rectae *VX* curvam aliquam pro loco puncti *X* adhibeas, vel supponas istud *X* datum esse, et locum ipsius *F* in densiori Medio indeterminatum esse putes, varia hujusmodi de magnitudine angulorum ad *F* et *X* enunciari possent, quorum determinationes non tanti videntur, ut ijs diutiùs incumberem. placuit tamen de illis modò commonuisse, siquidem ad apparentia spatia respiciunt, per quae colores (unicâ variè obliquâ superficie transpectâ) distendi videntur: quemadmodum et propositio praecedens ad spatia colorum in parietes, aut ejusmodi obstacula trajectorum referenda est.

[162 = I, 69]　　Caeterùm ut de mutuis angulorum *PGT* & *πHτ* (in fig [66]) proportionibus habeatur plenior determinatio, dico praeterea quod sunt inter se quàm proximè ut lineae *AB* et *AD* segmenta nempe basium triangulorum aequialtorum, quorum alterum *EGD* constituitur a radijs *GP* ac *GT* cum perpendiculo *AF* concurrentibus, et alterum *CGB* sit simile triangulo *MHL* a radijs *Hπ* et *Hτ* similiter constituto. Nam anguli *EGD*, & *CGB*, si essent infinitè parvi, forent inter se ut *AB* ad *AD*, per Lem 2. At isti ex Hypothesi sunt aequales angulis *PGT* et *πHτ*, Quare etiam illi *PGT* & *πHτ* modò essent infinitè parvi, forent itidem ut *AB* ad *AD*. Cùm itaque semper sint admodum (licèt non infinitè) parvi, sequitur quòd sunt quàm proximè ut *AB* ad *AD*. Et pari ratione constat quòd sunt quàm proximè ut *AC* ad *AE*. Scilicet eorum ratio has duas rationes semper intercedit; et ideò veritatem adhuc propiùs assequemur adhibendo rationem intermediam, nempe quòd sit *PGT* ad *πHτ* ut *AB* + *AC* ad *AD* + *AE*; vel ut $\sqrt{AB \times AC}$ ad $\sqrt{AD \times AE}$ proximè.[22]

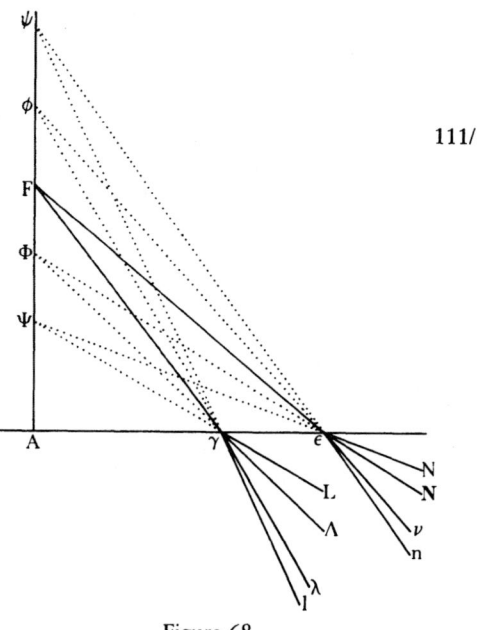

[163 = I, 78]　　/ Prop: 2. Heterogeneis radijs in superficiem quaecunque Media disterminantem incidentibus; quò magis Media differunt densitate[23] eò major erit inaequalitas rationis sinuum refractionis. In fig: [68] Sit *Fγ* radius e minime[24] refrangibilibus utcunque in superficiem, *Aγ* incidentibus, sitque refractus ejus *γλ*, qui retroactus secet perpendiculum *FA* in *φ*. Dein capiatur *Aε* ut sit *Fε* ad *Fγ* in datâ

(a) sect 117, 118, 122　　quâdam ratione qualem ante[a] descripsimus,[25] hâc scilicet conditione ut habito *Fε* pro radio maximè[26] refrangibili, refractus ejus *εν* ab eodem *φ* divergat.[27] Facto hoc, si pro

111/

Figure 68

(22) Since Lemma 2 applies to an infinitesimal angular dispersion, where $AB \approx AC$ and $AD \approx AE$, Newton here replaces them by their average value for mean refrangible rays, $\frac{1}{2}(AB + AC)$ and $\frac{1}{2}(AD + AE)$. Then since $AB >> BC$, he takes

more than the other angle at *F* increases. In the same way, if instead of the straight line *VX* you used some curve for the position of the point *X*, or if you assumed *X* to be given and considered the position of *F* in the denser medium to be indeterminate, various things of this kind could be related about the magnitude of the angles at *F* and *X*, the evaluation of which does not seem that important for me to be concerned with them any longer. Nevertheless, it seemed desirable simply to have called attention to them, since they concern the apparent spaces through which colors seen across a single, variously inclined surface appear to be spread out, just as the preceding proposition is to be related to the spaces of colors transmitted onto walls or such obstacles.

Furthermore, for a more complete determination of the mutual proportions [162 = I, 69] of the angles *PGT* and *πHτ* in Fig. 66, I say, in addition, that they are to one another very nearly as the lines *AB* and *AD*; namely, segments of the bases of equal-altitude triangles, one of which, *EGD*, is formed by the rays *GP* and *GT* meeting the perpendicular *AF*, while the other, *CGB*, is similar to the triangle *MHL* similarly formed by the rays *Hπ* and *Hτ*. For if the angles *EGD* and *CGB* were infinitely small, they would be to one another as *AB* to *AD*, by Lemma 2. But by hypothesis they are equal to the angles *PGT* and *πHτ*. Consequently, provided that those angles *PGT* and *πHτ* were infinitely small, they would likewise be as *AB* to *AD*. Since, therefore, they are always very (although not infinitely) small, it follows that they are very nearly as *AB* to *AD*, and for the same reason it is evident that they are very nearly as *AC* to *AE*. Namely, their ratio always lies between these two ratios, and thus we will approach still nearer the truth by using the intermediate ratio, specifically, that *PGT* is to *πHτ* as *AB* + *AC* is to *AD* + *AE*, or approximately as $\sqrt{[AB \times AC]}$ is to $\sqrt{[AD \times AE]}$.[22]

Proposition 2. When heterogeneous rays fall on a surface separating any [163 = I, 78] media, the more the media differ in density[23] the greater will be the inequality of the ratio of the sines of refraction.

In Fig. 68 let *Fγ* be a least[24] refrangible ray incident in any way on the surface *Aγ*, and let its refracted ray be *γλ*, which extended backward intersects the perpendicular *FA* in *φ*. Then take *Aε* such that *Fε* is to *Fγ* in a certain given ratio, as we described[25] earlier,[a] namely, with this condition, (a) §§117, 118, 122 that when *Fε* is taken to be a most[26] refrangible ray, its refracted ray *εν* diverges from the same point *φ*.[27] When this has been done, if another

---

$$AB + AC = 2AB(1 + BC/2AB) \approx 2AB\sqrt{[1 + BC/AB]} = 2\sqrt{[AB \times AC]};$$

and similarly for *AD* + *AE*.

(23) Heterogeneis ... densitate] **II**: Posito radiorum diversi generis communi sinu incidentiae quo magis diversa est Mediorum densitas (Assuming a common sine of incidence of rays of different kinds, the greater the difference in density of the media).

(24) Read "maximè" (most). The original "rubiformibus" was also mistaken.

(25) Originally: determinavimus (we determined).

(26) Read "minimè" (least). The initial "purpuriformi" was similarly in error.

(27) Originally continued: Id quod diximus eventurum circiter si dicta ratio *Aγ* ad *Aε* ponatur 39 ad 40. (We said this would just about occur if that ratio *Aγ* to *Aε* were assumed to be 39 to 40.) This should read "*Fγ* ad *Fε*."

posteriori Medio aliud utcunque densum rarumve substituatur, ejusmodi tamen duo radij secundum easdem rectas $F\epsilon$, $F\gamma$ incidentes semper debent ita refringi ut ab eodem aliquo perpendiculi istius puncto similiter divergant[b]; quemadmodum a ψ versus l et n, posito quod hoc Medium posterius sit densitatis ab anteriori magis diversae quàm alterum posterius Medium quod efficiebat divergentes a ϕ. Ostendendum est itaque quòd major sit inaequalitas rationis sinuum refractionis in posteriori quàm priori casu. Scilicet radij $F\gamma\lambda$

(b) sec 118 & 122

(c) sect 120 sinus incidentiae est ad sinum refractionis ut $\phi\gamma$ ad $F\gamma$[c], hoc est ut 1 ad $\dfrac{F\gamma}{\phi\gamma}$. Et

sic radij $F\epsilon\nu$ sinus isti sunt ut 1 ad $\dfrac{F\epsilon}{\phi\epsilon}$. Quare sinus refractionum eorundem

radiorum sunt inter se ut $\dfrac{F\gamma}{\phi\gamma}$ ad $\dfrac{F\epsilon}{\phi\epsilon}$. Et simili discursu constabit quòd radiorum

a ψ refractorum consimiles refractionum sinus sunt ut $\dfrac{F\gamma}{\psi\gamma}$ ad $\dfrac{F\epsilon}{\psi\epsilon}$. Restat itaque

probandum quod inter $\dfrac{F\gamma}{\psi\gamma}$ & $\dfrac{F\epsilon}{\psi\epsilon}$ major sit disproportio quàm inter $\dfrac{F\gamma}{\phi\gamma}$ et $\dfrac{F\epsilon}{\phi\epsilon}$.

(d) Lemma 3 Hoc est, (cùm sit $\dfrac{F\epsilon}{\psi\epsilon}\ \sqsubset\ /\ \dfrac{F\gamma}{\psi\gamma}$,[d]) probandum restat quod sit $\dfrac{F\epsilon}{\psi\epsilon}\cdot\dfrac{F\gamma}{\psi\gamma}\ \sqsubset\ \dfrac{F\epsilon}{\phi\epsilon}\cdot\dfrac{F\gamma}{\phi\gamma}$. **112/**

Scilicet est $\psi\epsilon\cdot\phi\epsilon\ \sqsupset\ \psi\gamma\cdot\phi\gamma$ per Lem 3, et sumendo reciproca rationum erit

$\dfrac{1}{\psi\epsilon}\cdot\dfrac{1}{\phi\epsilon}\ \sqsubset\ \dfrac{1}{\psi\gamma}\cdot\dfrac{1}{\phi\gamma}$ ducendoque priorem rationem in $F\epsilon$ et posteriorem in

$F\gamma$, orietur $\dfrac{F\epsilon}{\psi\epsilon}\cdot\dfrac{F\epsilon}{\phi\epsilon}\ \sqsubset\ \dfrac{F\gamma}{\psi\gamma}\cdot\dfrac{F\gamma}{\phi\gamma}$ et vicissim $\dfrac{F\epsilon}{\psi\epsilon}\cdot\dfrac{F\gamma}{\psi\gamma}\ \sqsubset\ \dfrac{F\epsilon}{\phi\epsilon}\cdot\dfrac{F\gamma}{\phi\gamma}$. Q.E.D.[28]

[164 = I, 79]   Schol. Demonstratio perinde se habet in literis majusculis (quibus refractiones designavi cùm posterius Medium sit anteriori rarius) si modò vice signi $\sqsubset$ ubique subintelligatur signum $\sqsupset$ et $\sqsubset$ vice $\sqsupset$. Notabis insuper quòd in hâc demonstratione posui densitatem posterioris tantum Medij variatam esse, sed eodem recidit si anteriora Media successivè varia adhiberi, posteriori non mutato, sive quod tantundem est si refractiones e posteriori Medio in anterius vicissim peragi concipias: siquidem radijs in superficiem alterutrinque incidentibus consimiles sunt sinuum rationes. Caeterùm de exactâ horum sinuum pro quibuslibet propositis Medijs ratione investigandâ disserui ante, et hanc utique propositionem haud attigissem,[29] nisi id in gratiam quartae secuturae fuisset factum.

[165 = I, 90]   / Prop: 3.[30] Heterogeneis radijs e densiori Medio in rarius secundum eandem datam lineam in superficiem positione datam incidentibus: quo rarius sit Medium in quod radij refringuntur eo major erit differentia refractionis. Sit FL [fig 69] linea secundum quam duo radij incidunt in superficiem AL quorum maximè refrangibilis refringatur ad P, et minimè refrangibilis[31] ad T: **106/[30]**

---

(28) This proposition states that $T'/P' > T/P$ or $n'_P/n'_T > n_P/n_T$, where the primes refer to the denser medium (that is, $n' > n$), or recasting it $\Delta n'/n'_T > \Delta n/n_T$, which is in general not physically valid; for instance, turpentine has a smaller index of refraction than common glass but is proportionally more dispersive, thus causing the inequality to be reversed. The proposition holds both for the quadratic and linear dispersion laws. On Newton's concept of optical density, see Lects. 16, 17, note (20).

medium, however dense or rare, is substituted for the second medium, two rays of this kind incident along the same lines $F\epsilon$ and $F\gamma$ must still always be so refracted that they likewise diverge from some one point of that perpendicular;[b] for instance, from ψ toward l and n, with the density of this second medium being assumed to differ more from the first one than the former second medium that caused them to diverge from ϕ. It must therefore be shown that the inequality of the ratio of the sines of refraction is greater in the latter than in the former case. Specifically, the sine of incidence of the ray $F\gamma\lambda$ is to the sine of refraction as $\phi\gamma$ to $F\gamma$,[c] that is, as 1 to $F\gamma/\phi\gamma$, and likewise those sines for the ray $F\epsilon\nu$ are as 1 to $F\epsilon/\phi\epsilon$. Consequently, the sines of refraction of the same rays are to one another as $F\gamma/\phi\gamma$ to $F\epsilon/\phi\epsilon$; and by a similar argument it will be clear that for the refracted rays from ψ the like sines of refraction are as $F\gamma/\psi\gamma$ to $F\epsilon/\psi\epsilon$. It therefore remains to be proved that there is a greater disproportion between $F\gamma/\psi\gamma$ and $F\epsilon/\psi\epsilon$ than between $F\gamma/\phi\gamma$ and $F\epsilon/\phi\epsilon$; that is (since $F\epsilon/\psi\epsilon > F\gamma/\psi\gamma$),[d] it remains to be proved that $F\epsilon/\psi\epsilon : F\gamma/\psi\gamma > F\epsilon/\phi\epsilon : F\gamma/\phi\gamma$. To be sure, by Lemma 3, $\psi\epsilon : \phi\epsilon < \psi\gamma : \phi\gamma$; then taking the reciprocal of the ratios, $1/\psi\epsilon : 1/\phi\epsilon > 1/\psi\gamma : 1/\phi\gamma$; and multiplying the first ratio by $F\epsilon$ and the second by $F\gamma$, there will result $F\epsilon/\psi\epsilon : F\epsilon/\phi\epsilon > F\gamma/\psi\gamma : F\gamma/\phi\gamma$, and by permutation $F\epsilon/\psi\epsilon : F\gamma/\psi\gamma > F\epsilon/\phi\epsilon : F\gamma/\phi\gamma$. As was to be demonstrated.[28]

(b) §§118, 122

(c) §120

(d) Lemma 3

Scholium. The demonstration is just the same with the capital letters by which I have designated the refracted rays when the second medium is rarer than the first, provided that the sign $<$ is everywhere understood for the sign $>$, and $>$ is understood for $<$. You will note, moreover, that in this demonstration I assumed only the second medium's density to vary, but it amounts to the same thing if you imagine the first medium to be treated as successively varying, while the second medium is unchanged, or equivalently if you imagine the refractions to proceed in turn from the second medium into the first, inasmuch as the ratios of the sines for rays incident on either side of a surface are completely similar. But I previously discussed investigating the exact ratio of their sines for any proposed media, and in any case I would not have undertaken[29] this proposition had it not been done for the sake of the fourth proposition to follow.

[164 = I, 79]

Proposition 3.[30] When heterogeneous rays are incident from a denser into a rarer medium along the same given line onto a surface given in position, the rarer the medium into which the rays are refracted the greater will be the difference of refraction.

[165 = I, 90]

Let FL (Fig. 69) be a line along which two rays fall upon the surface AL, and let the most refrangible of these be refracted to P and the least refrangible[31] to T. I say that if the rarer medium were still rarer, so that it

(29) Originally: composuissem (composed).

(30) Following Newton's instruction ("Pete ex pag 106"), Prop. 3 is here inserted from pp. 106–7; see note (9), this lecture.

(31) maximè refrangibilis . . . minimè refrangibilis] Originally: purpuriformis . . . rubriformis (purple-making . . . red-making).

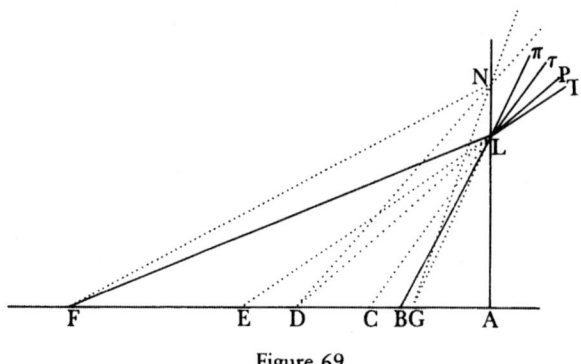

Figure 69

Dico quòd si Medium rarius foret adhuc magis rarum ut refringeret maximè refrangibilem radium ad π et minimè refrangibilem[32] ad τ, tunc angulus $\pi L\tau$ foret major angulo *PLT*. Demittatur enim *FA* ad refringentem superficiem normalis, quae secet refractos radios retrorsum ductos in *G, C, D*, et *E*. Deinde in refringente superficie quaeratur tale punctum *N*, ut sit *FN . DN ::*
FL . EL, ac *DN* productus erit refractus radij minimè refrangibilis[33] inciden-
(a) sec 120   tis ab *F* ad *N*.[a] Jam cùm talis supponatur positio linearum *FL* et *FN*, ut radij maximè refrangibilis secundum *FL* et minimè refrangibilis[31] secundum *FN* incidentis refracti *DL* ac *DN* divergant à puncto *D* quod situm est in perpendiculo *FA*: eâ de causâ, licèt raritas Medij in quod refractio peragitur foret alia quàm supponitur, tamen ejusmodi radiorum secundum easdem lineas *FN* et *FL* incidentium refracti semper divergerent ab aliquo puncto quod in ea-
(b) sec 122   dem *FA* sit positum: quemadmodum in praecedentibus ostensum est[b]. Sic cùm raritas dicti Medij talis esse supponitur, ut maximè refrangibilis radius secundum *FL* incidens refringatur a puncto quopiam *G*, tunc minimè refrangibilis[31] secundum *FN* incidens refringetur ab eodem *G*. Sed cùm maximè refrangibilis radius supponebatur a puncto *G* refringi tunc etiam minimè refrangibilis[31] secundum eandem lineam *FL* incidens supponebatur
(c) Sec 98 & 120   refringi a puncto *C*. Quare est *GN . FN :: CL . FL*.[c] / et praeterea cum antea 107/ posuerim esse *FN . DN :: FL . EL*,[34] ex aequo erit *GN . DN :: CL . EL*. Sed
(d) Lem 3   est *GN . DN* ⊏ *GL . DL*.[d] adeoque *CL . EL* ⊏ *GL . DL*. Quare si linea quaedam *BL* ita ducatur ut sit *CL . EL :: BL . DL* erit *BL* ⊏ *GL*, propter majorem rationem quam habet ad *DL*, et insuper erit *CL* ⊏ *BL*, eò quòd sit *EL* ⊏ *DL*. Et proinde punctum *B* cadet inter *G* et *C*, eritque ang *GLC* ⊏ ang *BLC*. Cùm verò sit *CL . EL :: BL . DL*, aut vicissim *BL . CL :: DL . EL*,
(e) Lem 1   erit ang *BLC* ⊏ ang *DLE*.[e] Et multo magis ang *GLC* ⊏ ang *DLE*.   Q.E.D.[35]
[166 = I, 91]   Prop: 4. Heterogeneis radijs e Medio densiori in rarius secundum eandem datam lineam in superficiem positione datam incidentibus: quo densius est Medium e quo radij incidunt, eo major erit differentia refractionis. Scilicet

---

(32) maximè refrangibilem . . . minimè refrangibilem] Originally: purpuriformem . . . rubiformem (purple-making . . . red-making).

(33) minimè refrangibilis] Originally: rubriformis (red-making).

(34) Note that *FL : EL = T : I*, whereas *GN : FN = CL : FL = I' : T'*.

refracted the most refrangible ray to π and the least refrangible[32] to τ, the angle $\pi L \tau$ would be greater than the angle PLT. For to the refracting surface drop the normal FA, which intersects the refracted rays extended backward at G, C, D, and E. Then in the refracting surface find a point N such that $FN : DN = FL : EL$, and the extension of DN will be the refraction of the least refrangible[33] ray falling from F to N.[a] Now, since such a position of the lines FL and FN is supposed that the refracted rays DL and DN of the most refrangible ray incident along FL and the least refrangible one[31] incident along FN diverge from the point D located on the perpendicular FA, hence even if the rarity of the medium in which the refraction is completed were different than supposed, still the refracted rays of rays of this sort incident along the same lines FN and FL would always diverge from some point located on the same FA, as was shown in the preceding.[b] Thus when the rarity of that medium is assumed to be such that the most refrangible ray incident along FL is refracted from some point G, then the least refrangible[31] ray incident along FN will be refracted from the same point G. But when the most refrangible ray was assumed to be refracted from G, then in addition the least refrangible[31] ray incident along the same line FL was assumed to be refracted from the point C. Consequently, $GN : FN = CL : FL$;[c] and moreover, since I assumed previously that $FN : DN = FL : EL$,[34] from the equality of the ratios there will be $GN : DN = CL : EL$. But $GN : DN > GL : DL$,[d] and thus $CL : EL > GL : DL$. Therefore, if some line BL is so drawn that $CL : EL = BL : DL$, then $BL > GL$, because it has a greater ratio to DL; and moreover $CL > BL$, because $EL > DL$. Consequently, the point B will fall between G and C, and $\widehat{GLC} > \widehat{BLC}$. But since $CL : EL = BL : DL$, or permuting, $BL : CL = DL : EL$, then $\widehat{BLC} > \widehat{DLE}$,[e] and even more so $\widehat{GLC} > \widehat{DLE}$. As was to be demonstrated.[35]

(a) §120

(b) §122

(c) §§98, 120

(d) Lemma 3

(e) Lemma 1

Proposition 4. When heterogeneous rays are incident from a denser into a rarer medium along the same given line upon a surface in a given position, the denser the medium from which the rays are incident the greater will be the difference of refraction.

[166 = I, 91]

---

(35) If we write the sine law in the form $n_1 \sin i = n_2 \sin r$, where $n = n_2/n_1$, then we find the angular dispersion to be

$$dr = \frac{\sin i(n_2 dn_1 - n_1 dn_2)}{n_2{}^2 \cos r} = \frac{\sin i(n_2 dn_1 - n_1 dn_2)}{n_2(n_2{}^2 - n_1{}^2 \sin^2 i)^{1/2}}.$$

In this and the next three propositions Newton investigates the variation of the angular dispersion in the four possible cases: In the first two, for refraction from a dense to a rare medium ($n_1 > n_2$), he lets the rarer medium (n_2) become rarer (Prop. 3), and then the denser medium (n_1) become denser (Prop. 4); and in the second two cases, for refraction from a rare to a dense medium ($n_1 < n_2$), he lets the rarer medium (n_1) become rarer (Prop. 5), and then the denser medium (n_2) become denser (Prop. 6). To evaluate the dispersion in these propositions Newton uses the refraction and dispersion laws set forth in Lect. 11, and generally invokes them in the alternative form that a homocentric bundle of heterogeneous rays remains homocentric after any refraction. Underlying these propositions is Newton's still more fundamental assumption that there is a law relating dn to n independent of the nature of the media, for without such an assumption even to formulate these propositions would be a futile endeavor.

(propter majores refractiones) eo majores erunt sinus refractionum respectu dati circuli, ad quem referuntur: & simul eo major erit inaequalitas rationis istorum sinuum (per prop 2 praecedentem) Adeoque eo major erit differentia angulorum quos subtendunt (per coroll 2 ad Lem 5.) Hoc est, eo major differentia refractionis. Q.E.O.

Lect 16 et 17
[167 = I, 92]

/ Prop: 5. Heterogeneis radijs e Medio rariori in densius secundum eandem datam lineam in superficiem positione datam incidentibus: quo rarius est Medium e quo radij incidunt, eo major erit differentia refractionis. Sit *AD* [fig 70] superficies in quam duo radij secundum eandem lineam *IX* incidunt, quorum alter maxime refrangibilis refringatur ad *P* et alter minimè refrangibilis[1] ad *T*: Dico quod si Medium ex quo radij incidunt foret adhuc rarius, ut dictos radios magis refringeret, puta maximè

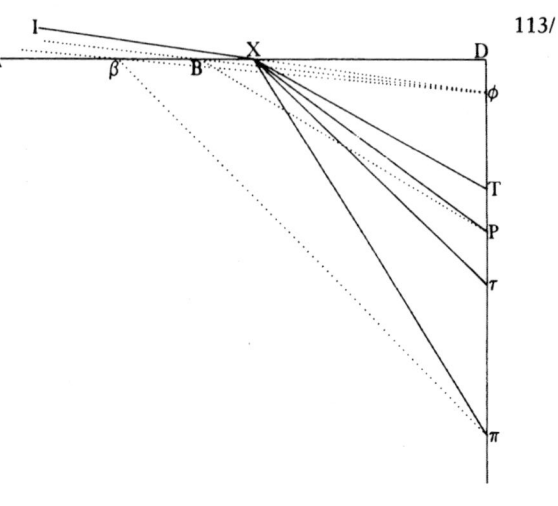

113/

Figure 70

refrangibilem versus *π*, et minimè refrangibilem[2] versus *τ*; tunc *πXτ* major angulus evaderet, quàm *PXT*. Id quod gradatim sic demonstro.

[168 = I, 93]   Cas 1. Ponamus primò quod recta *IX* secundum quam radij incidunt sit ad refringentem superficiem obliquissima. Ac ducatur quaelibet recta *PD* eidem superficiei normaliter insistens in *D*, et secans refractos radios in punctis *T, P, τ, π*. Et *IX* producatur donec istam *PD* secet in *φ*. Tum in linea *AD* quaeratur punctum quoddam *B*, hâc lege ut ductis *Bφ, BP*, fiat *Xφ . XT :: Bφ . BP*. Liquet ergo quod si minimè refrangibilis[3] radius incidat in *B* versus *φ* tendens, is debet versus *P* refringi: Quippe cùm ex Hypothesi sit *BP . Bφ :: XT . Xφ*, hoc est sinus incidentiae ejus et refractionis, sicut sinus incidentiae et refractionis alterius maximè[4] refrangibilis radij *IXT*. Quamobrem si supponamus hosce radios retrocedere, alterum nempe e minimè refrangibilibus a *T* ad *X* et alterum a *P* ad *B*, et maximè refrangibilem[5] a *P* ad *X*; eorum omnium refracti tendent a puncto *φ*, siquidem notum est Theorema[6] quod radij secundum refractum ejus retro incidentis, incidens vicissim fit refractus. Jam cùm radij difformes *PB PX* ab eodem puncto *P* manantes refringantur ab eodem *φ* quod situm est in perpendiculo *PD,* / pro- 114/ portione inter *PX* et *PB* semel cognitâ, si ab alio quovis ejusdem perpendiculi puncto ad refringentem superficiem duae ducantur lineae eandem ratio-

---

(1) maximè refrangibilis . . . minimè refrangibilis] Originally: purpuriformis . . . rubiformis (purple-making . . . red-making).

(2) maximè refrangibilem . . . minimè refrangibilem] Originally: purpuriformem . . . rubiformem (purple-making . . . red-making).

Namely, because of the greater refractions, the greater will be the sines of refraction with respect to the given circle to which they are referred, and at the same time the greater will be the inequality of the ratio of those sines (by the preceding Prop. 2). Hence, the greater will be the difference of the angles that they subtend (by Lemma 5, Corollary 2), that is, the greater will be the difference of refraction. As was to be shown.

Proposition 5. When heterogeneous rays are incident from a rarer into a denser medium along the same given line upon a surface in a given position, the rarer the medium from which the rays are incident the greater will be the difference of refraction. Lectures 16, 17 [167 = I, 92]

Let *AD* (Fig. 70) be a surface upon which two rays are incident along the same line *IX*, the most refrangible one being refracted to *P* and the other, least refrangible[1] one to *T*. I say that if the medium from which the rays are incident were still rarer, so that it would refract those rays more, for example, the most refrangible to π and the least refrangible[2] to τ, then the angle $\pi X \tau$ would become greater than the angle *PXT*. I will prove this stepwise.

Case 1. Let us first assume that the line *IX* along which the rays are incident is the most oblique to the refracting surface. Draw any line *PD* perpendicular to the same surface at *D* and intersecting the refracted rays at the points *T*, *P*, τ, and π; and extend *IX* until it intersects *PD* in ϕ. Next, in the line *AD* find a certain point *B*, subject to the condition that when *Bϕ* and *BP* are drawn, $X\phi : XT = B\phi : BP$. It is evident, therefore, that if a least refrangible[3] ray were incident at *B* tending toward ϕ, it ought to be refracted toward *P*, inasmuch as by hypothesis $BP : B\phi = XT : X\phi$; that is, its sines of incidence and refraction are as the sines of incidence and refraction of the other most[4] refrangible ray *IXT*. Consequently, if we suppose these rays to pass backward—namely, one of the least refrangible rays from *T* to *X* and the other from *P* to *B*, and the most refrangible[5] one from *P* to *X*—all of their refracted rays will tend from the point ϕ, since it is a known theorem[6] that for a ray incident backward along its refracted ray, the incident ray becomes interchangeably the refracted ray. Now, since the dissimilar rays *PB* and *PX* proceeding from the same point *P* are refracted from the same point ϕ located on the perpendicular *PD*, once the proportion between *PX* and *PB* is known, if from any other point of the same perpendicular to the refracting surface two lines having the same ratio are drawn, that is, so that one, [168 = I, 93]

---

(3) minimè refrangibilis]: Originally: rubiformis (red-making).

(4) Read (as II) "minimè" (least), although the original "rubiformis" was correct.

(5) alterum nempe e minimè refrangibilibus . . . et maximè refrangibilem] Originally: duos rubiformes nempe alterum . . . purpuriformem verò (namely, two red-making rays, the one . . . but the purple-making).

(6) See the *Optica*, Prop. 1, §I, 53.

nem habentes, hoc est ut altera designans maximè refrangibilem radium sit
ad alteram quae designet minimè refrangibilem,[2] ut *PX* ad *PB*: tunc istorum

(b)[7] Sect 118 refracti (ex ante monstratis[b]) divergent ab aliquo etiam puncto quod situm
est in eodem perpendiculo *PD*; utcunque Medium ex parte radij *IX* supponatur rarum, dummodo Mediorum alterum ex parte radij *PX* eandem densitatem retineat. Quemadmodum si maximè refrangibilis[8] radius incidat secundum *πX* [e]t refringatur a *φ*, Medio scilicet versus *IX* jam posito rariori
quàm ante; tunc rectâ *πβ* sic ductâ ut sit *PX . PB :: πX . πβ*, radius etiam
minimè refrangibilis[3] *πβ* refringetur ab eodem *φ*. Unde sequitur esse *πβ* ad
φβ sicut sinus incidentiae radiorum minimè refrangibilium[9] ad sinum

(c) sect 120 refractionis[c]. Ast in ratione istorum sinuum est etiam *τX* ad *φX*, eò quòd
inflexa *IXτ* designet radium aequaliter refrangibilem[10] cujus pars *IX* producta transit per idem *φ*. Quare est *πβ . φβ :: τX . φX*. Cùm verò radius *IX*
supponatur esse ad refringentem superficiem summè obliquus sive in angulo
infinitè parvo inclinatus, adeò ut recta *Dφ* pro infinitè parvâ sive nullâ
haberi debeat, sequitur esse *DX = Xφ, DB = Bφ*, ac *Dβ = βφ*: quos valores
pro *Xφ, Bφ*, et *βφ* substituendo in supra recensitas proportiones
BP . Bφ :: XT . Xφ & *πβ . φβ :: τX . φX*, emergent *BP . BD :: XT . XD*. et
πβ . Dβ :: τX . DX. Ex quibus pateat rectas *BP* ad *XT* & *βπ* ad *Xτ* parallelas esse, angulosque *BPX* ad *PXT* et *βπX* ad *πXτ* aequales. Sed ex Hypothesi

(d) Coroll 1[11] est *PX . PB :: πX . πβ*. Et proinde ang *βπX* ⊏ ang *BPX*.[d] Hoc est ang *πXτ*
Lem 5   ⊏ ang *PXT*.   Q.E.D.

[169 = I, 94]   Casus 2. Incidentibus verò radijs angulum definitè magnum cum refringente superficie constituentibus, propositum sic patebit. Sit *HX* [fig 71] recta
secundum quam incidunt, / et   115/
cum e Medio minùs raro adveniunt sit *XM* minimè refractus et
XN maximè[12] refractus. Cùm
verò adveniunt e magis raro, sit
Xμ minimè refractus et *Xν*
maximè[12] refractus. Adhibeantur etiam obliquissimè incidentes
radij *IX* cum eorum refractis *XT*,
XP, *Xτ* et *Xπ*, qualis jam descripsimus. Ita scilicet ut cùm
tanta sit anterioris Medij raritas
ut radios *HX* incurvari faciat
versus *M* et *N*, tunc etiam consimiles radios *IX* incurvet versus *T*
et *P*. Cùm verò tantò major sit
ejus raritas ut illos cogat versus
μ et *ν* tunc hosce simul cogat

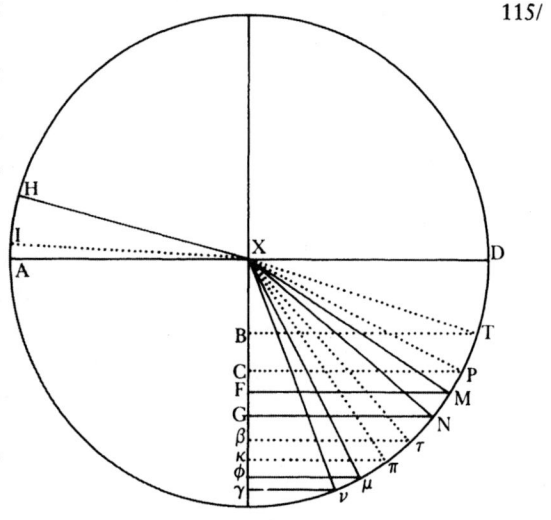

Figure 71

versus *τ* et *π*. Sit insuper *APD* circulus centro *X* et intervallo quolibet descriptus qui secet hosce refractos radios in *T*, *P*, *M*, *N*, *τ*, *π*, *μ*, & *ν*, a quibus ad
perpendiculum *BX* demittantur sinus refractionum *TB, PC, MF, NG, τβ, πκ*,

(a) Sect 99 & 110 *μφ, νγ*. Et ex lege refractionum[a] patebit esse *TB . PC :: MF . NG* et

representing the most refrangible ray, is to the other, representing the least refrangible[2] ray, as *PX* to *PB*, then their refracted rays (from what was shown before)[b] will also diverge from some point located on the same perpendicular *PD*—however rare the medium on the side of the ray *IX* is supposed—provided that the density of the other medium on the side of the ray *PX* remains the same. For instance, if a most refrangible[8] ray is incident along πX and refracted from ϕ (that is, assuming that the medium along *IX* is now rarer than before), then drawing the line $\pi\beta$ so that $PX : PB = \pi X : \pi\beta$, the least refrangible[3] ray $\pi\beta$ would also be refracted from the same point ϕ. Whence it follows that $\pi\beta$ is to $\phi\beta$ as the sine of incidence is to the sine of refraction of the least refrangible[9] rays.[c] But τX to ϕX is also in the ratio of those sines, because the bent line $IX\tau$ represents a ray equally as refrangible[10] whose part *IX*, when it is extended, passes through the same point ϕ; therefore $\pi\beta : \phi\beta = \tau X : \phi X$. Since the ray *IX* is supposed to be the most oblique to the refracting surface, or inclined at an infinitely small angle, so that the line $D\phi$ ought to be considered as infinitely small or nothing, it follows that $DX = X\phi$, $DB = B\phi$, and $D\beta = \beta\phi$. Substituting these values for $X\phi$, $B\phi$, and $\beta\phi$ in the proportions enumerated above—$BP : B\phi = XT : X\phi$, and $\pi\beta : \phi\beta = \tau X : \phi X$—there will result $BP : BD = XT : XD$ and $\pi\beta : D\beta = \tau X : DX$. It is clear from this that the lines *BP* and $\beta\pi$ are parallel to *XT* and $X\tau$, and that the angles *BPX* and $\beta\pi X$ are equal to *PXT* and $\pi X\tau$. But by hypothesis $PX : PB = \pi X : \pi\beta$. Consequently, $\widehat{\beta\pi X} > \widehat{BPX}$,[d] that is, $\pi\widehat{X}\tau > P\widehat{X}T$. As was to be demonstrated.

Case 2. When, however, the incident rays make a finitely large angle with the refracting surface, the proposition will be evident in this way: Let *HX* (Fig. 71) be the line along which the rays are incident; when they arrive from a less rare medium, let *XM* be the least refracted ray and *XN* the most[12] refracted; but when they arrive from a more rare medium, let $X\mu$ be the least refracted ray and $X\nu$ the most[12] refracted. Also add the most obliquely incident rays *IX* with their refracted rays *XT*, *XP*, $X\tau$, and $X\pi$, as we have already described, specifically, such that when the rarity of the first medium is so great that it makes the rays *HX* bend toward *M* and *N*, then the completely similar rays *IX* are also bent toward *T* and *P*; but when its rarity is so much greater that it drives the former toward μ and ν, then at the same time it drives the latter toward τ and π. Moreover, let *APD* be a circle described with center *X* and any radius [*AX*], intersecting these refracted rays at *T*, *P*, *M*, *N*, τ, π, μ, and ν, from which let the sines of refraction *TB*, *PC*, *MF*, *NG*, $\tau\beta$, $\pi\kappa$, $\mu\phi$, and $\nu\gamma$ be dropped to the perpendicular *BX*. From the law of refraction[a] it will be evident that $TB : PC = MF : NG$ and $\tau\beta : \pi\kappa = $

(b)[7] §118

(c) §120

(d) Lemma 5, Corollary 1[11]

[169 = I, 94]

(a) §§99, 100

---

(7) Newton eliminated note (a), when he put the reference (by hypothesis) in the text.

(8) *maximè refrangibilis*] Originally: *purpuriformis* (purple-making).

(9) *minimè refrangibilium*] Originally: *rubiformium* (red-making).

(10) *aequaliter refrangibilem*] Originally: *similiter rubiformem* (similarly red-making).

(11) The lemma itself rather than Corollary 1 is called for here.

(12) *minimè ... maximè*] Originally: *rubiformium ... purpuriformium* (red-making ... purple-making).

τβ . πκ :: μφ . νγ. Et insuper ex Hypothesi et constructione patebit esse *TB* sinuum istorum maximum et νγ minimum. Adeóque per coroll 3 Lem 6 est

ang *TXP* . ang *MXN* ⊏ ang τX*π* . ang μX*ν*.

Et permutando est ang *TXP* . ang τX*π* ⊏ ang *MXN* . ang μX*ν*. Verùm (ex ostensis in casu primo) est ang *TXP* ⊐ ang τX*π*. Quare et multò magis erit ang *MXN* ⊐ ang μX*ν*. Q.E.D.

[170 = I, 95]   Prop 6. Heterogeneis radijs e Medio rariori in densius secundum eandem lineam in superficiem positione datam incidentibus, quo densius sit Medium in quod radij incidunt eo major erit differentia refractionum ad certum usque terminum, et post eo minor perpetuò.[13] Nam si Medium posterius densitate suâ valdè parùm superet anterius, ita ut refractiones indefinitè parvas efficiat, differentia refractionum erit etiam indefinitè parva, et proinde minor quàm foret si Medium posterius / supponeretur densius ut refractiones evaderent 116/ majores. Quare aucta Medij posterioris densitate augebitur dicta refractionum differentia. Quod si densitas ejus in infinitum augeatur refractiones etiam quantum poterunt augebuntur, hoc est usque dum omnes refracti radij perpendiculariter emergant, angulis refractionum et eorum differentijs tunc prorsus evanescentibus. Quare differentia refractionum rursus diminuta est donec in nihilum evanuit.

[171 = I, 96]   Schol. Etsi limitis ejus determinatio ubi differentia refractionis evadit maxima plus taedij et laboris administr[ar]e possit[14] quam utilitatis cùm tamen alicujus fortè momenti censeatur densitatem Medij cognoscere quod radijs in se refractis colores maximè conspicuos efficiat, non pigebit hunc insuper designare. Idque primò cùm incidentia fit obliquissimè.

Cas: 1. Esto *IX* [fig 72] communis radiorum in superficiem *AX* quaecunque Media dirimentem obliquissimè incidentium via. Et eorum refracti ut ante sunto X*π* et X*τ*.[15] Et agatur recta quaevis *ππ* praefatae superficiei parallela, quae radijs istis occurrat in *π* ac *τ*; A quibus ad *AX* demissis perpendicularibus *πC*, *τE*, bisecetur *CE* in *D* et centro *D*, distantia *DX* circulus describatur secans C*π* in *P* et E*τ* in *T*, junganturque *XP* et *XT*.

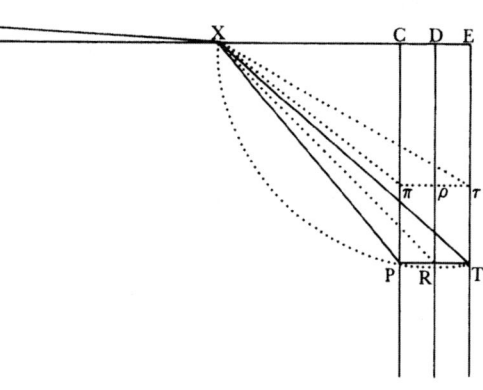

Figure 72

Dico quod cùm ea sit posterioris Medij densitas, ut radiorum secundum *IX* incidentium maximè refrangibiles ad *P* et minimè refrangibiles[16] ad *T* refringat, tunc angulus *PXT* quam maximus evadet.[17] Etenim utcunque Medium posterius ponatur densum, refracti radij ita lineas *CP* et [*E*]*T* in punctis *π* ac

---

(13) This proposition, except for §§172, 175, is included in *Mathematical Papers*, 3:464–75.

$\mu\phi : \nu\gamma$. Moreover, by hypothesis and construction it will be evident that *TB* is the greatest of those sines and $\nu\gamma$ the least. By Lemma 6, Corollary 3, then, $\hat{TXP} : \hat{MXN} > \hat{\tau X\pi} : \hat{\mu X\nu}$, and by permutation $\hat{TXP} : \hat{\tau X\pi} > \hat{MXN} : \hat{\mu X\nu}$. But from what has been shown in the first case, $\hat{TXP} < \hat{\tau X\pi}$, whence even more so $\hat{MXN} < \hat{\mu X\nu}$. As was to be demonstrated.

Proposition 6. When heterogeneous rays are incident from a rarer into a [170 = I, 95] denser medium along the same line upon a surface in a given position, the denser the medium into which the rays are incident the greater will be the difference of refraction up to a certain limit, and after that it will be continually smaller.[13]

For if the second medium very slightly exceeds the first one in its density so that it produces indefinitely small refractions, the difference of refraction will also be indefinitely small and consequently less than it would be if the second medium were supposed denser so that the refractions would prove to be greater. Therefore, increasing the density of the second medium will increase that difference of refraction. But if its density is increased infinitely, the refractions will also increase as much as they can, that is, until all the refracted rays emerge perpendicularly, with the angles of refraction and their differences then vanishing completely. Hence the difference of refraction has again decreased until it has vanished to nothing.

Scholium. Although determining the limit where the difference of refrac- [171 = I, 96] tion turns out greatest may afford[14] more tedium and effort than utility, still since it may perhaps be considered of some importance to find the density of the medium that produces the most conspicuous colors when rays are refracted in it, it will not be troublesome to designate this in addition—and first when the incidence is most oblique.

Case 1. Let *IX* (Fig. 72) be the common path of the rays most obliquely incident on the surface *AX* separating any media, and as before let their refracted rays be $X\pi$ and $X\tau$.[15] Parallel to that surface draw any line $\pi\tau$ meeting those rays in π and τ. Dropping the perpendiculars πC and τE from these points to *AX*, bisect *CE* in *D*; then with center *D* and radius *DX* describe a circle intersecting $C\pi$ in *P* and $E\tau$ in *T*; and join *XP* and *XT*. I say that when the density of the second medium is such that it refracts the most refrangible of the rays incident along *IX* to *P* and the least refrangible[16] to *T*, then the angle *PXT* will prove to be greatest.[17] For however dense the second medium is assumed, the refracted rays will intersect the lines *CP* and

---

(14) Etsi . . . possit] Originally: Caeterum istius limitis determinatio ubi dicta refractionis differentia fit maxima, metuo nè plus taedij et laboris administret (However, determining that limit where that difference of refraction becomes greatest, I fear may afford . . .).

(15) $X\pi$ et $X\tau$.] Originally: $X\pi$ purpuriformis et $X\tau$ rubiformis. ($X\pi$ purple-making and $X\tau$ red-making.)

(16) maximè refrangibiles . . . minimè refrangibiles] Originally: purpuriformes . . . rubiformes (purple-making . . . red-making).

(17) quam maximus evadet.] Originally: differentia refractionis erit omnium qui poterit maximum. (the difference of refraction will be the greatest of all possible.)

τ secabunt ut recta $\pi\tau$ ipsi *AX* parallela sit. Quare si ducatur linea $D\rho$, quae lineas omnes $\pi\tau$ bisecet, / centrum cujuscunque circuli per puncta π ac τ 117/ transeuntis semper jacebit in eâdem $D\rho$.[18] At ang: $\pi X\tau$ est ang: in segmento circuli per puncta π, τ, et *X* transeuntis; qui ideo erit maximus cùm ejusmodi circulus existit minimus, propterea quòd ratio subtensae $\pi\tau$ ad circuli dimensiones tunc evadit maxima. Verùm iste circulus fit omnium minimum cùm centrum ejus cadit in *D*, siquidem pro semidiametro tunc habet *XD* minimam rectarum quae ab *X* ad *RD* duci possunt. Est ergo ang $\pi X\tau$ tunc maximus cùm centrum circuli transeuntis per puncta π, τ, et *X* cadit in *D*. Adeóque cùm circulus *XPT* et angulus *PXT* ejusmodi sunt, liquet propositum.

[172 = I, 97]   Hinc obiter pateat hunc angulum *PXT* tunc etiam maximum evadere cùm talis est posterioris Medij densitas ut angulus refractionis mediocriter refrangibilium radiorum *XR*[19] obliquissimè secundum *IX* incidentium sit semirectus; et eo minorem perpetim fieri quò iste refractionis angulus a semirecto (excessu vel defectu) magis deviat. Quemadmodum si refractiones ex aere in aquam, in vitrum et in crystallum peractae conferantur, e calculo patebit quòd cùm angulus incidentiae sit 90^{gr} proximè, tunc angulus refractionis in aquam erit major semirecto, inque vitrum erit minor. Quamobrem aqua minùs densa est et vitrum magis densum quàm ut efficiant angulum *PXT* maximum. Et proinde cùm crystallum sit adhuc densius, efficiet istum *PXT* minorem quàm vitrum efficeret. Et sic vitrum etsi minùs refringat, in isthoc tamen casu heterogeneos radios in se refractos magis abinvicem dissipabit quàm crystallum, eóque pacto colores in oppositam ejus superficiem projiciet magis distinctos. Sed haec sunt expertu difficillima, quòd vitrum et crystallum densitate parùm differant, nec possint haberi satis crassa; et si possent, tum propter maximam crassitiem haud forent satis perspicua.[20]

[173 = I, 98]   Cas: 2. Quòd si linea secundum quam incidunt radij non sit maximè obliqua, Problema emerget solidum; sed lubet modum ostendere quo conditioni-

---

(18) By Barrow's *Euclidis Elementorum*, III, 1, Corollary, p. 51, or Porism in modern editions of the *Elements*.

(19) mediocriter ... radiorum *XR*] Originally: viridiformium *XR* (green-making *XR*). In the *Optica* "*XR*" is omitted, presumably because according to Newton's refraction model the point *R*, to which mean refrangible rays are refracted, lies on the chord *PT* rather than the arc *PT*. If, however, another mean refrangible ray, *XR'*, is taken, where *R'* is the intersection of the line $D\rho R$ and the arc *XPT*, then $DX = XR'$ and $D\widehat{X}R' = 45°$; nonetheless, the difference between the rays *XR* and *XR'* will be unobservable, since $RR' \approx \frac{1}{2}(CD^2/XD)$.

(20) While Newton's concept of density in these propositions is somewhat ambiguous, it is to be understood as optical density and not specific weight, or density. A few years earlier Hooke had defined optical density: "By density, I mean not the density in respect of gravity (with which the refractions or transparency of *mediums* hold no proportion) ... By *Density* and *Rarity*, I understand a property of a transparent body, that does either more or less refract a Ray of Light ... towards its perpendicular" (*Micrographia*, Observations IX, LVII, pp. 57, 219). Hooke published some data to support this concept (p. 219; see also the preface, sig. *f*), and Newton took notes on his measurements (Newton, *Unpublished Papers*, p. 411). Although the identity of Newton's "crystal" is unknown, he included it in his table of refractions on Add. 4000, f. 33^v and assigned it the unusually high index of refraction $n = 5/3$. If we calculate the mean angle of refraction, $\bar{r}$, and the difference of refraction, Δr, for glass and crystal using the values from the

[E]T in points π and τ such that the line $\pi\tau$ is parallel to AX. Consequently, if the line $D\rho$ is drawn bisecting all lines $\pi\tau$, the center of any circle passing through the points π and τ will always lie on the same line $D\rho$.[18] The angle $\pi X\tau$, however, is the angle in a segment of the circle passing through the points π, τ, and X, and thus it will be greatest when a circle of this kind turns out smallest, because the ratio of the chord $\pi\tau$ to the circle's dimensions then turns out to be greatest. But that circle becomes smallest of all when its center falls at D, since it then has for a radius XD, the smallest of the lines that can be drawn from X to RD. Therefore, the angle $\pi X\tau$ is greatest at the moment when the center of the circle passing through the points π, τ, and X falls at D. Hence, since the circle XPT and the angle PXT are of such a sort, the proposition is evident.

In passing, it should be clear from this that the angle PXT also proves to be [172 = I, 97] greatest just when the density of the first medium is such that the angle of refraction of the mean refrangible rays XR[19] incident most obliquely along IX is half a right angle, and that it becomes continually smaller the more that angle of refraction deviates (by excess or defect) from half a right angle. For example, if the refractions made from air into water, glass, and crystal are compared, it will be evident from a calculation that when the angle of incidence is about 90°, then the angle of refraction in water will be greater than half a right angle, and smaller in glass. Water is therefore not dense enough and glass too dense to make the angle PXT greatest, and consequently since crystal is still denser, it will make the angle PXT smaller than glass would. Thus although glass refracts less, yet in this case it will spread heterogeneous rays refracted in it more from one another than crystal will, and in this way it will project more distinct colors onto its opposite surface. But these are very difficult things to test, because glass and crystal differ little in density, nor can they be had sufficiently thick; and even if they could be, then because of their very great thickness they would be insufficiently transparent.[20]

Case 2. But if the line along which the rays are incident is not the most [173 = I, 98] oblique, the problem will emerge as a solid one; but I intend to show how by

---

table, we find for glass $\bar{r} = 40°14'$ and $\Delta r = 42'$, and for crystal $\bar{r} = 36°54'$ and $\Delta r = 40'$. This 2′ difference in their dispersion is fully consistent with Newton's assertion that they do not differ sufficiently in density—that is, index of refraction—to test. Incidentally, for water, $\bar{r} = 48°33'$, which is greater than the maximal 45° occurring at $n = 1/\sin 45° = \sqrt{2}$. Newton's crystal is probably an imaginary substance whose (very large) index of refraction was adopted for computational convenience. It is most unlikely that it is either rock crystal, whose index of refraction is not that great, or Iceland crystal, for Erasmus Bartholin's first account of it in his *Experimenta crystalli Islandici*, giving it an index of refraction of 5/3, was not published until 1669, well after Newton composed his table. In his *Lectiones XVIII* (Lect. XIV. §II, p. 120) Barrow adopted 5/3 for the index of refraction of glass and 4/3 for water for the sake of illustration. In the *Opticks* (Bk. II, Pt. III, Prop. X, pp. 270–7 = Add. 3970, ff. 191ʳ–5ʳ) Newton demonstrated that the refractive powers ($n^2 - 1$) of bodies are "very nearly" proportional to their densities, except for "unctuous & sulphureous bodies," and he included both "Crystall of yᵉ rock" ($n = 25/16$) and "Island Crystall" ($n = 5/3$) in his experimental test of the law; compare Whiteside in *Mathematical Papers*, 3:469, note 39.

bus ejus nonnihil mutatis, ad planum reduci poterit.[21] Sciendum est itáque quod cùm inter extremos seu maximè difformes[22] radios innumeri sint intermedij / qui gradibus continuò successivis et infinitè parvis alij magis alijs 118/ refringuntur: differentia refractionis extremorum[23] radiorum conflata erit ex consimilibus intermediorum differentijs numero et parvitate infinitis. Jam cognitis proprietatibus istarum infinitè parvarum differentiarum possumus exinde de omnibus simul aggregatis, sive de differentijs finitè parvis quales intercedunt extremorum[23] refractionibus, judicium proferre, praesertim cùm istae differentiae sint admodum exiguae. Sic cognito quòd infinitae parvae differentiae augentur, diminuuntur vel simul maximae evadunt aut minimae: concludendum erit quòd omnium summa perinde augetur, diminuitur, vel maxima fit aut minima. Quod si non sint omnes simul maximae vel minimae, tamen summa pro maxima aut minima haberi potest cùm id accidit intermediae parti. Sic omnium colorum latitudo tunc maxima censeri potest, cùm id accidit viriditati. Jam licèt Problema propositum cùm de differentijs finitè parvis agitur existat solidum, si tamen instituatur de differentijs infinitè parvis, ad planum reduci potèst. Verùm huic solvendo nolo obnixè incumbere, sed breviter tantùm ostendam quo pacto calculus in hoc et ejusmodi alijs sit ineundus, ut ad aequationem perveniatur, ex quâ maximus angulorum infinitè parvorum possit elici. Et insuper ex eodem fundamento determinabo proportiones differentiarum refractionis respectu diversorum Mediorum, quas in praecedentibus quatuor propositionibus generaliter tantùm descripsi.

[174 = I, 99]   Primò itaque investiganda est regula vel Aequatio, quâ ex uno utcunque refracto radio dato, refractus alter cum eo constituens angulum infinitè parvum cognosci poterit. Radijs e Medio densitate dato in Medium cujuslibet densitatis secundum obliquissimam lineam *IX* [fig 73] ut priùs incidentibus, sint *XR* et *Xρ* refracti duo, quorum alter *XR* sit altero / *Xρ* paulo magis 119/ refrangibilis, differentiâ tamen infinitè parvâ. Et agatur lineola quaevis *Rρ* his in *R* et *ρ* occurrens, et refringenti[24] superficiei parallela. Ad quam superficiem normales etiam *RD, ρδ* demittantur, quas datam finitamque distantiam

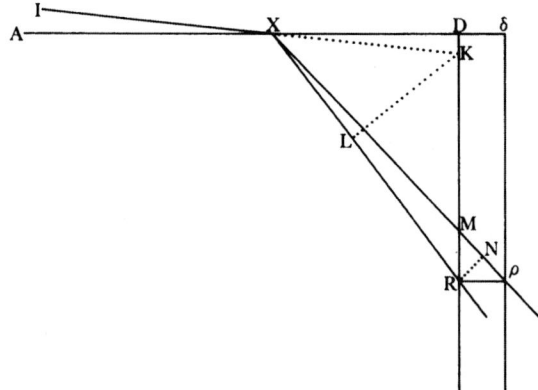

Figure 73

changing its conditions somewhat it can be reduced to a plane one.[21] It must therefore be understood that since between the extreme or most dissimilar[22] rays there are innumerable intermediate ones that are refracted, some more than others, in a continuous succession of infinitely small gradations; the difference of refraction of the extreme[23] rays will be made up of the quite similar differences of the intermediate ones, infinite both in number and smallness. Now when the properties of those infinitely small differences are known, we can then make a judgment about all of them taken together, or about the finitely small differences that lie between the refractions of the extreme[23] rays, especially since those differences are very small. Thus by knowing that the infinitely small differences increase, decrease, or together become greatest or least, it will have to be concluded that the sum of all of them likewise increases, decreases, or becomes greatest or least. But if they are not all simultaneously a maximum or minimum, their sum can still be considered a maximum or minimum when it occurs for an intermediate part. Thus the width of all the colors can be considered a maximum at the moment when it occurs for green. Now, although the proposed problem proves to be a solid one when it concerns finitely small differences, yet if it is undertaken for infinitely small differences, it can be reduced to a plane one. I certainly do not wish to apply myself obstinately to its solution, but I will only briefly show how the calculation in this and similar cases is to be begun in order to arrive at an equation from which the maximum of the infinitely small angles can be derived. In addition, on the same basis I will determine the proportions of the differences of refraction for different media, which I described only generally in the preceding four propositions.

In the first place, therefore, a rule or equation must be found whereby [174 = I, 99] given any one refracted ray, another refracted ray making an infinitely small angle with it can be determined. When, as before, rays are incident from a medium of given density into a medium of any density along the most oblique line *IX* (Fig. 73), let *XR* and *Xρ* be two refracted rays, with one, *XR*, being slightly more refrangible than the other, *Xρ*, although by an infinitely small difference. Draw any small line, *Rρ*, which meets these in *R* and *ρ* and is parallel to the refracting[24] surface. To this surface also drop the normals *RD* and *ρδ*, which you should imagine to be a given, finite distance from *X*

---

(21) According to the classical distinction, "plane" problems are those solvable by "plane" loci (straight lines and circles) alone, "solid" problems are those requiring in addition the use of "solid" loci (the three species of conic sections), and "linear" ones are those requiring "linear" (higher order and transcendental) curves; see Heath, *History of Greek Mathematics*, 2:117–8.

(22) extremos ... difformes] Originally: extremè rubiformes ac purpuriformes (extreme red-making and purple-making).

(23) Originally: purpuriformium et rubiformium (purple-making and red-making).

(24) Originally: refractivae (refractive).

ab *X*, ab invicem verò infinitè parvam habere fingeto, sed lineolam *R*ρ, cum radijs per *R* et ρ transeuntibus plus aut minùs ab *XD* vergere (quemadmodum in praecedentibus) concipito pro variâ posterioris Medij assumendâ densitate. Jam si recta *DR* secet radios *X*ρ in *M*, et *IX* in *K*, cùm infinitè parvum triangulum *RM*ρ sit simile triangulo *DMX* a quo triangulum *KRX* non nisi infinitè parvis differentijs *RXM* et *DXK* discrepat, quae dissimilitudinem non inferunt, triangula etiam *RM*ρ et *RDX* pro similibus haberi debent. Et proinde demissis perpendicularibus *KL* et *RN*, erit *XK . LR :: R*ρ *. MN*.[25] Adeoque cùm sit

$$LR = \frac{XR^q - XK^q}{XR} \text{ (nam est } XR . KR (= \sqrt{:XR^q - XK^q:}) :: KR . LR.)$$

erit etiam $MN = \frac{XR^q - XK^q}{XR \times XK}$ in *R*ρ. Quae differentia est inter *XN* sive *XR* et

XM, et inde erit $XM = XR - : \frac{XR^q - XK^q}{XR \times XK}$ in *R*ρ. Inventa est itaque relatio inter *XK*, *XM*, et *XR* cùm angulus *IXA* sit infinitè parvus: Quinetiam utcunque *IX* obliqua ponatur, illae *XK*, *XM*, et *XR* eandem relationem observabunt, siquidem reciprocè sunt ut sinus incidentiae et refractionis; et proinde inventa est etiam inter eas relatio pro quâvis obliquitate incidentis *IX*. Atque ita cognitis vel utcunque ad arbitrium assumptis *XK* et *XR*, inde *XM* simul cognoscitur. Quod primò determinandum proposui.

Quamobrem sit *IX* linea datum quemvis angulum *AXI* cum refringente superficie constituens; caeterisque stantibus, erit $MN = \frac{XR^q - XK^q}{XR \times XK}$ in *R*ρ. Insuper est $RD (= \sqrt{XR^q - XD^q}) . XD :: MN . NR$. Atque adeò est

$$NR = \frac{XR^q - XK^q \text{ in } R\rho \times XD}{XR \times XK \times \sqrt{XR^q - XD^q}}.$$

Quòd si *NR* dividatur[26] per *XR* prodibit sinus anguli *RXN* respectu circuli cujus semidiameter sit unitas. Quare cùm angulus iste et / sinus ejus sunt[27] simul maximi, ad maximum angulum determinandum quaerenda erit maxima quantitas $\frac{NR}{XR}$, hoc est maximum $\frac{XR^q - XK^q \text{ in } R\rho \times XD}{XR^q \times XK \times \sqrt{XR^q - XD^q}}$. Sive (factâ per datum $\frac{R\rho \times XD}{XK}$ divisione) quaerendum erit maximum

---

(25) See Lect. 12, note (22).
(26) Quòd . . . dividatur] Originally (and mistakenly): Vel dividendo.
(27) Originally continued: infinitè p[arvi] (infinitely small).

but an infinitely small distance from one another; whereas (as in the preceding) you should conceive the small line $R\rho$, together with the rays passing through R and ρ, to lie nearer or farther from XD according as various densities are assumed for the second medium. Now if the line DR intersects the rays $X\rho$ in M and IX in K, since the infinitely small triangle $RM\rho$ is similar to the triangle DMX, from which the triangle KRX differs only by the infinitely small differences RXM and DXK, without effect on the similarity, the triangles $RM\rho$ and RDX also ought to be considered as similar. Consequently, dropping the perpendiculars KL and RN, there will be $XK : LR = R\rho : MN$.[25] Hence, since

$$LR = (XR^2 - XK^2)/XR \text{ (for } XR : KR \ (= \sqrt{[XR^2 - XK^2]}) = KR : LR),$$

also

$$MN = \frac{XR^2 - XK^2}{XR \times XK} R\rho.$$

This is the difference between XN (or XR) and XM, so that

$$XM = XR - \frac{XR^2 - XK^2}{XR \times XK} R\rho.$$

Therefore, a relation has been found between XK, XM, and XR when the angle IXA is infinitely small. In fact, however oblique IX is assumed, XK, XM, and XR will keep the same relation, since they are inversely as the sines of incidence and refraction; consequently, a relation has also been found among them for any obliquity of the incident ray IX. Thus when XK and XR have been found or arbitrarily assumed, XM will be immediately known from them, as I first proposed to determine.

Therefore, let IX be a line making any given angle AXI with the refracting surface and, with other things remaining the same, there will be

$$MN = \frac{XR^2 - XK^2}{XR \times XK} R\rho.$$

Moreover, $RD \ (= \sqrt{[XR^2 - XD^2]}) : XD = MN : NR$, so that

$$NR = \frac{(XR^2 - XK^2)R\rho \times XD}{XR \times XK \times \sqrt{[XR^2 - XD^2]}}.$$

But if NR is divided by XR, the result will be the sine of the angle RXN with respect to a circle of unit radius. Hence, since that angle and its sine are[27] simultaneously a maximum, to determine the maximum angle the maximum quantity NR/XR will have to be found, that is, the maximum of

$$\frac{(XR^2 - XK^2)R\rho \times XD}{XR^2 \times XK \times \sqrt{[XR^2 - XD^2]}};$$

or rather (dividing through by the given quantity $R\rho \times XD/XK$), the maximum of

$$\frac{XR^q - XK^q}{XR^q \times \sqrt{XR^q - XD^q}}.$$

Id quod per Methodos de maximis et minimis satis notas fieri potest, et prodibit $XR^{qq} = 3XK^q \times XR^q - 2XK^q \times XD^q$.[28] Cujus aequationis constructio est ejusmodi.

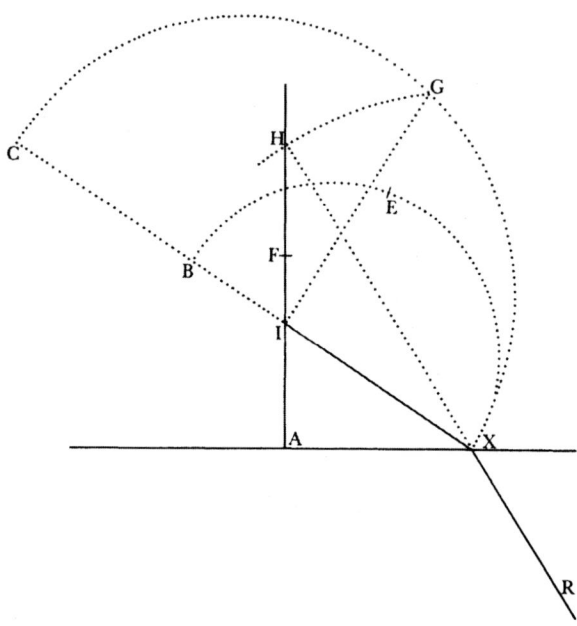

Figure 74

A puncto quolibet incidentis radij *IX* [fig 74] demitte perpendiculum *IA*, et in eo sume *AF* = *AX*. Et *XI* producto ad *B*, ut sit *IB* = $\frac{1}{2}$*IX*, super *BX* describe semicirculum *BEX* cui inscribe *XE* = *XF*. Dein *XB* producto ad *C* ut sit *BC* = *BE*, super *CX* describe semicirculum *CGX* quem in *G* secet perpendiculum *IG* super diametro ejus ad *I* erectum. Denique centro *X* et intervallo *GX* describatur arcus *GH* secans *AI* productum in *H*, ducatur *HX*, et producatur versus *R*, eritque *XR* ipsius *IX* refractus[29] cùm tanta sit posterioris Medij densitas ut differentia refractionis *RXM* fiat omnium maxima. Quo invento, densitas posterioris Medij talem refractionem efficientis facilè dabitur. Concipe ergo radios *XR* et *Xρ* esse mediocriter refrangibiles, diverso tamen gradu,[30] et posterius Medium sic inventum, non modò inter istos sed et inter extremos seu maxime difformes radios[31] maximam circiter[32] quam potest refractionis differentiam efficiet.

175 = I, 100]   Sin autem hujusmodi differentiarum proportiones ad variam raritatem vel densitatem Mediorum desiderentur, e jam ostensis facilè determinabuntur dummodo ponantur infinitè parvae. Sic raritate vel densitate posterioris

$$\frac{XR^2 - XK^2}{XR^2 \times \sqrt{[XR^2 - XD^2]}}$$

will have to be found. This can be done by the rather well known methods of maxima and minima, and the result will be

$$XR^4 = 3XK^2 \times XR^2 - 2XK^2 \times XD^2.^{(28)}$$

The construction of this equation is of the following sort.

From any point of the incident ray *IX* (Fig. 74) drop the perpendicular *IA* and on it take *AF* = *AX*. Then extending *XI* to *B* so that *IB* = $\frac{1}{2}$*IX*, on *BX* describe the semicircle *BEX*, and in it inscribe *XE* = *XF*. Next, extending *XB* to *C* so that *BC* = *BE*, on *CX* describe the semicircle *CGX*, which the perpendicular *IG*, erected on its diameter at *I*, intersects at *G*. Finally, with center *X* and radius *GX* describe the arc *GH* intersecting the extension of *AI* at *H*; draw *HX* and extend it toward *R*. *XR* will be the refracted ray of *IX*[29] when the density of the second medium is such that the difference of refraction *RXM* becomes the greatest of all. When this is found, the density of the second medium producing such a refraction will readily be given. Therefore, imagine the rays *XR* and *Xρ* to be intermediately refrangible, though to a different degree,[30] and the second medium found in this way will produce not only between those but also between the extreme or most dissimilar rays[31] approximately[32] the greatest difference of refraction possible.

If now the proportions of such differences are required at a different rarity [175 = I, 100] or density of the media, they will easily be determined from what has just been shown, provided that they are assumed to be infinitely small. Thus, when the rarity or density of only the second medium is varied, so that the

---

(28) In the intended companion treatise of the *Optical Lectures,* "De methodis serierum et fluxionum," Problem 3 (*Mathematical Papers,* 3:116–21), Newton explains how to determine maxima and minima. Taking the derivative with respect to *XR* and setting it equal to zero, we find $[XK^2(3XR^2 - 2XD^2) - XR^4]/XR^3(XR^2 - XD^2)^{3/2} = 0$, which immediately yields his result.

(29) To construct the solution of his equation $XR = \pm\sqrt{[XK(\frac{3}{2}XK \pm \sqrt{\{\frac{9}{4}XK^2 - 2XD^2\}})]}$, Newton sets $BX = \frac{3}{2}XI(= \frac{3}{2}XK \times XA/XD$, since $XI/XA = XK/XD)$, $EX/XA = \sqrt{2}$,

$$CB = EB = \sqrt{[BX^2 - EX^2]}(= \sqrt{[\frac{9}{4}XK^2 - 2XD^2]}XA/XD),$$

and $XG = \sqrt{[XI(CB + BX)]}$ (since *IG* is perpendicular to the hypoteneuse of the right triangle *CGX*), so that $XG(= XH)/XA = XR/XD$, and thus $\widehat{HXA} = \widehat{RXD}$; see Whiteside's note 47, which we have followed here, in *Mathematical Papers,* 3:475. Vavilov presents a modern version of Newton's derivation and shows that a maximum of dispersion occurs when the refraction law depends on two separate terms: one a function of color or index of refraction, and the other of the medium; Newton, *Lekcii po Optike,* p. 285, note 57. Both of Newton's dispersion laws satisfy this condition (see Lect. 11, note (6)), as does the Lorentz-Lorenz dispersion law.

(30) mediocriter . . . gradu] Originally: diverso gradu, viridiformes (green-making to a different degree).

(31) extremos . . . radios] Originally: radios extremam purpuram et viriditatem pingentes (the rays depicting extreme purple and green).

(32) Added.

tantùm Medij variatâ ut radij secundum *IX* [fig 75] incidentes nunc refringantur ad *M* et *R*, nunc ad μ et ρ: ductaque qualibet *DK* ipsi *DX* normali quae secet eos in *K, M, R,* μ et ρ: erit angulus infinitè parvus / *MXR* ad consimilem angulum $\mu X\rho$ sicut

$$\frac{XR^q - : XK^q}{XR^q \times RD} \text{ ad } \frac{X\rho^q - : XK^q}{X\rho^q \times \rho D}. \quad (33)$$

Quod si raritas vel densitas anterioris Medij varietur, non mutato posteriori Medio: Analysta facilè deprehendet quòd, (in fig [73]) sit

$$MN = \frac{XR^q - : XK^q}{XK^q} \text{ in } R\rho.$$ Et proinde quod, in fig [75], sit

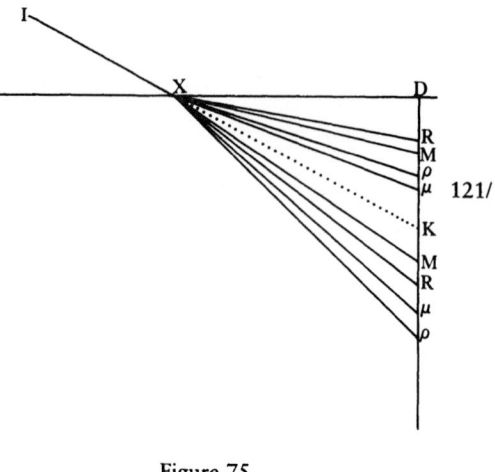

Figure 75

$$\text{ang } MXR \text{ . ang } \mu X\rho :: \frac{XR^q - : XK^q}{XR \times RD} \cdot \frac{X\rho^q - : XK^q}{X\rho \times \rho D}.$$

Non enim perinde est sive raritas vel densitas anterioris Medij, sive posterioris varietur, ut e praeostensis pateat.

    Propositiones praecedentes ad luminis e longinquo emanantis diffusionem spectant. In duabus sequentibus agitur de refractione luminis e propinquo manantis.[34]

[176 = I, 101]   Prop. 7. Heterogeneis radijs a dato puncto ad datum punctum per superficiem positione datam refractis: quo Medium densius sit magis densum eò major erit eorum ad invicem[35] inclinatio ex parte Medij utriusque ad certum usque terminum, et post erit eo minor. Scilicet cùm densitas ejus haud major sit quàm densitas alterius Medij ut refractio fiat infinitè parva, tum differentia refractionis erit etiam infinitè parva, et proinde augebitur ex auctâ densitate. Quod si densitas ejus in infinitum augeatur tum omnium radiorum in illud incidentium refracti perpendiculariter emergent,[a] et e contra soli perpendiculares possunt ingredi Medium rarius e densiori[a]. Unde omnes radij a puncto ad punctum refracti tunc pergent in ijsdem lineis, sive coincident, et sic differentia refractionis rursus in nihilum evanescet.

(a) Duplex haec assertio contemplanti sectiones 115 & 118 patebit.

[177 = I, 102]   Prop. 8. Heterogeneis radijs a dato puncto ad datum punctum per superficiem positione datam refractis: quo Medium rarius sit magis rarum eo major erit eorum ad invicem[35] inclinatio ex parte Medij utrius.que. Sit *AT* [fig 76] superficies ita refringens difformes radios *FTX* et *FPX*,[36] / ut manantes ab 122/ eodem puncto *F*, in idem rursus ad *X* conveniant. Dico si Medium posterius[37] esset rarius ut praefatos radios adhuc magis refringeret, puta *FTX* secundum

121/

---

    (33) Since the angles are assumed to be infinitesimal, $M\widehat{X}R \approx MR/XR$ and $\mu\widehat{X}\rho \approx \mu\rho/X\rho$.
    (34) This paragraph replaced: Propositiones praecedentes ad amplitudinem colorum in objectis lumen trans refringentem superficiem mutuantibus visorum spectant. Quae vero sequuntur duae, ad colores per interpositionem prismatis inter oculum et objectum generatos praesertim referi debent. (The preceding propositions concern the magnitude of colors visible in objects

rays incident along *IX* (Fig. 75) are at one time refracted to *M* and *R* and at another to μ and ρ; and drawing to *DX* any normal *DK*, which intersects the rays in *K*, *M*, *R*, μ, and ρ; the infinitely small angle *MXR* will be to the altogether similar angle $\mu X\rho$ as $(XR^2 - XK^2)/(XR^2 \times RD)$ is to $(X\rho^2 - XK^2)/(X\rho^2 \times \rho D)$.[33] If, however, the rarity or density of the first medium is varied, while the second is unchanged, the analyst will easily discover that (in Fig. 73)

$$MN = \frac{XR^2 - XK^2}{XK^2}R\rho,$$

and consequently that (in Fig. 75)

$$M\widehat{X}R : \mu\widehat{X}\rho = (XR^2 - XK^2)/(XR \times RD) : (X\rho^2 - XK^2)/(X\rho \times \rho D).$$

For it is not equivalent to vary the rarity or density of the first medium or of the second, as is obvious from what has been shown earlier.

The preceding propositions concern the diffusion of light flowing from afar. In the following two the refraction of light flowing from nearby is treated.[34]

Proposition 7. When heterogeneous rays are refracted from a given point [176 = I, 101] to a given point across a surface in a given position, the greater the density of the denser medium, the greater will be the rays' inclination to one anther[35] on the side of each medium up to a certain limit, and afterward it will be smaller.

Namely, when its density is not greater than the density of the other medium, so that the refraction becomes infinitely small, then the difference of refraction will also be infinitely small, and therefore it will increase with an increase in the density. But if its density is increased infinitely, then the refracted rays of all those incident on it will emerge perpendicularly,[a] and (a) This double conversely only perpendicular rays can enter the rarer medium from the assertion will be denser.[a] Consequently, all rays refracted from a point to a point will then clear to anyone proceed in the same lines—or rather coincide—and thus the difference of studying §§115, refraction will again vanish to nothing. 118.

Proposition 8. When heterogeneous rays are refracted from a given point [177 = I, 102] to a given point across a surface in a given position, the greater the rarity of the rarer medium the greater will be the rays' inclination to one another[35] on the side of each medium.

Let *AT* (Fig. 76) be a surface so refracting the dissimilar rays *FTX* and *FPX*[36] that when they proceed from the same point *F* they meet again at one point *X*. I say that if the second[37] medium were rarer, so that those rays would be refracted still more, for instance, *FTX* along *FτX*

---

getting their light across a refracting surface. But the following two are to be related particularly to colors generated by interposing a prism between the eye and object.) "Refringentem superficiem" replaced "prismata."

(35) ad invicem] Added.

(36) difformes . . . *FPX*] Originally: radios *FTX* rubiformem et *FPX* purpuriformem (the rays *FTX*, red-making, and *FPX*, purple-making). (37) Read (as II): prius (first).

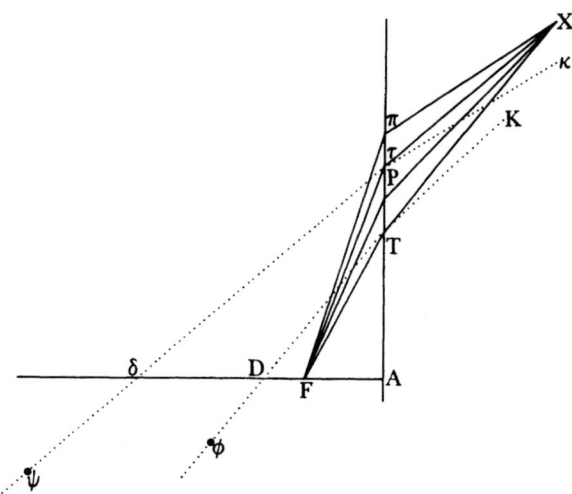

Figure 76

$F\tau X$ et $FPX^{(38)}$ secundum $F\pi X$; quòd angulus $\pi X\tau$ foret major angulo PXT, ut et angulus $\pi F\tau$ major angulo PFT.

Ad abbreviandam prioris casûs demonstrationem ponamus radios esse quàm minimè difformes ut propter infinitè parvam differentiam refractionis, angulos PXT & $\pi X\tau$ constituant infinitè parvos.* Tum ducatur TK refractus radij conformis ipsi FPX, ut infinitè parvus angulus KTX sit differentia refractionis radiorum secundum eandem lineam FT incidentium. Et pari modo ducatur $\tau\kappa$ refractus radij conformis ipsi $F\pi X$, ut angulus infinitè parvus $\kappa\tau X$ existat differentia refractionis radiorum secundum eandem $F\tau$ incidentium. Liquet ergo quòd cùm $F\tau$ sit obliquior quàm FT atque etiam in Medium densius incidat, erit angulus $\kappa\tau X$ major angulo KTX. Adhaec producantur XT et $X\tau$ donec in punctis D ac δ secent lineam FA quae sit plano AT perpendicularis: et ultra producantur ad ϕ et ψ ita ut sit

*Consule cas 2 scholij ad Prop 6 [§173].

$$\frac{FA^q}{FT} \cdot \frac{DA^q}{DT} :: TF \cdot T\phi, \text{ Et } \frac{FA^q}{F\tau} \cdot \frac{\delta A^q}{\delta\tau} :: \tau F \cdot \tau\psi.$$

Et erunt puncta sic inventa ϕ et ψ foci radiorum FTX et $F\tau X$.$^{(a)}$ Et inde

(a) Schol ad prop 3, sec [129].

$$X\phi \cdot T\phi :: \text{ang } KTX \cdot \text{ang } PXT;$$

(b) Schol ad prop 11 sect [143].

ut et $X\psi \cdot \tau\psi :: \text{ang } \kappa\tau X \cdot \text{ang } \pi X\tau.^{(b)}$ Istae quidem proportionalitates non sunt omninò verae ubi anguli praefati per differentiam refractionis effecti ponuntur esse definitae alicujus magnitudinis, sed ad veritatem eo magis accedunt quo anguli isti statuuntur minores, adeò ut in angulis infinitè parvis pro accuratè veris haberi debeant. Jam cùm ex Hypothesi sit $A\tau \sqsubset AT$ erit etiam $X\tau \sqsubset XT,^{(39)}$ ut et$^{(40)}$ $\tau\psi \sqsubset T\phi$ quemadmodum pateat ex determinatione / punctorum ψ et ϕ supra positâ. Quamobrem est $\tau\psi \cdot T\phi \sqsubset \tau X \cdot TX$, vel 123/ permutando $\tau\psi \cdot \tau X \sqsubset T\phi \cdot TX$, et componendo $\tau\psi \cdot X\psi \sqsubset T\phi \cdot X\phi$. Hoc est substituendo rationes his aequales, ang $\pi X\tau \cdot$ ang $\kappa\tau X \sqsubset$ ang $PXT \cdot$ ang KTX.

and FPX[38] along $F\pi X$, then the angle $\pi X\tau$ would be greater than the angle PXT, and also the angle $\pi F\tau$ greater than the angle PFT.

To shorten the demonstration of the first case, let us assume the rays to be the least dissimilar possible, so that because of the infinitely small difference of refraction, they make the angles PXT and $\pi X\tau$ infinitely small.* Then draw TK, the refraction of a ray identical to FPX, so that the infinitely small angle KTX is the difference of refraction of rays incident along the same line FT; likewise draw $\tau\kappa$, the refraction of a ray identical to $F\pi X$, so that the infinitely small angle $\kappa\tau X$ is the difference of refraction of rays incident along the same line $F\tau$. It is therefore clear that since $F\tau$ is more oblique than FT and also incident into a denser medium, the angle $\kappa\tau X$ will be greater than the angle KTX. Moreover, extend XT and $X\tau$ until they intersect the line FA, which is perpendicular to the plane AT, in the points D and δ; and extend them farther to ϕ and ψ such that $FA^2/FT : DA^2/DT = TF : T\phi$ and $FA^2/F\tau : \delta A^2/\delta\tau = \tau F : \tau\psi$. Then the points ϕ and ψ thus found will be the foci of the rays FTX and $F\tau X$.[a] Thus $X\phi : T\phi = \widehat{KTX} : \widehat{PXT}$, and also $X\psi : \tau\psi = \widehat{\kappa\tau X} : \widehat{\pi X\tau}$.[b] In fact these proportions are not strictly true when those angles made by the difference of refraction are assumed to be some definite magnitude, but they approach nearer the truth the smaller those angles are made, so much so that for infinitely small angles they are to be considered as exactly true. Now since by hypothesis $A\tau > AT$, it will also be $X\tau > XT$,[39] as well as[40] $\tau\psi > T\phi$, as is clear from the determination of the points ψ and ϕ presented above. Consequently, $\tau\psi : T\phi > \tau X : TX$, or by permutation $\tau\psi : \tau X > T\phi : TX$, and by composition $\tau\psi : X\psi > T\phi : X\phi$; that is, by substituting ratios equal to these, $\widehat{\pi X\tau} : \widehat{\kappa\tau X} > \widehat{PXT} : \widehat{KTX}$, and

*Consult Prop. 6, Scholium, Case 2 [§173]

(a) Prop. 3, Scholium, §[129]
(b) Prop. 11, Scholium, §[143]

---

(38) *FTX ... FPX*] Originally: rubiformem ... purpuriformem (red-making ... purple-making).

(39) Read: $X\tau \sqsupset XT$ ($X\tau < XT$).

(40) Originally continued: $\tau\delta \sqsubset TD$ ac $\delta A \sqsubset DA$. Quare $FA^q . DA^q \sqsubset FA^q . \delta A^q$. ($\tau\delta > TD$ and $\delta A > DA$. Therefore $FA^2 : DA^2 > FA^2 : \delta A^2$.)

et permutando ang $\pi X \tau$. ang $PXT \sqsubset$ ang $\kappa \tau X$. ang KTX. Verùm est ang $\kappa \tau X \sqsubset$ ang KTX ut dictum fuit; et ideò multò magis est ang $\pi X \tau \sqsubset$ ang PXT.   Q.E.D.

Exhinc verò de posteriori etiam casu, quòd semper sit ang $\pi F \tau \sqsubset$ ang PFT, fiat conjectura, siquidem demonstrationem longè difficiliorem postularet; et his tam multa impendisse verba jamdudum pertaesum est. Haec itáque de refractionibus solitariae superficiei sufficiant.

De radiorum bis refractorum affectionibus.

Lect 18
[178 = I, 103]   Quòd si gemina sit refractio perinde ut in Prismatibus contingit, quorum phaenomena praesertim explicare statui: radiorum sic refractorum passiones e praecedentibus ita manifestae sunt, ut circa illas parùm negotij superesse videatur. De parallelis quidem superficiebus nihil aliud occurrit observandum, quàm quòd posterior tantùm recurvat radios quantùm prior incurvat. De inclinatis verò sequentia notentur.

[179 = I, 104]   1 Homogenei radij ad Prisma divergentes, post utramque refractionem divergere pergent. Patet per Prop 3, sect [128].

Atque idem de parallelis, vel convergentibus radijs intellige quod nempe post utramque refractionem manebunt paralleli vel convergentes.

[180 = I, 105]   Schol Quòd si punctum a quo quilibet indefinitè[1] propinqui post utramque refractionem divergunt, sive locus imaginis trans prisma conspicuae desideretur, inventio ejus a Scholio ad praefatam Prop 3 manifesta est. Sed ut promptiùs fiat conjectura, juvabit adhibere Theorema hocce mechanicum; Quòd imago ad eandem illam circiter distantiam post prisma apparebit, quam habet objectum, cujus est imago, dummodò refractiones hinc, et indè non sint admodum inaequales.[2]

[181 = I, 106]   2. Ex heterogeneis radijs ad Prisma divergentibus aliqui post utramque refractionem convergent. Id quod constat e prop 5 & 8 sect [133, 137].[3]

---

(1) **II:** infinitè (infinitely).

(2) By the Scholium to Prop. 3 (see Lect. 12, note (23), this volume) the position of the primary image point ϕ after one refraction is $R\phi = nFR \cos^2 r / \cos^2 i$. Since the point ϕ then becomes the object for a second refraction, if FR is replaced by $R\phi + RR'$, and the ray is assumed to pass near the vertex so that $RR' \approx 0$ (though RR' is greatly exaggerated in the figure), and also if n is replaced by $1/n$, i by r' and r by i'—where r' and i' are the angles of incidence and refraction at the second face, CB—there results for the image point ϕ' after the

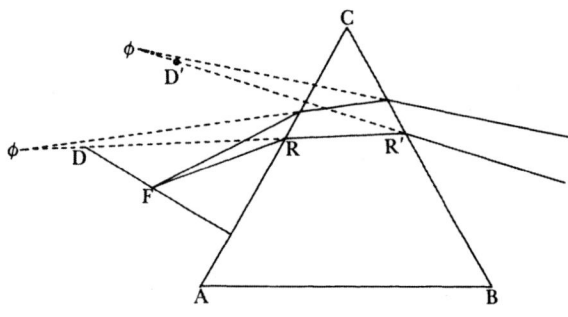

by permutation $\pi \widehat{X} \tau : P \widehat{X} T > \widehat{\kappa \tau X} : K \widehat{T} X$. But $\widehat{\kappa \tau X} > K \widehat{T} X$ (as was indicated), and thus even more so $\pi \widehat{X} \tau > P \widehat{X} T$. As was to be demonstrated.

But it may also be conjectured from this for the second case that $\pi \widehat{F} \tau > P \widehat{F} T$ always, as it would demand a far more difficult demonstration, and I am already weary from expending so many words on these matters. Therefore, let these things suffice for the refractions of a single surface.

The Properties of Rays Refracted Twice

If now the refraction is twofold, as happens in prisms, whose phenomena I particularly intended to explain, the effects of the rays thus refracted are so clear from the preceding that it may seem that little business concerning them remains. Indeed, for parallel surfaces nothing other arises to be observed than that the second surface bends back the rays as much as the first one bends them in, but for inclined surfaces the following may be noted. [Lecture 18 [178 = I, 103]

Proposition 1. When homogeneous rays diverge at a prism, they continue to diverge after both refractions. [179 = I, 104]

It is evident by Prop. 3, §128. The same thing is to be understood for parallel or convergent rays, namely, that they will remain parallel or convergent after both refractions.

Scholium. If, however, the point is required from which any indefinitely[1] close rays diverge after both refractions, that is, the position of the image seen through the prism, its determination is evident from the Scholium to the aforementioned Prop. 3. But to make a more rapid conjecture, it will help to use this mechanical theorem: The image will appear at approximately the same distance behind the prism as does the object whose image it is, provided that the refractions on both sides are not too unequal.[2] [180 = I, 105]

Proposition 2. Of heterogeneous rays diverging at a prism some will converge after both refractions. [181 = I, 106]

This is evident by Props. 5 and 8, §§133, 137.[3] Namely, the more refran-

---

second refraction,

$$R' \phi' = R \phi \cos^2 i'/n \cos^2 r' = FR \cos^2 r \cos^2 i'/\cos^2 i \cos^2 r'.$$

Thus near minimum deviation, where $i = i'$ and $r = r'$, the image will be located at $R' \phi' = FR$, which is Newton's result. Moreover, since the secondary image point D is located at $DR = nFR$ (see Lect. 12, note (19)), that image point, D', after the second refraction is similarly found by replacing FR by $DR + RR'$ (where again $RR' \approx 0$) and n by $1/n$, so that $D'R' = DR/n = FR$, and at minimum deviation the two image points coincide. After Newton touched here ever so briefly upon prismatic images the subject was virtually ignored for over a century and a half until Henry Coddington arrived at the present results, though independently of Newton, in his influential work, *A Treatise on the Reflexion and Refraction of Light, Being Part I of a System of Optics* (Cambridge, 1829), pp. 82–3; compare note (19), this lecture.

(3) In the *Optica* Newton cites the new Props. 10, 12.

Scilicet ex illis qui in plano ad utraque refringentia plana perpendiculari jacent, / magis refrangibiles ex incidentia paulo obliquiori convenient cum minùs re- 124/ frangibilibus. Atque idem in innumeris alijs ferè planis superficiebus continget.

[182 = I, 107]   3 E radijs itaque sic a puncto ad punctum sive ab objecto ad oculum refractis, alij ad verticem prismatis gradatim alijs propiores transibunt pro eo ut sint magis atque magis refrangibiles (per prop 5 et 11 sect [133, 140]).[4] Unde colorum ordines definiuntur.

[183 = I, 108]   4 Quo major est angulus verticalis Prismatis caeteris paribus, differentia refractionis fiet eo major, et inde colorum apparentia distinctior. Et hoc manifestum est e prop 1 sect [160].[5]

[184 = I, 109]   5. Quo densior est Prismatis materia, vel quo rarius est medium circumfluum caeteris paribus, eo major erit refractionis differentia, et inde colorum apparentia manifestior. Scilicet posterior casus e prop 3 & 5 sect [165, 167] patet. Priorem verò ne per prop 6 sect [170] in dubium revocetur, sic ostendo. Concipe [fig 77] magis refrangibilem radium *PD* et minùs refrangibilem[6] *TD* sic in Prisma ad idem quodvis punctum *D* incidere, ut refracti pergant in

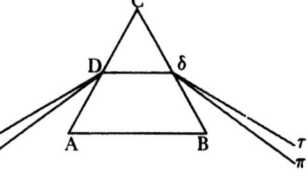

Figure 77

eadem linea *Dδ* ac denuò in *δ* refracti divergant versus *π* ac *τ*. Quo posito constat per Prop 4 sect [166] quod angulus *πδτ* ex auctâ Prismatis densitate augebitur. Deque angulo *PDT* par est ratio, si modò radij consimiles secundum easdem lineas retrocedere concipiantur. Patet itaque assertio de radijs in Prismate coincidentibus, et inde etiam de parallelis.[7]

[185 = I, 110]   Lemma 7. Radijs tribus homogeneis *βI, γI, δI* [fig 78] e Medio densiori in rarius per superficiem *IK* refractis; si differentiae incidentiarum *βIγ, γIδ* sint aequales, summa refractorum angulorum extremis radijs effectorum erit major duplo anguli refracti per intermedium radium effecti. Hoc est, refractis radijs retro-actis ad *B, G,* ac *D,* dico quòd angulus *βIB* + ang *δID* ⊏ 2ang *γIG*. Etenim descripto quovis circulo *ADI* tangente refringentem superficiem in *I,* cu-

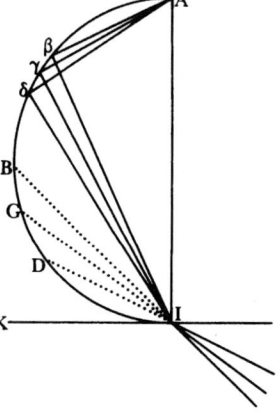

Figure 78

---

(4) In the *Optica* only the new Prop. 10 is invoked.

(5) For any ray passing through a prism, we have sin $i = n$ sin r, n sin r' = sin i', and $C = r + r'$, where i and r' are the angles of incidence at the first and second surfaces and r and i' are the angles of refraction at those same surfaces, and C is the vertex angle, so that

$$\cos i \, di = \sin r \, dn + n \cos r \, dr, \quad \cos i' \, di' = \sin r' \, dn + n \cos r' \, dr', \text{ and } dr = -dr'.$$

For parallel incident rays, where $di = 0$, the angular dispersion, dD, or the difference of refraction, becomes

$$dD = di' = \frac{\sin r' \cos r + \cos r' \sin r}{\cos r \cos i'} dn = \frac{\sin(r' + r)}{\cos r \cos i'} dn = \frac{\sin C}{\cos r \cos i'} dn.$$

Thus, as the prism's angle C increases, and n, i, and consequently r remain constant, since $\cos i' = \sqrt{[1 - n^2 \sin^2 (C - r)]}$, the angular dispersion also increases.

gible of those rays lying in a plane perpendicular to both refracting planes will, because of their slightly more oblique incidence, meet the less refrangible ones; and the same thing will happen in nearly innumerable other plane surfaces.

Proposition 3. Therefore, of rays thus refracted from a point to a point, or [182 = I, 107] from an object to an eye, some will gradually pass nearer to the vertex of the prism than others, according as they are more and more refrangible, by Props. 5 and 11, §§133, 140.[4]

Whence the order of the colors is defined.

Proposition 4. The greater a prism's vertex angle, with other things being [183 = I, 108] equal, the greater will the difference of refraction become and, consequently, the more distinct the colors' appearance.

This is manifest from Prop. 1, §160.[5]

Proposition 5. The denser the matter of a prism or the rarer its surrounding [184 = I, 109] medium, with other things being equal, the greater will be the difference of refraction and consequently the clearer the colors' appearance.

To be sure, the latter case is evident from Props. 3 and 5, §§165, 167, but so that the former case may not be cast into doubt by Prop. 6, §170, I show as follows. Conceive of a more refrangible ray, PD (Fig. 77), and a less refrangible[6] one, TD, to fall on the prism at any one point D so that their refracted rays continue in the same line $D\delta$ and after being refracted again at δ diverge toward π and τ. With this assumption, it is established by Prop. 4, §166, that the angle $\pi\delta\tau$ will increase when the prism's density is increased. The reason is the same for the angle PDT, provided that identical rays are understood to proceed backward along the same lines. Therefore, the assertion is evident for coincident rays in the prism, and thus also for parallel rays.[7]

Lemma 7. When three homogeneous rays βI, γI, and δI (Fig. 78) are [185 = I, 110] refracted from a denser medium into a rarer one by the surface IK, if the differences of the angles of incidence, $\beta I\gamma$ and $\gamma I\delta$, are equal, the sum of the deviations made by the exteme rays will be more than twice the deviation made by the intermediate ray. That is, drawing the refracted rays backward to B, G, and D, I say that $\widehat{\beta IB} + \widehat{\delta ID} > 2\widehat{\gamma IG}$.

For tangent to the refracting surface at I describe any circle ADI, which

---

(6) magis refrangibilem . . . minùs refrangibilem] Originally: purpuriformem . . . rubiformem (purple-making . . . red-making).

(7) Originally continued: Et sanè in alijs quibuscunque casibus ubi divergunt ante refractionem et post convergunt, non adeò multùm a parallelismo intra prisma recedunt unquam, quin ut pro parallelis, sine aliquo circa differentiam refractionis errore sensibili, haberi possint. (In fact, in any other cases whatever, where they diverge before refraction and converge afterward, they will never depart so much from parallelism within the prism that they cannot be considered as parallel without any sensible error in the difference of refraction.) "Ante refractionem" replaced "ante prisma" (before the prism). Newton deleted this when he decided to elaborate this (erroneous) justification into a more comprehensive Scholium in §189 = I, 114. Rather than treating the general case, if we consider minimum deviation, where $i = i'$ and $r = r'$, then from note (5), this lecture, together with the identity $\sin 2r = 2 \sin r \cos r$ and the sine law of refraction $\sin i = n \sin r$, we find $dD = 2 \tan i \, dn/n$. Evidently the angular dispersion increases with an increase in the index of refraction, at least according to Newton's assumed dispersion law $\Delta n'/n' > \Delta n/n$, where $n' > n$; see Prop. 2, §163, esp. note (28).

jus Diameter sit *AI,* quíque dictos radios secet in β, γ, δ; *B, D, G:* Quandoquidem anguli β*I*γ et γ*I*δ sint aequales, erunt etiam / arcus βγ, et γδ aequales. 125/
Sed ductis *A*β, *A*γ, &c: erunt *A*β, *A*γ, *A*δ sinus[8] incidentiarum, adeoque
inter se ut sunt *AB, AG, AD* sinus refractionum. Quare (per Lemma 6) est
arcus *GD* major arcu *BG.* Et inde

$$2\gamma G \sqsupset 2\gamma G + GD - GB = \gamma D + \gamma B = \gamma D - \gamma\delta + \gamma\beta + \gamma B = D\delta + B$$

Hoc est 2γ*G* ⊐ *D*δ + *B*β, sive ang β*IB* + ang δ*ID* 2ang γ*IG.*   Q.E.D.

[186 = I, 111]   6. Homogeneis radijs a Prismate
refractis, angulus quem incidentes
et emergentes comprehendunt tunc
maximus evadit cum aequalis est
hinc et inde refractio. Sit [fig
79] *ABC* Prisma, z0*y o*N* radius
utrinque ad *R* et *S* aequaliter refractus, et *IPQL* alius radius refractus
inaequaliter, magis quidem ad *P*
minùs ad *Q.* Et producantur hi
radij donec sibi occurrent *IP* et *QL*
in *T, GR* verò et *NS* in *V.* Dico
angulum *RVS* esse majorem angulo
PTQ. Quod ut pateat, concipe radios in lineis *PQ* et *RS* hinc inde

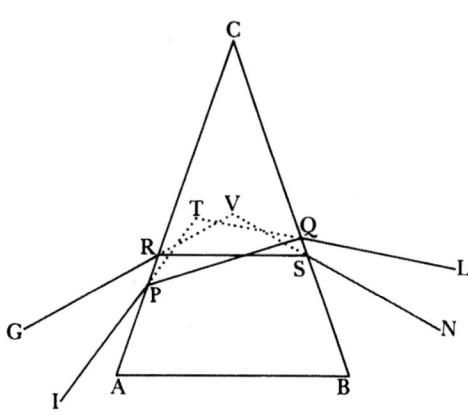

Figure 79

pergentes utrinque egredi Prismate, et sic e medio densiori in rarius refringi.
Jam in triangulis *CPQ* et *CRS,* cùm angulus *C* communis sit, caeterorum
angulorum summae erunt aequales. Et proinde cùm *CRS* sit Isosceles, duplum anguli *CSR* aequabitur angulis *CPQ* + *CQP.* Quamobrem radij *QP*
incidentia ad *P* tanto major est incidentiâ radij *RS* ad *S,* quanto eadem
incidentia sit major incidentia radij *PQ* ad *Q.* Trium itaque incidentiarum
differentiae sunt aequales, adeoque (juxta Lemma praemonstratum) summa
refractorum angulorum per incidentiam maximam et minimam effectorum
major erit duplo anguli refracti per incidentiam mediocrem effecti. Hoc est

ang *QPT* + ang *PQT* ⊏ 2ang *RSV,* sive ⊏ ang *RSV* + ang *SRV.*

Itaque cùm in triangulis *PTQ* et *RVS* summa angulorum ad Basin *PQ* sit
major summâ eorum ad basin *RS,* erit angulus verticalis *RVS* major angulo
verticali *PTQ.*[9]   Q.E.D.

[187 = I, 112]   Lemma 8. Si secundum tres lineas β*I,* γ*I,* δ*I* [fig 80] aequales angulos β*I*γ
et γ*I*δ continentes, tres radij minimè refrangibiles incidant ad *I* in superficiem
IK, et e Medio rariori in densius[10] refringantur, quorum refracti retrorsum
producti sint *IB, IG, ID*; et praeterea si trium maximè refrangibilium[11]
radiorum secundum easdem lineas β*I,* γ*I,* δ*I* incidentium refracti retrorsum
producti sint *Ib, Ig, Id*: Differentia / refractionis radiorum quorum incidentia 126/
est minima unà cum differentia refractionis eorum quorum incidentia est

---

(8) Originally continued: angulorum (of the angles).

has diameter *AI* and intersects those rays in β, γ, and δ, and *B*, *G*, and *D*. Since the angles $\beta I\gamma$ and $\gamma I\delta$ are equal, the arcs $\beta\gamma$ and $\gamma\delta$ will also be equal. But when *Aβ*, *Aγ*, and so forth are drawn, *Aβ*, *Aγ*, and *Aδ* will be the sines[8] of incidence, and so be to one another as *AB*, *AG*, and *AD*, the sines of refraction. Therefore, by Lemma 6 the arc *GD* is greater than the arc *BG*. Consequently,

$$2\widehat{\gamma G} < 2\widehat{\gamma G} + \widehat{GD} - \widehat{GB} = \widehat{\gamma D} + \widehat{\gamma B} = \widehat{\gamma D} - \widehat{\gamma \delta} + \widehat{\gamma \beta} + \widehat{\gamma B} = \widehat{D\delta} + \widehat{B\beta};$$

that is, $2\widehat{\gamma G} < \widehat{D\delta} + \widehat{B\beta}$, or $\widehat{\beta I B} + \widehat{\delta I D} > 2\widehat{\gamma I G}$. As was to be demonstrated.

Proposition 6. When homogeneous rays are refracted by a prism, the angle that the incident and emergent rays form will be a maximum at the moment when the refractions on each side are equal. [186 = I, 111]

Let *ABC* (Fig. 79) be a prism, *GRSN* a ray refracted equally on each side at *R* and *S*, and *IPQL* another ray refracted unequally, more in fact at *P* and less at *Q*. Then extend these rays until they meet each other, *IP* and *QL* at *T*, but *GR* and *NS* at *V*. I say that the angle *RVS* is greater than the angle *PTQ*. For this to be evident, imagine rays proceeding in each direction along the lines *PQ* and *RS* to leave the prism on each side and thus to be refracted from a denser into a rarer medium. Now, since in the triangles *CPQ* and *CRS* the angle *C* is common, the sums of the other angles will be equal. Thus, since *CRS* is isosceles, twice the angle *CSR* will equal the sum of the angles *CPQ* and *CQP*. Therefore, the angle of incidence of the ray *QP* at *P* is just as much greater than the angle of incidence of the ray *RS* at *S* as the same angle of incidence is greater than the angle of incidence of ray *PQ* at *Q*. Consequently, the differences of the three angles of incidence are equal, so that (by the lemma just demonstrated) the sum of the deviations made at the greatest and least angle of incidence will be greater than twice the deviation made at the intermediate angle of incidence, that is, $\widehat{QPT} + \widehat{PQT} > 2\widehat{RSV}$ or $> \widehat{RSV} + \widehat{SRV}$. Therefore, since in the triangles *PTQ* and *RVS* the sum of the angles at the base *PQ* is greater than their sum at the base *RS*, the vertex angle *RVS* will be greater than the vertex angle *PTQ*.[9] As was to be demonstrated.

Lemma 8. If along the three lines βI, γI, and δI (Fig. 80) containing the equal angles $\beta I\gamma$ and $\gamma I\delta$, three least refrangible rays are incident at *I* on the surface *IK* and are refracted from a rarer into a denser[10] medium, and their refracted rays extended backward are *IB*, *IG*, and *ID*; if, moreover, the refracted rays extended backward of three most refrangible[11] rays incident along the same lines, βI, γI, and δI, are *Ib*, *Ig*, and *Id*; the difference of refraction of the rays whose angle of incidence is the smallest together with the difference of refraction of those whose angle of incidence is the greatest [187 = I, 112]

---

(9) Since the angles *RVS* and *PTQ* are the supplements of what is now defined as the angle of deviation, Newton has at last demonstrated that the deviation is a minimum when a ray is refracted symmetrically. In the renaissance of geometric optics in the 1760s, partly stimulated by Newton's earlier investigations, several analytical proofs of the position of minimum deviation were presented; see Lect. 9, note (26).

(10) rariori in densius] Read: densiori in rarius (denser into a rarer).

(11) minimè refrangibiles ... maximè refrangibilium] Originally: rubiformes ... purpuriformium (red-making ... purple-making).

maxima, major erit quàm dupla differentia refractionis eorum quorum incidentia est mediocris. Hoc est, ang *BIb* + ang *DId* ⊏ 2ang *GIg*.
Etenim descripto quovis circulo *ADI* tangente
refringentem superficiem in *I*; cujus diameter sit
AI, quíque praefatos radios in punctis β, γ, δ; *B*,
b; *G*, *g*; *D*, *d* secet: concipe subtensas ab *A* ad
quodlibet istorum punctorum duci. Et erunt *Aβ*,
Aγ, *Aδ* inter se ut sunt *AB*, *AG*, *AD*, atque
etiam ut sunt *Ab*, *Ag*, *Ad*. Unde sequitur quod
AB, *AG*, *AD* inter se sunt ut *Ab*, *Ag*, *Ad*; et
praeterea (per Lemma 6, sect [155]) quòd sit arcus *GD* ⊏ arcu *BG*, et arcus *gd* ⊏ arcu *bg*. Jam
fiat arcus *GM* = *BG*, eritque *GD* ⊏ *GM*, et *AD*
⊏ *AM*. Item in peripheria *AD* sume punctum
quoddam *n* sub hac conditione, ut, si concipias
AM, *An* subtensas duci, sit *AB* . *Ab* :: *AM* . *An*.
Et erunt *AB*, *AG*, *AM* inter se ut sunt *Ab*, *Ag*,

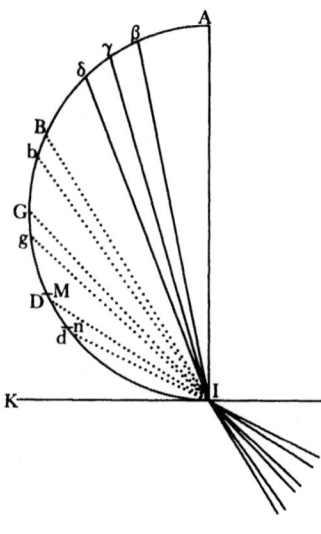

Figure 80

An. Adeoque cùm arcus *BG* ac *GM* sint aequales, erit summa arcuum *Bb* +
Mn (per Lemma [7]) major duplo arcu *Gg*. Sed cùm sit *AM* . *An*(:: *AB* . *Ab*)
:: *AD* . *Ad*,[12] vel conversè *AM* . *AD* :: *Mn* . *Dd*, propter *AD* ⊏ *AM* erit
arc *Dd* ⊏ arc *Mn*. Et utrobique addito arcu *Bb*, erit arc *Bb* + arc *Dd* ⊏
arc *Bb* + arc *Mn*. Et multò magis erit

arc *Bb* + arc *Dd* ⊏ duplo arcu *Gg*:

sive ang *BIb* + ang *DId* ⊏ 2ang *GIg*.   Q.E.O.

[188 = I, 113]   7. Heterogeneis radijs a Prismate refractis, differentia angulorum quos incidentes cum emergentibus constituunt, tunc minima evadit cùm aequales
sunt utrobique refractiones. In Prismate *ABC* [fig 81] sumatur *CR* = *CS*, et
RS ducatur, ut et alia quaevis linea *PQ* quae non sit parallela ad *RS*. Et
concipe radios in Prismate secundum has lineas *PQ* et *RS* hinc inde pergentes
ad puncta *P*, *Q*; *R*, et *S* egredi, et maximè refrangibiles versus *K*, *M*; *H*, et *O*
refringi, ac minimè refrangibiles[13] versus *I*, *L*; *G*, et *N*. Dico quòd refractio-

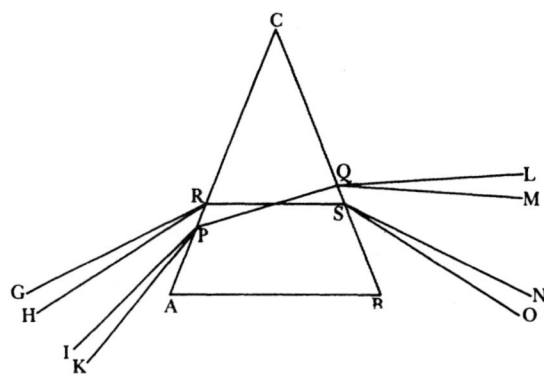

Figure 81

will be more than twice the difference of refraction of those at the mean angle of incidence, that is, $\widehat{B I b} + \widehat{D I d} > 2\widehat{G I g}$.

For tangent to the refracting surface at *I* describe any circle *ADI*, which has diameter *AI* and intersects those rays in the points β, γ, and δ; *B* and *b*; *G* and *g*; and *D* and *d*. Imagine the chords from *A* to each of those points to be drawn, and $A\beta$, $A\gamma$, and $A\delta$ will be to one another as are *AB*, *AG*, and *AD*, and also as *Ab*, *Ag*, and *Ad*. Whence it follows that *AB*, *AG*, and *AD* are to one another as are *Ab, Ag,* and *Ad,* and moreover (by Lemma 6, § 155) that $\widehat{GD} > \widehat{BG}$, and $\widehat{gd} > \widehat{bg}$. Now make $\widehat{GM} = \widehat{BG}$, and it will be $\widehat{GD} > \widehat{GM}$, and $\widehat{AD} > \widehat{AM}$. Also, in the circumference *AD* take some point *n* with this condition, namely, if you imagine the chords *AM* and *An* to be drawn, then *AB : Ab* = *AM : An*. Then *AB, AG,* and *AM* will be to one another as are *Ab, Ag,* and *An*. Hence since the the arcs *BG* and *GM* are equal, the sum of the arcs *Bb* and *Mn* (by Lemma 7) will be more than twice the arc *Gg*. But since

$$AM : An \ (= AB : Ab) = AD : Ad,^{(12)}$$

or by conversion $AM : AD = \widehat{Mn} : \widehat{Dd}$, and because $\widehat{AD} > \widehat{AM}$, it will be $\widehat{Dd} > \widehat{Mn}$. Adding the arc *Bb* to both sides, $\widehat{Bb} + \widehat{Dd} > \widehat{Bb} + \widehat{Mn}$, and even more so $\widehat{Bb} + \widehat{Dd} > 2\widehat{Gg}$, or $\widehat{BIb} + \widehat{DId} > 2\widehat{GIg}$. As was to be shown.

Proposition 7. When heterogeneous rays are refracted by a prism, the [188 = I, 113] difference of the angles that the incident rays make with the emergent ones proves to be a minimum at the moment when the refractions on both sides are equal.

In the prism *ABC* (Fig. 81) take *CR* equal to *CS*, and draw *RS* as well as any other line, *PQ*, that is not parallel to *RS*. Then imagine rays to proceed in the prism in each direction along these lines *PQ* and *RS* and to leave at the points *P* and *Q*, and *R* and *S*, with the most refrangible ones being refracted toward *K* and *M*, and *H* and *O*, and the least refrangible[13] ones toward *I* and *L*, and *G* and *N*. I say that the sum of the differences of refraction made

---

(12) Originally continued: et *AM* minor caeteris *An, AD,* vel *Ad*; erit arcus (and *AM* is smaller than the others, *An, AD,* or *Ad*; there will be arc . . .).

(13) maximè refrangibiles . . . minimè refrangibiles] Originally: purpuriformes . . . rubiformes (purple-making . . . red-making).

num inaequaliter ad *P* et *Q* / factarum differentiae simul sumptae *IPK* + 127/
LQM sint majores quàm *GRH* + *NSO* differentiae refractionum aequaliter
ad *R* et *S* factarum simul sumptae.[14] Nam incidentiarum ad *P Q* et *S* differentiae sunt aequales, ut ostensum erat in Propositione praecedenti. Atque
adeò (per Lemma 8) differentia refractionis radiorum difformium ad *P* ubi
maxima est incidentia, unà cum differentia consimili ad *Q* ubi minima est
incidentia, excedit[15] duplum consimilis differentiae ad *S* ubi incidentia est
mediocris. Hoc est ang *IPK* + ang *LQM* ⊏ 2ang *NSO*: Sive cùm *NSO* ac
GRH aequentur, ang *IPK* + ang *LQM* ⊏ ang *NSO* + ang *GRH*. Q.E.D.

[189 = I, 114]   Schol. Posui quidem radios e Prismate utrobique egredi; sin pergant ab *I* et
K per *P* et *Q* versus *L* et *M*, et a *G* et *H* per *R* et *S* versus *N* et *O*, linearum
positiones et quantitates angulorum non inde mutabuntur. Et proinde demonstratio praefata tunc etiam valebit. Et propter eandem rationem valebit
etiam cùm radij ad Prisma divergentes evadunt in Prismate paralleli. Quod
idem de Propositionum 5 et 6 demonstrationibus itidem intellige. Quinetiam
in alijs quibuscunque casibus ubi divergunt ante refractionem et post convergunt, vel in Prisma incidunt paralleli; non adeò multùm a parallelismo intra
Prisma recedunt unquam, quin ut anguli vel differentiae angulorum quos
incidentes cum emergentibus constituunt, pro ijsdem circiter haberi possint ac
si intus essent paralleli; adeóque dictas propositiones ad omnes omninò casus
extendi.[16]

[190 = I, 115]   8. Si denique, radijs a dato puncto *F* [fig 82] ad datum punctum *X*, per

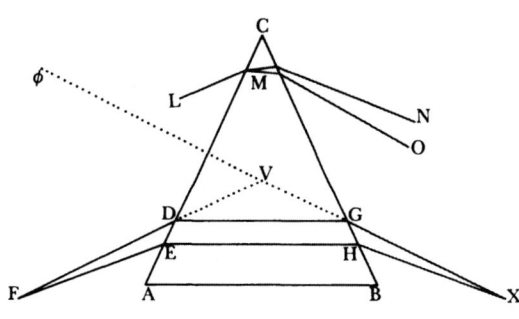

Figure 82

Prisma *ABC* positione datum refractis, desiderentur anguli *DFE*, *GXH* quos
heterogenei comprehendunt: Problema ex eorum numero est quae veteres
linearia dixêre. At sequens mechanica solutio, quantùm exigunt res practicae,
veritati appropinquat. Finge summam angulorum *DFE* + *GXH* aequalem
esse angulo *NMO* quem radij duo alteris *FD* et *FE* quoad refrangibilitatem
consimiles, ac juxta quamvis lineam *LM*, rectae angulum *DFE* bisecanti

---

(14) That is, Newton claims the minimum of dispersion occurs at the position of minimum
deviation. At least by the time he wrote the *Opticks*, he had experimentally discovered his error.
In a late addition in the manuscript of the *Opticks*, he for the first time clearly distinguished
between the spectrum's motion, or the deviation, and its length, or the dispersion. He observed
that near the position of minimum deviation, where the image is stationary, if the prism was

unequally at P and Q, $\hat{IPK} + \hat{LQM}$, is greater than $\hat{GRH} + \hat{NSO}$, the sum of the differences of refraction made equally at R and S.[14] For the differences of the angles of incidence at P, Q, and S are equal, as was shown in the preceding proposition, so that (by Lemma 8) the difference of refraction of the dissimilar rays at P, where the angle of incidence is greatest, together with the similar difference at Q, where the angle of incidence is least, exceeds[15] twice the identical difference at S, where the angle of incidence is intermediate; that is, $\hat{IPK} + \hat{LQM} > 2\hat{NSO}$, or since $\hat{NSO}$ and $\hat{GRH}$ are equal, $\hat{IPK} + \hat{LQM} > \hat{NSO} + \hat{GRH}$. As was to be demonstrated.

Scholium. I have, to be sure, assumed that the rays leave the prism on each [189 = I, 114] side, but if they proceed from I and K through P and Q toward L and M, and from G and H through R and S toward N and O, the positions of the lines and magnitudes of the angles will not on that account be changed. Consequently, that demonstration will then still be valid; and for the same reason, it will also be valid when the rays diverge at the prism and become parallel within it. The same thing must likewise be understood for the demonstrations of Props. 5 and 6. Moreover, in any other cases whatever—where they diverge before refraction and converge afterward, or are incident upon the prism parallel—they will never depart so much from parallelism within the prism but that the angles or the differences of the angles that the incident and emergent rays make can be considered nearly the same as if they were parallel within it, so that these propositions can be totally extended to all cases.[16]

Proposition 8. Finally, if rays are refracted from a given point F (Fig. 82) to [190 = I, 115] a given point X through a prism ABC in a given position, the angles DFE and GXH made by heterogeneous rays are required.

The problem belongs to that class that the ancients called linear, but the following mechanical solution approaches the truth as much as practical matters require. Imagine that the sum of the angles DFE and GXH equals the angle NMO made after two refractions by two rays that are identical in refrangibility to the other rays, FD and FE, and are incident along any line,

---

rotated one way about its axis "the image soon became an inch or two longer," whereas if it was rotated the opposite way "the Image soon became an inch or two shorter" (Bk. I, Pt. I, Prop. II, Expt. III, Add. 3970, f. 31ᵛ = *Opticks*, p. 19). In 1827, John Herschel, who was the first to determine the minimum of dispersion after Newton's unsuccessful and long-forgotten attempt, noted that "the position which gives a minimum of breadth to the spectrum is very different from that which gives a minimum of deviation" ("Light," *The Encyclopaedia of Mechanical Philosophy* (London, 1848), pp. 341–586, esp. §449, p. 419). In fact, with the large vertex angles of Newton's prisms the dispersion continually increases and there is no minimum; see R. A. Herman, *A Treatise on Geometrical Optics* (Cambridge, 1900), pp. 174–6.

(15) Originally: major est quam (is greater than).

(16) Newton assumes that the rays are parallel within the prism and that the total angular dispersion is the sum of the dispersion at each face. Contrary to his assumption, this sum is not equal to the angular dispersion of parallel incident rays refracted successively at each surface, except at minimum deviation. Newton, in fact, had used this assumption in §110 ≈ I, 37, to calculate the chromatic dispersion measured at minimum deviation, and its validity in that symmetric situation undoubtedly led him to generalize it.

quàm proximè parallelam incidentes,[17] post binam refractionem constituunt. / Et e radijs ad X refractis aliquem GX cum incidente radio FD convenientem in V, produc ad φ, ut sit φ locus imaginis quam objectum F oculo in X constituto exhibet. Dein angulo NMO, ac distantijs φX et φV mechanicè cognitis, dic esse φX . φV :: ang NMO . ang GXH. Et erit GXH quem quaeris proximè. Quemadmodum ex ostensis ad Schol Prop [11, §143] quodammodo manifestum est.[18]   128/

Cùm refractiones utrobique non sint admodum inaequales res expeditiùs absolvitur per Schol ad Prop [1, §162] fingendo esse

$$VX . FV :: \text{ang } DFE . \text{ang } GXH$$

vel compositè FV + VX . FV :: ang NMO . ang GXH.[19]

---

(17) quem ... incidentes] Originally: quem radij duo consimiles, juxta quamvis lineam LM, utrisque radijs FD et FE quàm proximè parallelam incidentes (by two rays that are identical and incident along any line LM very nearly parallel to both rays FD and FE).

(18) Et ... est.] Added.

(19) Namely, by the Scholium to Prop. 11, D̂FE + ĜXH is approximately equal to the total dispersion N̂MO in the prism, and by the Scholium to Prop. 1 the perceived angles are inversely proportional to the distances. Thus at minimum deviation, where VX ≈ FV, the angle DFE at the eye is one half of the total dispersion, or ½N̂MO. In a study of the measure of chromatic dispersion, directed particularly against an experimental investigation by Nicolas Béguelin, who was unaware of the factor of one half in measuring the dispersion, Giambattista Venturi demonstrated Newton's approximate result. To prove it he first showed that near minimum deviation,

LM, very nearly parallel to the line bisecting the angle *DFE*.[17] Next, of the rays refracted to *X* extend any one, *GX*, to ϕ meeting the incident ray *FD* in *V*, so that ϕ is the place of the image that the object *F* exhibits to an eye situated at *X*. Then when the angle *NMO* and the distances ϕX and ϕV are known mechanically, fix it so that $\phi X : \phi V = N\widehat{M}O : G\widehat{X}H$, and the angle *GXH* will be very nearly what you sought, as is clear in a way from what was shown in the Scholium to Prop. 11, §143.[18]

When the refractions on each side are not very unequal, the matter is resolved more readily by the Scholium to Prop. 1, §162, by imagining that $VX : FV = D\widehat{F}E : G\widehat{X}H$, or by composition

$$FV + VX : FV = N\widehat{M}O : G\widehat{X}H.^{[19]}$$

---

after two refractions the angle of divergence at ϕ of an infinitely narrow bundle of monochromatic rays is equal to their initial divergence from the source *F*; "Considerazioni ottiche," *Memorie di matematica e di fisica della società italiana*, 3 (1786):268–77, esp. 269–71. Although Venturi does not here cite the *Optical Lectures*, he does do so further on in his paper (see Lect. II, 14, note (14), this volume), and in a later paper ("Indagine fisica sui colori," *Memorie di matematica e di fisica della società italiana*, 8, ii (1799):699–754, esp. 703) he refers to Props. 20 and 22 in the *Optica*, Pt. I, which are equivalent to Props. 1 and 3 in this lecture. Newton applies this rule with a quantitative example in the *Optica*, §II, 106.

Opticae[(1)] pars 1ᵃ,
De radiorum Lucis Refractionibus.

SECTIO I

RADIORUM DIVERSAM ESSE REFRANGIBILITATEM.

Jan 1669
Lect 1
[1 = 1. Incepti
ratio.]

Inventio Telescopiorum nupera plerosque Geometras ita exercuit, ut nihil in opticâ non tritum, nullum inventioni praeterea locum alijs reliquisse videantur. Et insuper cum dissertationes quas hìc non ita pridem audivistis, tantâ rerum opticarum varietate novorum copiâ, et accuratissimis eorundem demostrationibus fuerint compositae: frustranei forte videantur conatus et labor inutilis, si ego scientiam hanc iterum tractandam suscepero. Verùm cum Geometras in quâdam lucis proprietate, quae ad refractiones spectat hucusque hallucinatos videam, demonstrationes suas in Hypothesi quadam Physica haud benè stabilita tacitè fundantes:[(2)] non ingratum me facturum judico, si principia scientiae hujus examini severiori subjiciam, et quae ego de ijs simul excogitavi, et experientiâ multiplici habeo comperta subnectam ijs, quae Reverendus meus Antecessor hic loci postrema dixit.

Imaginantur Dioptrices studiosi, quod Perspicilla ad quemlibet perfectionis gradum perduci possent, modò vitris dum perpoliuntur geometricam, quam vellent, figuram communicare concederetur. et in eum finem instrumenta varia fuerunt excogitata, quibus vitra in figuras Hyperbolicas, vel etiam Parabolicas contererentur; sed exacta figurarum istarum fabricatio nemini hucusque successit. Scilicet aratur littus, et ne labores suos in negotio desperato diutiùs insumant ijs audeo spondere, quòd licèt omnia fierent feliciter,[(3)] nihil minùs tamen quàm votis suis responderent: etenim vitra licet efformentur secundum figuras in istum finem optimas, quae possunt excogitari, tamen non duplo plus praestabunt quàm sphaerica, aequali politurâ perfecta. Haec autem non ideò loquor, quasi peccatum esse a scriptoribus Optices contenderem; illi enim omnia pro intentione demonstrationum suarum accuratè quidem et verissimè dixerunt, sed aliquid tamen idque maximi momenti reliquerunt posteris inveniendum. Scilicet in refractionibus irregularitatem quan-

---

(1) ULC, Dd.9.67, pp. 1–77, and ₂1–101; the figures are gathered together at the end of each part (pp. 79–84 and ₂103–7), and the remainder of this bound volume is blank. To distinguish the figures of Parts I and II of the *Optica* from one another and from those of the *Lectiones opticae*, we have added, as appropriate, a roman numeral I or II to all figure numbers. Newton's college roommate, John Wickins, transcribed the manuscript, but Newton himself corrected it and completed the last few pages. Article headings in square brackets have been inserted from the corresponding articles in the *Lectiones*. The history of the manuscript and its subsequent publication is discussed in the Introduction, §§3 and 4.

Optics,[1] Part 1
The Refractions of Light Rays

SECTION I
THE REFRANGIBILITY OF RAYS DIFFERS

The recent invention of telescopes has so occupied most geometers that they seem to have left to others nothing in optics untouched nor any room for further discovery. Moreover, since the lectures that you heard here not so long ago brought together such a great variety of optical topics and a vast quantity of discoveries with their very accurate demonstrations, it might perhaps seem a vain endeavor and futile effort for me to undertake to treat this science again. But since I observe that geometers have hitherto erred with respect to a certain property of light pertaining to its refractions by implicitly founding their demonstrations on a certain not well established physical hypothesis,[2] I judge it will not be unappreciated if I subject the principles of this science to a rather strict examination, adding what I have conceived concerning them and confirmed by numerous experiments to what my reverend predecessor last delivered in this place. Jan. 1669
Lecture 1
[1 = 1. The reason for this undertaking.]

Those knowledgeable in dioptrics imagine that telescopes may be brought to any degree of perfection, provided that it were possible to impart any desired geometrical figure to the lenses while they are being finely polished. For this purpose various instruments have been devised for grinding lenses into hyperbolic or even parabolic figures, but no one has yet succeeded in exactly forming those shapes. It is, to be sure, a futile endeavor, and lest they devote their efforts any longer to a hopeless occupation, I venture to promise them that even if everything were to turn out successfully,[3] it still would fall far short of their expectations. For even if lenses were fashioned according to the best shapes that can be conceived for that purpose, they will nonetheless perform no more than twice as well as spherical ones polished equally perfectly. I am not, however, pointing this out as if I maintained that those who write on optics are in error, for everything they have asserted with regard to the intent of their demonstrations is accurate and very true. However, they have left something—and of the greatest importance—to be discovered by their successors; namely, I find in refractions a certain irregularity that upsets

---

(2) demonstrationes . . . fundantes] Originally (as I): dum demonstrationibus suis Hypothesin quandam Physicam haud benè stabilitam tacitè supponunt (while they implicitly assume in their demonstrations a certain not well established physical hypothesis).

(3) I: perquam felicitèr (thoroughly successfully).

dam reperio, quae omnia perturbat, et non solum efficit, ut figurae Conicarum Sectionum[4] Sphaericas non multum superent, sed etiam ut sphaericae
multo minus praestent, quam praestarent, si dicta refractio esset uniformis.

/ Itaque in Dioptricâ pedem figo, non ut eam pertractarem de integro, sed   2/
tantùm, ut hanc de natura lucis proprietatem rimarer primò, deindè ut ostenderem quantum ex hâc proprietate perfectio Dioptrices impeditur, et quo
pacto incommodum istud, quatenus natura rei sinit, devitetur. Ubi nonnulla
proferam quae ad Telescopiorum, juxta et Microscopiorum, tum Theoriam
tum praxin spectant; ostendens quod Optices summa perfectio (praeter opinionem receptam) ex Dioptricâ et Catoptricâ mixtis petenda est. Ac interea
discrimen colorum et eorum genesin a prismatibus et corporibus etiam coloratis fusè explicabo.

2 [≈ 2]. Quòd   De luce itaque compertum habeo, quòd radij ejus quoad quantitatem re-
omnium radiorum fractionis ab invicem differant: Ex ijs qui omnes habent eundem angulum
non sit eadem   incidentiae, alij angulum refractionis aliquanto majorem alij minorem[5] habe-
refrangibilitas. bunt. Plenioris illustrationis gratiâ sit *EFG* superficies quaelibet refringens,
Fig. I, 1

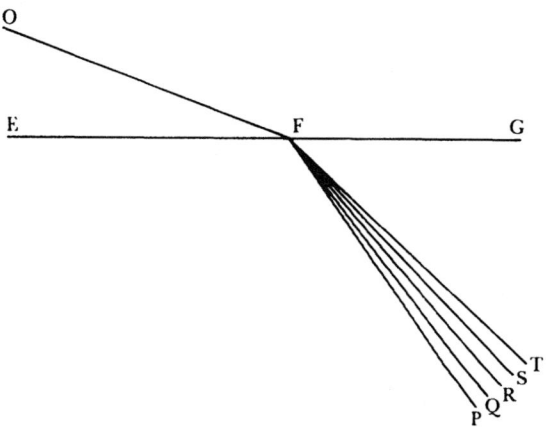

Figure I, 1

puta vitrea; et ducatur quaevis *OF* huic occurrens in *F,* et cum ea efficiens
angulum *OFE* acutum: concipe etiam radios solares per istam lineam *OF* sibi
continuo successivos fluere, ita ut alij post alios in punctum *F* impingant,
ibidemque in medium densius refringantur. ‖ Vel si mavis, finge parallelos
‖ radios indefinitè parùm distare ab *OF,* et incidere in puncta ipsi *F*
‖ vicinissima.[6] ‖ Jam ex opinione recepta hi radij eandem habentes incidentiam, eandem quoque omnes refractionem habere debent puta in lineam *FR.*
At contrarium compertum habeo; scilicet quod postquam refringuntur, divergant ab invicem; quasi quidam refringerentur in lineam *FP,* alij in lineam *FQ,*
et alij in lineas *FR, FS,* & *FT,* ac alij etiam innumeri per spatia intermedia, ut
et ultra citraque nonnulli pervagantes; prout radius quilibet ad refractionem
majorem minoremve patiendam sit aptus. Invenio praeterea quod radij *FP*
maximè refracti colores purpureos producant, et illi *FT* minimè refracti ru-

everything, which not only causes the figures of conic sections[4] to be little superior to spherical ones but also makes spherical ones perform far less well than they would if this refraction were uniform.

I therefore have set foot on dioptrics not to treat it systematically anew, but only, in the first place, to examine thoroughly this property in the nature of light, and then to show how much the perfection of dioptrics is impeded by this property and how that obstacle, insofar as its nature allows, may be avoided. Here I will set forth several things concerning both the theory and practice of telescopes as well as microscopes, showing that the ultimate perfection of optics (contrary to the received opinion) must be sought in dioptrics and catoptrics combined. In the meantime I will fully explain the difference of colors and their generation by prisms as well as colored bodies.

Concerning light, therefore, I have discovered that its rays differ from one another with respect to the quantity of refraction: Of those rays that all have the same angle of incidence, some will have an angle of refraction somewhat larger, and others a smaller one.[5] For the sake of a fuller illustration, let *EFG* be any refracting surface, for example, glass, and draw any line *OF* meeting it at *F* and making with it an acute angle *OFE*; also imagine that along this line *OF* solar rays flow in continual succession to one another, so that they strike the point *F* one after another and are there refracted into the denser medium; ‖ or, if you prefer, imagine parallel rays to be an indefinitely ‖ small distance from *OF* and to fall on the points in the immediate vicinity of *F*.[6] ‖ Now, according to the received view, since these rays have the same incidence, they should all also have the same refraction, suppose, in the line *FR*. But I have discovered the contrary; that is, after they are refracted, they diverge from one another, just as if some were refracted in the line *FP*, some in the line *FQ*, others in the lines *FR*, *FS*, and *FT*, and still innumerable others through the spaces between, while some extend on both sides as well, insofar as any ray is disposed to undergo a greater or smaller refraction. Moreover, I find that the most refracted rays *FP* produce purple colors and those least refracted *FT* produce red, but those that proceed along the inter-

2 [≈ 2]. The refrangibility of all rays is not the same.

Fig. I, 1

---

(4) Added.
(5) alij minorem] I: alijs (than others).
(6) Newton first introduced this broader concept of light ray in §20 = I, 21.

bros, qui autem hisce intermedij pergunt, *FQ, FR, FS,* ij colores intermedios nempe caeruleos, virides, et flavos generant. Et sic radij prout apti sunt, ut alij alijs magis atque magis refringantur, hos ordine colores, rubrum, flavum, viridem, caeruleum, et purpureum generant una cum omnibus intermedijs quos in Iride liceat conspicere. Unde productio colorum Prismatis et Iridis facilè patebit. Sed his jam perfunctoriè notatis, quae de coloribus dicenda sunt, in posterum differam.

3 [= 3]. Probatur experimento vulgari per longitudinem imaginis solaris refractae.[7]

Fig. I, 2

/ Sententia nostra de hac re sic breviter[8] explicata, ne putetis fabulas pro veris enarratas esse, rationes et experimenta quibus isthaec innituntur, continuò proferam, et quoniam experimentum quoddam Prismatis valde obvium mihi primò dedit occasionem excogitandi reliqua istùd primum explicabo. Sit *F* foramen aliquod in pariete, vel fenestra cubiculi, per quod radij

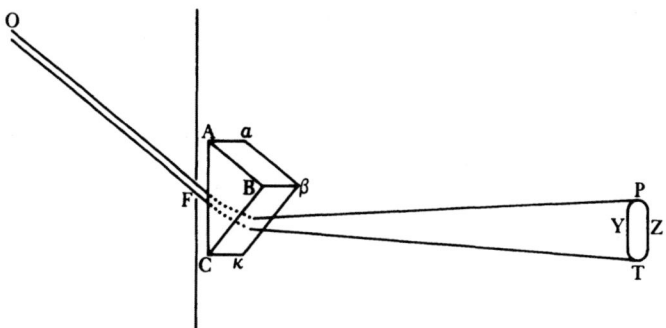

Figure I, 2

solares *OF* trajiciantur, reliquis ubique foraminibus diligenter obturatis, ne lux alibi ingrediatur. Ista autem obscuratio cubiculi non est necessaria,[9] sed efficit tantùm, ut experimentum evadat aliquanto evidentius. Deinde Prisma triangulare vitreum *AαBβCκ* ad foramen istud applicetur, quod radios *OF* per se trajectos refringat versus *PYTZ:* quos radios opposito pariete, vel papyro aliquâ ad distantiam a Prismate satis magnam objectâ terminatos, videbis in figuram *PYTZ* valde oblongam efformari: cujus nempe longitudo *PT* sit quadruplex latitudinis *YZ* et amplius. Et hinc evinci certò videtur, quod radiorum aequaliter incidentium alij refractionem majorem alijs patiuntur. Nam si contrarium esset verum, praedicta solis imago appareret ferè orbicularis et in quadam positione Prismatis omnino ad sensum orbicularis[10] conspiceretur: Id quod contra omnem experientiam est. Quocunque enim situ Prisma disposui, nunquam tamen potui efficere, quin longitudo imaginis esset latitudinis plusquam quadrupla. Angulo scilicet Prismatis *ACB* vel *ακβ* existente graduum plus minus sexaginta.

4 [= 4]. Casus in quo radij aeque refrangibiles faciunt imaginem orbicularem.

Fig. I, 3

Quod autem datur quaedam Prismatis positio in qua imago solis ex opinione de refractionibus receptâ, appareret orbicularis, sic ostendo. Juxta foramen in fenestra cubiculi factum, Prisma collocetur foras, vel quod eodem recidit, sit *EG* corpus aliquod opacum citra Prisma locatum in quo sit *F* foramen indefinite parvum et orbiculare, per quod radij refracti in parietem

mediate lines *FQ, FR,* and *FS* generate the intermediate colors, namely, blue, green, and yellow. Hence, insofar as the rays are so disposed that some are refracted more and more than others, they generate in order these colors, red, yellow, green, blue, and purple, together with all the intermediate ones that can be seen in the rainbow. The production of the colors of the prism and rainbow will be readily evident from this, but having now perfunctorily noted these things, I will postpone what I have to say about colors for later.

Having thus briefly[8] explained my view on this subject, I will at once present the reasoning and experiments that support these things, lest you think that I have set forth fables instead of the truth. Since a certain very commonly encountered experiment with a prism first presented me the opportunity to think out the rest, I will explain that first. In the wall or window of a room let *F* be some hole through which solar rays *OF* are transmitted, while other holes elsewhere have been carefully sealed off so that no light enters from any other place. This darkening of the room, however, is not necessary;[9] it only enables the experiment to turn out somewhat more clearly. Then place at that hole a triangular glass prism *AαBβCκ* that refracts the rays *OF* transmitted through it toward *PYTZ*. You will see these rays, terminated by the opposite wall or by some paper placed sufficiently far from the prism, formed into a very oblong figure *PYTZ*, specifically, one whose length *PT* is four times and more its breadth *YZ*. Hence, this definitely appears to establish that at equal incidence some rays undergo a greater refraction than others; for if the contrary were true, that solar image would seem almost circular, and in a certain position of the prism it would appear to the senses completely circular,[10] which is contrary to all experience. Indeed in whatever position I placed the prism, I nonetheless could never make it happen that the image's length was not more than four times its breadth, that is, with the angle of the prism *ACB* or *ακβ* being about 60°.

That there exists, however, a certain position of the prism in which, according to the received view of refractions, the sun's image should appear circular, I show in this way. Place the prism outside, next to the hole in the window of the room, or, what amounts to the same thing, let *EG* be some opaque body placed on this side of the prism; and let *F* be an indefinitely small circular hole in it through which the refracted rays are transmitted onto

3 [= 3]. A common experiment proves this by the length of the refracted solar[7] image.

Fig. I, 2

4 [= 4]. The case in which equally refrangible rays make a circular image.

Fig. I, 3

---

(7) solaris refractae] I: coloratae (colored).

(8) breviter] I: in genere (generally).

(9) Originally (as I): necessaria omninò (absolutely necessary). In fact, in the corresponding Fig. 2 in the *Lectiones opticae* Newton drew the room in which the experiment is performed and the spectrum projected onto its wall.

(10) See Lect. 1, note (17).

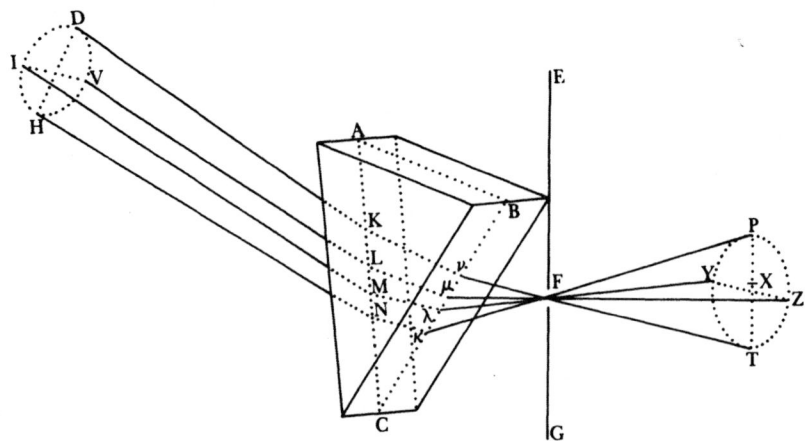

Figure I, 3

directè oppositum ad Imaginem *PYTZ* ibi depingendam trajiciantur. Et ponatur *ABC* esse planum secans *AC* et *BC* plana refringentia perpendiculariter
atque etiam transiens per foramen *F* ut et per centrum Solis *DIHV,* quem
bisecet secundum diametrum ejus *DH,* a cujus extremitatibus duo radij *DK* et
HN in eodem plano jacentes adveniant qui postquam refringuntur *DK* in *Kν*
& *νT,* atque *HN* in *Nκ* et *κP* utrique pergant per centrum foraminis *F.* Et
praeterea sit talis / inclinatio Prismatis ad istos radios, ut anguli *AKD* & *BκF* 4/
fiant aequales. Deinde sit *IV* alia solis diameter praedicto plano *ABC* perpendicularis a cujus extremitatibus alij duo radij *VL* et *IM* adveniant, alter *IM*
cis planum *ABC,* qui refringatur in *Mλ* & *λY,* alter vero *VL* ultra planum
istud qui refringatur in *Lμ* et *μZ.* Et praedicti quatuor radij sese omnes
decussent in medio foraminis *F.* Denique ponatur quod imago lucida *PYTZ*
foramen *F* directè respiciat, ita scilicet ut *FP* et *FT* item *FY* et *FZ* aequales
fiant. Dico jam quod in ista positione Prismatis, anguli *PFT* ac *YFZ* aequales
essent, supposito radios omnes aeque refringi qui eundem habent angulum
incidentiae: Et proinde quòd imago ista sensui saltèm, deberet esse orbicularis, utpote cujus diametri *PT* et *YZ* sese decussant perpendiculariter, et aequales istos angulos subtendunt.

5 [= 5].
Demonstratio istius
Casûs. ejus pars 1ᵐᵃ.   Angulos autem istos *PFT* et *YFZ* aequales esse sic demonstro[.] concipe
radium aliquem a *P* per *κ* et *N* retrocedere, dum alius radius pergit a *D* per *K*
et *ν*: Itaque cùm anguli *AKD* et *BκF* supponantur aequales, erunt etiam
anguli per primas refractiones facti *AKν* et *BκN* aequales. Unde triangula
CKν et *CκN* erunt similia et eorum anguli externi *κNA, KνB* aequales et
proinde anguli per secundas refractiones facti *ANH* et *BνF* etiam aequales.
Quare cùm anguli *AKD* et *BκF,* item *ANH* et *BνF* sint aequales eorum
differentiae erunt etiam aequales, hoc est angulus *νFκ* sive *PFT* aequalis
angulo quem radij *DK* et *HN* comprehendunt, sive diametro solari. Est
itaque angulus *PFT* aequalis diametro solari[:][(11)] Quare cum praeterea

the directly opposite wall depicting the image *PYTZ* there. Assume that *ABC* is a plane perpendicularly cutting the refracting planes *AC* and *BC* while also passing through both the hole *F* and the center of the sun *DIHV*, which it bisects along its diameter *DH*; and from its ends let there proceed two rays *DK* and *HN* lying in the same plane, which, after they are refracted (*DK* into *Kν* and *νT*, and *HN* into *Nκ* and *κP*), both pass through the center of the hole *F*. Moreover, let the prism's inclination to those rays be such that the angles *AKD* and *BκF* become equal. Next, let *IV* be another diameter of the sun, perpendicular to that plane *ABC*; and from its ends let there proceed two other rays *VL* and *IM*; letting one, *IM*, on this side of the plane *ABC*, be refracted into *Mλ* and *λY*, while letting the other, *VL*, on the other side of that plane, be refracted into *Lμ* and *μZ*; and let the specified four rays all cross one another in the middle of the hole *F*. Finally, let it be assumed that the bright image *PYTZ* directly faces the hole *F*, specifically, so that *FP* and *FT* as well as *FY* and *FZ* become equal. Now, I assert that in that position of the prism the angles *PFT* and *YFZ* are equal—supposing that all rays that have the same angles of incidence are equally refracted—and consequently that that image must be circular, at least to the senses, inasmuch as its diameters *PT* and *YZ* cross each other perpendicularly and subtend those equal angles.

That these angles *PFT* and *YFZ* are equal, however, I demonstrate as follows: Imagine some ray to proceed backward from *P* through *κ* to *N*, while another ray proceeds from *D* through *K* and *ν*. Therefore, since the angles *AKD* and *BκF* are assumed to be equal, the angles *AKν* and *BκN* made by the first refractions will also be equal. Hence the triangles *CKν* and *CκN* will be similar, and their external angles *κNA* and *KνB* will be equal; thus the angles *ANH* and *BνF* made by the second refractions will also be equal. Accordingly, since the angles *AKD* and *BκF* as well as *ANH* and *BνF* are equal, their differences will also be equal; that is, the angle *νFκ* or *PFT* will be equal to the angle contained by the rays *DK* and *HN* or to the solar diameter. The angle *PFT* is therefore equal to the solar diameter.[11] More-

5 [= 5]. The demonstration of that case. Part I.

---

(11) Est . . . solari] Deleted in *editio princeps*.

demonstratum[12] fuerit quod angulus *YFZ* aequatur eidem diametro, liquebit propositum. Istud autem ut fiat Theorema quoddam more Lemmatis praesternendum est.[13]

6 [= 6]. Lemma ad
secundam partem.

Fig. I, 4

Sunto duo plana *ABCD* et *EFGH* sibimet perpendicularia quorum communis intersectio sit *KL,* et sit *IP* radius quilibet qui in planum *ABCD* incidens ad punctum *P,* ab eo refringitur in *PR.* Dico quod sinus anguli quem radius

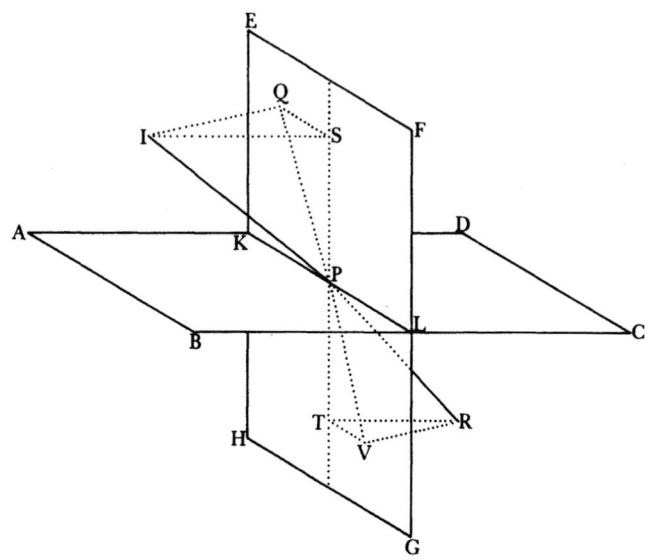

Figure I, 4

incidens *IP* efficit cum plano perpendiculari *FH,* sit ad sinum anguli quem radius refractus *PR* efficit cum eodem plano, sicut sinus incidentiae ad sinum refractionis; et proinde in ratione datâ. Sumptis enim radijs *IP* et *PR* aequalibus et demissis *IQ* et *RV* ad planum *FH* perpendicularibus, et praeterea ad punctum incidentiae *P* erectâ *SPT* perpendiculari ad planum refringens *BD,* (quae ideo cum altero plano *FH* coincidet,) et ad istam demissis *IS,* et *RT* iterum perpendicularibus: Erit *IPQ* angulus quem radius incidens *IP* efficit cum plano perpendiculari *FH,* et *RPV* angulus quem radius refractus *PR* efficit cum eodem plano: / Item *IPS* angulus incidentiae et *RPT* angulus 5/ refractionis. Quare si *IP* vel *PR* supponatur radius circuli, erunt *IQ, RV, IS,* et *RT* dictorum angulorum sinus. Sed *IQ* et *RV* sunt paralleli[a] propterea quod eidem plano *FH* sunt perpendiculares. Item *IS* et *RT* sunt paralleli[b], quia jacentes in eodem plano *ISPTR* eidem rectae *ST* perpendiculariter insistunt. Hoc est rectae *IQ, IS* quae angulum *QIS* comprehendunt sunt parallelae rectis *RV, RT* comprehendentibus angulum *VRT.* Quare isti anguli *QIS* et *VRT* sunt aequales[c]. Ductis autem *QS* et *VT* fient anguli *IQS* et *RVT* recti[d], quia rectae *IQ* et *RV* plano *FH* perpendiculariter insistunt. Ergo triangula *IQS* et *RVT* sunt similia[e]: et *IQ . RV :: IS . RT:* Hoc est sinus angulorum quos radius incidens et refractus efficiunt cum plano aliquo *FH* ad refringens planum *BD* perpendiculari, sunt ut sinus incidentiae et refractionis et proinde

(a) 6. 10[14] Elem.
(b) 28. 1 Elem.
(c) 10. 11 Elem.
(d) Def. 3. 11 Elem.
(e) 4. 6 Elem.

over, since it will be demonstrated[12] that the angle *YFZ* is equal to the same diameter, the proposition will be evident. However, in order to do that a certain theorem, in the form of a lemma, must be premised.[13]

Let there be two mutually perpendicular planes *ABCD* and *EFGH* the common intersection of which is *KL,* and let *IP* be any ray that falls on the plane *ABCD* at the point *P* and is refracted by it into *PR.* I say that the sine of the angle made by the incident ray *IP* with the perpendicular plane *FH* is to the sine of the angle made by the refracted ray *PR* with the same plane as the sine of incidence is to the sine of refraction, and thus is in a given ratio.

6 [= 6]. A Lemma for Part II.

Fig. I, 4

Set the rays *IP* and *PR* equal; drop *IQ* and *RV* perpendicular to the plane *FH;* moreover, at the point of incidence *P* and perpendicular to the refracting plane *BD* erect the line *SPT* that will thus lie in the other plane *FH;* and also drop *IS* and *RT* perpendicular to that line: *IPQ* will be the angle that the incident ray *IP* makes with the perpendicular plane *FH,* and *RPV* will be the angle that the refracted ray *PR* makes with the same plane; similarly, *IPS* will be the angle of incidence and *RPT* the angle of refraction. Consequently, if *IP* or *PR* is assumed to be the radius of a circle, *IQ, RV, IS,* and *RT* will be the sines of those angles. But *IQ* and *RV* are parallel,[a] since they are perpendiculars to the same plane *FH.* Likewise, *IS* and *RT* are parallel,[b] because they lie in the same plane *ISPTR* and are perpendicular to the same line *ST.* That is, the lines *IQ* and *IS* that contain the angle *QIS* are parallel to the lines *RV* and *RT* that contain the angle *VRT.* Therefore those angles *QIS* and *VRT* are equal.[c] Drawing *QS* and *VT,* the angles *IQS* and *RVT* will become right angles,[d] because the lines *IQ* and *RV* stand perpendicular to the plane *FH.* Therefore, the triangles *IQS* and *RVT* are similar,[e] and *IQ* : *RV* = *IS* : *RT.* That is, the sines of the angles that the incident ray and the refracted ray make with any plane *FH* perpendicular to the refracting plane *BD* are as the

(a) *Elements,* XI,[14] 6
(b) *Elements,* I, 28

(c) *Elements,* XI, 10
(d) *Elements,* XI, Def. 3
(e) *Elements,* VI, 4

---

(12) In §7 directly following.
(13) Originally continued (as I): quod cùm postea nobis forsan erit ex usu, jam facere non pigebit (since it will perhaps be useful for us afterward, it will not be troublesome to do this now).
(14) Newton mistakenly cited Bk. 10 rather than 11.

in ratione datâ. Quippe sinuum istorum rationem esse datam Cartesius edocuit et alij deinde fuerunt experti.

　　Quinetiam Theorematis jam demonstrati veritas manebit salva, licet planum *EF* plano refringenti *BD* alibi perpendiculariter insistat quam ad punctum refringens *P*. Exinde enim neque anguli cum radijs et plano *FH* effecti,
neque ideo sinus istorum angulorum immutabuntur.

7 [= 7]. Pars
　　secunda.　　Hisce ita praemonstratis ad propositum jam revertor demonstraturus scilicet angulum *YFZ* (in Fig. I, 3) diametro Solis ac proin angulo *PFT* aequari:
Ex supra positis liquet quod planum *KDHNκFν* bisecat angulum radijs *IM* et
VL utrinque jacentibus contentum. Itaque cùm iste angulus aequetur diametro solari, angulus quem radiorum alter puta *IM* cum dicto plano facit,
aequabitur semidiametro solari, cujus esto sinus α, et β sinus anguli quem
radius iste refractus *Mλ* facit cum eodem plano. Jam cum planum istud
supponatur perpendiculare ad prismatis refringens planum *AC*, erit ex praecedenti Lemmate sinus α ad sinum β, sicut sinus incidentiae ad sinum refractionis e medio rariori in medium densius. Vel e contra sicut sinus incidentiae
ad sinum refractionis e medio densiori in rarius, ita erit β ad α. Quare cum
dictum planum *DHF* etiam perpendiculare sit ad alterum Prismatis planum
BC, quod radios e medio densiori in rarius refringit; et insuper cum β supponatur anguli sinus, quem radius incidens *Mλ* facit cum plano isto perpendiculari *DHF*: erit (per Lemma praecedens) α sinus anguli quem radius refractus *λF* facit cum eodem plano *DHF*. Sed α ponitur sinus semidiametri
solaris, ergo ille angulus quem radius[15] facit cum plano *DHF* aequatur semidiametro solari: et ejus duplus *λFμ* sive *YFZ* toti diametro et cum suprà
fuerit ostensum, quòd angulus *PFT* sit eidem diametro aequalis, isti duo
anguli *YFZ* et *PFT* erunt aequales.　　Q.E.D.

　　/ Jam si planum *YFZ* esset perpendiculare plano imaginis *PYTZ* aeque ac 6/
planum *PFT*, istae quatuor lineae *FP, FT, FY*, & *FZ* quae angulos aequales
comprehendunt, essent omnes inter se aequales et proin subtensae *PT* et *YZ*
etiam aequarentur. Sed qui rem serió perpendet inveniet radios collaterales
VLμFZ et *IMλFY* duobus reliquis *DKνFT* et *HNκFP* paulo minus refringi et
idcirco planum *YFZ* paulo magis declinabit a radio *FP* quàm ab *FT* secans
lineam *PT* infra medium ejus punctum *X*. et sic divaricans a perpendiculari
FX (quam concipe ductam) erit aliquantulum obliquum ad planum imaginis
PYTZ. Et ea de causa lineae *FY*, et *FZ* erunt paulo majores quàm *FP* et *FT*, et
subtensa *YZ* paulo major quam subtensa *PT*. Sed hujus rei demonstrationem
utpote longiusculam et proposito meo non omnino necessariam praetermitto:
Etenim non multum refert utrum planum *YFZ* sit rectum ad planum imaginis
PYTZ, vel nonnihil obliquum, hoc est, utrum *YZ* sit aequalis vel major quàm
PT. sufficit quod nequit esse minor. Imo cùm propter ἰσοςκελέα *PFT* et *YFZ*

---

(15) I: refractus radius *λF* (the refracted ray *λF*).

sines of incidence and refraction, and are thus in a given ratio. Indeed, Descartes has taught that the ratio of these sines is given, and others have since verified it.

Moreover, the truth of the theorem just demonstrated will remain valid even if the plane *EF* is perpendicular to the refracting plane *BD* other than at the refracting point *P*, for neither the angles made by the rays with the plane *FH* nor consequently the sines of these angles will be changed on this account.

Thus having first shown these things, I now return to the proposition to be demonstrated, namely, that the angle *YFZ* (in Fig. I, 3) is equal to the sun's diameter and thus to the angle *PFT*. From what was considered above it is clear that the plane *KDHNκFν* bisects the angle formed by the rays *IM* and *VL* lying on each side. Since, therefore, that angle is equal to the solar diameter, the angle that one of the rays, for example, *IM*, makes with that plane will be equal to the sun's radius; and let α be its sine and β the sine of the angle that the refracted ray *Mλ* makes with the same plane. Now, since that plane is assumed to be perpendicular to the prism's refracting plane *AC*, from the preceding lemma the sine α will be to the sine β as the sine of incidence to the sine of refraction from a rarer into a denser medium; or, in the opposite case from a denser into a rarer medium, the sine of incidence will be to the sine of refraction as β to α. Accordingly, since that plane *DHF* is also perpendicular to the prism's other plane *BC*, which refracts the rays from a denser into a rarer medium, and besides since β is assumed to be the sine of the angle that the incident ray *Mλ* makes with that perpendicular plane *DHF*, by the preceding lemma α will be the sine of the angle that the refracted ray *λF* makes with the same plane *DHF*. But α is assumed to be the sine of the sun's radius, and so that angle that the ray[15] makes with the plane *DHF* is equal to the sun's radius, and its double, *λFμ* or *YFZ*, is equal to the whole diameter. Furthermore, since it has been shown above that the angle *PFT* is equal to the same diameter, those two angles *YFZ* and *PFT* will be equal. As was to be demonstrated.

If now the plane *YFZ*, just as the plane *PFT*, were perpendicular to the plane of the image *PYTZ*, those four lines, *FP*, *FT*, *FY*, and *FZ*, that contain equal angles would all be equal to one another, and thus the chords *PT* and *YZ* would also be equal. But whoever will carefully consider this will find that the rays on the two sides, *VLμFZ* and *IMλFY*, are refracted a little less than the other two, *DKνFT* and *HNκFP*, and that the plane *YFZ* will therefore deviate a little more from the ray *FP* than from *FT* and intersect the line *PT* below its midpoint *X*. Hence, inclining away from the perpendicular *FX* (which you should image to be drawn), it will be a little oblique to the plane of the image *PYTZ*, and thus the lines *FY* and *FZ* will be slightly greater than *FP* and *FT*, and the chord *YZ* slightly greater than the chord *PT*. But I am omitting the demonstration of this, as it is rather long and not completely necessary for my purpose. For it makes little difference whether the plane *YFZ* is normal to the plane of the image *PYTZ* or somewhat oblique, that is, whether *YZ* is equal to or greater than *PT*; it is sufficient that it cannot be

7 [= 7]. Part II

sit *FP . FY :: PT . YZ*, atque *FP* et *FY* sint quàm proximè aequales, tantilla erit inter *PT*, et *YZ* differentia, ut quoad sensum pro aequalibus haberi possint.

Ostensus itaque casus est in quo longitudo solaris imaginis per Prisma trajectae conspiceretur aequalis ejusdem latitudini; et proinde in quo imago ista quasi orbicularis appareret, modò vera esset opinio vulgaris. Quinimo licèt positio Prismatis alia sit atque descripsi, modò radij refractionem utrinque non valde inaequalem patiantur, figura tamen imaginis ea propter vix immutabitur. Nec multum interest an corpus opacum *EG*, foramine *F* ad radios transmittendos terebratum citra Prisma collocetur vel ultra: neque figura foraminis multum curanda est, modò sit exigua. etenim tam parvae variationes haud plus mutabunt imaginem quàm decimâ forte vel quintâ parte diametri suae sicut cogitanti patebit. Atque ita ut paucis tandem comprehendam omnia, liquet quod imago solis refracta utplurimum deberet esse sensui quasi orbicularis; si modò ejusdem incidentiae in idem medium refractio semper foret eadem. Sed priùs repugnat experientiae, longitudine scilicet ejus latitudinem plusquam quatuor vicibus, ut dictum fuit, excedente. Ergo posterius repugnat veritati: et ejusdem incidentiae refractio est varia.

Ex eodem experimento potui propositum sic breviùs indicasse: Nempe cum ita disposuissem Prisma, ut refractio radiorum tum ingredientium tum egredientium foret quasi aequalis; angulos *PFT* et *YFZ* (Fig I, 2 vel I, 3) dimensus sum, et inveni quidem angulum *YFZ* semissi gradûs / sive diametro solis 7/ aequalem, at angulum *PFT*[2] eandem diametrum quater et ampliùs superavit, cui tamen aequalis esse debuisset ex parte priori demonstrationis praecedentis: et inde planissimè liquet propositum. Verùm in eorum gratiam quae mox sequentur, oporteret demonstrasse quod illi radij quorum refrangibilitas non est dispar, efformabunt imaginem pene orbicularem: et eâ de re mihi visum fuit demonstrationem istam etiamsi longiusculam in illustrationem hujus experimenti hìc adduxisse.

Verum cum in experiendis praedictis eam esse positionem Prismatis supposuerim ut radij ad utramque faciem Prismatis aequalitèr refringantur: Conclusionis loco[3] dicam quâ ratione istud citò fiat et facilè. Si Prisma teneatur in luce solari et motu lento circa suum axin convertatur, videbis colores quos efficit, de loco in locum continuo motu translatos esse, ita quidem ut aliquando progredi, deindè verò regredi videantur. Observabis itaque medium inter istos contrarios motus, quando colores modò progressi, et statim regressuri, videntur quiescere.[4] Quod ubi vides siste Prisma, idque in eo situ fige. Dico factum. Scilicet in eo situ summa refractionum utrobique factarum, sive

---

(1) I: *contractior* (shorter).
(2) Originally (as I): *RFS*. See Lect. 1, note (34).
(3) That is, a conclusion to Lecture 1 of the *Lectiones opticae*.
(4) Originally (as I): *sistere* (stand still).

smaller. On account of the isosceles triangles *PFT* and *YFZ*, there is *FP* : *FY* = *PT* : *YZ*; and since *FP* and *FY* are very nearly equal, the difference between *PT* and *YZ* will be so small that they can be considered as sensibly equal.

The case has therefore been presented in which the length of the solar image transmitted through the prism would appear equal to its breadth, and consequently one in which that image would appear nearly circular, provided that the common opinion were true. Moreover, even if the prism's position were other than I have described, as long as the rays do not undergo a particularly unequal refraction on each side, the shape of the image will nonetheless hardly be changed because of that. Nor does it make much difference whether the opaque body *EG*, perforated with the hole *F* for transmitting the rays, is placed on the near or the far side of the prism; nor does the shape of the hole matter much, provided that it is small. For such small alterations will scarcely change the image more than a tenth or even a fifth of its diameter, as will be clear to one who considers it. Finally, to include everything in a few words, it is manifest that generally the sun's refracted image must be sensibly nearly circular, provided that in the same medium the refraction at the same incidence be always the same. But the former is contrary to experience, specifically, its length exceeded its breadth more than four times, as has been noted. Therefore the latter is contrary to the truth, and the refraction at the same incidence varies.

8 [= 8]. In this case the image's length is nevertheless more than four times its breadth: whence diverse refrangibility is proven.

From the same experiment I could have shown what was proposed more briefly as follows. Namely, when I had so placed the prism that the refraction of both the entering and emerging rays was almost equal, I measured the angles *PFT* and *YFZ* (Fig. I, 2 or I, 3) and found that the angle *YFZ* was indeed equal to half a degree, or the sun's diameter, but that the angle *PFT*[2] exceeded the same diameter by four times and more—which it ought to have nonetheless been equal to according to the first part of the preceding demonstration. Consequently what was proposed is most plainly evident. Yet for the sake of what will follow next, it ought to have been demonstrated that those rays whose refrangibility is not unequal will form a nearly circular image, and for this reason I considered it appropriate to have introduced that demonstration here, even if it was somewhat long, to illustrate this experiment.

Lecture 2
9 [= 9]. A briefer[1] demonstration of the same thing.

Since in doing these experiments I supposed the prism's position to be such that the rays would be refracted equally at each face of the prism, in place of a conclusion[3] I will explain how that may be done quickly and easily. If the prism is held in the sun's light and turned about its axis with a slow motion, you will see the colors that it makes shifted with a continuous motion from place to place, so that in fact they appear to move now forward now backward. Therefore you will observe a mean between those opposite motions when the colors, now having moved forward and just about to move backward, appear to come to rest.[4] When you see that, stop the prism and fix it in that position. I say you are done. Namely, in that position the sum of the refractions made on both sides, or the inclination of the emergent to the

10 [= 10]. How the prism can be easily placed in the position required to do that experiment.

radij emergentis ad incidentem inclinatio, evadit omnium minima: quod cum accidit, refractiones utrobique sunt aequales uti posthac demonstrabitur.[5]

11 [≈ 11]. Imaginis praefatae figura describitur, quae partim rectis partim semicirculis comprehensa est.[6]

Caeterum experimenti hujus varias circumstantias non minùs jucundas experienti, quàm propositi nostri indicativas prosequi jam animus est. ‖ Et primò notandum venit quod imaginis istius figura secundum longitudinem suam lineis rectis terminata fuit, et secundum latitudinem duobus (ut ex visu potui judicare) semicirculis: In figura I, 5 sit *PT* imago solis Prismate refracta: Hanc observabam ad latera duabus lineis *AB* et *CD* qu[o]ad sensum rectis et sibi parallelis terminari, ad extremitates autem duobus semicirculis *APC* et *BTD* cujus quidem eventus causa ex praemonstratis sic determinatur.

Figures I, 5 & I, 6

12 [= 12]. Quomodo talis evadit per circulares[7] imagines (quas unumquodque genus radiorum aequaliter[8] refrangibilium facit) in longum dispositas.

Semicirculi illi terminales in circulos compleantur ut vides in Fig I, 6 et alius inscribatur circulus *YZ* istis intermedius. Jam concipe radios quosdam a Sole provenientes qui apti sunt ut aequaliter incidentes, etiam aequaliter refringantur: Illi per Prisma trajecti, ex supra demonstratis imaginem quoad sensum (si sola posset videri) circularem depingent, puta *BD*. Deinde concipe alios ejusdem solis radios sibi etiam conformes, qui apti sunt, ut prioribus paulò magis refringantur, illi itaque aliam imaginem depingent circularem puta *YZ*: Et alios etiam radios adhuc magis refrangibiles concipe, qui tertiam circularem imaginem *AC* efficiant. Denique alios innumeros cogita praedictis plus et minus refrangibiles, et illi alias etiam innumeras circulares imagines prioribus tum intermedias tum extremas efformabunt, illuminantes oblongum spatium *PYTZ* rectis[9] lineis *AB*, et *CD*, / duobusque semicirculis contentum. Verùm cùm imagines illae 8/ sint omnes ejusdem penè magnitudinis et inter lineas *AB* et *CD* in directum dispositae, istae lineae *AB* et *CD* pro rectis sibi parallelis possunt haberi et ad sensum tales videbuntur. et sic totum spatium *PYTZ* radijs ex eadem incidentiâ variè refractis illuminatum, partim parallelis rectis, et partim semicirculis oppositis terminabitur; sicut experientiâ compertum est.

13 [= 13]. Exinde deducitur experimentum quo termini recti fiant distinctissimi.

Hanc autem conjecturam ut penitùs probarem, cogitabam de imagine solis per foramen aliquod sine ullâ refractione ad distantiam magnam trajectâ, scilicet quod malè definitur, termino existente inter lucem et tenebras minimè distincto: at si radij isti per lentem convexam transeant, cujus focus ad imaginem est, imago terminabitur distinctissimè. Simili modo de radijs aequè refrangibilibus intellexi quod si per Prisma trajicerentur ad distantiam magnam, depingerent imaginem circularem malè definitam, cujus tamèn terminus, mediante lente convexâ, distinctissimus evaderet. Itaque cum vidissem terminos imaginis refractae *PYTZ* non admodum distinctos, de imaginibus *BD*, *YZ*, *AC*, et reliquis circularibus oblongam istam formantibus conjiciebam quod multo distinctiùs terminarentur per lentem convexam trajectae quàm alitèr.

---

(5) In §I, 111 = 186.

(6) quae . . . comprehensa est.] I: quòd . . . terminatur. (which is bounded . . .)

(7) I: orbiculares.

incident ray, proves to be least of all; and when this happens, the refractions on both sides are equal, as will be demonstrated later.[5]

Now, however, I intend to pursue various features of this experiment that are no less pleasant for the experimenter than they are informative for our purpose. ‖ In the first place, it should be noted that the shape of that image was bounded lengthwise by straight lines and in its breadth by two semicircles (insofar as I have been able to judge by sight). In Fig. I, 5 let *PT* be the sun's image refracted by the prism. I observed this to be terminated on its sides by two lines, *AB* and *CD*, sensibly straight and parallel to each other, but at the ends by two semicircles *APC* and *BTD*. The cause of this result is determined from what has previously been described as follows.

Let those terminal semicircles be completed into circles, as you see in Fig. I, 6, and inscribe another circle *YZ* between them. Now imagine some rays to come from the sun that are disposed to be also equally refracted when they are equally incident; according to the above demonstration, after passing through the prism, these will depict a sensibly circular image, for example, *BD*—if it could be seen alone. Then imagine other rays from the same sun, also similar to one another, that are disposed to be refracted a little more than the first; those therefore will depict another circular image, for example, *YZ*. Also imagine other still more refrangible rays that make a third circular image *AC*. Finally, conceive innumerable others more and less refrangible than the preceding, and those will also form other innumerable circular images both in between and beyond the preceding ones, illuminating the oblong space *PYTZ* contained within the straight[9] lines *AB* and *CD* and the two semicircles. But since those images are all nearly the same size and disposed in a straight line between the lines *AB* and *CD*, the lines *AB* and *CD* can be considered to be straight lines parallel to one another, and to the senses they will appear as such. Thus the entire space *PYTZ*, illuminated by rays diversely refracted at the same incidence, will be terminated partly by parallel straight lines and partly by opposite semicircles, just as is found by doing the experiment.

So that I might fully confirm this conjecture, however, I thought about the sun's image cast through some hole to a great distance without any refraction, in particular, I thought about the fact that it is poorly defined, the border between light and darkness being very indistinct; but if those rays pass through a convex lens, whose focus is at the image, the image will be terminated very distinctly. Similarly, for equally refrangible rays I perceived that if they were transmitted through a prism to a great distance, they would depict a poorly defined circular image, whose border would nevertheless turn out to be very distinct by interposing a convex lens. When, therefore, I had seen that the borders of the refracted image *PYTZ* were not very distinct, I inferred that the images *BD, YZ, AC,* and all the other circular ones forming that oblong would be terminated much more distinctly when transmitted through a convex lens than they would otherwise be. And this was evident by doing

11 [≈ 11]. The shape of that image, which is contained[6] partly by straight lines and partly by semicircles, is described.

12 [= 12]. How this arises from circular[7] images (which each kind of equally[8] refrangible ray produces) disposed lengthwise.

13 [= 13]. From this an experiment is deduced whereby the straight edges are made very distinct.

(8) I: aequè. This common stylistic change henceforth will not be indicated.   (9) Added.

Et experienti res patuit: Nam rectas *AB* et *CD*, in quas imagines omnes istae circulares utrinque terminantur, vidi admodùm distinctas, quas antea confusas videram.

14 [≈ 14]. Quare termini circulares semper apparent confusi.

Sed quod notatu valdè dignum videtur, termini circulares *APC* ac *BTD* imaginis istius semper apparuere maximè confusi, luce paulatim deficiente donec tandem in tenebras desijt. Scilicet intermedij circuli ut *YZ* miscentur alijs circulis utrinque cadentibus, quibuscum, ex aliquâ sui parte, coincidunt: at extremi quidem circuli *AC* et *BD* ex unâ tantum parte cum alijs concurrunt, et eorum concursus continuò fit rarior et exinde lux usque remissior dum ad extremitates *P* ac *T* deventum est. Sed et alia prodit istius rei causa, scilicet quod radiorum maxima copia apta sit ut mediocrem refractionem patiatur, et sic in medium imaginis incidat. ‖ et quod eorum numerus ‖ continuò minor existat quibus competit gradus refrangibilitatis alterutrinque ‖ magis extremus.

15 [= 15]. Admonitio de figura et situ Lentium et Prismatum.

Caeterum ad isthaec experienda Lentes adhiberi vellem, quarum foci sunt longinqui, sex forté vel duodecim pedibus a lentibus distantes modò tales praesto sint: saltem non sint minùs distantes quam duobus. Atque etiam latera Prismatis debent esse accuratè plana. Sin latera ejus sint aliquatenus convexa tum praestat adhibere lentem cujus focus ad pedes tantùm duos, vel tres a se remotus est. Quibus paratis Lentem Prismati ex utravis parte vicinam colloca; ita scilicet ut radios per se trajectos directè respiciat. Deinde radij in papyrum aliquam excipiantur, quam ultrò citróque / transfer, donec imaginem coloratam utrinque rectis parallelis distinctissimè terminatam videas. 9/

16 [= 16]. Deque imagine quadam orbiculari.

Sed observandum est quòd cum Prisma collocatur ultra foramen *F* ut in figura I, 3, vel ipsi quàm proximè citrà; et lens magis distat ab isto foramine quàm focus lentis, quem radij in eam parallelōs incidentes efficerent, distat a lente: duplicem invenies casum in quo imago in papyrum projecta evadet distincta; alter quando radij omnes homogenei qui in lentem paralleli incidunt ita refringuntur ut ad papyrum istam in eodem puncto concurrant, quod fit cùm vides imaginem coloratam, oblongam, et parallelis rectis distinctè terminatam. Alter casus est quando radij omnes homogenei ab uno puncto foraminis *F* divergentes, postquam a lente refringuntur, ad unum iterum punctum dictae papyri convergunt. Id autem accidit cum imaginem albam, orbicularem et undique bene definitam vides: De quo fusè dicetur alibi;[10] sufficiat hoc monitum hic dedisse, ne quis proprijs oculis haec experturus, per ambiguitatem effectus incautè decipiatur, et exinde praedicta in dubium revocet.

17. Ac de umbris[11] nebularum intercedentium solem.

Juvat annotare praeterea quod cum nebulae aliquae tenuiores interceperunt discum solis, eum non penitus obscurantes umbras in hanc imaginem *PT* projecerunt, non sui similes sed in longum protensas et imaginis terminis rectilineis parallelas. Id quod ratiocinijs modò allatis accuratè convenit. Nam [Fig I, 7] concipe nebulam aliquam in disco solis ad instar maculae conspicuam esse, et ea, si radij maximè refrangibiles circu-

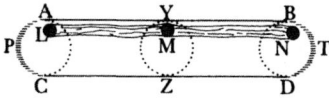

Figure I, 7

---

(10) In §§II, 43–7 = 54–8.   (11) Originally: figuris (shapes).

the experiment, for I saw the lines *AB* and *CD* that terminate all those circular images on both sides to be quite distinct, whereas previously I had seen them to be confused.

It seems especially noteworthy, however, that the circular ends of that image, *APC* and *BTD*, have always appeared extremely confused, with their light gradually decreasing until finally ending in darkness. Specifically, the intermediate circles, such as *YZ*, are mixed with other circles, which fall on each side and with which they partially coincide; but the outermost circles *AC* and *BD* in fact run into others on only one side, and their intersection becomes continually rarer and accordingly the light continually fainter until reaching the ends *P* and *T*. This also has another cause, namely, that the greatest quantity of rays is disposed to undergo a mean refraction and thus falls in the middle of the image, ‖ and their number becomes continually ‖ smaller, corresponding to the more extreme degree of refrangibility on each ‖ side. ‖

To do these experiments I suggest that lenses be used whose foci are far off, distant perhaps six or twelve feet from the lenses—if such lenses are readily available; in any case, they should not be distant less than two feet. In addition, the sides of the prism ought to be exactly flat; but if its sides are somewhat convex, then it is preferable to use a lens, the focus of which is only two or three feet away from it. Having procured these things, place the lens near either side of the prism such that it directly faces the rays transmitted through it. Then catch the rays on some paper while shifting it back and forth until you see a colored image terminated very distinctly on both sides with parallel straight lines.

It must be noted that when the prism is placed outside the hole *F* (as in Fig. I, 3) or else inside as close as possible to it, and the lens is farther from that hole than the lens's focus (which would be produced by parallel rays falling on it) is from the lens, you will find two cases in which the image projected onto the paper will turn out to be distinct. The first happens when all homogeneous rays falling parallel on the lens are refracted such that they meet at the same point on that paper; this occurs when you see a colored oblong image terminated distinctly by parallel straight lines. The second case occurs when all homogeneous rays diverging from one point of the hole *F*, after being refracted by the lens, converge again to one point on that paper; this occurs, however, when you see a white circular image well defined all around. I will discuss this case in detail elsewhere;[10] let it suffice to have given this warning here, so that anyone trying these things with his own eyes may not be unwarily deceived by the variability of the effect and then call the preceding in doubt.

It is, moreover, useful to note that when some rather thin clouds have intercepted the sun's disk without fully obscuring it, they have cast shadows onto this image *PT* not like themselves, but stretched out lengthwise and parallel to the straight edges of the image in exact agreement with the reasoning just set forth. For (Fig. I, 7) imagine some cloud to appear like a spot on the sun's disk, and if the most refrangible rays bounded by the circle *AC* were

14 [≈ 14]. Why the circular ends always appear confused.

15 [= 15]. Advice about the shape and position of the lenses and prisms.

16 [= 16]. And about a certain circular image.

17. And about the shadows[11] of clouds intercepting the sun.

loque *AC* circumscripti spectentur, umbram projiciat in locum *L*, ita ut circulus *AC* cum umbra *L* discum solis nebulâ deficientum referat. Quo posito, si radij minimè refrangibiles circuloque *BD* circumscripti spectentur umbra nebulae ab ijs projicietur in locum *N*, cujus talis erit situs in circulo *BD* qualis est ipsius *L* in circulo *AC*, quippe hic etiam discum solis nebulâ deficientem refert. Atque idem porro discursus de circulo quolibet intermedio cum umbrellâ ejus *M* intelligatur. Adeò ut propter indefinitam multitudinem circulorum spatium integrum *ABDC* occupantium, nebula suas umbras per totam longitudinem *LN* dispergat, eamque reddat obscuram. Et sic cum plures nebulae vel nubium sinus soli interveniant, imago ejus pluribus umbris in longum diffusis et parallelis obscurabitur.

18 [= 17]. Ab imaginis figura aliud etiam experimentum deducitur quo fiat multum oblongior. Ut dictas proprietates lucis quâ potui diligentiâ perscrutarer sequentem praeterea modum excogitavi quo illas examini subjicerem. / Nempe in Fig. I, 10/ 6 cum magnitudo circulorum *AC, YZ, BD*, dependeat a magnitudine solari: si diameter solis fieret aliquanto minor quam nunc reverà existit, tum illi etiam circuli fierent minores, distantia centrorum *H I K* non omninò mutatâ: ut videre est in fig I, 8. et sic latitudo imaginis ad ejusdem longitudinem comparata multo minor evaderet quàm antea, utrâque scilicet per eandem quantitatem diminutâ. Haec probaturus effeci ut

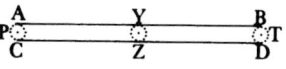

Figure I, 8

radij solis per duo parva foramina ab invicem longe distantia transirent antequam inciderent Prismati, quo pacto radij ab extremis partibus solis venientes excludebantur et res perinde successit quasi diameter solis reverà fuisset diminuta. Illustrationis gratiâ sit in fig I, 9 *εφγ* fenestra parvo foramine *φ* penetrata, per quod radij solares cubiculum alias obscuratum ingrediantur. Deinde sit *EFG* corpus aliquod opacum perforatum ad *F* et in medio cubiculo ita locatum ut radij iterum permeent foramen istud, antequam Prisma *ABC* pone locatum attingant. Jam foraminum istorum diametro existente $\frac{1}{8}$ digiti, et eorundem distantiâ *φF* 12 pedibus (ita scilicet ut maxima radiorum utrumque foramen permeantium inclinatio foret angulus ferè minutorum sex, hoc est, quasi quinta pars diametri solaris) atque etiam Imagine *PT* projectâ in papyrum decem pedes a Prismate distantem, prout angustia cubiculi tulit: inveni longitudinem imaginis esse plusquam quatuor digitorum cum semisse et latitudinem trientis digiti, hoc est longitudinem plusquam quatuordecim

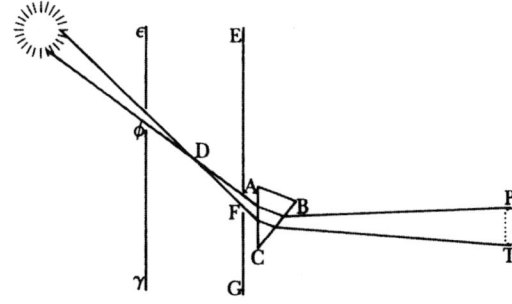

Figure I, 9

seen, it would cast a shadow onto the place *L*, so that the circle *AC* together with the shadow *L* would represent the sun's disk when eclipsed by the cloud. Having assumed this, if the least refrangible rays bounded by the circle *BD* were seen, the cloud's shadow would be cast by them to the place *N* whose position in the circle *BD* would be just like that of *L* in the circle *AC*; of course, this too represents the sun's disk when eclipsed by the cloud. Moreover, the same argument is to be understood for any intermediate circle with its little shadow *M*. Hence, because of the indefinitely large number of circles filling the entire space *ABDC*, the cloud would spread its shadows through the entire length *LN* and obscure it; and thus when several clouds or wisps of clouds intercept the sun, its image will be obscured by several shadows spread out lengthwise and parallel.

In order to investigate the said properties of light as diligently as I could, I thereafter conceived of the following way whereby I might put them to a test. Namely, in Fig. I, 6, since the size of the circles *AC*, *YZ*, and *BD* depends on the size of the sun, if the sun's diameter were to become somewhat smaller than it now actually is, then those circles would also become smaller without at all changing the distance of their centers *H, I,* and *K*, as can be seen in Fig. I, 8. Thus the breadth of the image compared to its length would turn out to be much smaller than before, that is, both decreased by the same amount. To prove this, I made the sun's rays pass through two small holes, very far from one another, before they fell upon the prism; in this way the rays coming from the outer parts of the sun were cut out, and it succeeded just as if the sun's diameter had really been diminished. For an illustration, in Fig. I, 9 let $\epsilon\phi\gamma$ be a window pierced with a small hole ϕ through which solar rays may enter an otherwise darkened room. Then let *EFG* be an opaque body perforated at *F* and so placed in the middle of the room that the rays may also pass through that hole before they arrive at the prism *ABC* placed behind it. Now, when the diameter of those holes was $\frac{1}{8}$ inch and their distance ϕF was twelve feet (so that, in particular, the greatest inclination of rays passing through both holes would be an angle of nearly six minutes, that is, about a fifth part of the solar diameter); and also when the image *PT* was projected onto a paper ten feet away from the prism, as the narrowness of the room allowed, I found the image's length to be more than $4\frac{1}{2}$ inches and its breadth $\frac{1}{3}$ inch; that is, the length was more than fourteen times greater than its breadth, just

18 [= 17]. From the shape of the image still another experiment is deduced whereby it becomes much more oblong.

vicibus majorem latitudine, sicut ex praedictis oportuit evenisse. Etenim cùm
isti tantùm radij mittuntur intrò, qui minùs quam quintâ parte solaris diametri ad se invicem inclinantur, diametri *AC, YZ,* et *BD* diminutae diametro
foraminis *F,* debent esse quintuplo minores quam secundum priora contingeret; ut videre est in fig I, 6 & I, [8].[12] Quasi a sole essent effectae cujus
diameter sit quinquies minor diametro solis nostri. Verùm si corpus opacum
[ϵ]$\phi\gamma$ (fig I, [9]) tolleretur, ut radij per unum solummodo foramen *F* ad
Prisma transirent, sicut in prioribus factum est, latitudo imaginis evaderet
$1\frac{1}{6}^{\text{dig.}}$ et longitudo plusquam $5^{\text{dig.}}$: angulo nempe Prismatis existente 60 grad,
vel paulo majori. Itaque diameter circulorum *AC, YZ,* et *BD,* qui eo quo
dictum est modo imaginem constituunt, esset $1\frac{1}{6}^{\text{dig.}}$ a quâ subducatur diameter
foraminis nempe $\frac{1}{8}$ et manebit $1\frac{1}{24}^{\text{dig.}}$ cujus quintae parti rursus adjungatur
eadem foraminis diameter sive $\frac{1}{8}^{\text{dig.}}$ et prodibit $\frac{1}{3}^{\text{dig.}}$ diameter circulorum *AC,*
YZ, et *BD* in / Fig I, 8: quae minor est quàm diameter circulorum istorum in   11/
fig I, 6, quantitate $\frac{5}{6}^{\text{dig.}}$. Quamobrem figura [8]^{ma} quaquaversum minor est
quàm sexta, quantitate $\frac{5}{6}^{\text{dig}}$. Atque ideo longitudo ejus fit plusquam 4 dig,
latitudo autem digiti triens. Id quod cum experientiâ modo recensitâ quadrat.
Ad eundem modum si foramina ϕ et *F* adhuc minora forent, vel si distantia
ϕF foret major, imago *PT* oblongior evaderet. Quod idem quoque quadantenus contingeret, ex imagine *PT* a Prismate longiùs dissitâ. Caeterum notandum est quod foramina ϕ et *F* ad radios directè respicientia supponam, licet
non multum refert an situs eorum sit parum obliquus ut in apposita figura 9^{nâ}
factum est.

19 [= 18].   Porrò si in hoc experimento convexam Lentem ut prius adhibueris cujus
Experimentum istud   focus ad imaginem cadit, foramine *F* si placet dilatato vel opaco corpore *EG*
promovetur.   prorsus ablato, ut radij per longinquum foramen ϕ solummodò transeant, et
si foramen istud ϕ effeceris angustius quàm anteà, caeteris ut priùs stantibus,
imaginem valde oblongam, et pro longitudine lucidiorem videbis quam in
casu priori. Exempli gratiâ, si diameter foraminis sit pars digiti vigessima, et
si pedibus abindè duodecim Prisma cum Lente disposueris, videbis longitudinem imaginis plusquam 80 vel 100 vicibus latitudine majorem. Sed in his
experiendis oportet cubiculum quaquaversus bene obturatum esse, ne lux
alibi quàm per foramen ϕ ingressa perturbet imaginem et juxta circulares ejus
extremitates obscuram reddat. et praeterea si superficies Prismatis sint
accuratè planae, praestat adhibere lentem, quae focum ad distantiam magnam projicit, puta ad 12 aut 20 pedes modo loci amplitudo sinat; quo pacto
de proportionibus imaginis meliùs[13] judicium proferas. Quod si latera prismatis sint aliquantulùm convexa ut ijs nonnunquam contingit quae vulgò
venduntur: licebit istud absque ulla lente solum adhibere, et ejus convexitas
radios vice lentis ad magnam distantiam congregabit. Quinimò si cum Prismate quolibet lentem parvam adhibeas cujus focus non sit duobus tribusve
pedibus longinquior, imaginem conspicies satis longam quidem, sed cujus
latitudo haud[14] sensibilis existit. Id quod proposito nostro non minùs inser-

---

(12) When in revision Newton added §17 and Fig. I, 7, he did not correct all the figure
numbers in this article.   (13) I: satiùs.   (14) I: tamen haud (yet not).

as it should turn out according to what was just described. For since only those rays are allowed to pass inside that are inclined to one another less than a fifth part of the solar diameter, the diameters *AC, YZ,* and *BD,* decreased by the diameter of the hole *F,* must be five times smaller than would result according to the former case, as is seen in Figs. I, 6 and I, [8].[12] It is just as if they were produced by a sun whose diameter was five times smaller than the diameter of our sun. But if the opaque body [ϵ]$\phi\gamma$ (Fig. I, [9]) were removed so that the rays would pass to the prism through only one hole *F,* as happened in the former experiments, then the image's breadth would turn out to be $1\frac{1}{6}$ inch and its length more than 5 inches, that is, when the prism's angle was 60° or a little more. Therefore, the diameter of the circles *AC, YZ,* and *BD,* which form the image in the way already described, would be $1\frac{1}{6}$ inch. From this subtract the diameter of the hole, namely, $\frac{1}{8}$ inch, and $1\frac{1}{24}$ inch will remain; to its fifth part in turn add the same diameter, or $\frac{1}{8}$ inch, and this will produce $\frac{1}{3}$ inch for the diameter of the circles *AC, YZ,* and *BD* in Fig. I, 8, which is less than the diameter of those circles in Fig. I, 6 by a quantity of $\frac{5}{6}$ inch. Consequently, the [eighth] figure is less than the sixth on all sides by $\frac{5}{6}$ inch. Hence its length becomes more than 4 inches but its breadth $\frac{1}{3}$ inch, in agreement with the experiment just now set forth. Likewise, if the holes ϕ and *F* were even smaller, or if the distance ϕF were greater, the image *PT* would become more oblong. The same thing would also occur to a certain extent when the image *PT* has been set farther away from the prism. It must be noted, however, that I am assuming that the holes ϕ and *F* are in a straight line with the rays, although it does not matter much whether their position is a little oblique, as is drawn in the adjoined Fig. I, 9.

Moreover, if in this experiment as before you use a convex lens whose focus falls at the image, with the hole *F* enlarged or, if you prefer, the opaque body *EG* completely removed so that the rays pass through only the far hole ϕ; and if you make that hole ϕ narrower than previously, with everything else remaining as before, you will see the image as very oblong and brighter in its length than in the former case. For example, if the diameter of the hole is $\frac{1}{20}$ inch, and if you place the prism with the lens twelve feet from there, you will see the length of the image eighty or one hundred times greater than its breadth. In these experiments it is necessary that the room be thoroughly closed off everywhere, so that light entering anywhere but through the hole ϕ does not disturb the image and make it indistinct near its circular ends. Moreover, if the prism's surfaces are exactly flat, it is preferable to use a lens that projects its focus to a great distance, for instance, to twelve or twenty feet if the room's size allows it. This way you can make a better[13] judgment of the proportions of the image. If, however, the sides of the prism are slightly convex (as sometimes happens in those commonly sold), it may be used alone without any lens, and its convexity will gather the rays at a great distance like a lens. In fact, if with any prism you use a small lens whose focus is not longer than two or three feet, you will, to be sure, observe an image sufficiently long, but with a width that is not[14] perceptible; this serves our purpose just as well as if you could accurately judge the proportion of the

19 [= 18]. That experiment is extended.

vit, quam si posses de proportione longitudinis ad latitudinem ejus accuratè judicare. In istis etiam experiendis notetur praeterea quod lens debet ita longe post Prisma locari quin possit ad omnes radios simul transmittendos extendi ne imaginem successivè per partes tantùm observare sis coactus: / Et notetur 12/ denique quòd si foramen *F* citra Prisma locaveris et lentem deinceps citra foramen istud, ad distantiam majorem ab eo, quam focus radiorum a foramine ϕ longinquiori manantium abest a lente: duplex erit casus in quo imago in papyrum projecta conspicietur distincta, prout radij venientes a singulis punctis foraminis *F,* aut a singulis punctis foraminis ϕ, in totidem iterum punctis papyri colliguntur. In uno casu imago erit alba et orbicularis uti prius commonui,[a] in altera autem oblonga et colorata, sicut praesens experimentum exigit.

(a) Numb. 16

Lect 3
20 [= 19]. Magis adhuc promovetur per imaginem Stellae Veneris.

Jam liquet ex praefatis quod Imaginis *PT* latitudo semper evadit eo minor, quo foramen longinquum ϕ factum est angustius; ut nihil dubitandum sit quin dicta latitudo prorsus evanesceret, si vice foraminis istius translucidi unum duntaxat punctum ibi lucidissimum existeret: atque istud sic futurum esse confirmatur ex observatione non dissimili quam habui quondam de stellâ Veneris. Cubiculo nempe quaquaversus obturato, excepto foramine paulò plusquam duos digitos lato, ut tenebrosissimum efficeretur. In isto foramine vitrum objectivum perspicilli septempedalis collocavi; latitudine ejus, ad sufficientem radiorum copiam transmittendam, duos digitos et ampliùs apertâ. Deinde ad distantiam septem pedum papyro transversè positâ, in eam vidi sideris imaginem ad instar puncti lucidi projectam. et interposito Prismate ad distantiam pedis unius duorumvé ab ista papyro, per quod radij trajecti aliò refringerentur: pro puncto illo lucido ad distantiam inde plusquam pedalem, vidi lineolam licèt non valdè lucidam facilè conspicuam tamen, et cujus longitudo semissem digiti superavit; latitudo autem fuit, quoad sensum nulla, saltèm haud major quàm ut sentiretur. Atque idem credo de stellis primae magnitudinis, uti de Sirio, liceat observare, praesertim si lens adhibeatur quatuor vel sex digitos lata, ut plures radios transmittat.

21 [= 20]. Et applicatur descriptioni refractionis ad fig I, 1 traditae.

Hoc experimentum quam bene convenit cum explicatione nostrâ quam de refractione radiorum ad eundem angulum incidentium variâ, sub initio dedi, operae pretium videatur adnotare. In figura prima supposui complures radios per eandem rectam in superficiem aliquam refringentem successivè delatos esse, ibidemque alios alijs paulò magis, gradatìm, / refringi. Quod si fieri 13/ concipiatur, abinde sequeretur quod radij sic refracti, si corpore deinceps opaco quovis, ut papyro, interciperentur, lineolam ibi lucidam depingerent.[1] Jam licèt radij a stellâ aliquâ venientes, non omnes in eadem rectâ pergant; tamen, quod tantundem est, pro parallelis possunt haberi; et quòd a lente convexâ effecti sunt convergentes antequam attingant prisma, hoc adeò non destruit Analogiam, ut eam maxime confirmet. Etenim pro singulis in eadem rectâ pergentibus, debes tantùm concipere tot radiorum penicillos qui omnes habent eundem axem, et idem punctum concursus; et quòd istorum penicillo-

---

(1) Originally (as I): expingerent.

length to its breadth. It should also be noted that in doing these experiments the lens must not be placed so far behind the prism that it cannot transmit all the rays at one time, lest you be forced to observe the image only successively and piecemeal. Finally, it should be noted that if you place the hole *F* beyond the prism, and then you place the lens beyond that hole at a distance from the hole greater than that of the focus of the rays flowing from the farther hole φ from the lens, there will be two cases in which the image projected onto the paper will be seen distinctly, according as the rays coming from each point of the hole *F,* or from each point of the hole φ, are gathered again in as many points on the paper. In the one case the image will be white and circular, as I urged earlier,[a] but in the other it will be oblong and colored, just as the present experiment requires.

(a) §I, 16

Now, from the preceding it is evident that the breadth of the image *PT* always becomes smaller as the farther hole φ is made narrower, so that it cannot be doubted that the breadth would completely vanish if instead of that transparent hole there were only one extremely bright point there. That this will happen is confirmed by a not dissimilar observation I once made of the planet Venus. The room was of course closed off everywhere, except for a hole slightly more than two inches wide, in order to make it exceedingly dark. In this hole I placed the object glass of a seven-foot telescope with an aperture more than two inches wide, in order to transmit a sufficient number of rays. Then placing a paper transversely at a distance of seven feet, I saw the planet's image projected on it like a bright point. When one or two feet away from that paper a prism was inserted that refracted the transmitted rays elsewhere; instead of that bright point, distant more than a foot from there, I saw a small line which, though not very bright, I nonetheless easily perceived. Its length exceeded half an inch, whereas it had no sensible breadth, or at least none greater than could be perceived. I believe that the same thing might be observed for first magnitude stars, such as Sirius, especially if a lens four or six inches wide is used, so that it transmits many rays.

Lecture 3
20 [= 19]. This is extended still further by the image of the planet Venus.

I might usefully observe how well this experiment agrees with my explanation that I presented at the beginning concerning the different refraction of rays incident at the same angle. In the first figure I supposed that very many rays were conveyed successively along the same straight line onto some refracting surface and that there some of them were refracted gradually a little more than the others. But if this were imagined to occur, it would consequently follow that if the rays thus refracted were then intercepted by any opaque body, such as a paper, they would depict[(1)] a small bright line there. Now, although the rays coming from the planet do not all proceed in the same straight line, they can nonetheless be considered parallel, which is equivalent; and because they are made to converge by the convex lens before they reach the prism, this does not destroy the analogy but rather greatly strengthens it. For instead of individual rays proceeding in the same straight line, you should imagine just as many pencils of rays all having the same axis and the same point of intersection; and because some of those pencils are

21 [= 20]. And it is applied to the description of refraction related at Fig. I, 1.

rum alij magis alijs a Prismate refringuntur, ita ut eorum puncta concursus sive foci qui priùs coincidere, jam singuli cadant seorsìm, lineam rectam conficientes. Ac proinde quòd axes penicillorum, qui radijs, puta succ[e]ssivis, eousque coincidebant donec attigere Prisma, ibi per variam refractionem sint effecti divergentes, ut ad focos penicillorum in lineâ rectâ jacentes pergant.

Si Prisma stellae Veneris vicinius quam lentem collocaveris, ut radij per illud trajiciantur primò, et a lente deinde convergentes fiant: eandem lineolam ut priùs videbis licet minùs conspicuam et inventu difficilliorem. Jam in hoc specimine cum radij omnes adveniant paralleli, si aequaliter refringerentur transientes Prisma, manerent postea paralleli usque dum lenti inciderent; et in eâ proinde sic refringerentur, ut omnes deinceps ad idem punctum pergerent, et sic punctum lucidum conspiceretur. Quare cùm vice puncti istius apparet linea, concludendum est quod omnes radij non aequalitèr refringuntur.

Si jam objiciat aliquis quod in refractionibus detur quidem irregularitas, sed eam esse contingentem et non ex praevia radiorum dispositione, vel ullis certis legibus ortam. Respondeo, quod imago solis praefata si radijs nullâ certâ lege refractis fieret oblonga, non posset in lineas rectas secundum longitudinem suam distincte terminari; sicut ad figuram 5^{tam} ostensum est: Quin-etiam non omnino deberet esse oblonga, sed parte ejus mediâ, et magìs splendidâ in morem orbis effingi, sensibilique termino distingui ab erratica luce debiliori quaquaversum dispersâ. Perinde ut Sol apparet cum nubibus penè obscuratur; vel ut ejus / imago cernitur cum trajicitur per laminam vitream parallelis planis terminatam et halitu vel fumo leviter obductam, ut lux inter refringendum paululum conturbetur. Adhaec [Fig I, 10] si duo Prismata similia *ABC* et *αβκ* juxtaponantur secundum longitudines suas parallela, cum lateribus planis *AC* et *ακ*, ut et *BC* & *βκ* parallelis: et si sol transluceat utrumque in locum *Z*, ubi corpus opacum luci directè opponitur:[(3)] radijs tamèn ejus per orbiculare foramen *F* prius trajectis: Lux incidens in dictum *Z* apparebit distinctè orbicularis, non secus quàm si directè tenderet ab *F*, Prismatibus

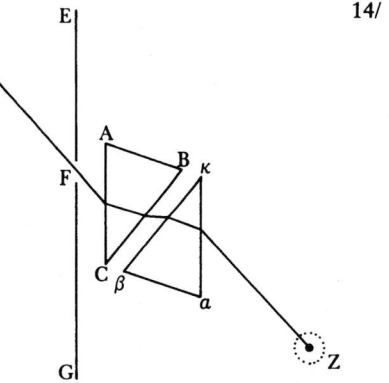

14/

Figure I, 10

non omninò interpositis. Fatendum est itaque quod utriusque Prismatis conjunctim refractiones sunt regulares, et proinde etiam refractiones alterutrius. Scilicet radij illi similiter incidentes non omnes aeque refringuntur in primo Prismate *ABC*, ut neque in secundo *αβκ*: tamen cum ea refractionis inaequalitas non contingens sit; sed oriatur ex praevia radiorum dispositione: ideo licet varij radij variè refringantur tamen ejusdem radij eadem erit refractionis quantitas in utroque Prismate, et quantum incurvatur a priori *ABC* tantum recurvabitur a posteriori *αβκ*. Unde radius quilibet utcunque sit refrangibilis postquam ex utroque Prismate emerserit, sibimet ipsi cum nondum ijs inci-

refracted by the prism more than others, their points of intersection, or foci, that previously coincided, now each fall separately and form a straight line. Consequently, because the axes of the pencils coincided with the rays, which are assumed to be successive, until they reached the prism, where they were made to diverge by the varying refraction, they proceed to the foci of the pencils that lie in a straight line.

If you have placed the prism rather than the lens nearer to the planet Venus, so that the rays are transmitted through that prism first and then are made to converge by the lens, you will see the same small line as before, although less visible and harder to find. Now, since in this example all rays arrive parallel, if they were equally refracted in crossing the prism, they would afterward remain parallel until they fell on the lens. Consequently, they would be so refracted in it that they would then all proceed to the same point, and thus a bright point would be seen. Therefore, since a line appears instead of such a point, it must be concluded that all rays are not equally refracted.

22 [= 21]. A changed feature is again introduced to the same description.

If someone now objects that there is indeed an irregularity in refractions, but that it is contingent and does not arise from a previous disposition of the rays or from any definite laws, I respond that if that image of the sun became oblong by rays that are refracted according to no definite law, it could not be distinctly terminated by straight lines along its length, as is shown in Fig. I, 5. Moreover, it ought not to be oblong at all, but in its middle and brighter part it ought to be shaped circularly with a boundary perceptibly distinguished from the scattered weaker light dispersed all around, just as the sun appears when it is almost obscured by clouds, or as its image is seen when it is transmitted through a glass plate bound by parallel planes and lightly covered with breath or smoke, so that the light is somewhat disturbed while being refracted. Furthermore (Fig. I, 10), if two similar prisms ABC and $\alpha\beta\kappa$ are placed next to one another, parallel along their lengths with their plane sides AC and $\alpha\kappa$ as well as BC and $\beta\kappa$ parallel; and if the sun shines through both to the place Z where an opaque body directly faces[3] the light, with its rays first having passed through the circular hole F, the light falling upon Z will appear distinctly circular, just as if it had traveled directly from F with no prisms at all placed in between. It must therefore be admitted that the refractions of both prisms together are regular and consequently also the refractions of each one. That is to say, those similarly incident rays are not all equally refracted in the first prism ABC nor in the second one $\alpha\beta\kappa$; yet, since this inequality of refraction is not contingent but arises from a previous disposition of the rays, then although different rays are differently refracted, the quantity of refraction of the same ray will nevertheless be the same in each prism, and as much as it is bent by the first prism ABC it will be bent back by the second one $\alpha\beta\kappa$. Consequently, any ray, however refrangible, after emerging from both prisms will become parallel to itself when it has not

23 [≈ 22]. The refractions in the experiments brought forward do not become unequal by chance nor by any other cause than[2] unequal refrangibility.

---

(2) neque . . . quàm] Originally (as I): sed ex (but by).
(3) I: objicitur.

derat fiet parallelus. Atque ideò cum omnes ad easdem plagas tendant ad quas liberè tenderent, si Prismatibus non interciperentur; necesse est, ut eandem orbicularem imaginem ad Z exhibeant quas illuc liberè tendentes exhiberent. Quod si imago oblonga per refractionem unici Prismatis (ut dictum est) effecta, figuram suam a radijs nullâ certâ lege divaricantibus, sed forte fortuna huc illuc vagè refractis acquireret; cùm refractiones binis Prismatibus geminentur, errores etiam radiorum duplo plures evaderent, ut et duplo majores. et exinde imago ad Z fieret multo oblongior, quae tamen, experientia teste, in orbem contrahitur.

Nonnullis fortè in suspicionem veniet quod terminatio lucis, sive quiescentis medij confinium diversitatem refractionis efficiat; sed huic dubitationi in promptu est remedium; efficiendo nempe ut lux a posticâ parte Prismatis (sicut ad Fig I, 3) solummodo terminetur ne fiat umbrae confinis priusquam fuerit refracta.[4]

Et praeterea ne suspicio sit de variâ crassitie vitri, potest refractio ejus ad varias crassities tentari promovendo Prisma / transversè juxta lucis ingressum parallelo motu, ita ut lux primò ad aciem ejus trajiciatur, deinde ad partes crassiores; et in quovis casu persimilis erit colorum apparitio.[5] Neque multum interest, si foramen per quod lux ingreditur sit latius vel angustius nam exinde nihil aliud eveniet quàm lucis colores exhibentis augmentatio vel diminutio, ac tanta dilat[at]io vel contractio imaginis quanta est foraminis.[6]

Experimento duorum parallelorum Prismatum jam antè descripto, constat etiam quod haec imaginis in longitudinem distractio non oritur ex ejusdem cujusque radij diffusione vel diffractione[7] in complures divergentes radios: Siquidem illi per iteratam diffusionem vel diffractionem[7] in transitu per secundum Prisma tunc resolvi deberent in longè plures et magis divergentes radios.

Quin et ijsdem omnibus objectionibus adversatur experimentum ubi posterius Prisma non statuitur parallelum anteriori sed perpendiculariter transversum: Nam in isto casu si anterius Prisma distraheret imaginem in longitudinem ob aliam quamcunque causam quàm diversam refrangibilitatem diversorum radiorum, tunc posterius Prisma per transversam refractionem distrahere deberet illam oblongatam imaginem in latitudinem, et sic quadrilateram efficeret. sed experimentum tentanti res secùs evenit, imagine scilicet non secundum latitudinem dilatatâ, sed solùm obliquata per majorem refractionem

---

(4) Newton probably had in mind here Descartes and Hooke for whom the formation of the spectrum depended on an adjacent shadow; on Descartes see the Introduction, note (23). Hooke's "split or rarifyd" rays spread into the adjacent quiescent or unilluminated medium to produce both the elongation and colors of the spectrum; see *Micrographia*, Observation IX, pp. 59, 63; and Hooke's critique of Newton's theory in his letter to Oldenburg on 15 February 1671/2, *Correspondence*, 1:110–4. Pardies's theory likewise depended on the boundary of light and shadow; see Pardies to Oldenburg for Newton, 11 May 1672, *Correspondence*, 1:156–7 = *Phil. Trans.*, 7 (1672):5012. Newton returned to reject this common explanation of the origin of spectral colors in §§II, 32–3 = 45–6 and §II, 89 ≈ 92.

(5) Newton perhaps learned of this popular explanation from Charleton, who, in turn, apparently adopted this idea from Kircher's *Ars magna*, Bk. I, Pt. III, Ch. IV, p. 75. Charleton held that colors arise from a mixture of light and shadow, with the blues containing a greater and the reds a lesser mixture of shadow. Blue appears toward the base of the prism, because the rays

yet fallen on the prisms. Hence, since all rays tend to the same places to which they would tend freely if they were not intercepted by the prisms, it necessarily follows that they exhibit the same circular image at Z that they would exhibit had they freely tended there. If, however, the oblong image made by the refraction of one prism (as was said) acquired its shape from rays that are spread about not by a definite law but are refracted scatteredly here and there by mere chance, then when the refractions were repeated by two prisms, the straying of the rays would also be doubled and doubly great, and so the image at Z would become much more oblong; yet by the evidence of experience it is contracted into a circle.

Some may perhaps come to suspect that the termination of the light, or the boundary of the quiescent medium, may cause the diversity of refraction, but the cure for this doubt is at hand, namely, by causing the light to be terminated only by the rear part of the prism (as in Fig. I, 3), so that it does not become bounded by shadow before being refracted.[4]

Moreover, lest the varying thickness of the glass be suspected, its refraction may be tried at various thicknesses by moving the prism with a parallel motion across the entrance of the light so that the light is first transmitted at its sharp edge and then at its thicker parts, and in every case the appearance of its colors will be altogether similar.[5] Nor does it matter much if the hole through which the light enters is wider or narrower, for nothing else will result from this other than an increase or decrease of the light exhibiting the colors, and the image expands or contracts as much as the hole.[6]

From the experiment with two parallel prisms just described, it is also evident that this lengthwise expansion of the image does not arise from a diffusion or splitting[7] of one and the same ray into very many divergent rays, since they then ought to be separated into more numerous and more divergent rays by the repeated diffusion or splitting[7] in their passage through the second prism.

Moreover, all these same objections are countered by an experiment where the second prism is not placed parallel to the first one but perpendicularly across it. For in that case, if the first prism expanded the image lengthwise from any other cause than the different refrangibility of the different rays, then the second prism by its transverse refraction ought to expand that oblong image in its breadth and thus make a quadrilateral one. But when the experiment is tried, it turns out differently; that is, the image is not dilated in its breadth but is only inclined by the greater refraction of the red end than

---

"are trajected through it by a longer tract or way, than those arriving at or nearer to the *Top* thereof: and therefore, the Glass being in that part most crass, there must be more impervious particles obsistent to the Rayes of Light; each one whereof ... causeth that the number of shadowes is multiplyed ..." (*Physiologia*, p. 194). Newton briefly considers and rejects this explanation in the opening paragraphs of the "New theory."

(6) Diffraction will of course appreciably affect the size of the image when the hole is sufficiently small, but Newton does not seem to have learned of Grimaldi's discovery of diffraction, which was published in 1665, until 1672 or somewhat later.

(7) *Editio princeps*: distractione, distractionem.

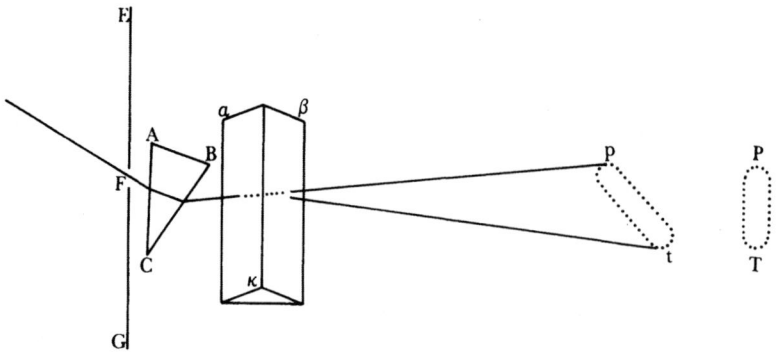

Figure I, 11

extremitatis rubrae quàm violaceae.[8] Quemadmodum videre est ad Fig I, 11, ubi Imago *PT* per secundi Prismatis refractionem transfertur ad *pt*.[9]

24 [≈ 22, 23].
Perstringuntur alia
experimenta
praecedentibus[10]
affinia.

Ex dictis opinor satis superéque constat id quod initio proposui commonstrandum; quoniam verò jucunditatem intellectui et assensum plerunque firmiorem harmonia rerum plurium affert quam unici licèt maximè scientifici argumenti testimonium: non erit abs re si in aliud experimentorum genus praecedentibus affinium experturos breviter introducam.

In fig I, 12[11] sit *F* foramen valde exiguum per quod lumen solis trajiciatur: deinde ad distantiam pro lubitu magnam statuatur Prisma *ABC* per quod radij transeant refracti; pro ut in prioribus explicui: Tum oculo ponè admoto, circularis foraminis *F* videbis imaginem *TP* oblongam: cujus longitudo ad latitudinem collata tanto major erit quanto foramen *F* fi[e]t angustius. / Et exinde pateat quod ra-

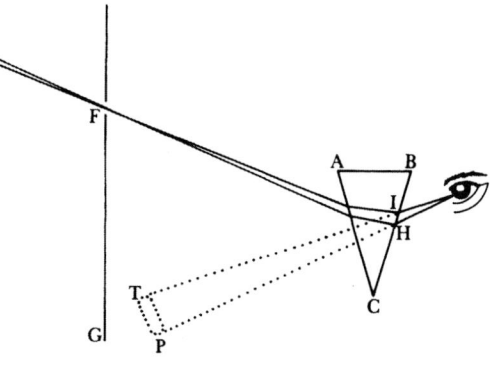

Figure I, 12

16/

diorum alij tendentes ad oculum per *H* quasi manassent a *P*, sunt magis refracti quàm alij tendentes per *I* quasi a *T* venissent. Et radijs sic in oculum non secus ingressis quam si profluxissent ab oblongo spatio *PT*, necesse est ut spatium istud longum appareat luminosum.

Sed cavendum est ne foraminis *F* tanta sit apertura ut nimiae lucis introitu laedatur oculus: imò ne tanta sit quin ut possis nudo oculo particulam solis per foramen istud quasi punctum lucidum distinctè et absque ullâ circumradiatione transpicere. Verum si lumen solis censeatur nimium, huic experiendo lumen a nubibus transmissum sufficiat; modò talis sit oculi tui dispositio ut foramen sine radijs circumcirca superfluis distinctum cernas antequam interponas Prisma: aliàs imaginem ejus non cernes distinctam neque debitâ longitudine diductam. Adhaec liceat tantundem observare, si filum albens inter-

the violet[8] one, as is seen in Fig. I, 11 where the image *PT* is shifted to *pt* by the refraction of the second prism.[9]

From what I have already related I believe that I have more than suffi-ciently established what I initially proposed to demonstrate. Since, however, the agreement of several things imparts an intellectual pleasure and a gener-ally more assured acceptance than the evidence of a single, though highly scientific, argument, it will not be without benefit if I briefly introduce inves-tigators to another kind of experiment related to the preceding ones.

24 [≈ 22, 23]. Other experiments related to the preceding ones[10] are briefly considered.

In Fig. I, 12[11] let *F* be a very small hole through which the sun's light is transmitted; then at a large distance chosen arbitrarily place a prism *ABC* through which the rays pass and are refracted, as I explained earlier. Then putting your eye right behind it, you will see the oblong image *TP* of the circular hole *F*, and its length compared to its breadth will be greater as the hole *F* becomes narrower. From this it is evident that those rays tending to the eye through *H*, as if they had proceeded from *P*, are refracted more than the others tending through *I*, as if they had come from *T*. Thus, since the rays entered the eye no differently than if they had flowed from the oblong space *PT*, it necessarily follows that that long space appears illuminated.

You must be careful, however, that the opening of the hole *F* is not so large that your eye is hurt by the admission of too much light; in fact, it should be just large enough that with your naked eye you can see through that hole a small part of the sun like a bright point distinctly and without any surround-ing radiation. If you consider the sun's light too great, light transmitted from clouds is sufficient for doing this experiment, provided that the location of your eye is such that before you insert the prism you can see the hole dis-tinctly without any superfluous rays all around; otherwise you will see its image neither distinctly nor drawn into its proper length. Moreover, you may

---

(8) rubrae . . . violaceae] Read (as *editio princeps*): violaceae . . . rubrae (violet . . . red).

(9) Quemadmodum . . . *pt*.] Added. The crossed-prism experiment together with the parallel-prism experiment described in the preceding paragraph are Newton's two major weapons in his attack against Hooke's and Pardies's "diffusion theories"—theories in which both the elongation and colors of the spectrum are conceived to arise from a diffusion or spreading on the sides of the rays—as well as any other explanation that does not admit unequal refrangibility. In the *Lectiones opticae* the crossed-prism experiment was directed only against the general idea that the elongation of the spectrum resulted from a "contingent cause" rather than unequal refrangi-bility; but in his letters to Pardies and Hooke Newton directed it, just as in this addition, specifically against their diffusion theories. Similarly in the *Lectiones opticae* (§32 ≈ II, 7, 8) the crossed-prism experiment was used simply to demonstrate that rays of different colors are refracted unequally, but in the revision (§II, 11) and in his letter to Hooke it was directly opposed to the diffusion theories. See Newton's letters to Oldenburg for Pardies on 10 June, and for Hooke on 11 June 1672, *Correspondence*, 1:165–6, 178–9 = *Phil. Trans.*, 7 (1672):5015–16, 5092–3. Since both the long addition to this article and §II, 11 parallel Newton's disputes with Hooke and Pardies, it is quite possible that they were added after the *Optica* was composed but before it was transcribed for deposit in the University Library in 1674. See the Introduction, §3; Shapiro, "Newton's definition of a light ray"; and Mamiani, *Newton*, p. 26, note 24.

(10) I: prioribus (the previous ones).

(11) In the corresponding Fig. 10 in the *Lectiones* there is a complete eye, where the rays cross and proceed to the retina.

posito Prismate aspicias, etenim filum multò latius apparebit cùm in situ ad longitudinem Prismatis parallelo quam cùm in transverso statuitur. Caeterùm ut in uno comprehendam omnia, si stellam fixam primae magnitudinis mediante Prismate intuearis, ejus etiam imago conspicietur longa: At cùm radij stellarum pro parallelis habeantur, si omnes aequè refringerentur, manerent etiam paralleli postquam egrediuntur Prismate et oculum sic ingressi efficerent imaginem omninò similem stellae vel puncto lucido et nullatenus oblongam; perinde ut fit cùm stella parallelos radios in oculum directè mittit. Videtis itaque quod radij paralleli superficiebus planis refracti fiunt inclinati, unde necesse est ut inaequalem refractionem patiantur. In transitu autem notetur quod Telescopio si placeat primùm adhibito, tum ut copia lucis ad oculum transmittatur tum ut scintillatio qua fixae quasi coronâ solent cingi minuatur; et Prismate deinceps interposito videbis albicantem / lineam dis- 17/ tinctiorem quàm priùs, cum latitudine vix aut ne vix quidem conspicua. His paucis de radiorum diversâ refrangibilitate narratis, quorum sensus plenior in sequentibus ubi coloribus agitur elucescet: restat ut refractionum quantitates et mensurae jam determinentur.

SECTIO 2da
DE MENSURÂ REFRACTIONUM.

Lect 4
25 [≈ 98]. De mensura refractionis dati generis radiorum e quavis incidentia data.

Refractiones ope angulorum quos incidentes et refracti radij cum perpendiculo refringentis plani constituunt, quasi datam rationem habentium a Veteribus determinatae fuerunt. Quemadmodùm si in fig I, 13 *IH* sit planum refringens cui linea *DCE* ad aliquod ejus punctum *C* perpendiculariter insistit. Et in illud *C* radius quilibet *AC* incidat et refringatur ad *R*: Posito refractum radium *CR* in plano *ACI* jacere quod refringenti plano perpendiculare est; supposuêre Veteres quòd angulus incidentiae *ACD*, angulus refractionis *RCE*, et angulus refractus *RCF* semper sint in datâ quâdam ratione, vel potiùs Hypothesin credidere satìs accuratam esse ubi radij a perpendiculo non multùm divaricant. Sic

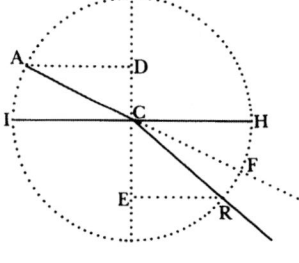

Figure I, 13

in vitro statuerunt angulum refractionis[1] quasi triplum esse anguli refracti. ‖ At illa refractionum aestimatio minùs exacta deprehenditur, quàm ut pro fundamento dioptrices debeat statui, et Cartesius aliam regulam primus[2] excogitavit, quâ istud exactiùs determinaretur ponendo dictorum angulorum sinus esse in ratione datâ. In fig I, 13 si centro *C* et distantiâ quâlibet *AC* circulus describatur secans radios praefatos in *A* et *R*, et ab istis punctis ad plani perpendiculum *DCE* demittantur normales *AD* et *RE*; ipsarum *AD* et *RE* proportio erit eadem perpetuò. Cujus rei veritatem Author non inele-

---

(1) Read (as I): incidentiae (of incidence).

observe the equivalent if you look at a white thread through an interposed prism, for the thread will appear much broader when it is placed in a position parallel to the length of the prism than when it is transverse. However, to encompass everything in one, if you look through a prism at a fixed star of the first magnitude, you will also perceive its image to be long. Since the rays from stars can be considered parallel, if they were all equally refracted they would also remain parallel after they left the prism, and so entering the eye would produce an image totally like the star, or a bright point, and not at all oblong, just as what happens when the star sends parallel rays directly into the eye. You therefore see that parallel rays refracted at plane surfaces become inclined; consequently, it necessarily follows that they experience an unequal refraction. In passing, however, it may be noted that if it seems appropriate to use a telescope first, both because it transmits an abundance of light to the eye and because the twinkling that generally encircles the fixed stars like a crown is diminished, and then the prism is inserted, you will see a whitish line that is more distinct than before with its breadth hardly or not even visible. ‖ Having related these few things about the diverse refrangibility of rays, the fuller meaning of which will become apparent in the following when colors are considered, it now remains to determine the quantities and measures of refractions.

SECTION 2

THE MEASURE OF REFRACTIONS

Men of old determined refractions by means of the angles that the incident and refracted rays make with the perpendicular to the refracting plane, as if they had a given ratio. For instance (Fig. I, 13), if *IH* is the refracting plane to which the line *DCE* stands perpendicular at some point *C* of it; and any ray *AC* is incident at *C* and refracted to *R*; and assuming that the refracted ray *CR* lies in the plane *ACI* that is perpendicular to the refracting plane; men of old supposed the angle of incidence *ACD*, the angle of refraction *RCE*, and the angle of deviation *RCF* to be always in a certain given ratio, or rather, they believed the hypothesis to be sufficiently accurate when the rays did not deviate much from the perpendicular. Thus in glass they considered the angle of refraction[1] to be about triple the deviation. ‖ Yet to be considered as a foundation of dioptrics, that estimate of refractions is found to be less exact than it ought to be. Descartes first[2] conceived another rule whereby that could be more exactly determined by setting the sines of those angles to be in a given ratio. If (Fig. I, 13) with center *C* and any radius *AC* a circle is described cutting the specified ray in *A* and *R*, and the normals *AD* and *RE* are dropped from these points to the [refracting] plane's perpendicular *DCE*, then the ratio of *AD* and *RE* will always be the same. The author would have

Lecture 4
25 [≈ 98]. The measure of refraction of rays of a given kind at any given incidence.

---

(2) The *editio princeps* notes here that in the *Principia* Newton attributes this discovery to Snel; see Lect. 9, note (5).

ganter demonstrasset, modò de causis physicis quas assumpsit nullum dubitandi locum reliquisset. / Ut ut, quoniam instrumentis in istum finem 18/
accuratè constructis[3] examinarunt aliqui, et veritati (quoad sensum) exactè
convenientem adinvenerunt, non dubitabimus pro fundamento statuere; hoc
solùm adhibito moderamine, quod cùm is de quibuslibet radijs indifferenter
affirmarit quàsi omnium persimilis fuisset refractio; nos tantùm affirmamus
de singulis eorum generibus seorsim spectatis,[4] ponendo quòd radiorum
aequè refrangibilium sinus refractionis sunt ut sinus incidentiae. Concipiamus aliquot genera radiorum secundum lineam *AC* in fig. I, 14 allapsa esse
ad punctum *C* ibique refracta per superficiem *IH*
puta mediocriter refrangibiles radios in *CR*, minimè
refrangibiles in *CT*, et maximè refrangibiles[5] in
CP, ac innumeros alios gradibus intermedijs plus
minùs refrangibiles per totum spatium *TCP* diffusos esse. Jam si ducatur *DCG* perpendicularis ad
planum refringens *IH*, et centro *C*, distantiâ quâvis
AC circulus ut priùs describatur secans radios dictos in *A*, *P*, *R*, *T*; atque ex istis punctis demittantur
perpendiculares *AD*, *PG*, *RE*, *TF* pro sinubus angulorum *ACD*, *PCG*, *RCE*, *TCF*: pono quod ut-

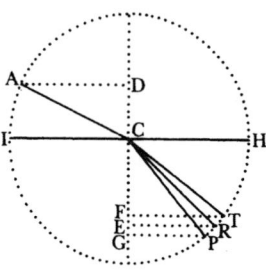

Figure I, 14

cunque radij incidant tamèn semper erit *AD* ad *PG* in eadem ratione; quâ
semel cognitâ regulam habes pro refractione radiorum maximè refrangibilium in eandem superficiem ad angulum quemvis incidentium mensurandâ.
Et sic semper erit *AD* ad *TF* in eadem ratione; quâ cognitâ regulam habes
quâcum refractio minimè refrangibilium in quavis incidentiâ determinabitur.
Atque idem de ratione ipsius *AD* ad *RE* et ad sinum cujusvis intermedij
generis concipiatur.

26 [= 99]. De   Porro autem cùm sinus *PG*, *RE*, *TF* caeterique datam habeant rationem ad
conferendis   sinum *AD*, datam quoque rationem inter sese habebunt, atque adeò si ex
refractionibus   unica observatione / proportionem sinuum *PG*, *RE*, *TF*, et reliquorum ad 19/
radiorum diversi   radios ex eadem incidentiâ refractos pertinentium cognoveris, regulam exinde
generis.[6]   habebis quacum ex sinu refractionis cujusvis generis radiorum, et in istam
superficiem utcunque incidentium dato, caeterorum omnium ex eâdem
incidentiâ prolabentium sinus elicias: licèt quaenam sit eorum incidentia non
innotuerit. Quinimò si omnium *AD*, *TF*, *RE*, *PG*, &c proportiones inter se
semel cognoscantur, habito respectu ad eadem media refringentia, regulam
habes pro caeteris omnibus exquirendis ex unico quovis unquam dato. Itaque
quo rationes istorum sinuum investigentur convenit ut in aliquo radiorum
genere proportio sinus incidentiae ad sinum refractionis primùm exquiratur:
deinde ut proportiones sinuum refractionis pro radijs diversorum generum ad
eundem angulum incidentium determinentur.

---

(3) *Editio princeps*: instructis.

(4) seorsim spectatis] Added.

(5) mediocriter refrangibiles ... minimè refrangibiles ... maximè refrangibiles] I originally:
maximè viridiformes ... intensissimè rubriformes ... intensissimè purpuriformes (the most

demonstrated not inelegantly the truth of this, if only he had not left room for doubt concerning the physical causes that he assumed. Since, however, some have tested this with instruments accurately constructed[3] for that purpose and have found it to agree exactly with the truth (at least as to sense), we will not hesitate to consider it as a foundation, subject only to this qualification: Whereas he had asserted this about every ray without distinction, as if the refraction of all of them was identical, we assert this about only the individual kinds of them considered separately[4] and assume that the sines of refraction of equally refrangible rays are proportional to the sines of incidence. Let us conceive several kinds of rays to flow along the line *AC* (Fig. I, 14) to the point *C* and to be refracted there by the surface *IH*, namely, the mean refrangible rays into *CR*, the least refrangible into *CT*, and the most refrangible[5] into *CP*, and innumerable others more or less refrangible by intermediate gradations to be spread through the entire space *TCP*. Now, if *DCG* is drawn perpendicular to the refracting plane *IH*; and if with center *C* and any radius *AC* a circle is described, as before, cutting the said rays in *A*, *P*, *R*, and *T*; and from those points the perpendiculars *AD*, *PG*, *RE*, and *TF* are dropped for the sines of the angles *ACD*, *PCG*, *RCE*, and *TCF*; I assume that however the rays are incident, *AD* to *PG* will nevertheless always be in the same ratio. Once this is known, you have a rule for measuring the refraction of the most refrangible rays incident upon the same surface at any angle. Similarly *AD* to *TF* will always be in the same ratio, and knowing this you have a rule whereby the refraction of the least refrangible rays will be determined at any incidence. The same thing is to be understood for the ratio of *AD* to *RE*, and for the sine of any intermediate kind.

Moreover, since the sines *PG, RE, TF*, and the others have a given ratio to the sine *AD*, they will also have a given ratio to one another. Hence, if from a single observation you know the proportion of the sines *PG, RE, TF*, and the other belonging to rays refracted at the same incidence, you will then have a rule whereby given the sine of refraction of any kind of ray, whatever its incidence on that surface, you may derive the sines of all the others proceeding at the same incidence, even if their incidence is not known. In fact, if the proportions of all of them, *AD, TF, RE, PG*, and so forth, to one another are once known with respect to the same refracting medium, you have a rule for finding all the others whenever any one is given. Therefore, to investigate the ratios of these sines it is proper first to find the ratio of the sine of incidence to the sine of refraction for some kind of ray, so that then the proportions of the sines of refraction for rays of different kinds incident at the same angle may be determined.

26 [= 99].
Comparing the refractions of rays of different kinds.[6]

---

green-making . . . most intense red-making . . . most intense purple-making). In revising this part of the *Lectures*, Newton consistently replaced the color of the rays by their degree of refrangibility, and henceforth we will note these changes only in the *Lectiones*; see Lect. 9, note (9).

(6) De . . . generis.] I: De mensurâ refractionum radiorum genere differentium ex eâdem quavis incidentiâ. (The measure of refractions of rays of different kinds at any identical incidence.)

27 [= 100]. Ad
sinus incidentiae
et refractionis
conferendos
adhibetur mediocre
genus radiorum.

Ad sinus incidentiae cum sinubus refractionis conferendos commodum erit, ut medium genus eligatur, puta genus illud radiorum qui viriditatem vel potiùs colorem viridi et caeruleo intermedium exhibent. Credo enim illos qui refractiones antehàc mensuravêre (sive id factum sit, ut jam dicta hypothesis Cartesij probaretur sive alijs de causis) credo illos, inquam, mensuram institutisse ad medietatem refractae lucis: hoc est, si spatium a coloribus occupatum spectemus, ad confinium viridis et caerulei; aut si spectemus quantitatem lucis, ad medietatem viridis. Et praeterea punctum istud pro principali foco lentium habendum esse videtur in quod intermedium genus radiorum convergit: atque etiam siquando de radijs indistinctè disserendum est, ut hactenus apud Opticae peritos[7] consueverit, genus mediocre commodiùs quàm extremorum aliquod pro omnibus haberi potest.

28 [= 101].
Modus explorandi
sinuum istorum
rationes.

Porrò cum fortè desideretur accuratius examen dictae regulae Cartesianae, quàm antehàc instituebatur, dum varia radiorum refrangibilitas experientes latuit: / primo dicam quo pacto id non incommode fiat. Quoniam fluidi 20/ pellucidi superficies refringentes facilè possint inclinari ad quemvis datum angulum, quod solido non est concessum, fluida in hunc finem fuerunt adhibita sed instrumento magis laborioso quàm opus erat, et erroribus fortè magis obnoxio quàm si omni apparatu privaretur, demptâ trabe cui vasculum aquae plenum affigitur. Sit itaque *HK* in fig I, 15 vectis ligneus duas tresvè ulnas longus aut campliùs, satis crassus ne ob longitudinem et pondus

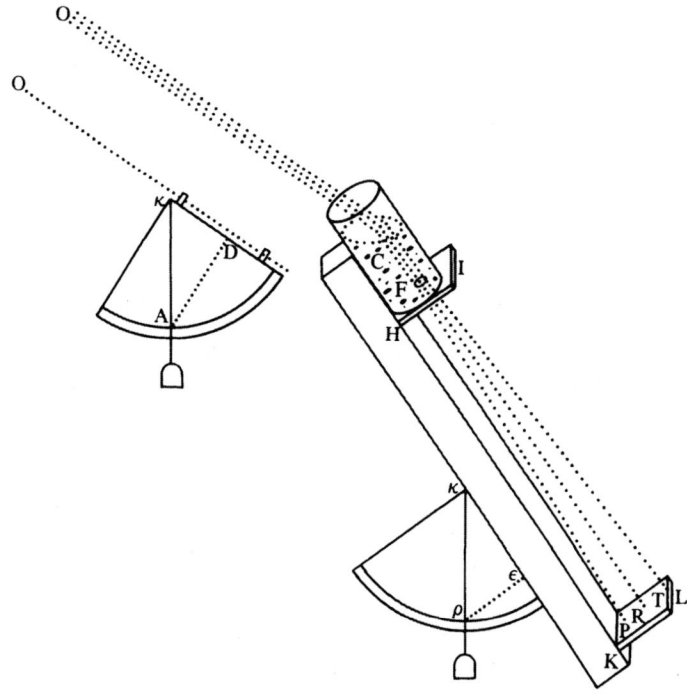

Figure I, 15

To compare the sines of incidence with the sines of refraction it will be appropriate to choose a middle kind, for example, that kind of ray that exhibits green, or rather, the color between green and blue. For I believe that those who have hitherto measured refractions (whether it was done to prove Descartes's hypothesis just described, or for other reasons), I believe, I say, that they determined their measure at the middle of the refracted light; that is, if we consider the space occupied by the colors, at the boundary of green and blue, or, if we consider the quantity of light, at the middle of the green. Moreover, it appears that that point ought to be taken as the principal focus of lenses at which the intermediate kind of rays converge; and also whenever rays must be discussed indiscriminately—as has hitherto been the custom with those skilled [7] in optics—the middle kind, rather than any of the extreme ones, can be more properly taken for all of them.

27 [= 100]. To compare the sines of incidence and refraction a mean kind of ray is used.

Moreover, since a more accurate investigation of that Cartesian rule may perhaps be desired than was previously undertaken while the diverse refrangibility of rays was unknown to experimenters, I first describe how that can be conveniently done. Since the refracting surfaces of a transparent fluid can easily be inclined at any given angle (something that is impossible with a solid), fluids were used for this purpose, but with an instrument more troublesome than was necessary and perhaps more prone to errors than if it were freed from the entire apparatus, except for a beam to which a vessel full of water is attached. Accordingly, let *HK* in Fig. I, 15 be a wooden beam, two, three, or more yards long, sufficiently thick (so that it is not at all able to

28 [= 101]. A method for investigating the ratios of those sines.

---

(7) I: scriptores (writers).

minimè inflecti queat, quadrilaterus, rectangulus, et rectus, cum lateribus oppositis exactè parallelis. Tum lamellae duae *HI,* et *KL* super unum ejus latus ad angulos rectos erigantur *KL* proximè ad unam extremitatem, et *HI* quasi quatuor digitos ab alterâ distans, quarum longitudo sit trium digitorum quatuorvè, latitudo autem duorum vel trium. Deinde sumatur vasculum aliquod cylindricum vel prismiforme *CF* duos tresve digitos latum, longum verò quatuor, vel quinque: ejus basis super lamellam *HI* figatur[8] cemento aliquo duro et tenaci ac in eo situ firmetur ope trabis *HK* ultra dictam lamellam *HI* productae. Tum trajiciatur ejus fundum in medietate, et lamella simul, parvo foramine *F* puta decimâ parte digiti lato: et juxta foramen istud in altera lamellâ notetur punctum *R,* quod aequè distat a trabe ac dicti foraminis centrum; ita scilicet ut linea *FR* per centrum foraminis ad *R* ducta sit parallela longitudini trabis. Denique sumatur lamella vitrea plana polita et uniformiter crassa eaque applicetur ad planitiem lamellae *HI* vasculo *CF* obversam super foramen *F* et cemento figatur ita ut vasculum istud aquae (quâ repleatur) non sit pervium. et cum normâ aliquâ fiat periculum, an illa vitrea lamella perpendicularitèr insistat trabi. Quod si non contingat, corrigatur situs donec sit exactè perpendicularis. In cujus rei gratiam convenit ut dicta lamella vitrea sit / tres vel quatuor digitos longa et lata, quò de situ ejus 21/ meliùs judicare liceat. Instrumento hoc sic fabricato et aquâ vasi *CF* plusquam ad medietatem ejus infusâ, illud in radijs solaribus ita statuatur, ut in superiori superficie aqueâ refracti perpendicularitèr emergant ad foramen *F,* rectàque progrediantur versus laminam *KL*; rubedine ad *T,* purpurâ ad *P,* et viridi vel confinio caerulei et viridis ad *R* incidentibus. Convenit autem ut dicta lamina *KL* dealbetur, aut albente papyro vestiatur, quò de coloribus judicium certius feras. Interea verò cum Quadrante aliquo amplo et exactè fabricato εκρ quaeratur inclinatio trabis *HK* ad horizontem et habebis angulum refractionis εκρ et ejus sinum ερ. Tum solis altitudo statim inquiratur ejusque complementum ad 90^{grad} AκD erit angulus incidentiae, et *AD* sinus. Quibus sinibus ad invicem collatis et experimento ad diversas solis altitudines repetito, constabit an sinuum ratio semper sit eadem. Quòd si velis ut experimenta varia simul fiant, aut ad minorem incidentiam quàm sit complementum maximae altitudinis solaris: vice radiorum a sole directè manantium possis adhibere reflexos.

Lect. 5
29 [= 102]. Modus
explorandi vim
refractivam Solidi
cujusvis aere
circundati.[1]

Cùm eandem sinuum incidentiae et refractionis rationem alicui radiorum generi utcunque in eandem quamvis superficiem incidenti perpetuò competere sat exploratum fuerit, proponatur exquirere rationem illam ad superficiem data quaelibet media disterminantem; idque unico experimento. Si aer sit unum ex datis medijs, et liquor quilibet alterum, Instrumentum novissimè descriptum non incommodè potest adhiberi. Sin mediorum alterum sit solidum, res expeditè perficitur ad diagramma I, 16. In cujus explicationem praemittantur sequentia duo Lemmata.

---

(8) Omitted in *editio princeps.*

(1) Solidi . . . circundati.] I: Medij cujusvis aëre circundati, praesertim verò Solidi. (of any medium surrounded by air, but especially of a solid.)

bend on account of its length and weight), quadrilateral, rectangular, and straight, with its opposite sides exactly parallel. Then erect two small plates, *HI* and *KL*, at right angles to one of its sides, with *KL* being nearly at the one end and *HI* about four inches from the other; and let their length be three or four inches, but their width two or three. Next take some small cylindrical or prismatic vessel, *CF*, two or three inches wide but four or five inches long; attach[8] its base over the plate *HI* with some hard and gripping cement, and let the vessel be supported in that position by means of the beam *HK* extending beyond that plate *HI*. Then pierce both its base and the plate in the middle with the small hole *F*, say, a tenth of an inch wide; and opposite that hole in the other plate mark the point *R*, which is just as far from the beam as the center of that hole is, namely, so that the line *FR* drawn through the center of the hole to *R* is parallel to the length of the beam. Finally, take a flat, polished, and uniformly thick glass plate and attach it to the side of plate *HI* facing the vessel *CF* above the hole *F*, and fasten it with cement, so that the vessel is impervious to the water with which it is filled. With any square, test whether that glass plate is perpendicular to the beam, and if it is not, correct its position until it is exactly perpendicular. For this purpose it is convenient that that glass plate be three or four inches long and wide, so that its position may be better judged. After thus constructing this instrument and filling the vessel *CF* more than halfway with water, place it in the sun's rays so that the rays refracted at the upper surface of the water emerge perpendicular to the hole *F* and proceed in a straight line toward the plate *KL*, with red falling at *T*, purple at *P*, and green, or else the boundary of blue and green, falling at *R*. It is convenient, moreover, to paint that plate *KL* white or to cover it with white paper, so that you may more accurately judge the colors. Meanwhile, with some large and exactly constructed quadrant $\epsilon\kappa\rho$ determine the inclination of the beam *HK* to the horizon, and you will have the angle of refraction $\epsilon\kappa\rho$ and its sine $\epsilon\rho$. Then immediately find the sun's altitude, and its complement to 90°, *AκD*, will be the angle of incidence and *AD* its sine. By comparing these sines to one another and repeating the experiment at different altitudes of the sun, it will be ascertained whether the ratio of the sines is always the same. But if you wish to make various experiments at the same time, or at a smaller [angle of] incidence than the complement of the greatest solar altitude, you can use reflected rays instead of rays flowing directly from the sun.

Since it has been sufficiently tested that the same ratio of the sines of incidence and refraction perpetually applies to any one kind of ray, however it is incident on any one surface, it is proposed to discover that ratio at a surface separating any given media—and indeed by a single experiment. If air is one of the given media, and any fluid the other, the instrument just described can be conveniently used. But if one of the media is a solid, the matter is readily carried out according to Fig. I, 16. For the explanation of this, I premise the following two lemmas:

Lecture 5
29 [= 102]. A way to investigate the refractive force of any solid surrounded by air.[1]

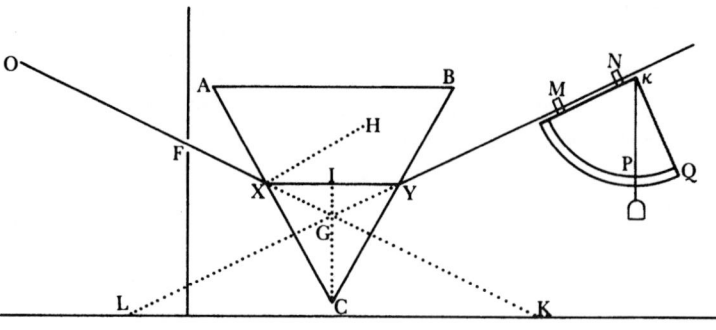

Figure I, 16

Lem: 1. In fig I, 16 sit *ABC* Prisma ex materiâ quavis pellucidâ confectum, cujus axis sit horizonti parallelus et / perpendicularis ad radios solis, et prae· 22/
terea sit ejus positio talis ut dictos radios *OX* aeque refringat ingredientes ad *X* et egredientes ad *Y*; istud autem quo pacto debet fieri ostensum fuit ad Num 10. Jam dico quòd angulus refractionis ad alterutram refringentem superficiem, ut *AC*, factae, sit aequalis dimidio verticalis anguli prismatici *ACB*. Scilicet ad punctum incidentiae *X* erigatur perpendicularis *HX*, eritque *HXY* angulus refractionis ad superficiem *AC*. Porrò demittatur *CI* perpendicularis in radium *XY*; et ista bisecabit angulum *YCX* propterea quod triangulum *YCX* (ob aequalitatem refractionis in *X* et *Y*) sit isosceles. Dico itaque quod anguli *HXY* et *ICX* aequantur. Nam ang. *AXY* = ang. *XIC + ICX* (per 32. 1. Elem.) Sed anguli *AXH* et *XIC* sunt recti; Ergo residui *HXY* et *ICX* aequantur.   Q.E.D.

Lem: 2. Ad hoc si radius incidens *OX* et emergens *YN* indefinitè producantur occurrentes in *G*, et praeterea si recta quaevis *KL* horizonti parallela radijs istis interjiciatur constituens triangulum *GKL*; Et cùm refractus radius *YN* tendit sursum si summa angulorum *LKX* et *KLY* sumatur, aut eorum differentia cùm iste *YN* tendit deorsùm: Dico quòd illius summae vel differentiae dimidium unà cum angulo refractionis *HXY* aequabitur angulo incidentiae *HXG*. Nam dicta summa vel differentia aequatur angulo *NGK* (per 32 1 elem) hoc est angulis *GXY + GYX*. et cùm triangulum *GYX* sit isosceles, dictae summae vel differentiae dimidium aequabitur angulo refracto *GXY* qui cum angulo refractionis *YXH* constituit angulum incidentiae. Q.E.D.

His praemissis problema propositum sic perficitur. Primò mensuretur angulus verticalis Prismatis *ACB* et ejus dimidium erit angulus refractionis. Dein Prismate in positione praefata disposito, per quod radij trajiciantur ingressi foramen *F*, ope Quadrantis *MNPQ* ampli et accurati (puta cujus pinnarum *M* et *N* distantia sit pedis unius minimùm) exploretur angulus *YLK* vel *PκQ* quem refracti radij *YMN* cum horizonte / constituunt: faciendo ut medio· 23/
criter refrangibiles per pinnas *M* et *N*, ad distantiam decem aut viginti pedum a prismate trajiciantur, et simul observetur solis altitudo *XKL*. Qui duo anguli addantur si refracti radij *YMN* sursum tendant, sicut in schemate

Lemma 1. In Fig. I, 16 let *ABC* be a prism made from any transparent material; let its axis be parallel to the horizon and perpendicular to the sun's rays; and moreover let its position be such that it equally refracts the specified rays *OX* upon entering at *X* and emerging at *Y*—and how that should be done was shown in § I, 10. Now, I assert that the angle of refraction made at either refracting surface, such as at *AC*, is equal to half the prism's vertex angle *ACB*.

Specifically, at the point of incidence *X*, erect the perpendicular *HX*, and *HXY* will be the angle of refraction at the surface *AC*. Furthermore, drop *CI* perpendicular to the ray *XY*, and it will bisect the angle *YCX*, because the triangle *YCX* is isosceles (on account of the equality of the refractions at *X* and *Y*). I say, therefore, that the angles *HXY* and *ICX* are equal. For $A\widehat{X}Y = X\widehat{I}C$ + $I\widehat{C}X$ (*Elements*, I, 32), but the angles *AXH* and *XIC* are right angles; therefore the remainders *HXY* are *ICX* are equal. As was to be demonstrated.

Lemma 2. Moreover, if the incident ray *OX* and the emergent ray *YN*, meeting in *G*, are extended indefinitely; if, besides, any line *KL* parallel to the horizon is placed between those rays and forms the triangle *GKL*; and if the sum of the angles *LKX* and *KLY* is taken when the refracted ray *YN* tends upward, or their difference when *YN* tends downward; I assert that half of that sum or difference together with the angle of refraction *HXY* will equal the angle of incidence *HXG*.

For the specified sum or difference equals the angle *NGK* (*Elements*, I, 32), that is, $G\widehat{X}Y + G\widehat{Y}X$. Since the triangle *GYX* is isosceles, half of the specified sum or difference will equal the deviation *GXY*, which together with the angle of refraction *YXH* constitutes the angle of incidence. As was to be demonstrated.

With these lemmas having been premised, the proposed problem is carried out as follows: First, measure the prism's vertex angle *ACB*, and half of it will be the angle of refraction. Then after the prism that transmits the rays entering the hole *F* has been arranged in the specified position, using a large and accurate quadrant *MNPQ* (for example, one whose sights *M* and *N* are at least one foot apart), determine the angle *YLK* or *PκQ*, which the refracted rays *YMN* make with the horizon, by causing the mean refrangible rays to pass through the sights *M* and *N* ten or twenty feet away from the prism. Also, at the same time observe the sun's altitude *XKL*. Add these two angles if the refracted rays *YMN* tend upward (as is drawn in the figure);

describitur; alias minor subtrahatur de majori: Et summae vel differentiae dimidium unà cum angulo refractionis priùs invento erit angulus incidentiae ut pateat per Lemma 2^{dum}. Denique ex angulis incidentiae et refractionis sic datis, dantur eorum sinus.   Q.E.F.

30 [≈ 103].
Exemplum in refractione cujusdam generis vitri.

Sic in prismate quodam vitreo dimensus sum angulum ejus maximum *ACB* et inveni 63^{grad}, 12′: Cujus dimidium *HXY* est 31^{grad} 36′: ejusque sinus 5240, posito sinu 90$^{grad.}$ 10000. Deinde cum altitudo solis *OKL* observabatur esse 14^{grad} 4′, alter angulus *MLK* a radio *YN* ad medium viriditatis tendente conflatus erat 30^{grad} 52′: quorum summa est 44^{grad} 56′ ejusque dimidium *YXK* 22^{grad} 28′: quod unà cum angulo refractionis *HXY* facit 54^{grad} 4′, angulum incidentiae, cujus sinus est 8097. Denique conferendo sinus jam inventos, ut eorum proportio in minimis terminis haberetur; inveni esse ut 11 ad 17 ferè. Quare pro regula generali statuendum est, quod radiorum viriditatem exhibentium sinus incidentiae ex aere in vitrum quodvis aequè refractivum ac illud prisma, sit ad sinum refractionis ut Septendecim ad undecim.

Haud secus dimetiendo refractionem radiorum colorem inter viridem et caeruleum exhibentium, investigatur 45^{gr} 8^{min} pro duplo anguli refracti[2] cujus dimidium 22^{gr} 34^{min} unà cum angulo refractionis 31^{gr} 36^{min} dat angulum incidentiae 54^{gr} 10^{min}. Ejusque sinus 8107 est ad sinum refractionis 5240 ut 82 ad 53 proximè.[3]

31 [= 104]. Modi praefati commoditas.

Hujus autem modi commoditas in mensurandis refractionibus ex eo conjicietur, quòd instrumento nullo hic opus sit, dempto Quadrante, et Prismate cujus refractio desideratur: quòd refractionem, dum geminatur, facta ad *X* et *Y*, exinde certiùs metiri possis: et quod facillimum sit Prisma in desiderato

Num 10.

situ disponere, ut supra ostenditur. Imò quòd parvus error a situ / desiderato   24/ ferè nihili sit, dum quoad sensum haud inde mutabitur angulus refractus *MGK*, ut experienti patebit. Quippe angulus iste tunc minimus est; et quantitatum per motum generatarum, cùm maximae existant vel minimae, hoc est in momento regressûs, motus utplurimùm sunt infinitè parvi. Sic verbi gratiâ in fig I, 17 si centro *C* describatur circulus *λLλ* et extra eum sumatur punctum quoddam *G*, ducaturque *GC*, et erigatur normalis *GK*; Deinde si concipiatur quòd punctum *λ* moveatur uniformiter in illius circuli circumferentiâ, per quod punctum recta quaedam *Gλ* circa centrum *G* rotata perpetuò transeat: manifestum est quòd quò major sit angulus *CGλ* sive quò minor angulus *KGλ*, eò minor erit motus angularis ipsius *Gλ*: et cùm angulus *CGλ* sit maximus, sive angulus *KGλ* minimus, hoc est in momento regressûs (rectâ *Gλ* tunc circulum in *L* tangente) motus ejus erit infinitè parvus, et quoad sensum nullus, parvusque error a

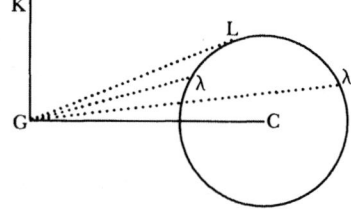

Figure I, 17

---

(2) That is, the deviation after refraction at a single face of the prism.

(3) Newton arrived at this approximation by converting to a unit continued fraction, where

$$\frac{8107}{5240} \approx 1 + \frac{1}{1+} \Big/ \frac{1}{1+} \Big/ \frac{1}{4+} \Big/ \frac{1}{1+} \Big/ \frac{1}{4} = \frac{82}{53},$$

otherwise subtract the smaller from the larger. Half the sum or difference together with the angle of refraction found earlier will be the angle of incidence, as is evident by Lemma 2. Finally, from the angles of incidence and refraction having been thus given, their sines will be given. As was to be done.

In this way I measured in a certain glass prism its largest angle *ACB* and found it to be 63°12', half of which, *HXY*, is 31°36', and its sine is 5240, assuming the sine of 90° to be 10000. Then since the sun's altitude *OKL* was observed to be 14°4', and the other angle, *MLK*, made by the ray *YN* tending to the middle of the green, was 30°52', their sum is 44°56', and its half, *YXK*, is 22°28', which together with the angle of refraction *HXY* makes the angle of incidence 54°4', whose sine is 8097. Finally, comparing the sines just found so that their proportion may be in the least terms, I found it to be about 11 to 17. Accordingly, it must be considered as a general rule that for rays exhibiting green the sine of incidence from air into any glass equally as refractive as that prism is to the sine of refraction as 17 to 11.

In the same way, by measuring the refraction of rays exhibiting the color between green and blue, 45°8' was found for double the deviation,[2] whose half, 22°34', together with the angle of refraction, 31°36', gives the angle of incidence, 54°10'. Its sine 8107 is to the sine of refraction 5240 approximately as 82 to 53.[3]

The advantage of this method of measuring refractions will be inferred from the following: No instrument is needed here except a quadrant and the prism whose refraction is sought; you may measure the refraction more precisely, because it is doubled at *X* and *Y*; and the prism may be very easily placed in the required position, as was shown above. Indeed, a small departure from the required position is almost inconsequential, insofar as the deviation *MGK* will not be perceptibly changed from this, as will be clear to anyone trying it. Namely, that angle is then a minimum, and when quantities generated by motion are either a maximum or minimum—that is, in the moment of regression—their motions are generally infinitely small. Thus for example (Fig. I, 17), if the circle λ*L*λ is described with the center *C*, some point *G* is chosen outside of it, *GC* is drawn, and the normal *GK* is erected; then if one imagines the point λ to move uniformly in the circumference of that circle, and a certain straight line *G*λ turning about its center *G* always to pass through that point, it is evident that the greater the angle *CG*λ, or the smaller the angle *KG*λ, the smaller will be the angular motion of *G*λ. When the angle *CG*λ is a maximum, or the angle *KG*λ a minimum, that is, in the moment of regression (the line *G*λ then being tangent to the circle at *L*), its motion will be infinitely small and imperceptible; and a small departure from

30 [≈ 103]. An example of the refraction of a certain kind of glass.

31 [= 104]. The advantage of the preceding method.

§I, 10

---

see Lect. 9, note (29), this volume. This is not a new measurement made after the composition of the *Lectiones opticae*, since it was already used in §110 ≈ I, 37. In the "Fundamentum Opticae" (Prop. 9, Expt. 27, Add. 3970, f. 411ʳ) and initially in the manuscript of the *Opticks* (Bk. I, Pt. II, Prop. III, Expt. VII, Add. 3970, f. 98ʳ), Newton reported new measurements on this prism for the rays between blue and green with the deviation differing by 23' and giving $\bar{n} = 1.5435 \approx$ 65/42. In §II, 116, he uses 65/42, but without specifying the kind of ray.

puncto contactûs *L* nullam sensibilem variationem in angulis istis *KGL* et *CGL* producet. Et ad eundem ferè modum parva convolutio prismatis haud omninò mutabit angulum *MGK*, cùm iste sit minimus, sive complementum[(4)] ejus maximum. Quod si prisma disponeretur in quovis alio situ quàm hic describitur (puta cùm radij perpendiculariter ingressi ad egressum duntaxat refringuntur) minimus error ab isto desiderato situ multùm mutaret angulum refractum, et sic experientia foret incertitudini et erroribus multò magis obnoxia.

32 [= 105]. Regula de investiganda refractione mediorum sibi ipsis contiguorum quorum aeri contiguorum refractiones cognoscuntur.

In majorem hujus rei copiam, quia dantur aliqui casus ubi refractiones per modos jam descriptos haud possint mensurari, (ut cum refractio fit ex vitro in Crystallum, ex aquâ in vitrum; vel ex uno liquore in alium), et nequa omninò sit refringens superficies cujus refractio nequit investigari, problema sequens lubet proponere.

Datis refractionibus quas duo media alicui tertio contigua conficiunt, illorum sibi ipsis contiguorum refractionem invenire.

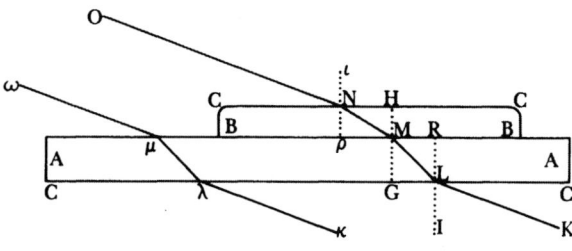

Figure I, 18

In fig I, 18 sunto duo media proposita *A* et *B,* quorum / superficiei disterminantis refractio quaeritur, et sit *C* medium tertium cujus superficiei ipsis *A* et *B* contiguae refractiones dantur, sitque sinus incidentiae ad sinum refractionis ex medio *C* in medium *A* sicut *I* ad *R,* et sinus incidentiae ad sinum refractionis ex eodem medio *C* in alterum medium *B* sicut *ι* ad *ρ.* Dico quod sit $I \times \rho \,.\, R \times \iota$ :: sinus incidentiae ad sinum refractionis ex medio *B* in medium *A.*

25/

Verbi gratiâ proponatur investigatio refractionis ex aquâ in vitrum, datâ refractione ex aere in utrumque, sitque sinus incidentiae ex aere in vitrum ad sinum refractionis ut 17 ad 11, et sinus incidentiae ex aere in aquam ad sinum refractionis ut 4 ad 3. Quare sinus istos multiplicando reciprocè, erit ut 17×3 ad 11×4, sive ut 51 ad 44, ita sinus incidentiae ex aquâ in vitrum ad sinum refractionis. Et sic cognitâ refractione ex aere in quaevis alia media proposita possis adipisci eorum refractionem inter se; et e contra.

33 [= 106]. Ejus regulae demonstratio.

Caeterùm demonstratio hujus non est omittenda, in quem finem praesternatur Lemma sequens. Si media duo proposita *A* et *B* in fig. I, 18 concipiantur esse planis parallelis terminata, contigua, et dicto Medio tertio (puta aere) circundata, et radius quilibet *ON* obliquè incidens ad *N* refringatur primò ad *M,* ac deinde ad *L* et emergens pergat ad *K*: Dico radium incidentem *ON* sibi

the point of contact *L* will produce no perceptible variation in those angles *KGL* and *CGL*. In almost the same way, a small rotation of the prism will not at all change the angle *MGK*, since it is a minimum, or its complement[4] a maximum. But if the prism were placed in any position other than that described here (as when rays entering perpendicularly are refracted only upon leaving), the least departure from that required position would greatly change the deviation, and the experiment would thus be much more liable to uncertainty and errors.

For greater facility in this matter, because some cases are given where the refractions cannot be measured by the methods already described (as when the refraction is made from glass into crystal, from water into glass, or from one fluid into another), and lest there be any refracting surface whose refraction cannot be investigated, it is desirable to propose the following problem:

32 [= 105]. A rule for investigating the refraction of media contiguous to one another, when their refractions contiguous to air are known.

> Given the refractions that two media make when contiguous to any third one, to find their refraction when they are contiguous to each other.

In Fig. I, 18 let *A* and *B* be the two proposed media, and the refraction of the surface separating them is to be found; let *C* be the third medium, and let the refractions of its surface when contiguous to *A* and *B* be given; and let the sine of incidence be to the sine of refraction from the medium *C* into the medium *A* as *I* to *R,* and let the sine of incidence be to the sine of refraction from the same medium *C* into the other medium *B* as ι to ρ. I say that $I \times \rho$ is to $R \times \iota$ as the sine of incidence is to the sine of refraction from the medium *B* into the medium *A*.

For example, the investigation of the refraction from water into glass may be proposed, given the refraction from air into each. Let the sine of incidence from air into glass be to the sine of refraction as 17 to 11, and the sine of incidence from air into water be to the sine of refraction as 4 to 3. Consequently by multiplying those sines reciprocally, then as 17×3 to 11×4, or as 51 to 44, so will the sine of incidence from water into glass be to the sine of refraction. Thus knowing the refraction from air into any other proposed media, you can obtain their mutual refraction, and conversely.

The demonstration of this, however, must not be omitted, and to this end the following lemma is provided beforehand. If the two proposed media *A* and *B* in Fig. I, 18 are conceived to be bounded by parallel planes, contiguous, and surrounded by the specified third medium (for instance, air); and if any ray *ON* incident obliquely at *N* is refracted first to *M* and then to *L* and upon emerging continues to *K*; I say that the incident ray *ON* is parallel to

33 [= 106]. The demonstration of this rule.

---

(4) Read: *supplementum* (supplement).

emergenti *LK* parallelum esse. Cujus quidem assertionis veritas experientiâ patet.[5] Etenim ponatur Medium *A* esse Vitrum, et Medium *B* esse aquam, Mediumque tertium circundans esse äera: Et laminae vitreae *A* superficies *ρMR* tenuitèr illinatur aquâ *B*, et statuatur parallela ad Horizontem, ut aqua consistat uniformiter crassa. Quo facto videbis, quòd radij per utrumque medium *A* et *B* trajecti tendent ad easdem plagas, versus quas tenderent a sole directi.

Praemisso hoc, erigantur *ιNρ*, *HMG*, et *RLI* perpendiculares ad refringentia puncta *N*, *M*, et *L*. Est ergo *ι* ad *ρ* ut sinus anguli *ONι* ad sinum anguli *MNρ*, sive *NMH*. Et multiplicando rationem antecedentem per *I* fiet *I* × *ι* ad *I* × *ρ* ut sinus ipsius *ONι* ad sinum ipsius *NMH*. Porrò est *I* ad *R*, / ut sinus 26/ anguli *KLI* sive *ONι*, ad sinum anguli *MLR* sive *LMG*. Et multiplicando rationem antecedentem per *ι*, fiet *I* × *ι* ad *R* × *ι* ut sinus ang: *ONι* ad sinum ipsius *LMG*. Jam permutando terminos utriusque proportionis fiet

$$I \times \iota . \text{sin: } ON\iota :: I \times \rho . \text{sin } NMH,$$

et *I* × *ι* . sin *ONι* :: *R* × *ι* . sin *LMG*. Quare ex aequalitate rationis est[6] *I* × *ρ* . *R* × *ι* :: sin: *NMH* . sin: *LMG*.   Q.E.D.

34 [= 107]. Modus dimetiendi refractiones solidorum ad fluida accommodantur.

Ex hisce sic ostensis problema non inutile proficiscitur, quo refractiones fluidorum eodem modo metiri possis, ac de solidis ostensum est ad fig I, 16, non adhibito instrumento *HILK* quod in fig I, 15 describitur.[7] Scilicet ex laminis vitreis in morem cunei connexis vasculum prismiforme conficiatur, cujus acies sive angulus verticalis sit 80$^{\text{grad.}}$ circiter vel 90. Istius autem anguli quantitatem exactissimâ mensurâ cognitam habebis, ejusque dimidij sinum pro sinu refractionis semper statues.[8] Quo peracto, cùm liquoris alicujus vis refractiva desideratur, vasculum cum illo liquore impleatur, et in tali situ disponatur, ut acies ejus a concursu refringentium planorum constituta, sit parallela ad Horizontem, et perpendicularis ad radios solares, atque ut illi radij per praefata refringentia plana trajecti refractiones ad ingressum et egressum aequales patiantur. Et ope Quadrantis, ut ostensum erat ad fig I, 16, exploretur angulus incidentiae, cujus sinus ad praefatum sinum refractionis erit ut sinus incidentiae ad sinum refractionis ex aere in liquorem propositum.

35 [= 108]. Refractio aquae, prout ipse dimensus sum, in specimen ejus rei adducitur.

Instantiae gratiâ, quo aquae refractionem cognoscerem curavi ut Prisma ligneum conficeretur, quale est *ABκ* in fig I, 19, cujus ille angulus *ACB*, quem pro verticali designabam, foret rectus, caeterique duo semirecti, et effeci ut refringentia plana *Aκ* et *Bκ* per Meditullium trajicerentur foramine *F* parallelo ad basem *Aβ*, per quod foramen lux itura esset; et ut tertium planum *Aβ* foderetur in *G* usque dum aditus ad foramen *F* transversè pertingeret. Dein sumptis duabus ex vitro lamellis, quas / speculum confractum mihi subminis-

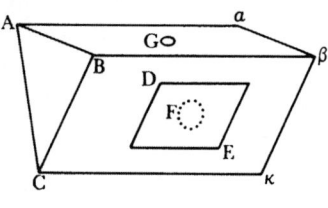

Figure I, 19   27/

---

(5) Originally (as I): patebit (will be evident).

(6) I continues with the intermediate step: *I* × *ρ* . sin *NMH* :: *R* × *ι* . sin *LMG*. Et permutando (*I* × *ρ* : sin $\widehat{NMH}$ = *R* × *ι* : sin $\widehat{LMG}$, and by a permutation).

itself emerging as *LK*. The truth of this assertion is certainly evident[5] from experience. For let it be assumed that the medium *A* is glass, the medium *B* is water, and the third, surrounding medium is air; and cover the surface *ρMR* of the glass plate *A* lightly with the water *B* and place it parallel to the horizon, so that the water is uniformly thick. Having done this, you will see that rays transmitted through both media *A* and *B* will tend in the same direction to which they would tend directly from the sun.

With this premised, erect the perpendiculars *ιNρ*, *HMG*, and *RLI* at the refracting points *N*, *M*, and *L*. Therefore,

$$\iota : \rho = \sin \widehat{ON\iota} : \sin \widehat{MN\rho} \text{ (or } \widehat{NMH});$$

and multiplying the preceding ratio by *I*, it will become

$$I \times \iota : I \times \rho = \sin \widehat{ON\iota} : \sin \widehat{NMH}.$$

In addition, $I : R = \sin \widehat{KLI}$ (or $\widehat{ON\iota}$) : $\sin \widehat{MLR}$ (or $\widehat{LMG}$); and multiplying the preceding ratio by *ι*, it will become $I \times \iota : R \times \iota = \sin \widehat{ON\iota} : \sin \widehat{LMG}$. Now permuting the terms of each proportion there will result $I \times \iota : \sin \widehat{ON\iota} = I \times \rho : \sin \widehat{NMH}$, and $I \times \iota : \sin \widehat{ON\iota} = R \times \iota : \sin \widehat{LMG}$. Therefore from the equality of the ratios,[6] $I \times \rho : R \times \iota = \sin \widehat{NMH} : \sin \widehat{LMG}$. As was to be demonstrated.

From these things thus demonstrated a rather useful problem arises whereby you can measure the refractions of fluids in the same way as that demonstrated for solids at Fig. I, 16 without using the instrument *HILK* illustrated in Fig. I, 15.[7] Namely, make a prismatic vessel from glass plates joined in the shape of a wedge whose edge or vertex angle is about 80° or 90°. You will, however, have to know the quantity of that angle from the most exact measurement, and always set the sine of half of it for the sine of refraction. Having carried this out, when the refractive force of any fluid is required, fill the vessel with that fluid and set it in a position so that its edge, formed by the intersection of the refracting planes, is parallel to the horizon and perpendicular to the solar rays, and also so that those rays transmitted through these refracting planes undergo equal refractions upon entering and emerging. Then using a quadrant, as was shown at Fig. I, 16, find the angle of incidence, the sine of which will be to the specified sine of refraction as the sine of incidence is to the sine of refraction from air into the proposed fluid.

For example, to learn the refraction of water I arranged for the construction of a wooden prism such as *ABκ* (Fig. I, 19) whose angle *ACB* (which I designated as the vertex) was a right angle and the other two half a right angle. I caused the refracting planes *Aκ* and *Bκ* to be bored through the middle with a hole *F* parallel to the base *Aβ* for the light to pass through; and I also caused the third plane *Aβ* to be pierced at *G* until a passage extended across to the hole *F*. Then taking two glass plates, which a shattered mirror

34 [= 107]. The method of measuring the refractions of solids is adapted to fluids.

35 [= 108]. The refraction of water, as I myself have measured it, is set forth as an example of this.

---

(7) That is, in §§I, 28–9 = 101–2.
(8) habebis . . . statues.] I: habeto . . . statue.

travit, unam *DE* super meditullium plani *Bκ* cum caemento fixi, et alteram super meditullium alterius plani *Aκ* ut meatus *F* utrinque clauderetur. Tum aquam pluvialem per orificium *G* in excavatum spatium infudi; et cum oper-culo ex subere conciso clausi. Atque adeò aqua duabus vitreis lamellis ad angulum rectum inclinatis interjecta vices subibat aqueae prismatis habentis angulum rectum. Eas autem laminas rectum angulum exactè comprehendere ex applicatione normae cognovi, cujus ideò dimidium 45^{grad} pro angulo re-fractionis habendum est. Hoc prisma dein ita statuebam ad ingressum lucis in obscurum cubiculum, ut eadem foret utrinque refractionis quantitas. et ex altitudine solis, et refractorum radiorum viriditatem exhibentium[9] inclina-tione ad Horizontem, inveni angulum refractum esse 51^{grad} 16^{min}, cujus dimi-dium 25^{grad} 38^{min}, unà cum angulo refractionis 45^{gr} dabit angulum incidentiae 70^{grad} 38^{min}. Horum verò angulorum 70^{grad} 38^{min} et 45$^{grad.}$ sinus sunt 9434 et 7071, respectu sinus 90^{graduum} 10000. Quorum quidem numerorum ratio est paulò minor quàm Cartesiana 250 ad 187, & paulo major quàm 4 ad 3 nempe 4,002 ad 3, quae tamen a ratione $\frac{4}{3}$ tam parvâ differentiâ recedit; ut error fuerit insensibilis, si posuerim esse ut 4 ad 3, idque maximè cùm aquae refractio non perpetim eadem maneat, sed a caloris vicissitudine nonnihil patiatur, variosque densitatis gradus induat. Quod idem et aeri circundanti contingit, qui a vaporibus etiam non solùm variè incrassatur, sed et arctiùs (auctâ Atmosphaerae gravitate) vel laxiùs comprimitur. Adde quòd aquarum ex diversis terrarum regionibus scaturientium aut vi solis in vapores et plu-viam deinde conversarum diversae sint densitates, et internae dispositiones ad refringendum ortae ex varijs mineralium tincturis, quas e locis subterraneis extrahunt, et exhalationibus variè crassis et[10] / copiosis quae simul cum 28/ vaporibus in altum attolluntur.

Lem 1. Num 29. (margin)

Lect. 6
36 [= 109].
Praefatorum demonstratio. (margin)

Problematis hujus de refractionis fluidorum mensurâ sic soluti veritas constabit ex ostenso, quod refractionis in hoc prismate ex aqua et vitris com-posito eadem sit quantitas, quae foret si vitrum tolleretur, et aqua sola ma-neret aere ci[r]cundata. Sit itaque *ABC* prisma, in fig: I, 20,[1] confectum ex laminis vitreis *ACφδ* et *BCφε* (ut dic-tum est) et aqua *δφε* repletum: et con-cipiatur quod *DEF* sit aqueum prisma immediatè circundatum aere, et om-ninò simile aquae *δεφ* circumclusae vi-tro, similiterque positum, et incidant radij paralleli *ON OX* in utrumque

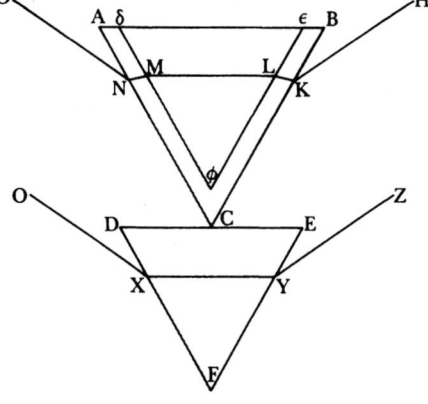

Figure I, 20

quorum alter *ON* refractus in *N, M, L,* et *K* tendit ad *H*, alter vero *OX* refractus in *X* et *Y* tendit ad *Z*. Dico jam quod emergentes *KH* et *YZ* erunt paralleli, atque adeò quod in utroque prismate tota refractionum quantitas

provided to me, I fastened the one plate *DE* with cement over the middle of
the plane *Bκ* and the other over the middle of the other plane *Aκ* in order to
seal the passage *F* on both sides. Then I poured rain water through the
opening *G* into the hollowed space and sealed it with a stopper cut from
cork. Hence the water, placed between the two glass plates inclined at a right
angle, assumed the role of a right-angled aqueous prism. But by applying a
square I determined that those plates comprehended exactly a right angle,
half of which, 45°, must therefore be taken for the angle of refraction. I then   Lemma 1, §I, 29
so placed this prism at the light's entrance into a dark room that the quantity
of refraction would be the same on both sides. From the sun's altitude and
the inclination to the horizon of the refracted green-exhibiting[9] rays, I found
the deviation to be 51°16′, half of which, 25°38′, together with the angle of
refraction, 45°, will give the angle of incidence, 70°38′. The sines of these
angles, 70°38′ and 45°, are 9434 and 7071 with respect to sin 90° as 10000.
The ratio of these numbers is in fact a little smaller than the Cartesian 250 to
187 and a little more than 4 to 3, namely, 4.002 to 3. Yet this deviates from
the ratio 4/3 by so small a difference that the error would be insensible if I
assumed it to be 4 to 3; particularly since the refraction of water does not
always remain the same, but it is affected somewhat by the change of heat
and assumes various degrees of density. The same thing happens to the
ambient air, which in addition is not only variably thickened by vapors but is
also compressed more closely (by the increased weight of the atmosphere) or
more loosely. Moreover, the density of the water flowing from different
regions of the earth or transformed by the force of the sun into vapors and
then into rain differs, and its internal disposition to refract also differs, because of various mineral tinctures extracted from subterranean locations and
variably thick and[10] abundant exhalations that are carried aloft together
with vapors.

    The validity of this problem of the measurement of the refraction of fluids   Lecture 6
solved in this way will be established by showing that the quantity of refrac-   36 [= 109]. The
tion in this prism composed of water and of glass is the same as it would be if   demonstration of
the glass were removed and only water surrounded by air remained. There-   the preceding.
fore (Fig. I, 20),[1] let *ABC* be a prism made (as was described) from the glass
plates *ACφδ* and *BCφε* and filled with water *δφε*; and imagine that *DEF* is a
prism of water immediately surrounded by air, entirely similar to the water
δεφ enclosed by glass, and similarly positioned. Let there fall upon each
prism the parallel rays *ON* and *OX*, of which the one ray *ON*, refracted at
N, *M*, *L*, and *K*, tends to *H*, whereas the other *OX*, refracted at *X* and *Y*,
tends to *Z*. I say now that the emergent rays *KH* and *YZ* will be parallel, and
hence that the total quantity of the refractions in each prism will be the same.

---

    (9) viriditatem exhibentium] I: viridiformium (green-making).
    (10) I: aut (or).
    (1) In the *Lectiones opticae*, Fig. 48, the prisms are right angled rather than equilateral.

erit eadem. Etenim in fig I, 18 si radius ωμ ipsi ON parallelus incidat in vitream laminam A, emergatque in λκ, notum est quod radius λκ erit parallelus ipsi ωμ hoc est ipsis ON et LK et cùm λκ et LK sint paralleli, erunt etiam μλ et ML paralleli. Unde liquet propositio quod quantitas refractionis ex aere in medium quodvis propositum sit eadem, sive radij immediatè ingrediantur istud medium ex aere (ut fit ad ωμλ) sive priùs permeent aliud medium interpositum, et parallelis planis terminatum (uti fit ad ONML) et e contra. Atque idem intellige, cùm vice aeris aliud quodpiam adhibetur medium. Quare in fig I, 20 cùm paralleli radij OX et ON incidant in Prismata DFE et ACB similia, et similitèr posita, refractionis quantitas ex aere in aquam erit eadem, sive radij immediatè intrent, ut videre est, ad DEF, sive priùs permeent lamellam vitream AδφC. hoc est radius XY semel / refractus erit paral- 29/ lelus ipsi ML bis refracto, et ob eandem rationem, cum XY et ML sint paralleli, radij emergentes YZ et KH erunt etiam paralleli. Quare cùm radij et incidentes et emergentes sint paralleli, refractio tota Prismatis utriusque erit eadem. Atque adeo cum aqueum Prisma aeri contiguum propter fluiditatem aquae fabricari nequeat ejus vice liceat adhibere vitreum Prisma cum aquâ repletum. Q.E.O.

Et sic modus generalis, quo refractiones ex aere in quaelibet media proposita determinentur, ostensus est; facillimus quidem et erroribus minimè obnoxius, praesertim si angulus Prismatis sit magnus et exactè cognitus, Quadrans magnus et accuratus, et observatio facta longè post Prisma ubi colores multum dilatati faciliùs distinguuntur. Et praeterea, cùm refractiones inter aerem et media proposita sic experientijs determinantur: indicata est regula, qua mediorum eorundem sibi ipsis contiguorum refractiones eliciantur. Quod satis est in gratiam primi casûs de refractionibus dimetiendis, cum in eodem quopiam radiorum genere proportio sinûs incidentiae et refractionis quaeritur, ostendisse.

Prosequendus est jam alter casus, ubi Heterogeneorum radiorum refractiones conferendae sunt. ‖ Quod autem sinus refractionis cujusque radiorum generis sit ad sinum incidentiae in data quadam ratione experiri possis dimetiendo refractiones singulorum insigniorum generum juxta varias obliquitates in medium aliquod refringens seorsim procidentium veluti in aquam (ad fig I, 15)[2] in vase stagnantem vel in Prismata vitrea, quorum diversae sint quantitates angulorum verticalium. Nam per unum Prisma proportiones / sinuum 30/ ad singula radiorum genera investigari possis prout ostenditur ad fig I, 16[3] deinde per alia Prismata (vel ejusdem Prismatis alios seu minores seu majores angulos) exquirere an eaedem proportiones in alijs obliquitatibus obveniant. Atque ita (observationibus exactissimè factis) simul constabit refractiones cujusque generis radiorum secundum certas rationes sinuum peragi, et istorum sinuum rationes innotescent. Impraesentia verò cùm eandem esse cujusque radij refractionem cognoverim sive heterogeneis radijs (ut in lumine solis nondum refracto) commistus incidat sive ab heterogeneis priùs separetur: Ostendam quomodo per refractionem immediati luminis solaris hae pro-

Num 32

37 [≈ 110].
Radiorum diversi generis refractiones conferuntur et maxima refrangibilitatis differentia investigatur.

---

(2) In §I, 28 = 101.   (3) In §I, 29 = 102.

For in Fig. I, 18, if the ray ωμ parallel to *ON* falls upon the glass plate *A* and emerges in λκ, it is known that the ray λκ will be parallel to ωμ, that is, to *ON* and *LK*; and since λκ and *LK* are parallel, μλ and *ML* will also be parallel. Whence the proposition is evident: The quantity of refraction from air into any proposed medium is the same whether the rays immediately enter that medium from air (as occurs with ωμλ) or first cross another interposed medium bounded by parallel planes (as occurs with *ONML*), and conversely. The same thing must be understood when in place of air any other medium is used. Accordingly, in Fig. I, 20, since the parallel rays *OX* and *ON* fall upon the similar and similarly positioned prisms *DFE* and *ACB*, the quantity of refraction from air into water will be the same whether the rays enter immediately, as can be seen at *DEF*, or first cross the glass plate *AδφC*; that is, the ray *XY* refracted once will be parallel to *ML* refracted twice. For the same reason, since *XY* and *ML* are parallel, the emergent rays *YZ* and *KH* will also be parallel. Therefore, since the incident and emergent rays are parallel, the total refraction of each prism will be the same. Hence, since an aqueous prism contiguous to air cannot be constructed because of the fluidity of water, one may use in its place a glass prism filled with water. As was to be shown.

Thus a general method has been presented whereby the refractions from air into any proposed medium may be determined, and indeed very easily and hardly subject to errors; especially if the angle of the prism is large and exactly known, the quadrant is large and accurate, and the observation is made far behind the prism, where the colors being greatly spread out are more easily distinguished. Moreover, when the refractions between air and the proposed media are thus determined by experiments, a rule has been disclosed whereby the refractions of the same media when contiguous to one §I, 32 another may be derived. It suffices to have shown this for the first case of measuring refractions, when the ratio of the sine of incidence and of refraction for any one kind of ray is sought.

The other case must now be pursued, where the refractions of heterogeneous rays are to be compared. ‖ That, however, the sine of refraction of each kind of ray is to the sine of incidence in a certain given ratio you may test by measuring the refractions of each of the more prominent kinds of rays falling separately upon some refracting medium at various inclinations, as upon water standing in a vessel (Fig. I, 15),[(2)] or upon glass prisms whose vertex angles are of different sizes. For with one prism you can investigate the ratios of the sines for the individual kinds of rays, as is shown in Fig. I, 16,[(3)] and then with other prisms (or other angles, either larger or smaller, of the same prism) you can examine whether the same ratios turn out at other inclinations. Thus (having made extremely exact observations) it will simultaneously be evident that the refractions of each kind of ray occur according to definite ratios of the sines, and also the ratios of those sines will become known. But for the present, since I have learned that the refraction of each ray is the same whether it is incident mixed with heterogeneous rays (as in sunlight not yet refracted) or is first separated from heterogeneous rays, I will show how these

37 [≈ 110]. The refractions of different kinds of rays are compared, and the greatest difference of refrangibility is investigated.

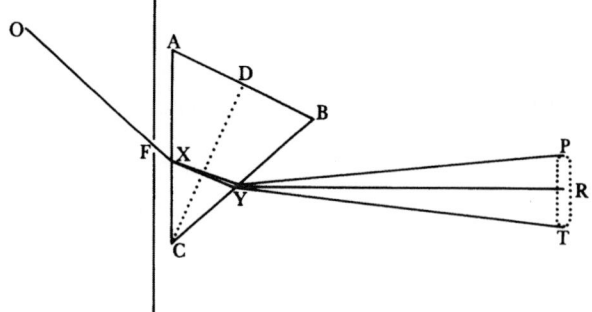

Figure I, 21

portiones obtineri possint,[4] imprimis determinando proportiones sinuum refractionis inter se respectu ejusdem incidentiae, ac deinde cum communi sinu
incidentiae conferendo. ‖ Et quoniam de intermedijs radiorum generibus facile esset judicium ferre si modo refractiones extremorum essent cognitae:
satisfecero si radios maxime omnium refrangibiles cum minimè refrangibilibus comparavero. Itaque in fig. I, 21 sit *ABC* Prisma vitreum ita positum ut
radij tum ingredientes tum egredientes eandem quantitatem refractionis ut
priùs patiantur. Dies autem seligatur splendidus, et cubiculum esto valdè
obscurum, ut colores usque ad ultima quae occupent spatia distinctè satis
videri possint. Tunc ad distantiam viginti pedum autampliùs a Prismate radij
excipiantur in papyrum aliquam directè obversam, et spatij a coloribus illuminati (ut *PT*) longitudo et latitudo mensuretur. Sic Prismate adhibito cujus
angulus verticalis *ACB* fuit 63^{grad} 12$^{min.}$, et latitudine foraminis radios intromittentis existente $\frac{1}{4}^{ta}$ parte digiti, ad distantiam *XP* vel *XT* 22^{pedum} inveni
maximam longitudinem / imaginis *PT* esse 13$\frac{1}{4}^{dig}$ circiter et latitudinem 2$\frac{5}{8}^{dig}$.   31/
Jam si latitudo hujus imaginis ab ejus longitudine subtrahatur, manebit
10$\frac{5}{8}^{digiti}$ pro longitudine quam habere debuisset si solis discus et foraminis *F*
diameter fuissent infinitè parvus. Hoc est si radij advenissent omnes in eadem
recta *OF*. Ista itaque linea 10$\frac{5}{8}^{digitorum}$ subtendit angulum quem radij duo similiter incidentes per inaequalitatem refractionis constituunt, quorum alter
maximè omnium similiter incidentium, et alter minimè omnium refringitur.
qui proinde angulus ex calculo reperietur 2^{grad} 18^{min}. Verùm cùm angulus iste
bina refractione ad *X* et *Y* conficiatur, et praeterea cùm utraque supponatur
aequalis: calculus ad hoc negotium satis accuratus ex unicâ tantùm refractione poterit institui, puta quae conficitur ad latus *BC*. Etenim si verticalis
angulus *ACB* plano *DC* bisecetur, et alterum Prismatis dimidium *DCB* vel
DCA concipiatur tolli, refractio ad alterum dimidium facta, radijs *OF*
obliquè incidentibus in latus *AC*, et perpendiculariter emergentibus e latere
DC, vel perpendiculariter incidentibus in latus *DC* secundum unicam quandam lineam *XY* et obliquè emergentibus e latere *BC*, refractio, inquam, sic ad
alterum dimidium facta foret semissis refractionis ad integrum Prisma, si
modò unicum quodpiam radiorum[5] mediocriter refrangibilium genus specte-

ratios can be obtained by the refraction of the sun's immediate light,[4] first by ‖ determining the proportions of the sines of refraction to each other with ‖ respect to the same incidence, and then by comparing them with their com- ‖ mon sine of incidence. ‖ Since it would be easy to make a judgment about the intermediate kinds of rays, provided that the refractions of the extreme ones were known, it will be sufficient for me to have compared the most refrangible of all rays with the least refrangible. Therefore, in Fig. I, 21 let *ABC* be a glass prism so placed that, as earlier, both the entering and emerging rays undergo the same quantity of refraction. Choose a bright day, however, and let the room be very dark, so that the colors can be seen sufficiently distinctly to the farthest space they occupy. Then at a distance of twenty feet or more from the prism receive the rays on a paper directly opposite them, and measure the length and breadth of the space, such as *PT*, illuminated by the colors. In this way, using a prism whose vertex angle *ACB* was 63°12′, and with the breadth of the hole admitting the rays being $\frac{1}{4}$ inch, at a distance *XP* or *XT* of twenty-two feet I found the greatest length of the image *PT* to be approximately $13\frac{1}{4}$ inches and its breadth $2\frac{5}{8}$ inches. Now if the breadth of this image is subtracted from its length, $10\frac{5}{8}$ inches will remain for the length it ought to have had if the sun's disk and the diameter of the hole *F* had been infinitely small; that is, if all the rays had arrived in the same straight line *OF*. That $10\frac{5}{8}$ inch line thus subtends an angle that two similarly incident rays make because of the inequality of the refraction, one ray of which is refracted the most of all similarly incident ones and the other the least of all. This angle, consequently, will be found by calculation to be 2°18′. But since that angle is made by a twofold refraction at *X* and *Y*, and moreover since both are assumed to be equal, a calculation sufficiently accurate for this purpose can be carried out from only one refraction, for instance, from that made at the side *BC*. For if the vertex angle *ACB* were bisected by the plane *DC*, and one half of the prism, either *DCB* or *DCA*, were imagined to be removed, then the refraction made at the other half (when the rays *OF* fall obliquely onto the side *AC* and emerge perpendicularly from the side *DC*, or fall perpendicularly along some one line *XY* onto the side *DC* and emerge obliquely from the side *BC*), the refraction, I repeat, thus made in the other half would be one half of the refraction in the whole prism, if only some one kind of ray,[5] the mean refrangible, were considered. Moreover, if all

---

(4) Newton has once again rejected experimentally verifying the sine law of refraction for each color and remained content to assume its truth, as he had in §II, 12 = 34. Not until the *Opticks* (Bk. I, Pt. I, Prop. VI, Expt. XV) did he attempt to demonstrate it experimentally. But rather than directly testing the law by measuring the angle of refraction at different angles of incidence, as sketched here, he attempted to prove it indirectly by his hapless crossed-prism experiment; see Lect. II, 2, note (4).

(5) I continues: *puta* (suppose).

tur. Quinetiam si caetera omnia radiorum genera simul spectentur, assertio illa licèt non ampliùs sit absolutè vera, tamen veritati tam proximè accedet, ut quoad sensum et calculum mechanicum pro verâ habeatur. Quamobrem cum refractionis utriusque[6] ad X et Y peractae computatio geometrica aegriùs institui possit, istud more ad praxim màgis accommodato, ut ut mechanico perficere non verebor, confisus id mihi vitio verti non debere, si dum computationes rebus physicis adhibeo minutias quae operam molestè et sine fructu producerent, missas faciam. Refractionem itaque ex unicâ tantùm parte Prismatis perpendam; et quoniam omnes radij, demptis mediocritèr refrangibi-/libus,[7] a dimidio *ACD* bis deberent refringi, et semel tantum ab altero di- 32/
midio *DCB*, perpendiculariter ingressi latus planum *DC* secundum lineam *XY*: itaque in dimidio *DCB* fiat calculus, hoc est ad latus planum *BC*: supposito quod omnibus radijs secundum eandem lineam *XY* allapsis, angulus quem maximè refrangibiles cum minimè refrangibilibus, postquam refringerentur a latere *BC*, constituerent, foret dimidium anguli *PYT*, hoc est 1^{grad} 9^{min}. Jam

Num 30 cùm angulus incidentiae radij *XY* ex praemonstratis sit 31^{grad} 36^{min} et angulus refractionis mediocris 54^{grad} 10^{min}, tran[s]ferantur haec omnia in schema I, 22 ponendo quod *CB* sit superficies disterminans

Medium vitreum versus *A* et aereum versus *F*, et quod angulus incidentiae *XYH* sit 31^{grad} 36^{min}; eritque angulus refractionis *RYF* 54^{grad} 10^{min}, et angulus *PYT* 1^{gr} 9^{min}, differentia nempe refractionis inter maximè refrangibiles *YP* et minime refrangibiles *YT*. Qui angulus a radio *YR* mediocriter refracto et confinium caerulei ac viridis occupante, bisecatur: Ac proin ang. *PYR* vel *RYT* erit $34\frac{1}{2}^{min}$ dimidium totius *PYT*. Adeoque angulus *PYE* $54^{gr.}$ $44\frac{1}{2}^{min}$, et angulus *TYE* 53^{gr} $35\frac{1}{2}^{min}$, et eorum sinus *PG*, ac *TF* erunt 81656 et 80481; quorum proportione ad simpliciores numeros redactâ, erit *TF* ad *PG*[8] ut $69\frac{1}{2}$ ad $68\frac{1}{2}$ circiter.[9] Ad hunc modum experimenta et calculum cùm saepius in-

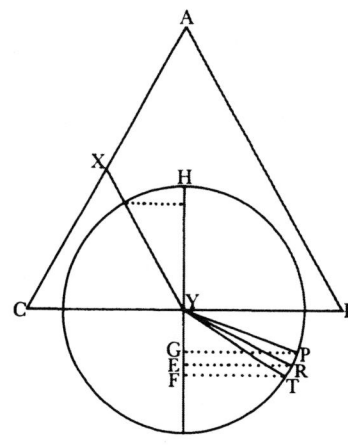

Figure I, 22

stituerim horum sinuum proportiones inter terminos 67 ad 66 et 72 ad 71 semper obvenerunt. Sed ut plurimum incidi in proportiones 69 ad 68, $69\frac{1}{2}$ ad $68\frac{1}{2}$ et 70 ad 69 quarum tantilla est differentia, ut parvi intersit quaenam adhibeatur.

38 [= 111]. Illarum refractionum sinus ad communem sinum incidentiae conferuntur. Ratione sinuum refractionis pro extremis radiorum similitèr incidentium generibus sic inventâ, eorum comparatio ad sinum incidentiae simul innotescit, quippe qui paulo ante inventus est 52400. Et conferendo hunc 52400 ad sinus 81656 et 80481 eorum ratio in minoribus numeris reperietur $44\frac{1}{2}$ ad $69\frac{1}{2}$, et $68\frac{1}{2}$: aut $44\frac{1}{4}$ ad 69 & 68 ferè. Refractionibus nempe ex vitro in aerem peractis.

---

(6) I: *refractiones utrinque* (the refraction at each side).

the other kinds of rays were considered simultaneously, that assertion, although no longer absolutely true, will still so closely approach the truth that it may be taken as true with respect to sense and a mechanical calculation. Consequently, since a geometric calculation of each refraction[6] made at X and Y can be undertaken rather difficultly, I will not be afraid to do it in a way which, however mechanical, is more suited to practice, being confident that no fault ought to be attributed to me if, when I perform computations in physical matters, I omit the minutiae that entail burdensome and fruitless work. I will accordingly consider the refraction from only a single side of the prism. And since all rays, except the mean refrangible[7] ones, must be refracted twice by the half ACD, and only once by the other half DCB (since they enter the plane side DC perpendicularly along the lines XY), therefore let the calculation be made in the half DCB, that is, at the plane side BC, while supposing that for all rays flowing along the same line XY, the angle that the most refrangible rays would make with the least refrangible ones after they were refracted by the side BC would be one half of the angle PYT, that is, 1°9'. Now, since the angle of incidence of the ray XY is 31°36' (from what has been shown before), and the angle of the mean refraction is 54°10', §I, 30 transfer all these to Fig. I, 22, while assuming that CB is the surface separating the glass medium toward A and the air toward F, and that the angle of incidence XYH is 31°36'. The angle of refraction RYF will be 54°10' and the angle PYT will be 1°9', namely, the difference of refraction between the most refrangible rays YP and the least refrangible YT. Let this be bisected by the mean refracted ray YR, occupying the boundary of blue and green, and then the angle PYR or RYT will be $34\frac{1}{2}'$, half of the whole angle PYT. Hence the angle PYE will be $54°44\frac{1}{2}'$; the angle TYE will be $53°35\frac{1}{2}'$; and their sines PG and TF will be 81656 and 80481. And reducing their ratio to simpler numbers, TF to PG[8] will be approximately[9] as $69\frac{1}{2}$ and $68\frac{1}{2}$. When I had in this way frequently carried out the experiments and calculations, the ratios of their sines always turned out between the limits of 67 to 66 and 72 to 71; but most often I came upon the ratios of 69 to 68, $69\frac{1}{2}$ to $68\frac{1}{2}$, and 70 to 69—whose difference is so small that it matters little which is used.

After the ratio of the sines of refraction for the extreme kinds of rays similarly incident has thus been found, their comparison to the sine of incidence becomes known at the same time, inasmuch as a little earlier it was found to be 52400. Comparing this 52400 to the sines 81656 and 80481, their ratio in smaller numbers will be found to be $44\frac{1}{2}$ to $69\frac{1}{2}$ and $68\frac{1}{2}$, or nearly $44\frac{1}{4}$ to 69 and 68, that is, when the refractions are made from glass into air.

38 [= 111]. The sines of those refractions are compared to the common sine of incidence.

(7) mediocritèr refrangibilibus] I: viridiformibus (green-making).
(8) *TF* ad *PG*] Read (as *editio princeps*): *PG* ad *TF*.
(9) Added.

39 [= 112].
Radiorum ad
oppositas partes
ejusdem[9]
refringentis
superficiei
incidentium sinus
sunt reciproce
proportionales.

Quod si radij è contra ex aere in vitrum similiter incidant: proportiones sinuum nullo negotio ex jam inventis eruuntur, utpote quae sunt reciprocae. Sit I communis[9] sinus incidentiae e vitro in aerem P sinus refractionis maximè refrangibilium radiorum, R mediocriter / refrangibilium, et T minimè 33/ refrangibilium. Dico quod ex horum reciprocè proportionalibus si $\frac{1}{I}$ ponatur esse sinus incidentiae ex aere in vitrum erit $\frac{1}{P}$ sinus refractionis maximè refrangibilium radiorum $\frac{1}{R}$ sinus refractionis mediocriter refrangibilium ac $\frac{1}{T}$ minimè refrangibilium. Nam cùm sinus incidentiae radij maxime refrangibilis è vitro in aerem sit I, et sinus refractionis P, radij ejus ex aere in vitrum, per easdem lineas retroacti sinus incidentiae erit P, et sinus refractionis I, siquidem jam radius est incidens qui prius fuerit refractus.[10] Est ergo sinus incidentiae radij maximè refrangibilis ex aere in vitrum utcunque incidentis ad sinum refractionis ut P ad I hoc est (applicando terminos rationis[11] ad P) ut 1 ad $\frac{I}{P}$, hoc est (applicando ad I denuò) ut $\frac{1}{I}$ ad $\frac{1}{P}$. Et simili argumento constabit ejusmodi sinus radij mediocritèr refrangibilis esse ut $\frac{1}{I}$ ad $\frac{1}{R}$ et sinus minimè refrangibilis ut $\frac{1}{I}$ ad $\frac{1}{T}$. Liquet ergo quod posito $\frac{1}{I}$ communi sinu incidentiae erunt $\frac{1}{P}$, $\frac{1}{R}$, et $\frac{1}{T}$ singulorum generum respectivè sinus.

Rem numeris illustro. Cum $44\frac{1}{4}$ ad 69 et 68 sit ratio sinus communis incidentiae ad sinus maximè discrepantium refractionum e vitro in aerem: sinus incidentiae communis ad sinus refractionum ex aere in vitrum erit ut $\frac{1}{44\frac{1}{4}}$ ad $\frac{1}{69}$ et $\frac{1}{68}$ sive $\frac{69 \times 68}{44\frac{1}{4}}$ (= 106 ferè) ad 68 et 69. Hoc est pro radijs maximè refrangibilibus sinus incidentiae ad sinum refractionis ut 106 ad 68, et pro minimè refrangibilibus ut 106 ad 69.

41 [= 114].
E refractionibus
extremorum
generum facile est
de intermedijs
conjecturam facere.

Hisce sic determinatis rationes sinuum pro radijs intermedijs facilè determinantur ex cognitis colorum distantijs, quas in imagine coloratâ observant. Sic radij qui ad caeruleum paulo[12] magis quam flavum vergunt cum in mediam imaginem cadant intermediam rationem sinuum $44\frac{1}{4}$ ad $68\frac{1}{2}$[13] vel 106 ad $68\frac{1}{2}$ circiter[12] habebunt. et sic de alijs.

Lect. 7
42 [= 115].
Theoremate
ostenditur quomodo
e refractionibus
heterogeneorum ad
vitrum vel quodvis
medium inter se
determinatis,

Ad eundem modum quo refractiones ad vitrum determinatae sunt, id ipsum posset fieri ad alia media. Sed e re erit ut regulam jam ostendam, quâ refractionum istarum mensurae ex sinubus earum sic ad vitrum determinatis, possunt determinari ad quodlibet aliud medium propositum, idque licèt istud sit alij medio quam aeri contiguum. In fig. I, 23 sit AB superficies terminans aerem ex parte F, et vitrum ex parte G, ad cujus aliquod punctum X ducatur linea FXG ei perpendiculariter insistens, et praeterea concipiatur rectam IX ad angulum IXA infinitè parvum duci secundum quam omnes omnium for-

---

(10) By Prop. 1, §I, 53 to follow.　　(11) terminos rationis] I: rationem.

(12) Added.　　(13) $44\frac{1}{4}$ ad $68\frac{1}{2}$] Read: $\frac{1}{44\frac{1}{4}}$ ad $\frac{1}{68\frac{1}{2}}$.

If now, conversely, the rays are similarly incident from air into glass, the proportions of the sines are derived with little difficulty from what has already been found since they are reciprocals. Let I be the common[(9)] sine of incidence from glass into air, P the sine of refraction of the most refrangible rays, R that of the mean refrangible ones, and T that of the least refrangible. I say that from the reciprocal proportions of these, if $\frac{1}{I}$ is assumed to be the sine of incidence from air into glass, $\frac{1}{P}$ will be the sine of refraction of the most refrangible rays, $\frac{1}{R}$ that of the mean refrangible ones, and $\frac{1}{T}$ that of the least refrangible. For since the sine of incidence of a most refrangible ray from glass into air is I and the sine of refraction P, then the sine of incidence of its ray going back along the same lines from air into glass will be P and the sine of refraction I, because the incident ray now was the refracted ray before.[(10)] Therefore the sine of incidence of a most refrangible ray from air into glass, however incident, is to the sine of refraction as P to I, that is (dividing the terms of the ratio[(11)] by P), as 1 to $\frac{I}{P}$, that is (dividing again by I), as $\frac{1}{I}$ to $\frac{1}{P}$. By a similar argument it will be evident that the sines of this sort for a mean refrangible ray are as $\frac{1}{I}$ to $\frac{1}{R}$, and the sines of a least refrangible one as $\frac{1}{I}$ to $\frac{1}{T}$. It is clear therefore that setting $\frac{1}{I}$ as the common sine of incidence, $\frac{1}{P}$, $\frac{1}{R}$, and $\frac{1}{T}$ will be the sines of the individual kinds of rays respectively.

I will illustrate this with numbers. Since $44\frac{1}{4}$ to 69 and 68 is the ratio of the common sine of incidence to the sines of the most different refractions from glass into air, the common sine of incidence to the sines of refraction from air into glass will be as $1/44\frac{1}{4}$ to $1/69$ and $1/68$, or $(69 \times 68)/44\frac{1}{4}$ (≈ 106) to 68 and 69. That is, for the most refrangible rays the sine of incidence to the sine of refraction is as 106 to 68, and for the least refrangible as 106 to 69.

After these have been thus determined, the ratios of the sines for the intermediate rays are easily determined from the known distances that the colors maintain in the colored image. Thus the rays that incline slightly[(12)] more to blue than to yellow, since they fall in the middle of the image, will have the intermediate ratio of sines, $44\frac{1}{4}$ to $68\frac{1}{2}$,[(13)] or approximately[(12)] 106 to $68\frac{1}{2}$; and similarly for the others.

In the same way that the refractions were determined at glass, it could have been done at other media; but it will be useful, having determined their sines in this way at glass, if I now show a rule whereby the measures of those refractions can be determined at any other proposed medium, and even if that proposed medium is contiguous to a medium other than air. Let AB (Fig. I, 23) be a surface bounding air on the side F and glass on the side G, and at any point X of the surface draw the line FXG perpendicular to it. In addition, imagine that at the infinitely small angle IXA there is drawn the line IX,

39 [= 112]. The sines of rays incident on opposite sides of the same[(9)] refracting surface are reciprocally proportional.

40 [= 113]. It is illustrated by the refraction of glass.

41 [= 114]. From the refractions of the extreme kinds of rays, it is easy to infer the intermediate ones.

Lecture 7
42 [= 115]. It is shown by a theorem how having determined the refractions of heterogeneous rays with respect to one another at glass or any other medium,

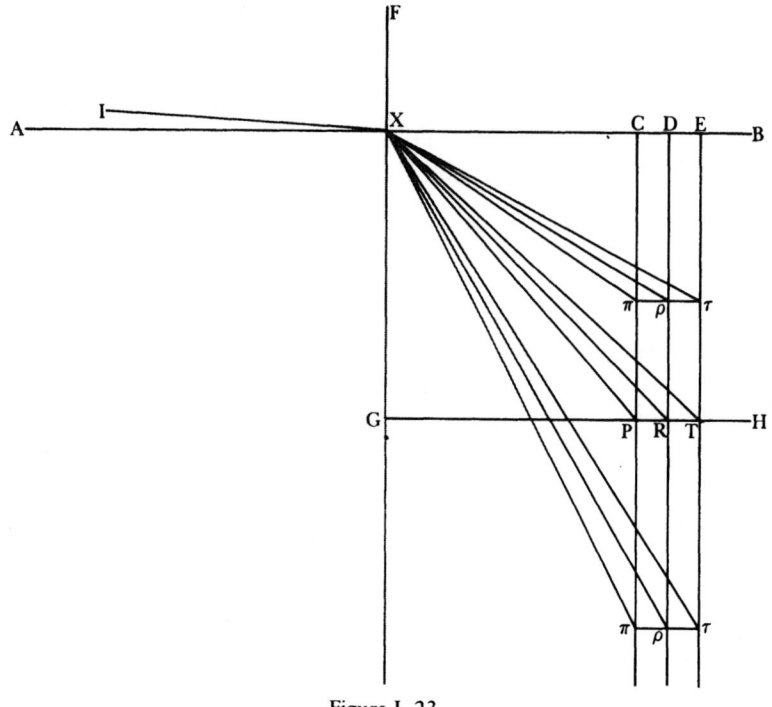

Figure I, 23

possunt etiam ad
alia quaelibet media
aëri contigua
refractiones,
sine novis
experimentorum
molestijs, inter se
determinari.

marum radij supponantur incidere et in *X* refringi, puta / mediocriter refran-   34/
gibiles versus *R* maximè refrangibiles versus *P* et minimè refrangibiles versus
T aliosque intermedios versus intermedias plagas. Porrò ducatur linea quaevis
GH parallela ad lineam incidentiae *IX* hoc est perpendicularis ad *FG*. ea verò
secet radios in punctis *P, R* ac *T*, a quibus demittantur *PC, RD*, ac *TE*
perpendiculares ad refringentem superficiem *AB*. His ad vitrum sic determinatis ac descriptis, si aliud quodvis medium in locum vitri jam concipiatur
substitui, caeteris stantibus; et radij alicujus mediocriter refrangibilis secundum lineam *IX* incidentis ad *X*, refractus *Xρ* ducatur secans rectam *DR* in *ρ*.
Quod fieri suppono, siquidem modum quo mediocriter refrangibilium refractiones ad media quaelibet investigari possint, antehac exposui.[1] Dein per
punctum *ρ* ducatur recta *ππ* secans lineas *CP* et *ET* in *π* ac *τ* perpendiculariter, junganturque *πX* et *τX*: Dico quod radij maximè refrangibiles secundum
dictam lineam *IX* incidentes refringentur in lineam *Xπ* et minimè refrangibiles in lineam *Xτ* radijque cujusvis speciei quos vitrum refringebat ad quodlibet punctum rectae[2] *PT*, illi ad correspondens punctum rectae *ππ* per alterum dictum Medium refringentur quod pro vitro supponitur substitui; istis
punctis linearum *PT* et *ππ* habitis pro correspondentibus per quae recta quaevis parallela ipsi *DR* transit. Patet itaque modus quo refractiones quorumvis
radiorum ex aere in quodlibet medium propositum obliquitate maximâ incidentium determinari poterunt, cognitâ unici tantùm[3] radiorum generis in
istud medium refractione; et proportionibus sinuum ex obliquissima ista refractione determinatis eorundem radiorum refractiones dabuntur ad quamlibet aliam datam incidentiam.

along which it is supposed that all rays of every nature are incident and refracted at *X,* suppose, mean refrangible rays toward *R,* most refrangible ones toward *P,* least refrangible ones toward *T,* and other, intermediate ones toward the intermediate regions. Moreover, draw any line *GH* parallel to the line of incidence *IX,* that is, perpendicular to *FG;* but let it intersect the rays at the points *P, R,* and *T,* from which the perpendiculars *PC, RD,* and *TE* are dropped to the refracting surface *AB.* After these things have been so determined and described at glass, if any other medium is now conceived to be substituted for glass, with everything else remaining the same, and the refracted ray *Xρ* of some mean refrangible one incident along the line *IX* at *X* is drawn intersecting the line *DR* in *ρ* (I assume this to be done, since I previously explained how the refractions of mean refrangible rays can be found at any medium);[1] then if through the point *ρ* the line *ππ* is drawn cutting the lines *CP* and *ET* perpendicularly at *π* and *τ,* and *πX* and *τX* are joined; I say that most refrangible rays incident along the line *IX* will be refracted into the line *Xπ,* least refrangible ones into the line *Xτ,* and those rays of any sort that glass refracted to some point of the line[2] *PT* will be refracted to a corresponding point of the line *ππ* by that other medium that is assumed to be substituted for glass—those points of the lines *PT* and *ππ* through which any straight line parallel to *DR* passes are considered to be corresponding points. It is therefore evident how the refractions of any rays incident with the greatest obliquity from air into any proposed medium can be determined by knowing the refraction of only[3] a single kind of ray in that medium; and when the ratios of the sines have been determined from that most oblique refraction, the refractions of the same ray at any other given incidence will be given.

the refractions can also be determined with respect to one another at any other medium contiguous to air without bothering anew with experiments.

(1) In §§I, 28–9 = 101–2.
(2) Omitted in *editio princeps.*
(3) **I** continues: cujusvis (any).

Hujus quidem Theorematis certitudinem ab experimentis nondum habeo depromptam, sed cùm a veritate vix multùm discrepare videatur, nil veritus sum impraesentiâ gratis assumere. Posthac fortè vel experientia confirmabo, vel si falsum invenero corrigam.

Calculum quod attinet is facilè potest institui ex hac proportionalitate, quòd sinus incidentiae radij *IX* (hoc est sinus 90^{grad}) sit ad sinum refractionis (puta quae facta sit in lineam *XR*) sicut *XR* ad *RG*. Sic ad vitrum erit *XR . RG* :: 106 . 68½, et *XP . PG* [::] / 106 . 68, et *XT . TG* :: 106 . 69 et inde deducetur quod *GP . GR . GT* :: 39 . 39½ . 40. Quae proportiones semel inventae possunt in eum finem asservari ut earum ope refractiones ad alia media quàm vitra determinentur. Nam quolibet medio proposito sumatur *XE* = 40, *DE* = ½, et *CD* = ½ atque perpendicula *CP, DR*, et *ET* erigantur. Tum ex datâ sinuum refractionis radiorum mediocriter refrangibilium proportione, hoc est ex datâ proportione ipsius *Xρ* ad *XD*, dabitur punctum *ρ* et longitudo *Dρ*, cui aequales sunt *Cπ* et *Eτ*. Punctisque *π* ac *τ* sic datis dantur rationes ipsarum *Xπ* et *XC* hoc est sinuum incidentiae et refractionis pro radijs maximè refrangibilibus, ut et rationes ipsarum *Xτ* et *XE*, hoc est sinuum incidentiae et refractionis pro radijs minimè refrangibilibus. Sic pro superficie aquam et aerem disterminante sinus isti sunt, ut 68 ad 90 pro minimè refrangibilibus, et ut 68 ad 91 pro maximè refrangibilibus proximè.[4]

Proportionibus linearum *XC, XD* et *XE* sic inventis mensura refractionum ex aere in Medium quodvis propositum et ad quamlibet incidentiam factarum per aliud insuper Theorema non inelegans determinari potest. In linea *FX* (fig I, 24) ad refringens planum *AB* perpendiculari sumatur punctum aliquod *F*

35/

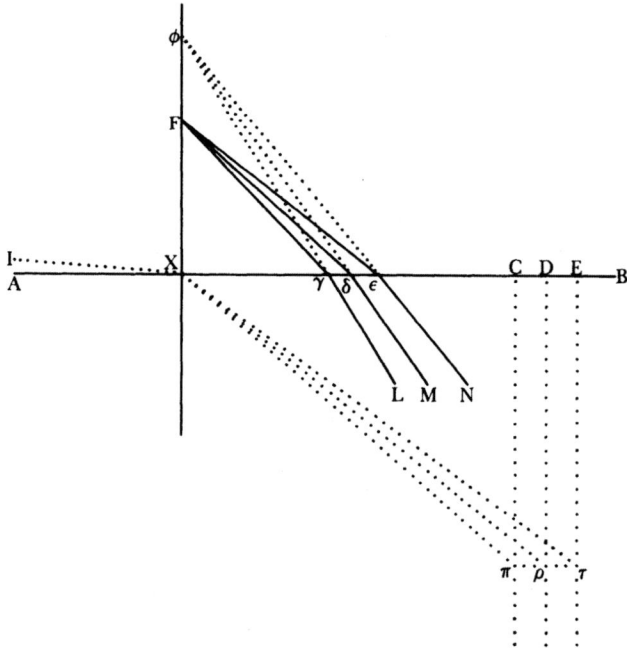

Figure I, 24

I have not yet, to be sure, derived the certainty of this theorem from experiments, but since it appears scarcely to differ much from the truth, for the present I have not hesitated to assume it gratuitously. In the future perhaps I will either confirm it by experiment or, if I will find it to be false, correct it.

43 [→ 116]. The certainty of this theorem.

As for the calculation, it can easily be carried out from this proportionality, the sine of incidence of the ray IX (that is, sin 90°) is to the sine of refraction (suppose that one made in the line XR) as XR to RG; thus in glass $XR : RG = 106 : 68\frac{1}{2}$, $XP : PG = 106 : 68$, and $XT : TG = 106 : 69$. From this it will be deduced that $GP : GR : GT = 39 : 39\frac{1}{2} : 40$. Once these proportions have been found, they can be maintained for the purpose of determining with their help the refractions in media other than glass. For in any proposed medium take $XE = 40$, $DE = \frac{1}{2}$, and $CD = \frac{1}{2}$, and erect the perpendiculars CP, DR, and ET. Then from the given ratio of the sine of refraction of the mean refrangible rays, namely, from the given ratio of $X\rho$ to XD, the point ρ and the length $D\rho$, which equals $C\pi$ and $E\tau$, will be given. The points π and τ thus being given, there is given the ratio of $X\pi$ and XC, that is, of the sines of incidence and refraction for the most refrangible rays, as well as the ratio of $X\tau$ and XE, that is, of the sines of incidence and refraction for the least refrangible rays. Thus for a surface separating water and air those sines are approximately[4] as 68 to 90 for the least refrangible rays and 68 to 91 for the most refrangible.

44 [= 117]. The proportion of certain lines that serve for carrying out the calculation by this theorem.

With the proportions of the lines XC, XD, and XE having thus been found, the measure of the refractions made from air into any proposed medium and at any incidence can be determined by still another, not inelegant theorem. In the line FX (Fig. I, 24) perpendicular to the refracting plane AB

45 [= 118]. Another theorem for accomplishing the same thing.

---

(4) Added.

quod lucidum fingatur ac ducatur quaelibet $F\delta$ secans AB in δ eaque concipiatur esse mediocriter refrangibilis radius cujus refractus ex aere in medium
propositum esto δM, qui retro-ductus secet FX in ϕ. Porro fiat

$$F\delta \cdot FE :: X\delta \cdot XE^{(5)} :: (39\tfrac{1}{2} \cdot 40) \text{ et } F\delta \cdot F\gamma :: XD \cdot XC \,(:: 39\tfrac{1}{2} \cdot 39)$$

centroque F et intervallis Fe et $F\gamma$ describantur circuli secantes AB in ϵ et γ
junganturque $F\epsilon$, $F\gamma$, $\phi\epsilon$, $\phi\gamma$ et producantur $\phi\epsilon$ et $\phi\gamma$ indefinitè versus N et L.
Dico jam si radius minimè refrangibilis incidat secundum lineam $F\epsilon$ quod iste
refringetur in lineam ϵN: et si maximè refrangibilis incidat secundum $F\gamma$
quod iste refringetur secundum γL.[6] Et sic radij quorumlibet intermediorum
generum manantes a puncto F et in puncta sibi correspondentia inter γ et ϵ
incidentes, ita refringentur a medio proposito quasi manassent omnes à
/ puncto ϕ. Istis punctis inter C et E atque γ et ϵ habitis pro correspondenti- 36/
bus, quorum distantiae ab X et F respectivè sunt in eadem ratione cum DX
ac δF.

Cujus Theorematis demonstrationi praesternantur duo Lemmata sequentia.
   1. Duobus punctis γ, δ in lineâ quâpiam AB (fig I, 24) sumptis, et alijs
duobus ϕ et F in ejus perpendiculo FX; junctisque $\phi\delta$, $F\delta$ $\phi\gamma$ et $F\gamma$: differentia quadratorum a duobus $\phi\delta$ et $F\delta$ concurrentibus ad δ aequabitur differentiae quadratorum ab alijs duobus $\phi\gamma$ et $F\gamma$ concurrentibus ad γ. Nam cùm
$\phi\delta^q = \phi X^q + X\delta^q$, et $F\delta^q = FX^q + X\delta^q$; erit differentia $\phi\delta^q - F\delta^q = \phi X^q -$
FX^q et ob eandem rationem est differentia $\phi\gamma^q - F\gamma^q = \phi X^q - FX^q$. Quare
dictae differentiae sic aequales eidem tertio sunt aequales inter se.   Q.E.D.

   2. Si radius aliquis FG (fig I, 25) incidat in
superficiem AB et refringatur versus H, lineâ
GH retro-ductâ ut secet perpendiculum FX in ϕ,
dico quod $\phi G \cdot FG ::$ sinus incidentiae ad sinum
refractionis. et e contra si $\phi G \cdot FG ::$ sin: incid .
sin: refract. erit ϕGH refractus ipsius FG. Etenim sumatur $\phi K = FG$ et demittatur KL perpendicularis ad FX[,] quo facto cùm ang: GFX
aequetur angulo incidentiae et ang: $G\phi X$ angulo
refractionis, erit GX sinus incidentiae et KL sinus refractionis, habito respectu ad circulum cu-
jus semidiameter sit FG vel ϕK. Sed $\phi G \cdot \phi K :: GX \cdot KL$, hoc est $\phi G \cdot FG ::$
$GX \cdot KL$.   Q.E.D.

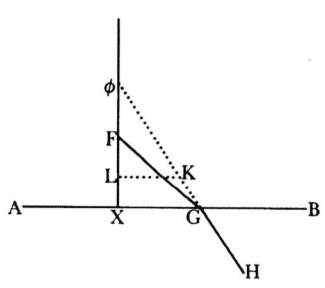

Figure I, 25

His praemissis Theorema propositum sic demonstratur. In Fig. I, 24 ducatur IX obliquissima linea secundum quam radij omnium formarum ex aere
ad X incidere ponantur et in medium propositum refringi[8] maximè refrangibiles versus π, et minimè refrangibiles versus τ[,] eosque lineae ad puncta D,
C, et E normaliter erectae secent in punctis ρ, π, ac τ, ut explicabatur ad Fig

(5) Read (as *editio princeps*): $F\delta \cdot F\epsilon :: XD \cdot XE$.

(6) secundum γL.] I: in ipsam γL. (into γL.)

(7) Ad . . . demonstrationem.] I: Ad ejus demonstrationem duo Lemmata praelibantur. (For its
demonstration two lemmas are prefaced.)

take any point *F*, and imagine it to be luminous. Also draw any line *Fδ* intersecting *AB* in *δ*, and conceive it to be a mean refrangible ray, whose refracted ray from air into the proposed medium is *δM*, which being drawn backward intersects *FX* in *φ*. Moreover, let

$$Fδ : FE = Xδ : XE^{(5)} (= 39\tfrac{1}{2} : 40),$$

and *Fδ* : *Fγ* = *XD* : *XC* (= 39½ : 39); and with center *F* and radii *Fε* and *Fγ* describe circles intersecting *AB* in *ε* and *γ*; join *Fε*, *Fγ*, *φε*, and *φγ*; and extend *φε* and *φγ* indefinitely toward *N* and *L*. I now say that if a least refrangible ray is incident along the line *Fε*, it will be refracted into the line *εN*; and if a most refrangible one is incident along *Fγ*, it will be refracted along *γL*.[6] Similarly, rays of any intermediate kind flowing from the point *F* and falling upon their corresponding points between *γ* and *ε* will be so refracted by the proposed medium as if they had all flowed from the point *φ*—those points between *C* and *E* and *γ* and *ε* whose distances from *X* and *F*, respectively, are in the same ratio as *DX* and *δF*, being considered as corresponding points.

To demonstrate this theorem the following two lemmas are premised.

1. Taking two points, *γ* and *δ* (Fig. I, 24), in any line *AB*, and another two, *φ* and *F*, in its perpendicular *FX*, and joining *φδ*, *Fδ*, *φγ*, and *Fγ*, the difference of the squares of the two [lines] *φδ* and *Fδ* meeting at *δ* will equal the difference of the squares of the other two, *φγ* and *Fγ*, meeting at *γ*.

For since *φδ²* = *φX²* + *Xδ²*, and *Fδ²* = *FX²* + *Xδ²*, their difference will be *φδ²* − *Fδ²* = *φX²* − *FX²*; and for the same reason there will be the difference *φγ²* − *Fγ²* = *φX²* − *FX²*. Consequently, those differences, since they are equal to the same third one, are equal to each other. As was to be demonstrated.

2. If any ray, *FG* (Fig. I, 25), falls on the surface *AB* and is refracted toward *H*, and the line *GH* is drawn backward to intersect the perpendicular *FX* in *φ*, I say that *φG* is to *FG* as the sine of incidence is to the sine of refraction; and conversely, if *φG* is to *FG* as the sine of incidence is to the sine of refraction, then *φGH* will be the refracted ray of *FG*.

For take *φK* to be equal to *FG*, and drop *KL* perpendicular to *FX*. When this has been done, since the angle *GFX* is equal to the angle of incidence and the angle *GφX* to the angle of refraction, *GX* will be the sine of incidence and *KL* the sine of refraction with respect to the circle whose radius is *FG* or *φK*. But *φG* : *φK* = *GX* : *KL*, that is, *φG* : *FG* = *GX* : *KL*. As was to be demonstrated.

With these premises, the proposed theorem is demonstrated in this way. In Fig. I, 24 draw *IX*, the most oblique line along which rays of every nature are assumed to be incident from air at *X* and to be refracted into the proposed medium—the most refrangible ones toward *π* and the least refrangible toward *τ*; and let the lines erected perpendicularly at the points *D*, *C*, and *E* intersect them at the points *ρ*, *π*, and *τ*, as was explained at Fig. I, 23. Now

46 [= 119]. For the demonstration of this theorem.[7] Lemma 1.

47 [= 120]. Lemma 2.

48 [= 121]. The demonstration.

(8) *Editio princeps* continues: mediocriter refrangibiles versus [*ρ*].

I, 23. Jam cum istorum radiorum sinus incidentiae et refractionis statuantur / esse ut $X\rho$ ad XD, $X\pi$ ad XC et $X\tau$ ad XE respectivè si propterea[9] demon- 37/ stratum fuerit, quod $\phi\delta$ ad $F\delta$, $\phi\gamma$ ad $F\gamma$, et $\phi\epsilon$ ad $F\epsilon$ respectivè sint in eadem ratione (hoc est, quod $\phi\delta \, . \, F\delta :: X\rho \, . \, XD ::$ sinus incidentiae ad sinum refractionis radiorum mediocriter refrangibilium et $\phi\gamma \, . \, F\gamma :: X\pi \, . \, XC ::$ sinus incidentiae ad sinum refractionis radiorum maximè refrangibilium,) constabit propositum ex Lemmate secundo. Et ad mediocriter refrangibiles quod attinet, cum $\phi\delta$ supponatur refractus ipsius $F\delta$, erit (per Lem 2^{dum}) $\phi\delta$ ad $F\delta$ ut sinus incidentiae ad sinum refractionis, hoc est, ut $X\rho$ ad $X[D]$. Sed eadem proportionalitas in caeteris radiorum generibus jam demonstranda proponitur, puta quod sit $\phi\gamma \, . \, F\gamma :: X\pi \, . \, XC$. Scilicet est $F\gamma \, . \, F\delta :: XC \, . \, XD$ ut et $F\delta \, . \, \phi\delta :: XD \, . \, X\rho$, per Hypothesin. Quare permutando et connectendo rationes aequales, est $F\gamma \, . \, XC :: F\delta \, . \, XD :: \phi\delta \, . \, X\rho$. et quadrando, $F\gamma^{\text{q}} \, . \, XC^{\text{q}} :: F\delta^{\text{q}} \, . \, XD^{\text{q}} :: \phi\delta^{\text{q}} \, . \, X\rho^{\text{q}}$, diminuendoque per terminos aequalis rationis

$$F\gamma^{\text{q}} \, . \, XC^{\text{q}} :: \phi\delta^{\text{q}} - F\delta^{\text{q}} \; (\text{Sive per Lem 1}, \; \phi\gamma^{\text{q}} - F\gamma^{\text{q}}) \, . \, X\rho^{\text{q}} - XD^{\text{q}} \; (\text{sive } C\pi^{\text{q}})$$

et augendo per terminos aequalis rationis

$$F\gamma^{\text{q}} \, . \, XC^{\text{q}} :: \phi\gamma^{\text{q}} \, . \, C\pi^{\text{q}} + XC^{\text{q}}(X\pi^{\text{q}}.).$$

Denique terminorum radices extrahendo, permutandoque, est $\phi\gamma \, . \, F\gamma :: X\pi \, . \, XC$. Quare $\phi\gamma$ sive γL est refractus ipsius $F\gamma$ per Lemma 2^{dum}. Q.E.D. Et eodem argumento patebit quod ϵN sit refractus radij $F\epsilon$. Deque alijs radijs pro varijs refrangibilitatis gradibus intermedia spatia variè occupantibus, idem intelligendum est.

49 [= 122].
Heterogeneorum refractiones a superficiebus aëri neutra ex parte contiguis Theoremate etiam determinantur.

De refractionibus superficierum aeri contiguarum mensurandis haec satis. Quod si desideretur id ipsum ad alias superficies aeri ex neutra parte contiguas fieri, sunto (in fig I, 26) $AB\beta H$ et $\alpha\beta\nu\mu$ duo quaelibet media secundum planam superficiem $H\beta$ contigua, et aere circundata. Sitque AB planum ipsi $H\beta$ parallelum, et in eo sumatur punctum X, ad quod ducatur XV perpendicularis, et IX obliquissima linea secundum quam (ut jam antè) radij omnium formarum incidant, et pro gradu refrangibilitatis refringantur ad P, R, ac T, aliaque intermedia loca. Horum radiorum in propositam superficiem $\alpha\beta$ sic incidentium refractiones jam quaeruntur, et cùm refractiones mediocritèr refrangibilium[10] ad quaslibet superficies fuerint antehac expositae;[11] radij XR sit refractus RM, et is retro-ducatur donec secet / perpendiculum XV in 38/ ϕ. Et insuper ducantur ϕP, ϕT, et producantur ad L et N. Dico quod PL erit refractus ipsius XP, ac TN ipsius XT, atque omnes aliarum formarum radij incidentes inter P ac T ita refringentur, ut postea divergant a puncto ϕ. Concipiatur enim quòd medium $\alpha\beta\nu\mu$ longiùs versus $\alpha\mu$ producitur quàm medium $AB\beta H$, ita ut ejus plani $\alpha H\beta$ pars inter H & α sit aeri contigua et ad aliquod in eo punctum F ducatur perpendicularis $F\gamma$ necnon obliquissima linea EF,[12] secundum quam radij omnium formarum incidant, et pro gradu

---

(9) I: praetera (moreover).

(10) et . . . refrangibilium] I: Atque equidem cùm viridiformium refractiones (Since indeed the refractions of green-making rays).

(11) Namely, in §I, 32 = 105.   (12) In I this ray is designated ιF.

since the sines of incidence and refraction of those rays are determined to be
as $X\rho$ to XD, $X\pi$ to XC, and $X\tau$ to XE, respectively; if therefore[9] it were
demonstrated that $\phi\delta$ to $F\delta$, $\phi\gamma$ to $F\gamma$, and $\phi\epsilon$ to $F\epsilon$, respectively, are in the
same ratio (that is, $\phi\delta : F\delta = X\rho : XD$ is as the sine of incidence to the sine
of refraction of mean refrangible rays, and $\phi\gamma : F\gamma = X\pi : XC$ is as the sine
of incidence to the sine of refraction of most refrangible ones); the proposition will be established by the second lemma. As for the mean refrangible
rays, since $\phi\delta$ is supposed to be the refracted ray of $F\delta$, then (by Lemma 2)
$\phi\delta$ to $F\delta$ will be as the sine of incidence to the sine of refraction, that is, as
$X\rho$ to $X[D]$. It is now proposed to demonstrate the same proportionality for
the other kinds of rays, that is, that $\phi\gamma : F\gamma = X\pi : XC$. Namely, by hypothesis $F\gamma : F\delta = XC : XD$, and also $F\delta : \phi\delta = XD : X\rho$. Consequently, by
permuting and equating the equal ratios, $F\gamma : XC = F\delta : XD = \phi\delta : X\rho$; and
by squaring, $F\gamma^2 : XC^2 = F\delta^2 : XD^2 = \phi\delta^2 : X\rho^2$; and by diminishing the
terms of the equal ratios,

$$F\gamma^2 : XC^2 = \phi\delta^2 - F\delta^2 \text{ (or, by Lemma 1, } \phi\gamma^2 - F\gamma^2): X\rho^2 - XD^2 \text{ (or } C\pi^2);$$

and by augmenting the terms of the equal ratios,

$$F\gamma^2 : XC^2 = \phi\gamma^2 : C\pi^2 + XC^2 \text{ (or } X\pi^2).$$

Finally, by extracting the roots of the terms and permuting, it is $\phi\gamma : F\gamma =
X\pi : XC$. By Lemma 2, therefore, $\phi\gamma$, or rather γL, is the refracted ray of $F\gamma$.
As was to be demonstrated. By the same argument it will be clear that ϵN is
the refracted ray of $F\epsilon$. The same thing must also be understood for the other
rays, which according to their various degrees of refrangibility variously occupy the intermediate spaces.

Let this suffice for measuring the refractions of surfaces contiguous to air.
If now it is required to do this at other surfaces contiguous to air on neither
side, let (Fig. I, 26) $AB\beta H$ and $\alpha\beta\nu\mu$ be any two media contiguous along the
plane surface $H\beta$ and surrounded by air. Let the plane AB be parallel to $H\beta$;
in it take a point X at which the perpendicular XV is drawn and also the
most oblique line IX along which (as just before) rays of every form are
incident and refracted according to their degree of refrangibility to P, R, T,
and other, intermediate places. The refractions of these rays thus incident
upon the proposed surface $\alpha\beta$ are now to be found. Since the refractions of
mean refrangible rays[10] at any surface have been explained earlier,[11] let RM
be the refraction of the ray XR, and draw it backward until it intersects the
perpendicular XV in ϕ. Moreover, draw ϕP and ϕT and extend them to L
and N. I say that PL will be the refracted ray of XP, TN that of XT, and all
rays of other forms falling between P and T will be so refracted that afterward they diverge from the point ϕ. For conceive the medium $\alpha\beta\nu\mu$ to be
extended farther toward $\alpha\mu$ than the medium $AB\beta H$ so that the part of its
plane $\alpha H\beta$ between H and α is contiguous to air; and at any point F in it
draw the perpendicular $F\gamma$ and also the most oblique line EF[12] along which
rays of every form are incident and according to their degree of refrangibility

49 [= 122]. The refractions of heterogeneous rays by surfaces contiguous to air on neither side are also determined by a theorem.

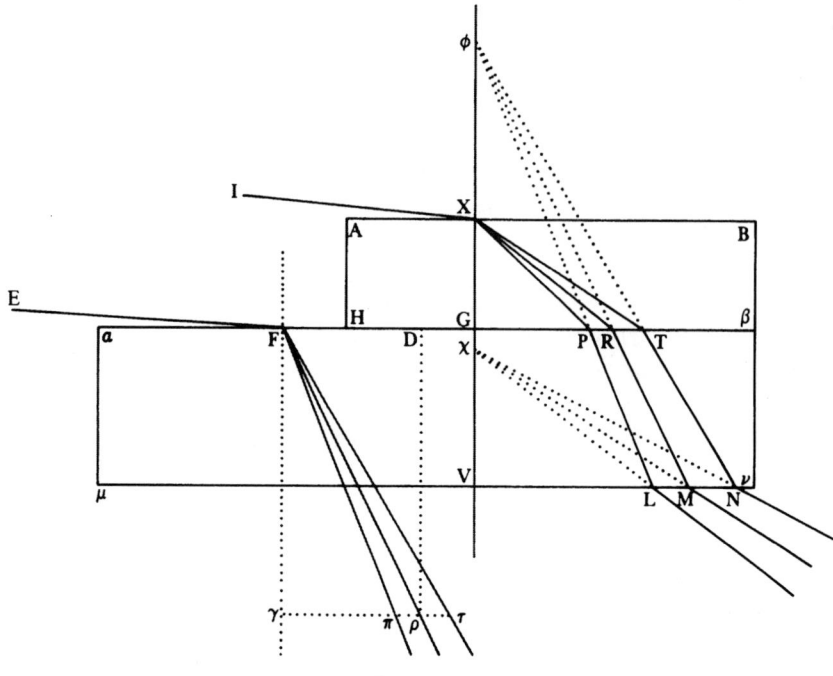

Figure I, 26

refrangibilitatis refringantur ad π, ρ, τ, locaque intermedia, perinde ut effectum erat ad alterius medij superficiem *AB*. Praeterea sumatur *FD* = *GR* et ducatur *Dρ* ipsi *Fγ* parallela, ut secet radium *Fρ* in *ρ*, unde *ργ* demittatur ad *Fγ* normalis, aliosque radios *Fπ* et *Fτ* secans in *π* ac *τ*. Jam cùm sit *γρ* = *GR*, erit etiam *γπ* = *GP*, et *γτ* = *GT* ex ostensis ad fig I, 23;[13] et insuper ex ostensis ad fig I, 18,[14] cùm radiorum secundum *IX* et *EF* lineas parallelas incidentium eadem sit refractio in Medium *αβνμ*, sive immediatè ingrediantur ex aere sicut fit ad *F*, sive priùs permeent aliud medium ut *ABβH* parallelis planis terminatum: sequitur quòd radij alterutro modo refracti in dictum Medium *αβνμ* sunt paralleli radijs homogeneis altero modo in idem medium refractis: hoc est, quod *Fπ* ad *PL*, *Fρ* ad *RM*, et *Fτ* ad *TN* sunt paralleli. Quapropter si refracti radij *PL*, *RM*, ac *TN* retroducantur donec singuli occurrunt perpendiculo *GX*: cum eo et basibus *GP*, *GR* ac *GT* constituent triangula similia triangulis *γπF*, *γρF*, & *γτF*, imò et ipsis aequalia, siquidem eorum bases *γπ* et *GP*, *γρ* et *GR*, *γτ* et *GT* sibimet respectivè sint aequales. Quare cùm horum triangulorum vertices conveniant ad idem punctum *F*, illorum etiam vertices ad idem aliquod punctum *φ* convenient. Hoc est radij *PL*, *RM*, ac *TN*, ipsorum *XP*, *XR* et *XT* refracti divergent omnes ab eodem puncto *φ*. Q.E.D.

50 [≈ 123].
Theorema illud
notis quibusdam
promovetur.

Ostenso hoc, sequentia obveniunt notanda. 1. Quod proportiones sinuum incidentiae et refractionis ad superficiem *Hβ* factae, ex his facilè determinantur. Nam pro radijs maximè refrangibilibus sinus isti sunt ut *φP* ad *XP*; et pro minimè refrangibilibus ut *φT* ad *XT*; &c.

refracted to π, ρ, τ, and the intermediate places, just as was done at the surface *AB* of the other medium. Moreover, take *FD* equal to *GR*; draw *Dρ* parallel to *Fγ* so that it intersects the ray *Fρ* in ρ; and from here drop $\rho\gamma$ normal to *Fγ*, intersecting the other rays, *Fπ* and *Fτ*, in π and τ. Now since $\gamma\rho = GR$, then also $\gamma\pi = GP$ and $\gamma\tau = GT$, from what was shown at Fig. I, 23.[13] Furthermore, from what has been shown at Fig. I, 18,[14] since the refraction of the rays incident along the parallel lines *IX* and *EF* is the same in the medium $\alpha\beta\nu\mu$, whether they enter immediately from air, as happens at *F*, or they first pass through another medium, such as *ABβH*, bounded by parallel planes: It follows that the rays refracted in the specified medium $\alpha\beta\nu\mu$ in the first way are parallel to the homogeneous rays refracted in the same medium in the second way, that is, that *Fπ* is parallel to *PL*, *Fρ* to *RM*, and *Fτ* to *TN*. Therefore, if the refracted rays *PL*, *RM*, and *TN* are drawn backward until each one meets the perpendicular *GX*, they will form with it and the bases *GP*, *GR*, and *GT* triangles similar to the triangles $\gamma\pi F$, $\gamma\rho F$, and $\gamma\tau F$, indeed, equal to them, since their bases $\gamma\pi$ and *GP*, $\gamma\rho$ and *GR*, and $\gamma\tau$ and *GT* are respectively equal to one another. Consequently, since the vertices of the latter triangles meet at the same point *F*, the vertices of the former will also meet at some one point ϕ. That is, the rays *PL*, *RM*, and *TN*, the refracted rays of *XP*, *XR*, and *XT*, will all diverge from the same point ϕ. As was to be demonstrated.

With this having been shown, the following things occur to us to be noted:

1. The ratios of the sines of incidence and refraction made at the surface *Hβ* are easily determined from this. For instance, for the most refrangible rays those sines are as ϕP to *XP*, and for the least refrangible as ϕT to *XT*, and so forth.

50 [$\approx$ 123]. That theorem is extended by some notes.

---

(13) In §I, 42 = 115.
(14) In §I, 33 = 106.

/ 2. Hinc si proportiones sinuum refractionis ex aere in duo quaelibet 39/
media proposita, paribus incidentijs, dentur; proportiones sinuum refractionis ex altero mediorum in alterum facilè dabuntur; dividendo nempe sinus
posterioris medij per correspondentes sinus anterioris. Sic cùm refractio fit ex
aëre in vitrum dicti sinus sunt ut 68, $68\frac{1}{2}$, 69; et cùm fit ex aere in aquam sunt
ut 90, $90\frac{1}{2}$, 91: Ergo cùm fit ex aquâ in vitrum erunt ut, $\frac{68}{90}\ \frac{68\frac{1}{2}}{90\frac{1}{2}}\ \frac{69}{91}$, hoc est
ut 281, $281\frac{1}{2}$, 282 ferè.

3. Si tertium aliquod medium aere densius postponatur medio $\alpha\beta\nu\mu$, contingens illud in superficie $\mu\nu$, quae concipiatur plana, ipsisque AB et $\alpha\beta$
parallela; et si radij divergentes a puncto ϕ (sicut modò ostensum erat) in
illud incidant ad puncta L, M, et N; postquam in ijsdem refringuntur, divergent rursus ab alio quodam puncto χ quod situm est in perpendiculo XG: Et
sic praeterea in infinitum, quotcunque licet media parallelis planis ab invicem
discreta sese ordine subsequantur. Quod si aer immediatè succedat Medio
$\alpha\beta\nu\mu$, punctum istud χ a quo emergentes radij tendunt situm erit ad V in ipsâ
refringente superficie, propterea quod emergent paralleli ad summè obliquam
lineam IX secundum quam primùm incidebant ex aere, si modò emergere
dicantur, qui nunquam divaricabant a refringenti superficie.

4. Si radij ab aliquo puncto F in aere sito divergentes, tendant ad puncta,
γ, δ, ϵ, eo more quem ad schema I, 24 explicui, et per varia deinde plana
refringentia ipsisque AB parallela transeant semper divergent omnes ab eodem aliquo puncto quod situm est in perpendiculo planorum per punctum F
transeunte, non secus quàm si incidissent in planum AB advenientes in
obliquissimâ lineâ IX. Et longitudines radiorum punctis refringentibus dic-
toque perpendiculo interceptorum sunt ut sinus incidentiae et refractionis ad
singula plana quae respiciunt.[15] Quarum assertionum demonstrationes cùm
facilè eruantur e praedictis, praetermitto, ne nimius in hac re videar.

Sectio 3ᵃ
De Planorum Refractionibus.

Lect 8   Positis refractionum legibus, radiorum per diversa media trajectorum
[51 → 124] affectiones[1] aliae jam tradendae sunt. Et primo refractiones / planorum in 40/
gratiam doctrinae de coloribus post explicandae describam; Deinde sphae-
ricarum et aliarum superficierum proprietates enarrabo, tum ut colorum exinde ortorum[2] phaenomena detegantur, tum ut instrumentorum Opticis usibus inservientium constructio rectiùs innotescat. Imprimis autem plani solita-
rij refractiones, deinde planorum refractiones iteratas considerabo.[3]

---

(15) Originally (as I): respectant.
(1) Originally: passiones.
(2) Originally: scaturientium (flowing). Spherical surfaces are treated in Section 4, Lects. I,
13–15.

2. Hence, if the proportions of the sines of refraction from air into any two proposed media at the same incidence are given, the proportions of the sines of refraction from one of the media into the other will be easily given, namely, by dividing the sines of the second medium by the corresponding sines of the first. Thus when the refraction is made from air into glass, those sines are as $68 : 68\frac{1}{2} : 69$, and when it is made from air into water, they are as $90 : 90\frac{1}{2} : 91$. Therefore, when it is made from water into glass, they will be as $68/90 : 68\frac{1}{2}/90\frac{1}{2} : 69/91$, that is, approximately as $281 : 281\frac{1}{2} : 282$.

3. If any third medium denser than air is placed behind the medium $\alpha\beta\nu\mu$ in contact with it at the surface $\mu\nu$, conceived to be plane and parallel to AB and $\alpha\beta$; and if rays diverging from the point ϕ (as was just shown) fall upon it at the points L, M, and N; then after they are refracted in the same medium, they will diverge again from some other point χ located in the perpendicular XG. And so on in infinitum, no matter how many media separated from each other by parallel planes may follow in turn. But if air immediately follows the medium $\alpha\beta\nu\mu$, that point χ from which the emergent rays tend will be located at V in the refracting surface itself, because they will emerge parallel to the most oblique line IX along which they were initially incident from air—if rays may be said to emerge that never diverge from the refracting surface.

4. If rays diverging from any point F located in air tend to the points γ, δ, and ϵ in the way that I explained at Fig. I, 24 and then pass through various refracting planes parallel to AB, they will all always diverge from some one point located in the planes' perpendicular that passes through the point F, just as if they were incident upon the plane AB and arrived in the most oblique line IX. The lengths of the rays intercepted between the refracting points and the specified perpendicular are as the sines of incidence and refraction at each plane they confront. Since the demonstrations of these assertions may be easily derived from the preceding, I am omitting them lest I seem to linger too long on this matter.

SECTION 3
THE REFRACTIONS OF PLANES

With the laws of refractions set forth, other properties[1] of rays transmitted through diverse media must now be treated. First, I will describe the refractions of planes for the sake of the doctrine of colors to be explained afterward, and then I will expound on the properties of spherical and other surfaces, so that both the phenomena of colors arising[2] from there may be disclosed and the construction of instruments for optical uses may become more properly known. In the first place, however, I will consider the refractions of a single plane and then the repeated refractions of planes.[3]

Lecture 8
[51 → 124]

---

(3) Repeated refractions in prisms is considered in Lect. I, 12.

De Plani Solitarij refractionibus.

[52]   Quod ad radios ejusdem cujuscunque generis attinet passiones in Lectionibus D^ris Barrow (his fundamentis, quod radij lucis in similari medio directi
sunt, quod eorum refractio fit in superficie ad medij refringentis superficiem
perpendiculari, et quod sinus incidentiae perpetuo sunt proportionales sinubus refractionum in aliud quodpiam medium similare factarum.)[4] traduntur,
et idcircò sufficiet aliquas sub formâ lemmaticarum propositionum sine
demonst[r]ationibus hic recensuisse.

[53]   Prop. 1. Radij cujusvis refracti incidens incidentis vicissim fit refractus.[5]

[54 → 125, 126]   Prop. 2. Angulo incidentiae aequali aequalis, et majori major convenit tum
angulus refractionis tum refractus. Et contra.[6]

[55]   Prop. 3. Incidentium radiorum refractos exhibere. Instantiam in radijs ad
medium densius e rariori divergentibus accipe. In fig I, 27 sit *F* punctum

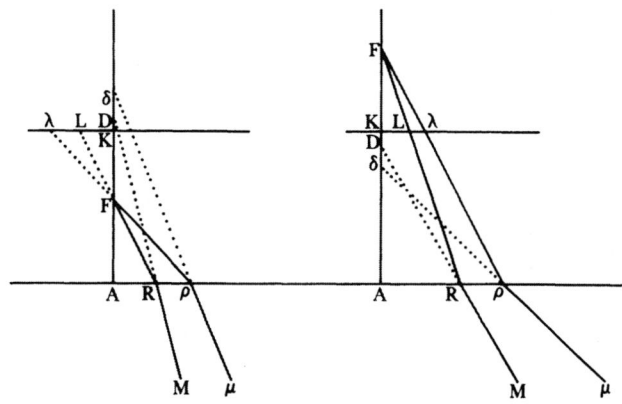

Figure I, 27

radios *FR*, *Fρ* aliosque innumeros versus refringentem superficiem *AR* ejaculans, sitque *FA* radius perpendicularis, quem produc ad *K* ut sit *AF* ad *AK*
sicut sinus refractionis ad sinum incidentiae. Et ad *K* Erige perpendiculum
KL. Quo facto radios quoslibet incidentes *FR*, *Fρ* retrorsum produc donec
praefatae *KL* occurrant in *L* et *λ*, et in angulo *FAR* inscribe *RD* = *RL* et *ρδ*
= *ρλ*. Quibus versus *M* et *μ* productis habebis re-
fractos radios *RM* et *ρμ*. Et eadem ratione refractos quamplurimos confestim duces.[7]

[56]   Prop. 4. Radium datae rectae parallelum designare cujus refractus per datum punctum transibit.
In fig I, 28, sit *AB* superficies refringens, *M* punctum datum, et *GH* recta cui radius incidens debet
esse parallelus. Et imprimis radij secundum *GH*
incidentis duc refractum *HI* per prop 3. eique parallelum age *MR* et *FR* datae *GH* parallelè ductus
erit radius incidens.

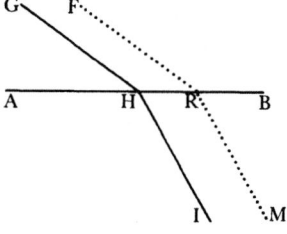

Figure I, 28

THE REFRACTIONS OF A SINGLE PLANE

As for rays of any one kind, Dr. Barrow in his *Lectures* has treated their [52] properties with these principles: Rays of light in a uniform medium are straight; their refraction is made in a surface perpendicular to the surface of the refracting medium; and the sines of incidence are always proportional to the sines of refraction made into any other uniform medium.[4] Consequently, it will suffice to recount some of them here in the form of lemmatical propositions without demonstrations.

Proposition 1. The incident ray of any refracted one becomes interchange- [53] ably the refracted ray of an incident one.[5]

Proposition 2. To an equal angle of incidence there corresponds both an [54 → 125, 126] equal angle of refraction and deviation, and to a greater angle of incidence, greater ones; and conversely.[6]

Proposition 3. To display the refracted rays of incident ones. [55]

Consider the example of rays diverging from a rarer into a denser medium. In Fig. I, 27 let F be a point casting out the rays FR, $F\rho$, and innumerable others toward the refracting surface AR; let FA be a perpendicular ray extended to K so that AF is to AK as the sine of refraction is to the sine of incidence; and at K erect the perpendicular KL. Having done this, extend any incident rays, FR and $F\rho$, backward until they meet the specified KL in L and λ; and in the angle FAR inscribe $RD = RL$ and $\rho\delta = \rho\lambda$. When these are extended toward M and μ, you will have the refracted rays RM and $\rho\mu$; and by the same method you may immediately draw very many rays.[7]

Proposition 4. To describe a ray that is parallel to a given line and whose [56] refracted ray will pass through a given point.

In Fig. I, 28 let AB be the refracting surface, M the given point, and GH the line to which the incident ray must be parallel. First draw the refracted ray HI of the ray incident along GH (by Prop. 3), then draw MR parallel to it, and FR being drawn parallel to the given GH will be the incident ray.

---

(4) Barrow, in fact, founded his *Lectiones XVIII* on six hypotheses, of which Newton has invoked the first, fourth, and sixth; Lect. I, §§VIII, XI, and Lect. II, §IV, pp. 18, 21–2, 27–9. His other hypotheses are: (2) light rays are directed from every point of a luminous body to any point of the medium, (3) a light ray falling perpendicularly on a surface either proceeds in a straight line or is reflected backward along the same line, and (5) the angles of incidence and reflection are equal; Lect. I, §§IX, X, and Lect. II, §II, pp. 18–21, 25–6.

(5) Ibid., Lect. III, §III, pp. 37–8.

(6) Ibid., Lect. III, §§IV, VI, pp. 38, 39; see also, Lect. 9, note (4).

(7) See *Lectiones XVIII*, Lect. IV, §V, pp. 47–8. Ray DRM evidently satisfies the sine law; since the triangles ARF and KLF are similar, then $RF/FL = AF/FK$, or $RF/RL = RF/RD = AF/AK = R/I$, thus fulfilling the condition of §120 = I, 47.

[57]   Prop. 5. Radium e dato puncto prodientem designare, cujus refractus evadet rectae positione datae parallelus. Absolvitur ad modum 4^{tae} propositionis denominatione radiorum secundum prop 1 permutatâ.

[58 → 141]   / Prop. 6. Radium e dato puncto *F* prodientem designare, cujus refractus per aliud punctum datum *M* transibit. Per *F* et *M* (fig I, 29) ducantur refringenti perpendiculares, et (radio in Medium densius incidente) fiat *AE* ad *AF* ut sinus incidentiae ad radicem differentiae quadratorum a sinubus incidentiae et refractionis. Item *T* ad *MI* ut sinus refractionis ad eandem radicem. Anguloque *AIM* per *E* transiens ipsamque *T* adaequans, inscribatur recta *RH,* et connectantur *FR, RM*; nam ipsae *FR, RM,* erunt radij quaesiti.[8]

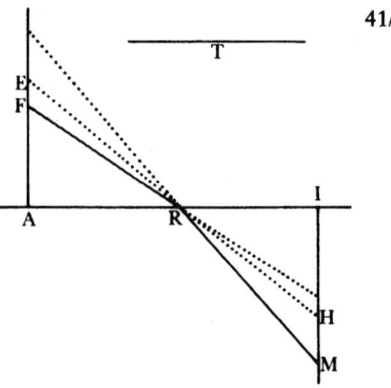

41/

Figure I, 29

Cùm radius incidit in medium rarius, appellatione (secundum prop. 1.) commutatâ, absolvitur ut ante.

Caeterùm quo pacto data recta angulo recto interserenda sit, quae per punctum datum transibit in Lect 5 D^{ris} Barrow per Hyperbolae et circuli intersectionem ostenditur.[9]

[59]   Prop. 7. Radiorum ad planam superficiem divergentium, parallelorum vel convergentium refracti itidem divergent, paralleli erunt vel convergent, et e contrà.[10]

[60]   Prop. 8. Punctum a quo refracti illi radij divergunt, vel ad quod convergunt invenire.

Cas. 1. Cùm radiorum definita est inclinatio duc refractos per prop. 3, 4, 5, vel 6, et intersectionem habebis.

[61 → 130]   Cas. 2. At cùm inclinatio existit quâvis datâ indefinitè minor eodem recidit problema ac si punctum in radio obliquo refracto quaeras quod radiorum alterutrinque jacentium intersectiones disterminat, et intercedit, quodque pro radiationis centro seu loco imaginis respectu oculi per cujus pupillae centrum radius ille transigitur, haberi debet. Ejus autem inventio ejusmodi est. In fig I, 30[11]

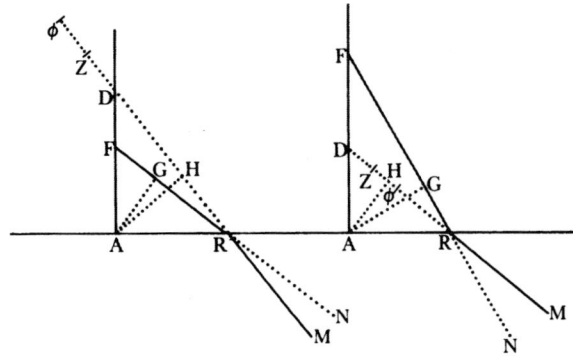

Figure I, 30

Proposition 5. To describe a ray that proceeds from a given point and [57] whose refracted ray will become parallel to a line in a given position.

It is solved in the manner of Prop. 4 by interchanging the rays' designations according to Prop. 1.

Proposition 6. To describe a ray that proceeds from a given point *F* and [58 → 141] whose refracted ray will pass through another given point *M*.

Through *F* and *M* (Fig. I, 29) draw perpendiculars to the refracting surface; (when the ray falls upon a denser medium) make *AE* be to *AF* as the sine of incidence is to the root of the difference of the squares of the sines of incidence and refraction; and also make *T* be to *MI* as the sine of refraction is to the same root. In the angle *AIM* inscribe the line *RH* passing through *E* and equal to *T*, and join *FR* and *RM*, for *FR* and *RM* will be the rays sought.[8]

When the ray falls upon a rarer medium, it is solved as before by changing its name (by Prop. 1).

But how a given line that will pass through a given point may be inserted in a right angle Dr. Barrow shows in Lecture V by the intersection of a hyperbola and a circle.[9]

Proposition 7. The refracted rays of rays that are divergent, parallel, or [59] convergent at a plane surface will likewise be divergent, parallel, or convergent; and conversely.[10]

Proposition 8. To find the point from which those refracted rays diverge or [60] to which they converge.

Case 1. When the inclination of the rays is finite, draw the refracted rays by Props. 3, 4, 5, or 6, and you will have the intersection.

Case 2. But when their inclination is indefinitely smaller than any given [61 → 130] one, the problem proves to be the same as if in an oblique refracted ray you sought the point that separates and falls between the intersections of the rays lying on each side and that must be considered as the center of radiation or the place of the image with respect to an eye through the center of whose pupil that ray passes. It is found, moreover, in this way. In Fig. I, 30[11] let

---

(8) Newton has abandoned his own solution to the anaclastic problem from §141 in favor of Barrow's, *Lectiones XVIII*, Lect. V, §XII, pp. 58–9. It is necessary to show that the rays determined by this construction satisfy the sine law of refraction. By construction $AE/AF = I/\sqrt{[I^2 - R^2]}$ and $RH/MI = R/\sqrt{[I^2 - R^2]}$, so that

$$I^2/R^2 = (IM^2 + RH^2)/RH^2 = AE^2/(AE^2 - AF^2).$$

Since the triangles *AER* and *IHR* are similar, and also $IM^2 + RH^2 = MR^2 + IH^2$ and $AE^2 - AF^2 = ER^2 \times FR^2$, it follows that $AE^2 - AF^2 = IH^2(FR^2/MR^2)$, whence

$$I^2/R^2 = (AE^2/IH^2)/(FR^2/MR^2) = (AR^2/FR^2)/(IR^2/MR^2),$$

which yields the sine law $\sin \widehat{AFR} : \sin \widehat{IMR} = I : R = AR/FR : IR/MR$. See *Mathematical Papers*, 3:453, note 6.

(9) *Lectiones XVIII*, Lect. V, §VII, pp. 56–7; and also in the *Lectiones geometricae* (1670), Lect. VI, §III, *Mathematical Works of Isaac Barrow*, 2:210.

(10) Newton demonstrated this for divergent rays in the *Lectiones opticae*, §128.

(11) We have changed the line *DR* in the figure on the left to a dotted one in agreement with the figure on the right.

sit *DRM* refractus cujusvis incidentis *FRN* sitque *F* centrum radiationis incidentium (sive divergentium sive convergentium) radiorum et *FA* refringenti normaliter insistat, secetque *RM* in *D*. Jam ab *A* demitte ad hos radios perpendicula
AG et *AH* et fac esse *RF . Rϕ :: FG . DH,* et ipsius *RM,* aliorumque refractorum proximè *RM* utrinque jacentium centrum radiationis erit ϕ.[12]

[62 ≈ 129]   Schol. Caeterùm hoc ϕ radiorum in plano *FAR* jacentium concursus
solummodò existit. Nam aliorum extra planum *FAR* jacentium refracti ‖ nec
in puncto ϕ nec ullibi omninò radium *Rϕ* secabunt, si eas solummodò excipias, quorum incidentes jacent in superficie conicâ, cujus axis est *AF,* vertex *F,*
et semiangulus *AFR;* utpotè qui omnes praefatum *Rϕ* in puncto *D* secabunt,
quod in axe *FA* sit positum. Et hujus itaque *Rϕ* centra radiationis praecipuè
sunt duo, / alterum ϕ a refractis jacentium in plano *FAR* effectum, et alterum 42/
a refractis jacentium in conicis superficiebus axe *FA,* angulisque *AFR, ADR*
descriptis. Ad reliquos autem radios quod attinet aliter circa *FR* quaquaversum positos, eorum refracti maximè appropinquant radio *Rϕ* alicubi inter *D*
et ϕ. Adeo ut respectu oculi per cujus pupillae centrum radius *RM* transit,
locus imaginis per totum spatium ϕ*D* diffundi debeat. Vel potiùs cùm spatium ϕ*D* sit unici tantùm puncti *F* imago, debemus unicum aliquod in eo
punctum quod omnis lucis ab eo versus oculum pergentis meditullium occupet, inter puncta *D* et ϕ in mediâ circiter distantiâ interjacens pro sensibili
imagine statuere. Puncti vero illius accurata determinatio, cùm omnium radiorum ab *F* versus oculi pupillam refractorum habenda sit aestimatio, problema solutu difficillimum praebebit nisi Hypothesi alicui saltem verisimili si
non accuratè verae innitatur assertio. Quemadmodum cùm radij aequè multi
a termino *D,* alijsque vicinis punctis ac a termino ϕ alijsque punctis similiter
sibi vicinis versus oculum videantur profluere; locus imaginis ita debet in
medio istorum terminorum statui, ut angulus quem radij duo a *D* et ϕ ad
idem quodpiam pupillae punctum convergentes includant, a radio ab illo
visionis loco ad idem pupillae punctum pergente quàm proximè semper bisecetur. Quâ Hypothesi admissâ, nihil aliud agendum est quàm ut fiat

$$Mϕ + MD . MD :: ϕD . DZ,$$

et erit *Z* locus visionis puncti *F* quaesitus, posito nempe quod *M* sit locus
oculi. Nam cùm ponatur *Mϕ + MD . MD :: ϕD . DZ,* erit divisìm *Mϕ . MD
:: ϕZ . DZ:* et proinde ductis tribus lineis a ϕ, *D,* et *Z* ad *M* vel potiùs ad
punctum quodpiam huic *M* indefinitè vicinum, angulus quem externae duae
continent, ab interjacente lineâ (per 3. 6 Elem) quam proximè semper bise-
cabitur.

[63 = 131]   His[13] paucis circa radios homogeneos in gratiam sequentium obiter notatis, ut eorum penitior cognitio habeatur, Lectiones quas Vir Reverendus D^r

---

(12) Newton has again abandoned his own solution to a problem and, by implication, referred the reader to Barrow's *Lectiones XVIII;* see Lect. 12, note (19), this volume. In the
"General Remarks" to his *Compleat System* (§§490–4, 2:81–2), Smith gave his own demonstration of this remark and translated the following Scholium.

DRM be the refraction of any incident ray *FRN*; let *F* be the center of radiation of the incident rays (whether divergent or convergent); and let *FA* be normal to the refracting surface and intersect *RM* in *D*. Now, from *A* drop the perpendiculars *AG* and *AH* to these rays, and make $RF : R\phi = FG : DH$. The center of radiation of *RM* and the other refracted rays lying near *RM* on each side will be ϕ.[12]

Scholium. This ϕ, however, is the intersection only of rays lying in the plane *FAR*. For the refracted rays of the others lying outside the plane *FAR* will intersect the ray $R\phi$ neither in the point ϕ nor anywhere else, except only those whose incident rays lie in the surface of the cone whose axis is *AF*, vertex is *F*, and half angle is *AFR*, inasmuch as all those will intersect the specified $R\phi$ in the point *D* located on the axis *FA*. Therefore, there are principally two centers of radiation of $R\phi$: the one, ϕ, made by refracted rays lying in the plane *FAR*, and the other, [*D*], made by refracted rays lying in the surfaces of the cones described with the axis *FA* and angles *AFR* and *ADR*. As for the other rays otherwise situated all around *FR*, their refractions come nearest to the ray $R\phi$ somewhere between *D* and ϕ, so that with respect to an eye through the center of whose pupil the ray *RM* passes, the image's position must be diffused through the entire space ϕD. Or rather, since the space ϕD is the image of only the single point *F*, we ought to take for the sensible image some single point in it that occupies the middle of all the light proceeding from there toward the eye, and that lies approximately midway between the points *D* and ϕ. Since an estimate must be made of all rays refracted from *F* toward the pupil of the eye, the accurate determination of that point will in fact present a very difficult problem to solve, unless the claim is based on some hypothesis that is at least probable, if not strictly true. For instance, since equally many rays appear to flow toward the eye from the end point *D* and other nearby points as from the end point ϕ and other points similarly nearby, the image's position ought to be set in the middle of those end points so that the angle contained by two rays converging from *D* and ϕ to any one point of the pupil is always bisected as closely as possible by the ray proceeding from that place of vision to the same point of the pupil. Once this hypothesis is admitted, nothing else needs to be done other than to make $M\phi + MD : MD = \phi D : DZ$, and *Z* will be the required place of vision of the point *F*, assuming, that is, that *M* is the eye's position. For since it is assumed that $M\phi + MD : MD = \phi D : DZ$, upon dividing it will be $M\phi : MD = \phi Z : DZ$; consequently, three lines having been drawn from ϕ, *D*, and *Z* to *M*, or rather to any point indefinitely near *M*, the angle contained by the outer two will always be bisected as closely as possible by the line lying between them (*Elements*, VI, 3).

After[13] these few things about homogeneous rays have been noted in passing for the sake of what follows, in order to acquire a more profound

‖ [62 ≈ 129]
‖

[63 = 131]

---

(13) I: *Sed videor actum agere, et his itaque* (But I seem to be wasting your time, and therefore after . . .).

Barrow de ijsdem fusiùs[14] composuit consulendas esse hortor,[15] deque Heterogeneis sive dissimiliter refrangibilibus radijs pergo actutùm disserere.

[64 → 138]   Prop. 9 E radijs diversi generis a puncto lucido fluentibus, soli possunt ad
aliud commune punctum refringi, que jacent in plano / per utrunque punctum   43/
transeunte, et ad planum refringens perpendiculari. Utpote cùm radij cujusque refractio semper fiat in plano ad medij refringentis superficiem perpendiculari, et ejusmodi duo plana per utrumque punctum transire nequeant.

[65]   Prop. 10. E radijs diversorum generum a dato puncto fluentibus, quorum
refracti ad aliud punctum datum convergunt, illis magis a lineâ rectâ punctis
concursuum sive radiationum centris interja-
cente divaricant, qui sunt magis refrangibiles.
Sint *FPφ*, *FQφ* (fig I, 31) radij dissimiles hinc et
inde convenientes in *F* et *φ* et manifestum est
quòd non penitùs coincident, quia sic par esset
refractio contra Hypothesin. Neque radius mâ-
gis refrangibilis potest esse rectae *Fφ* propior;
Sic enim propter obliquitatem ex parte Medij
densioris majorem, major esset ejus refractio per
prop 2 et Hypoth. Hoc est, angulus *Fπφ* esset
minor angulo *FQφ* contra 21. 1. Elem.[16] Restat
itaque ut sit magis refrangibilis *FPφ* qui a rectâ
Fφ magis divaricat.

Figure I, 31

[66 = 146.
Lemmata quaedam
ad prosequendam
de difformium
radiorum
affectionibus
doctrinam
ponuntur.]

Lem. 1. Quatuor lineis *GB*, *GC*,
GD, *GE* (fig I, 32) a dato puncto
G ad datam lineam *EB* ita ductis
ut sit *GB . GC :: GD . GE*: angulus *BGC* quem minima *GB* cum
alterutrâ intermediarum *GC* constituit, major est quam angulus
DGE ab alterâ intermediâ *GD* et
maximâ *GE* constitutus. Nam centro *G* radio *GE* describatur circulus *EK* et radius *GK* ducatur con-
stituens angulum *DGK* aequalem angulo *BGC*, et puncta *K D* jungantur;

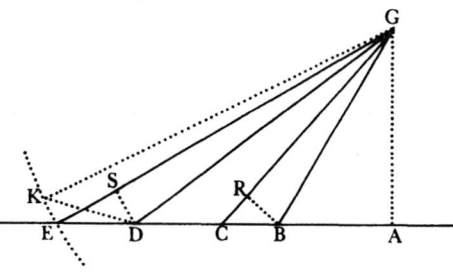

Figure I, 32

(a) 6. 6. Elem.
Et Hypoth.
(b) 16. 1. Elem.
(c) 7. 3. Elem.
(d) 25. 1. Elem.

eruntque triangula *GDK*, *GBC* similia propter aequales angulos ad *G* et
latera circa illos proportionalia[a], nempe *GB . GC :: GD . (GE)GK*. Quare
angulus *KDG* = ang *CBG*. Sed ang *EDG* ⊏ ang *CBG*[b]. Ergo linea *KD* ⊏
ED,[c] et ang *KGD* ⊏ ang *EGD*[d] hoc est ang. *CGB* ⊏ ang. *EGD*.   Q.E.D.

[67 = 147]   Lemma 2. Positis istis angulis infinitè parvis, ac *GA* perpendiculari ad
lineam *EB* demissâ; erit ang: *EGD* . ang: *CGB :: BA . DA*. A punctis enim *B*
& *D* ad lineas *GC*, *GE*, demittantur normalia *BR* ac *DS*, et erunt anguli

praefati ad se invicem ut est $\frac{DS}{DG}$ ad $\frac{BR}{BG}$ ponendo nempe lineas istas *BR* ac

DS aequipollentes esse arcubus infinitè parvis, quibus anguli isti subtenduntur. Est autem *BG . CG :: DG . EG* ex Hypoth: et divi[dendo] *BG . CR*
:: *DG . ES*.[17] Item propter similia triangula *BAG*, *CRB*, est *BA . AG*

knowledge of them I urge[15] you to consult the *Lectures* where Rev. Dr. Barrow has written on the same things at greater length;[14] and I immediately proceed to lecture on heterogeneous or unequally refrangible rays.

Proposition 9. Of rays of different kinds flowing from a luminous point, only those can be refracted to another common point that lie in the plane passing through both points and perpendicular to the refracting plane. [64 → 138]

Namely, the refraction of every ray is always made in a plane perpendicular to the surface of the refracting medium, and two such planes cannot pass through both points.

Proposition 10. Of rays of different kinds that flow from a given point and whose refracted rays converge to another given point, those that are more refrangible deviate more from the straight line lying between the points of intersection or the centers of radiation. [65]

Let $FP\phi$ and $FQ\phi$ (Fig. I, 31) be dissimilar rays meeting on both sides at F and ϕ, and it is evident that they will not wholly coincide, because then their refraction would be equal, contrary to the hypothesis. Nor can the more refrangible ray be nearer to the line $F\phi$; for because of the greater obliquity on the denser side of the medium, its refraction would be greater by Prop. 2 and hypothesis, that is, the angle $F\pi\phi$ would be smaller than the angle $FQ\phi$, contrary to *Elements*, I, 21.[16] It remains, therefore, that it is the more refrangible ray $FP\phi$ that deviates more from the line $F\phi$.

Lemma 1. When four lines, GB, GC, GD, and GE (Fig. I, 32), are drawn from a given point G to a given line EB so that $GB : GC = GD : GE$, the angle BGC made by the smallest one, GB, with either of the intermediate ones, GC, is greater than the angle DGE made by the other intermediate line, GD, and the largest one, GE. [66 = 146. Some lemmas are proposed for pursuing the doctrine of the properties of dissimilar rays.]

For with center G and radius GE describe the circle EK; then draw the radius GK making the angle DGK equal to the angle BGC; and join the points K and D. The triangles GDK and GBC will be similar because of the equal angles at G and the proportional sides around them,[a] namely, $GB : GC = GD : (GE$ or$) GK$. Consequently $\widehat{KDG} = \widehat{CBG}$, but $\widehat{EDG} > \widehat{CBG}$.[b] Therefore $KD > ED$,[c] and $\widehat{KGD} > \widehat{EGD}$;[d] that is, $\widehat{CGB} > \widehat{EGD}$. As was to be demonstrated.

(a) *Elements*, VI, 6 & Hypothesis
(b) *Elements*, I, 16
(c) *Elements*, III, 7
(d) *Elements*, I, 25

Lemma 2. Assuming that those angles are infinitely small and dropping GA perpendicular to the line EB, there will be $\widehat{EGD} : \widehat{CGB} = BA : DA$. [67 = 147]

For from the points B and D drop the normals BR and DS to the lines GC and GE, and the specified angles will be to one another as DS/DG is to BR/BG; namely by considering those lines BR and DS to be equivalent to the infinitely small arcs that those angles subtend. By hypothesis, however, $BG : CG = DG : EG$, and by dividing $BG : CR = DG : ES$.[17] Also, because

---

(14) ijsdem fusiùs] I: ijs fusè (these things at length).   (15) I: moneo (I recommend).
(16) If two lines are drawn from the ends of one side of a triangle and meet within it, they will contain a greater angle than that made by the other two sides.
(17) I continues: permutandoque $BG . DG :: CR . ES$ (and then by permuting $BG : DG = CR : ES$).

:: CR . BR et pari ratione EA, vel DA . AG :: ES . DS. Sive AG . DA ::
DS . ES. Quamobrem addendo rationes aequales est

$$BA . AG + AG . DA \; (:: BA . DA) :: CR . BR + DS . ES$$

(et permutatis terminis / posteriorum rationum) :: CR . ES + DS . BR (et 44/
aequipollenti ratione pro CR . ES substitutâ) :: BG . DG + DS . BR (termi-
nisque ad invicem vicissim[18] applicatis) :: $\frac{DS}{DG} \cdot \frac{BR}{BG}$. Est itaque BA . DA ::
$\frac{DS}{DG} \cdot \frac{BR}{BG}$, hoc est ut ang. EGD ad ang. CGB. Q.E.D.

[68 = 160]   Prop. 11. Heterogeneis radijs secundum eandem lineam incidentibus, quo
obliquior est eorum incidentia caeteris paribus eo major erit differentia refractionis. In fig. I, 33 sit FG linea secundum quam duo radij incidunt quo-

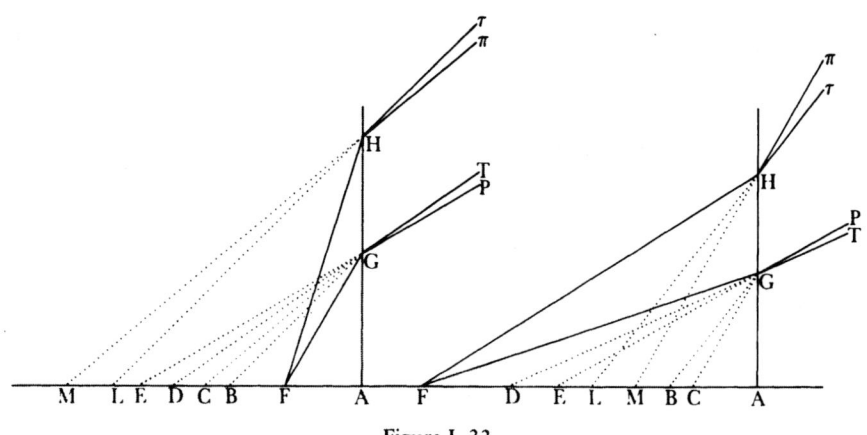

Figure I, 33

rum unus maximè refrangibilis refringitur[19] versus P et alter minimè
refrangibilis[20] versus T, eritque angulus PGT differentia refractionis. Item sit
FH linea obliquior quàm FG et secundum hanc alij duo ejusmodi radij incidant quorum maximè refrangibilis versus π et minimè refrangibilis[20] versus τ
refringitur; et similiter erit angulus πHτ eorum differentia refractionis. Dico
jam quòd sit ang: πHτ ⊏ ang. PGT. Demittatur enim FA ad refringens
planum linea normalis, quae refractos radios retro-actos in D, E, L, et M
secet. Et ad hanc a puncto G ducantur lineae duae GB, GC ipsis HL, HM
parallelae. Jam cùm tres lineae, GF, GD, GE, (ex naturâ refractionis antè
Num 25, 26 & sequ. descriptâ) sint in ratione datâ, et alterae tres HF, HL, HM in eadem ratione:
proportionales erunt HL . HM :: GD . GE. Sed est HL . HM :: GB . GC
propter sim: tri: LMH et BCG. Quare GB . GC :: GD . GE. Adeoque
ang. BGC ⊏ ang. DGE per Lem. 1. hoc est ang. LHM ⊏ ang. DGE. Sive
angulus πHτ ⊏ ang. PGT. Q.E.D.

[69 = 162]   Caeterùm ut de mutuis angulorum PGT et πHτ (in fig. I, 33) proportionibus habeatur plenior determinatio dico praeterea quòd sunt inter se quam

of the similar triangles *BAG* and *CRB*, there is *BA* : *AG* = *CR* : *BR*; and for the same reason *EA* (or *DA*) : *AG* = *ES* : *DS*, or *AG* : *DA* = *DS* : *ES*. Consequently, when the equal ratios are multiplied

$$BA : AG \times AG : DA \ (= BA : DA) = CR : BR \times DS : ES$$

(and permuting the terms of the last ratio) = *CR* : *ES* × *DS* : *BR* (and substituting the equivalent ratio for *CR* : *ES*) = *BG* : *DG* × *DS* : *BR* (and alternately[18] dividing the terms by one another) = *DS*/*DG* : *BR*/*BG*. Therefore, it is *BA* : *DA* = *DS*/*DG* : *BR*/*BG*, that is, as the angle *EGD* to the angle *CGB*. As was to be demonstrated.

Proposition 11. When heterogeneous rays are incident along the same line, [68 = 160] with other things being equal, the more oblique their incidence, the greater will be their difference of refraction.

In Fig. I, 33 let *FG* be the line along which two rays fall. One of these, the most refrangible, is refracted[19] toward *P*, while the other, the least refrangible,[20] is refracted toward *T*; and the angle *PGT* will be the difference of refraction. Likewise, let the line *FH* be more oblique than *FG*, and along this line let there fall two other such rays, of which the most refrangible one is refracted toward π and the least refrangible[20] toward τ; and similarly the angle $\pi \widehat{H} \tau$ will be their difference of refraction. I say now that $\pi \widehat{H} \tau > P\widehat{G}T$. For normal to the refracting plane drop the line *FA*, intersecting the refracted rays extended backward at *D*, *E*, *L*, and *M*; and to this line from the point *G* draw two lines, *GB* and *GC*, parallel to *HL* and *HM*. Now, since the three lines *GF*, *GD*, and *GE* are in a given ratio (from the nature of refraction previously described), and the other three, *HF*, *HL*, and *HM*, are in the same §§I, 25, 26, ff. ratio, they will be proportional, *HL* : *HM* = *GD* : *GE*. But *HL* : *HM* = *GB* : *GC*, because of the similar triangles *LMH* and *BCG*, and therefore *GB* : *GC* = *GD* : *GE*. By Lemma 1 then $B\widehat{G}C > D\widehat{G}E$, that is, $L\widehat{H}M > D\widehat{G}E$, or $\pi \widehat{H} \tau > P\widehat{G}T$. As was to be demonstrated.

Furthermore, for a more complete determination of the mutual proportions [69 = 162] of the angles *PGT* and $\pi H \tau$ in Fig. I, 33, I say, in addition, that they are to

---

(18) Omitted in *editio princeps*.

(19) *Editio princeps*: pergat.

(20) maximè refrangibilis ... minimè refrangibilis] **I** : purpuriformis ... rubiformis (purple-making ... red-making).

proximè ut lineae *AB* et *AD*, segmenta nempe basium triangulorum aequialtorum quorum alterum *EGD* constituitur a radijs *GP* ac *GT* cum perpendiculo *AF* concurrentibus, et alterum *CGB* sit simile triangulo *MHL* a radijs
Hπ et *Hτ* similitèr constituto. Nam anguli *EGD* et *CGB* si essent infinitè
parvi, forent inter se ut *AB* ad *AD*, per Lemma 2. At isti ex Hypothesi sunt
aequales angulis *PGT* et *πHτ*. Quare etiam illi *PGT* et *πHτ* modò essent
infinite parvi, forent itidem [ut *AB* ad *AD*. Cùm itaque semper sint admodum
(licèt non infinitè) parvi, sequitur quòd sunt quàm proximè][21] ut *AB* ad *AD*,
et pari ratione constat quod sunt quàm proximè ut *AC* ad *AE*. Scilicet eorum
ratio has duas rationes semper intercedit; et ideò veritatem adhuc propiùs
assequemur adhibendo rationem intermediam, nempe quod sit *PGT* ad *πHτ*
ut *AB* + *AC* ad *AD* + *AE*, vel ut √ *AB* × *AC* ad √ *AD* × *AE* proximè.

Lect. 9.
Octob: 1670
[70]
/ Prop. 12 Radios diversorum generum à dato puncto profluos exhibere 45/
quorum refracti per aliud punctum datum transibunt. Cùm punctorum alterutrum infinitè distat ut radij ex ea parte existant paralleli, res per prop 4 et
5 absolvitur. et per prop. 6 cùm utrumque finitè distat.

[71 → 134]
Schol. Caeterùm e re erit ut ostendam quomodo ex data alicujus radij
positione caeteri omnes expeditiùs determinentur.

Cas 1. Sint (in fig I, 34) *FT, FR, FP* radij ab *F* prodeuntes quorum refracti
TO, RM, PK paralleli sunt futuri. Et radij *FT* esto sinus incidentiae ad sinum
refractionis sicut *I* ad *T*, quemadmodum et radiorum *FR* et *FP* sinus isti sicut
I ad *R* et *P*. Jam ut horum quovis *FT* positione data caeteri confestim designentur, demitte *FA* refringenti normalem et in angulo *FAT* inscribe *TE, TD*,
hac lege ut sit *T . R . P :: TF . TE . TD*. Et ipsis *TE, TD* age parallelas *FR*,
FP. Dico factum. Scilicet, refractis *TO, RM* cum perpendiculo *DA* occurrentibus in *G* et *H*,[1] erit *I . T :: TG . TF* (Art. [47]).[2] Et praeterea cùm sit

$$T . R :: TF . TE$$

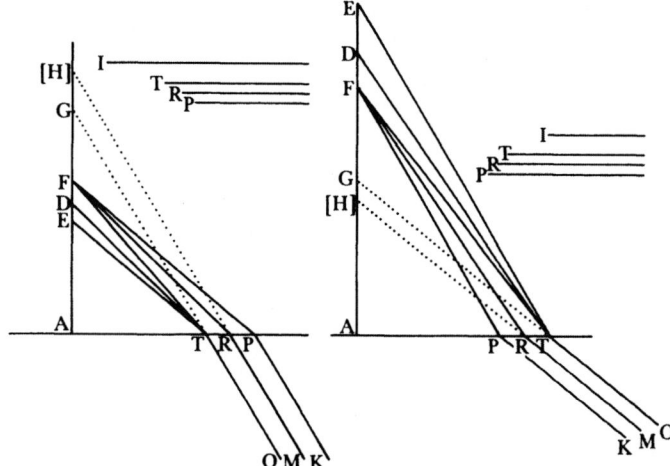

Figure I, 34

one another very nearly as the lines *AB* and *AD*; namely, segments of the bases of equal-altitude triangles, one of which, *EGD*, is formed by the rays *GP* and *GT* meeting the perpendicular *AF*, while the other, *CGB*, is similar to the triangle *MHL* similarly formed by the rays *Hπ* and *Hτ*. For if the angles *EGD* and *CGB* were infinitely small, they would be to one another as *AB* to *AD*, by Lemma 2. But by hypothesis they are equal to the angles *PGT* and *πHτ*. Consequently, provided that those angles *PGT* and *πHτ* were infinitely small, they would likewise be [as *AB* to *AD*. Since, therefore, they are always very (although not infinitely) small, it follows that they are very nearly][21] as *AB* to *AD*, and for the same reason it is evident that they are very nearly as *AC* to *AE*. Namely, their ratio always lies between these two ratios, and thus we will approach still nearer the truth by using the intermediate ratio, specifically, that *PGT* is to *πHτ* as *AB* + *AC* is to *AD* + *AE*, or approximately as $\sqrt{[AB \times AC]}$ is to $\sqrt{[AD \times AE]}$.

Proposition 12. To display rays of different kinds that flow from a given point and whose refracted rays will pass through another given point.

Lecture 9
October 1670
[70]

When either of the points is infinitely distant so that the rays on that side are parallel, the matter is accomplished by Props. 4, and 5, and when both are infinitely distant by Prop. 6.

Scholium. It will in addition be worthwhile for me to show how from the given position of any ray all the others may be rather readily determined.

[71 → 134]

Case 1. In Fig. I, 34 let *FT*, *FR*, and *FP* be rays that proceed from *F* and whose refracted rays *TO*, *RM*, and *PK* are to be parallel. Let the sine of incidence to the sine of refraction of the ray *FT* be as *I* to *T*, and also let those sines of the rays *FR* and *FP* be as *I* to *R* and *P*. Now given the position of any one of these, *FT*, to describe immediately the others drop *FA* normal to the refracting surface; in the angle *FAT* inscribe *TE* and *TD* subject to the condition that *T* : *R* : *P* = *TF* : *TE* : *TD*; and parallel to *TE* and *TD* draw *FR* and *FP*. I say it is done. Namely, since the refracted rays *TO* and *RM* meet the perpendicular *DA* in *G* and *H*,[1] then *I* : *T* = *TG* : *TF* (§I, 47).[2] Furthermore, since *T* : *R* = *TF* : *TE* (by hypothesis), there will be from the

---

(21) ut . . . proximè] An accidental skip restored from I.

(1) We have added to Fig. I, 34, the inadvertently omitted line *RH*.

(2) The equivalent §120 of the *Lectiones* was mistakenly cited and then deleted.

(Hyp.) erit ex aequo *I . R* :: *TG . TE*. Sed est *I . R* :: *RH . RF* (Art. [47]).[2] Ergo *TG . TE* :: *RH . RF*. Atque adeò cùm *TE* et *RF* parallelae sint (ex Hypoth) erunt etiam *TG* et *RH* parallelae. Q.E.O. Deque radij *PK* parallelismo consimile est ratiocinium.

[72]   Cas. 2. Si, parallelis incidentibus, refracti ad datum punctum convergant, propositum nihil secus exequaris ut e prop: 1 patet.

[73 → 142]   Cas 3. Si denique divergant incidentes et refracti convergant, problema solidum est, sed ad planum[3] quodam[m]odo reducetur fingendo differentiam refrangibilitatis indefinitè parvam esse. Quae cùm semper sit admodum exigua, solutionem ex istâ Hypothesi haud gravatim exhibebo.

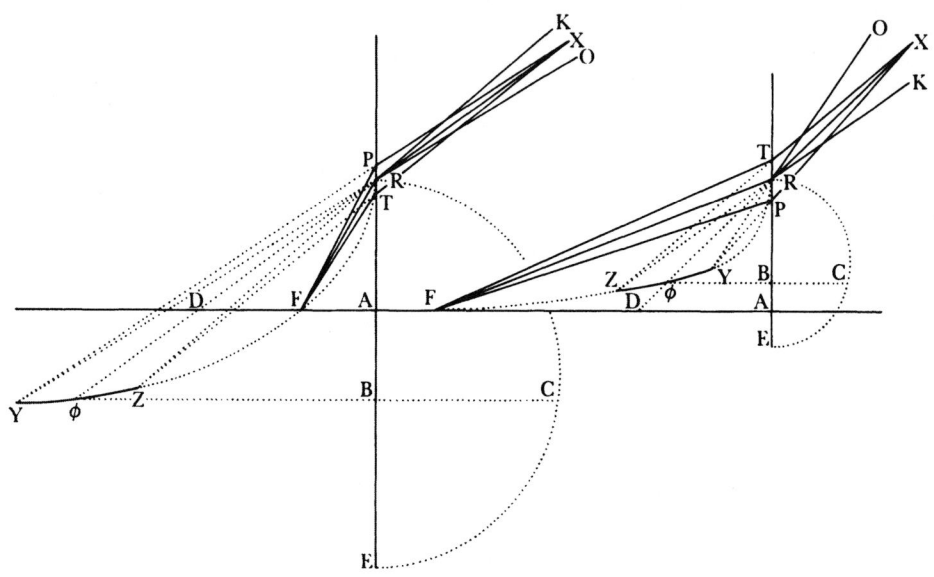

Figure I, 35

Pone *FRX* (fig I, 35) radium esse positione datum, et radios *F[P]X, FTX* (quorum datae sunt sinuum incidentiae et refractionis rationes) punctis *F* et *X* interserendos esse. Jam alios etiam radios aequè refrangibiles ac radios *FP, FT* finge secundum lineam *FR* incidere, et refractos eorum *RO, RK* (ope prop 3) describe. Centraque radiationum *Y* et *Z* (per prop 8) quaere. Ac junge *YX*, et *ZX* refringenti occurrentes in *P* ac *T*. Dico factum; nempe *FPX, FTX* esse radios quos oportuit designare. Nam cùm ex Hypothesi differentia refrangibilitatis, adeòque distantia / punctorum *T R* et *P* sit indefinitè parva; constat 46/ homogeneos radios *RO, PX* sibi mutuò vicinissimos esse et inde ab eodem radiationis puncto *Y* divergere. Rectè itáque determinavi radium *PX* per radij *RO* centrum radiationis transiturum esse. Deque radio *TX* par est ratio.

[74 ≈ 142, 143]   Verum enim verò cùm anguli *PXT* determinatio eò spectet ut noscatur quanta sit objectorum mediante refractione visorum propter inaequabiles absimilium radiorum refractiones confusio perque quantum spatium colores inde emergentes extenduntur, quemadmodum pateat concipiendo *F* esse

equality of the ratios $I : R = TG : TE$. But $I : R = RH : RF$ (§I, 47),[2] and therefore $TG : TE = RH : RF$. Hence, since TE and RF are parallel (by hypothesis), TG and RH will also be parallel. As was to be shown. As to the parallelism of the ray PK, the reasoning is completely similar.

Case 2. If the incident rays are parallel and the refracted rays converge to a [72] given point, you may accomplish what was proposed no differently, as is clear by Prop. 1.

Case 3. Finally, if the incident rays diverge and the refracted ones converge, [73 → 142] the problem is a solid one but will in a way be reduced to a plane one[3] by imagining the difference of refrangibility to be indefinitely small. Since this difference is always extremely small, I will freely present the solution according to that hypothesis.

Suppose the ray FRX (Fig. I, 35) to be given in position, and the rays $F[P]X$ and FTX (whose ratios of the sines of incidence and refraction are given) must be inserted between the points F and X. Now also imagine other rays, equally as refrangible as the rays FP and FT, to be incident along the line FR, and draw their refracted rays RO and RK (by means of Prop. 3). Find the centers of radiation Y and Z (by Prop. 8), and join YX and ZX, which meet the refracting surface at P and T. I say it is done; namely, FPX and FTX are the rays that it was required to describe. For, since by hypothesis the difference of refrangibility and hence the distance of the points $T, R,$ and P is indefinitely small, it is evident that the homogeneous rays RO and PX are exceedingly close to one another and therefore diverge from the same point of radiation Y. I have therefore properly determined that the ray PX will pass through the center of radiation of the ray RO; and the reason is the same for the ray TX.

But to be sure, since the determination of the angle PXT aims at knowing [74 ≈ 142, 143] the extent of the confusion of objects seen by an intervening refraction because of the unequal refractions of dissimilar rays, and the extent of the space through which the colors emerging from there are spread out (as may be evident by imagining F to be a luminous point which to an eye located at X

(3) On plane and solid problems see Lects. 16, 17, note (21).

punctum lucidum, quod oculo in X constituto per totum angulare spatium
PXT quod radijs PX ac TX maximè minimèque omnium refrangibilibus
comprehenditur,[4] dilatatum ac diffusum appareat; de magnitudine ejus pau‑
cula adjiciam.[5] Finge lineam curvam YφZ descriptam esse,[6] in qua radiatio‑
num centra[7] radiorum omnigenorum jacent, secundum lineam FR inciden‑
tium et ita refractorum in puncto R ut per totum angulum KRO divaricent;
Et ista Curva non malè assimiliabitur objecto lucido cujus angulus visibilis
sive apparens magnitudo[8] ad oculum in X situm sit YXZ, ac distantia ab
eodem oculo ad meditullium ejus aestimata, φX. ‖ Et hinc consectatur,

　1 Quod (cùm rei visibilis apparens magnitudo pene sit reciprocè ut distan‑
tia ejus,) stante puncto F et puncto X in lineâ RX ubicunque sumpto angulus
PXT sive YXZ pene erit reciprocè ut longitudo φX. Et hinc intervallo RX
diminuto angulus PXT augetur, ejusque quantitas in qualibet puncti X
distantiâ dabitur, si modò data fuerit unquam in quâpiam distantiâ.

　2 Quinetiàm, angulo ORK cognito, cognoscetur angulus quilibet PXT su‑
mendo in ratione ad ORK quam habet Rφ ad Xφ, quippe cùm YRZ (cui
ORK aequatur) sit objecti YφZ in distantiâ φR apparens magnitudo.

　Cùm itaque angulus ORK pro qualibet obliquitate radiorum juxta FR
incidentium supra in Schol: ad prop: 11 determinatus habeatur, et punctum
φ haud difficilè inveniatur fa[c]iendo (juxta prop 8) ut sit

$$RF . R\phi :: \frac{AF^q}{RF} . \frac{AD^q}{RD};$$

satis constat anguli PXT inventio.

[75 = 144] At ex abundanti subnoto praedictam curvam YφZ[9] in qua radiorum om‑
nis generis in puncto R refractorum / radiationum centra locantur esse Cissoi‑ 47/
dem vulgarem sive Diocleam, circulo accommodatam cujus diameter RE sit
ad AR ut FR^q ad AF^q. Nam super diametro RE descripto circulo isto RCE,
agatur quaevis recta φBC normalis ad RE, circuloque in C, et curvâ in φ
terminata. Et propter analoga latera similium triangulorum RAD, RBφ erit
$AD^q . AR \times DR :: B\phi^q . BR \times \phi R$. et applicando posteriorem rationem ad
BR fiet $AD^q . AR \times DR :: \frac{B\phi^q}{BR} . \phi R$. rursusque ducendo consequentes ratio‑
num in Rφ, et applicando ad AR orietur $AD^q . DR \times R\phi :: \frac{B\phi^q}{BR} . \frac{R\phi^q}{AR}$. Est
autem $\frac{AF^q}{FR} . \frac{AD^q}{DR} :: RF . R\phi$ ut priùs, et consequentibus in DR, et antece‑
dentibus in FR ductis, oritur $AF^q . AD^q :: FR^q . DR \times R\phi$ et vicissim

$$AF^q . FR^q :: AD^q . DR \times R\phi.$$

Quamobrem rationes eidem tertiae congruentes connectendo, habebitur

$$\frac{B\phi^q}{BR} . \frac{R\phi^q}{AR} :: AF^q . FR^q;$$

ducendoque antecedentes rationum in BR et consequentes in AR prodibit

appears extended and diffused through the entire angular space *PXT* contained by the rays *PX* and *TX*, the most and least refrangible of all),[4] I will now add a few words concerning its magnitude.[5] Imagine that there is described the[6] curved line *YϕZ* on which lie the centers of radiation[7] of rays of every kind incident along the line *FR* and refracted at the point *R* so that they spread through the whole angle *KRO*: That curve will not improperly be compared to a luminous object whose visible angle, or apparent magnitude,[8] to an eye located at *X* is *YXZ* and whose estimated distance from the same eye to its middle is *ϕX*. Hence it follows:

1. Since the apparent magnitude of a visible object is nearly inversely as its distance, when the point *F* is fixed and the point *X* is taken anywhere in the line *RX*, the angle *PXT* or *YXZ* will be nearly inversely as the length *ϕX*. Consequently, as the interval *RX* is diminished, the angle *PXT* increases, and its size will be given for any distance of the point *X*, provided it was once given at some distance.

2. Moreover, when the angle *ORK* is known, any angle *PXT* will be known by taking it in the ratio to the angle *ORK* that *Rϕ* has to *Xϕ*, since the angle *YRZ* (which is equal to the angle *ORK*) is the apparent magnitude of the object *YϕZ* at the distance *ϕR*.

Therefore, since the angle *ORK* has been determined above (in Prop. 11, Scholium) for any inclination of rays incident along *FR*, and the point *ϕ* may be found with no difficulty by making $RF : R\phi = AF^2/RF : AD^2/RD$ (according to Prop. 8), the discovery of the angle *PXT* is satisfactorily known.

In addition, I observe that that curve *YϕZ*[9] on which are located the centers of radiation of rays of every kind refracted at the point *R* is a common or Dioclean cissoid adapted to a circle whose diameter *RE* is to *AR* as FR^2 is to AF^2. For having described that circle *RCE* on the diameter *RE*, draw normal to *RE* any line *ϕBC*, which is bounded by the circle at *C* and the curve at *ϕ*. Because of the corresponding sides of the similar triangles *RAD* and *RBϕ*, there will be $AD^2 : AR \times DR = B\phi^2 : BR \times \phi R$. Dividing the latter ratio by *BR*, it will become $AD^2 : AR \times DR = B\phi^2/BR : \phi R$; and again, multiplying the denominators of the ratios by *Rϕ* and dividing by *AR*, there will arise $AD^2 : DR \times R\phi = B\phi^2/BR : R\phi^2/AR$. However, as before, $AF^2/FR : AD^2/DR = RF : R\phi$; and multiplying the denominators by *DR* and the numerators by *FR*, there arises $AF^2 : AD^2 = FR^2 : DR \times R\phi$; and by permuting, $AF^2 : FR^2 = AD^2 : DR \times R\phi$. Consequently, by equating ratios equal to the same third ratio, there will be had $B\phi^2/BR : R\phi^2/AR = AF^2 : FR^2$; when the numerators of the ratios are multiplied by *BR* and the

[75 = 144]

---

(4) quod . . . comprehenditur] Added.

(5) de . . . adjiciam.] Added to replace a long passage in §142 and to serve as a transition to §143 which follows.

(6) Finge . . . esse] I: Adhaec si describatur linea quaedam curva *YϕZ* (If, moreover, there is described some . . .).

(7) radiationum centra] I: foci.

(8) sive apparens magnitudo] Added.

(9) At . . . *YϕZ*] I: Caeterùm curva praedicta (Moreover, that curve).

$$B\phi^q . R\phi^q :: AF^q \times BR . FR^q \times AR$$

et insuper applicando posteriorem rationem ad AF^q fiet

$$B\phi^q . R\phi^q :: BR . \frac{FR^q \times AR}{AF^q}.$$

Sed cùm posuerim $RE . AR :: FR^q . AF^q$, erit $\dfrac{FR^q \times AR}{AF^q} = RE$ et proindè $B\phi^q . R\phi^q :: BR . RE$, ac divisim $B\phi^q . R\phi^q - B\phi^q(BR^q) :: BR . BE$. Atqui ex natura circuli est BC media proportionalis inter BR et BE, adeoque est $BR . BE :: BR^q . BC^q$ et proinde $B\phi^q . BR^q :: BR^q . BC^q$, sive $B\phi . BR :: BR . BC$.[10] Quod indicat curvam esse Cissoidem sicut ostendendum proposui.

[76 → 145]   Refractionibus ad superficiem data duo media disterminantem transactis, ad explorandum quid ex auctâ alterutrius Medij raritate vel densitate consequetur, sive ad diversorum Mediorum effectus inter se conferendum, jam animum adjicio.

[77 = 148]   Lemma. 3. Si a duobus punctis D, G (Fig I, 36) in lineâ quâpiam AD sitis, ad alia duo puncta L, N in ejus perpendiculo[11] sita, ducantur quatuor rectae DN, DL, GN, GL; ratio ductarum ad punctum remotius N magis accedit ad aequalitatem quàm ratio ductarum ad vicinius punctum L. Sive est $GN . DN$ ⊏ $GL . DL$. Sit enim $GN . DN :: GL . R$, et erit

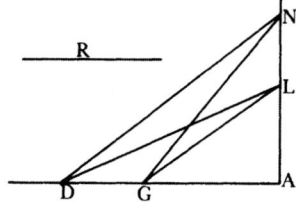

Figure I, 36

$$GN^q . DN^q :: GL^q . R^q :: GN^q - GL^q . DN^q - R^q.$$

Quare cum sit DN ⊏ GN sive DN^q ⊏ GN^q erit $DN^q - R^q$ ⊏ $GN^q - GL^q$. Verùm est $GN^q - GL^q = DN^q - DL^{q(a)}$ et ideo $DN^q - R^q$ ⊏ $DN^q - DL^q$. Hoc est DL^q ⊏ R^q sive DL ⊏ R. Atque adeò cum supponatur $GN . DN :: GL . R$, erit $GN . DN$ ⊏ $GL . DL$. Q.E.D.

(a) Num 46

[78 = 163]   / Prop 13. Posito radiorum diversi generis communi sinu incidentiae quo magis diversa est Mediorum densitas[12] eo major erit inaequalitas rationis sinuum refractionis. In fig: I, 37 sit $F\gamma$ radius e minimè[13] refrangibilibus utcunque in superficiem $A\gamma$ incidentibus, sitque refractus ejus $\gamma\lambda$ qui retroactus secet perpendiculum FA in ϕ. Dein capiatur $A\epsilon$ ut sit $F\epsilon$ ad $F\gamma$ in datâ quâdam ratione qualem ante descripsimus,[a] hâc scilicet conditione ut habito $F\epsilon$ pro radio maximè[14] re-

(a) Num 44, 45 & 49

48/

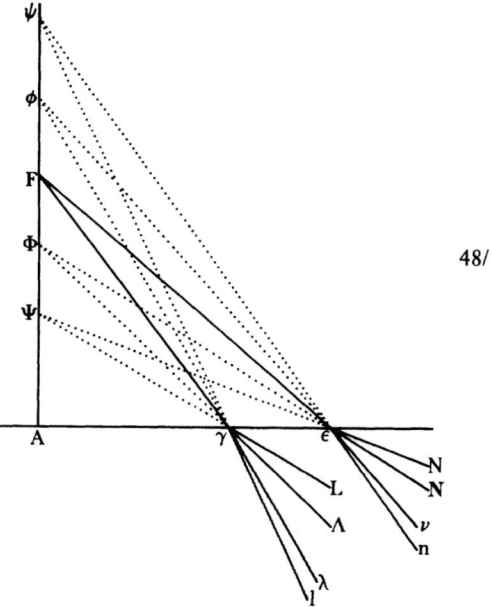

Figure I, 37

denominators by AR, there will appear $B\phi^2 : R\phi^2 = AF^2 \times BR : FR^2 \times AR$; and furthermore, dividing the latter ratio by AF^2, it will become $B\phi^2 : R\phi^2 = BR : FR^2 \times AR/AF^2$. But since I set $RE : AR = FR^2 : AF^2$, it will be

$$FR^2 \times AR/AF^2 = RE,$$

and consequently $B\phi^2 : R\phi^2 = BR : RE$, and by division,

$$B\phi^2 : R\phi^2 - B\phi^2 \text{ (or } BR^2) = BR : BE.$$

However, from the nature of a circle, BC is the mean proportional between BR and BE, and hence $BR : BE = BR^2 : BC^2$; and consequently $B\phi^2 : BR^2 = BR^2 : BC^2$, or $B\phi : BR = BR : BC$,[10] which indicates that the curve is a cissoid, as I proposed to show.

Having completed refractions at a surface separating two given media, I [76 → 145] now turn to explore the consequences of an increase of the rarity or density of either medium, or to compare to one another the effects of different media.

Lemma 3. If from two points, D and G (Fig. I, 36), situated on any line, [77 = 148] AD, four straight lines, DN, DL, GN, and GL, are drawn to two other points, L and N, situated on its perpendicular,[11] the ratio of the lines drawn to the more distant point N approaches closer to an equality than the ratio of those drawn to the nearer point L, or $GN : DN > GL : DL$.

For let $GN : DN = GL : R$, and it will be

$$GN^2 : DN^2 = GL^2 : R^2 = GN^2 - GL^2 : DN^2 - R^2.$$

Consequently, since $DN > GN$, or $DN^2 > GN^2$, it will be $DN^2 - R^2 > GN^2 - GL^2$. But $GN^2 - GL^2 = DN^2 - DL^2$,[a] and thus $DN^2 - R^2 > DN^2 - DL^2$, (a) §I, 46 that is, $DL^2 > R^2$ or $DL > R$. Hence, since it is assumed that $GN : DN = GL : R$, it will be $GN : DN > GL : DL$. As was to be demonstrated.

Proposition 13. Assuming a common sine of incidence of rays of different [78 = 163] kinds, the greater the difference in density of the media[12] the greater will be the inequality of the ratio of the sines of refraction.

In Fig. I, 37 let $F\gamma$ be a least[13] refrangible ray incident in any way on the surface $A\gamma$, and let its refracted ray be $\gamma\lambda$, which extended backward intersects the perpendicular FA in ϕ. Then take $A\epsilon$ such that $F\epsilon$ is to $F\gamma$ in a certain given ratio, as we described earlier,[a] namely, with this condition, that (a) §§I, 44, 45, 49 when $F\epsilon$ is taken to be a most[14] refrangible ray, its refracted ray [$\epsilon\nu$] diverges

---

(10) $B\phi . BR :: BR . BC$] I: $B\phi . BR . BC \div$ ($B\phi$, BR, and BC are in continued proportion).

(11) I: perpendiculo AN.

(12) Posito . . . densitas] I: Heterogeneis radijs in superficiem quaecunque Media disterminantem incidentibus; quò magis Media differunt densitate (When heterogeneous rays fall on a surface separating any media, the more the media differ in density).

(13) Read : maximè (most).

(14) Read : minimè (least).

frangibili, refractus ejus [$\epsilon\nu$] ab eodem ϕ divergat.[15] Facto hoc, si pro posteriori medio aliud utcunque densum rarumve substituatur, ejusmodi duo radij
secundum easdem rectas $F\epsilon$, $F\gamma$ incidentes semper debent ita refringi, ut ab
eodem aliquo perpendiculi istius puncto similiter divergant.[b] Quemadmodum a ψ versus l et n, posito quod hoc medium posterius sit densitatis ab
anteriori magis diversae quàm alterum posterius medium quod efficiebat divergentes a ϕ. Ostendendum est itaque quod major sit inaequalitas rationis
sinuum refractionis in posteriori quam priori casu. Scilicet radij $F\gamma\lambda$ sinus

(b) Num 45 & 49

(c) Num 47   incidentiae est ad sinum refractionis ut $\phi\gamma$ ad $F\gamma$[c], hoc est, ut 1 ad $\dfrac{F\gamma}{\phi\gamma}$. Et sic

radij $F\epsilon\nu$ sinus isti sunt ut 1 ad $\dfrac{F\epsilon}{\phi\epsilon}$. Quare sinus refra[c]tionum eorundem

radiorum sunt inter se ut $\dfrac{F\gamma}{\phi\gamma}$ ad $\dfrac{F\epsilon}{\phi\epsilon}$. Et simili discursu constabit quod radio-

rum a ψ refractorum consimiles refractionum sinus sunt ut $\dfrac{F\gamma}{\psi\gamma}$ ad $\dfrac{F\epsilon}{\psi\epsilon}$. Restat

itaque probandum quod inter $\dfrac{F\gamma}{\psi\gamma}$ & $\dfrac{F\epsilon}{\psi\epsilon}$ major sit disproportio, quam inter $\dfrac{F\gamma}{\phi\gamma}$

(d) Lem 3   et $\dfrac{F\epsilon}{\phi\epsilon}$. hoc est, (cùm sit $\dfrac{F\epsilon}{\psi\epsilon} \sqsubset \dfrac{F\gamma}{\psi\gamma}$[d]) probandum restat, quod sit $\dfrac{F\epsilon}{\psi\epsilon}$. $\dfrac{F\gamma}{\psi\gamma} \sqsubset$

$\dfrac{F\epsilon}{\phi\epsilon}$. $\dfrac{F\gamma}{\phi\gamma}$. Scilicet est $\psi\epsilon$. $\phi\epsilon \sqsupset \psi\gamma$. $\phi\gamma$ per Lem 3 et sumendo reciproca

rationum erit $\dfrac{1}{\psi\epsilon}$. $\dfrac{1}{\phi\epsilon} \sqsubset \dfrac{1}{\psi\gamma}$. $\dfrac{1}{\phi\gamma}$ ducendoque priorem rationem in $F\epsilon$ et

posteriorem in $F\gamma$, orietur $\dfrac{F\epsilon}{\psi\epsilon}$. $\dfrac{F\epsilon}{\phi\epsilon} \sqsubset \dfrac{F\gamma}{\psi\gamma}$. $\dfrac{F\gamma}{\phi\gamma}$ et vicissim $\dfrac{F\epsilon}{\psi\epsilon}$. $\dfrac{F\gamma}{\psi\gamma} \sqsubset \dfrac{F\epsilon}{\phi\epsilon}$. $\dfrac{F\gamma}{\phi\gamma}$.
Q.E.D.

[79 = 164]   Schol. Demonstratio perinde se habet in literis majusculis (quibus refractiones designavi cùm posterius Medium sit anteriori rarius) si modò vice signi
$\sqsubset$ ubique subintelligatur signum $\sqsupset$, et $\sqsubset$ vice $\sqsupset$. Notabis insuper quod in hâc
demonstratione posui densitatem posterioris tantùm Medij variatam esse. Sed
eodem recidit si anteriora Media successivè varia adhiberi, posteriori non
mutato, sive quod tantundem est, si refractiones e posteriori medio in
anteriùs vicissim peragi concipias: Siquidem radijs in superficiem alterutrinque incidentibus consimiles sunt sinuum / rationes. Caeterùm de exactâ   49/
horum sinuum pro quibuslibet propositis
Medijs ratione investiganda disserui ante,
et propositionem haud attigissem si non
exegisset prop 15 mox tradenda.[16]

[80 = 149]   Lemma 4. Centro *A*, distantia quavis
AD (in fig. I, 38) describatur circulus
DGγ. Deinde centro quolibet *C* distantiâ
AC describatur alius circulus secans rectam *AD* in *B* et circulum priùs descriptum
in *G*. Tum arcus *BG* bisecetur in *F*, et *FK*
demittatur ad *BD* perpendicularis. His ita
constitutis, dico quòd *FK* sic perpendiculariter demissa dictam *BD* bisecabit. Junc-

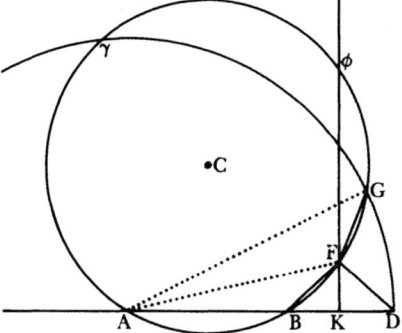

Figure I, 38

from the same point ϕ.[15] When this has been done, if another medium, however dense or rare, is substituted for the second medium, two rays of this kind incident along the same lines $F\epsilon$ and $F\gamma$ must always be so refracted that they likewise diverge from some one point of that perpendicular;[b] for in- (b) §§I, 45, 49 stance, from ψ toward l and n, with the density of this second medium being assumed to differ more from the first one than the former second medium that caused them to diverge from ϕ. It must therefore be shown that the inequality of the ratio of the sines of refraction is greater in the latter than in the former case. Specifically, the sine of incidence of the ray $F\gamma\lambda$ is to the sine of refraction as $\phi\gamma$ to $F\gamma$,[c] that is, as 1 to $F\gamma/\phi\gamma$, and likewise those sines for (c) §I, 47 the ray $F\epsilon\nu$ are as 1 to $F\epsilon/\phi\epsilon$. Consequently, the sines of refraction of the same rays are to one another as $F\gamma/\phi\gamma$ to $F\epsilon/\phi\epsilon$; and by a similar argument it will be clear that for the refracted rays from ψ the like sines of refraction are as $F\gamma/\psi\gamma$ to $F\epsilon/\psi\epsilon$. It therefore remains to be proved that there is a greater disproportion between $F\gamma/\psi\gamma$ and $F\epsilon/\psi\epsilon$ than between $F\gamma/\phi\gamma$ and $F\epsilon/\phi\epsilon$; that is (since $F\epsilon/\psi\epsilon > F\gamma/\psi\gamma$),[d] it remains to be proved that (d) Lemma 3 $F\epsilon/\psi\epsilon : F\gamma/\psi\gamma > F\epsilon/\phi\epsilon : F\gamma/\phi\gamma$. To be sure, by Lemma 3, $\psi\epsilon : \phi\epsilon < \psi\gamma : \phi\gamma$; then taking the reciprocal of the ratios, $1/\psi\epsilon : 1/\phi\epsilon > 1/\psi\gamma : 1/\phi\gamma$; and multiplying the first ratio by $F\epsilon$ and the second by $F\gamma$, there will result $F\epsilon/\psi\epsilon : F\epsilon/\phi\epsilon > F\gamma/\psi\gamma : F\gamma/\phi\gamma$, and by permutation $F\epsilon/\psi\epsilon : F\gamma/\psi\gamma > F\epsilon/\phi\epsilon : F\gamma/\phi\gamma$. As was to be demonstrated.

Scholium. The demonstration is just the same with the capital letters by [79 = 164] which I have designated the refracted rays when the second medium is rarer than the first, provided that the sign $<$ is everywhere understood for the sign $>$, and $>$ is understood for $<$. You will note, moreover, that in this demonstration I assumed only the second medium's density to vary, but it amounts to the same thing if you imagine the first medium to be treated as successively varying, while the second medium is unchanged, or equivalently if you imagine the refractions to proceed in turn from the second medium into the first, inasmuch as the ratios of the sines for rays incident on either side of a surface are completely similar. But I previously discussed investigating the exact ratio of their sines for any proposed media, and I would not have undertaken the proposition had not Prop. 15, soon to be related, required it.[16]

Lemma 4. With center A (Fig. I, 38) and any radius AD describe the circle [80 = 149] $DG\gamma$. Next with any center C and radius AC describe another circle intersecting the line AD in B and the previously described circle in G. Then bisect the arc BG at F, and drop the perpendicular FK to BD. Having so constructed these things, I say that FK, dropped perpendicularly in this way, will bisect that line BD.

---

(15) I originally continued: Id quod diximus eventurum circiter si dicta ratio $A\gamma$ ad $A\epsilon$ ponatur 39 ad 40. (We said this would just about occur if that ratio $A\gamma$ to $A\epsilon$ were assumed to be 39 to 40.) This should read "$F\gamma$ ad $F\epsilon$."

(16) si . . . tradenda.] I: nisi id in gratiam quartae secuturae fuisset factum. (had it not been done for the sake of the fourth proposition to follow.)

tis enim *AF, AG, BF, FG* et *FD*; in triangulis *AFG* et *AFD* anguli ad *A* sunt aequales propter aequales arcus *BF, FG* quibus subtenduntur; Item latera circa istos angulos *AD* et *AG* sunt aequalia, quippe radij ejusdem circuli. Et aliud latus *AF* habent commune. Quare etiam tertia latera *FG* et *FD* sunt aequalia. Sed est *BF* = *FG* propter aequalitatem arcuum quos subtendunt, adeòque *BF* = *FD* et triangulum *FKB* = triangulo *FKD* et inde *BK* = *KD*.

[81 = 150]   Coroll: 1. Hinc recta *KF*, quae bisecat *BD* insistens ei normaliter, bisecabit etiam arcus *BG* circulorum omnium per data duo puncta *A* et *B* transeuntium, et alicubi in *G* secantium datum circulum *DG* centro *A* intervallo *AD* descriptum. Imò et bisecabit arcus *BGγ* in altero intersectionis puncto *φ*.

[82 = 151]   Coroll: 2. Idem eveniet cùm *A* et *B* coincidunt, hoc est cùm circuli *AFG* tangunt rectam *AD* in puncto *AB*. Potest etiam *B* sumi ad alteras partes ipsius *A*. In transcursu etiam notetur, quod anguli *BFK, BGD*, quos circulus *ABF* cum rectâ *FK* et arcu *GD* efficit, sint aequales.

Lect 10   Lemma 5. Lineis quatuor *Aβ, AB, Aγ*, et *AG* (fig I, 39) circulo alicui ab
[83 = 152]   eodem circumferentiae puncto ita inscriptis ut sit *Aβ . AB :: Aγ . AG* quarum omnium *Aβ* sit minima: Dico angulum *BAG* majorem esse angulo *βAγ*. Describatur enim alius circulus *ABgd* secans priorem in punctis *A* et *B* cujus Diameter sit ad ejus *ABG* diametrum sicut *AB* ad *Aβ*, centris utrisque ad easdem partes ipsius *AB* jacentibus. Dein centro *A* distantia *AG* describe tertium circulum *GH* secundo occurrentem in *g*, et istud *g* ex constructione jacebit alicubi inter *G* et *H*, atque adeò si *Ag* ducatur, erit angulus *BAG* major angulo *BAg*. Est autem angulus *BAg* = ang: *βAγ* propterea quòd *AB* et *Ag* similiter inscriptae sunt circulo *ABg* ac *Aβ* et *Aγ* ipsi *Aβγ*, habentes nempe easdem rationes / et inter se (*Aβ . Aγ :: AB . AG* vel *Ag*.) et ad diame-   50/

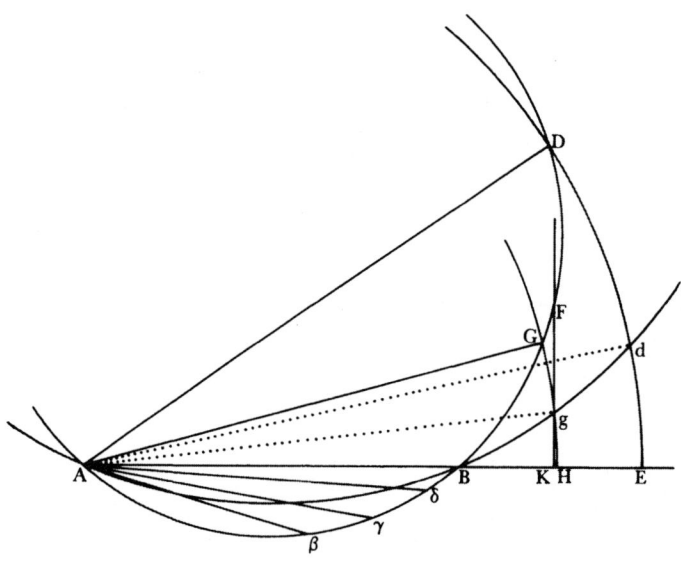

Figure I, 39

For when *AF, AG, BF, FG,* and *FD* are joined, the angles at *A* in the triangles *AFG* and *AFD* are equal because they subtend equal arcs, *BF* and *FG*. Likewise, the sides *AD* and *AG* about those angles are equal, inasmuch as they are radii of the same circle; and they have the other side, *AF*, in common. Consequently, the third sides, *FG* and *FD*, are also equal. But *BF* = *FG*, because they subtend equal arcs, so that *BF* = *FD*; and the triangle *FKB* is equal to the triangle *FKD*, and thus *BK* = *KD*.

Corollary 1. Hence the line *KF*, which bisects *BD* and is normal to it, will [81 = 150] also bisect the arcs *BG* of all circles that pass through the two given points *A* and *B* and intersect, somewhere at *G*, the given circle *DG* described with center *A* and radius *AD*. In fact, it will also bisect the arc *BGγ* in the other point of intersection *φ*.

Corollary 2. It will turn out to be the same when *A* and *B* coincide, that is, [82 = 151] when the circles *AFG* are tangent to the line *AD* at the point *AB*. *B* can also be taken on the other side of *A*. In passing, it may be noted that the angles *BFK* and *BGD* that the circle *ABF* makes with the line *FK* and the arc *GD* are equal.

Lemma 5. Inscribing four lines, *Aβ, AB, Aγ,* and *AG* (Fig. I, 39), in any **Lecture 10** circle from the same point of the circumference so that *Aβ : AB* = *Aγ : AG*, [83 = 152] and letting *Aβ* be the smallest of all, I say that the angle *BAG* is greater than the angle *βAγ*.

For describe another circle, *ABgd*, intersecting the first in the points *A* and *B*, and let its diameter be to the diameter of *ABG* as *AB* to *Aβ*, with both centers lying on the same side of *AB*. Then with center *A* and radius *AG* describe a third circle, *GH*, meeting the second one at *g*. By construction *g* will lie somewhere between *G* and *H*, so that if *Ag* is drawn, the angle *BAG* will be greater than the angle *BAg*. The angle *BAg*, however, equals the angle *βAγ*, because *AB* and *Ag* are inscribed in the circle *ABg* similarly to *Aβ* and *Aγ* in *Aβγ*, namely, they have the same ratios both to each other (*Aβ : Aγ* = *AB : AG* or *Ag*) and to the diameters of the circles in which they are in-

tros circulorum quibus inscribuntur. Cùm ergo sit *BAG* ⊏ *BAg* = *βAγ* erit *BAG* ⊏ *βAγ*.   Q.E.D.

[84 = 153]   Coroll 1. Hinc in eodem quovis radiorum genere quo major est refractio eo major erit angulus refractus. In Fig I, 27[1] ubi est *FR . RD* :: *Fρ . ρδ*, erit ang *Fρδ* ⊏ ang *FRD*.[2]

[85 = 154]   Coroll. 2. Hinc etiam si sit *AG . AB* ⊏ *Aγ . Aβ*, multo magis erit angu. *BAG* ⊏ ang. *βAγ*. Hoc est in genere, quo majores sunt subtensae, et simul quo major est inaequalitas rationis earum, eo major erit differentia angulorum quos subtendunt. Atque idem de sinubus et eorum angulis, utpote subtensarum et earum angulorum dimidijs intellige.

[86 = 155]   Lemma 6. Insuper si arcus *γδ* ipsi *βγ* capiatur aequalis et *AD* inscribatur circulo *ABD* quae sit ad *Aδ* sicut *AG* ad *Aγ*, caeteris stantibus: dico quod arcus *DG* erit arcu *GB* major. Nam centro *A*, radio *AD* describe circulum *DdE* circulo *ABg* occurrentem in *d* et rectae *AB* in *E*, et *Ad* ducatur. Jam cùm *Ad, Ag,* et *AB* circulo *ABgd* similiter inscribantur atque *Aδ, Aγ,* et *Aβ* ipsi *Aβγ*, erit arcus *gd* = arcui *Bg*. Quare demissâ *gK* ad *BE* perpendiculari, et productâ donec secet arcum *BD* in *F*, ista [*g*]*K* (per Lem. 4) bisecabit tum rectam *BE*, tum arcum *BD*. At quoniam *gF* ex constructione jacet extra circulum *gG* punctum *F* cadet inter *G* et *D*. Quare *DG* ⊏ *DF* Sive ⊏ *FB* et multo magis ⊏ *GB*.   Q.E.D.

[87 = 156]   Coroll. 1. Hinc si arcus *βδ* non tantùm duabus sed quotcunque partibus aequalibus constet, correspondentes partes arcus *BD* a termino *B* ad terminum *D* sese gradatim superabunt longitudine. Adeóque si arcus *βγ* ad arc: *γδ* habeat quamcunque rationem commensurabilem; erit arc. *GD* . arc. *BG* ⊏ arc *γδ* . arc *βγ*, siquidem numeris aequalium partium mensurantium arcus *βγ* et *γδ* correspondent consimiles numeri partium inaequalium constituentium arcus *BG* ac *GD*, quarum illae in *GD* sunt omnes parte maximâ ipsius *BG* majores. Quinetiam si *βγ* ad *γδ* habeat quamcunque rationem incommensurabilem, erit itidem *GD* . *BG* ⊏ *γδ* . *βγ*. Nam rationum similitudines quae quantitatibus commensurabilibus indefinitè conveniunt eo nomine conveniunt etiam incommensurabilibus similiter affectis; quemadmodum ex Euclideâ definitione similium rationum ostendi potest, sed faciliùs deprehenditur imaginando / quantitates quas vocant incommensurabiles posse mensurari[3]   51/ per partes indefinitè parvas, et sic ad naturam commensurabilium praesertim quoad rationum habitudines quodammodò reduci. Con[c]ipias itaque arcum *βγ* in aequales et indefinitè multas partes dividi, et ejusmodi tot sumi quae minùs quàm unâ parte, (hoc est indefinite parùm) differunt ab arcu *γδ*, atque adeò ipsi pro more consueto censeantur aequales. Concipe etiam *BD* in partes quales antè definivi correspondentes partibus ipsius *βδ* dividi; et propter tot inaequales partes majores quidem in *GD* et minores in *BG* quot sunt aequales in *γδ* et *βγ*, erit *GD* . *BG* ⊏ *γδ* . *βγ*.

---

(1) That is, in §I, 55.
(2) This corollary is a late, marginal addition; see Lects. 14, 15, note (14).

scribed. Therefore, since $\widehat{BAG} > \widehat{BAg} = \widehat{\beta A\gamma}$, it will be $\widehat{BAG} > \widehat{\beta A\gamma}$. As was to be demonstrated.

Corollary 1. Hence, for any one kind of ray the greater the [angle of] [84 = 153] refraction, the greater will be the deviation. In Fig. I, 27[1] where $FR : RD = F\rho : \rho\delta$, it will be $\widehat{F\rho\delta} > \widehat{FRD}$.[2]

Corollary 2. Hence also, if $AG : AB > A\gamma : A\beta$, even more so will it be [85 = 154] $\widehat{BAG} > \widehat{\beta A\gamma}$. That is, in general the greater the chords, as well as the greater the inequality of their ratio, the greater will be the difference of the angles they subtend. The same is to be understood for the sines and their angles, as for the halves of the chords and their angles.

Lemma 6. Moreover, if the arc $\gamma\delta$ is taken to be equal to $\beta\gamma$, and in the [86 = 155] circle ABD there is inscribed AD that is to $A\delta$ as AG to $A\gamma$, with other things remaining the same, I say that the arc DG will be greater than the arc GB.

For with center A and radius AD describe the circle DdE meeting the circle ABg in d and the line AB in E; and draw Ad. Now, since $Ad, Ag,$ and AB are inscribed in the circle $ABgd$ similarly to $A\delta, A\gamma,$ and $A\beta$ in the circle $A\beta\gamma$, then $\widehat{gd} = \widehat{Bg}$. Consequently, when gK is dropped perpendicular to BE and extended until it intersects the arc BD in F, gK will bisect both the line BE and the arc BD (Lemma 4). But since by construction gF lies outside of the circle gG, the point F will fall between G and D. Therefore $\widehat{DG} > \widehat{DF}$ or $\widehat{FB}$ and even more so $\widehat{DG} > \widehat{GB}$. As was to be demonstrated.

Corollary 1. Hence if the arc $\beta\delta$ consists not only of two but of any [87 = 156] number of equal parts, the corresponding parts of the arc BD from the end B to the end D will gradually exceed one another in length. Thus if the arc $\beta\gamma$ has any commensurable ratio to the arc $\gamma\delta$, then $\widehat{GD} : \widehat{BG} > \widehat{\gamma\delta} : \widehat{\beta\gamma}$, since to the number of equal parts measuring the arcs $\beta\gamma$ and $\gamma\delta$ there corresponds a like number of unequal parts forming the arcs BG and GD, and those in GD are all greater than the greatest part of BG. In fact, if $\beta\gamma$ has any incommensurable ratio to $\gamma\delta$, then likewise $\widehat{GD} : \widehat{BG} > \widehat{\gamma\delta} : \widehat{\beta\gamma}$. For the similarity of ratios, which applies without limitation to commensurable quantities, for that reason also applies to incommensurables similarly treated, as can be shown from the Euclidean definition of similar ratios. But this is more easily understood by imagining that quantities that are called incommensurable can be measured[3] by indefinitely small parts and so in a certain way be reduced to the nature of commensurables, particularly as to the condition of ratios. You may conceive, therefore, the arc $\beta\gamma$ to be divided into indefinitely many equal parts and so many of these to be taken that they differ by less than one part (that is, indefinitely little) from the arc $\gamma\delta$, and thus in the usual way they are to be considered equal. Conceive the arc BD also to be divided into parts (such as I defined before) corresponding to the parts of $\beta\delta$; and because there are as many unequal parts, indeed greater in the arc GD and smaller in the arc BG, as there are equal parts in the arcs $\gamma\delta$ and $\beta\gamma$, it will be $\widehat{GD} : \widehat{BG} > \widehat{\gamma\delta} : \widehat{\beta\gamma}$.

(3) *Editio princeps*: numerari.

[88 = 157]   Coroll. 2. Hinc praeterea componendo sequitur esse $BD . BG ⊏ βδ . βγ$, nec non $GD . BD ⊏ γδ . βδ$.

[89 = 158]   Coroll 3. Consectatur denique quod ductis utcunque quatuor subtensis $Aβ, Aγ, Aδ, Aε$; (in fig I, 40) et alijs quatuor AB, AG, AD, AE quarum singulae ad priorum singulas eandem rationem observant, (nempe $AB . Aβ :: AG . Aγ :: AD . Aδ :: AE . Aε$:) si AE sit omnium maxima et $Aβ$ minima erit arcus ED . arc: $GB ⊏$ arc: $εδ$. arc: $γβ$. Nam per Coroll 1 hujus est $ED . DG ⊏ εδ . δγ$, et $DG . GB ⊏ δγ . γβ$. Et multò magis $ED . GB ⊏ εδ . γβ$. Haud secùs pateat esse arc EG . arc $DB ⊏$ arc $εγ$. arc $δβ$. Scilicet ex coroll. 2 hujus est $EG . DG ⊏ εγ . δγ$ ac $DG . DB ⊏ δγ . δβ$ et multò magis $EG . DB ⊏ εγ . δβ$.

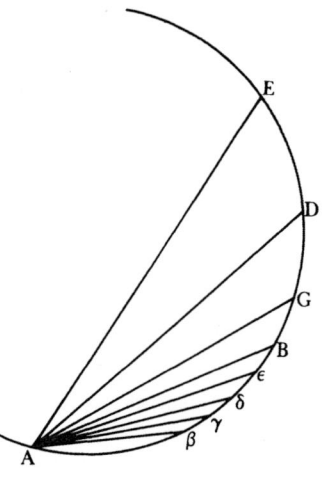

Figure I, 40

Denique quae de subtensis et earum arcubus dicta sunt, possunt etiam de sinubus et eorum arcubus aut angulis intelligi.

[90 = 165]   Prop. 14. Heterogeneis radijs e densiori Medio in rarius secundum eandem datam lineam in superficiem positione datam incidentibus, quo rarius sit medium in quod radij refringuntur, eo ma-

Fig I, 41[4]   jor erit differentia refractionis. Sit FL linea secundum quam duo radij incidunt in superficiem AL quorum maximè refrangibilis refringatur ad P, et minimè refrangibilis ad T: Dico quod si Medium rarius foret adhuc magis rarum ut refringeret maximè refrangibilem radium ad $π$ et minimè refrangibilem ad $τ$, tunc angulus $πLτ$ foret major angulo PLT. demittatur enim FA ad refringentem superficiem normalis, quae secet refractos radios retrorsum ductos in G, C, D, et E. Deinde in refringente superficie quaeratur tale punctum N ut sit $FN . DN :: FL . EL$, ac DN productus

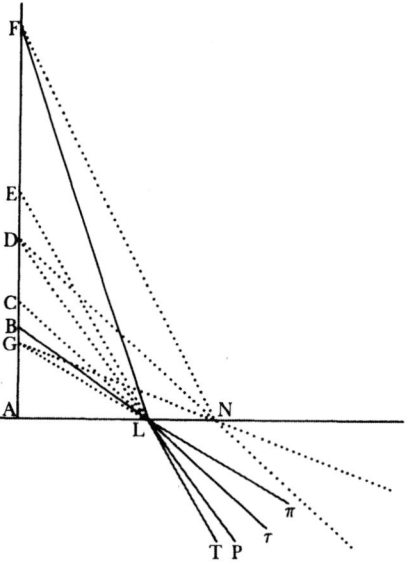

Figure I, 41

(a) Num 47.   erit refractus radij minimè refrangibilis incidentis ab F ad N.[a] Jam cum talis supponatur positio linearum FL et FN, ut radij maxime refrangibilis secundum FL et minimè refrangibilis secundum FN incidentis refracti DL ac DN divergant / a puncto D quod situm est in perpendiculo FA: eâ de causâ, licèt 52/ raritas Medij in quod refractio peragitur foret alia quàm supponitur, tamen ejusmodi radiorum secundum easdem lineas FN et FL incidentium refracti semper divergerent ab aliquo puncto quod in eadem FA sit positum; quemad-

Corollary 2. Hence furthermore, by composition it follows that $\widehat{BD} : \widehat{BG}$ [88 = 157]
$> \widehat{\beta\delta} : \widehat{\beta\gamma}$, and also $\widehat{GD} : \widehat{BD} > \widehat{\gamma\delta} : \widehat{\beta\delta}$.

Corollary 3. It follows, finally, that when four chords, $A\beta$, $A\gamma$, $A\delta$, and $A\epsilon$ [89 = 158]
(Fig. I, 40), are arbitrarily drawn, together with another four, AB, AG, AD,
and AE, each of which maintains the same ratio to each of the former ones
(namely, $AB : A\beta = AG : A\gamma = AD : A\delta = AE : A\epsilon$), if AE is the greatest
and $A\beta$ the smallest of all, then $\widehat{ED} : \widehat{GB} > \widehat{\epsilon\delta} : \widehat{\gamma\beta}$. For by Corollary 1 of
this lemma, $\widehat{ED} : \widehat{DG} > \widehat{\epsilon\delta} : \widehat{\delta\gamma}$, and $\widehat{DG} : \widehat{GB} > \widehat{\delta\gamma} : \widehat{\gamma\beta}$, and even more so
$\widehat{ED} : \widehat{GB} > \widehat{\epsilon\delta} : \widehat{\gamma\beta}$. In the same way it is clear that $\widehat{EG} : \widehat{DB} > \widehat{\epsilon\gamma} : \widehat{\delta\beta}$.
Specifically, from Corollary 2 of this lemma, $\widehat{EG} : \widehat{DG} > \widehat{\epsilon\gamma} : \widehat{\delta\gamma}$, and
$\widehat{DG} : \widehat{DB} > \widehat{\delta\gamma} : \widehat{\delta\beta}$, and even more so $\widehat{EG} : \widehat{DB} > \widehat{\epsilon\gamma} : \widehat{\delta\beta}$.

Finally, what has been said about chords and their arcs can also be understood about sines and their arcs or angles.

Proposition 14. When heterogeneous rays are incident from a denser into a [90 = 165]
rarer medium along the same given line onto a surface given in position, the
rarer the medium into which the rays are refracted the greater will be the
difference of refraction.

Let FL be a line along which two rays fall upon the surface AL, and let the Fig. I, 41[4]
most refrangible of these be refracted to P and the least refrangible to T. I say
that if the rarer medium were still rarer, so that it refracted the most refrangible ray to π and the least refrangible to τ, the angle $\pi L\tau$ would be greater
than the angle PLT. For to the refracting surface drop the normal FA, which
intersects the refracted rays extended backward at G, C, D, and E. Then in
the refracting surface find a point N such that $FN : DN = FL : EL$, and the
extension of DN will be the refraction of the least refrangible ray falling from
F to N.[a] Now, since such a position of the lines FL and FN is supposed that (a) §I, 47
the refracted rays DL and DN of the most refrangible ray incident along FL
and the least refrangible one incident along FN diverge from the point D
located on the perpendicular FA, hence even if the rarity of the medium in
which the refraction is completed were different than supposed, still the
refracted rays of rays of this sort incident along the same lines FN and FL
would always diverge from some point located on the same FA, as was

---

(4) The corresponding Fig. 69 in the *Lectiones opticae* is essentially identical, except that it is
oriented with the line FA horizontal.

(b) Num 49. modum in praecedentibus ostensum est.[b] Sic cum raritas dicti Medij talis
esse supponitur ut maximè refrangibilis radius secundum *FL* incidens refringatur a puncto quopiam *G*, tunc minimè refrangibilis secundum *FN* incidens
refringetur ab eodem *G*. Sed cùm maximè refrangibilis radius supponebatur a
Puncto *G* refringi tunc etiam minimè refrangibilis secundum eandem lineam
FL incidens supponebatur refringi a puncto *C*. Quare est

(c) Num 25 & 47 $GN \cdot FN :: CL \cdot FL:$[c]

et praeterea cum antea posuerium esse *FN . DN :: FL . EL*, ex aequo erit
GN . DN :: CL . EL. Sed (per Lem. 3) est *GN . DN* ⊏ *GL . DL*, adeoque
CL . EL ⊏ *GL . DL*. Quare si linea quaedam *BL* ita ducatur ut sit
CL . EL :: BL . DL, erit *BL* ⊏ *GL* propter majorem rationem quam habet ad
DL, et insuper erit *CL* ⊏ *BL*, eò quòd sit *EL* ⊏ *DL*. et proinde punctum *B*
cadet inter *G* et *C*, eritque ang. *GLC* ⊏ ang. *BLC*. Cùm verò sit
CL . EL :: BL . DL, aut vicissim *BL . CL :: DL . EL*, erit ang *BLC* ⊏
ang *DLE* (Lem. 1) et multò magis ang. *GLC* ⊏ ang *DLE*. Q.E.D.

[91 = 166]   Prop. 15. Heterogeneis radijs e Medio densiori in rarius secundum eandem
datam lineam in superficiem positione datam incidentibus: quo densius est
medium e quo radij incidunt eo major erit differentia refractionis. Scilicet
(propter majores refractiones) eo majores erunt sinus refractionum respectu
dati circuli, ad quem referuntur; et simul eo major erit inaequalitas rationis
istorum sinuum (per prop 13) Adeoque eo major erit differentia angulorum
quos subtendunt (per Coroll. 2 ad Lem. 5.), hoc est eò major differentia
refractionis. Q.E.O.

[92 = 167]   Prop. 16. Heterogeneis radijs e
Medio rariori in densius secun-
dum eandem datam lineam in superficiem positione datam incidentibus; quo rarius est Medium
e quo radij incidunt eo major erit
Fig I, 42 differentia refractionis. Sit *AD*
superficies, in quam duo radij secundum eandem datam lineam
IX incidunt quorum alter maximè refrangibilis refringa[tur] ad
P, et alter minimè refrangibilis
ad *T*: Dico quòd si / medium ex
quo radij incidunt foret adhuc
rarius, ut dictos radios magis refringeret, puta maximè refran-

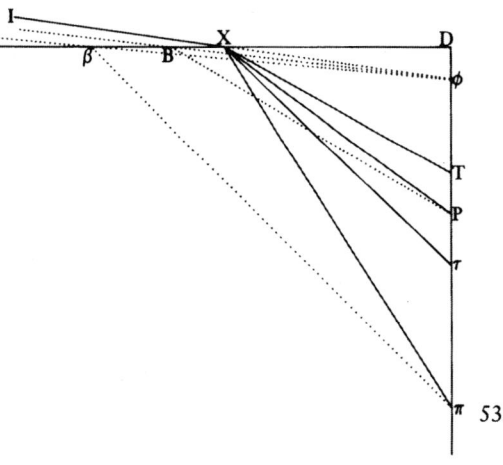

Figure I, 42

gibilem versus *π* et minimè refrangibilem versus *τ*: tunc *πXτ* major angulus
evaderet quàm *PXT*: Id quod gradatim sic demonstro.

[93 = 168]   Cas. 1. Ponamus primò quòd recta *IX* secundum quam radij incidunt sit ad
refringentem superficiem obliquissima. Ac ducatur quaelibet recta *PD* eidem
superficiei normaliter insistens in *D*, et secans refractos radios in punctis *T, P,
τ, π*; et *IX* producatur donec istam *PD* secet in *φ*. Tum in linea *AD* quaeratur

shown in the preceding.[b] Thus when the rarity of that medium is assumed to   (b) §I, 49
be such that the most refrangible ray incident along FL is refracted from
some point G, then the least refrangible ray incident along FN will be refracted from the same point G. But when the most refrangible ray was
assumed to be refracted from G, then in addition the least refrangible ray
incident along the same line FL was assumed to be refracted from the point
C. Consequently, $GN : FN = CL : FL$;[c] and moreover, since I assumed   (c) §§I, 25, 47
previously that $FN : DN = FL : EL$, from the equality of the ratios there will
be $GN : DN = CL : EL$. But $GN : DN > GL : DL$ (Lemma 3), and thus
$CL : EL > GL : DL$. Therefore, if some line BL is so drawn that $CL : EL =
BL : DL$, then $BL > GL$, because it has a greater ratio to DL; and moreover
$CL > BL$, because $EL > DL$. Consequently, the point B will fall between G
and C, and $\widehat{GLC} > \widehat{BLC}$. But since $CL : EL = BL : DL$, or permuting,
$BL : CL = DL : EL$, then by Lemma 1, $\widehat{BLC} > \widehat{DLE}$, and even more so
$\widehat{GLC} > \widehat{DLE}$. As was to be demonstrated.

Proposition 15. When heterogeneous rays are incident from a denser into a   [91 = 166]
rarer medium along the same given line upon a surface in a given position,
the denser the medium from which the rays are incident the greater will be
the difference of refraction.

Namely, because of the greater refractions, the greater will be the sines of
refraction with respect to the given circle to which they are referred, and at
the same time the greater will be the inequality of the ratio of those sines (by
Prop. 13). Hence, the greater will be the difference of the angles that they
subtend (by Lemma 5, Corollary 2), that is, the greater will be the difference
of refraction. As was to be shown.

Proposition 16. When heterogeneous rays are incident from a rarer into a   [92 = 167]
denser medium along the same given line upon a surface in a given position,
the rarer the medium from which the rays are incident the greater will be the
difference of refraction.

Let AD be a surface upon which two rays are incident along the same line   Fig. I, 42
IX, the most refrangible one being refracted to P and the other, least refrangible one to T. I say that if the medium from which the rays are incident were
still rarer, so that it would refract those rays more, for example, the most
refrangible to π and the least refrangible to τ, then the angle $\pi X \tau$ would
become greater than the angle PXT. I will prove this stepwise.

Case 1. Let us first assume that the line IX along which the rays are   [93 = 168]
incident is the most oblique to the refracting surface. Draw any line PD
perpendicular to the same surface at D and intersecting the refracted rays at
the points T, P, τ, and π; and extend IX until it intersects PD in ϕ. Next, in

punctum quoddam *B* hac lege ut ductis *Bφ, BP,* fiat *Xφ . XT :: Bφ . BP.* Liquet ergo quod si minimè refrangibilis radius incidat in *B* versus *φ* tendens, is debet versus *P* refringi: Quippe cùm ex Hypoth. sit *BP . Bφ :: XT . Xφ*, hoc est sinus incidentiae ejus et refractionis, sicut sinus incidentiae et refractionis alterius minimè refrangibilis radij *IXT.* Quamobrem si supponamus hosce radios retrocedere, alterum nempè e minimè refrangibilibus a *T* ad *X* et alterum a *P* ad *X*, et maximè refrangibilem a *P* ad *X*; eorum omnium refracti tendent a puncto *φ*: Siquidem notum est Theorema[5] quod radij secundum refractum ejus retro incidentis, incidens vicissim fit refractus. Jam cùm radij difformes *PB, PX* ab eodem puncto *P* manantes refringantur ab eodem *φ* quod situm est in perpendiculo *PD*, proportione inter *PX* et *PB* semel cognitâ, si ab alio quovis ejusdem perpendiculi puncto ad refringentem superficiem duae ducantur lineae eandem rationem habentes, hoc est, ut una designans maximè refrangibilem radium sit ad alteram quae designet minimè refrangibilem, ut

(d) Num 45   *PX* ad *PB*: tunc istorum refracti (ex antè monstratis[d]) divergent ab aliquo etiam puncto quod situm est in eodem perpendiculo *PD*; utcunque Medium ex parte radij *IX* supponatur rarum, dummodo Mediorum alterum ex parte radij *PX* eandem densitatem retineat. Quemadmodum si maximè refrangibilis radius incidat secundum *πX* et refringatur a *φ*, Medio scilicet versus *IX* jam posito rariori quam ante; tunc rectâ *πβ* sic ductâ ut sit *PX . PB :: πX . πβ* radius etiam minimè refrangibilis *πβ* refringeretur ab eodem *φ*. Unde sequitur esse *πβ* ad *φβ* sicut sinus incidentiae radiorum minimè refrangibilium

(e) Num 47.   ad sinum refractionis.[e] Ast in ratione istorum sinuum est etiam *τX* ad *φX*, eò quòd inflexa *IXτ* designet radium aequaliter refrangibilem cujus pars *IX* producta transit per idem *φ*. Quare est *πβ . φβ :: τX . φX.* Cum verò radius *IX* supponatur esse ad refringentem superficiem summè obliquus sive in angulo infinite parvo inclinatus, adeò ut recta *Dφ* pro / infinitè parvâ 54/ sive nullâ haberi debeat, sequitur esse *DX = Xφ, DB = Bφ,* ac

Dβ = βφ: quos valores pro *Xφ Bφ* et *βφ* substituendo in supra recensitas proportiones *BP . Bφ :: XT . Xφ*, & *πβ . φβ :: τX . φX*, emergent *BP . BD :: XT . XD*, et *πβ . Dβ :: τX . DX.* Ex quibus pateat rectas *BP* ad *XT* et *βπ* ad *Xτ* parallelas esse, angulosque *BPX* ad *PXT*, et *βπX* ad *πXτ* aequales. Sed ex Hypothesi est *PX . PB :: πX . πβ*, et proinde ang: *βπX* ⊏ ang: *BPX* (per Coroll 1. Lem 5.) hoc est ang: *πXτ* ⊏ ang: *PXT*. Q.E.D.

[94 = 169]   Casus. 2. Incidentibus verò radijs angulum definitè magnum cum refringente superficie constituentibus, propositum sic patebit. Sit *HX* recta secundum

Fig I, 43

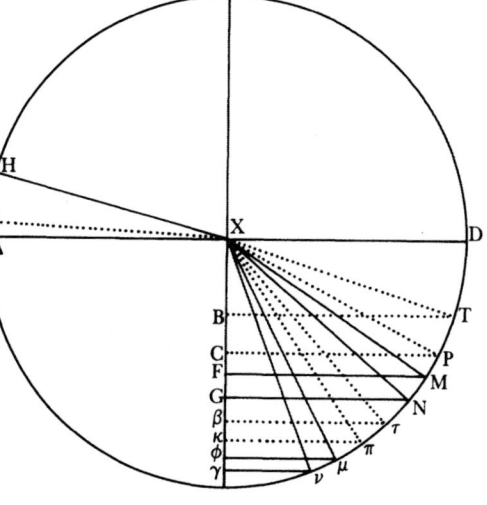

Figure I, 43

the line AD find a certain point B, subject to the condition that when $B\phi$ and BP are drawn, $X\phi : XT = B\phi : BP$. It is evident, therefore, that if a least refrangible ray were incident at B tending toward ϕ, it ought to be refracted toward P, inasmuch as by hypothesis $BP : B\phi = XT : X\phi$; that is, its sines of incidence and refraction are as the sines of incidence and refraction of the other least refrangible ray IXT. Consequently, if we suppose these rays to pass backward—namely, one of the least refrangible rays from T to X and the other from P to B, and the most refrangible one from P to X—all of their refracted rays will tend from the point ϕ, since it is a known theorem[5] that for a ray incident backward along its refracted ray, the incident ray becomes interchangeably the refracted ray. Now, since the dissimilar rays PB and PX proceeding from the same point P are refracted from the same point ϕ located on the perpendicular PD, once the proportion between PX and PB is known, if from any other point of the same perpendicular to the refracting surface two lines having the same ratio are drawn, that is, so that one, representing the most refrangible ray, is to the other, representing the least refrangible ray, as PX to PB, then their refracted rays (from what was shown before)[d] will also diverge from some point located on the same perpendicular (d) §I, 45 PD—however rare the medium on the side of the ray IX is supposed—provided that the density of the other medium on the side of the ray PX remains the same. For instance, if a most refrangible ray is incident along πX and refracted from ϕ (that is, assuming that the medium along IX is now rarer than before), then drawing the line $\pi\beta$ so that $PX : PB = \pi X : \pi\beta$, the least refrangible ray $\pi\beta$ would also be refracted from the same point ϕ. Whence it follows that $\pi\beta$ is to $\phi\beta$ as the sine of incidence is to the sine of refraction of the least refrangible rays.[e] But τX to ϕX is also in the ratio of (e) §I, 47 those sines, because the bent line $IX\tau$ represents a ray equally as refrangible whose part IX, when it is extended, passes through the same point ϕ; therefore $\pi\beta : \phi\beta = \tau X : \phi X$. Since the ray IX is supposed to be the most oblique to the refracting surface, or inclined at an infinitely small angle, so that the line $D\phi$ ought to be considered as infinitely small or nothing, it follows that $DX = X\phi$, $DB = B\phi$, and $D\beta = \beta\phi$. Substituting these values for $X\phi$, $B\phi$, and $\beta\phi$ in the proportions enumerated above—$BP : B\phi = XT : X\phi$, and $\pi\beta : \phi\beta = \tau X : \phi X$—there will result $BP : BD = XT : XD$ and $\pi\beta : D\beta = \tau X : DX$. It is clear from this that the lines BP and $\beta\pi$ are parallel to XT and $X\tau$, and that the angles BPX and $\beta\pi X$ are equal to PXT and $\pi X\tau$. But by hypothesis $PX : PB = \pi X : \pi\beta$. Consequently, $\widehat{\beta\pi X} > \widehat{BPX}$ (Lemma 5, Corollary 1), that is, $\widehat{\pi X\tau} > \widehat{PXT}$. As was to be demonstrated.

Case 2. When, however, the incident rays make a finitely large angle with [94 = 169] the refracting surface, the proposition will be evident in this way: Let HX be Fig. I, 43

---

(5) Namely, Prop. 1, §I, 53.

quam incidunt, et cum e Medio minùs raro adveniunt sit *XM* minimè refractus et *XN* maximè refractus, Cum verò adveniunt e magis raro sit *Xμ*
minimè refractus, et *Xν* maximè refractus. Adhibeantur etiam obliquissimè
incidentes radij *IX* cum eorum refractis *XT, XP, Xτ,* et *Xπ* quales jam
descripsimus. Ita scilicet ut cùm tanta sit anterioris Medij raritas ut radios
HX incurvari versus *M* et *N* faciat, tunc etiam consimiles radios *IX* incurvet
versus *T* et *P*; Cùm verò tanto major sit ejus raritas ut illos cogat versus *μ* et
ν tunc hosce simul cogat versus *τ* et *π*. Sit insuper *APD* circulus centro *X* et
intervallo quolibet *AX* descriptus qui secet hosce refractos radios in *T, P, M,
N, τ, π, μ,* & *ν* a quibus ad perpendiculum *BX* demittantur sinus refractio-

Num 26 & 37. num *TB, PC, MF, NG, τβ, πκ, μφ, νγ,* et ex lege refractionum patebit esse
TB . PC :: MF . NG, et *τβ . πκ :: μφ . νγ* et insuper ex Hypothesi et constructione patebit esse *TB* sinuum istorum maximum et *νγ* minimum. Adeóque per Coroll 3 Lem 6 est ang: *TXP* . ang: *MXN* ⊏ ang: *τXπ* . ang: *μXν*.
Et permutando est ang: *TXP* . ang: *τXπ* ⊏ ang: *MXN* . ang: *μXν*. Verùm (ex
ostensis in primo Casu) est ang *TXP* ⊐ ang. *τXπ*. Quare et multo magis erit
ang *MXN* ⊐ ang *μXν*. Q.E.D.

Lect 11   Prop. 17. Heterogeneis radijs e Medio rariori in densius secundum eandem
[95 = 170] lineam in superficiem positione datam incidentibus, quo densius sit Medium
in quod radij incidunt eo major erit differentia refractionum ad certum usque
terminum, et post eo minor perpetuò. Nam si medium Posterius densitate suâ
valdè parùm superet anterius ita ut refractiones indefinitè parvas efficiat,
differentia refractionum erit etiam indefinitè parva, et proinde minor quàm
foret si / Medium posterius supponeretur densius, ut refractiones evaderent 55/
majores. Quare aucta Medij posterioris densitate augebitur dicta refractionum differentia. Quod si densitas ejus in infinitum augeatur refractiones
etiam quantùm poterunt, augebuntur, hoc est usque dum omnes refracti radij
perpendiculariter emergant, angulis refractionum et eorum differentijs tunc
prorsus evanescentibus. Quare differentia refractionum rursus diminuta est,
donec in nihilum evanuit.

[96 = 171]   Schol. Etsi limitis ejus determinatio ubi differentia refractionis evadit maxima, plus taedij et laboris admi-
nistrare possit, quàm utilitatis:
Cum tamen alicujus fortè momenti censeatur densitatem medij
cognoscere quod radijs in se refractis colores maxime conspicuos efficiat, non pigebit hunc
insuper designare. Idque primo
cum incidentia sit obliquissimè.

Fig I, 44   Cas: 1. Esto *IX* communis radiorum in superficiem *AX* quaecunque Media dirimentem obliquissimè incidentium via. et
eorum refracti ut antè sunto *Xπ* et

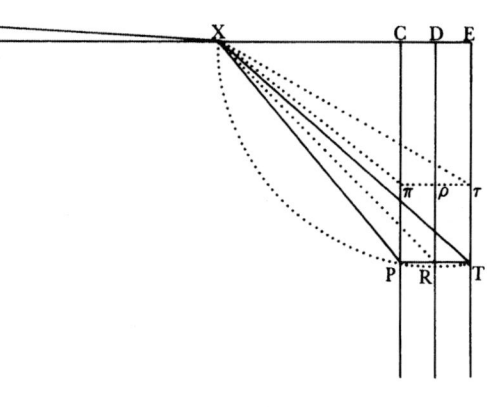

Figure I, 44

the line along which the rays are incident; when they arrive from a less rare medium, let *XM* be the least refracted ray and *XN* the most refracted; but when they arrive from a more rare medium, let *Xμ* be the least refracted ray and *Xν* the most refracted. Also add the most obliquely incident rays *IX* with their refracted rays *XT, XP, Xτ,* and *Xπ,* as we have already described, specifically, such that when the rarity of the first medium is so great that it makes the rays *HX* bend toward *M* and *N,* then the completely similar rays *IX* are also bent toward *T* and *P;* but when its rarity is so much greater that it drives the former toward *μ* and *ν,* then at the same time it drives the latter toward *τ* and *π.* Moreover, let *APD* be a circle described with center *X* and any radius *AX,* intersecting these refracted rays at *T, P, M, N, τ, π, μ,* and *ν,* from which let the sines of refraction *TB, PC, MF, NG, τβ, πκ, μϕ,* and *νγ* be dropped to the perpendicular *BX.* From the law of refraction it will be §§I, 26, 37 evident that *TB : PC* = *MF : NG* and *τβ : πκ* = *μϕ : νγ.* Moreover, by hypothesis and construction it will be evident that *TB* is the greatest of those sines and *νγ* the least. By Lemma 6, Corollary 3, then, $T\hat{X}P : M\hat{X}N >$ $\tau\hat{X}\pi : \mu\hat{X}\nu$, and by permutation $T\hat{X}P : \tau\hat{X}\pi > M\hat{X}N : \mu\hat{X}\nu$. But from what has been shown in the first case, $T\hat{X}P < \tau\hat{X}\pi$, whence even more so $M\hat{X}N <$ $\mu\hat{X}\nu$. As was to be demonstrated.

Proposition 17. When heterogeneous rays are incident from a rarer into a **Lecture 11** denser medium along the same line upon a surface in a given position, the [95 = 170] denser the medium into which the rays are incident the greater will be the difference of refraction up to a certain limit, and after that it will be continually smaller.

For if the second medium very slightly exceeds the first one in its density so that it produces indefinitely small refractions, the difference of refraction will also be indefinitely small and consequently less than it would be if the second medium were supposed denser so that the refractions would prove to be greater. Therefore, increasing the density of the second medium will increase that difference of refraction. But if its density is increased infinitely, the refractions will also increase as much as they can, that is, until all the refracted rays emerge perpendicularly, with the angles of refraction and their differences then vanishing completely. Hence the difference of refraction has again decreased until it has vanished to nothing.

Scholium. Although determining the limit where the difference of refrac- [96 = 171] tion turns out greatest may afford more tedium and effort than utility, still since it may perhaps be considered of some importance to find the density of the medium that produces the most conspicuous colors when rays are refracted in it, it will not be troublesome to designate this in addition—and first when the incidence is most oblique.

Case 1. Let *IX* be the common path of the rays most obliquely incident on **Fig. I, 44** the surface *AX* separating any media, and as before let their refracted rays be

$X\tau$. et agatur recta quaevis $\pi\tau$ praefatae superficiei parallela, quae radijs istis occurrat in π ac τ; a quibus ad AX demissis perpendicularibus πC, τE, bisecetur CE in D, et centro D distantia DX circulus describatur secans $C\pi$ in P et $E\tau$ in T, junganturque XP et XT. Dico quod cùm ea sit posterioris Medij densitas ut radiorum secundum IX incidentium maximè refrangibiles ad P, et minimè refrangibiles ad T refringat, tunc angulus PXT quam maximus evadet. Etenim utcunque Medium posterius ponatur densum refracti radij ita lineas CP et $[E]T$ in punctis π ac τ secabunt ut recta $\pi\tau$ ipsi AX parallela sit. Quare si ducatur linea $D\rho$ quae lineas omnes $\pi\tau$ bisecet, centrum cujuscunque circuli per π ac τ transeuntis semper jacebit in eâdem $D\rho$. At ang $\pi X\tau$ est ang: in segmento circuli per puncta π, τ, et X transeuntis, qui ideò erit maximus cum ejusmodi circulus existit minimus, propterea quòd ratio subtensae $\pi\tau$ ad circuli dimensiones tunc evadit maxima. Verùm iste circulus fit omnium minimus, cum centrum ejus cadit in D, siquidem pro semidiametro tunc habet XD minimam refractarum[1] quae ab X ad RD duci possint. Est ergo ang $\pi X\tau$ tunc maximus cum centrum circuli transeuntis per puncta π, τ, et X cadit in D. Adeoque cum circulus XPT et ang PXT ejusmodi sunt, liquet propositum.

[97 = 172] / Hinc obiter pateat hunc angulum PXT tunc etiam maximum evadere cum 56/ talis sit posterioris Medij densitas ut angulus refractionis mediocriter refrangibilium radiorum[2] obliquissimè secundum IX incidentium sit semirectus; et eo minorem perpetim fieri quò iste refractionis angulus a semirecto (excessu vel defectu) magis deviat. Quemadmodum si refractiones ex aere in aquam, in vitrum et in Crystallum peractae conferantur, e calculo patebit quod cum angulus incidentiae sit 90^{gr} proximè tunc angulus refractionis in aquam erit major semirecto, inque vitrum erit minor. Quamobrem aqua minùs densa est, et vitrum magis densum quàm ut efficiant angulum PXT maximum. Et proinde cùm Crystallum sit adhuc densius efficiet istum PXT minorem quàm vitrum efficeret. Et sic vitrum etsi minùs refringat, in isthoc tamen casu heterogeneos radios in se refractos magis ab invicem dissipabit quàm Crystallum, eoque pacto colores in oppositam ejus superficiem projiciet magis distinctos. Sed haec sunt expertu difficillima, quòd vitrum et Crystallum densitate parùm differant, nec possint haberi satis crassa: et si possent, tunc propter maximam crassitiem haud forent satis perspicua.

[98 = 173] Cas: 2. Quod si linea secundum quam incidunt radij non sit maxime obliqua Problema emerget solidum; sed lubet modum ostendere quod conditionibus ejus nonnihil mutatis, ad planum reduci poterit. Sciendum est itaque quod cùm inter extremos seu maximè difformes radios innumeri sint intermedij qui gradibus continuò successivis, et infinitè parvis alij magis alijs refringuntur; differentia refractionis[3] extremorum radiorum conflata erit ex consimilibus intermediorum differentijs numero et parvitate infinitis. Jam cognitis proprietatibus istarum infinite parvarum differentiarum possumus exinde de omnibus simul aggregatis, sive de differentijs finite parvis quales intercedunt extremorum refractionibus, judicium ferre, praesertim cum istae

---

(1) Read (as I): rectarum (lines).
(2) I continues: XR. See Lects. 16, 17, note (19).
(3) Omitted in *editio princeps*.

$X\pi$ and $X\tau$. Parallel to that surface draw any line $\pi\tau$ meeting those rays in π and τ. Dropping the perpendiculars πC and τE from these points to AX, bisect CE in D; then with center D and radius DX describe a circle intersecting $C\pi$ in P and $E\tau$ in T; and join XP and XT. I say that when the density of the second medium is such that it refracts the most refrangible of the rays incident along IX to P and the least refrangible to T, then the angle PXT will prove to be greatest. For however dense the second medium is assumed, the refracted rays will intersect the lines CP and $[E]T$ in the points π and τ such that the line $\pi\tau$ is parallel to AX. Consequently, if the line $D\rho$ is drawn bisecting all lines $\pi\tau$, the center of any circle passing through the points π and τ will always lie on the same line $D\rho$. The angle $\pi X\tau$, however, is the angle in a segment of the circle passing through the points π, τ, and X, and thus it will be greatest when a circle of this kind turns out smallest, because the ratio of the chord $\pi\tau$ to the circle's dimensions then turns out to be greatest. But that circle becomes smallest of all when its center falls at D, since it then has for a radius XD, the smallest of the lines[1] that can be drawn from X to RD. Therefore, the angle $\pi X\tau$ is greatest at the moment when the center of the circle passing through the points π, τ, and X falls at D. Hence, since the circle XPT and the angle PXT are of such a sort, the proposition is evident.

In passing, it should be clear from this that the angle PXT also proves to be [97 = 172] greatest just when the density of the first medium is such that the angle of refraction of the mean refrangible rays[2] incident most obliquely along IX is half a right angle, and that it becomes continually smaller the more that angle of refraction deviates (by excess or defect) from half a right angle. For example, if the refractions made from air into water, glass, and crystal are compared, it will be evident from a calculation that when the angle of incidence is about 90°, then the angle of refraction in water will be greater than half a right angle, and smaller in glass. Water is therefore not dense enough and glass too dense to make the angle PXT greatest, and consequently since crystal is still denser, it will make the angle PXT smaller than glass would. Thus although glass refracts less, yet in this case it will spread heterogeneous rays refracted in it more from one another than crystal will, and in this way it will project more distinct colors onto its opposite surface. But these are very difficult things to test, because glass and crystal differ little in density, nor can they be had sufficiently thick; and even if they could be, then because of their very great thickness they would be insufficiently transparent.

Case 2. But if the line along which the rays are incident is not the most [98 = 173] oblique, the problem will emerge as a solid one; but I intend to show how by changing its conditions somewhat it can be reduced to a plane one. It must therefore be understood that since between the extreme or most dissimilar rays there are innumerable intermediate ones that are refracted, some more than others, in a continuous succession of infinitely small gradations; the difference of refraction[3] of the extreme rays will be made up of the quite similar differences of the intermediate ones, infinite both in number and smallness. Now when the properties of those infinitely small differences are known, we can then make a judgment about all of them taken together, or about the finitely small differences that lie between the refractions of the

differentiae sint admodum exiguae. Sic cognito quod infinitae parvae differentiae augentur diminuuntur, vel simul maximae evadant aut minimae. Concludendum erit quod omnium summa perinde augetur, diminuitur, vel maxima fit, vel minima. Quod si non sint omnes simul maximae vel minimae tamen summa pro maxima vel minima haberi potest, cum id accidit intermediae parti. Sic omnium colorum latitudo tunc maxima censeri possit cum id accidit viriditati. Jam licèt problema propositum cùm de differentijs finitè parvis agitur, existat solidum si tamen instituatur de differentijs infinitè parvis ad planum reduci potest. Verùm huic solvendo nolo obnixè incumbere / sed breviter tantùm ostendam quo pacto calculus in hoc et ejusmodi alijs sit 57/ ineundus, ut ad aequationem perveniatur, ex quâ maximus angulorum infinitè parvorum possit elici. Et insuper ex eodem fundamento determinabo proportiones differentiarum refractionis respectu diversorum Mediorum quas in praecedentibus quatuor propositionibus generaliter tantùm descripsi.

[99 = 174]   Primò itaque investiganda est regula vel Aequatio, qua ex uno utcunque refracto radio dato, refractus alter cum eo constituens angulum infinitè parvum cognosci poterit. Radijs e Medio densitate dato in Medium cujuslibet

Fig I, 45   densitatis secundum obliquissimam lineam *IX* ut prius incidentibus, sint *XR*

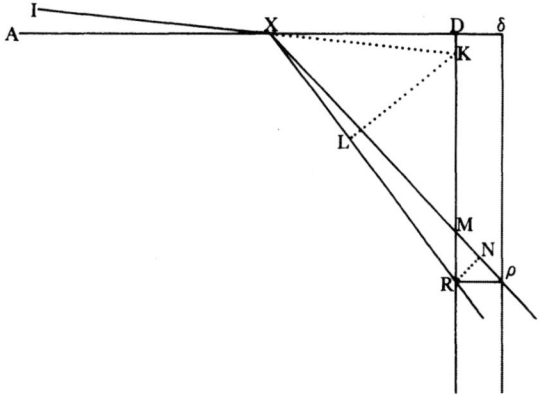

Figure I, 45

et *Xρ* refracti duo, quorum alter *XR* sit altero *Xρ* paulo magis refrangibilis, differentiâ tamen infinite parvâ. Et agatur lineola quaevis *Rρ* his in *R* et *ρ* occurrens, et refringenti superficiei parallela. Ad quam superficiem normales etiam *RD*, *ρδ* demittantur, quas datam finitamque distantiam ab *X*, ab invicem verò infinitè parvam habere fingeto sed lineolam *Rρ* cum radijs per *R* et *ρ* transeuntibus plus aut minus ab *XD* vergere (quemadmodum in praecedentibus) concipito pro variâ posterioris Medij assumenda densitate. Jam si recta *DR* secet radios *Xρ* in *M*, et *IX* in *K*, cùm infinitè parvum triangulum *RMρ* sit simile triangulo *DMX* a quo triangulum *KRX* non nisi infinitè parvis differentijs *RXM* et *DXK* discrepat, quae dissimilitudinem non inferunt, triangula etiam *RMρ* et *RDX* pro similibus haberi debent. Et proinde demissis perpendicularibus *KL* et *RN* erit *XK . LR :: Rρ . MN*. Adeoque cùm sit

$$LR = \frac{XR^q - XK^q}{XR}$$ (nam est *XR . KR* $(= \sqrt{:\ XR^q - X[K]^q :})$ :: *KR . LR*.) erit

extreme rays, especially since those differences are very small. Thus by knowing that the infinitely small differences increase, decrease, or together become greatest or least, it will have to be concluded that the sum of all of them likewise increases, decreases, or becomes greatest or least. But if they are not all simultaneously a maximum or minimum, their sum can still be considered a maximum or minimum when it occurs for an intermediate part. Thus the width of all the colors can be considered a maximum at the moment when it occurs for green. Now, although the proposed problem proves to be a solid one when it concerns finitely small differences, yet if it is undertaken for infinitely small differences, it can be reduced to a plane one. I certainly do not wish to apply myself obstinately to its solution, but I will only briefly show how the calculation in this and similar cases is to be begun in order to arrive at an equation from which the maximum of the infinitely small angles can be derived. In addition, on the same basis I will determine the proportions of the differences of refraction for different media, which I described only generally in the preceding four propositions.

In the first place, therefore, a rule or equation must be found whereby [99 = 174] given any one refracted ray, another refracted ray making an infinitely small angle with it can be determined. When, as before, rays are incident from a medium of given density into a medium of any density along the most oblique line IX, let XR and $X\rho$ be two refracted rays, with one, XR, being Fig. I, 45 slightly more refrangible than the other, $X\rho$, although by an infinitely small difference. Draw any small line, $R\rho$, meeting these in R and ρ and parallel to the refracting surface. To this surface also drop the normals RD and $\rho\delta$, which you should imagine to be a given, finite distance from X but an infinitely small distance from one another; whereas (as in the preceding) you should conceive the small line $R\rho$, together with the rays passing through R and ρ, to lie nearer or farther from XD according as various densities are assumed for the second medium. Now if the line DR intersects the rays $X\rho$ in M and IX in K, since the infinitely small triangle $RM\rho$ is similar to the triangle DMX, from which the triangle KRX differs only by the infinitely small differences RXM and DXK, without effect on the similarity, the triangles $RM\rho$ and RDX also ought to be considered as similar. Consequently, dropping the perpendiculars KL and RN, there will be $XK : LR = R\rho : MN$. Hence, since

$$LR = (XR^2 - XK^2)/XR \text{ (for } XR : KR \ (= \sqrt{[XR^2 - X[K]^2]}) = KR : LR),$$

etiam $MN = \dfrac{XR^q - XK^q}{XR \times XK}$ in $R\rho$. Quae differentia est inter XN sive XR et

XM. Et inde erit $XM = XR - : \dfrac{XR^q - XK^q}{XR \times XK}$ in $R\rho$. Inventa est itaque relatio

inter XK, XM, et XR cùm angulus IXA sit infinitè parvus: Quinetiam utcunque IX obliqua ponatur, illae XK, XM, et XR eandem relationem observabunt, siquidem reciprocè sunt ut sinus incidentiae et refractionis; et proinde inventa est etiam inter eas relatio pro quavis obliquitate incidentis IX.
Atque ita cognitis vel utcunque ad arbitrium assumptis XK et XR, inde XM
simul cognoscitur. Quod primò determinandum proposui.

Quamobrem sit IX linea datum quemvis angulum AXI cum refringente

superficie constituens; caeterisque stantibus, erit $MN = \dfrac{XR^q - XK^q}{XR \times XK}$ in $R\rho$.
Insuper est RD $(= \sqrt{XR^q - XD^q})$. $XD :: MN . NR$. Atque adeo est

$$NR = \frac{XR^q - XK^q \text{ in } R\rho \times XD}{XR \times XK \times \sqrt{XR^q - XD^q}}.$$

Quod si NR dividatur per XR / prodibit sinus anguli RXN respectu circuli 58/
cujus semidiameter sit unitas. Quare cùm angulus iste et sinus ejus sint simul
maximi, ad maximum angulum determinandum quaerenda erit maxima

quantitas $\dfrac{NR}{XR}$,[4] hoc est maximum $\dfrac{XR^q - XK^q \text{ in } R\rho \times XD}{XR^q \times XK \times \sqrt{XR^q - XD^q}}$[5]   Sive

(factâ per datum $\dfrac{R\rho \times XD}{XK}$ divisione) quaerendum erit maximum

$$\frac{XR^q - XK^q}{XR^q \times \sqrt{XR^q - XD^q}}.$$

Id quod per Methodos de maximis et minimis satis notas fieri potest, et
prodibit $XR^{qq} = 3XK^q \times XR^q - 2XK^q \times XD^q$. Cujus aequationis constructio est ejusmodi.

Fig I, 46   A puncto quolibet incidentis radij IX demitte perpendiculum IA, et in eo
sume $AF = AX$. Et XI producto ad B ut sit $IB = \frac{1}{2}IX$, super BX describe
semicirculum BEX, cui inscribe $XE = XF$. Dein XB producto ad C ut sit BC
$= BE$, super CX describe semicirculum CGX quem in G secet perpendiculum
IG super diametro ejus ad I erectum. Denique centro X et intervallo GX

---

(4) *Editio princeps*: NR.
(5) The *editio princeps* omits XK from the denominator.

also

$$MN = \frac{XR^2 - XK^2}{XR \times XK} R\rho.$$

This is the difference between XN (or XR) and XM, so that

$$XM = XR - \frac{XR^2 - XK^2}{XR \times XK} R\rho.$$

Therefore, a relation has been found between XK, XM, and XR when the angle IXA is infinitely small. In fact, however oblique IX is assumed, XK, XM, and XR will keep the same relation, since they are inversely as the sines of incidence and refraction; consequently, a relation has also been found among them for any obliquity of the incident ray IX. Thus when XK and XR have been found or arbitrarily assumed, XM will be immediately known from them, as I first proposed to determine.

Therefore, let IX be a line making any given angle AXI with the refracting surface and, with other things remaining the same, there will be

$$MN = \frac{XR^2 - XK^2}{XR \times XK} R\rho.$$

Moreover, $RD \ (= \sqrt{[XR^2 - XD^2]}) : XD = MN : NR$, so that

$$NR = \frac{(XR^2 - XK^2)R\rho \times XD}{XR \times XK \times \sqrt{[XR^2 - XD^2]}}.$$

But if NR is divided by XR, the result will be the sine of the angle RXN with respect to a circle of unit radius. Hence, since that angle and its sine are simultaneously a maximum, to determine the maximum angle the maximum quantity NR/XR[4] will have to be found, that is, the maximum of

$$\frac{(XR^2 - XK^2)R\rho \times XD}{XR^2 \times XK \times \sqrt{[XR^2 - XD^2]}};\quad {}^{(5)}$$

or rather (dividing through by the given quantity $R\rho \times XD/XK$), the maximum of

$$\frac{XR^2 - XK^2}{XR^2 \times \sqrt{[XR^2 - XD^2]}}$$

will have to be found. This can be done by the rather well known methods of maxima and minima, and the result will be

$$XR^4 = 3XK^2 \times XR^2 - 2XK^2 \times XD^2.$$

The construction of this equation is of the following sort.

From any point of the incident ray IX drop the perpendicular IA and on it take $AF = AX$. Then extending XI to B so that $IB = \frac{1}{2}IX$, on BX describe the semicircle BEX, and in it inscribe $XE = XF$. Next, extending XB to C so that $BC = BE$, on CX describe the semicircle CGX, which the perpendicular IG, erected on its diameter at I, intersects at G. Finally, with center X and radius

Fig. I, 46

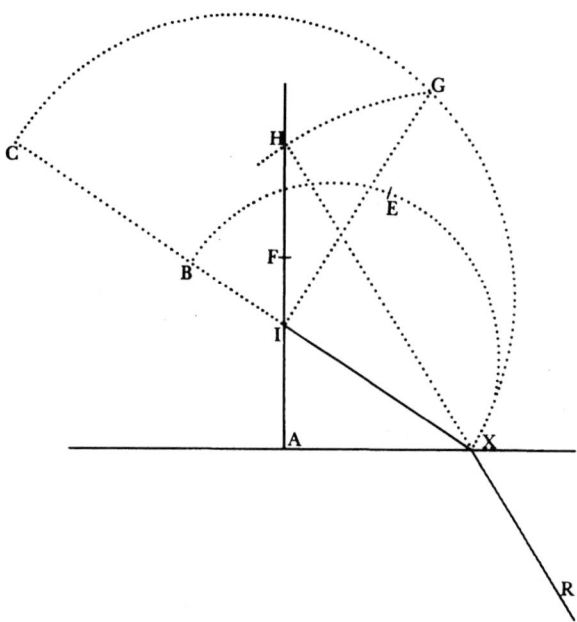

Figure I, 46

describatur arcus *GH* secans *AI* productum in *H*, ducatur *HX* et producatur versus *R*, eritque *XR* ipsius *IX* refractus cum tanta sit posterioris Medij densitas ut differentia refractionis *RXM* fiat omnium maxima. Quo invento densitas posterioris Medij talem refractionem efficientis facile dabitur. Concipe ergo radios *XR* et *Xρ* esse mediocriter refrangibiles, diverso tamen gradu, et posterius Medium sic inventum non modo inter istos, sed et inter extremos seu maximè difformes radios maximam circiter quam potest refractionis differentiam efficiet.

[100 = 175]   Sin autem hujusmodi differentiarum proportiones ad variam raritatem vel densitatem Mediorum desiderentur, e jam ostensis facilè determinabuntur dummodò ponantur infinitè parvae. Sic raritate vel densitate posterioris Medij tantùm

Fig I, 47 variatâ ut radij secundum *IX* incidentes nunc refringantur ad *M* et *R*, nunc ad *μ* et *ρ*: ductaque qualibet *DK* ipsi *DX* normali quae secet eos in *K*, *M*, *R*, *μ*, et *ρ*: erit angulus infinitè parvus *MXR* ad consimilem angulum *μXρ* sicut $\dfrac{XR^q - : XK^q}{XR^q \times RD}$ ad $\dfrac{X\rho^q - : XK^q}{X\rho^q \times \rho D}$. Quòd si raritas vel

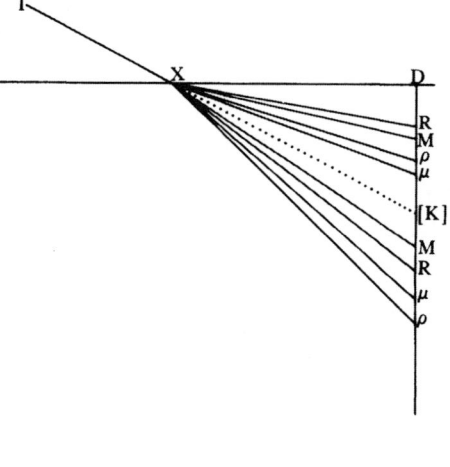

Figure I, 47

GX describe the arc *GH* intersecting the extension of *AI* at *H*; draw *HX* and extend it toward *R*. *XR* will be the refracted ray of *IX* when the density of the second medium is such that the difference of refraction *RXM* becomes the greatest of all. When this is found, the density of the second medium producing such a refraction will readily be given. Therefore, imagine the rays *XR* and *Xρ* to be intermediately refrangible, though to a different degree, and the second medium found in this way will produce not only between those but also between the extreme or most dissimilar rays approximately the greatest difference of refraction possible.

If now the proportions of such differences are required at a different rarity [100 = 175] or density of the media, they will easily be determined from what has just been shown, provided that they are assumed to be infinitely small. Thus, when the rarity or density of only the second medium is varied, so that the rays incident along *IX* are at one time refracted to *M* and *R* and at another to *μ* and *ρ*; and Fig. I, 47 drawing to *DX* any normal *DK*, which intersects the rays in *K, M, R, μ,* and *ρ*; the infinitely small angle *MXR* will be to the altogether similar angle *μXρ* as $(XR^2 - XK^2)/(XR^2 \times RD)$ is to $(X\rho^2 - XK^2)/(X\rho^2 \times \rho D)$. If, however, the

densitas anterioris Medij varietur non mutato posteriori Medio: Analysta facile deprehendet quòd (in fig I, 45) sit $MN = \dfrac{XR^q - : XK^q}{XK^q}$ in $R\rho$, et proinde quod in fig I, 47 sit

$$\text{ang. } MXR . \text{ ang. } \mu X\rho :: \frac{XR^q - : XK^q}{XR \times RD} \cdot \frac{X\rho^q - : XK^q}{X\rho \times \rho D}.$$

Non enim perinde est sive raritas sive densitas anterioris Medij sive posterioris varietur, ut e praeostensis pateat.

Propositiones praecedentes ad luminis e longinquo mananantis diffusionem spectant. In duabus sequentibus agitur de refractione luminis e propinquo mananantis.

[101 = 176]　/ Prop. 18. Heterogeneis radijs a dato puncto ad datum punctum per 59/ superficiem positione refractis: quo Medium densius sit magis densum eo major erit eorum ad invicem inclinatio ex parte medij utriusque ad certum usque terminum, et post erit eo minor. Scilicet cum densitas ejus haud minor[6] sit quàm densitas alterius Medij ut refractio fiat infinitè parva, tum differentia refractionis erit etiam infinitè parva, et proinde augebitur ex auctâ densitate. Quod si densitas ejus in infinitum augeatur, tum omnium radiorum in illud incidentium refracti perpendiculariter emergent.[a] et e contra soli perpendiculares possunt ingredi Medium rarius e densiori.[a] Unde omnes radij a puncto ad punctum refracti tunc pergent in ijsdem lineis, sive coincident; et sic differentia refractionis rursus in nihilum evanescet.

(a) Num 42 & 45

[102 = 177]　Prop 19. Heterogeneis radijs a dato puncto ad datum punctum per superficiem positione datam refractis; quo Medium rarius sit magis rarum eò major erit eorum ad invicem inclinatio ex parte Medij utriusque. Sit *AT* superficies

Fig I, 48[7]

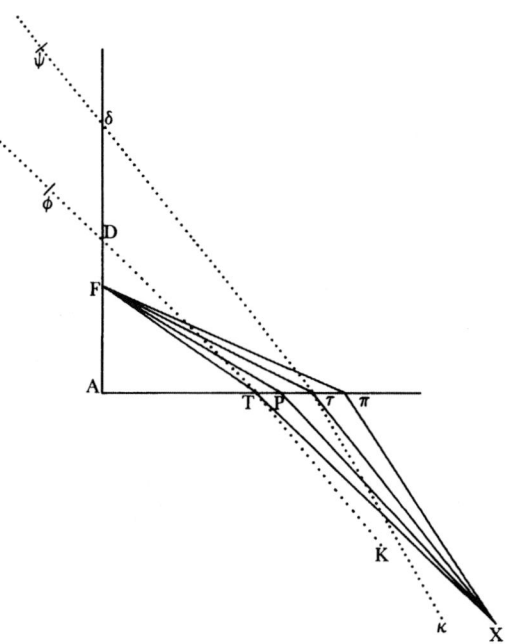

Figure I, 48

rarity or density of the first medium is varied, while the second is unchanged, the analyst will easily discover that (in Fig. I, 45)

$$MN = \frac{XR^2 - XK^2}{XK^2}R\rho,$$

and consequently that (in Fig. I, 47)

$$M\widehat{X}R : \mu\widehat{X}\rho = (XR^2 - XK^2)/(XR \times RD) : (X\rho^2 - XK^2)/(X\rho \times \rho D).$$

For it is not equivalent to vary the rarity or density of the first medium or of the second, as is obvious from what has been shown earlier.

The preceding propositions concern the diffusion of light flowing from afar. In the following two the refraction of light flowing from nearby is treated.

Proposition 18. When heterogeneous rays are refracted from a given point [101 = 176] to a given point across a surface in a given position, the greater the density of the denser medium, the greater will be the rays' inclination to one another on the side of each medium up to a certain limit, and afterward it will be smaller.

Namely, when its density is not smaller[6] than the density of the other medium, so that the refraction becomes infinitely small, then the difference of refraction will also be infinitely small, and therefore it will increase with an increase in the density. But if its density is increased infinitely, then the refracted rays of all those incident upon it will emerge perpendicularly,[a] and (a) §§I, 42, 45 conversely only perpendicular rays can enter the rarer medium from the denser.[a] Consequently, all rays refracted from a point to a point will then proceed in the same lines—or coincide—and thus the difference of refraction will again vanish to nothing.

Proposition 19. When heterogeneous rays are refracted from a given point [102 = 177] to a given point across a surface in a given position, the greater the rarity of the rarer medium the greater will be the rays' inclination to one another on the side of each medium.

Let *AT* be a surface so refracting the dissimilar rays *FTX* and *FPX* that Fig. I, 48[7]

---

(6) Read (as I and originally) : major (greater).

(7) In the *Lectiones opticae* the corresponding Fig. 76 is drawn with the line *FA* horizontal but is otherwise the same.

ita refringens difformes radios *FTX* et *FPX* ut manantes ab eodem puncto *F* in idem rursus ad *X* conveniant. Dico si medium prius[8] esset rarius, ut praefati radij adhuc magis refringerentur, puta *FTX* secundum *FτX* et *FPX* secundum *FπX*; quòd angulus $\pi X\tau$ foret major angulo *PXT*, ut et angulus $\pi F\tau$ major angulo *PFT*.

(b) Consule cas: 2 Scholij ad Prop 17

Ad abbreviandam prioris casûs demonstrationem ponamus radios esse quam minimè difformes ut propter infinitè parvam differentiam refractionis, angulos *PXT* et $\pi X\tau$ constituant infinitè parvos.[b] Tum ducatur *TK* refractus radij conformis ipsi *FPX*, ut infinitè parvus angulus *KTX* sit differentia refractionis radiorum secundum eandem lineam *FT* incidentium. Et pari modo ducatur *τκ* refractus radij conformis ipsi *FπX*, ut angulus infinitè parvus *κτX* existat differentia refractionis radiorum secundum eandem *Fτ* incidentium. Liquet ergo quod cùm *Fτ* sit obliquior quàm *FT* atque etiam in Medium densius incidat, erit angulus *κτX* major angulo *KTX*. Adhaec producantur *XT* et *Xτ*[9] donec in punctis *D* ac *δ* secent lineam *FA* quae sit plano *AT* perpendicularis: et ultra producantur ad *φ* et *ψ* ita ut sit

$$\frac{FA^q}{FT}\cdot\frac{DA^q}{DT} :: TF . T\phi. \text{ et } \frac{FA^q}{F\tau}\cdot\frac{\delta A^q}{\delta\tau} :: \tau F . \tau\psi.$$

Et erunt puncta sic inventa *φ* et *ψ* foci radiorum *FTX* et *FτX* per prop. 8 cas. 2, et inde *Xφ* . *Tφ* :: ang. *KTX* . ang *PXT*; ut et *Xψ* . *τψ* :: ang. *κτX* . ang *πXτ* (cas 3 Schol. ad prop. 12). Istae quidem proportionalitates non sunt omninò verae ubi anguli praefati per differentiam refractionis effecti ponuntur esse definitae alicujus magnitudinis sed ad veritatem eo magis accedunt, quo anguli isti statuuntur minores / adeò ut in angulis infinitè parvis pro accuratè veris 60/ haberi debeant. Jam cùm ex Hypothesi sit *Aτ* ⊏ *AT*, erit etiam *Xτ* ⊏ *XT*,[10] ut et *τψ* ⊏ *Tφ*, quemadmodum pateat ex determinatione punctorum *ψ* et *φ* supra positâ. Quamobrem est *τψ* . *Tφ* ⊏ *τX* . *TX*, vel permutando *τψ* . *τX* ⊏ *Tφ* . *TX* et componendo *τψ* . *Xψ* ⊏ *Tφ* . *Xφ*. Hoc est substituendo rationes his aequales ang *πXτ* . ang. *κτX* ⊏ ang *PXT* . ang *KTX*. et permutando

ang *πXτ* . ang *PXT* ⊏ ang *κτX* . ang *KTX*.

Verùm est ang *κτX* ⊏ ang *KTX* ut dictum fuit. et ideò multò magis est ang *π[X]τ* ⊏ ang *PXT*.   Q.E.D.

Exhinc verò de posteriori casu, quod semper sit ang *πFτ* ⊏ ang *PFT* fiat conjectura, siquidem demonstrationem longè difficiliorem postularet; et his tam multa impendisse verba jamdudum pertaesum est. Haec itaque de refractionibus solitariae superficiei sufficiant.

DE RADIORUM BIS REFRACTORUM AFFECTIONIBUS.

Lect 12 [103 = 178]

Quod si gemina sit refractio perinde ut in Prismatibus contingit quorum phaenomena praesertim explicare statui; radiorum sic refractorum passiones e praecedentibus ita manifestae sunt, ut circa illas parùm negotij superesse[1] videatur. De parallelis quidem superficiebus nihil aliud occurrit observan-

when they proceed from the same point F they meet again at one point X. I say that if the first[8] medium were rarer, so that those rays would be refracted still more, for instance, FTX along $F\tau X$ and FPX along $F\pi X$, then the angle $\pi X\tau$ would be greater than the angle PXT, and also the angle $\pi F\tau$ greater than the angle PFT.

To shorten the demonstration of the first case, let us assume the rays to be the least dissimilar possible, so that because of the infinitely small difference of refraction, they make the angles PXT and $\pi X\tau$ infinitely small.[b] Then draw TK, the refraction of a ray identical to FPX, so that the infinitely small angle KTX is the difference of refraction of rays incident along the same line FT; likewise draw $\tau\kappa$, the refraction of a ray identical to $F\pi X$, so that the infinitely small angle $\kappa\tau X$ is the difference of refraction of rays incident along the same line $F\tau$. It is therefore clear that since $F\tau$ is more oblique than FT and also incident into a denser medium, the angle $\kappa\tau X$ will be greater than the angle KTX. Moreover, extend XT and $X\tau$[9] until they intersect the line FA, which is perpendicular to the plane AT, in the points D and δ; and extend them farther to ϕ and ψ such that $FA^2/FT : DA^2/DT = TF : T\phi$ and $FA^2/F\tau : \delta A^2/\delta\tau = \tau F : \tau\psi$. Then the points ϕ and ψ thus found will be the foci of the rays FTX and $F\tau X$ (Prop. 8, Case 2). Thus $X\phi : T\phi = \widehat{KTX} : \widehat{PXT}$, and also $X\psi : \tau\psi = \widehat{\kappa\tau X} : \widehat{\pi X\tau}$ (Prop. 12, Scholium, Case 3). In fact these proportions are not strictly true when those angles made by the difference of refraction are assumed to be some definite magnitude, but they approach nearer the truth the smaller those angles are made, so much so that for infinitely small angles they are to be considered as exactly true. Now since by hypothesis $A\tau > AT$, it will also be $X\tau > XT$,[10] as well as $\tau\psi > T\phi$, as is clear from the determination of the points ψ and ϕ presented above. Consequently, $\tau\psi : T\phi > \tau X : TX$, or by permutation $\tau\psi : \tau X > T\phi : TX$, and by composition $\tau\psi : X\psi > T\phi : X\phi$; that is, by substituting ratios equal to these, $\widehat{\pi X\tau} : \widehat{\kappa\tau X} > \widehat{PXT} : \widehat{KTX}$, and by permutation $\widehat{\pi X\tau} : \widehat{PXT} > \widehat{\kappa\tau X} : \widehat{KTX}$. But $\widehat{\kappa\tau X} > \widehat{KTX}$ (as was indicated), and thus even more so $\widehat{\pi X\tau} > \widehat{PXT}$. As was to be demonstrated.

But it may be conjectured from this for the second case that $\widehat{\pi F\tau} > \widehat{PFT}$ always, as it would demand a far more difficult demonstration, and I am already weary from expending so many words on these matters. Therefore, let these things suffice for the refractions of a single surface.

The Properties of Rays Refracted Twice

If now the refraction is twofold, as happens in prisms, whose phenomena I particularly intended to explain, the effects of the rays thus refracted are so clear from the preceding that it may seem that little business concerning them remains.[1] Indeed, for parallel surfaces nothing other arises to be observed

(b) Consult Prop. 17, Scholium, Case 2

---

(8). Originally (as I) : post[erius] (the second).
(9) The *editio princeps* has "KT et $\kappa\tau$," and the figure is correspondingly misdrawn.
(10) Read (as *editio princeps*): $X\tau \sqsupset XT$ ($X\tau < XT$).   (1) *Editio princeps*: interesse.

dum, quàm quòd posterior tantùm recurvat radios quantùm prior incurvat. De inclinatis verò sequentia notentur.

[104 = 179]   Prop 20. Homogenei radij ad Prisma divergentes, post utramque refractionem divergere pergent. Patet per prop. 7.

Atque idem de parallelis vel convergentibus radijs intellige, quòd nempe post utramque refractionem manebunt paralleli vel convergentes.

[105 = 180]   Schol. Quod si punctum a quo quilibet infinitè[2] propinqui post utramque refractionem divergunt, sive locus imaginis trans Prisma conspicuae desideretur, inventio ejus a Scholio ad praefatam Prop. 8 manifesta est. Sed ut promptiùs fiat conjectura, juvabit adhibere Theorema hocce mechanicum; Quod imago ad eandem illam circiter distantiam post prisma apparebit, quam habet objectum cujus est imago, dummodò refractiones hinc et inde non sint admodum inaequales.

[106 = 181]   Prop: 21. Ex Heterogeneis radijs ad Prisma divergentibus aliqui post utramque refractionem convergent. Id quod constat / e prop 10 & 12.[3] 61/ Scilicet ex illis qui in plano ad utraque refringentia plana perpendiculari jacent, magis refrangibiles ex incidentia paulo obliquiori convenient cum minùs refrangibilibus atque idem in innumeris alijs ferè planis superficiebus continget.

[107 = 182]   Pr: 22. E radijs itaque sic a puncto ad punctum sive ab objecto ad oculum refractis, alij ad verticem prismatis gradatim alijs propiores transibunt, pro eo ut sint magis atque magis refrangibiles (per prop. 10).[4] Unde colorum ordines definiuntur, de quibus posthac.[5]

[108 = 183]   Pr: 23. Quo major est angulus verticalis Prismatis caeteris paribus, differentia refractionis fiet eo major, et inde colorum apparentia distinctior. Et hoc manifestum est e prop. 2.[6]

[109 = 184]   Pr: 24. Quo densior est Prismatis materia, vel quo rarius est medium circumfluum caeteris paribus, eo major erit refractionis differentia, et inde colorum apparentia manifestior. Scilicet posterior casus e prop 14 et 16 patet. Priorem verò nè per prop 17 in dubium revocetur, sic ostendo. Concipe magis refrangibilem radium Fig I, 49 *PD* et minimè[7] refrangibilem *TD* sic in Prisma ad idem quodvis punctum *D* incidere ut refracti pergant in eadem linea *Dδ* ac denuò in δ refracti divergant versus π ac τ. Quo posito constat per prop. 15 quod angulus πδτ ex aucta prismatis densitate augebitur. Deque angulo *PDT* par est ratio, si modo radij consimiles secundum easdem lineas retrocedere concipiantur. Patet itaque assertio de radijs in prismate coincidentibus, et inde etiam de parallelis.

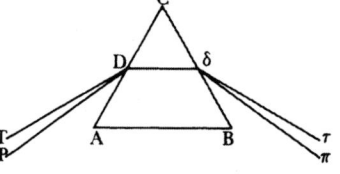

Figure I, 49

(2) I: indefinitè (indefinitely).
(3) In the *Lectiones* Newton cites the omitted Props. 5, 8, §§133, 137.
(4) Newton invokes the omitted Props. 5, 11, §§133, 140, in the *Lectiones*.
(5) de quibus posthac] Added.

than that the second surface bends back the rays as much as the first one bends them in, but for inclined surfaces the following may be noted.

Proposition 20. When homogeneous rays diverge at a prism, they continue [104 = 179] to diverge after both refractions.

It is evident by Prop. 7. The same thing is to be understood for parallel or convergent rays, namely, that they will remain parallel or convergent after both refractions.

Scholium. If, however, the point is required from which any infinitely[2] [105 = 180] close rays diverge after both refractions, that is, the position of the image seen through the prism, its determination is evident from the Scholium to the aforementioned Prop. 8. But to make a more rapid conjecture, it will help to use this mechanical theorem: The image will appear at approximately the same distance behind the prism as does the object whose image it is, provided that the refractions on both sides are not too unequal.

Proposition 21. Of heterogeneous rays diverging at a prism some will [106 = 181] converge after both refractions.

This is evident by Props. 10 and 12.[3] Namely, the more refrangible of those rays lying in a plane perpendicular to both refracting planes will, because of their slightly more oblique incidence, meet the less refrangible ones; and the same thing will happen in nearly innumerable other plane surfaces.

Proposition 22. Therefore, of rays thus refracted from a point to a point, [107 = 182] or from an object to an eye, some will gradually pass nearer to the vertex of the prism than others, according as they are more and more refrangible, by Prop. 10.[4]

Whence the order of the colors is defined, which is to be discussed afterward.[5]

Proposition 23. The greater a prism's vertex angle, with other things being [108 = 183] equal, the greater will the difference of refraction become and, consequently, the more distinct the colors' appearance.

This is manifest from Prop. 2.[6]

Proposition 24. The denser the matter of a prism or the rarer its surround- [109 = 184] ing medium, with other things being equal, the greater will be the difference of refraction and consequently the clearer the colors' appearance.

To be sure, the latter case is evident from Props. 14 and 16, but so that the former case may not be cast into doubt by Prop. 17, I show as follows. Conceive of a more refrangible ray, PD, and a least[7] refrangible one, TD, to Fig. I, 49 fall on the prism at any one point D so that their refracted rays continue in the same line $D\delta$ and after being refracted again at δ diverge toward π and τ. With this assumption, it is established by Prop. 15 that the angle $\pi\delta\tau$ will increase when the prism's density is increased. The reason is the same for the angle PDT, provided that identical rays are understood to proceed backward along the same lines. Therefore, the assertion is evident for coincident rays in the prism, and thus also for parallel rays.

(6) Read: prop. 11.   (7) I: minùs (less).

[110 = 185]　　Lemma. 7. Radijs tribus homogeneis βI, γI, δI e medio densiori in rarius
Fig I, 50　per superficiem *IK* refractis; si differentiae inciden-
tiarum $\beta I\gamma$, $\gamma I\delta$ sint aequales, summa refractorum
angulorum extremis radijs effectorum erit major du-
plo anguli refracti per intermedium radium effecti.
Hoc est, refracti[s] radijs retro-actis ad *B*, *G* ac *D*,
dico quod sit angulus βIB [+] ang δID ⊏ 2ang γIG.
Etenim descripto quovis circulo *AD[I]* tangente refringentem superficiem in *I*, cujus diameter sit *AI*,
quique dictos radios secet in β, γ, δ; *B*, *D*, *G*: Quandoquidem anguli $\beta I\gamma$ et $\gamma I\delta$ sint aequales, erunt
etiam arcus $\beta\gamma$ et $\gamma\delta$ aequales. Sed ductis *A*β, *A*γ,　K
&c: erunt *A*β, *A*γ, *A*δ sinus incidentiarum adeoque
inter se ut sunt *AB*, *AG*, *AD*, sinus refractionum.
Quare (per Lem. 6) est arcus *GD* major arcu *BG*; et
inde

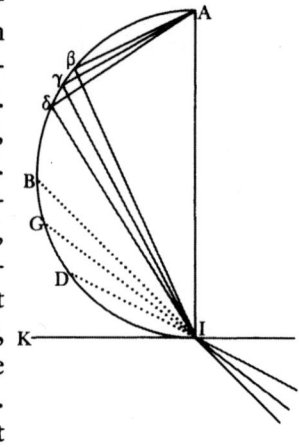

Figure I, 50

$$2\gamma G \ \beth \ 2\gamma G + GD - GB = \gamma D + \gamma B = \gamma D - \gamma\delta + \gamma\beta + \gamma B = D\delta + B\beta.$$

Hoc est $2\gamma G \ \beth \ D\delta + B\beta$, sive ang: βIB + ang: δID ⊏ 2ang: γIG.　Q.E.D.

[111 = 186]　　/ Pr: 25. Homogeneis radijs a Pris-　　　　　　　　　　　　　　62/
mate refractis, angulus quem inci-
dentes et emergentes comprehendunt
tunc maximus evadit, cum aequalis est
Fig I, 51　hinc et inde refractio. Sit *ABC* Prisma
GRSN radius utrinque ad *R* et *S* aequaliter refractus, et *IPQL* alius radius
refractus inaequaliter, magis quidem
ad *P*, minùs ad *Q*. Et producantur hi
radij donec sibi occurrant, *IP* et *QL* in
T, *GR* vero et *NS* in *V*. Dico angulum
RVS esse majorem angulo *PTQ*. Quod
ut pateat, concipe radios in lineis *PQ*
et *RS* hinc inde pergentes utrinque

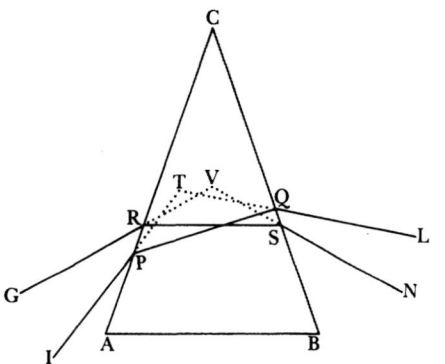

Figure I, 51

egredi Prismate, et sic e Medio densiori in rarius refringi. Jam in triangulis
CPQ et *CRS*, cum angulus *C* communis sit, caeterorum angulorum summae
erunt aequales. Et proinde cùm *CRS* sit Isosceles, duplum anguli *CSR* aequabitur angulis *CPQ* + *CQP*. Quamobrem radij *QP* incidentia ad *P* tantò
major est incidentia radij *RS* ad *S*, quanto eadem incidentia sit major incidentia radij *PQ* ad *Q*. Trium itaque incidentiarum differentiae sunt aequales
adeòque (juxta Lemma praemonstratum) summa refractorum angulorum per
incidentiam maximam et minimam effectorum major erit duplo anguli refracti per incidentiam mediocrem effecti. Hoc est ang *QPT* + ang *PQT* ⊏
2ang *RSV*, sive ⊏ ang *RSV* + ang *SRV*. Itaque cùm in triangulis *PTQ* et *RVS*
summa angulorum ad Basin *PQ* sit major summâ eorum ad basin *RS* erit
angulus verticalis *RVS* major angulo verticali *PTQ*.　Q.E.D.

Lemma 7. When three homogeneous rays βI, γI, and δI are refracted from
a denser medium into a rarer one by the surface *IK*, if the differences of the
angles of incidence, $\beta I\gamma$ and $\gamma I\delta$, are equal, the sum of the deviations made
by the extreme rays will be more than twice the deviation made by the
intermediate ray. That is, drawing the refracted rays backward to *B*, *G*, and
D, I say that $\widehat{\beta IB} + \widehat{\delta ID} > 2\widehat{\gamma IG}$.

[110 = 185]
Fig. I, 50

For tangent to the refracting surface at *I* describe any circle *ADI*, which
has diameter *AI* and intersects those rays in β, γ, and δ, and *B*, *G*, and *D*.
Since the angles $\beta I\gamma$ and $\gamma I\delta$ are equal, the arcs $\beta\gamma$ and $\gamma\delta$ will also be equal.
But when $A\beta$, $A\gamma$, and so forth are drawn, $A\beta$, $A\gamma$, and $A\delta$ will be the sines
of incidence, and so be to one another as *AB*, *AG*, and *AD*, the sines of
refraction. Therefore, by Lemma 6 the arc *GD* is greater than the arc *BG*.
Consequently,

$$2\widehat{\gamma G} < 2\widehat{\gamma G} + \widehat{GD} - \widehat{GB} = \widehat{\gamma D} + \widehat{\gamma B} = \widehat{\gamma D} - \widehat{\gamma \delta} + \widehat{\gamma \beta} + \widehat{\gamma B} = \widehat{D\delta} + \widehat{B\beta};$$

that is, $2\widehat{\gamma G} < \widehat{D\delta} + \widehat{B\beta}$, or $\widehat{\beta IB} + \widehat{\delta ID} > 2\widehat{\gamma IG}$. As was to be demonstrated.

Proposition 25. When homogeneous rays are refracted by a prism, the
angle that the incident and emergent rays form will be a maximum at the
moment when the refractions on each side are equal.

[111 = 186]

Let *ABC* be a prism, *GRSN* a ray refracted equally on each side at *R* and *S*,
and *IPQL* another ray refracted unequally, more in fact at *P* and less at *Q*.
Then extend these rays until they meet each other, *IP* and *QL* at *T*, but *GR*
and *NS* at *V*. I say that the angle *RVS* is greater than the angle *PTQ*. For this
to be evident, imagine rays proceeding in each direction along the lines *PQ*
and *RS* to leave the prism on each side and thus to be refracted from a denser
into a rarer medium. Now, since in the triangles *CPQ* and *CRS* the angle *C* is
common, the sums of the other angles will be equal. Thus, since *CRS* is
isosceles, twice the angle *CSR* will equal the sum of the angles *CPQ* and
CQP. Therefore, the angle of incidence of the ray *QP* at *P* is just as much
greater than the angle of incidence of the ray *RS* at *S* as the same angle of
incidence is greater than the angle of incidence of ray *PQ* at *Q*. Consequently, the differences of the three angles of incidence are equal, so that (by
the lemma just demonstrated) the sum of the deviations made at the greatest
and least angle of incidence will be greater than twice the deviation made at
the intermediate angle of incidence, that is, $\widehat{QPT} + \widehat{PQT} > 2\widehat{RSV}$ or $> \widehat{RSV}$
$+ \widehat{SRV}$. Therefore, since in the triangles *PTQ* and *RVS* the sum of the angles
at the base *PQ* is greater than their sum at the base *RS*, the vertex angle *RVS*
will be greater than the vertex angle *PTQ*. As was to be demonstrated.

Fig. I, 51

[112 = 187]   Lemma 8. Si secundum tres lineas *βI*, *γI*, *δI*
Fig I, 52 aequales angulos *βIγ* et *γIδ* continentes, tres
radij minimè refrangibiles incidant ad *I* in superficiem *IK*, et e medio rariori in densius[(8)] refringantur, quorum refracti retrorsum producti sint
IB, *IG*, *ID* et praeterea si trium maximè refrangibilium radiorum secundum easdem lineas *βI*,
γI, *δI* incidentium refracti retrorsum producti
sint *Ib*, *Ig*, *Id*: Differentia refractionis radiorum
quorum incidentia est minima, unà cum differentia refractionis eorum quorum incidentia est
maxima, major erit quam dupla differentia refractionis eorum quorum incidentia est mediocris. Hoc est, ang *BIb* + ang *DId* ⊏ 2ang *GIg*.
Etenim descripto quovis circulo *ADI* tangente
refringentem superficiem in *I*; cujus diameter sit
AI, quique praefatos radios in punctis *β*, *γ*, *δ*; *B*,
b; *G*, *g*; *D*, *d* secet; Concipe subtensas ab *A* ad

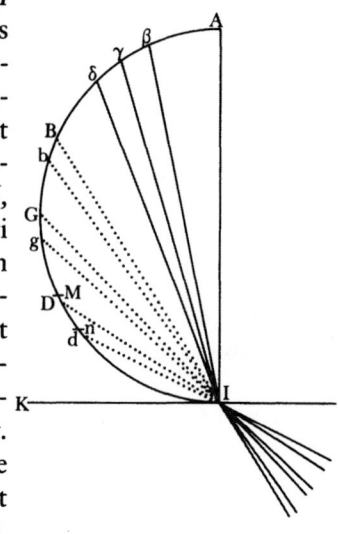

Figure I, 52

quodlibet istorum punctorum duci. Et erunt *Aβ*, *Aγ*, *Aδ* inter se ut sunt *AB*,
AG, *AD*, atque etiam ut sunt *Ab*, *Ag*, *Ad*. Unde sequitur quod *AB*,
AG, *AD*, / inter se sunt ut *Ab*, *Ag*, *Ad*: et praeterea (per Lemma 6) quod sit 63/
arcus *GD* ⊏ arcu *BG*, et arcus *gd* ⊏ arcu *bg*. Jam fiat arcus *GM* = *BG*
eritque *GD* ⊏ *GM*, et *AD* ⊏ *AM*. Item in peripheria *AD* sume punctum
quoddam *n* sub hac conditione, ut, si concipias *AM*, *An* subtensas duci, sit
AB . Ab :: *AM . An*. et erunt *AB*, *AG*, *AM* inter se ut sunt *Ab*, *Ag*, *An*.
Adeoque cùm arcus *BG* ac *GM* sint aequales, erit summa arcuum *Bb* + *Mn*
(per Lem [7]) major duplo arcu *Gg*. Sed cùm sit

$$AM . An \ (:: AB . Ab) :: AD . Ad,$$

vel conversè *AM . AD* :: *Mn . Dd* propter *AD* ⊏ *AM* erit arc *Dd* ⊏ arc *Mn*.
Et utrobique addito arcu *Bb*, erit arc *Bb* + arc *Dd* ⊏ arc *Bb* + arc *Mn*. Et
multo magis erit arc *Bb* + arc *Dd* ⊏ duplo arcu *Gg*: sive ang *BIb* + ang. *DId*
⊏ 2ang *GIg*.   Q.E.O.

[113 = 188]   Pr: 26. Heterogeneis radijs a Prismate refractis, differentia angulorum quos
incidentes cum emergentibus constituunt, tunc minima evadit, cum aequales
Fig I, 53 sunt utrobique refractiones. In Prismate *ABC* sumatur *CR* = *CS*, et *RS* ducatur, ut et alia quaevis linea *PQ* quae non sit parallela ad *RS*. Et concipe
radios in Prismate secundum has lineas *PQ* et *RS* hinc inde pergentes ad
puncta *P*, *Q*; *R*, et *S* egredi, et maxime refrangibiles versus *K*, *M*; *H*, et *O*
refringi, ac minimè refrangibiles versus *I*, *L*; *G*, et *N*. Dico quod refractionum
inaequaliter ad *P* et *Q* factarum differentiae simul sumptae *IPK* + *LQM* sint
majores quàm *GRH* + *NSO* differentiae refractionum aequaliter ad *R* et *S*
factarum simul sumptae. Nam incidentiarum ad *P*, *Q*, et *S* differentiae sunt
aequales, ut ostensum erat in propositione praecedenti. Atque adeò (per Lem.
8.) differentia refractionis radiorum difformium ad *P* ubi maxima est incidentia, unâ cum differentia consimili ad *Q* ubi minima est incidentia, excedit

Lemma 8. If along the three lines βI, γI, and δI containing the equal angles [112 = 187]
$\beta I \gamma$ and $\gamma I \delta$, three least refrangible rays are incident at I on the surface IK Fig. I, 52
and are refracted from a rarer into a denser[8] medium, and their refracted
rays extended backward are IB, IG, and ID; if, moreover, the refracted rays
extended backward of three most refrangible rays incident along the same
lines, βI, γI, and δI, are Ib, Ig, and Id; the difference of refraction of the rays
whose angle of incidence is the smallest together with the difference of refrac-
tion of those whose angle of incidence is the greatest will be more than twice
the difference of refraction of those at the mean angle of incidence, that is,
$\widehat{BIb} + \widehat{DId} > 2\widehat{GIg}$.

For tangent to the refracting surface at I describe any circle ADI, which
has diameter AI and intersects those rays in the points β, γ, and δ; B and b;
G and g; and D and d. Imagine the chords from A to each of those points to
be drawn, and $A\beta$, $A\gamma$, and $A\delta$ will be to one another as are AB, AG, and
AD, and also as Ab, Ag, and Ad. Whence it follows that AB, AG, and AD
are to one another as are Ab, Ag, and Ad, and moreover (by Lemma 6) that
$\widehat{GD} > \widehat{BG}$, and $\widehat{gd} > \widehat{bg}$. Now make $\widehat{GM} = \widehat{BG}$, and it will be $\widehat{GD} > \widehat{GM}$,
and $\widehat{AD} > \widehat{AM}$. Also, in the circumference AD take some point n with this
condition, namely, if you imagine the chords AM and An to be drawn, then
$AB : Ab = AM : An$. Then AB, AG, and AM will be to one another as are
Ab, Ag, and An. Hence since the arcs BG and GM are equal, the sum of the
arcs Bb and Mn (by Lemma 7) will be more than twice the arc Gg. But since
$AM : An$ ($= AB : Ab$) $= AD : Ad$, or by conversion $AM : AD = Mn : Dd$,
and because $\widehat{AD} > \widehat{AM}$, it will be $\widehat{Dd} > \widehat{Mn}$. Adding the arc Bb to both sides,
$\widehat{Bb} + \widehat{Dd} > \widehat{Bb} + \widehat{Mn}$, and even more so $\widehat{Bb} + \widehat{Dd} > 2\widehat{Gg}$, or $\widehat{BIb} + \widehat{DId}$
$> 2\widehat{GIg}$. As was to be shown.

Proposition 26. When heterogeneous rays are refracted by a prism, the [113 = 188]
difference of the angles that the incident rays make with the emergent ones
proves to be a minimum at the moment when the refractions on both sides
are equal.

In the prism ABC take CR equal to CS, and draw RS as well as any other Fig. I, 53
line, PQ, that is not parallel to RS. Then imagine rays to proceed in the prism
in each direction along these lines PQ and RS and to leave at the points P
and Q, and R and S, with the most refrangible ones being refracted toward K
and M, and H and O, and the least refrangible ones toward I and L, and G
and N. I say that the sum of the differences of refraction made unequally at P
and Q, $\widehat{IPK} + \widehat{LQM}$, is greater than $\widehat{GRH} + \widehat{NSO}$, the sum of the differences
of refraction made equally at R and S. For the differences of the angles of
incidence at P, Q, and S are equal, as was shown in the preceding proposition, so that (by Lemma 8) the difference of refraction of the dissimilar rays
at P, where the angle of incidence is greatest, together with the similar difference at Q, where the angle of incidence is least, exceeds twice the indentical

---

(8) rariori in densius] Read (as *editio princeps*): densiori in rarius (denser into a rarer)

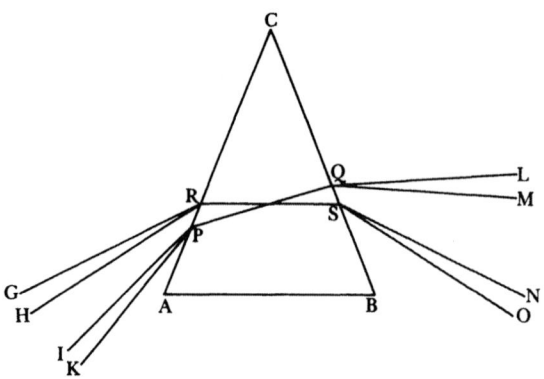

Figure I, 53

duplum consimilis differentiae ad *S* ubi incidentia est mediocris. Hoc est ang. *IPK* + ang. *LQM* ⊏ 2ang. *NSO*: Sive cùm *NSO* ac *GRH* aequentur ang. *IPK* + ang *LQM* ⊏ ang *NSO* + ang *GRH*.   Q.E.D.

[114 = 189]   Schol. Posui quidem radios e Prismate utrobique egredi; sin pergant ab *I* et *K* per *P* et *Q* versus *L* et *M* et a *G* et *H* per *R* et *S* versus *N* et *O*, linearum positiones et quantitates angulorum non inde mutabuntur; et proinde demonstratio praefata tunc etiam valebit, et propter eandem rationem valebit etiam, cum radij ad Prisma divergentes evadunt in Prismate paralleli. Quod idem de Prop[o]sitionum 24 et 25 demonstrationibus itidem intellige. Quinetiam in / alijs quibuscunque casibus ubi divergunt ante refractionem et post conver- 64/ gunt, vel in Prisma incidunt paralleli; non adeo multum a parallelismo intra Prisma recedunt unquam, quin ut anguli vel differentiae angulorum quos incidentes cum emergentibus constituunt, pro ijsdem circiter haberi possint ac si intus essent paralleli; adeoque dictas propositiones ad omnes omninò casus extendi.

[115 = 190]   Prop. 27. Si denique radijs a dato puncto *F* ad datum punctum *X*, per
Fig I, 54 prisma *ABC* positione datum refractis, desiderentur anguli *DFE*, *GXH* quos Heterogenei comprehendunt: Problema ex eorum numero est quae veteres linearia dixêre. At sequens mechanica solutio quantum exigunt res practicae veritati appropinquat. Finge summam angulorum *DFE* + *GXH* aequalem esse angulo *NMO* quem radij duo alteris *FD* et *FE* quoad refrangibilitatem consimiles, ac juxta quamvis lineam *LM* rectae angulum *DFE* bisecanti quàm proximè parallelam incidentes, post binam refractionem constituunt. Et e radijs ad *X* refractis aliquem *GX* cum incidente radio *FD* convenientem in *V*, produc ad *φ*, ut sit *φ* locus imaginis quam objectum *F* oculo in *X* constituto exhibet. Dein angulo *NMO* ac distantijs *φX* et *φV* mechanicè cognitis, dic esse *φX* . *φV* :: ang. *NMO* . ang. *GXH*. Et erit *GXH* quem quaeris proximè. Quemadmodum ex ostensis ad Schol Prop 12 quodammodo manifestum est.

difference at *S*, where the angle of incidence is intermediate; that is, $\widehat{IPK}$ + $\widehat{LQM} > 2\widehat{NSO}$, or since $\widehat{NSO}$ and $\widehat{GRH}$ are equal, $\widehat{IPK}$ + $\widehat{LQM} > \widehat{NSO}$ + $\widehat{GRH}$. As was to be demonstrated.

Scholium. I have, to be sure, assumed that the rays leave the prism on each [114 = 189] side, but if they proceed from *I* and *K* through *P* and *Q* toward *L* and *M*, and from *G* and *H* through *R* and *S* toward *N* and *O*, the positions of the lines and magnitudes of the angles will not on that account be changed. Consequently, that demonstration will then still be valid; and for the same reason, it will also be valid when the rays diverge at the prism and become parallel within it. The same thing must likewise be understood for the demonstrations of Props. 24 and 25. Moreover, in any other cases whatever—where they diverge before refraction and converge afterward, or are incident upon the prism parallel—they will never depart so much from parallelism within the prism but that the angles or the differences of the angles that the incident and emergent rays make can be considered nearly the same as if they were parallel within it, so that these propositions can be totally extended to all cases.

Proposition 27. Finally, if rays are refracted from a given point *F* to a given [115 = 190] point *X* through a prism *ABC* in a given position, the angles *DFE* and *GXH* Fig. I, 54 made by heterogeneous rays are required.

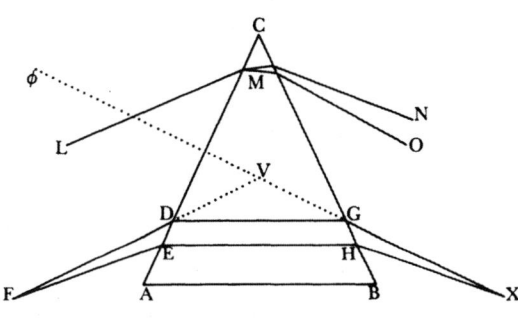

Figure I, 54

The problem belongs to that class that the ancients called linear, but the following mechanical solution approaches the truth as much as practical matters require. Imagine that the sum of the angles *DFE* and *GXH* equals the angle *NMO* made after two refractions by two rays that are identical in refrangibility to the other rays, *FD* and *FE*, and incident along any line, *LM*, very nearly parallel to the line bisecting the angle *DFE*. Next, of the rays refracted to *X* extend any one, *GX*, to ϕ meeting the incident ray *FD* in *V*, so that ϕ is the place of the image that the object *F* exhibits to an eye situated at *X*. Then when the angle *NMO* and the distances ϕX and ϕV are known mechanically, fix it so that $\phi X : \phi V = \widehat{NMO} : \widehat{GXH}$, and the angle *GXH* will be very nearly what you sought, as is clear in a way from what was shown in the Scholium to Prop. 12. When the refractions on each side are not

Cum refractiones utrobique non sint admodum inaequales res expeditiùs absolvitur per Schol: ad Prop: 1,[9] fingendo esse $VX . FV :: $ ang DFE . ang GXH; vel compositè $FV + VX . FV :: $ ang NMO . ang GXH.

Sectio 4ta
De Refractionibus Curvarum Superficierum.

Lect 13 Haec de refractionibus planorum. De curvis et praesertim sphaericis super-
[116] ficiebus jam agendum est, quarum doctrinam respectu homogeneorum radiorum sequentibus propositionibus complecti conabimur.[1]

[117]   Prop. 28. Radij in curvam superficiem incidentis refractum ducere. Nempe eadem est refractio radij a Curva ac est a plano contingente Curvam in puncto refractionis. Quaere ergo refractum a contingente plano per Prop: 3.

[118]   / Prop. 29. Si radij seu paralleli seu ad punctum aliquod contermini se 65/ sphaerae objiciant refringendos, refractorum axi quamproximorum concursum sive focum determinare.

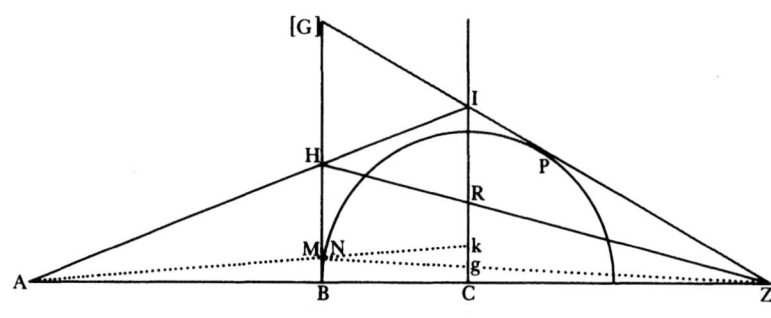

Figure I, 55

Sit A punctum radios ejaculans versus sphaericam superficiem BNP centro C (in fig I, 55) descriptam: E vertice et centro erige ad axem AC perpendiculares BH et CI; ipsisque occurrentem in H et I age quamlibet AI per punctum A. Tum a puncto C versus I cape CR quae sit ad CI ut sinus refractionis ad sinum incidentiae, et age rectam HR occurrentem AC in Z et erit Z concursus refractorum quem determinare oportuit.

Sit enim AN radius axi vicinissimus incidens ad N, et occurrens CI in k. Age NZ occurrentem CI in g. Et, ut mos est, concipe infinitè parvum arcum BN aequalem esse BM segmento rectae BH ad radium Ak terminato, et erit $CI . BH :: Ck . BN$, ac $BH . CR :: BN . Cg$. et ex aequo $CI . CR :: Ck . Cg$. Hoc est Ck ad Cg ut est sinus incidentiae ad sinum refractionis. Et proinde cùm anguli CAk et CZg ex Hypoth sint infinitè parvi, adeóque Ck ad AN et Cg ad NZ perpendiculares vel saltem aequipollentes perpendiculis, erit NZ refractus ipsius AN.   Q.E.D.

[119]   Coroll. 1. Posito I ad R ut est sinus incidentiae ad sinum refractionis erit

very unequal, the matter is resolved more readily by the Scholium to Prop. 1,[9] by imagining that $VX : FV = D\widehat{F}E : G\widehat{X}H$, or by composition

$$FV + VX : FV = N\widehat{M}O : G\widehat{X}H.$$

SECTION 4
THE REFRACTIONS OF CURVED SURFACES

This will suffice for the refractions of planes. We must now consider curved and, especially, spherical surfaces, and we will attempt to comprehend their doctrine with respect to homogeneous rays in the following propositions.[1]   **Lecture 13** [116]

Proposition 28. To draw the refraction of a ray incident on a curved surface.   [117]

The refraction of a ray by a curve is certainly the same as that of a plane tangent to the curve at the point of refraction. Therefore, find the ray refracted by the tangent plane by Prop. 3.

Proposition 29. If rays that are either parallel or terminate at some common point are cast upon a sphere to be refracted, to determine the intersection or focus of the refracted rays immediately near the axis.   [118]

Let A (Fig. I, 55) be a point casting rays toward the spherical surface BNP described with center C. From its vertex and its center erect the perpendiculars BH and CI to the axis AC, and through the point A draw any line AI meeting them at H and I. Next, from the point C in the direction of I take CR such that it is to CI as the sine of refraction is to the sine of incidence. Draw the line HR meeting AC at Z, and Z will be the intersection of the refracted rays that it was required to determine.

For let the ray AN, very close to the axis, fall at N and meet CI in k; and draw NZ meeting CI in g. Then, as is customary, imagine the infinitely small arc BN to be equal to the segment BM of the line BH terminated by the ray Ak. There will then be $CI : BH = Ck : BN$ and $BH : CR = BN : Cg$, and from the equality of the ratios, $CI : CR = Ck : Cg$; that is, Ck is to Cg as the sine of incidence is to the sine of refraction. Consequently, since the angles CAk and CZg are by hypothesis infinitely small, so that Ck is perpendicular to AN and Cg to NZ (or at least equivalent to perpendiculars), NZ will be the refracted ray of AN. As was to be demonstrated.

Corollary 1. On setting I to R as the sine of incidence to the sine of   [119]

---

(9) Read: Prop. 12.

(1) Section 4 is included in its entirety, together with extensive commentary, in *Mathematical Papers*, 3:474–513.

$\frac{I}{R} AB . AC :: BZ . CZ$. Est enim $\frac{I}{R} AB . AB(:: I . R) :: CI . CR$ et $AB . AC$

$:: BH . CI$, et ex aequo perturbate $\frac{I}{R} AB . AC(:: BH . CR) :: BZ . CZ$.[2]

[120]　　Coroll. 2. Si quando punctum A infinitè distet, seu parallelos radios ejaculetur, tum propter aequales BH et CI erit $I . R :: BZ . CZ$.[3] Atque ita si refracti radij paralleli sint tum propter aequales BH et CR erit $I . R :: AC . AB$.

[121]　　Coroll. 3. Si e quatuor punctis A, B, C, et Z tria quaevis dentur, potest quartum inveniri, ut e sequentibus exemplis patebit.

Exempl: 1. Dentur A, B, C, et quaeratur Z. Scilicet est

$$\frac{I}{R} AB . AC :: BZ . CZ$$

adeoque divisim $\frac{I}{R} AB - AC . AC :: BC . CZ$.

Exempl. 2. Si datis A, B, et Z, quaeratur C. Cùm sit $\frac{I}{R} AB . AC :: BZ . CZ$ vicissim erit $\frac{I}{R} AB . BZ :: AC . CZ$ et composite $\frac{I}{R} AB + BZ . BZ :: AZ . CZ$.

Exempl. 3. Si datis A, C, et Z quaerantur B, cùm sit $\frac{I}{R} AB . AC :: BZ . CZ$, sive $AB . \frac{R}{I} AC :: BZ . CZ$. vicissim erit $\frac{R}{I} AC . CZ :: AB . BZ$ et composite $\frac{R}{I} AC + CZ . CZ :: AZ . BZ$.

/ Possunt eadem determinari per ductum linearum,[4] veluti si datis A, B, et 66/ Z quaeratur C, erige ad AZ normalem BH, cujusvis longitudinis et in ea cape BG quae sit ad BH ut I ad R. Junge AH et GZ occurrentes in I, et IC normaliter demissa ad AZ incidet in punctum quaesitum C.

[122]　　Nota 1. Quod Z sit locus imaginis objecti A per refractionem exhibitae cum spectatoris oculus in ipso axe ultra Z constituitur.[5]

2. Siquando refracti radij divergant, vel incidentes convergant, vel sint paralleli, similis erit problematis constructio mutatis tantum suo modo mutandis.

3. Si lux e puncto A emissa per plures sphaericas superficies eundem Axem AC retinentes successivè trans-mittatur; Ad concursum post omnes refractiones determinandum, quaere primò concursum radiorum post primam refractionem, deinde concursum eorundem post secundam refractionem, juxta ac si primariò emissi fuissent e puncto praecedentis concursûs. Et sic deinceps, donec ad ultimum concursum deventum sit.[6] Atque hoc pacto locus imaginis Objecti cujusvis per Telescopium vel Microscopium visi determinari potest.

4. Ope Coroll 3. Lentes ex sphaericis superficiebus confici possunt quae Telescopijs modo quolibet designato constituendis inservient. Patet enim ex

---

(2) If we set $AB = s$, $BZ = s'$, $BC = \rho$, and $I/R = n'/n$, then Newton's expression for the refraction of paraxial rays, $(I/R)AB : AC = BZ : CZ$, becomes $n's/n(s + \rho) = s'/(s' - \rho)$, or in "Gaussian" form, $n/s + n'/s' = (n' - n)/\rho$. In a calculation appended to his "Of Refractions" Newton already deduced this result; *Mathematical Papers*, 1:573–4. When he composed the *Opticks*, however, he reformulated it in "Newtonian" form, that is, measuring the distances of

refraction, it will be $(I/R)AB : AC = BZ : CZ$. For $(I/R)AB : AB (= I : R) = CI : CR$, and $AB : AC = BH : CI$, and equating the equal ratios and inverting, $(I/R)AB : AC (= BH : CR) = BZ : CZ$.[2]

Corollary 2. Whenever the point A is infinitely distant or casts out parallel [120] rays, then because of the equality of BH and CI, it will be $I : R = BZ : CZ$.[3] Similarly, if the refracted rays are parallel, then because of the equality of BH and CR, it will be $I : R = AC : AB$.

Corollary 3. If any three of the four points A, B, C, and Z are given, the [121] fourth can be found, as will be evident from the following examples.

Example 1. Let A, B, and C be given and Z sought. Specifically, $(I/R)AB : AC = BZ : CZ$, and by dividing $(I/R)AB - AC : AC = BC : CZ$.

Example 2. If A, B, and Z are given and C is sought, since $(I/R)AB : AC = BZ : CZ$, and by permutation $(I/R)AB : BZ = AC : CZ$, and by composition $(I/R)AB + BZ : BZ = AZ : CZ$.

Example 3. If A, C, and Z are given and B is sought, since $(I/R)AB : AC = BZ : CZ$, or $AB : (R/I)AC = BZ : CZ$, by permutation $(R/I)AC : CZ = AB : BZ$, and by composition $(R/I)AC + CZ : CZ = AZ : BZ$.

The same thing can be determined by a linear construction.[4] For example, if A, B, and Z are given and C is sought, normal to AZ erect BH of any length, and in it take BG in proportion to BH as I is to R. Join AH and GZ meeting in I, and IC dropped normal to AZ will fall at the required point C.

Note 1. Z is the position of the image of the object A produced by refraction when the observer's eye is located on the axis itself beyond Z.[5] [122]

2. Whenever the refracted rays diverge, or the incident rays converge or are parallel, the construction of the problem will be similar, by making only the necessary changes.

3. If light emitted from the point A is transmitted successively through several spherical surfaces having the same axis AC, to determine the intersection after all refractions, first find the rays' intersection after the first refraction; then find their intersection after the second refraction, just as if they had been originally emitted from the point of the previous intersection; and so on until the last intersection is reached.[6] In this way the position of the image of any object seen through a telescope or microscope can be determined.

4. With the aid of Corollary 3 lenses can be made from spherical surfaces to serve for the construction of telescopes of any design. For it is clear from

---

the object and image, x and x', from the primary and secondary foci, f and f', he found $ff' = xx'$; see Bk. I, Axiom VI, Case 3. Barrow derived this corollary for the general case of rays incident off the axis; *Lectiones XVIII*, Lect. XIII, §II, p. 109.

(3) Barrow deduced this using a clever construction that establishes that Z is the point where paraxial rays intersect the axis, whereas rays incident farther from the axis intersect it closer to B; *Lectiones XVIII*, Lect. XI, §§II, IV, pp. 96, 97.

(4) We have added the missing lines GIZ to Fig. I, 55, as does the *editio princeps*.

(5) Barrow rigorously demonstrates this in Lect. XI of his *Lectiones XVIII*.

(6) In "an elegant and convenient" geometrical method published by Barrow but "communicated by a friend," Newton applies his linear construction twice to determine the image point of a lens; *Lectiones XVIII*, Lect. XIV, pp. 127–8.

illo Corollario quòd non tantùm refractiones datarum Lentium investigari possunt, Sed et Lentes delineari quae datas refractiones peragent.

[123]   Lemma 9. Ad datam quamvis Curvam, concursum Axis et vicinissimi perpendiculi determinare.

In fig I, 56 sit *BNn* Curva, et ad quodvis ejus punctum *n* indeterminatè spectatum quaere perpendiculum *nc* per notas methodos ducendi perpendicula Curvarum;[7] et simul invenies longitudinem *Bc*. Tum (demisso ad *Bc* normali *nt*) finge *Bt* vel *nt* infinitè parvam esse, seu nullam, et emerget longitudo *BC* cujus terminus est ad concursum axis cum vicinissimo perpendiculo.

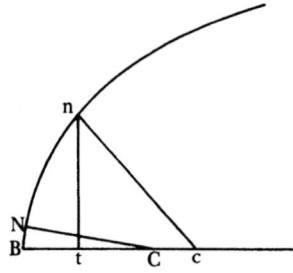

Figure I, 56

Exempl. 1. Sit *BNn* Parabola cujus latus rectum *r*; et *Bt* dic *x*, erit $BC = x + \frac{1}{2}r$ ut notum est.[8] Pone jam $x = 0$ et restabit $\frac{1}{2}r$ pro longitudine *BC* ad verticem.

Exempl. 2. Sit *BNn* Ellipsis cujus latus rectum *r* et transversum *q*; eritque, ut notum est, $BC = x - \frac{rx}{q} + \frac{1}{2}r$. Jam pone $x = 0$ et restabit iterum $\frac{1}{2}r$ pro longitudine *BC* ad verticem. Nec secus in Curvis magis compositis procedendum est.

[124]   Prop. 30. Radijs in curvam quamvis superficiem quàm / proxime perpendiculariter incidentibus, refractorum concursum seu focum determinare.   67/

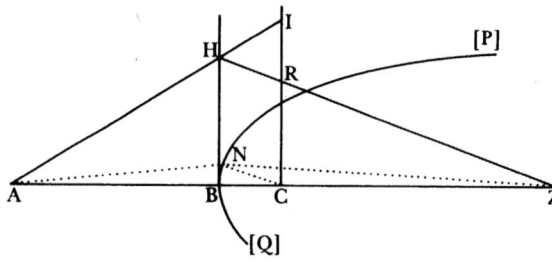

Figure I, 57

Esto *PBQ* (fig I, 57)[9] curva quaevis, *A* commune punctum, seu concursus incidentium radiorum, *AB* radius perpendicularis sive axis: et *AN* radius quàm proximè perpendicularis sive axi proximus. Sitque *NC* ad curvam perpendicularis axique *A[B]* occurrens ad *C*. Et puncto *C* per Lem. 9 invento, Erige ad *B* et *C* perpendicula *BH* et *CI*, quibus in *H* et *I* occurrentem age quavis *AI*: versus *I* cape *CR* quae sit ad *CI* ut sinus refractionis ad sinum incidentiae et recta *HR* occurret *AB* in quaesito refractorum concursu *Z*.

Probatur ad modum praecedentis Propositionis.[10] Et huic etiam consimilia Corollaria et Notae competunt.

[125]   Prop. 31. Parallelis radijs in sphaeram incidentibus refractorum ab Axe remotorum errorem a principali foco determinare.

that corollary that not only can the refractions of given lenses be found but also lenses can be designed that will produce given refractions.

Lemma 9. For any given curve to determine the intersection of the axis and [123] the nearest normal.

In Fig. I, 56 let *BNn* be the curve, and at any arbitrarily chosen point *n* of it find the perpendicular *nc* by the known methods of drawing normals to curves,[7] and at the same time you will find the length *Bc*. Then dropping *nt* perpendicular to *Bc*, imagine *Bt* or *nt* to be infinitely small or zero, and there will emerge the length *BC* whose end is at the intersection of the axis with the nearest normal.

Example 1. Let *BNn* be a parabola with *latus rectum r*, and call *Bt x*; then, as is known,[8] $BC = x + \frac{1}{2}r$. Now set $x = 0$, and $\frac{1}{2}r$ will remain for the length of *BC* at the vertex.

Example 2. Let *BNn* be an ellipse with *latus rectum r* and main axis *q*; and then, as is known, $BC = x - (r/q)\,x + \frac{1}{2}r$. Now set $x = 0$, and again $\frac{1}{2}r$ will remain for the length of *BC* at the vertex. One should proceed equivalently for more complex curves.

Proposition 30. When rays are incident upon any curved surface very [124] nearly perpendicularly, to determine the intersection or focus of the refracted rays.

Let *PBQ* (Fig. I, 57)[9] be any curve, *A* the common point or intersection of the incident rays, *AB* the perpendicular ray or axis, and *AN* a ray very nearly perpendicular, that is, exceedingly close to the axis. Let *NC* be normal to the curve and meet the axis *A[B]* in *C*. Then after finding the point *C* by Lemma 9, at *B* and *C* erect the perpendiculars *BH* and *CI*, and draw any line *AI* meeting them in *H* and *I*. In the direction of *I* take *CR* in proportion to *CI* as the sine of refraction is to the sine of incidence, and the line *HR* will meet *AB* in the required intersection *Z* of the refracted rays.

This is proved in the manner of the preceding proposition,[10] and wholly similar corollaries and notes also apply to it.

Proposition 31. When parallel rays are incident upon a sphere, to deter- [125] mine the error from the principal focus of refracted rays remote from the axis.

(7) See Whiteside's "Historical note," *Mathematical Papers*, 1:213.

(8) See Apollonius, *Conics*, V, 58 and, for the ellipse in Example 2, Prop. 59. Newton drew these examples from Descartes's *La Geometrie*, Bk. II, *Oeuvres*, 6:414–18.

(9) Newton's figure lacks the point *P* and the arc *BQ*, which we have added.

(10) By substituting the center of curvature in the immediate vicinity of the point *N*, Prop. 29 for spherical surfaces can be applied to any curve that is the cross-section through the axis of a surface of revolution.

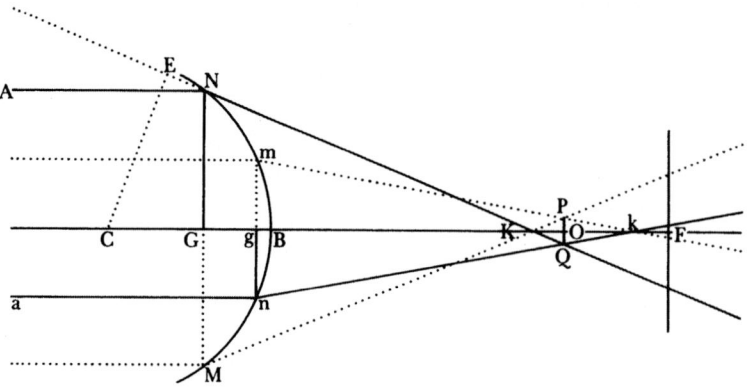

Figure I, 58

In Schem I, 58 sit *NBn* sphaera, *C* centrum ejus, *CB* semi-diameter incidentibus radijs parallela, *AN* radius incidens et *NK* refractus ejus occurrens axi
seu semi-diametro *CB* in *K*. Et posito *F* principali foco, i.e. in quem radij
prope axem jacentes congregantur, quaerendus erit error *FK*. Demitte ergo
perpendiculares *CE* in *NK* et *NG* in *CK*, et dic *CB* = *a*, *GB* = *x*, et *CK* = *z*,
atque ex natura circuli erit $NG^q = 2ax - xx$,[11] cui adde GK^q, hoc est $zz +
2x[z] - 2az + xx - 2ax + aa$, et prodibit $NK^q = zz + 2xz - 2az + aa$. Jam
cùm *NG* sit ad *CE* ut sinus incidentiae ad sinum refractionis, sive ut *I* ad *R* et
propter similitudinem triangulorum *CEK* et *NGK*, *NK* et *CK* sint in eadem
ratione; erit *II* . *RR* (:: NK^q . CK^q) :: $zz + 2xz - 2az + aa$. zz; Adeoque
$IIzz = RRzz + 2RRxz - 2RRaz + RRaa$. et facta reductione

$$zz = \frac{2RRaz - 2RRxz - RRaa}{RR - II}.$$

Extractaque radice

$$z = \frac{RRa - RRx + R\sqrt{IIaa - 2RR\, ax + RRxx}}{RR - II}.$$

Ac radicali in infinitam seriem redacta[12]

$$z = \frac{Ra}{R - I} - \frac{RR}{IR - II}\, x - \frac{R^3}{2I^3a}\, xx - \frac{R^5}{2I^5a[a]}\, x^3 \ \&c.$$

Jam cum per Coroll 2 vel 3 ad Prop 29 sit $\dfrac{Ra}{R - I}$ = *CF* (id quod etiam
innotescit ex valore *z* jam invento, fingendo esse *x* = 0) ex hoc *CF* subduc
inventum valorem *z* et restabit $\dfrac{RR}{IR - II}\, x + \dfrac{R^3}{2I^3a}\, xx + \dfrac{R^5}{2I^5a[a]}\, x^3$ &c pro
valore erroris *KF*, quem quaerimus.[13]

---

(11) If *NG* = *y*, the circle has the equation $(x - a)^2 + y^2 = a^2$, or $y^2 = 2ax - x^2$.
(12) Specifically, on computing this root,

$$\sqrt{[I^2a^2 - 2R^2ax + R^2x^2]} = Ia - \frac{R^2}{I}x - \frac{R^2(R^2 - I^2)}{2I^3a}x^2 - \frac{R^4(R^2 - I^2)}{2I^5a^2}x^3 \cdots$$

For Newton's general method of extracting roots, see his "De methodis serierum et fluxionum";
Mathematical Papers, 3:42–64.

In Fig. I, 58 let *NBn* be the sphere, *C* its center, *CB* the radius parallel to the incident rays, *AN* an incident ray, and *NK* its refracted ray meeting the axis, or the radius *CB*, in *K*. Then on taking *F* as the principal focus, that is, the point at which rays lying near the axis are gathered, the error *FK* will have to be found. Accordingly, drop the perpendiculars *CE* to *NK* and *NG* to *CK*, and call $CB = a$, $GB = x$, and $CK = z$. From the nature of a circle $NG^2 = 2ax - x^2$,[11] and adding to it GK^2, that is, $z^2 + 2xz - 2az + x^2 - 2ax + a^2$, there will result $NK^2 = z^2 + 2xz - 2az + a^2$. Now, since *NG* is to *CE* as the sine of the incidence is to the sine of refraction, or as *I* to *R*, and, because of the similarity of the triangles *CEK* and *NGK*, *NK* and *CK* are in the same ratio, there will be

$$I^2 : R^2 \; (= NK^2 : CK^2) = (z^2 + 2xz - 2az + a^2) : z^2,$$

and hence $I^2z^2 = R^2z^2 + 2R^2xz - 2R^2az + R^2a^2$. Making a reduction,

$$z^2 = \frac{2R^2z(a - x) - R^2a^2}{R^2 - I^2},$$

and extracting the root,

$$z = \frac{R^2a - R^2x + R\sqrt{[I^2a^2 - 2R^2ax + R^2x^2]}}{R^2 - I^2},$$

and also bringing the radical to an infinite series,[12]

$$z = \frac{Ra}{R - I} - \frac{R^2}{I(R - I)}x - \frac{R^3}{2I^3a}x^2 - \frac{R^5}{2I^5a^2}x^3 \cdots$$

Now, since by Corollary 2 or 3, Prop. 29, $Ra/(R - I) = CF$ (which also is determined from the value of *z* just found by supposing $x = 0$), from *CF* subtract the value of *z* that was found, and there will remain,

$$\frac{R^2}{I(R - I)}x + \frac{R^3}{2I^3a}x^2 + \frac{R^5}{2I^5a^2}x^3 \cdots$$

for the value of the error *KF* that we seek.[13]

---

(13) Newton previously determined the longitudinal spherical aberration in "Of Refractions"; ibid., 1:572. He stated only the final results, but he probably calculated it just as here; although he expressed the aberration in terms of *NG* (as in Corollary 2, below) rather than *GB*. In his reply to Hooke on 11 June 1672, Newton used this proposition to compare the spherical aberration of a concave mirror and a plano-convex lens of the same focal length, while making the necessary changes for the case when the lens's spherical surface faces the incident rays and also keeping higher order terms; *Correspondence*, 1:172–3 = *Phil. Trans.*, 7 (1672):5085. To apply this proposition to reflection, Newton simply set $I = -R$, or $n = -1$; see "Theoremata Optica," Add. 4004, f. 71ʳ = *Mathematical Papers*, 3:516.

Kepler made the imperfect focal properties of spherical refracting surfaces a central concern of optical research, but a calculation, now lost, by Huygens in 1653 seems to be the first attempt to quantify the magnitude of this aberration. In 1665, at approximately the same time as Newton's first calculations, Huygens undertook a study of spherical aberration in lenses—not just single spherical surfaces—that was far more systematic than Newton's and also had the practical goal of correcting for it. Smith in his *Compleat System* (Bk. II, Ch. XIII, Prop. II, 1:254–7) adopted Huygens's approach and extended it to the general case of nonparallel rays incident upon any lens, although he also utilized some of Newton's results (see note (16), this lecture). For a thorough account of Huygens's, Smith's, and other contemporary investigations of spherical

[126]   / Coroll. 1. Si *BG* sive x ponatur valde exigua erit $\dfrac{RRx}{IR - II}$ quàm proximè

aequalis *KF*. Tunc enim quantitates $\dfrac{R^3xx}{2I^3a} + \dfrac{R^5x^3}{2I^5a^2}$ &c propter ascendentes

potestates ejusdem x evadunt admodum exiguae ut respectu termini $\dfrac{RRx}{IR - II}$

pro nullis haberi possint.

[127]   Coroll. 2 Quinetiam si statuas $NG = y$, erit $\dfrac{RRyy}{2IRa - 2IIa} = KF$ circiter.

Etenim est $NG^q = BG \times \overline{BC + CG}^{(14)}$ sive $= BG \times 2BC$ proximè, hoc est $yy =$

$2ax$ ferè vel $\dfrac{yy}{2a} = x$. et substituto $\dfrac{yy}{2a}$ pro x in valore ipsius *KF* emerget

$\dfrac{RRyy}{2IRa - 2IIa} = KF$.

[128]   Coroll. 3. Hinc errores *KF* sunt ut sagittae *GB* vel ut quadrata semichorda-
rum *NG*.

[129]   Coroll 4. Si radius *ANK* detur positione et paralleli alicujus Axique
prop[r]ioris et ad alteras axis partes incidentis radij *an* refractus *nk* ducatur
secans Axem in *k* et hunc refractum *NK* in *Q*, et ad Axem demittatur normalis *QO*: linea *KO* evadet omnium maxima ubi radius *an* duplo minùs distat
ab Axe circiter quàm radius alter *AN*. Demissa enim ad Axem normalis *ng*
ponatur $= v$, $KO = s$, $GK = f$, et $KF = h$. Et per Coroll. 3 hujus erit

$yy . vv :: KF . kF$, adeoque $kF = \dfrac{hvv}{yy}$, quo a *KF* subducto restat $Kk =$

$\dfrac{hyy - hvv}{yy}$. Praeterea est $GK . GN :: KO . QO$. Adeoque $QO = \dfrac{ys}{f}$. Item

$gn . gk \, (= GK$ proximè$) :: QO . Ok$. Quare $Ok = \dfrac{ys}{v}$. Huic adde *KO* et

iterum prodit $Kk = \dfrac{vs + ys}{v}$. Quamobrem est $\dfrac{vs + ys}{v} = \dfrac{hyy - hvv}{yy}$, factaque

divisione per $v + y$ et reducta aequatione prodit $s = \dfrac{hvy - hvv}{yy}$.

Jam ut maximum s inveniatur multiplica terminos juxta Methodum
Huddenij[15] per dimensiones quantitatis indeterminatae v, et emerget

$0 = \dfrac{hvy - 2hvv}{yy}$, sive $y = 2v$. Hoc est $NG = 2ng$.

[130]   Coroll 5. Et hinc *KO* ubi maximum est, aequatur quartae parti ipsius *KF*
circiter. Nam in valore ipsius s jam antè invento si scribas $2v$ pro y exoritur
$\tfrac{1}{4}h = s$.

[131]   Coroll 6. Est etiam $OQ = \dfrac{Ry^3}{8Iaa}$ [16]. Nam est

$$GK \, (= BF \text{ proximè}) \, . \, GN :: KO . OQ,$$

hoc est $\dfrac{Ra}{R - I}$ [17] $. \, y :: \dfrac{RRyy}{8IRa - 8IIa} \, (= \tfrac{1}{4}KF) \, . \, \dfrac{Ry^3}{8Iaa}$ [16].

aberration see Huygens, *Oeuvres complètes*, 13, i; lii–lxxviii; see also in that volume the "Pars
secunda. De aberratione a foco" of Huygens's *Dioptrica*, which was first (posthumously) published in 1703. On Newton's investigation see Karl Stiegler, "Das Problem der sphärischen
Aberration und seine Lösung durch Isaac Newton. Ein Beitrag zur Geschichte der Theorie der
optischen Instrumente," *Technikgeschichte*, **44** (1977):121–52.

Corollary 1. If *BG*, that is, x, is assumed to be extremely small,   [126]

$$[R^2/I(R - I)]x$$

will be very nearly equal to *KF*. For then the quantities $(R^3/2I^3a)x^2 + (R^5/2I^5a^2)\, x^3 + \dots$ turn out to be exceedingly small because of the increasing powers of x, so that with respect to the term $[R^2/I(R - I)]x$ they can be considered as zero.

Corollary 2. Moreover, if you set $NG = y$, then $R^2y^2/2Ia(R - I) = KF$ [127] approximately. For $NG^2 = BG(BC + CG)^{(14)} \approx BG \times 2BC$, that is, $y^2 \approx 2ax$ or $y^2/2a \approx x$; and substituting $y^2/2a$ for x in the value of *KF*, there will result $R^2y^2/2Ia(R - I) = KF$.

Corollary 3. Hence the errors *KF* are as the versed sines *GB* or as the [128] squares of the half chords *NG*.

Corollary 4. If the ray *ANK* is given in position and for some parallel ray [129] *an*, incident nearer to the axis but on its other side, its refraction *nk* is drawn, intersecting the axis in *k* and this refracted ray *NK* in *Q*; and the normal *QO* to the axis is dropped; the line *KO* will turn out to be the greatest of all when the ray *an* is about half as distant from the axis as the other ray *AN*. For drop the normal *ng* to the axis; set it equal to v, and also $KO = s$, $GK = f$, and $KF = h$. Then by the present Corollary 3, $y^2 : v^2 = KF : kF$, so that $kF = hv^2/y^2$; subtracting this from *KF* there remains $Kk = h(y^2 - v^2)/y^2$. Moreover, $GK : GN = KO : QO$, so that $QO = ys/f$. Likewise $gn : gk\ (\approx GK) = QO : Ok$, whence $Ok = ys/v$. Add *KO* to this, and it once more produces $Kk = s(v + y)/v$. Consequently, $s(v + y)/v = h(y^2 - v^2)/y^2$, and dividing by $v + y$ and reducing the equation, there results $s = h(vy - v^2)/y^2$.

Now to find the maximum of s according to Hudde's method,[15] multiply the terms by the dimensions of the variable quantity v, and there will emerge $0 = h(vy - 2v^2)/y^2$, or $y = 2v$, that is, $NG = 2ng$.

Corollary 5. Hence when *KO* is a maximum, it is approximately equal to [130] one quarter of *KF*. For, if in the value of s found just before you write $2v$ for y, there arises $\frac{1}{4}h = s$.

Corollary 6. Also $OQ = Ry^3/8Ia^2$.[16] For $GK\ (\approx BF) : GN = KO : OQ$, [131] that is, $Ra/(R - I)^{(17)} : y = [R^2y^2/8Ia(R - I)]$ (or $\frac{1}{4}KF) : Ry^3/8Ia^2$.[16]

---

(14) By Euclid, *Elements*, III, 35; see also note (11), this lecture.

(15) Hudde's method is equivalent to setting the derivative $ds/dy = 0$; see also Lects. 16, 17, note (28).

(16) Read (as *editio princeps*): $\dfrac{RRy^3}{8IIaa}$. Newton mistakenly substituted $CF = Ra/(R - I)$ instead of $BF = (I/R)CF = Ia/(R - I)$. He carried over this mistake into Prop. 37 below and thence into the first English (1704) and Latin (1706) editions of the *Opticks* (Bk. I, Pt. I, Prop. VII). He finally caught the error and noted it in the margin of his copy of the *Optice* (1706), pp. [xii], 79 (Harrison, *Library of Isaac Newton* no. 1162), but the slip was not corrected in print until the second English edition (1717). Smith incorporated Newton's derivation of the circle of least confusion in his *Compleat System*, Bk. II, Ch. VI, Prop. IV, 1:137–8. Just a few years earlier than Newton, Huygens arrived at the same result by a similar method; *Oeuvres complètes*, **13**, i: 390–1.

(17) Read (as *editio princeps*): $\dfrac{Ia}{R - I}$.

[132]   Coroll 7. Si arcus *BM* sumatur aequalis *BN*, et *Bm* = *Bn*, ac radij ad puncta *M* et *m* refracti ducantur sibi occurrentes in *P*, constat esse spatium $PQ = \dfrac{Ry^{3\,(18)}}{4Iaa}$ duplum nempe ipsius *OQ*. et praeterea constat refractos omnium radiorum in sphaericam superficiem inter *N* et *M* cadentium convergere in spatium hocce *PQ*, et idem *PQ* esse minimum circulare spatium in quod possent omnes congregari, adeoque focum esse seu locum imaginis objecti parallelos radios in lentem / ad usque limites *M* et *N* apertam ejaculantis. 69/ Scilicet nulli radij possunt transilire hoc spatium, quia cùm *OQ* sit in data ratione ad *KO*, erit *OQ* simul maximum adeoque punctum *Q* omnium versus *F* jacentium remotissimum ab Axe in quo radius quisquam concurrit cum externo radio *NK*. Neque possunt in minus spatium congregari quia radij *nk* et *mk* secant externos radios in ipsissimis punctis *P* et *Q* quibus spatium *PQ* terminatur.

[133]   Coroll 8. Si circuli *NBM* apertura augeatur vel minuatur Error lateralis *PQ* erit ut y^3, sive ut cubus latitudinis aperturae *NM*. Item si immutata apertura mutetur circuli magnitudo error *PQ* erit reciprocè ut *aa* sive ut *CB*q, adeoque ut *BF*q siquidem *CB* et *BF* sint in datâ ratione. Sin vero et circuli magnitudo et apertura mutetur erit error ille *PQ* ut $\dfrac{y^3}{aa}$, sive ut $\dfrac{NM^{cub}}{BF^q}$ quemadmodum ex $\dfrac{Ry^{3\,(18)}}{4Iaa}$ valore istius *PQ* constare potest.

[134]   Schol. Eodem ferè modo quo radiorum parallelè incidentium errores *KF* et *PQ* determinavimus consimiles divergentium vel convergentium errores, licèt calculo difficiliori, determinari possunt.[19]

Lect 14   Prop. 32. Si radij seu paralleli, seu versus commune aliquod punctum
[135]   inclinati[1] se sphaerae objiciant refringendos: refractorum extra axem sibi quam proximorum et in eodem plano cum incidentibus jacentium concursum designare.

   In fig I, 59 sit *AN* incidens radius, *NK* refractus ejus et *NV* in plano

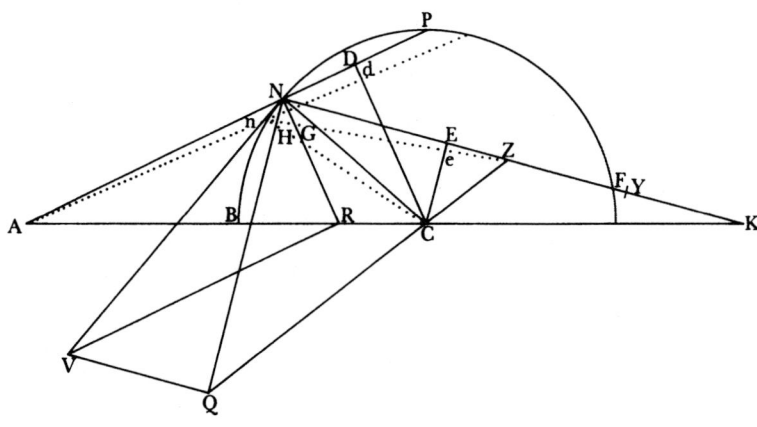

Figure I, 59

Corollary 7. If the arc *BM* is set equal to *BN*, and *Bm* equal to *Bn*, and the   [132]
rays refracted at the points *M* and *m* are drawn meeting one another at *P*, it
is established that the space $PQ = Ry^3/4Ia^2$,[18] that is, twice *OQ*. Furthermore, it is established that the refractions of all rays falling on the spherical
surface between *N* and *M* converge in this space *PQ*, and also that *PQ* is the
smallest circular space into which all rays can be gathered and is consequently the focus or place of the image of an object casting parallel rays onto
a lens opened as far as the limits *M* and *N*. Specifically, no rays can pass
outside this space, because since *OQ* is in a given ratio to *KO*, *OQ* will
simultaneously be a maximum and so the point *Q* is the farthest from the
axis of all those lying toward *F* in which any ray meets the outer ray *NK*.
Nor can they be gathered into a smaller space, because the rays *nk* and *mk*
intersect the outer rays in the very points *P* and *Q* that bound the space *PQ*.

Corollary 8. If the aperture of the circle *NBM* increases or decreases, the   [133]
lateral error *PQ* will be as y^3, or as the cube of the aperture's width *NM*.
Likewise, when the aperture remains unchanged, if the size of the circle
changes, the error *PQ* will be inversely as a^2, that is, as CB^2, and hence as
BF^2, since *CB* and *BF* are in a given ratio. If indeed both the circle's size and
the aperture change, that error *PQ* will be as y^3/a^2, that is, as NM^3/BF^2, as
can be established from $Ry^3/4Ia^2$,[18] the value of *PQ*.

Scholium. In almost the same way as we have determined the errors *KF*   [134]
and *PQ* of parallel incident rays, the quite similar errors of divergent or
convergent rays can be determined, although the calculation is more
difficult.[19]

Proposition 32. If rays that either are parallel or inclined toward some   **Lecture 14**
common point[1] are cast upon a sphere to be refracted, to specify the inter-   [135]
section of the refracted rays lying outside the axis immediately near one
another and in the same plane as the incident ones.
In Fig. I, 59 let *AN* be an incident ray, *NK* its refracted ray, and *NV* a

---

(18) Read (as *editio princeps*): $\dfrac{RRy^3}{4Ilaa}$.

(19) In "Of Refractions," Newton set out without demonstration the measure of the aberration of a single spherical surface for rays incident from a point on the axis, and about 1670 he
returned to revise this calculation a number of times; *Mathematical Papers*, 1:573–4; 3:516–21,
534–9.

(1) versus ... inclinati] Originally: ad punctum aliquod contermini (terminate at some
point). Whiston's extracts from the *Optica* begin with this proposition but omit the concluding
Notes 2 and 3; *Praelectiones*, Problem, pp. 232–4 = *Mathematick Philosophy*, pp. 273–5; see
the Introduction, note (74).

trianguli *ANK* recta linea tangens sphaeram ad *N*. Ad *AN* duc *NR* perpendicularem et occurrentem Axi *AC* in *R*, nec non *RV* parallelam et occurrentem tangenti *NV* in *V*. Item ad *NK* duc *NQ* perpendicularem et *VQ* parallelam convenientes in *Q*. Et age *QC* occurrentem *NK* in *Z*; eritque *Z* concursus radiorum ipsi *AN* vicinissimorum.

Sit enim *An* alius ex incidentibus priori *AN* infinitè vicinus et occurrens *NR* in *G*. Age *nZ* occurrentem *NQ* in *H* et ad *AN* et *NK* e *C* centro sphaerae demitte normales *CD* et *CE* occurrentes *An* et *nZ* in *d* et *e*. Jam cum *AN* supponatur infinitè vicinus *An* arcus infinitè parvus *Nn* pro recta coincidente cum tangente *NV* haberi potest, ac triangula *NGn*, *NRV*, ut et *NHn*, *NQV* pro similibus. Quare est

$$DC . Dd(:: NR . NG :: NV . Nn :: NQ . NH) :: EC . Ee,$$

et converse *DC* . (*DC* − *Dd*) *dC* :: *EC* . (*EC* − *Ee*) *eC*. et vicissim *DC* . *EC* :: *dC* . *eC*. Est autem *DC* ad *EC* ut sinus incidentiae ad sinum refractionis propterea quod *NK* sit refractus ipsius *AN*: adeoque etiam *dC* ad *eC* est ut sinus / incidentiae ad sinum refractionis. Et proinde cum anguli *DAd* et *EZe* 70/ sint infinite parvi atque adeo *Cd* ad *An* et *Ce* ad *nZ* perpendiculares vel saltem perpendiculis aequipollentes erit *nZ* refractus ipsius *An*. Q.E.D.[2]

[136]   Coroll. 1. Est *ND* . *NE* (sive *NP* . *NF*) :: *NR* . *NQ*. Nam actâ *NC* propter triang. *NDC* sim triang. *NRV* et triang *NEC* sim triang *NQV*, est *ND* . *NR* (:: *NC* . *NV*) :: *NE*. *NQ* et inversè *ND* . *NE* :: *NR* . *NQ*.

Hinc promptior emergit Problematis resolutio: nempe ad radios *AN*, *NK* erige normales *NR*, *NQ*, quorum *NR* axi *AC* occurrat, et *NQ* sit ad *NR* ut *NF* ad *NP*. Dein Age *QC* quae cum *NK* in quaesito puncto *Z* conveniet.[3]

[137]   Coroll 2. Est etiam $AN \times DC \times NE . AD \times EC \times ND :: NZ . EZ$. nam

est $AD . AN :: DC . NR$. et inde $NR = \dfrac{AN \times DC}{AD}$. Item

$$ND . NE :: NR . NQ$$

et inde $NQ = \dfrac{AN \times DC \times NE}{AD \times ND}$. Adeoque

$$AN \times DC \times NE . AD \times ND \times EC(:: NQ . EC) :: NZ . EZ.^{[4]}$$

---

(2) Newton's straightforward determination of the primary image point for rays in the plane of incidence was preceded by Barrow's equivalent solution in *Lectiones XVIII*, Lect. XIII, §XXIV, pp. 116–7. Finding the image point is equivalent to determining the caustic, for the refracted rays are tangent to the caustic curve at *Z* where infinitely close rays intersect. In 1693 Jakob Bernoulli published without demonstration his construction of the diacaustic locus (*Z*), and claiming full priority for himself, he unfairly minimized the earlier contributions of Huygens (*Traité de la lumière*, Ch. VI, *Oeuvres complètes*, 19:534–7) and Barrow as only special cases— limited to circles and, for Huygens, also to parallel rays—of his own general solution; "Curvae dia-causticae, earum relatio ad evolutas, aliaque nova his affinia," *Acta eruditorum* (1693):244– 9. Three years later L'Hospital published a clear and direct demonstration of Bernoulli's solution using the differential calculus and arrived independently at Barrow's and Newton's result; *Analyse des infiniment petits, pour l'intelligence des lignes courbes*, [1696], 2nd ed. (Paris, 1715), Pt. I, Sect. VII, §133, pp. 121–3; on early investigations of caustics see *Mathematical Papers*, 3:490, note 25. Smith, in his *Compleat System*, Bk. II, Ch. IX, preserved and extended Barrow's and Newton's approach, while clearly establishing the relation of the study of caustics to the determination of the primary image point.

straight line in the plane of the triangle *ANK* tangent to the sphere at *N*. To *AN* draw the perpendicular *NR* meeting the axis *AC* in *R*, and also parallel to it draw *RV* meeting the tangent *NV* in *V*. Likewise to *NK* draw the perpendicular *NQ* and the parallel *VQ* meeting each other in *Q*. Then draw *QC* meeting *NK* in *Z*, and *Z* will be the intersection of the rays closest to *AN*.

For let *An* be another incident ray infinitely close to the first one *AN* and meeting *NR* in *G*. Draw *nZ* meeting *NQ* in *H*, and to *AN* and *NK* from the sphere's center *C* drop the normals *CD* and *CE* meeting *An* and *nZ* in *d* and *e*. Now since *AN* is assumed to be infinitely close to *An*, the infinitely small arc *Nn* can be considered as a straight line coincident with the tangent *NV*, and the triangles *NGn* and *NRV* as well as *NHn* and *NQV* as similar. Therefore, *DC* : *Dd* (= *NR* : *NG* = *NV* : *Nn* = *NQ* : *NH*) = *EC* : *Ee*, and by conversion *DC* : (*DC* − *Dd* or) *dC* = *EC* : (*EC* − *Ee* or) *eC*, and by permutation *DC* : *EC* = *dC* : *eC*. *DC* is to *EC*, however, as the sine of incidence is to the sine of refraction because *NK* is the refracted ray of *AN*, and so also *dC* is to *eC* as the sine of incidence is to the sine of refraction. Consequently, since the angles *DAd* and *EZe* are infinitely small, so that *Cd* is perpendicular to *An*, and *Ce* to *nZ* (or at least equivalents to perpendiculars), *nZ* will be the refracted ray of *An*. As was to be demonstrated.[2]

Corollary 1. It is *ND* : *NE* (or *NP* : *NF*) = *NR* : *NQ*. For when *NC* is [136] drawn, because the triangle *NDC* is similar to the triangle *NRV*, and the triangle *NEC* to the triangle *NQV*, it is *ND* : *NR* (= *NC* : *NV*) = *NE* : *NQ*, and permuting, *ND* : *NE* = *NR* : *NQ*.

Hence a handier solution of the problem emerges: Namely, to the rays *AN* and *NK* erect the normals *NR* and *NQ*, letting *NR* meet the axis *AC* and *NQ* be to *NR* as *NF* to *NP*. Then draw *QC*, which will meet *NK* in the required point *Z*.[3]

Corollary 2. It is also *AN* × *DC* × *NE* : *AD* × *EC* × *ND* = *NZ* : *EZ*. [137] For, *AD* : *AN* = *DC* : *NR*, and thus *NR* = *AN* × *DC*/*AD*. Likewise, *ND* : *NE* = *NR* : *NQ*, and thus *NQ* = *AN* × *DC* × *NE*/(*AD* × *ND*). Consequently,

$$AN \times DC \times NE : AD \times ND \times EC (= NQ : EC) = NZ : EZ.^{(4)}$$

(3) In the *Lectiones XVIII* (Lect. XIII, §XXVI, p. 117) Barrow presented Newton's construction which, he relates, was "communicated by a friend and discovered by him by another method and elegantly demonstrated."

(4) Let $D\hat{N}C = i$, $E\hat{N}C = r$, $NC = \rho$, $AN = s$, and $NZ = s'$; and take *B* as the origin of the coordinate system, so that $DC/EC = \sin i/\sin r$, $NE/ND = \cos r/\cos i$, $AN/AD = s/(s - \rho \cos i)$, and $NZ/EZ = s'/(s' - \rho \cos r)$; then Newton's determination of the primary image point becomes

$$\cot r (1 - \frac{\rho}{s'} \cos r) = \cot i (1 - \frac{\rho}{s} \cos i),$$

or

$$1 = \frac{\rho \sin r \cos^2 i}{s \sin(i - r)} - \frac{\rho \sin i \cos^2 r}{s' \sin(i - r)}.$$

Newton's preceding analysis for these two corollaries survives; Add. 4004, insert between ff. 72, 73 = *Mathematical Papers*, 3:532.

[138]   Coroll 3. Si punctum radians *A* infinitè distet, sive parallelos radios ejaculetur: posito *I . R* :: sin: incid: . sin: refract: erit *I* × *NF . R* × *NP* :: *NZ . EZ*. In hoc enim casu *AN* et *AD*, cum sint infinitè longae pro aequalibus haberi debent. Atque adeò per Coroll 2 hujus erit

$$DC \times NE . EC \times ND :: NZ . EZ.$$

Sed ex Hypothesi est *DC . EC* :: *I . R*. et proinde

$$I \times NE . R \times ND (:: NZ . EZ) :: NP . NF.^{(5)}$$

Caeterùm de his vide plura in Lectionibus D^{ris} Barrow.

[139]   Notetur autem 1. Quod mutatis mutandis resolutio Problematis cuicunque casui facilè accommodatur, sive radij incidentes divergant a puncto aliquo vel ad idem convergant vel incidant paralleli.

2. Cum e radijs huic *ANK* proximis, qui jacent in plano *ANR* conveniant in *Z*, qui vero in Conicâ superficie per re[v]olutionem trianguli *ANK* circa latus *AK* generatâ jacent, conveniant in *K*; erit maxima radiorum ipsi *ANK* undique proximorum constipatio circa medium spatij *KZ*; puta ad *Y*. Et proinde oculo in linea *NK* ultra *K* constituto, sensibilis imaginis objecti *A* per refractionem sphaericae superficiei *BN* visi locus erit ad *Y* vel saltem intra limites *K* et *Z*. Nam locus ille non praecisè definitur.$^{(6)}$

3. Cum radij pluribus superficiebus successive refringuntur ut vicinorum post omnes refractiones concursum determines; / primo quaere concursum 71/ post primam refractionem, deinde concursum eorundem post secundam tanquam si primariò effluxissent e puncto praecedentis concursus: Et sic deinceps ut ad Prop 29 dictum fuit.$^{(7)}$

[140]   Prop. 33. Radijs in quamcunque curvam superficiem incidentibus refractorum sibi quàm proximorum, et in eodem plano cum incidentibus jacentium concursum designare.

In fig I, 59 finge *BNP* jam non sphaeram sed aliam quamcunque curvam referre. Sitque *A* commune punctum seu concursus incidentium radiorum, *AN* aliquis ex incidentibus, *NK* refractus ejus, et *NC* perpendicularis Curvae ad punctum refringens. In hac *NC* quaere intersectionem proximi alicujus perpendicularis (qualis *nC*) ad aliud proximum punctum refringens insistentis. Id quod alibi docebitur:$^{(8)}$ Sitque ista intersectio *C*. Jam ductâ *AC*, demitte

(5) *NP.NF.*] Read (as *editio princeps*): *I* × *NF . R* × *NP*. For parallel incident rays, or *s* = ∞, this defines the primary focal point *f'* = *ρ* sin *i* cos²*r*/sin (*i* − *r*); for Barrow's derivation see his *Lectiones XVIII*, Lect. XII, §VII, pp. 103–4.

(6) Newton has extended his insight into the location of the image for plane surfaces to spherical surfaces; see Prop. 3, Scholium, §129 ≈ I, 62, especially note (19). Although Smith did not include this passage in his *Compleat System*, in his translation of the equivalent passage for plane surfaces, he noted that it is "manifestly applicable" to spherical surfaces ("General Remarks," ibid., §494, 2:82). The subject of astigmatism remained essentially untouched until it was taken up again by Young, who protested—to no avail—that the work of Newton and Smith had been "too little noticed" ("On the mechanism of the eye," *Phil. Trans.*, **91** (1801):23–88, esp. 30). In his investigation of the eye's astigmatism, among other significant contributions Young first recognized that the focal points were lines and described the changing cross-section of an incident cylindrical pencil of rays in the region of *Z* and *K*; "On the mechanism of the

Corollary 3. If the radiant point *A* is infinitely distant, that is, if it casts [138] parallel rays, on setting *I* to *R* to be as the sine of incidence to the sine of refraction, then $I \times NF : R \times NP = NZ : EZ$. For in this case since *AN* and *AD* are infinitely long, they must be considered as equal. By the present Corollary 2 then, $DC \times NE : EC \times ND = NZ : EZ$. But by hypothesis $DC : EC = I : R$, and accordingly, $I \times NE : R \times ND (= NZ : EZ) = NP : NF$.[5] However, for more about these matters see Dr. Barrow's *Lectures*.

But let it be noted that: [139]

1. By making the necessary changes, the solution of the problem is easily accommodated to any case, whether the incident rays diverge from some point or converge to the same point or are incident parallel to one another.

2. Since of rays closest to this one, *ANK*, those that lie in the plane *ANR* meet in *Z*, but those that lie in the conical surface generated by the revolution of the triangle *ANK* around its side *AK* meet in *K*; the greatest crowding of rays closest to *ANK* on each side will be near the middle of the space *KZ*, that is, at *Y*. Consequently, when an eye is placed on the line *NK* beyond *K*, the place of the sensible image of the object *A* seen by the refraction of the spherical surface *BN* will be at *Y*, or at least within the bounds of *K* and *Z*, for that place is not precisely defined.[6]

3. When rays are successively refracted by several surfaces so that you may determine the intersection of the neighboring rays after all the refractions, first seek the intersection after the first refraction and then their intersection after the second, just as if they had flowed originally from the point of the previous intersection, and so on, as was described in Prop. 29.[7]

Proposition 33. When rays are incident on any curved surface, to specify [140] the intersection of the refracted rays lying immediately near one another and in the same plane as the incident rays.

In Fig. I, 59 imagine *BNP* no longer to represent a sphere but any curve whatsoever. Let *A* be the common point or intersection of the incident rays, *AN* any one of the incident rays, *NK* its refracted ray, and *NC* the normal to the curve at the refracting point. In this normal *NC* find its intersection with some other very close normal (such as *nC*) constructed at another very close refracting point (this construction will be explained elsewhere),[8] and let that intersection be *C*. Now after drawing *AC*, drop the normals *CD* and *CE* to

---

eye," Prop. IV, Scholium 4, ibid., p. 30. Young omitted his eight "dioptrical propositions" in the reprint of this paper in his *A Course of Lectures on Natural Philosophy and the Mechanical Arts*, 2 vols. (London, 1807), but in the "Mathematical elements of natural philosophy" appended to his *Lectures* he added a new series of theorems extending the study of astigmatism still further (ibid., 2:73–6). See Marius H. E. Tscherning's excellent annotated translation, *Oeuvres ophtalmologiques de Thomas Young* (Copenhagen, 1894); and John R. Levene, "Sir George Biddell Airy, F.R.S. (1801–1892) and the discovery and correction of astigmatism," *Notes and Records of the Royal Society of London*, 21(1966):180–99; and *Clinical Refraction and Visual Science* (London/Boston, 1977).

(7) Smith successfully applied this approach to determining the image point after any number of refractions; *Compleat System*, Bk. II, Props. VII–X, 1:167–71.

(8) Namely, in Problem 5 of his treatise "De methodis serierum et fluxionum" (*Mathematical Papers*, 3:150), which he intended to publish together with his *Optical Lectures*.

ad radios *AN, NE* normales *CD, CE*; ac erige *NR, NQ*; quorum *NR* occurrat *AC* in *R*. Sitque *NQ* ad *NR* ut *NE* ad *ND*, et acta *QC* conveniet cum refracto *NK* in desiderato proximorum refractorum concursu *Z*.

Probatur ad modum praecedentis Propositionis: et huic etiam consimilia Corollaria et Notae competunt.[9]

[141]   Prop. 34. Figuram determinare quae radios Homogeneos sive parallelos sive ad commune aliquod punctum terminatos, ita refringet ut refracti omnes ad aliud datum punctum accuratè conveniant.[10]

In fig I, 60, Sit *A* concursus incidentium radiorum, et *Z* refractorum, ac punctum aliquod *B* in recta *AZ* pro vertice curvae ad arbitrium sumatur. Ab illo *B* capiantur in linea *BZ* versus medium densius *BI* cujusvis longitudinis et *BR* in ratione ad *BI* quam habet sinus incidentiae ad sinum refractionis.[11] Centrisque *A* et *Z* et intervallis *AI* et *ZR* describantur circuli se intersecantes in *N* et ipsius *N* locus erit curva quae desideratam refractionem peraget.[12]

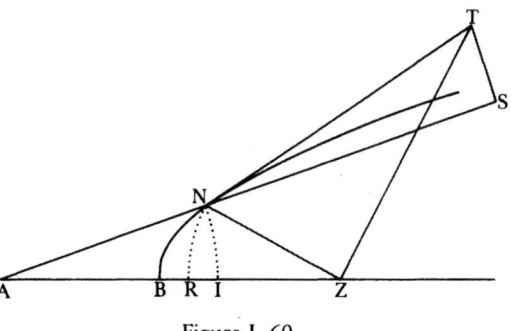

Figure I, 60

Quod ut pateat producatur *AN* ad *S* ut sit *NS . NZ :: BI . BR*. et ad *NS* et *NZ* erigantur perpendiculares *ST* et *ZT* concurrentes in *T* et acta *NT* curvam tanget in *N* ut ex methodo ducendi tangentes alibi exposita constabit.[13] Jam cum *NS* et *NZ* sint ut *BI* et *BR* hoc est ut sinus incidentiae et refractionis; et respectu sinus totiûs sive semidiametri *NT*, sit *NS* sinus anguli *NTS* qui aequatur angulo incidentiae radij *AN*, et *NZ* sinus anguli *NTZ* qui aequatur angulo refractionis radij *NZ* patet esse *NZ* refractum ipsius *AN*.   Q.E.D.[14]

[142]   / Nota 1. Potest etiam Curva huic usui inserviens describi quae per datum quodvis punctum *B* extra axem *AZ* positum transibit. Scilicet in fig I, 61 agantur *AB* et *ZB* et in ipsis capiantur *BI* et *BR* in ratione sinuum incidentiae et refractionis. Et centris *A* et *Z* ac intervallis *AI* et *ZR* describantur circuli concurrentes in *N* eritque *N* ad curvam quam oportet describere.[15]

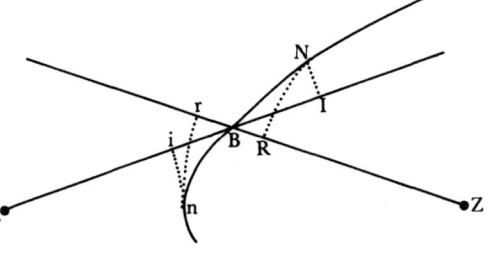

Figure I, 61

72/

---

(9) See Lecture I, 13, note (10).

(10) In Bk. II of his *Geometrie* (*Oeuvres*, 6:424–41) Descartes posed and solved this problem with what we now—following Newton—call a Cartesian oval, that is defined by the equation $(AN - AB)/(BZ - NZ) = I/R$. Descartes identified four varieties of oval, depending on whether the vertex *B* lies within or outside the points *A* and *Z*, and whether the index of refraction is positive or negative; see Joseph F. Scott, *The Scientific Work of René Descartes (1596–1650)*

the rays *AN* and *NE*, and erect *NR* and *NQ*, letting *NR* meet *AC* in *R*, and *NQ* be to *NR* as *NE* to *ND*. Then when *QC* is drawn, it will meet the refracted ray *NK* in *Z*, the required intersection of the nearest refracted rays.

This is proved in the manner of the preceding proposition, and quite similar notes and corollaries also apply to it.[9]

Proposition 34. To determine the figure that will refract homogeneous [141] rays, whether they are parallel or terminate at some common point, such that all the refracted rays meet exactly at some given point.[10]

In Fig. I, 60 let *A* be the intersection of the incident rays and *Z* that of the refracted ones, and arbitrarily choose some point *B* in the line *AZ* as the vertex of the curve. From that point *B* in the line *BZ* toward the denser medium take any length *BI* and also *BR* to *BI* in the ratio of the sine of incidence to the sine of refraction.[11] Then with centers *A* and *Z* and radii *AI* and *ZR* describe circles intersecting each other at *N*, and the locus of *N* will be the curve that brings about the desired refraction.[12]

For this to be evident extend *AN* to *S* such that *NS* : *NZ* = *BI* : *BR*; and to *NS* and *NZ* erect the perpendiculars *ST* and *ZT* meeting in *T*; then when *NT* is drawn it will be tangent to the curve at *N*, as will be evident from the method of drawing tangents set forth elsewhere.[13] Now, since *NS* and *NZ* are as *BI* and *BR*, that is, as the sines of incidence and refraction; and since with respect to the whole sine (that is, the radius *NT*), *NS* is the sine of the angle *NTS*, which is equal to the angle of incidence of the ray *AN*, and *NZ* is the sine of the angle *NTZ*, which is equal to the angle of refraction of the ray *NZ*; it is evident that *NZ* is the refraction of *AN*. As was to be demonstrated.[14]

Note 1. A curve can also be described for this purpose that will pass [142] through any given point *B* located off the axis *AZ*. Namely, in Fig. I, 61 draw *AB* and *ZB*, and in them take *BI* and *BR* to be in the ratio of the sines of incidence and refraction. Then with centers *A* and *Z* and radii *AI* and *ZR*, describe circles meeting in *N*, and *N* will be on the curve that it is required to describe.[15]

(London, [1952]), pp. 121–33. Newton had begun his investigations of Cartesian ovals by September 1664; see *Mathematical Papers*, 1:551–8.

(11) incidentiae . . . refractionis] Read (as *editio princeps*): refractionis . . . incidentiae (of refraction . . . of incidence).

(12) It is evident that the point *N* satisfies Descartes's defining condition, for by construction *BI/BR* = (*AN* − *AB*)/(*BZ* − *NZ*) = *I/R*.

(13) In Problem 4, Mode 3 of his "De methodis serierum"; *Mathematical Papers*, 3:136–9.

(14) Newton published a variant of this solution in the *Principia* (Bk. I, Sect. XIV, Prop. XCVII), where he formulated it in terms of the emission theory of light. In 1676 or 1677, a little later than Newton, Huygens proposed his own solution to this problem based on his wave theory of light: If a wave front diverges from *A* and converges to *Z*, then the time for its rays to traverse *AN* + *NZ* and *AB* + *BZ* must be equal, or—introducing the concept of optical path— *AN* + (*I/R*)*NZ* = *AB* + (*I/R*)*BZ*, which defines a Cartesian oval; see *Traité de la lumière*, Ch. VI, *Oeuvres complètes*, **19**:524–30.

(15) The point *N* clearly satisfies Descartes's defining condition, for by construction *BI/BR* = (*AN* − *AB*)/(*BZ* − *NZ*) = *I/R*.

2. Praefata Problematis resolutio mutatis mutandis se ad omnes casus extendit, sive incidentes aut refracti radij convergant, divergant vel existant paralleli, sive refractio fiat e rariori Medio in densius, vel e densiori in rarius. Et quidem si radij ex neutra parte paralleli sint i.e. si punctorum A et Z neutrum sit ad infinitam distantiam, Curva BN erit aliqua quatuor Ellipsium quas Cartesius in hunc usum in Geometria descripsit. Sin alterutrum infinitè distet, ita ut radij punctum illud respicientes evadant paralleli, Curva erit Conica sectio, uti notum est.[16] Et in hoc casu circulus RN vel IN propter infinitam centri distantiam evadet recta linea ipsi AZ ad R vel I perpendicularis.

[143]   Lemma 10. E parallelis radijs ad circulum refractis, radium illum determinare, cujus pars circulo inclusa datam habeat rationem ad partem refracti ejus eidem circulo inclusam.[17]

In fig I, 62 sit AN radius incidens NK refractus NP et NF partes eorum circulo inclusae, CD et CE perpendicula ad istas partes e centro circuli demissa, et BC semidiameter acta parallela ipsi AN. Sitque $CD \cdot CE :: I \cdot R$. et $NP \cdot NF$ :: $p \cdot q$. His positis ut innotescat punctum N quod radios AN & NK determinat, erige ad BC normalem BX cujus quadratum sit ad BC qua-

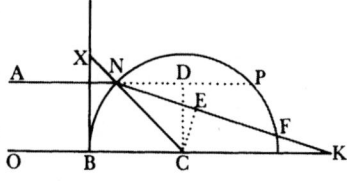

Figure I, 62

dratum ut $\dfrac{qq - pp}{pp}$ ad $\dfrac{II - RR}{II}$ [18] et acta CX secabit circulum in desiderato N.

Est enim ex Hypoth: $p \cdot q (:: NP \cdot NF) :: ND \cdot NE$. et $I \cdot R :: CD \cdot CE$; Quare

$$\frac{q}{p} ND = NE, \text{ et } \frac{R}{I} CD = CE.$$

Porro cùm sit $ND^q + CD^q (= NC^q) = NE^q + CE^q$, aufer hinc inde $ND^q + CE^q$ et restabit $CD^q - CE^q = NE^q - ND^q$. Hoc est substituendo valores CE et NE modo inventos, $CD^q - \dfrac{RR}{II} CD^q = \dfrac{qq}{pp} ND^q - ND^q$: et facta reductione

$\dfrac{II - RR}{II} CD^q = \dfrac{qq - pp}{pp} ND^q$. Quo in proportionalitatem resoluto fit

$\dfrac{qq - pp}{pp} \cdot \dfrac{II - RR}{II} (:: CD^q \cdot ND^q) :: BX^q \cdot BC^q$.   Q.E.D.

[144]   Prop. 35. Sole sphaeram pellucidam illustrante radiorum ejus post unam reflexionem emergentium maximam ad axem inclinationem determinare.

/ In fig I, 63 sit BNK sphaera proposita, $BC[Q]$ diameter sive axis incidentibus radijs parallelus, AN aliquis ex incidentibus, NF refractus ejus, FG reflexus et GR denuò refractus, et quaerendus erit maximus angulorum quos RG cum axe BQ potest conficere.[19]   73/

---

(16) In Discourse VIII of his *Dioptrique*, Descartes considered parallel incident rays that yield the simplest of his ovals, elliptical and hyperbolical lenses.

(17) This lemma and the following two propositions on the rainbow complete Whiston's extracts from the *Optica*; *Praelectiones*, Problems 2–4, pp. 234–8 = *Mathematick Philosophy*, pp. 275–80.

Note 2. When the necessary changes are made, this solution of the problem extends to all cases, whether the incident or refracted rays converge, diverge, or are parallel, or whether the refraction is made from a rarer medium into a denser one, or from a denser medium into a rarer one. In fact, if the rays are not parallel on either side, that is, if neither of the points *A* and *Z* is infinitely distant, the curve *BN* will be one of the four ovals that Descartes described for this purpose in the *Geometry*. But if either point is infinitely distant, so that the rays with respect to that point turn out parallel, the curve will be a conic section, as is known.[16] In this case, because of the infinite distance to its center, the circle *RN* or *IN* will turn out to be a straight line perpendicular to *AZ* at *R* or *I*.

Lemma 10. Of parallel rays refracted at a circle, to determine that ray [143] whose part included by the circle has a given ratio to the part of its refracted ray included by the same circle.[17]

In Fig. I, 62 let *AN* be the incident ray, *NK* its refracted ray, *NP* and *NF* their parts included by the circle, *CD* and *CE* perpendiculars to these parts dropped from the circle's center, and *BC* a radius drawn parallel to *AN*. Then let $CD : CE = I : R$ and $NP : NF = p : q$. With these assumptions, in order to discover the point *N* that determines the rays *AN* and *NK*, to *BC* erect the normal *BX*, and let its square be to *BC* squared as $(q^2 - p^2)/p^2$ to $(I^2 - R^2)/I^2$;[18] and when *CX* is drawn, it will intersect the circle in the desired point *N*. For by hypothesis $p : q \; (= NP : NF) = ND : NE$ and $I : R = CD : CE$; consequently, $(q/p)ND = NE$ and $(R/I)CD = CE$. Moreover, since $ND^2 + CD^2 \; (= NC^2) = NE^2 + CE^2$, subtract $ND^2 + CE^2$ from both sides, and $CD^2 - CE^2 = NE^2 - ND^2$ will remain. That is, upon substituting the values of *CE* and *NE* just found,

$$CD^2 - (R^2/I^2)CD^2 = (q^2/p^2)ND^2 - ND^2$$

and, making a reduction,

$$[(I^2 - R^2)/I^2] \; CD^2 = [(q^2 - p^2)/p^2] \; ND^2.$$

Resolving this into a proportion, it becomes

$$(q^2 - p^2)/p^2 : (I^2 - R^2)/I^2 \; (= CD^2 : ND^2) = BX^2 : BC^2.$$

As was to be demonstrated.

Proposition 35. When the sun shines on a transparent sphere, to determine [144] the greatest inclination to the axis of its rays emerging after one reflection.

In Fig. I, 63 let *BNK* be the proposed sphere, *BCQ* its diameter or axis parallel to the incident rays, *AN* one of these incident rays, *NF* its refraction, *FG* its reflection, and *GR* its second refraction; and the greatest angle that *RG* can make with the axis *BQ* will have to be found.[19]

---

(18) Note that $NP/NF = \cos i/\cos r = p/q$, $CD/CE = \sin i/\sin r = I/R$ and $BX/BC = \tan i$, where $D\widehat{N}C = i$ and $E\widehat{N}C = r$.

(19) Both Harriot and Descartes discovered that the primary or interior rainbow is seen when the inclination of the emergent ray *RG* with the incident one *AN* is a maximum. Whereas Harriot went on to deduce from this condition that the tangent of the angle of incidence is then

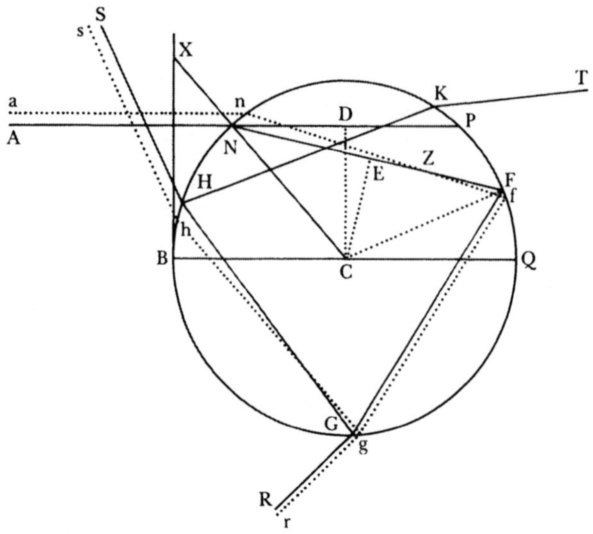

Figure I, 63

In quem finem advertendum est quod in eo solo casu ubi RG maximè inclinatur ad BQ, radij ipsi AN vicinissimi possunt emergere paralleli ad RG.[20] Nam in alijs casibus ex emergentibus radijs sibi vicinissimis alij magis alijs continuò inclinantur ad BQ, adeoq́ue aliquantulum inclinantur ad se invicem.

Advertendum est praeterea quod radij emergent paralleli qui conveniunt ad punctum reflexionis. Duc enim radium an ipsi AN parallelum et quàm proximum sitque ejus refractus nf reflexus fg ac iterum refractus gr. Et punctis F et f coincidentibus, cùm anguli NFn & GFg sint aequales et refractiones ad Nn et Gg similes, emergentes radij GR et gr aequè paralleli erunt ac incidentes AN et an.[21]

Quaerendus est itaque radius AN cújus refractus cum refracto vicinissimi radij an concurrit ad F. Et quidem per Coroll. 3 Prop. 32 (demissis a centro sphaerae ad radios normalibus CD et CE, positoque $I \,.\, R :: CD \,.\, CE$) si radij isti ad quodvis punctum Z concurrant, erit $I \times NF \,.\, R \times NP$ ($:: NZ \,.\, EZ :: NF \,.\, EF$, puncto nempe Z ad ipsum F juxta Hypoth. cadente) $:: 2 \,.\, 1$. Quare $I \times NF = 2R \times NP$, et $I \,.\, 2R :: NP \,.\, NF$. Datur itaque ratio NP ad NF et inde per Lemma 10 dabitur punctum N. Scilicet ad verticem circuli ducatur tangens BX cujus quadratum sit ad quadratum semidiametri BC ut $4RR - II$ ad $II - RR$ et agatur CX: haec enim circulo occurret in N. et ex invento N caetera nullo negotio determinantur.

[145] Coroll 1. Hinc fit $3RR \,.\, II - RR :: CN^q \,.\, ND^q$. cùm enim sit

$$4RR - II \,.\, II - RR :: BX^q \,.\, BC^q$$

componendo erit $3RR \,.\, II - RR$ ($:: CX^q \,.\, BC^q$) $:: CN^q \,.\, ND^q$.

[146] Coroll. 2. Est et $I \,.\, 2R :: ND \,.\, NE$. nam supra fuit $I \,.\, 2R :: NP \,.\, NF$. et ex his e[x]peditior evadit problematis resolutio.[22]

For this purpose it must be noted that in that case alone, when *RG* is most inclined to *BQ*, can the rays closest to *AN* emerge parallel to *RG*.[20] For in other cases of rays emerging closest to each other some are always more inclined to *BQ* than others, so that they are somewhat inclined to each other.

It must be noted in addition that those rays will emerge parallel that meet at the point of reflection. For draw the ray *an* parallel to *AN* and immediately near it, and let *nf* be its refraction, *fg* its reflection, and *gr* its second refraction. When the points *F* and *f* coincide, since the angles *NFn* and *GFg* are equal and the refractions at *Nn* and *Gg* are similar, the emergent rays *GR* and *gr* will be just as parallel as the incident ones *AN* and *an*.[21]

Therefore the ray *AN* whose refraction meets the refraction of an immediately neighboring ray *an* at *F* must be found. In fact, by Corollary 3, Prop. 32 (after dropping the normals *CD* and *CE* from the center of the sphere to the rays and setting $I : R = CD : CE$), if these rays meet at any point *Z*, then $I \times NF : R \times NP$ ($= NZ : EZ = NF : EF$, namely when by hypothesis the point *Z* falls at *F* itself) $= 2 : 1$. Consequently, $I \times NF = 2R \times NP$ and $I : 2R = NP : NF$. The ratio of *NP* to *NF* is therefore given, and thus by Lemma 10 the point *N* will be given. Specifically, at the circle's vertex draw the tangent *BX*, letting its square be to the square of the radius *BC* as $4R^2 - I^2$ is to $I^2 - R^2$, and draw *CX*; for this line will meet the circle in *N*. Once *N* has been found, the rest is determined with no effort.

Corollary 1. Hence, $3R^2 : (I^2 - R^2) = CN^2 : ND^2$. For since [145]

$$(4R^2 - I^2) : (I^2 - R^2) = BX^2 : BC^2,$$

by composition $3R^2 : (I^2 - R^2) (= CX^2 : BC^2) = CN^2 : ND^2$.

Corollary 2. Also $I : 2R = ND : NE$, for above there was $I : 2R = NP : NF$. [146] From these the solution of the problem proves to be more convenient.[22]

---

twice the tangent of the angle of refraction, Descartes ponderously carried out twenty-seven separate calculations at different angles of incidence on a spherical raindrop to "learn at what angles light could come to our eyes after two refractions and one or two reflections; I found that after one reflection and two refractions very many more could be seen at an angle of 41° to 42° than at any smaller one, and that none at all were visible at a larger angle" (*Meteora*, Ch. VIII, §9, *Oeuvres* 6:705, 336). Newton's introduction to the mathematical theory of the rainbow came directly from Descartes's *Meteora*, as is evident from his notes on page 60 of Hooke's *Micrographia* (Add. 3958, f. 1ᵛ = Newton, *Unpublished Papers*, p. 403) and his very sketchy account in "Of Colours" (§§51–3, Add. 3975, f. 8ᵛ = *Mathematical Papers*, 3:543–4). Since Newton's angle of inclination $\widehat{FCQ} = \widehat{FQ} = 2r - i$ (or $\frac{1}{2}\widehat{AXR}$ in Fig. II, 61), where $\widehat{DNC} = i$ and $\widehat{ENC} = r$, is the supplement of half the deviation, $D = \pi - 2(2r - i)$, he requires that it be a maximum rather than a minimum. On the history of the rainbow see Boyer, *Rainbow*; and, especially for Harriot, Lohne, "Regenbogen und Brechzahl," *Sudhoffs Archiv für Geschichte der Medizin und der Naturwissenschaften*, **49** (1965):401–15.

(20) The infinitesimally close rays emerge effectively parallel when their inclination to *BQ* is a (local) extremum, since the variation of that inclination is then momentarily nil.

(21) Newton's demonstration that at maximum inclination the rays emerge parallel depends on the symmetric situation of the refracted and reflected rays when *Z*, *F*, and *f* coincide and does not go significantly beyond Barrow's *Lectiones XVIII*, Lect. XII, §§XII–XVI, pp. 105–7. Newton's argument, however, can be made rigorous and general; see *Mathematical Papers*, 3:506, note 55.

(22) Since $CN/ND = 1/\cos i$ and $ND/NE = \cos i/\cos r$, these corollaries, deduced earlier by Barrow (*Lectiones XVIII*, Lect. XII, §XI, p. 105), readily yield the rainbow's radii; see §II, 137.

[147]   Schol. Una cum maxima inclinatione radij *RG* datur maximus arcuum *FQ* ad refractos *NF* terminatorum. Nam angulus *FCQ* quem *FQ* subtendit est aequalis angulo quem *CF* et *AN* comprehendunt, hoc est aequalis dimidio anguli, quem *RG* et *AN* vel *BQ* comprehendunt. et proinde arcuum *FQ* aequè ac angulorum ab *RG* et *BQ* comprehensorum maximus est qui radio *AN* in punctum jam inventum incidente definitur.[23]

Lect 15   / Prop. 36. Sole sphaeram pellucidam illustrante, radiorum ejus post duas 74/
[148]   reflexiones emergentium minimam ad axem inclinationem determinare.[1]

Sint *AN* et *an* (in fig I, 63) radij duo incidentes sibi quam proximi qui post duas reflexiones in *Ff* et *Gg* emergant secundum *HS* et *hs*. Et manifestum est quod in eo solo casu ubi acutus angulus quem *BQ* et *SH* comprehendunt minimus est, radij illi *HS* et *hs* possunt esse paralleli, uti supra de radijs *GR* et *gr* dictum fuit.[2] Et ubi hoc accidit radius etiam *FG* ad *fg* parallelus erit. Unde 2arc *Ff* (= arc *Ff* + *Gg* = arc *FG* − *fg* = arc *NF* − *nf*) = arc *Nn* − *Ff*, adeoque 3arc *Ff* = arc *Nn*. et cùm *NF* dividatur in *Z* in ratione istorum arcuum ut patet, erit *NZ* = 3*ZF*, seu 3*EZ*.[3] Cùm itaque per Coroll 3 Prop 32 sit $I \times NF \cdot R \times NP :: NZ \cdot EZ$ sive :: 3.1, erit $I \times NF = 3R \times NP$, sive $I \cdot 3R :: NP \cdot NF$. Datur itaque ratio *NP* ad *NF* et inde per Lem. 10 dabitur punctum *N*, ducendo nempe *BX* quae circulum tangat in vertice *B* et cujus quadratum sit ad *BC*^{quadr.} ut 9*RR* − *II* ad *II* − *RR*; et agendo *CX* quae occurret periferiae in *N*. Invento autem *N* caetera facile determinantur.

[149]   Coroll. 1. Hinc est $8RR \cdot II - RR :: CN^q \cdot ND^q$. Nam

$$9RR - II \cdot II - RR :: BX^q \cdot BC^q$$

et componendo $8RR \cdot II - RR \ (:: CX^q \cdot BC^q) :: CN^q \cdot ND^q$.

[150]   Coroll. 2. Est etiam $I \cdot 3R :: ND \cdot NE$ utpote cùm supra fuerit $I \cdot 3R :: NP \cdot NF$.[4]

[151]   Schol. Ad eundem modum maxima radij *KT* post tres reflexiones emergentis inclinatio ad axem, juxta ac maximus arcuum *QG* investigabitur. Scilicet in eo casu *FG* et *fg* convenient ad *G* eritque

$$\text{arc } Ff \ (= \text{arc } FG - fG = \text{arc } NF - nf) = Nn - Ff.$$

Et inde 2arc *Ff* = arc *Nn*, et *NZ* = 2*ZF*. Adeoque

$$4 \cdot 1 :: NZ \cdot EZ :: (\text{per Coroll 3 ad Prop 32}) \ I \times NF \cdot R \times NP.$$

Sive $I \cdot 4R :: NP \cdot NF$. Et proinde per Lem 10

---

(23) Unbeknownst to Newton, Harriot had determined the half-radius of the primary bow by finding the maximum of $\widehat{FQ}$; see Lohne, "Thomas Harriot als Mathematiker," *Centaurus,* 11 (1965):19–45, esp. 34–9.

(1) The condition that the secondary bow occurs at the least inclination of the emergent ray *HS* to the incident one *AN* (the angle *AYS* in Fig. II, 61) was determined numerically by Descartes, much as for the primary bow (see Lect. I, 14, note (19), this volume): "I also found that after two reflections and just as many refractions very many more rays proceed to the eye at an angle of 51° or 52° than at any greater one, and that none at all are perceived at a smaller one" (*Meteora,* Ch. VIII, §9, *Oeuvres,* 6:705, 336).

(2) See Lect. I, 14, note (20).

Scholium. Together with the greatest inclination of the ray *RG* there is [147] given the greatest of the arcs *FQ* bounded by the refracted rays *NF*. For the angle *FCQ* subtended by the arc *FQ* is equal to the angle made by *CF* and *AN*; that is, it is equal to half the angle made by *RG* and *AN* or *BQ*. Consequently, the greatest of the arcs *FQ*, just as that of the angles made by *RG* and *BQ*, is determined by the ray *AN* incident at the point just found.[23]

Proposition 36. When the sun shines on a transparent sphere, to determine **Lecture 15** the least inclination to the axis of its rays emerging after two reflections.[1] [148]

Let *AN* and *an* (Fig. I, 63) be two incident rays immediately near one another, which after two reflections at *F* and *f* and *G* and *g* emerge along *HS* and *hs*. It is evident that in that case alone, when the acute angle made by *BQ* and *SH* is the least, can those rays *HS* and *hs* be parallel, as was described above for the rays *GR* and *gr*.[2] When this occurs the ray *FG* will also be parallel to *fg*. Hence,

$$2\widehat{Ff} \ (= \widehat{Ff} + \widehat{Gg} = \widehat{FG} - \widehat{fg} = \widehat{NF} - \widehat{nf}) = \widehat{Nn} - \widehat{Ff},$$

so that $3\widehat{Ff} = \widehat{Nn}$. Then since *NF* is divided at *Z* in the ratio of those arcs, as is clear, it will be $NZ = 3ZF = 3EZ$.[3] Since therefore by Corollary 3, Prop. 32, $I \times NF : R \times NP = NZ : EZ = 3 : 1$, then $I \times NF = 3R \times NP$, or $I : 3R = NP : NF$. Consequently, the ratio of *NP* to *NF* is given, and thus by Lemma 10 the point *N* will be given; namely, by drawing *BX* tangent to the circle at its vertex *B*, letting its square be to BC^2 as $9R^2 - I^2$ is to $I^2 - R^2$, and then drawing *CX* that will meet the circumference in *N*. Once *N* has been found, however, the other things are easily determined.

Corollary 1. Hence, $8R^2 : (I^2 - R^2) = CN^2 : ND^2$. For [149]

$$(9R^2 - I^2) : (I^2 - R^2) = BX^2 : BC^2,$$

and by composition $8R^2 : (I^2 - R^2) \ (= CX^2 : BC^2) = CN^2 : ND^2$.

Corollary 2. Also, $I : 3R = ND : NE$, inasmuch as there was above $I : 3R$ [150] $= NP : NF$.[4]

Scholium. In the same way the greatest inclination to the axis of the ray *KT* [151] emerging after three reflections, as well as the greatest of the arcs *QG*, will be investigated. Namely, in this case *FG* and *fg* will meet at *G*, and

$$\widehat{Ff} \ (= \widehat{FG} - \widehat{fG} = \widehat{NF} - \widehat{nf}) = \widehat{Nn} - \widehat{Ff}.$$

Hence, $2\widehat{Ff} = \widehat{Nn}$ and $NZ = 2ZF$, so that $4 : 1 = NZ : EZ = I \times NF : R \times NP$ (by Corollary 3, Prop. 32), or $I : 4R = NP : NF$. Consequently, by Lemma 10,

---

(3) Since the arcs subtend equal angles at *Z*, $\widehat{Nn} : \widehat{Ff} = NZ : ZF = 3:1$. Moreover, since $NE = \frac{1}{3}NF = 2ZF$, then $EZ = NZ - NE = ZF$.

(4) In determining the condition for the secondary rainbow Newton has moved beyond Barrow, who left the problem untouched, and Huygens, who in 1667 only partially solved it; see *Oeuvres complètes*, 13, i:163–8; and also *Mathematical Papers*, 3:506, note 53. These two corollaries, which Newton uses in §II, 138 to calculate the radii of the secondary bow, together with those of the preceding proposition, are set out without demonstration in the *Opticks*, Bk. I, Pt. II, Prop. IX.

$$16RR - II \cdot II - RR :: BX^q \cdot BC^q.$$

Unde consectatur esse $15RR \cdot II - RR :: CN^q \cdot ND^q$, et $I \cdot 4R :: ND \cdot NE$.

Atque ita si radij post quatuor reflexiones emergentis inclinatio minima desideretur determinabis faciendo ut sit $25RR - II \cdot II - RR :: BX^q \cdot BC^q$. Vel $24RR \cdot II - RR :: CN^q \cdot ND^q$, et $I \cdot 5R :: ND \cdot NE$. Et sic praeterea in infinitum.[5]

[152]   Transactis refractionibus Homogeneorum radiorum. Jam restat ut Hetero-geneos conferamus. De horum ad plana / refractionibus paulo fusiùs ageba-   75/ mus eo ut Prismatum (quorum usus in faciendis experimentis posthac erit frequentissimus) affectiones innotescerent. Praecipuum vero quod circa cur-vas superficies jam determinandum occurrit, est quantitas erroris radiorum a quo oritur confusio, sive indistincta visio objectorum quae in Telescopijs per nimiam vitri objectum respicientis aperturam evenire solet: Et in hunc finem cum praemissa sit Prop 31 unde errores innotescunt qui in sphaericis superfi-ciebus per ineptitudinem figurae efficiuntur: Sequentem jam subjungimus quâ errores ex inaequali refrangibilitate diversorum radiorum[6] orti determinari possunt.

[153]   Prop 37. Heterogeneis radijs in sphaeram incidentibus erro[r]es ex inae-qualibus radiorum similiter incidentium refractionibus progenitos determi-nare.

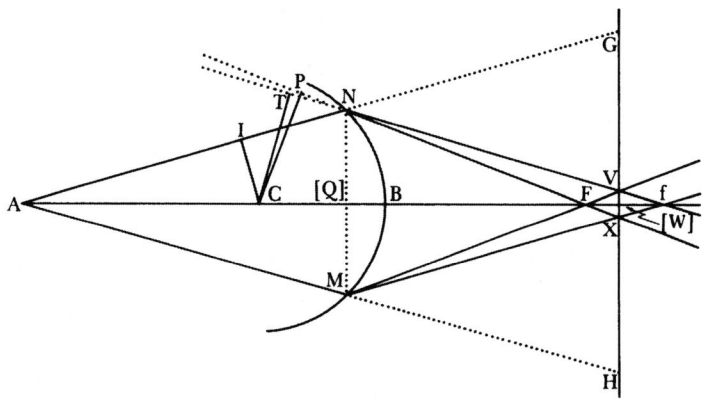

Figure I, 64

E puncto A (fig I, 64) in sphaeram NBM centro C descriptam incidant secundum lineam aliquam AN radij duo maximè difformes quorum refracti sint NF et Nf Axi occurrentes in F et f. et in illos demittantur perpendicula CI, CP et CT. Jam si accurata resolutio desideretur, refractiones radiorum NF et Nf seorsim computandae sunt. Sed cùm arcus NM ponatur admodum exigua portio circuli, veritatem quam proximè assequemur assumendo angu-los CNI, CNP et CNT ferè esse ut eorum sinus. Sit ergo I communis sinus

---

(5) After recapitulating these results in the *Opticks* (Bk. I, Pt. II, Prop. IX), Newton set forth a rule for any number of internal reflections: "the numbers 3, 8, 15, 24, &c. being gathered by

$$(16R^2 - I^2) : (I^2 - R^2) = BX^2 : BC^2.$$

Hence it follows that $15R^2 : (I^2 - R^2) = CN^2 : ND^2$, and $I : 4R = ND : NE$.

Similarly if you desire the least inclination of a ray emerging after four reflections, you will determine it by making

$$(25R^2 - I^2) : (I^2 - R^2) = BX^2 : BC^2,$$

or $24R^2 : (I^2 - R^2) = CN^2 : ND^2$, and $I : 5R = ND : NE$. And so on indefinitely.[5]

Having completed the refractions of homogeneous rays, it now remains for us to discuss heterogeneous rays. We have treated their refractions at planes somewhat comprehensively, so that we would come to know the properties of prisms, which will hereafter be used very frequently for making experiments. But the principal thing now presenting itself to be determined about curved surfaces is the quantity of the error of rays, from which there arises the confusion or indistinct vision of objects that usually occurs in telescopes through too large an aperture of the object glass. To this end, since Prop. 31 was premised in order to know the errors produced in spherical surfaces by their unsuitable shape, we now add the following proposition by which the errors arising from the unequal refrangibility of different rays[6] can be determined. [152]

Proposition 37. When heterogeneous rays are incident on a sphere, to determine the errors produced by the unequal refractions of similarly incident rays. [153]

From the point A (Fig. I, 64) let two most dissimilar rays be incident along any line AN onto the sphere NBM described with center C; let their refracted rays NF and Nf meet the axis in F and f; and to those rays drop the perpendiculars CI, CP, and CT. Now, if an accurate solution is required, the refractions of the rays NF and Nf must be calculated separately; but since the arc NM is assumed to be a very small portion of a circle, we may approach extremely close to the truth by assuming the angles CNI, CNP, and CNT to be nearly as their sines. Therefore, let I be the common sine of incidence, P

continuall addition of y^e arithmeticall progression 3, 5, 7, 9, &c." (Add. 3970, f. 120^r = *Opticks*, p. 128). Whereas Newton's derivation of the higher order bow was unique when the *Lectures* were composed, by the time the *Opticks* was published he had been forestalled by Halley. Halley simplified his derivation by maximizing the bows' half-radius $\frac{1}{2}\rho = (k + 1)r - i$, where the bows' radius is measured from the point opposite the sun for odd-numbered bows and from the sun itself for even ones, and k is the order of the bow or number of reflections. Since $\rho = (k\pi - D)$, where the deviation $D = k(\pi - 2r) + 2(i - r)$, his procedure is equivalent to maximizing the deviation for k odd and minimizing it for k even; see Halley, "De iride, sive de arcu coelesti, dissertatio geometrica, qua methodo directâ iridis utriusque diameter, data ratione refractionis, obtinetur: Cum solutione inversi problematis, sive inventione rationis istius ex data arcus diametro," *Phil. Trans.*, **22** (1700–1) [1702]:714–25, esp. 716. Jakob Hermann also derived the conditions for higher order bows independently of Halley, though he granted him priority, in a paper published the same year as the *Opticks*; Hermann, "Méthode géométrique & générale de déterminer le diamétre de l'arc-en-ciel, quelque hypothèse de la refraction qu'on suppose dans l'eau, ou dans toute autre liqueur transparente. Et le diamétre de l'arc-en-ciel étant donné par observation, en trouver la raison de la refraction," *Nouvelles de la republique des lettres*, **32** (1704):658–71.

(6) Originally continued: similiter incidentium (similarly incident).

incidentiae, P sinus refractionis radiorum maximè refrangibilium ac T sinus ille minimè refrangibilium, eritque ang: CNI. ang: $CNP :: I . P$, et ang: CNP. ang: $CNT :: P . T$. ac divisim ang: INP. ang: $CNP :: P - I . P$, et ang: CNP. ang: $PNT :: P . P - T$. et ex aequo ang: INP. ang $PNT :: P - I . P - T$.

Sume jam arcum BM aequalem arcui BN, et radiorum secundum AM incidentium duc refractos MF, $M[f]$, prioribus occurrentes in V et X, age VX, et produc donec occurrat incidentibus radijs ad G et H. Et patet VX esse latitudinem minimi spatij in quod omnes radij congregari possunt. Estque $GX . VX$ ($::$ ang GNX. ang VNX proximè $::$ ang INP. ang PNT) $:: P - I . P - T$. Et $GH + VX$ ($2GX$) . $VX :: 2[P] - 2I . P - T$. ac divisim $GH . VX :: P + T - 2I . P - T$.[7] Unde datis P, T, et I dabitur ratio GH ad VX.

E.G. Cùm supra[8] determinaverim quòd ad vitrum aeri conterminum sit $I . P :: 44\frac{1}{2} . 69\frac{1}{2}$, & $I . T :: 44\frac{1}{2} . 68\frac{1}{2}$. Si assumatur $I = 44\frac{1}{2}$, erit $P = 69\frac{1}{2}$, ac $T = 68\frac{1}{2}$, et $P + T - 2I = 49$, & $P - T = 1$. adeoque $GH . VX :: 49 . 1$ circiter.

[154]   / Schol. Ope hujus et Prop 31 errores homogeneorum radiorum quae in 76/ sphaericis superficiebus per figurae ineptitudinem eveniunt cum heterogeneorum erroribus conferi possunt et constabit hosce longe majores esse in parvis Sphaerarum portionibus:[9] Atque adeo heterogeneitatem Lucis et non ineptitudinem figurae Sphaericae in causa esse quod Telescopia in majorem perfectionis gradum nondum promota habeamus.[10]

Concipiamus e.g. quod NMB in figuris I, 58 et I, 64 referat objectivum Vitrum Telescopij, cujus anterior superficies NM plana sit eo ut radios in posteriori seu sphaerica superficie NBM solummodo refringat. Et ponamus CB semidiametrum hujus sphaerae esse 10 pedes ut Telescopium (fere pedes 20 sive 240 digitos longum) conficiat. Sitque Apertura NM 2^{dig} quanta maxima cum visione satis distincta adhibetur in hujusmodi telescopijs quae objectum quasi 70 vel 80 vicibus ampliant. Et sinus incidentiae sit ad sinum refractionis in confinio vitri et aeris peractae, ut 11 ad 17 circiter prout supra[11] determinavimus. His positis scribendum est 120 pro a, 1 pro y, 11

---

(7) That is, $GH/VX = 2(n - 1)/\Delta n$. If we assume a thin lens, or the points Q and B to coincide, and let $AQ = s$, $QW = s'$, and $NM = 2y$ (where we have added the points Q and W to Newton's figure), then since the triangles AGW and ANQ are similar, $\frac{1}{2}GH/(s + s') = y/s$, and the lateral chromatic aberration $VX = y(s + s')\Delta n/s(n - 1)$. Moreover, since the triangles NQF and VWF are similar, the longitudinal chromatic aberration $\Delta s' = Ff$ is

$$\Delta s'/s' \approx VX/y = (s + s')\Delta n/s(n - 1).$$

For parallel incident rays, where $(s + s')/s = 1$ and $s' = f$, the lens's focal length, then $VX = y/\nu$ and $\Delta s' = \Delta f = f/\nu$, where $\nu = (n - 1)/\Delta n$. In the *Opticks* (Bk. I, Pt. I, Prop. VII, pp. 61–2 = Add. 3970, f. 63), Newton laid down without demonstration—"they that are skilled in Opticks will easily understand"—the conclusion of this proposition and its immediate consequences, that is, the magnitudes of the lateral and longitudinal chromatic aberration for an object point both on the axis and infinitely distant, taking $\nu = 27\frac{1}{2}$. He then described in great detail a "nice and troublesome Experiment [16]" to measure the longitudinal aberration, in which he chose the advantageous case where the object and image are equidistant from the lens, or $(s + s')/s = 2$.

the sine of refraction of the most refrangible rays, and T that sine of the least refrangible rays. Then it will be $C\widehat{N}I : C\widehat{N}P = I : P$ and $C\widehat{N}P : C\widehat{N}T = P : T$, and by dividing $I\widehat{N}P : C\widehat{N}P = (P - I) : P$ and $C\widehat{N}P : P\widehat{N}T = P : (P - T)$, and from the equality of the ratios, $I\widehat{N}P : P\widehat{N}T = (P - I) : (P - T)$.

Now take the arc BM to be equal to the arc BN, and of the rays incident along AM draw the refracted rays MF and Mf meeting the previous ones at V and X. Draw VX and extend it until it meets the incident rays at G and H. Then it is evident that VX is the breadth of the smallest space into which all rays can be gathered. Hence

$$GX : VX \ (\approx G\widehat{N}X : V\widehat{N}X = I\widehat{N}P : P\widehat{N}T) = (P - I) : (P - T)$$

and

$$GH + VX \ (\text{or } 2GX) : VX = (2P - 2I) : (P - T),$$

and by dividing $GH : VX = (P + T - 2I) : (P - T)$.[7] Consequently, given P, T, and I, the ratio of GH to VX will be given.

For example, since I determined above[8] that for glass bounded by air, $I : P = 44\frac{1}{2} : 69\frac{1}{2}$ and $I : T = 44\frac{1}{2} : 68\frac{1}{2}$, if it is assumed that $I = 44\frac{1}{2}$, then $P = 69\frac{1}{2}$, $T = 68\frac{1}{2}$, $P + T - 2I = 49$, and $P - T = 1$, so that $GH : VX \approx 49 : 1$.

Scholium. With the aid of Prop. 31 and this one the errors of homogenous rays occurring in spherical surfaces because of their unsuitable figure can be [154] compared with the errors of heterogeneous rays, and it will be manifest that the latter are far greater in small portions of spheres.[9] Consequently, the heterogeneity of light and not the unsuitability of a spherical shape is the reason why we have not yet advanced telescopes to a greater degree of perfection.[10]

For example, let us imagine that NMB in Figs. I, 58 and 64 represents a telescope's object glass whose first surface NM is plane so that it refracts the rays solely in the second or spherical surface NBM. Then let us assume that this sphere's radius CB is 10 feet, thus making a telescope about 20 feet, or 240 inches, long; and let the aperture NM be 2 inches, the largest that is used with sufficiently distinct vision in such telescopes that magnify an object roughly 70 to 80 times. Moreover, let the sine of incidence to the sine of refraction occurring at the boundary of glass and air be approximately as 11 to 17, as we determined above.[11] With these assumptions we must write 120

Smith incorporated the present proposition and the following Scholium in his *Compleat System*, Bk. II, Ch. VI, Props. I, V, 1:134–5, 138–9. (8) In §I, 38 = 111.

(9) In the "New theory," Newton strongly implies that he had arrived at this conclusion—at least for parallel rays—in the beginning of 1666 and then abandoned refracting telescopes for reflectors; *Correspondence*, 1:95 = *Phil. Trans.*, 6 (1671/2):3079. If he had concluded this as early as 1666, it is possible that he derived the chromatic aberration by the simple physical argument he sets forth in §II, 128, and repeated for Huygens in his letter to Oldenburg, 8 July 1672 (*Correspondence*, 1:212–13). Newton had calculated the spherical aberration by the winter of 1665/6; see Lect. I, 13, note (13).

(10) Appropriately enough, Newton concludes Part I by returning to his opening remarks in §I, 1 = 1. (11) In §I, 30 ≈ 103.

pro *I*, & 17 pro *R* in valore ipsius *PQ* quem exhibuimus in Coroll 7 Prop 31 hoc est in termino $\frac{Ry^{3}}{4Iaa}$ [12] et emergit $\frac{17}{4, 11, 120, 120}$ dig sive $\frac{17}{633600}$ dig = *PQ*.[13] Estque hic error lateralis homogeneorum radiorum ortus ab ineptitudine figurae sphaericae.

Praeterea concipiamus radios *AN* et [*A*]*M* in fig I, 64 esse parallelos Axi et erit *GH* = *NM* = 2^{dig} adeoque *VX*, error nempe lateralis heterogeneorum ab invicem in eodem loco concursûs, erit $\frac{2}{49}$ dig. Confer jam hos errores et patebit *VX* esse ad *PQ* (seu $\frac{2}{49}$ ad $\frac{17}{633600}$ [13]) ut 1267200 ad 833 sive ut 1521 ad 1 circiter.[14] Adeoque *VX* esse quasi mille et quingentibus vicibus majorem quam *PQ*; Tanta sane disproportio, ut *PQ* respectu *VX* pro nullo haberi possit. Error quidem *VX* cum sit $\frac{2}{49}$ dig: tantus est ut miror quod objecta per ejusmodi Telescopia tam distinctè videri possint.[15] Sed alterius generis error *PQ* sive $\frac{17}{633600}$ dig i.e. $\frac{1}{37271}$ dig [13] circiter longè minor est quàm qui potest esse sensibilis, et proinde negligendus et indistincta visio erroribus ex heterogeneitate Lucis exortis solummodo tribuenda. Et hinc patet perfectionem Telescopiorum non e Conicis Sectionibus petendam esse, sed figuras sphaericas huic usui aequè inservire posse. In Microscopijs quidem errores homogeneorum radiorum ex sphaerica superficie vitri objectivi propter aperturam bene magnam enormes / oriuntur et admodum sensibiles. Adeo ut illa vitra si secundum conicam aliquam sectionem debitè formarentur, paulò perfectiora evaderent. Sed methodus tamen me non latet corrigendi errores illos absque conicis sectionibus et efficiendi ut vitra e sphaericis superficiebus formari possint quae radios homogeneos satis accuratè refringent,[16] ne dicam quae longè accuratius refringent obliquos radiorum penicillos[17] quam vitra alijs quibuscunque figuris terminata. Adeo ut Sphaericas superficies usibus dioptricis prae caeteris omnibus accommodatas esse censeam.

---

(12) Read: $\frac{RRy^{3}}{4IIaa}$. See Lect. I, 13, note (18).

(13) In accordance with the preceding note *PQ* should be 289/6,969,600 ≈ 26/633,600 ≈ 1/24,116.

(14) That is, approximately 984 to 1, using the correct value for *PQ*. In the "New theory," Newton simply noted that chromatic aberration is "some hundreds of times greater" than spherical (*Correspondence*, 1:95 = *Phil. Trans.*, 6 (1671/2):3079). However, in the *Opticks* (1721) (Bk. I, Pt. I, Prop. VII, p. 84), using a new lens with a 4-inch aperture, a 50-foot radius, and $(n - 1)/\Delta n = 27\frac{1}{2}$, where $n = R/I = 31/20$, he found the ratio of the aberrations to be a much larger 5449 to 1; in the first English and Latin editions (see Lect. I, 13, note (16)) he had 8151 to 1, although—because of yet another slip in the calculation—it should have been 8446 to 1.

(15) Huygens repeatedly rejected Newton's claim that the chromatic aberration is so large and argued that "Experience agrees not with what Mr. *Newton* holds," for a twelve-foot lens projects an image "too distinct and too well defined to be produced by rays that should stray the 50th. part of the Aperture" (Oldenburg to Newton, 18 January 1672/3, *Correspondence*, 1:256; translated in *Phil. Trans.*, 8 (1673):6087). Newton essentially conceded that chromatic aberration is not as large and serious as he had initially claimed, since "the rays, whose error is

for *a*, 1 for *y*, 11 for *I*, and 17 for *R* in the value of *PQ* that we presented in Corollary 7, Prop. 31, that is, in the term $Ry^3/4Ia^2$;[12] and there results $17/(4 \times 11 \times 120 \times 120)$ inches = $17/633,600$ inches = *PQ*.[13] This then is the lateral error of homogeneous rays arising from the unsuitability of a spherical figure.

Furthermore, let us conceive the rays *AN* and *AM* in Fig. I, 64 to be parallel to the axis, and then *GH* = *NM* = 2 inches; and consequently *VX*, that is, the lateral error of heterogeneous rays from one another at the same place of intersection will be 2/49 inches. Now compare these errors and it will be clear that *VX* is to *PQ* (or 2/49 to 17/633,600[13]) as 1,267,200 to 833, or approximately as 1521 to 1.[14] Hence *VX* is about fifteen hundred times greater than *PQ*—so exceedingly great a disproportion that *PQ* can be considered as nothing compared to *VX*. Indeed the error *VX*, 2/49 inches, is so great that I wonder how objects can be seen so distinctly through telescopes of this kind.[15] But the other type of error *PQ*, that is, 17/633,600 or about 1/37,271[13] inches, is far smaller than can be perceived and is consequently to be neglected; and indistinct vision must be attributed solely to errors arising from the heterogeneity of light. Thus it is evident that the perfection of telescopes ought not to be sought in conic sections, but that spherical shapes can serve equally for this purpose. In microscopes, in fact, the errors of homogeneous rays arising from the spherical surface of the object glass, because of its extremely large aperture, are so enormous and quite perceptible that if those glasses were duly shaped into some conic section, they would turn out scarcely more perfect. Yet I am not unaware of a method of correcting those errors without conic sections and bringing it about that glasses can be made from spherical surfaces that will refract homogeneous rays sufficiently accurately,[16] not that I say that they will refract oblique pencils of rays[17] far more accurately than glasses bounded by any other shapes. Therefore, I judge spherical surfaces above others to be appropriate for dioptrical purposes.

---

so great, are but very few in comparison to those wch are refracted more justly . . . And these are yet so much further weakened by the greater space through wch they are scattered" (Newton to Oldenburg, 3 April 1673, *Correspondence*, 1:266 = *Phil. Trans.*, 8 (1673):6111). Huygens, in turn, grudgingly accepted Newton's argument, for in about 1685 he incorporated it in his *Dioptrica* (Pt. III, Prop. VII) while jotting in the margin of his manuscript, "Newtoni diffusio. nimiam ponit. Sed longe major aberratione altera" (*Oeuvres complètes*, 13, ii:484, note 1). In the *Opticks* (Bk. I, Pt. I, Prop. VII) Newton mathematically developed his argument and added the further factor that "the yellow & Orange [rays] . . . affect the senses more strongly then all the rest together" (Add. 3970, f. 72^r = *Opticks*, p. 71). These two causes reduced the chromatic aberration to about 1/250 of the aperture, or about one fifth of his initial claim.

(16) Newton is undoubtedly alluding to the compound glass-water lens which he first proposed as a remedy for spherical aberration in one of his draft replies to Hooke (Add. 3970, f. 447^v = *Correspondence*, 1:191–2), and eventually set forth in the *Opticks* (Bk. I, Pt. I, Prop. VII, pp. 74–5 = Add. 3970, f. 74^r). See *Mathematical Papers*, 3:553–5; H. Boegehold, "Zur Vor- und Frühgeschichte der achromatischen Fernrohrobjektive," *Forschungen zur Geschichte der Optik*, 3 (1943):81–114; and also Bechler, "A less agreeable matter."

(17) obliquos radiorum penicillos] Added.

Opticae pars 2da
De Colorum Origine.

Lect 1
[1 ≈ 24. Dissertatio de coloribus inita.]

Qui in fabricandis Telescopijs occupati sunt, de coloribus conqueruntur quibus objecta dum vitris istis mediantibus aspiciuntur, tingi solent, quique eo magis augentur et apparent quo vitrum oculare ex sphaeris minoribus efformatur, vel etiam quo vitrum objectivum majori latitudine radijs intrantibus patet. Unde duplici incommodo implicati, impediuntur ne perspicilla ad optatum perfectionis gradum perducant: tum quod oculare vitrum ultra certos gradus parvum ad objecta magis amplianda nequeant adhibere, tum quod vitrum objectivum ultra certos limites aperire nequeant ad objecta magis lucida et perspicua reddenda. Qui gradus vel limites si non probe observentur, objecta coloribus involuta reddentur et multò minùs distincta, quam si vel minora cernerentur, ope vitri ocularis minus convexi; vel minus lucida, diminutâ perspicilli aperturâ. Jam cum istae perfectiones praecipuae sint, quae in perspicillis desiderantur, nempe ut objecta magis amplient et reddant lucidiora: operae pretium videtur in naturam colorum inquirere, ut investigemus tandem quid in causâ sit quod ita appareant et objecta reddant indistincta. Hujus enim ignorantia quamplurimos labore non exiguo sed inani tamen exercuit dum imperfectionem Telescopiorum a vitiosis vitrorum figuris ortam esse credentes, in istis meliori figura perpoliendis navarunt operam. At si causam horum colorum satis exploratam habuissent, simul innotuisset inaequalis diversorum radiorum refrangibilitas, et inde vitia Telescopiorum non ab ineptitudine figurae Sphaericae ad refractiones rite peragendas originem ducere constitisset. Quo bene intellecto, conatus suos procul dubio mutassent, ‖ et laboribus istis secundum aliam methodum dispositis Opticam in gradum multò perfectiorem jam promotam haberemus.

[2 = 25. De opinionibus Philosophorum et imprimis Peripateticorum]

₂2/ / Qui de coloribus hucusque disseruere vel id nomine tenus fecerunt ut Peripatetici, vel in eorum naturam et causas inquirere conabantur ut Epicurei et alij recentiores. Quae Peripatetici de hisce tradidere, etsi vera forent tamen ad nostrum propositum nihil valerent: quippe dum modum quo generantur, et causas unde fiunt tam varij, non[1] attingant. Etenim illi de originibus et varijs rerum speciebus disputantes pro causis ex quibus ipsarum existentiam

---

(1) I: non omninò (not at all).

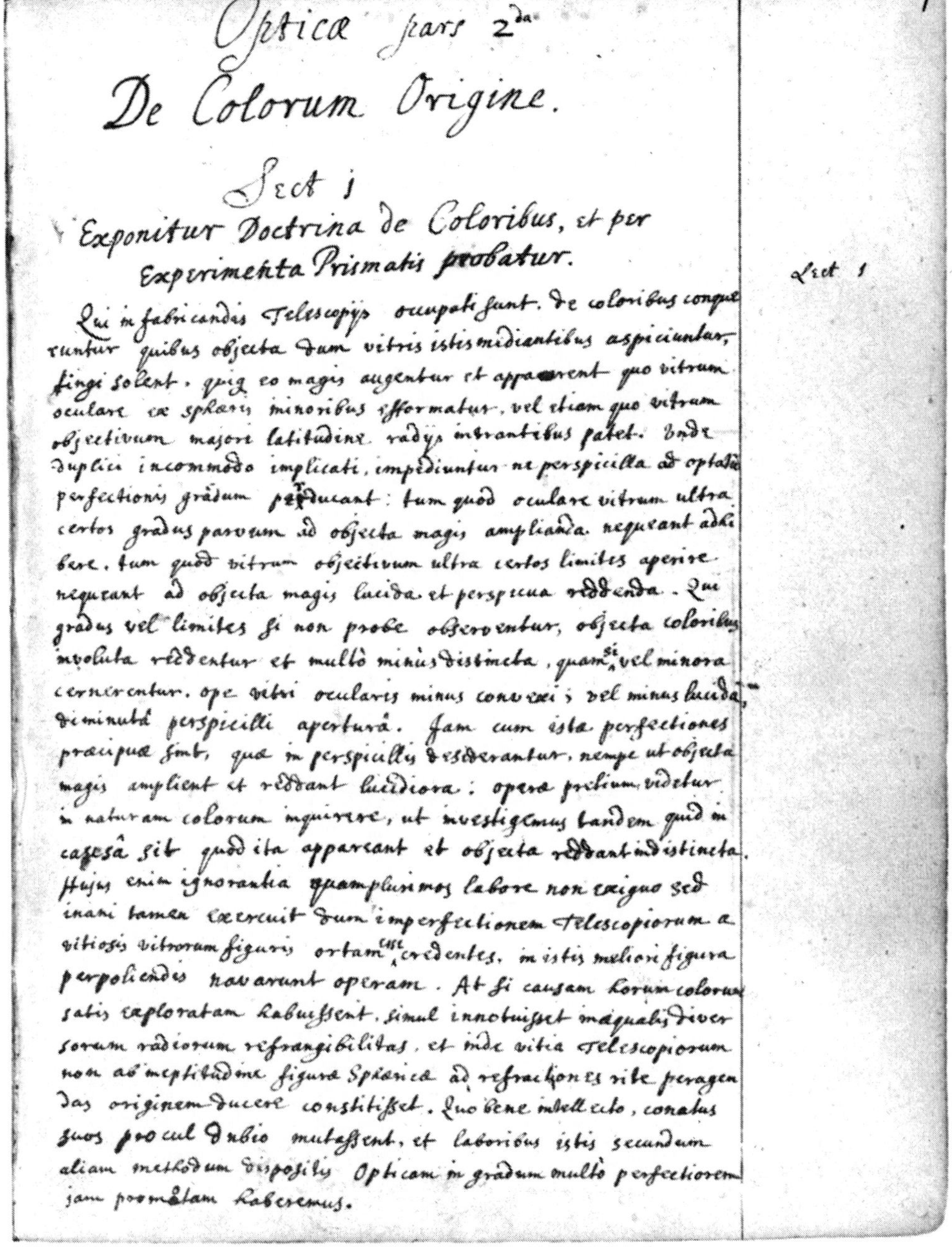

Optica pars 2ᵈᵃ

De Colorum Origine.

Sect 1

Exponitur Doctrina de Coloribus, et per Experimenta Prismatis probatur.

Qui in fabricandis Telescopijs occupati sunt, de coloribus conqueruntur quibus objecta dum vitris istis mediantibus aspiciuntur, tingi solent. quipe eo magis augentur et apparent quo vitrum oculare ex sphæris minoribus efformatur, vel etiam quo vitrum objectivum majori latitudine radijs intrantibus patet. Unde duplici incommodo implicati, impediuntur ne perspicilla ad optatæ perfectionis gradum perducant: tum quod oculare vitrum ultra certos gradus parvum ad objecta magis amplianda nequeant adhibere. tum quod vitrum objectivum ultra certos limites aperire nequeant ad objecta magis lucida et perspicua reddenda. Qui gradus vel limites si non probe observentur, objecta coloribus involuta reddentur et multò minùs distincta, quàmsi vel minora cernerentur. ope vitri ocularis minus convexi; vel minus lucidâ diminutâ perspicilli aperturâ. Jam cum ista perfectiones præcipuæ sint, quæ in perspicillis desiderantur, nempe ut objecta magis ampliant et reddant lucidiora: operæ pretium videtur in naturam colorum inquirere, ut investigemus tandem quid in causâ sit quod ita appareant et objecta reddant indistincta. Hujus enim ignorantia quamplurimos labore non exiguo sed inani tamen exercuit dum imperfectionem Telescopiorum a vitiosis vitrorum figuris ortam esse credentes, in istis meliori figurâ perpoliendis navarunt operam. At si causam horum colorum satis exploratam habuissent, simul innotuisset inæqualis diversorum radiorum refrangibilitas, et inde vitia Telescopiorum non ab ineptitudine figuræ Sphæricæ ad refractiones rite peragendas originem ducere constitisset. Quo bene intellecto, conatus suos procul dubio mutassent, et laboribus istis secundum aliam methodum dispositis Opticam in gradum multò perfectiorem jam promotam haberemus.

Plate II. The opening of Part II, "The origin of colors," of the *Optica* (§II, 1 ≈ 24)

Optics, Part II
The Origin of Colors

SECTION I
THE DOCTRINE OF COLORS IS SET FORTH AND PROVED BY PRISMATIC EXPERIMENTS

Those who are occupied with constructing telescopes complain about the colors that usually tinge objects when they are viewed through those glasses; the colors increase and are more evident as the eyeglass is made from smaller spheres, or also as the object glass is opened wider to the entering rays. Consequently, beset by this double obstacle they are hindered from bringing telescopes to their desired degree of perfection: both because they cannot use an eyeglass beyond certain degrees of smallness to magnify objects more; and because they cannot open the object glass beyond certain limits to make objects brighter and clearer. If these degrees or limits are not properly observed, objects will become enveloped with colors and be much less distinct than if they were seen either smaller, by using a less convex eyeglass, or less bright, by decreasing the aperture of the object glass. Now, since these are the principal perfections desired in telescopes, namely, that they magnify objects more and make them brighter, I consider it worthwhile to investigate the nature of colors, so that we may finally discover the cause that is responsible for their appearing in this way and rendering objects indistinct. Ignorance of this has indeed taxed quite a few with not a slight, but nonetheless vain, effort; while believing the imperfection of telescopes to arise from the defective shape of their glasses, they zealously directed their efforts toward finely polishing them to a better shape. ‖ But had they sufficiently examined the cause of these colors, the unequal refrangibility of different rays would have immediately become known, and hence it would have been established that the defects of telescopes do not derive from the unsuitability of a spherical shape to perform refractions properly. Had this been thoroughly understood, they would undoubtedly have altered their efforts, ‖ and by directing their labors according to a different method, we would already have advanced optics to a much more perfect state.

Those who have treated colors until now either have done it in name only, as the Peripatetics, or have endeavored to investigate their nature and causes, as the Epicureans and more recently others. What the Peripatetics have taught about these things, even if true, would nevertheless be of no value for our purpose, since as yet they do not[1] consider how they are generated and the causes whereby they become so diverse. Indeed, disputing about the

Lecture 1
[1 ≈ 24. The dissertation on colors begins.]

[2 = 25. The opinions of philosophers, and first the Peripatetics.]

433

et discrimen mutuantur varias quasdam formas assignarunt; verum de particulari cujusvis formae causâ et ratione ob quam differt ab alijs haud unquam quicquam disseruere. Et sic ea fecerunt missa quorum explicatio videtur summum Philosophorum officium, imo quae sola mentem scientiae naturalis avidam explere possint.

Attamen ne mancam tradidisse Philosophiam viderentur effecerunt ut ejusmodi disquisitiones' pro maxime absurdis et ridendis habeantur, utpote quae
supponunt formarum esse alias formas, et qualitates qualitatum. Itaque cum
lux definiatur esse qualitas vel forma quae dat esse lucidum, non expectandum est, ut aliquid de ejus causis audiamus, vel qua ratione ad varios colores
producendos fiat varia. Dicunt equidem quod plus luminis quibusdam coloribus immiscetur, quam alijs: at hoc non sufficit ad eorum productionem tum
quod nullus omnino color ex albedine et nigredine solummodò mixtis praeter
fuscos intermedios generatur: tum quod quantitas lucis non mutat speciem
coloris. Corpus enim rubrum, verbi gratiâ, semper apparet rubrum sive aspiciatur in crepusculo sive in meridie lucidissimâ. Porro autem ipsa definitio
quam attribuunt coloribus adeo non pandit eorum naturam ut eos ne nomine
tenus exprimat. Ait Aristoteles χρῶμα δέ ἐστι τοῦ διαφανοῦς ἐν σώματι
ὡρισμένῳ πέρας.[2] Quae superficiei coloratae potiùs quàm coloris descriptio
est. Illa enim dici potest extremitas perspicui in corpore terminato, at color
plerunque videtur ubi nulla talis datur extremitas: ut in Iride; et Prismate; in
vitris vel liquoribus perspicuis et aliquo colore leviter tinctis; in aquâ marinâ
quae viridis ut plurimum apparet, qui tamen color non in extremitate aquae
sed per totam ejus crassitiem generatur; in aere qui licèt maximè perspicuus
et nullo corpore denso terminatus, serenâ tamen nocte caeruleus apparet; et
in flammâ, quae non minus perspicua est, et luci pervia quàm ipse aer. Sic
cùm humores oculi colore aliquo tinguntur omnia videntur eodem colore
tincta, licet extremitas perspicui sit alijs coloribus praedita. Et cùm Solem
nudis oculis modo aspexeris, luminosa omnia deinceps videntur rubra, et
/ nigra plerumque apparent caerulea, qui color erit magis conspicuus si clausis ₂3/
oculis te in locum aliquem tenebrosissimum statim conferas. Imo premendo
oculum colores in tenebris excitare liceat: quis autem vocabit illos extremitatem perspicui? Caeterum non opus est ut has opiniones enixè refutem quae
etsi verae essent tamen non sunt sufficientes,[3] neque proposito meo adversantur. Esto enim lux qualitas corporis lucidi esto lumen actus perspicui, et
color ejus extremitas, et quicquid de istis dixerunt, esto; abinde tamen haud
concipi poterit quo pacto lux refringatur, unde colores sint varij, quid in
causâ sit quod in perspicillis apparent, et qua ratione incommodum istud
devitari possit.

[3 = 26. De　　　Ad opiniones aliorum Philosophorum quod attinet, dixerunt colores vel ex
opinionibus aliorum　umbra luceque variè mixtis; vel ex contortione globulorum aut eorum varijs
Philosophorum.]　pressionibus generari; vel denique ex varijs modis quibus Medium quoddam

---

(2) *De sensu*, 439 b 12–13.

(3) etsi . . . sufficientes] Originally (as I): non videntur tanti (seem of no great value).

origins and various species of things, they specify that certain different forms are the causes from which they derive their existence and difference, but hardly ever has anyone treated the particular cause of any form and how it differs from other forms. Thus they dismiss those things the explanation of which seems the highest function of philosophy and, indeed, alone can satisfy the eager mind of natural science.

But lest they seem to have handed down an imperfect philosophy, they have brought it about that investigations of this kind are considered highly absurd and ridiculous, inasmuch as they assume there are other forms of forms and qualities of qualities. Therefore, since light is defined to be a quality or form that allows a luminous thing to exist, we must not expect to learn anything about its causes or how it becomes different in order to produce different colors. To be sure, they say that more light is mixed with some colors than with others; but this is insufficient for their production, both because no color at all besides intermediate grays is generated from mixing white and black alone, and also because the quantity of light does not change the species of the color. For a red body, for example, will always appear red whether it is viewed at twilight or at brightest midday. Moreover, the very definition that they attribute to colors so little reveals their nature that it does not even represent them in name. Aristotle says, "Color then is the extremity of the transparent in a determinately bounded body"[2]—this is a description of a colored surface rather than of color. The extremity of the transparent can of course be spoken of in a bounded body, but color frequently appears where no such extremity exists: in a rainbow; in a prism; in transparent glasses or liquids lightly tinctured with some color; in sea water, which often appears green, which color is nonetheless produced not in the extremity of the water but in its entire thickness; in air, which although exceedingly transparent and not bounded by a dense body nevertheless appears blue on a clear night; and in a flame, which is no less transparent and pervious to light than air itself. Thus when the humors of the eye are tinged with some color, everything appears tinged with the same color, although the extremity of the transparent is endowed with other colors. Also when you have looked at the sun with only your naked eyes, afterward all bright things seem red, and black things often appear blue, which will be more conspicuous if with your eyes shut you move immediately into some very dark place. In fact, by pressing your eye you may produce colors in the dark, but who will call those an extremity of the transparent? There is no need, however, for me vigorously to refute these opinions that, even if true, are nevertheless insufficient[3] and not opposed to my purpose. For let light [*lux*] be a quality of a luminous body; let light [*lumen*] be the action of the transparent and color its extremity; and whatever else they have said about those things, let it be. Yet from this it can hardly be conceived how light is refracted, whence colors are different, why they appear in telescopes, and how that inconvenience can be avoided.

As for the opinions of other philosophers, they have said that colors are generated from shadow and light mixed in various ways, or from a spinning of little balls or their various pressures, or, finally, from the various ways in

[3 = 26. The opinions of other philosophers.]

aethereum vibratur, statuentes lucem productam esse ex impulsu vibrantis aetheris in retiformem tunicam delato. Extra oleas nimis evagarer, si has opiniones sigillatim refutandas[4] adortus essem; nec opus est ut faciam cum omnes in communi quodam errore consentiant: Scilicet quod modificatio lucis, quâ singulos colores exhibet, ei non sit insita ab origine suâ, sed inter reflectendum vel refringendum acquiritur. Inter radios lucis nullum contemplantur discrimen priusquam incidant in corpus aliquod colorificum; opinati tantùm quòd pro variâ dispositione corporis istius varijs modis reflectuntur vel refringuntur, et pro specie modificationis quam sic acquirunt, varia deinde colorum phantasmata spectantibus exhibent. Mixtura lucis et umbrae gyratio globulorum vel varia vibratio medij non supponuntur inesse radijs antecedenter ad eorum reflexiones vel refractiones, sed per istas actiones generari creduntur. Quemadmodum et Peripatetici statuunt colores a corporibus originem sumere,[5] quorum dicunt esse qualitates. Attamen contrarium esse verum ex sequentibus abunde patebit. Invenio scilicet quod modificatio lucis unde colores originem sumunt, luci connata sit, et non oriatur a reflexione neque a refractione neque a qualitatibus corporum, aut modis quibuslibet, nec ab ijs vel destrui potest vel ullo modo mutari.

[4 = 27. Colorum origines et fundamenta generalia describuntur.]

[5 ≈ 28. Idque quinque[6] propositionibus]

Verùm ut sententiam meam distinctius proferam: ‖ Invenio primo quòd radijs diverse refrangibilibus competant diversi colores. Maxime refrangibilibus purpura, sive violarum color competit, et rubor minime refrangibilibus, atque mediocribus viriditas vel potius confinium viridis et virescentis caerulei. Caeruleus autem purpurae intercedit et viriditati, flavusque viriditati et rubori. Adeoque radij prout sunt plus plusque refrangibiles, apti sunt ad hos ordine colores / rubrum, flavum, viridem, caeruleum et violaceum generandos una cum omnibus eorum successivis gradibus et coloribus intermedijs.[7]   ₂4/

Invenio praeterea quod nullius radiorum generis forma sive dispositio colorifica vel refractione vel alia quacunque (quam potuerim animadvertere) causa mutari potest, sed unicum tantùm sibi proprium colorem unumquodque semper conservat et exhibet, si modò a radiorum diversi generis misturâ non conturbetur. Nam colores qui refractionibus generari videntur, non nisi difformium radiorum misturâ variâ, vel separatione fiunt.[8]

Tertiò invenio quod color albus et niger una cum cinereis seu fuscis intermedijs fiunt ex radijs cujusque speciei confusè mistis; et similiter quod caeteri omnes colores qui non sunt ex primitivis, per varias horum radiorum misturas producuntur. Et inde non mirum est si, difformibus radijs per inaequalem refractionem segregatis, diversi colores ex his de novo emergere videantur.[9]

Quinetiam invenio quod primitívi colores per misturam radiorum alterutrinque confinium exhiberi possunt. Viridis nempe ex flavo et caeruleo, flavus ex adjacente viridi citrioque, et sic de alijs. Per colores autem primiti-

---

(4) I: confutandas.

(5) *Editio princeps*: ducere.

(6) Changed from "quatuor" (four) in the *Lectiones*.

(7) In §II, 7 → 30, 31, 32, Newton undertakes to prove this proposition along with its converse, even though in the revision he had eliminated the converse as a proposition.

(8) The demonstration of this new proposition begins in §II, 19.

which a certain aetherial medium is vibrated, assuming that light is produced by an impulse of the vibrating aether carried to the retina. I would stray too far from the path were I to attempt to refute[4] these views individually. Nor is it necessary that I do so, since they all agree in a certain common error; namely, the modification of light by which it exhibits individual colors is not innate to it from its source but is acquired by being reflected or refracted. They consider that there is no difference between the rays of light before they fall upon some color-making body and believe only that according to the varied disposition of that body they are reflected or refracted in various ways, and according to the kind of modification so acquired they thereafter exhibit to observers various sensations of color. The mixture of light and shadow, the gyration of the little balls, or the various vibrations of the medium are not assumed to be in the rays prior to their reflections or refractions, but are believed to be produced through those actions; likewise, the Peripatetics consider colors to originate in bodies, saying they are their qualities. Nonetheless, it will be manifestly evident from the following that the contrary is true. Namely, I find that the modification of light whereby colors originate is connate to light and arises neither from reflection nor from refraction, nor from the qualities or any modes whatsoever of bodies, and it cannot be destroyed or changed in any way by them.

[4 = 27. The origins of colors and their general principles are described.]

But to present my idea more distinctly: ‖ First, I find that to differently refrangible rays there correspond different colors. To the most refrangible ones there corresponds purple or violet, to the least refrangible red, and to the intermediate ones green, or rather the boundary of green and greenish blue. Blue, however, falls between purple and green, and yellow between green and red. Hence as the rays are more and more refrangible, they are disposed to generate these colors in order: red, yellow, green, blue, and violet, together with all of their successive gradations and intermediate colors.[7]

‖[5 ≈ 28. And that by five[6] propositions.]

Moreover, I find that the form or color-making disposition of no kind of ray can be changed either by refraction or by any other cause whatever that I have been able to observe; but each one always conserves and exhibits only a unique color proper to it, provided it is not disturbed by a mixture of rays of a different kind. For the colors that appear to be generated by refractions are made only by a diverse mixture of unlike rays, or rather by their separation.[8]

Third, I find that the colors white and black, together with intermediate ashens or grays, are made by rays of every sort confusedly mixed, and similarly, that all other colors that are not primitive are produced by various mixtures of these rays. Consequently, it is not surprising if, when unlike rays are separated by an unequal refraction, various colors seem to emerge from them de novo.[9]

In addition, I find that primitive colors can be exhibited by a mixture of the neighboring rays on each side; namely, green from yellow and blue, and yellow from the adjacent green and orange, and similarly for the others. By

(9) The proof of this proposition begins in §II, 27 → 40.

vos non tantum quinque praedictos intelligo sed et quoslibet alios quibus exhibendis aptum datur aliquod uniforme radiorum genus.[10]

Invenio denique[11] quòd omnes omnium corporum colores non aliunde generantur quam e dispositione quadam qua apta sunt ut alios radios reflectant et intromittant alios. Sic corpus rubrum est quod radios ad rubedinem aptos reflectit maxime, et plerosque caeteros intromittit; purpureum quod radios isti colori generando proprios reflectit, et intromittit alios: album verò quod fere omnes reflectit; et nigrum quod omnes intromittit, paucissimis, sed omnium tamen specierum, radijs repercussis.[12]

[6 ≈ 29, 30. De quibus non hypotheticè et probabiliter, sed ab experimentis aut demonstrativè disserendum esse promittitur.]

Verùm ne videar officij limites excessisse, dum naturam colorum pertrectare aggrediar, qui nihil ad Mathesin attinere censeantur: Non abs re erit, si de ratione incepti hujus iterum commonefaciam. Nimirum tanta est inter proprietates refractionum et colorum affinitas, ut seorsim explicari nequeant. Qui alterutras rite velit cognoscere, ut alteras cognoscat necesse est. Et praeterea si de refractionibus non agerem, et earum disquisitio non esset in causâ quòd negotium de coloribus simul explicandis inceptarem: tamen generatio colorum tantam Geometriam complectitur et eorum cognitio tantâ firmatur evidentiâ, ut vel ipsorum gratiâ possem aggredi, sic limites Mathesis nonnihil ampliaturus. Quemadmodum enim Astronomia, Geographia, Navigatio, Optica, et Mechanica pro Scientijs Mathematicis habentur, licet in ijs agatur de rebus physicis, Caelo, Terra, Navibus, luce, et motu locali: Sic etiamsi / colores ad Physicam pertineant, eorum tamen scientia pro mathematicâ ₂5/ habenda est, quatenus ratione mathematica tractantur. Imò verò cum horum accurata scientia videatur ex difficillimis esse, quae Philosophus desideret; spero me quasi exemplo monstraturum quantum Mathesis in Philosophia naturali valeat; et exinde ut Geometras[13] ad examen Naturae strictius aggrediendum, et avidos scientiae naturalis ad Geometriam prius addiscendam horter; ut ne priores suum omnino tempus in speculationibus humanae vitae nequaquam profuturis absumant, neque posteriores operam praeposterâ methodo usque navantes a spe sua decidant.[14] Verùm ut Geometris philosophantibus & Philosophis Geometriam exercentibus, pro conjecturis et probabilibus quae venditantur ubique scientiam naturae summis tandem evidentijs firmatam nanciscamur. Itaque ad institutum redeo de coloribus secundum praecedentes quinque[15] propositiones explicatis disceptaturus.

---

(10) This new proposition is demonstrated in §II, 70. The definition of primitive color is new, as neither it nor an equivalent for it occurs in the *Lectiones opticae*. Newton seems to have largely adopted Boyle's terminology for pigment mixing, which was an unfortunate choice, because Newton's concept of primary color is fundamentally different from that of pigment mixing. In *Touching Colours*, Pt. III, Expt. XII, pp. 219–20, Boyle explained that "there are but few Simple and Primary Colours (if I may so call them) from whose Various Compositions all the rest do as it were Result," and in Expt. XVII, p. 232, he calls them "Primitive" and "Simple." In the "New theory," Prop. 5, such colors are called "original," "simple," and "primary." In his response to Hooke, 11 June 1672, Newton clarified this concept and no longer defined simple colors with respect to their color, as he does here—for all colors, both simple and compound, appear to be simple—but rather with respect to their refrangibility, as those colors whose rays are all refracted equally; *Correspondence*, 1:180–1 = *Phil. Trans.*, 7 (1672):50[95–

primitive colors, however, I understand not only the five specified ones, but also any others to which a disposition to exhibit some uniform kind of rays is attributed.[10]

Finally,[11] I find that all colors of all bodies are produced in no other way but from a certain disposition whereby they are disposed to reflect some rays and to let in others. Thus a red body is one that mostly reflects rays disposed to redness and lets in most of the others; a purple body is one that reflects rays proper to producing that color and lets in the others; but a white body is one that reflects almost all rays; and a black body is one that lets in all rays but reflects very few, though ones of every kind.[12]

But lest I seem to have exceeded the bounds of my position while I under- [6 ≈ 29, 30. It is take to treat the nature of colors, which are thought not to pertain to mathe- affirmed that these matics, it will not be useless if I again recall the reason for this pursuit. The propositions are to relation between the properties of refractions and those of colors is certainly hypothetically and so great that they cannot be explained separately. Whoever wishes to investi- probably, but by gate either one properly must necessarily investigate the other. Moreover, if I experiments or were not discussing refractions, my investigation of them would not then be demonstratively.] responsible for my undertaking to explain colors; nevertheless, the generation of colors includes so much geometry, and the understanding of colors is supported by so much evidence, that for their sake I can thus attempt to extend the bounds of mathematics somewhat, just as astronomy, geography, navigation, optics, and mechanics are truly considered mathematical sciences even if they deal with physical things: the heavens, earth, seas, light, and local motion. Thus although colors may belong to physics, the science of them must nevertheless be considered mathematical, insofar as they are treated by mathematical reasoning. Indeed, since an exact science of them seems to be one of the most difficult that philosophy is in need of, I hope to show—as it were, by my example—how valuable mathematics is in natural philosophy. I therefore urge geometers[13] to investigate nature more rigorously, and those devoted to natural science to learn geometry first. Hence the former shall not entirely spend their time in speculations of no value to human life, nor shall the latter, while working assiduously with an absurd method, fail[14] to reach their goal. But truly with the help of philosophical geometers and geometrical philosophers, instead of the conjectures and probabilities that are being bla-zoned about everywhere, we shall finally achieve a natural science supported by the greatest evidence. I now return to my original design to discuss colors treated according to the preceding five[15] propositions.

6]. This new approach was incorporated in the reformulation of his theory for Huygens, 23 June 1673, where such colors are also called "homogeneal" (*Correspondence*, 1:292–3 = *Phil. Trans.*, 8 (1673):6090), and adopted in the *Opticks*, Bk. I, Pt. I, Defs. VII, VIII.

(11) I: Quarto (Fourth).

(12) Newton establishes this proposition in §§II, 71–6.

(13) Originally (as I): homines Geometras.

(14) I: perpetuò decidant (perpetually fail).

(15) I: quatuor (four). This final sentence was transferred in revision from the opening of §30, and the rest of that article was incorporated into §II, 7.

PROP: i. Radijs diversè refrangibilibus diversi colores competunt.

Lect 2
[7 → 30, 31, 32] Quò primam comprobem repetamus experimentum prismatis sub initio
propositum, nempe[1] [fig II, 1] radij solares obtenebratum cubiculum ad

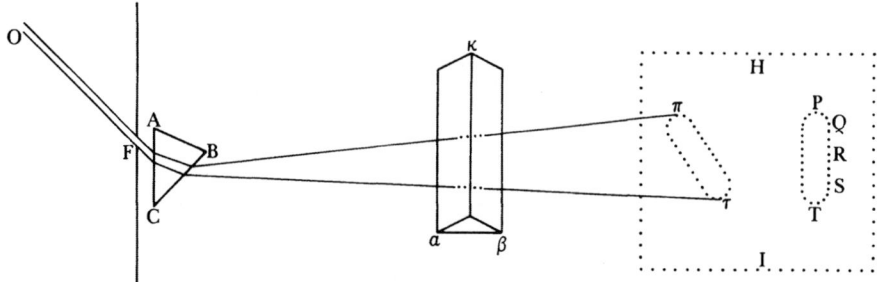

Figure II, 1

foramen *F* ingressi, a Prismate *ABC* quam proxime foramen istud intus disposito refringantur, tendentes deinde versus oppositum parietem *HI* ad
imaginem *PT* ibi depingendam et imago illa ut vulgo notum est coloribus
tingetur ‖ quorum rubeus ad extremitatem *T* a recto curso minùs deviantem,
et purpureus ad alteram procliviorem extremitatem *P* procidet, caeruleus
autem viridisque et flavus ad *Q, R,* et *S* intermedia loca cernentur. Constat
itaque quod radij maxime refracti purpuram exhibent et minimè refracti
ruborem, caeterique, intermediam refractionem passi colores in ordine praefi-
nito intermedios. Sed in majorem evidentiam tum doctrinae de radiorum
diversa refrangibilitate sub initio propositae, tum hujus doctrinae quod certis
refrangibilitatis gradibus certi conveniant colores; videamus e contra an diversi coloris radij diversam refractionem patientur, hoc est an radij versus *P*
tendentes refractionem iterum majorem patientur, quam qui tendunt versus
T. ‖ Id quod varijs modis tentare liceat, quorum facillimum et maximè per-
spicuum sequentem existimo.

[8 ≈ 32. Cui
probandae
adducitur
experimentum.] / Sume aliud Prisma *αβκ* (fig II, 1) et illud alicubi inter primum Prisma ₂₆/
ABC et imaginem *PT* ita colloca ut sit illi Prismati *ABC* transversum sive
parallelum imagini *PT*, radiosque versus *PT* tendentes intercipiat et aliover-
sum refringat, puta versus *ππ*. Hoc facto imaginem *ππ* refractionibus utriusque Prismatis sic effectam videbis ut prius coloratam, sed in alio tamen situ
dispositam: non parallelam imagini *PT*, sed secundum extremitates rubras
manifestò convergentem. Jam cùm radij ad utrosque colores rubeum *T* et
purpureum *P* pertinentes similiter incidant in Prisma secundum *αβκ*, si eandem praeterea refractionem paterentur, imagines *PT* et *ππ* deberent esse parallelae. Et ideo cùm non existant parallelae, sed imaginis *ππ* extremitas purpurea *π* longiùs ab alterâ imagine *PT* transferatur quàm extremitas rubea *τ*:
necessariò concedendum est quod radij ad extremitatem purpuream *P* tendentes magis refringantur quàm qui tendunt ad extremitatem rubeam *T*. hoc
est, quod radij generantes purpuram apti sint ut magis refringantur quàm
ruborem efficientes. Atque idem quoque de coloribus intermedijs eadem ratione constabit, sicut ostendendum proposui.

PROPOSITION 1. TO DIFFERENTLY REFRANGIBLE RAYS THERE CORRE-
SPOND DIFFERENT COLORS.

That I may prove the first, let us repeat the experiment with a prism   Lecture 2
proposed at the beginning. Namely[1] (Fig. II, 1), after solar rays enter a   [7 → 30, 31, 32]
darkened room at the hole *F*, they are refracted by the prism *ABC*, which is
placed inside exceedingly close to that hole, and then tend toward the opposite wall *HI*, depicting the image *PT* there. That image, as has been commonly observed, will be imbued with colors: ‖ Of those, the red will fall ‖
toward the end *T*, deviating less from a straight path, and the purple toward
the other more inclined end *P*; while the blue, green, and yellow will be
perceived at the intermediate positions *Q*, *R*, and *S*. It is therefore evident
that the most refracted rays exhibit purple, the least refracted red, and the
others undergoing an intermediate refraction exhibit the intermediate colors
in the prescribed order. But for greater clarity both of the doctrine initially
proposed concerning the different refrangibility of rays, and of this doctrine,
that to definite degrees of refrangibility there correspond definite colors, let
us see whether conversely rays of a different color will undergo a different
refraction; that is, whether the rays tending toward *P* will again undergo a
greater refraction than those that tend toward *T*. ‖ This may be tested in
various ways, of which I consider the following to be the easiest and clearest.

Take another prism *αβκ* (Fig. II, 1) and place that somewhere between the   [8 ≈ 32. To prove
first prism *ABC* and the image *PT*, so that it is transverse to that prism *ABC*,   this an experiment is
or parallel to the image *PT*, and intercepts the rays tending toward *PT* and   introduced.]
refracts them in another direction, for example, toward *ππ*. Having done this,
you will see the image *ππ*, produced in this way by the refractions of both
prisms, colored as before, but situated in still another position, not parallel to
the image *PT* but manifestly converging along its red ends. Now, since the
rays corresponding to each color, the red *T* and the purple *P*, are similarly
incident upon the second prism *αβκ*, if they then experienced the same refraction, the images *PT* and *ππ* ought to be parallel. Hence, since they are not
parallel, but rather the purple end *π* of the image *ππ* is shifted farther from
the other image *PT* than the red end *τ*, it must necessarily be concluded that
the rays tending to the purple end *P* are refracted more than those that tend
to the red end *T*; that is, the rays producing purple are disposed to be more
refracted than those making red. For the same reason the same thing will also
be evident for the intermediate colors, as I proposed to show.

---

(1) Quò ... nempe] I(§32): Hoc autem ut pateat, iterum repetatur experimentum Prismatis
quod in prioribus adduxeram. Nempe ponatur quòd (To make this clear I again repeat the
experiment with a prism that I introduced earlier. Namely, assume that). In the corresponding
Fig. 11 in the *Lectiones opticae* Newton drew the room in which the experiment is performed,
just as he did in Fig. 2 ≈ Fig. I, 2 and Fig. 13 ≈ Fig. II, 3.

[9 ≈ 33.
Experimenti praefati
circumstantia
notatur.]

In experiendis hisce notari poterit quod quò vicinius anteriori Prismati *ABC*, sive quo remotius a pariete *HI* collocetur Prisma posterius *αβκ*; imagines *ππ* ac *PT* eo magis ab invicem distantes, etiam ad se magis inclinabuntur. Adeò ut angulum semirectum vel paulo mi-norem eo contineant cum Prismata collocantur ad invicem vicinissima. Cujus rei ratio facillima est consi-deranti quod distantiae *Pπ* ac *Tτ* sunt in data quadam ratione. Sic in fig II, 2 si parallelae *Pπ* ac *Tτ* sint in ra-tione datâ, quo majores existant eo major erit inclinatio linearum *PT* ac *ππ*. ‖ Et hinc patet axes imaginum omnium *ππ* productos convenire ad commune aliquod punctum cum axe ipsius *PT*.[2]

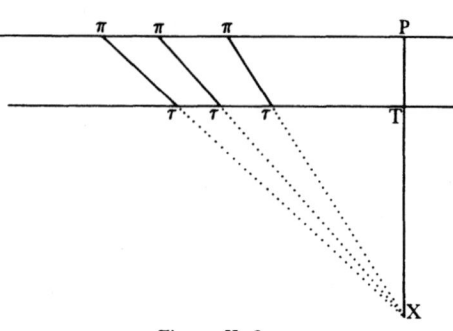

Figure II, 2

[10] Si forte concursus Imaginum desideretur, radij *OF* a Sole / directi eousque   ₂7/ producantur, donec occurrant cum plano *HI* in quod dictae imagines projici-untur, quemadmodum videre est ad *X*: Id quod fiet auferendo prisma *ABC*, ut[3] jubar per *F* trajectum rectà tendat ad *X*: et erit *X* locus ad quem imagines *PT* et *ππ* convergunt. Nam quemadmòdum radij maximè refrangibiles cadunt in *P* ac *π* et minime refrangibiles in *T* ac *τ*, conficientes imagines oblongas, *PT* ac *ππ*: si alij praeterea radij darentur minùs adhuc refrangi[bi]les illi citra punctum *T* ac *τ* in papyrum *IH* caderent: quo pacto imagines illae paulò longiores evaderent, auctae scilicet ad extremitates *T* ac *τ*. Atque ita si fin-geremus radios gradatim minùs atque adhuc minùs refrangibiles dari, usque dum deventum esset ad radios adeò pertinaces ut non possent omninò re-fringi, illi radij prismata sine aliquâ refractione pertranseuntes incidere debe-rent in ipsissimum punctum *X*, ad quod posuimus radios a sole directè ve-nientes tendere. Imagines itaque *PT* ac *ππ*, sic productae convenirent ad *X* et proinde ad idem *X* convergunt. Caeterùm dubitari potest an imagines illae, si eousque producerentur dum convenirent ad *X* forent accuratè rectae vel pau-lulum incurvatae; neque istud (cum multi foret laboris parvique momenti) jam lubet determinare. Sufficit quod ex observatione quàm proxime conver-gant ad *X*.[4]

---

(2) This claim is readily demonstrated, for from the similar triangles *XPπ* and *XTτ*, there follows *Pπ/Tτ = PX/TX*, or *(Pπ − Tτ)/Tτ = PT/TX*. Since *Pπ/Tτ*, and so also *(Pπ − Tτ)/Tτ*, is constant (see Lecture 3, note (32), this volume), as is *PT*, then *TX* is constant, and all the spectra will intersect at a common point *X*. In the corresponding Fig. 12 of the *Lectiones opticae* the projections of the spectra *ππ* intersecting at *X* are not drawn.

The next step is to determine the point of intersection. Before setting this forth in the next article, Newton began an alternative, incomplete account that he then deleted, the only para-graph in the entire revision to have been deleted. The figure is missing, but Fig. II, 2 with *ξ* being understood for *X* should suffice: "Nam omnes lineae *ππ* productae convenient cum *PT* in eodem puncti *ξ* ut pateat, atque adeo quo major sit *Pπ* vel *Tτ* eo majorem angulum *Pξπ* subtendet. Caeterùm cum dixi lineas omnes *ππ* cum ipsâ *PT* in eodem puncto *ξ* concurrere, non secus

In doing these experiments one will be able to observe that as the second [9 ≈ 33. A feature prism *αβκ* is placed closer to the first prism *ABC* or farther from the wall *HI,* of the preceding the images *ππ* and *PT* will become more distant from one another as well as experiment is more inclined to one another, until they make half a right angle or a little less noted.] when the prisms are placed as close as possible to each other. This is ex-plained very easily if one considers that the distances *Pπ* and *Tτ* are in a certain given ratio. Thus in Fig. II, 2, if the parallels *Pπ* and *Tτ* are in a given ratio, the greater they are, the greater will be the inclination of the lines *PT* and *ππ*. ‖ Consequently it will be evident that the axes of all the images *ππ* ‖ when produced will meet the axis of *PT* at some common point.[2]

If perhaps the intersection of the images is desired, let the sun's direct rays [10] *OF* be extended until they meet the plane *HI* onto which those images are projected, as is seen at *X.* This will be done by removing the prism *ABC,* so that[3] the sun beam transmitted through *F* tends directly to *X*; and *X* will be the place at which the images *PT* and *ππ* converge. For just as the most refrangible rays fall at *P* and *π,* and the least refrangible ones at *T* and *τ,* forming the oblong images *PT* and *ππ,* if there be granted in addition other rays still less refrangible than those, they would fall below the points *T* and *τ* in the paper *IH;* in this way those images would become a little longer, namely, expanding at the ends *T* and *τ.* If we imagine rays to be given gradually less and still less refrangible, until rays so persevering were reached that they could not be refracted at all, those rays passing through the prisms without any refraction ought to fall upon the very point *X* to which we have assumed the rays coming directly from the sun tend. Therefore the images *PT* and *ππ* thus produced would meet at *X* and consequently converge to the same *X.* It can be questioned, though, whether those images, if they were produced until they met at *X,* would be exactly straight or slightly curved; but now that is unpleasant to determine, since it would be a major effort with little significance. It is sufficient that by observation they converge as close as possible to *X.*[4]

imagines omnes *ππ* (in fig——) cum ipsa *PT* ad eundem quendam locum convergunt, et ibidem concurrerent si modo fingerentur eousque produci: cujus quidem concursûs locum determinasse propositio fortè non praetermittenda censeatur. Ad ejus autem determinationem nihil aliud agen-dum est quam ut Radij *OF* a sole . . . " (For all the lines *ππ* when produced will meet *PT* in the same point *ξ,* as should be obvious; and, moreover, the larger *Pπ* or *Tτ* is the greater will be the subtended angle *Pξπ.* When, however, I said that all the lines *ππ* intersect *PT* in the same point *ξ,* I meant nothing other than that all the images *ππ* (in fig.——) converge toward *PT* at some one place and would intersect there, provided they were imagined to be extended to there. It might of course be thought that to determine the place of their intersection, a proposition, ought perhaps not to be passed over. To do so, however, nothing other need be done than [to extend] the rays *OF* from the sun . . .).

(3) Id . . . ut] Originally: hoc est fiet ablato prismate *ABC.*

(4) Newton's assumption that the axes of the spectra *ππ* intersect the axis of spectrum *PT* at the position of the sun's direct image *X* is mistaken. He later made a similar erroneous premise in a related experiment in the *Opticks* (Bk. I, Pt. I, Prop. VI, Expt. XV), where to establish the sine law of refraction for rays of each color he used inclined spectra formed by varying the angle of the second, transverse prism that, however, always remained contiguous to the first one. Lohne's analysis of the inclined spectra in the *Opticks* ("Newton's 'proof' of the sine law," pp. 389–91)

[11]
Num 23

De hoc experimento sub initio observabam quòd omnibus adversatur objectionibus quae contra doctrinam de inaequali refrangibilitate traditam proponi possunt, ex eo quòd per transversam refractionem secundi Prismatis constat inaequales refractiones non esse fortuitas et irregulares neque ex radij cujusque diffusione vel dilatatione ortas esse, aut aliâ quavis causâ praeter dispositionem cujusque radij ad refractionem in gradu aliquo certo et constante patiendam, quandoquidem cujusque refractio in utroque Prismate secundum illam legem peragitur.[5] Addo jam quod exhinc etiam constat refractiones singulorum radiorum secundum easdem leges peragi sive commisceantur cum radijs aliorum generum, ut fit in albâ luce, sive separatim refringantur, luce prius in colores conversâ. Nam experiri est quod similes sunt refractiones posterioris Prismatis cùm proxime collocatur post alterum Prisma antequam lux per id trajecta transmigret in colores, atque cùm longius post illud Prisma statuitur ubi lux evasit colorata.[6]

[12 = 34. Idem
instrumentis
refractiones
dimetientibus posse
probari. Tamen
evidentiam
experimenti jam
descripti sufficere.]

/ Si cui in potestate est instrumentum aliquod ad quantitates refractionum ₂8/ accuratè mensurandas paratum, nullus dubito quin istius etiam ope seorsim dimetiendo refractiones diversorum generum radiorum facile observabit earum differentias: licet ego praedictis tanquam manifestissimis acquiescens haud operae pretium duxerim rem alijs modis experiri. Verum ut cuique magis pateat quanta sit praedictorum evidentia, quaedam quae exinde scaturiunt notatu dignissima proferre non pigebit.

can be applied here with some modifications. In the figure, the first prism alone at F casts a spectrum PT in the vertical plane that contains the incident ray OF, the prism's principal plane, and the refracted rays FP, FQ, and FT, with their deviation D_x being the angle made with the ray $OFAX$. When the spectrum is observed perpendicularly to the refracted rays, it forms an arc PTA on a sphere with center F and radius R equal to the distance of the prism to the wall, assuming, as Newton does, that the rays' re-

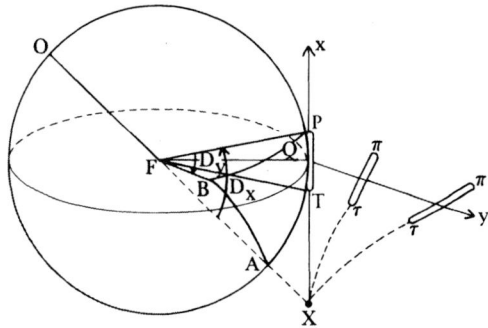

frangibility extends beyond the visible region PT continuously to zero refrangibility. The projection of this spectrum onto a screen tangent to the sphere at Q is the straight line PTX, where PT is the visible portion and X is the place of the sun's direct image. When the second prism is inserted transverse to the first one, its spectrum is the skewed arc BA, and the deviation D_y of the rays refracted by this prism is the angle they make with the rays, such as FP and FQ, refracted by the first prism. The stereographic projection of the arc BA onto the wall will be a circle $\pi\tau X$ whose small visible portion $\pi\tau$ will appear as an inclined straight line. If we take a coordinate system with the x-axis along the vertical and its origin at the green Q, and let $\overline{D}$ be the mean deviation for the green rays and ρ the distance of the second prism to the wall, then the equation for the spectra are

$$x = R \tan (D_x - \overline{D}_x) \text{ and } y = \rho \tan D_y / \cos (D_x - \overline{D}_x).$$

If now we ignore the cosine that is very nearly 1; take $D_x \approx D_y$, since the prisms are identical and the refractions in them nearly so; and also take $\rho = R$ when the spectra are most inclined; then effectively $x = R \tan (D - \overline{D})$ and $y = R \tan D$. The common point of intersection x_c is the point where the most inclined spectrum intersects the x-axis, and from similar triangles we have $x_T/(\overline{y} - y_\tau) = x_c/\overline{y}$, or

I observed at the beginning concerning this experiment that it counters all [11] objections that can be presented against the doctrine of unequal refrangibility §I, 23 that I taught, because it is evident from the transverse refraction of the second prism that the unequal refractions are not fortuitous and irregular, nor do they arise from a diffusion or expansion of every ray or from any other cause except the disposition of every ray to undergo a refraction at some definite and constant degree, since the refraction of every one occurs in each prism according to that law.[5] I add that it is also evident from this that the refractions of the individual rays occur according to the same laws, whether they are mixed with other sorts of rays, as in white light, or are refracted separately after the light has previously been converted into colors. For it is possible to prove that the refractions of the second prism are alike, both when it is placed close behind the other prism before the light passing through it has changed into colors, and when it is placed farther beyond that prism where the light has become colored.[6]

If anyone possesses some instrument equipped for accurately measuring the [12 = 34. The same quantity of refractions, I do not doubt that also with its aid in measuring the thing can be proved refractions of the different kinds of rays individually he will easily observe with instruments their differences; although, being content with the preceding as very clear, I refractions, yet the considered that it was not at all worthwhile to test this by other ways. But to evidence of the make it clearer to everyone how great the evidence of the preceding is, I will experiment just now not be displeased to reveal some very noteworthy things that flow from it. sufficient.]

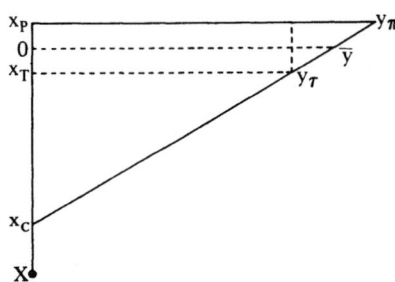

$x_c = R \tan (D_T - \overline{D}) \tan \overline{D}/(\tan \overline{D} - \tan D_T)$
$\quad = R \tan \overline{D}/(1 - \tan D_T \tan \overline{D}) \approx -R \tan \overline{D}/\sec^2 \overline{D},$

assuming $D_T \approx \overline{D}$. The point of common intersection therefore falls much higher than the sun's direct image, which is at $X = -R \tan \overline{D}$, where $\overline{D}$ is about 41°. Finally, since the tangent of the angle that the most inclined spectrum makes with the vertical axis equals $\overline{y}/x_c = \sec^2 \overline{D}$, that angle is approximately 60°, far from the 45° claimed by Newton.

A number of causes may be responsible for Newton's erroneous results: aligning the two prisms imperfectly; using prisms with refracting angles that are unequal, as is the case for most of his prisms (see his letter for Lucas, 18 August 1676, *Correspondence*, 2:77–8 = *Phil. Trans.*, 11 (1676):699–701); or using the hollow prisms filled with water that he suggests in the related experiment in the *Opticks*, p. 55 = Add. 3970, f. 58ʳ.

(5) Although Newton does not even specify that law here, in the *Opticks* he derived the sine law of refraction as well as his partial dispersion law (that the relative refraction or degree of refrangibility of the rays is the same in all substances) from inclined spectra. He correctly deduced the latter from his observation that the edges of the spectra remain straight lines even after three or four transverse refractions (*Opticks*, Bk. I, Pt. I, Prop. II, Expt. V, p. 29 = Add. 3970, f. 39ʳ); for if the partial dispersions of the prisms were different, the spectra would be curved. Since Newton's prisms were (physical imperfections apart) always of the same material, either the same glass or water, he ought to have observed straight spectra. August Kundt describes and illustrates the shape of inclined spectra in his "Ueber anomale Dispersion," *Annalen der Physik und Chemie*, ser. 2, **144** (1871):128–37.

(6) For a suggested dating of this article, see Lect. I, 3, note (9).

[13 = 35. Illud
promovetur
aliquantum, idque
vel tribus
prismatibus
adhibitis,]

Sit *Fφ* (fig II, 3) paries vel operculum fenestrae duobus foraminibus *F* et *φ* luci pervium, ijsque digitos duos ab invicem distantibus. et intus disponantur duo Prismata *ABC DEG* in situ sibi invicem parallelo et perpendiculari ad lineam *Fφ* per centra foraminum ductam: quae duo lucem ingressam refrin-

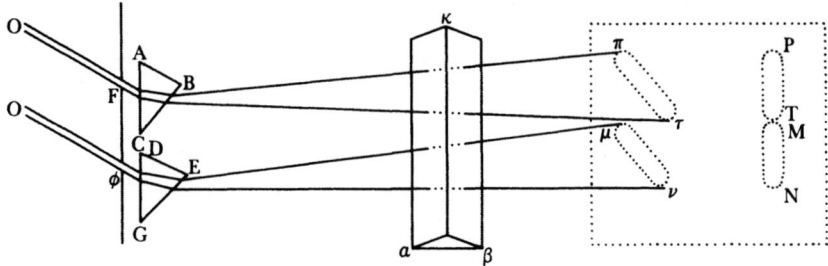

Figure II, 3

gant ad imagines duas *PT* et *MN* in oppositum parietem projiciendas, simili prorsus modo quo factum fuit experimento priori. Et praeterea sint anguli Prismatum *ACB, DGE* (comprehensi planis refringentibus) aequales. Quibus ita constitutis videbis imagines *PT* et *MN* in directum jacentes cum extremitatibus earum *T* et *M* contiguis. Quod si non eveniat, situs unius e Prismatibus parùm mutandus est donec extremitates contiguas esse cernas, vel fortè nonnihil coincidentes. Purpura *M* et rubore *T* sic juxta positis, adhibeatur Prisma tertium *αβκ* quod primis Prismatibus et eorum imaginibus interponatur in situ ad lineam *Fφ* sive ad imagines dictas *PT, MN,* parallelo; ita nempe ut radios utriusque Prismatis *ABC, DEG* tendentes versus *PT* et *MN* pariter intercipiat, eosque refringens aliò projiciat, quemadmodum ad *ππ* ac *μν*. Adeo ut quae duobus Prismatibus in priori specimine facta sunt hic videas facta tribus. His ita paratis et constitutis, videbis imagines *ππ* et *μν* ab invicem disjunctas esse, quae prius apud *PT* et *MN* fuerunt contiguae, et in directum positae: ita quidem ut purpura *μ* in extremitate imaginis *μν* magis distet ab imaginibus primis *PT* et *MN* quam rubor *τ* in extremitate imaginis *ππ*. Id quod nullo[7] modo potuisset accidisse nisi radij ad purpuram generandam apti aliquantò magis refringerentur ex incidentia pari quàm radij generantes rubedinem. Etenim cùm radij coloris utriusque pariter incidant in Prisma posterius *αβκ*; pariter etiam emergerent si aequaliter refringerentur, et exinde depingerent imagines *ππ* et *μν* prioribus *PT* et *MN* parallelas et in directum jacentes. Dixi radios utriusque coloris (purpurei rubeique) pariter incidere in Prisma posterius *αβκ*: Quod ne moram / injiciat alicui, concipien- ₂9/ dum est quod radij *FT* tantum inclinantur versus extremitatem ejus *K*[8] quantum alteri *φM* versus extremitatem alteram *αβκ*:[8] et sic incident pariter sive ad eosdem angulos, licet non paralleli. Siquis tamen velit efficere ut incidant etiam paralleli, nihil aliud agendum est, quam ut alterum e Prismatibus anterioribus *ABC* vel *DEG* circa suum axem paululum convertatur donec inter *T* et *M*, interiores imaginum extremitates, tanta intercedat distantia, quanta est inter foramina *F* et *φ*, sive quanta isti rei sufficiens videatur,[9] Imaginibus ad

Let $F\phi$ (Fig. II, 3) be a wall or a window shutter with two holes F and ϕ, which transmit light and are two inches from one another; inside place two prisms, ABC and DEG, in a position parallel to one another and perpendicular to the line $F\phi$, which is drawn through the center of the holes; and let these two prisms refract the admitted light, projecting two images, PT and MN, onto the opposite wall in a way wholly similar to what was done in the previous experiment. Moreover, let the prisms' angles, ACB and DGE, made by the refracting planes be equal. After this is so arranged, you will see the images PT and MN lying in a straight line with their ends T and M touching. If this does not occur, the position of one of the prisms must be changed a bit until you perceive that the ends touch or perhaps coincide somewhat. When the purple M and the red T are so juxtaposed, add a third prism, $\alpha\beta\kappa$, which is placed between the first prisms and their images in a position parallel to the line $F\phi$ or to those images PT and MN; that is, so that it may simultaneously intercept the rays tending toward PT and MN from each prism, ABC and DEG, and refract and project them elsewhere, as to $\pi\tau$ and $\mu\nu$. And here you shall see done with three prisms what was done with two prisms in the previous example. Having thus prepared and arranged these things, you will see the images $\pi\tau$ and $\mu\nu$, which earlier touched at PT and MN and lay in a straight line, separated from one another; so that, in fact, the purple μ at the end of the image $\mu\nu$ is farther from the first images PT and MN than the red τ at the end of the image $\pi\tau$. This could in no way[7] have happened, unless at equal incidence rays disposed to generating purple were refracted somewhat more than rays generating red. Since the rays of each color fall upon the second prism $\alpha\beta\kappa$ similarly, they would also emerge similarly if they were refracted equally, and then they would depict images, $\pi\tau$ and $\mu\nu$, parallel to the first ones, PT and MN, and lying in a straight line. I have said that the rays of each color, purple and red, fall upon the prism $\alpha\beta\kappa$ similarly. Lest this hinder anyone, it ought to be understood that the rays FT are inclined as much toward their end K[8] as the other ones, ϕM, are inclined toward the other end $\alpha\beta\kappa$;[8] thus they are incident similarly, or at the same angles, even if not parallel. Yet if anyone wants to make them also fall parallel, nothing other need be done than to rotate one of the first prisms, ABC or DEG, a little about its axis until as great a distance lies between the images' inner ends T and M as between the holes F and ϕ, or as great as seems sufficient[9]

[13 = 35. The former experiment is extended somewhat, either by using three prisms.]

---

(7) I: nullo prorsus (in absolutely no way).

(8) Read κ and $\alpha\beta$ respectively for K and $\alpha\beta\kappa$. The text was not correspondingly altered when in revision Fig. II, 3 was relabeled.

(9) quanta . . . videatur] Originally (as I): quanta intersit foraminibus F et ϕ, sive quantam isti rei sufficientem judicaverit (as between the holes F and ϕ, or as great as is judged sufficient).

istam distantiam in directum jacentibus. et Prismate $\alpha\beta\kappa$ deinceps interposito, facilè percipiet quod incidentes parallelè emergunt inclinati, tum quod imagines non amplius in directum jacebunt, tum quòd purpura M ad majorem distantiam transferetur quam rubedo T.

[14 = 36. Vel contractiùs duobus.] Si tria Prismata non praesto sint experimentum jam recitatum duobus experiri possis idque modo magis expedito et facili. Sit $ABCDE$ (fig II, 4)

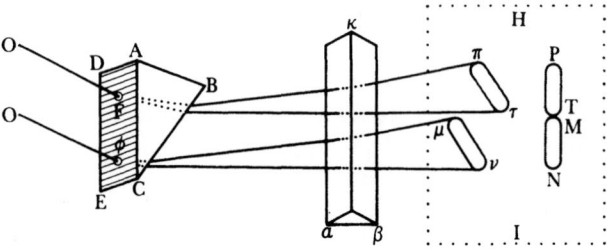

Figure II, 4

Prisma cujus unum latus planum $A[C]DE$ papyro denigratâ tegatur duobus parvis foraminibus F et ϕ luci perviâ, quorum foraminum situs esto ad longitudinem Prismatis transversus. Tum Prismate hoc ita disposito, ut radij permeantes ista foramina terminentur in oppositum quoddam planum, puta papyrum HI: transferatur ista papyrus ultra citraque donec videas imagines duas PT et MN contiguis extremitatibus in directum conjunctas, ut prius. Deinde altero Prismate $\alpha\beta\kappa$ interposito in situ ad alterum transverso, videbis imagines illas PT et MN ad $\pi\tau$ et $\mu\nu$ ita translatas esse ut non amplius jaceant in directum, rubedine τ a PN minùs remotâ quàm purpura μ, sicut in prioribus contingebat.

[15 = 37. Idem aliter promovetur] Est et aliud ex eodem fonte derivatum specimen haud expertu difficilius aut minoris evidentiae. Prismate ABC (Fig II, 5) juxta foramen F ut prius collocato; ad distantiam convenientem (veluti duodecim pedum) statuatur aliud Prisma $\alpha\beta\kappa$ in situ transverso respectu prioris, vel forte parallelo, aut alio quovis pro arbitrio: ita tamen ut anterius Prisma ABC lucem refractam et coloratam projiciat in aliquod ex ejus planis lateribus $\alpha[\kappa]$. Quod quidem latus obducatur papyro denigratâ et exiguo foramine G per medium transfossâ, per quod aliqui ex radijs ab anteriore Prismate refractis transeant in hoc Prisma posterius: ubi cùm rursus refracti fuerint pergant ad papyrum HI abinde decem pedibus vel pluribus distantem. Quibus ita constructis et dispositis / in situ illo figatur papyrus HI et prisma posterius $\alpha\beta\kappa$. Denique ₂10/ prae manibus sumatur anterius Prisma ABC non ut moveatur a loco ejus, sed ut motu tantùm angulari nunc huc nunc illuc paululum inclinetur, ut alios atque alios colores successive trajiciat per foramen G in oppositam papyrum HI. Et videbis quòd color quilibet diversus ad locum diversum perget. Veluti cùm ea sit positio Prismatis ABC ut rubeum colorem projiciat in G, si ponatur quod ille color ab altero Prismate $\alpha\beta\kappa$ refringatur ad T tum positione

for the images to lie in a straight line at that distance. Inserting the prism $\alpha\beta\kappa$ again, you will easily perceive that the rays falling parallel emerge inclined, both because the images no longer lie in a straight line and because the purple *M* is shifted a greater distance than the red *T*.

If three prisms are not available, you can try the experiment just related with two, and in a more convenient and easy way. Let (Fig. II, 4) *ABCDE* be a prism, and cover one of its plane sides *A[C]DE* with a blackened paper having two small holes *F* and *φ* to transmit light; and let the holes' position be transverse to the prism's length. When this prism has been so placed that the rays passing through those holes are outlined on some opposite plane, for example, the paper *HI,* shift that paper back and forth until you see the two images *PT* and *MN* joined in a straight line with their ends touching, as before. Then inserting the second prism $\alpha\beta\kappa$ in a position transverse to the other one, you will see those images *PT* and *MN* translated to $\pi\tau$ and $\mu\nu$, so that they no longer lie in a straight line, with the red τ being less remote from *PN* than the purple μ, just as it occurred in the previous ones. [14 = 36. Or more simply by two.]

There is another example drawn from the same source that is not at all more difficult to try or less evident. As before, having placed the prism *ABC* (Fig. II, 5) next to the hole *F,* at an appropriate distance (for instance, twelve [15 = 37. The same thing is accomplished differently.]

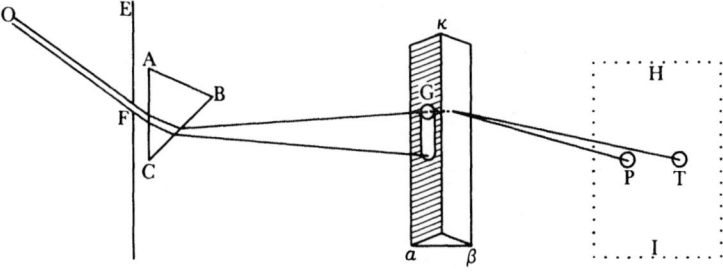

Figure II, 5

feet) set another prism $\alpha\beta\kappa$ in a position transverse to the first, or else perhaps parallel, or in any arbitrary way, but such that the first prism *ABC* projects the refracted colored light onto one of its plane sides $\alpha[\kappa]$. However, cover this side with a blackened paper pierced through the middle with a small hole *G* through which some of the rays refracted by the first prism may pass into this second prism where, when they have been refracted again, they continue to the paper *HI* ten or more feet away from it. After these have been thus constructed and set up, fix the paper *HI* and the second prism $\alpha\beta\kappa$ in that position. Finally, grasp the first prism *ABC,* not so it may be moved from its place, but only so it may be turned back and forth a bit with an angular motion to cast one and then another color successively through the hole *G* onto the facing paper *HI.* You will see that every different color proceeds to a different place. For example, when the position of the prism *ABC* is such that it projects the red color to *G,* if it is assumed that that color is refracted to *T* by the second prism $\alpha\beta\kappa$, then when the position of the

Prismatis *ABC* paululum mutatâ inclinando circa axem donec purpura cadat in *G*, videbis quod ille color juxta obliquiorem tramitem refringetur, puta ad *P*. Et pari modo si color aliquis intermedius incidat in *G*, idem refringetur ad locum ipsis *P* ac *T* interjacentem. Quamobrem cum radij cujuslibet generis pergentes a foramine *F* positione dato ad foramen *G* positione datum, et ideo similiter incidentes in Prisma posterius *αβκ* refringantur ad loca diversa *P*, *T*, caeteraque intermedia: constat quod inaequaliter refringantur et cum refractus *GP* observetur magis deflectere ab incidente *FG* quàm refractus *GT*: constat quòd radij purpuram exhibentes magis refringantur quàm exhibentes ruborem, caeterique deinceps in ordine intermedio.

[16 = 38. Quod specimen, circumstantiâ variatâ, fit maximè scientificum.]

Siqua forsan oboritur suspicio, quòd ex motu Prismatis *ABC* foraminibus *F* et *G* interpositi incidentia radiorum diversos colores efficientium tantum varietur quantum sufficiat ad varietatem efficiendam locorum *P*, *T*, &c. ad quos refringuntur: quamvis motus iste sit exiguus et ineptus huic effectui, tamen ut suspicio illa prorsus eximatur anterius Prisma *ABC* ad alteras partes foraminis *F* solem versus collocandum est, ut radij incidentes in foramen *G* directè veniant a dicto foramine *F*. Eo enim pacto cùm foramina *F* ac *G* positione determinentur, positio radiorum per utrumque trajectorum determinabitur, eademque accuratè erit omnium incidentia, quoscunque colores exhibentium; et tamen diversicolorum refractio non secus peragetur ad loca diversa *P*, *T*, &c quàm modò explicui.

[17] / Ex abundanti denique placet alium recensere modum quo haec eadem ₂11/ tentari possint, nè copia desit experturis. Nimirum radijs ut priùs per prisma *ABC* (fig II, 6) trajectis: ad distantiam quamlibet puta viginti pedum adhibea-

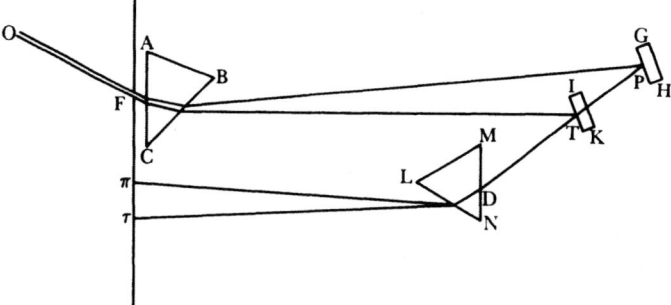

Figure II, 6

tur speculum planum quale *IK* vel *GH* quod eosdem versus locum quemvis *D* reflectat, ubi per aliud prisma *LMN* transmittuntur denuò versus *π* vel *τ*. His positis si speculum istud ita collocetur ad *IK* ut rubrum colorem reflectat, et notetur locus *τ* ad quem hi radij tendunt postquam transiere per prisma *LMN*, Deinde speculum statuatur ad *GH* ut violaceum vel caeruleum colorem ad idem prisma *LMN* secundum eandem lineam *PTD* reflectat, et notetur locus *π* ad quem isti etiam radij a dicto prismate *LMN* refringuntur; invenietur quòd caeruleus color versus *π* refractus longiùs divaricabit ab

prism *ABC* is changed slightly by turning it about its axis until the purple falls at *G,* you will see that color refracted along a more oblique path, say to *P.* Similarly, if some intermediate color falls on *G,* it will likewise be refracted to a place between *P* and *T.* Consequently, since rays of any sort whatsoever that proceed from the hole *F* in a given position to the hole *G* in a given position, and that therefore are similarly incident upon the second prism *αβκ,* are refracted to different places *P* and *T,* and all the others in between, it is certain that they are refracted unequally. Since the refracted ray *GP* is observed to be deflected more from the incident one *FG* than the refracted ray *GT,* it also is certain that rays exhibiting purple are refracted more than those exhibiting red and the others in sequence are refracted in an intermediate order.

If any suspicion perhaps arises that because of the motion of the prism *ABC,* placed between the holes *F* and *G,* the incidence of the rays producing the different colors is changed sufficiently to cause the diversity of the places *P, T,* and so forth to which they are refracted, that motion, however, is too small and unsuitable for this effect. Nonetheless, to remove that suspicion altogether the first prism *ABC* should be placed on the other side of the hole *F* toward the sun, so that the rays falling upon the hole *G* come directly from that hole *F.* Since in this way the holes *F* and *G* are fixed in position, the position of the rays transmitted through both holes will be fixed, and the incidence of all rays, whatever color they exhibit, will be exactly the same; yet the refraction of the different colors will proceed to the different positions *P, T,* and so forth, no differently than I already have explained.

[16 = 38. This example becomes most scientific when a feature is varied.]

Finally, in addition it seems appropriate to examine another way whereby these same things can be tried, lest experimenters lack a supply of them. To be sure, after the rays as before have passed through the prism *ABC* (Fig. II, 6), at any distance, perhaps twenty feet, put a plane mirror, such as *IK* or *GH,* that reflects them toward some place *D,* where they are again transmitted through another prism *LMN* toward *π* or *τ.* With these arrangements, if that mirror is placed at *IK* so that it reflects the color red; and if one notes the place *τ* to which these rays tend after having passed through the prism *LMN*; and then if the mirror is set at *GH* so that it reflects the violet or blue color to the same prism *LMN* along the same line *PTD*; and if one notes the place *π* to which those rays are also refracted by that prism *LMN,* it will be found that the blue color refracted to *π* will deviate farther from the incident

[17]

incidentibus radijs *PTD* quàm rubeus refractus versus τ: atque adeò quòd radij caeruleum generantes majorem refractionem patiantur quàm generantes rubeum.

Cùm veritatem propositam sic fecerim stabilitam, hanc propositionem concludam annotando connexionem et affinitatem quam coloribus et refractionibus interesse dixeram: Nempe ex ostensis non solùm pateat, quod diversa colorum genera cum definitis gradibus refrangibilitatis reciprocantur: Sed et ijsdem[10] experimentis probatur dari radios diversè refrangibiles, et radios diversè refrangibiles esse diversi coloris, ijsdemque[11] probatur e contra radios diversicolores esse diversè refrangibiles, et inde radios diversè refrangibiles dari. Et scopus eorum quae in primis lectionibus de dispari refrangibilitate radiorum edocui, quoad causas colorum intelligendas multùm illustratur; ut pateat quod una absque alijs dilucidè tractari nequeant.

PROP: 2. RADIORUM FORMAE SIVE DISPOSITIONES COLORIFICAE NON SUNT REFRACTIONE MUTABILES.[1]

Lect 3
[19]
Transacta assertione quod diversicolores radij sint diversè refrangibiles et contra; videamus jam an cujuscunque radiorum / seorsim[2] spectati generis ₂12/ color a refractione mutari possit; et hoc a novissimè tradito[3] experimento quadantenus decernitur. Scilicet cum extrema purpura incidebat in foramen *G,* radij secunda vice ad *P* refracti purpuram iterum exhibuere sine aliqua flavedine rubore aut viriditate exinde generata, et cum extrema rubedo in *G* projiciebatur, eadem rubedo in *T* absque violaceo caeruleo aut viridi emergente apparuit.

Sed experimentum nondum omnibus numeris absolutum est, nam ubi Prisma αβκ non transversum sed alteri Prismati *ABC* parallelum statuebatur,[4] e purpura caeruleus et e rubedine flavus eliciebatur, praesertim

---

(10) A marginal note in I refers to §3 [= I, 3] and §30 [→ II, 6, 7].

(11) A marginal note in I refers to §32, ff. [≈ II, 7, ff].

(1) Newton first invoked this proposition and recognized its significance in the midst of his argument that colors are innate to the sun's direct light in the *Lectiones,* §76 → II, 67, where he promised to prove it "afterward" but did not. Because of its fundamental significance for his theory and the intrinsic difficulty in demonstrating it in its greatest generality (see note (18), this lecture), Newton reformulated this proposition in each statement of his theory. In the "New theory," Prop. 3, he coupled the immutability of color with that of degree of refrangibility, though he justified the proposition only for color; *Correspondence,* 1:97 = *Phil. Trans.,* 6 (1671/2):3081–2. When he sketched his theory for Huygens on 23 June 1673, he resolved the principle into three distinct propositions whereby the rays are not changed by one another (Prop. 6), by refraction (Prop. 7), or by a bordering shadow (Prop. 8); *Correspondence,* 1:293–4 = *Phil. Trans.,* 8 (1673):6091. In the *Opticks,* Bk. I, Pt. II, Prop. II, he coupled the principle of color immutability with the correspondence between refrangibility and color; see Lect. 3, note (26), this volume.

A limited form of the principle of color immutability had been recognized by at least two of Newton's predecessors. Boyle had found "that the Prismatical Iris . . . might be Reflected without losing any of its several *Colours* . . . not onely from a plain Looking-glass and from the calm Surface of Fair Water, but also from a Concave Looking-glass; and that Refraction did as little Destroy those Colours as Reflection" (*Touching Colours,* Pt. III, Expt. V, p. 193). Before Boyle, Marcus Marci had even more fully stated and demonstrated this principle: "Theorem XIX. A

rays *PTD* than the red refracted toward τ, and hence that rays generating blue experience a greater refraction than those generating red.

Since I have thus established the proposed truth, I shall conclude this proposition by commenting on the connection and relation that I had said belong to colors and refractions. Namely, it is not only evident from what has been shown that different kinds of colors correspond to definite degrees of refrangibility, but also it is proved by the same experiments[10] that there exist differently refrangible rays, and that differently refrangible rays are differently colored; and conversely it is proved from the same experiments[11] that differently colored rays are differently refrangible, and consequently there exist differently refrangible rays. The aim of what I taught in the first lectures about the unequal refrangibility of rays with respect to understanding the causes of colors is made much more evident, so that it is obvious that the one cannot be clearly discussed without the other.

[18 = 39. Conclusion of the examination of the relation of colors and refractions.]

PROPOSITION 2. THE FORMS OR COLORIFIC DISPOSITIONS OF RAYS ARE NOT MUTABLE BY REFRACTION.[1]

Having concluded with the assertion that differently colored rays are differently refrangible, and conversely, let us now see whether the color of any kind of ray when observed separately[2] can be changed by refraction. This is partially settled by the experiment recently related.[3] Namely, when the extreme purple fell on the hole *G*, the ray refracted a second time to *P* again exhibited purple without any yellow, red, or green generated from there; and when the extreme red was projected to *G*, the same red appeared at *T* without violet, blue, or green emerging.

Lecture 3 [19]

But the experiment is not yet perfect in all respects. For when the prism αβκ was placed not transverse but parallel to the other prism *ABC*,[4] blue was drawn out from the purple, and yellow from the red, particularly if the

---

following reflection of a colored ray does not change the nature of its color . . . Theorem XX. A following refraction of a colored ray does not change the species of its color" (*Thaumantias*, p. 100). He had demonstrated the former principle, just as Boyle had, by reflecting colored light from specular, colorless bodies. Boyle, however, had demonstrated that colors are mutable by reflection from colored bodies, and undoubtedly Marci would have held to the same view, since it was nearly universally accepted that the color of the light and of the body would mix together and form a compound color just as with pigments; see Lect. II, 9, note (4).

(2) The statement of this proposition is flawed. Since compound rays are mutable, it should have been restricted to monochromatic rays. Newton, though, shows that he is aware of this restriction by this phrase, "when observed separately," that is, when the monochromatic rays are separated from one another.

(3) That is, in §II, 15 = 37.

(4) The parallel prisms should have their vertices in the same direction (as in the *Opticks*, Bk. I, Pt. I, Fig. 18) in order further to disperse and separate the colors. In the arrangement of the *experimentum crucis* sketched in his letter for Pardies on 10 June 1672, Newton, on the contrary, placed the prisms with their vertices in opposite directions in order to minimize the second dispersion and separation of the colors; *Correspondence*, 1:166–7 = *Phil. Trans.*, 7 (1672): 5016–7. In the "New theory" Newton noted that after passing through a single prism the colors are imperfectly separated and then brusquely observed that "how to make such further separations, will scarce be difficult to them, that consider the discovered laws of Refractions" (*Correspondence*, 1:102 = *Phil. Trans.*, 6 (1671/2):3087).

si non summae colorum extremitates per *G* trajiciebantur; cum autem viriditas trajecta fuit colores utrinque proximi (caeruleus nempe et flavus)
emersêre, et sic flavus citriusque ruborem et viriditatem, ac caeruleus viriditatem et purpuram praebuerunt. Eorum itaque reminisci oportet quae sub
initio[5] de more quo oblonga haec imago *PT* ex circulis in directum positis
formatur explicui: et inde constabit hosce colores non simplices esse sed e
plurium mistura componi. Nam concipe genus radiorum aequaliter refrangibilium et intensam purpuram gene-
rantium ab integro solari disco
profluere, et per Prisma versus
imaginem *PT* (fig II, 7)[6] trajectos
incidere in circulum *AC*. Deinde
aliud concipe radiorum paulò minus refrangibilium genus in alium
circulum *YZ* (qui priorem in *G*
contingat) incidere, et manifestum

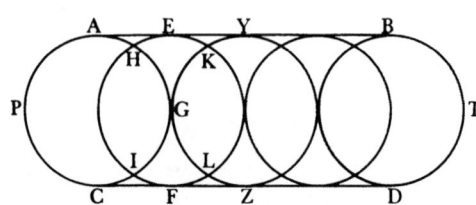

Figure II, 7

est quod nulli istorum generum radij commiscebuntur, quippe cum circulos
AC et *YZ* ex nulla parte coincidentes occupent. Quod si tertium radiorum
intermediam refractionem passorum genus in circulum [*EF*] quasi in medio
positum incidere fingas, patebit aliquos ex istis cum utrisque prioribus in
spatijs *HI* et *KL* misceri in quibus nempe circuli ab illis illuminati coincidunt.
Atque ita si concipias imaginem totam *PT* ex innumeris circulis in / longum ₂13/
dispositis componi quorum quilibet a diversis radiorum generibus illuminatur, constabit quod in omni ejus parte radij heterogenei commisceantur. Quibus proinde per iteratam refractionem magis segregatis, color quilibet in
simpliciores resolvi debet. Sic in viridi latet flavus et caeruleus qui tamen non
conspiciuntur, tum quod viriditatem generantes, sive (ut perspicuitatis gratiâ
voces fingam) viridiformes[7] radij propter copiam praepollent, tum quod
flavus et caeruleus viridem componunt.[8] Sed quatenus per secundam refractionem secernuntur, unusquisque sub propria forma videbitur. Et sic in alijs.

[20]   His perspectis periclitatus sum quid e pluribus refractionibus eveniret, hoc
fretus consilio quod colores iteratis refractionibus plus plusque mutari deberent, si modo a singulis quamlibet internam mutationem paterentur; contra
vero si non intrinsecus mutati sed per divergentiam difformium radiorum e
misturis tantùm educti et segregati essent, tum apparentes mutationes iteratis
refractionibus minores fieri propterea quod colores qualibet vice simpliciores
evaderent. Et experienti posterior casus evenit.[9] Scilicet cum coloris per posterius Prisma *αβκ* trajecti partem aliquam tertio Prismate ad distantiam aliquot pedum disposito exceperim, color ille denuò trajectus adeò perdurabat,

---

(5) In §I, 12 = 12.

(6) Appropriately enough, in the *editio princeps* this figure, as well as the next one, was
redrawn with seven rather than five circles.

(7) Newton had initially introduced his new terminology, such as "viridiformes," in §56 = II,
45, but in revision he decided to advance its introduction.

(8) Although the adjacent yellow and blue compound white or a very pale green, their admix-
ture with the predominant green would not change its hue, and consequently they would remain

extreme ends of the colors were not cast through G; whereas when green was transmitted, the adjoining colors on each side (namely, blue and yellow) emerged, and similarly yellow and orange, red and green, and blue, green, and purple appeared. It is therefore necessary to recall what I explained at the beginning[5] about how this oblong image PT is formed from circles set out in a straight line; and accordingly it will be evident that these colors are not simple but composed from a mixture of many colors. For, conceive the sort of equally refrangible rays that generate intense purple to flow from the entire solar disk and, after passing through the prism toward the image PT, to fall in the circle AC (Fig. II, 7).[6] Next conceive another, somewhat less refrangible sort of ray to fall in another circle YZ that touches the first one at G, and it is manifest that no rays of those sorts will be mixed together, since, to be sure, they occupy the circles AC and YZ that nowhere coincide. Now if you imagine a third sort of ray, which experiences an intermediate refraction, to fall in the circle [EF] situated approximately in the middle, it will be clear that some of these will mix with each of the preceding sorts in the spaces HI and KL, where indeed the circles illuminated by those coincide. Thus, if you conceive the entire image PT to be composed of innumerable circles disposed lengthwise, each of which is illuminated by a different sort of ray, it will be evident that heterogeneous rays are mixed together in every part of it. Consequently, when these have been further separated by repeated refraction, every color ought to be resolved into simpler ones. Yellow and blue, for example, are latent in green, but they are not discerned, both because the rays generating green, or (if I may invent the phrase for clarity's sake) green-making[7] rays, predominate because of their abundance, and also because yellow and blue compose green.[8] But to the extent that they are separated by a second refraction, each one will be seen according to its proper nature, and similarly for the others.

After I had observed these things, I examined what would happen from [20] several refractions. I relied on this consideration: that colors ought to be more and more changed by repeated refractions if each of them was subject to any internal change whatsoever; but, on the contrary, if they were not intrinsically changed but only drawn out of the mixtures and separated by the divergence of the dissimilar rays, then the apparent changes would become smaller by repeated refractions, because simpler colors would arise at every step. And by doing the experiment the latter case results.[9] Specifically, when I received any portion of color transmitted through the second prism αβκ, on a third prism placed several feet away, that color transmitted again

imperceptible. Newton's argument here is a special case of his fourth proposition (§II, 70), that a mixture of neighboring colors yields the intermediate one.

(9) Newton has here sketched his approach to demonstrating color immutability: first, to separate the colors; and then to show that as the colors are more separated their changes in subsequent refractions become continually smaller as they approach immutable, perfectly pure colors. In the *Opticks* he consequently divided this demonstration into two distinct propositions: the problem of separating the colors, in Bk. I, Pt. I, Prop. IV, Problem I; and a demonstration of color immutability, in Bk. I, Pt. II, Prop. II.

ut si non ratione constitisset mutationem aliquam eventuram fuisse, sensu
judice haud mutari percepissem. Tentabam deinde siquam quartâ refractione
mutationem sensibilem inducere potuerim: sed frustra. Interea verò cavendum est nè foramina *F* ac *G* caeteraque per quae lux transit majora statuantur quàm exigunt colores ut evadant perspicui.

[21]   Est et alia methodus qua diversi colores ab invicem segregari possunt, ut in
segregatis examen statuatur. Scilicet experimentum sub initio[10] traditum est,
quo solis imago *PT* per contractionem cujusque circularis imaginis oblongam
illam efformantis multò oblongior quàm aliàs evaserit. / Nam in contracta ₂14/
imagine $\pi\tau$ (fig II, 8) quae totidem circulis eadem centra retinentibus quot
sunt in majori *PT*, circuli minus coincidunt. Sic enim *AC* et *EF* ex parte *HI*
coincidunt. At cùm in minores $\alpha\gamma$ et $\epsilon\phi$ contrahuntur, videre est quòd ex
omni parte ab invicem distant: et sic de alijs. Quamobrem cùm circuli a
diversis radiorum generibus illuminati jam minus confundantur, colores evadent minus commixti; utut non fient omnino simplices propterea quod circuli
inter $\alpha\gamma$ $\epsilon\phi$ caeterosque positi, cum illis ex aliqua sui parte possint coincidere.
Sed hac de causa plures ejusdem cujusque coloris gradus tantum commisceri
possunt, ut cyaneus et indicus in caeruleo; coccineus, minius, et fortasse
citrius in rubro, et sic de alijs: Quae quidem mistura semper fiet eò minor
quo imago $\pi\tau$ in angustiorem contrahitur.[11]

[22]   Disposui itaque Prisma *ABC* (fig II, 9) unà cum Lente *LM* ad distantiam

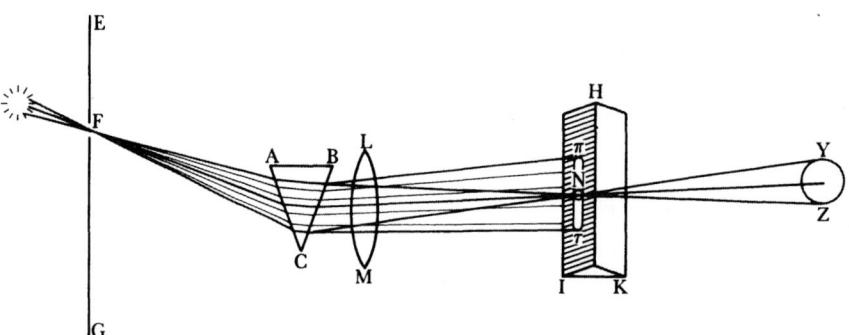

Figure II, 9

quasi decem pedum a foramine *F* per quod sol illuxit cubiculum. et radij per
haec duo vitra trajecti desideratam imaginem contractam $\pi\tau$ ad pedes exinde
decem circiter formabant, Lente *LM* existente tali ut radios parallelos ad
focum quinque pedibus a se distantem cogeret. Dein, aliud adhibui prisma
HIK, cujus latus planum *HI* velamine nigro ad *N* (ut dictum est) transfosso
tegebatur, et ad imaginem $\pi\tau$ statui ubi colores secundum latitudinem
maximè contractos ac distinctè terminatos vidi, ut eorum aliquis pro arbitrio
transmitteretur per *N* in parietem vel papyrum *YZ*. Quibus positis observabam deinde quod colores hoc modo multò minùs a repetitis refractionibus
mutati fuerint quàm in praecedentibus. Cùm rubor per *N* transmissus est,
idem rubor ad *YZ* apparuit, et non alius color quisquam demptis varijs

so persisted that if it had not been established by reason that some change would occur, judging by sense I would not have perceived it to have changed. Next I tested whether I could induce any sensible change by a fourth refraction, but it was in vain. One must be careful, though, that the holes *F* and *G* and the others through which the light passes are not made greater than is required for the colors to be evident.

There is yet another method by which the various colors can be separated [21] from one another so that they may be examined separately: namely, the experiment related at the beginning[10] whereby the sun's image *PT* became much more oblong than it would otherwise be by the contraction of each circular image forming that oblong one. In the contracted image *ππ* (Fig. II, 8), which has just as many circles keeping the same center as the larger one *PT*, the circles coincide less. For instance, *AC* and *EF* coincide in the part *HI*,

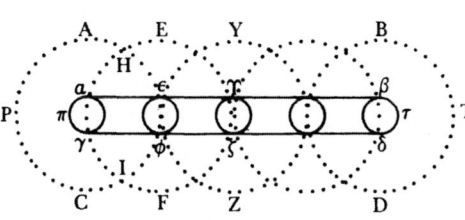

Figure II, 8

but when they are contracted into the smaller circles *αγ* and *εφ*, one can see that they are separated from one another in all parts; and the same is true for the others. Consequently, since the circles that are illuminated by different kinds of rays are now less confused, the colors turn out less mixed together, although they will not become completely simple because the circles situated between *αγ*, *εφ*, and the others can coincide with them in some part. But because of this only several gradations of any one color can be mixed together, such as cyan and indigo in blue; scarlet, vermillion, and perhaps orange in red; and similarly for the others. This mixture will of course always diminish as the image *ππ* is contracted into a narrower one.[11]

Accordingly, I placed the prism *ABC* (Fig. II, 9) together with the lens *LM* [22] at a distance of about ten feet from the hole *F*, through which the sun shined into the room. After having passed through these two glasses, the rays formed the desired contracted image *ππ* about ten feet from there; the lens *LM* being such that it gathered parallel rays to a focus distant about five feet from itself. Next I used another prism, *HIK*, whose plane side *HI* was covered with a black sheet pierced at *N* (as has been described), and placed it at the image *ππ*, where I saw the colors extremely contracted in their breadth and distinctly terminated so that any of them could be transmitted at will through *N* onto the wall or the paper *YZ*. Then having arranged these things, I observed that in this way the colors were changed much less by the repeated refractions than in the preceding ones. When red was transmitted through *N*, the same red appeared at *YZ* with no other color except various gradations

---

(10) See §I, 19 = 18.

(11) Newton gives a quantitative rule for determining the purity of the spectrum in the *Opticks*, Bk. I, Pt. I, Prop. IV, p. 46 = Add 3970, f. 51ʳ.

ejusdem gradibus, ut coccineo et / minio; et sic viriditas in varios solummodo ₂15/
gradus discreta fuit: ex una parte vergens ad flavescentem viriditatem, et ex
altera ad thalassinum: Sed in flavum aut caeruleum aliumvè colorem quemvis
ex nulla sui parte transformari potuit. Atque idem in alijs coloribus contigit.
Observabam praeterea quod cùm foramen *F* factum erat angustius, ut per
imaginis *ππ* majorem contractionem colores evaderent simpliciores, colores
versus *YZ* trajecti minùs adhuc mutati fuerint, et vix aliquam mutationem
sensibilem passi fuisse videbantur,[12] cùm foramina non latiùs duodecimà
parte digiti patuêre. hoc tantùm excepto discrimine quòd lux apud *ππ* fortior
erat (quia magis contracta) quàm apud *YZ*. Atque adeò nil dubitandum esse
censeo quin colores evaderent prorsus immutabiles, si modo (per indefinitam
parvitatem foraminum *F* et *N*) in simplices discerni possent. Et hoc ex eo
etiam confirmatur quòd cùm tegi Lentem *LM* juxta perimetrum ejus velamine nigro per medium ad latitudinem ferè semissis digiti circulariter pertuso,
figura imaginis *YZ* pene orbicularis evasit et eo magis orbicularis, quo magis
foramen *F* contraxi.[13] Id quod notari vellem cùm plurimùm illustret causam
imaginis *PT* in longitudinem diductae non aliam fore quàm radiorum coloribus dissimilium diversam refrangibilitatem.

[23]   Caeterùm quò propositum adhuc magis pateat, et ex abundanti ut constet,
quinam sint colores primitivi, adverto circulos *AC, EF, YZ* caeterosque in
alternas partes juxta lineam quae per omnium centra transit maximè extendi,
et ab invicem recedere antequam attingant parallelas rectas *AB* et *CD* quibuscum imago illa utrinque terminatur. Sic *AC* et *EF* se mutuo secantes in *H* et *I*
recedunt postea, non omnino coincidentes in triangulis *AHE* et *CIF*. Colores
itaque juxta ipsissimas extremitates *AB* et *CD* sunt omninò simplices. Et ex
hoc fundamento propositum assequi possem; sed cùm circuli illi statim ut
/ ab extremitatibus istis recedunt inter se mutuo nimis interserantur, quàm ut ₂16/
colores per aliquam sensibilem latitudinem satis ad experimenta commodè
instituenda segregentur; rem potiùs ad hunc modum assequor.

[24]   E praeostensis constat figuras ex quibus in longum dispositis imago *PT*
componitur, circulares esse propter solis discum circularem: et inde si discus
ille triangularis esset vel alio quocunque non circulari perimetro terminatus,
illae etiam figurae vel triangulares vel alio quovis modo, ad instar solis,
terminatae evaderent. Et par est ratio de foramine *F* et figuris ad instar istius
foraminis formatis ex quibus in longum similiter dispositis imago *ππ* constituitur. His animadversis vice orbicularis foraminis *F*, triangulare substitui cujus
altitudo verbi gratia sit plusquam digiti, basis tertiae quartaevè partis digiti,
et crura aequalia. Et prismate *ABC* ad trianguli hujus perpendiculum exis-

---

(12) In the *Opticks* (Bk. I, Pt. II, Prop. II, Expt. V, p. 88 = Add. 3970, f. 95ʳ) Newton
reported that "the colour of yᵉ light was never changed in the least." In general, in the *Opticks*
he obscures the subtle limit argument invoked here to demonstrate the immutability of monochromatic colors; only at the conclusion of his argument there does he note that the light could
not be made "absolutely homogeneal."

(13) Newton observed in the *Opticks*, Bk. I, Pt. I, where the argument of this experiment was
expanded to demonstrate the fifth proposition—"Homogeneal Light is refracted regularly with-

of the same, such as scarlet and vermillion. Similarly, green was separated only into its various gradations, inclining on the one side to yellowish green, and on the other to sea green, but it could not be transformed on either side into yellow, blue, or any other color. The same thing happened with the other colors. I observed, moreover, that when the hole *F* was made narrower so that the colors would become simpler by the greater contraction of the image $\pi\tau$, the colors projected to *YZ* were changed still less and appeared to have suffered scarcely any sensible change[12]—the hole being opened no wider than $\frac{1}{12}$ inch—except for only this difference, the light at $\pi\tau$, because it was more contracted, was stronger than at *YZ*. Hence I think that it must not be doubted that colors would certainly prove to be immutable if only they could be separated into simple ones through indefinitely small holes *F* and *N*. This is also confirmed because when I covered the lens *LM* up to its periphery with a black sheet perforated in the middle with a circle about half an inch wide, the shape of the image *YZ* turned out to be almost circular, and all the more circular the more I diminished the hole *F*.[13] I wanted this to be noted since it particularly illustrates that the cause of the image *PT* being dispersed lengthwise is none other than the diverse refrangibility of rays differing in color.

Moreover, so that the proposition may be still clearer and in addition it [23] may be evident which are the primitive colors, I observe that the circles *AC*, *EF*, *YZ*, and the others in every other part are most expanded along the line that passes through all the centers and that they recede from one another until they reach the parallel lines *AB* and *CD* that bound that image on each side. Thus *AC* and *EF* cut each other at *H* and *I* and then recede, not at all coinciding in the triangles *AHE* and *CIF*. Consequently the colors along the very edges themselves, *AB* and *CD*, are entirely simple. On this basis I could pursue the proposition; but since those circles, as soon as they recede from the edges, intermingle with one another too much for the colors to be sufficiently separated by some sensible width to carry out experiments conveniently, I prefer to pursue the matter in the following way.

From what has already been shown, it is evident that the figures disposed [24] lengthwise and composing the image *PT* are circular, because the sun's disk is circular. Therefore, if that disk were triangular or terminated by any other noncircular perimeter, those figures would also turn out to be triangular or terminated in any other way resembling the sun. The reason is the same for the hole *F* and the figures that are formed resembling that hole and, which disposed lengthwise, constitute the image $\pi\tau$. Having taken note of these things, in place of the circular hole *F* I substituted a triangular one whose altitude was, for example, more than 1 inch and the base of which was $\frac{1}{3}$ or $\frac{1}{4}$ inch, with the legs equal. When the prism *ABC* was parallel to the perpen-

---

out any dilatation splitting or shattering of yᵉ rays . . . "—that the image was "perfectly circular," and that without the additional covering on the lens used here. Earlier, in the "Fundamentum Opticae" (Prop. 5, Expt. 16, Add. 3970, f. 406ʳ), before he apparently recognized that his argument was erroneous and thus deleted it, he attempted to demonstrate the immutability of degree of refrangibility from the circularity of the image.

tente parallelo imago $\pi\tau$ (fig II, 10)
quadrilatera ex triangulis $\gamma\alpha\zeta$, $\epsilon\mu\theta$, $\eta\nu\iota$,
caeterisque infinitè multis efformata est,
quibus juxta bases in linea $\gamma\delta$ positas
cum se mutuò partes maximè commu-
nicantibus, exinde ad usque ipsorum
vertices recessio gradatim facta est,
donec in verticibus ad rectam $\alpha\beta$ sitis

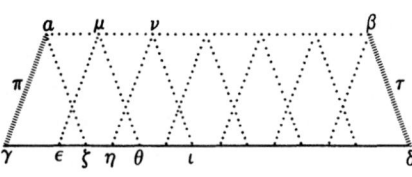

Figure II, 10

penitus dissocientur; Adeoque colores ibi simplices evadant.[14]

[25]   ˙ Jam verò observabam quòd simplices sive primitivi colores juxta terminum
$\alpha\beta$, etsi longè debiliores tamen (sensu judice) ejusdem speciei apparuêre ac
compositi juxta terminum $\gamma\delta$. Cujus rei ratio est quòd color quilibet primiti-
vus per commisturam colorum utrinque confinium exhiberi potest ut in
sequentibus[15] patebit. Colorum verò quos hoc pacto primitivos esse constitit,
gradus sunt insigniores Coccineus, sive purpureus,[16] Minius, Citrius, Luteus,
sive Heliocryseus[,] Subflava[,] viriditas, Gramineus, Thalassinus, Cyaneus,
Indicus, et ejusmodi violaceus[17] qui ad usque extremitatem imaginis inten-
debatur, sed absque immista rutili alicujus fulgoris tinctura apparuit, si modo
cubiculum factum fuerit valdè tenebrosum.

[26]   / Observabam praeterea quòd coloribus hisce juxta terminum $\alpha\beta$ conspic- ₂17/
uis non potui sensibilem speciei mutationem refractionibus utcunque repetitis
inducere. Quinetiam tentabam si quo alio pacto mutare possem, quemadmodum reflectendo a corporibus diversimodò coloratis, sed in eo frustra fui,
nam (superfluâ luce quaquaversum penitus exclusâ) si caeruliformes radij in
aurum incidêre, illud aurum caerulei coloris evasit, si flaviformes in Indicum
incidere, flavescebat indicum: et sic in caeteris. Adeò ut hanc Propositionem
satis superque stabilitam esse censeam.[18]

---

(14) The base of the triangular slit used in the *Opticks,* Bk. I, Pt. I, Prop. IV, Expt. XI, was a
tenth of an inch, and Newton also suggested using a parallelogram a tenth or twentieth of an
inch wide and an inch or two long. P. J. Bouma has shown that with these slits and lenses
Newton would have attained nearly perfect spectral purity, in *Colour,* pp. 173–4.

(15) In §II, 70.

(16) Except for this passage, in the *Optical Lectures* Newton consistently placed "purpureus"
at the violet end of the spectrum. It is possible that he meant to write "puniceus," a purple-red
color that was frequently placed at the red end of the spectrum; see, for example, Marci's and
Kircher's ordering of the spectral colors in Lect. 1, note (10), this volume. Aristotle called the
upper or red end of the rainbow—in Latin transliteration—"puniceus"; see Newton's notes from
Magirus in the Introduction, note (3).

(17) Although the translation of the names of colors can be elusive, Newton gives sufficient
data in various passages to place the spectral colors in a reasonably definite order, thus making
their precise English translation of secondary importance. In §II, 70, he presents the principal
spectral colors and their gradations in a reconstructible sequence, though for a different
purpose—to demonstrate Prop. 4, on mixing neighboring spectral colors; and so the two do not
agree in every nuance. His designations for the red, or least refrangible, end of the spectrum are
consistent in both enumerations: "coccineus" (scarlet) at the extreme end of the red (here
equated with "purpureus"), followed by "minius" (vermillion). Newton considers both of these
in §II, 21 to be gradations of "rubor" (red). In §II, 70 "minius" is followed by "croceus"
(saffron) rather than "citrius" (orange). The yellows, "luteus sive heliocryseus" (golden yellow or
marigold) and "subflavus" (pale yellow), are the same in both sequences. In place of two of the

dicular of this triangle, a quadrilateral image $\pi\tau$ (Fig. II, 10) was formed from the triangles $\gamma\alpha\zeta$, $\epsilon\mu\theta$, $\eta\nu\iota$, and infinitely many others. The triangles have the most parts in common along their bases situated on the line $\gamma\delta$ and thereafter they gradually recede as far as their vertices, until at their vertices located on the straight line $\alpha\beta$ they are completely separated. Hence the colors prove to be simple there.[14]

Now I observed that the simple or primitive colors along the edge $\alpha\beta$, even [25] if far weaker, nevertheless appeared (judging by sense) to be of the same sort as those compounded along the edge $\gamma\delta$. The reason for this is that any primitive color can be produced by a mixture of the colors bordering it on each side, as will be evident in the following.[15] The more prominent gradations of the colors established in this way as primitive are scarlet or purple,[16] vermillion, orange, golden yellow or marigold, pale yellow, green, grass green, sea green, cyan, indigo, and violet[17] of the kind that extended up to the end of the image but appeared without a tincture of any red effulgence mixed in, provided that the room was made completely dark.

I observed, moreover, that I could not bring about a sensible change of [26] species in these colors visible along the edge $\alpha\beta$ by repeating the refractions. In addition, I tested if I could change them in some other way, as by reflection from differently colored bodies, but I was frustrated in that. For after everywhere carefully shutting out superfluous light, if blue-making rays fell upon gold, that became gold of a blue color; if yellow-making ones fell upon indigo, the indigo became yellow; and similarly for the others. Hence I think that this proposition is more than sufficiently established.[18]

---

greens here, "viriditas" (green) and "gramineus" (grass green), in §II, 70 Newton invokes only one, "porraceus" (leek green); in §II, 94 he describes "porraceus" as "floridissima viriditas" (the most florid green). The remainder of the spectrum is identical in the two sequences except for the usual confusion of violet and purple: "thalassinus" (sea green), "cyaneus" (cyan), "indicus" (indigo), and "violaceus" or "purpureus." Newton in §II, 21 considers "cyaneus" and "indicus" to be gradations of "caeruleus" (blue), but in §II, 94 "thalassinus" to be the border of blue and green. A valuable, contemporary aid to translating Newton's color terms is Charleton's brief essay "On the differences and names of colors." Charleton classifies colors according to the primaries, red, yellow, and blue, together with white and black, and their principal compounds, and gives the Latin names of the various gradations with their English translations; *Exercitationes de differentiis & nominibus animalium*, 2nd ed. (Oxford, 1677), "Appendicula de *colorum* differentiis & nominibus," pp. 361–73. Except for such difficult cases as purple and violet, Newton's and Charleton's color terminology are surprisingly consistent.

(18) As Newton acknowledges in §II, 67, he has proved here that colors are immutable only after they have been refracted once and separated from one another—or for what he calls "secondary" refractions—and not that they are so in the case of the sun's direct light—or for the "primary" refraction. To be sure, at this time he considered it a simple matter to extend the proposition to the primary refraction, but as he came to recognize, it is impossible to demonstrate experimentally that the colors of sunlight are unchanged by the primary refraction. For if the colors of a beam of sunlight are compared before and after the first refraction, it seems to the contrary that they have changed, since before refraction the beam appears white and afterward it displays all the colors of the spectrum. Newton first recognized this problem in the version of his theory that he outlined for Huygens on 23 June 1673, where he decomposed the principle of color immutability into three propositions; see note (1), this lecture. In this formulation the burden of establishing the principle of color immutability for the sun's direct light fell upon his

PROP 3. COLORES ALBI AC NIGRI UNÀ CUM CINEREIS SIVE FUSCIS[1]
INTERMEDIJS EX RADIJS UNIUSCUJUSQUE SPECIEI CONFUSÈ GENERANTUR.

Lect. 4.
Octob. 1671
[27 → 40.
Transitur ad
propositionem
tertiam.]

Assertionis veritas e praecedenti propositione manifesta est. Nam colores qui non sunt ex primitivis (quales non reperiuntur jam recensiti)[2] per compositionem generari necesse est. At non gravabor tamen fusius probare: Idque potissimùm ut lucem, cui color albus competit, ex radijs quoad qualitates colorificas aequè ac refrangibilitatem heterogeneis componi, eâque de causâ albere, certissimè constat. Proponitur itaque jam monstrandum esse quòd cùm omnes omninò colores quos Prismata generant, debitè commiscentur, albedo exinde resultabit.[3] Deque tali misturâ perfectè componendâ plures modos eo quo cogitabam ordine recensere animus est.

[28 ≈ 40, 41.
Modus componendi
albedinem ex
coloribus
Prismatum.]

Ac primò rem aggressus sum cum pluribus Prismatibus ita dispositis ut colores eorum in eundem locum inciderent et sic inter se miscerentur.[4] Sint *ABC, DEF,* ac *GHI* (in fig II, 11) tria prismata juxta se situ parallelo ita

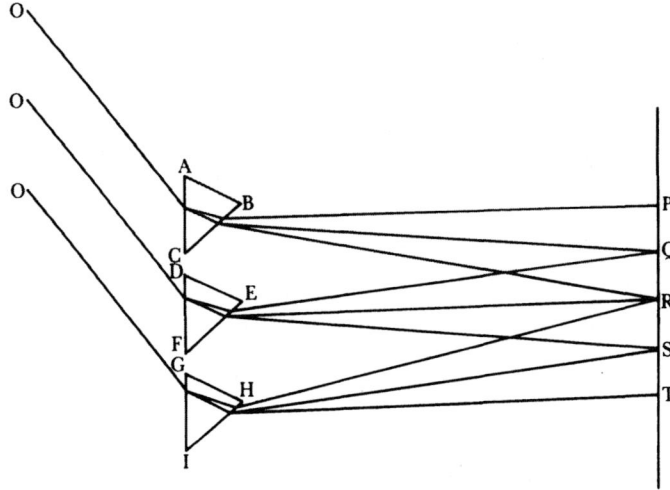

Figure II, 11

disposita ut alterum *DEF* sit alteris duobus *ABC* et *GHI* utrinque vicinissimis intermedium, in morem trium linearum conficientium capitalem literam graecam Ξ. Et lux per unumquodque prisma liberè transiens excipiatur in papyrum *PT* pede uno vel duobus postpositam. Coloribus omnium Prismatum sic in ipsam *PT* projectis, convertantur Prismata circa / proprios axes et videbis ₂18/ colores istos sibi invicem accedere vel recedere. Quare convertantur donec talis sit eorum situs ut unius Prismatis *ABC* rubor, et alterius *GHI* purpura vel color Indicus cum viriditate tertij *DEF* coincidant sicut vides factum ad *R.* Et ex istis coloribus ita sibi commixtis albedinem generari cernes, colore purpureo et caeruleo juxta *P* conspecto, rubeo verò et flavo juxta *T,* et albo juxta *R* caeteros intercedente.

---

sixth proposition, the principle of superposition, that rays do not act upon one another, and, in particular, that the rays composing sunlight do not alter each other's color, but only appear to compose a new color. Though Newton later pursued this and alternative proofs of the strong

PROPOSITION 3. THE COLORS WHITE AND BLACK TOGETHER WITH IN-
TERMEDIATE ASHENS OR GRAYS[1] ARE GENERATED FROM RAYS OF EVERY
SORT CONFUSEDLY MIXED.

The truth of the assertion is manifest from the preceding proposition, for
the colors that are not one of the primitives (those not among the ones just
now enumerated)[2] are necessarily generated by composition. Nevertheless I
will not be reluctant to prove it more fully, especially as it may be most
certainly established that light to which the color white corresponds is composed of heterogeneous rays with respect to colorific qualities as well as
refrangibility, and for that reason it is white. Therefore it is now proposed to
prove that when absolutely all the colors that prisms generate are duly mixed
together whiteness will consequently result.[3] I intend to relate several ways
of perfectly compounding such a mixture in the order in which I conceived of
them.

First I attempted this with several prisms so arranged that their colors fell
on the same place and thus were mixed among themselves.[4] Let *ABC, DEF,*
and *GHI* (Fig. II, 11) be three prisms placed near each other in a parallel
position so that one, *DEF,* is in between and exceedingly close to the other
two, *ABC* and *GHI,* on each side of it in the form of the three lines making
the Greek capital letter Ξ. Let the light passing freely through each of the
prisms be received on the paper *PT* placed one or two feet beyond them.
When the colors of all the prisms are thus projected onto *PT,* turn the prisms
around their axes, and you will see these colors approach or recede from each
other. Accordingly, rotate them until they are so situated that the red of the
one prism, *ABC,* and the purple or indigo color of the other, *GHI,* coincide
with the green of the third, *DEF,* as you see done at *R.* From these colors
thus mixed together you will see whiteness produced, with the colors purple
and blue visible at *P,* but red and yellow at *T,* and white falling in between
the others at *R.*

Lecture 4
October 1671
[27 → 40. Passing
to the third
proposition.]

[28 ≈ 40, 41. A way
of compounding
whiteness from
prismatic colors.]

principle of color immutability for the primary refraction in the "Fundamentum Opticae" (Prop.
12, Add. 3970, f. 418ʳ), by the time he completed the *Opticks,* he recognized that it could not be
proved and so he omitted a proof for that case. He did, however, contemplate raising the
principle of color immutability to axiomatic status in the form of the principle of superposition,
but he abandoned that idea too, as can be seen from a draft for a projected Axiom 4: "The effect
of reflexion or refraction upon any light is the same whether that light be alone or whether it be
mixed wᵗʰ any other light. Tis as much refracted alone as in a mixture & its colour and
properties are the same whether any other light do any where cross it or not" (Add. 3970, f.
392ʳ). Newton's attempts to establish color immutability are fully described in Shapiro, "Evolving structure."

(1) Newton explains elsewhere (§§25 = II, 2 and 51 = II, 39) that "fuscus" (gray) results
from a mixture of white and black, and whatever the exact translation of "cinereus" (ashen),
which he here equates with "fuscus," he followed common, contemporary usage in considering
both to be achromatic; see, for example, Scaliger's and De Boodt's terminology cited in Lect. 3,
note (10), this volume. Kircher, however, represented an alternative tradition in considering
neither to be achromatic, for he held that "cinereus" arises from a mixture of white and blue,
and "fuscus" from black and yellow; *Ars magna,* Bk. I, Pt. I, Ch. II, p. 67.

(2) Namely, in §II, 25.

(3) This sentence alone, somewhat revised, remains from §40.

(4) In the *Lectiones opticae* this sentence concluded §40.

[29 = 42. Notanda
quaedam quò
satiùs fiat.] Caeterum in istis experiendis convenit[5] observare sequentia.

Primo si anguli Prismatum planis refringentibus contenti *ACB, DFE,* et *GIH* sint inaequales; praestat ut illud Prisma cujus angulus *GIH* maximus est, ponatur versus exteriorem partem anguli contenti radijs incidentibus et refractis, et istud versus interiorem cujus angulus *ACB* est minimus.

Secundò, aperturae per quas lux transmittitur trans Prismata debent esse magnae. Imò convenit ut transitus luci per tota Prismata pateat, obstaculo nullo adhibito. Neque opus est ut experimentum in tenebris peragatur sicut in alijs quamplurimis requiritur.

Tertiò, papyrus *PT* in quam colores incidunt non nimis distare debet a Prismatibus. Sufficit distantia pedum plus minùs duorum. has autem aperturas et distantiam statuo ut colores eo meliùs commisceantur ad albedinem perfectiorem componendam.

Quartò ut colores ad *R* faciliùs etiam et satiùs commisceantur, Prisma *ABC* statuatur imprimis in situ quocunque tali ut radij tum ingredientes tum emergentes refractionem praeter propter aequalem patiantur: et in eo situ figatur. Et colores ejus ad distantiam duorum pedum excipiantur, vel ad eam potiùs ubi vides flavum ejus et caeruleum modò contiguos, albedine intermediâ tum evanescente: Postea figatur aliud Prisma *GHI* in tali situ ut purpura ejus contingat ruborem alterius *ABC,* non autem coincidat illi: et linea contactûs notetur. Deinde tertium Prisma *DEF* sic fige ut ejus colorum medietas cadat in dictam lineam contactûs, quod ubi contingit, facilè cognosces intercipiendo lucem caetera Prismata ingressuram. Denique papyrus *PT* / ultra citraque transferatur paululùm, donec videas albedinem perfectam in medio colorum ad *R* generari. Quam quidem albedinem ex varijs coloribus compositam esse constabit intercipiendo colores unius duorumve Prismatum priusquam attingant papyrum. Nam loco albedinis eos quos non intercipis colores intueberis. ₂19/

Denique si velis ut colores cujusque Prismatis[6] perfectius misceantur, possis adhibere plura prismata modò praesto sint: tamen eventus non deerit expectationi si tria tantum adhibeas. Etenim colores cujusque Prismatis seorsim spectati non sunt omninò simplices, sed viridis et rubeus nonnihil miscentur in flavo: et purpureus ac viridis in caeruleo; et sic de reliquis: quemadmodum in sequentibus ostendetur. Et inde fit quòd cùm tria tantùm Prismata adhibentur, non solùm tres colores rubeus, viridis, et indicus commisceantur in *R,* sed etiam caeruleus et flavus unà cum omnibus eorum gradibus intermedijs istam albedinis compositionem ingrediantur.

[30 = 43. Alius
ejusdem rei
perficiendae modus] Verùm cùm tot prismata in situ tam accurato disponere propter motum solis et alia incommoda difficile forsan et laboriosum simul inveniatur, nisi adhibeatur Machina quaedam ea de causâ fabricata ut ejus ope Prismata in desiderato situ figantur: alium propterea modum profero, quo ista negotio leviori, idque unico prismate periclitari poteris. Sumatur papyrus vel aliud opacum corpus attenuatum in morem laminae. Et in eo confodiantur oblongae rimae sex aut plures parallelae, quarum latitudines sint aequales distantijs

---

(5) Originally (as I): juvabit (it will help).   (6) cujusque Prismatis] I: adhuc (still).

Moreover, in doing these experiments it is appropriate[5] to note the fol- [29 = 42. Some things to note so that it occurs more satisfactorily.]
lowing:

First, if the prisms' angles, *ACB, DFE,* and *GIH,* made by the refracting planes are unequal, it is better to place that prism whose angle, *GIH,* is the largest toward the exterior side of the angle made by the incident and refracted rays, and that prism whose angle, *ACB,* is the smallest toward the interior.

Second, the apertures through which the light is transmitted across the prisms must be large. Indeed, it is fitting that the light's path through all the prisms be open with no obstacle being used. Nor is it necessary that the experiment be performed in darkness as is required in very many others.

Third, the paper *PT* on which the colors fall must not be too far from the prisms; a distance of two feet more or less is satisfactory. I fixed these apertures and the distance, however, so that the colors would be better mixed together to compound a more perfect white.

Fourth, to mix the colors at *R* more easily and also more satisfactorily, place the prism *ABC* initially in any position such that both the entering and emerging rays undergo a nearly equal refraction, and fix it in that position. Receive its colors at a distance of two feet, or rather at that distance where you see its yellow and blue just touching with the white in between then vanishing. Then fix the other prism, *GHI,* in a position such that its purple touches but does not coincide with the red of the first one, *ABC,* and note the line of contact. Next, fix the third prism, *DEF,* so that the middle of its colors falls on that line of contact, and you will readily identify where this occurs by blocking the light that enters by the other prisms. Finally, shift the paper *PT* back and forth slightly until you see a perfect white produced in the middle of the colors at *R.* Indeed, it will be established that this white is compounded of various colors by intercepting the colors of one or two prisms before they reach the paper. For, instead of white you will see those colors that you did not intercept.

Finally, if you want the colors of each prism[6] more perfectly mixed, you can use more prisms provided they are available; yet the result will not fall short of your expectations if you use only three prisms. For the colors of each prism when observed separately are not entirely simple; in fact green and red are mixed somewhat in the yellow, and purple and green in the blue, and similarly for the others, as will be shown in the following. Consequently, it turns out that when only three prisms are used, not only are three colors, red, green, and indigo, mixed together at *R,* but also blue and yellow together with all their intermediate gradations enter into that composition of white.

But it may be found both difficult and laborious to place so many prisms [30 = 43. Another way of accomplishing the same thing.]
so accurately in position, because of the sun's motion and other inconveniences, unless some device constructed for that purpose is used, so that with its help the prisms may be fixed in the required position. Accordingly, I offer another way—and with a single prism—whereby you will be able to try that with less trouble. Take a paper or other thin opaque body in the form of a sheet, and in it cut six or more parallel, oblong slits whose widths are equal

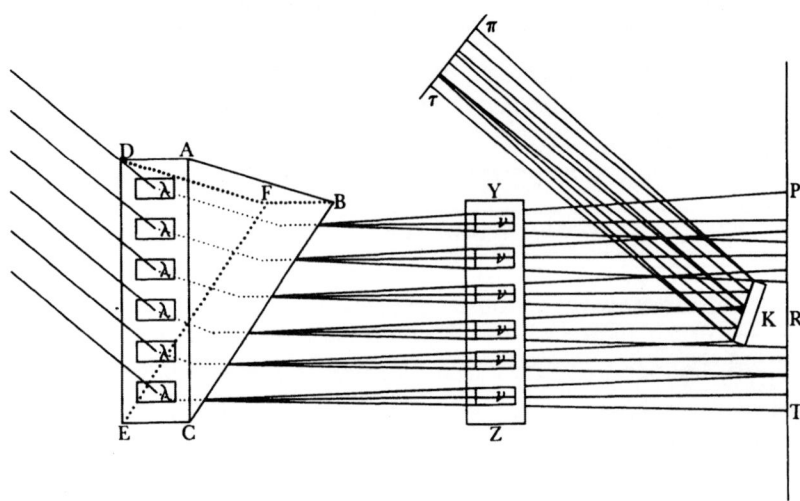

Figure II, 12

aut ijs paulo majores. Deinde papyrus ista figatur alicui ex planis lateribus Prismatis: sit istud latus papyro obductum *ACED* (fig II, 12) et rimae in papyro excisae literis λ designentur; quarum situs esto parallelus ad *EC* concursum laterum refringentium prismatis, sive ad verticem ejus. Papyrus autem debet toti isti plano *ADEC* superinduci, nequa lux alibi transmissa quàm per praedictas rimas perturbet experimentum. Tum prisma statuatur in luce solis ut radij ejus vel per dictas rimas id ingrediantur, vel postquam refracti fuerint per eas egrediantur: et in isto situ figatur. Quo facto sumatur alia / papyrus *PT* quae sic teneatur a postica parte Prismatis ad distantiam $_2$20/ duorum triumve digitorum ut in eam lux terminetur et videbis tot lineas colorum quot sunt oblongae rimae λ: quarum linearum cuique tot competent colores quot solent apparere virtute Prismatum. Nempe quaelibet rima subit officium unius e Prismatibus in experimento priori adhibitis et proprios colores caeruleum rubrum caeterosque generat quasi tot essent prismata quot sunt rimae. Porro si papyrus *PT* longiùs differatur a Prismate, coloratas istas lineas paululùm[7] dilatari cernes, et interjecta spatia minui, donec absorbeantur a coloribus tandem factis contiguis. Et si papyrus adhuc longiùs differatur colores a diversis rimis effecti (rubri cum caeruleis primò, deinde alij cum alijs) incipient plus plusque misceri. Et sic sese paulatim diluent, donec cum mistura satis absoluta est, convertantur in albedinem; praeterquam in eorum extremitatibus *P* ac *T* ubi mixtura et confusio ferè nulla est. Et isthaec accidunt cùm papyrus *PT* quasi ad distantiam decem vel duodecim vicibus majorem ipsa *AC* vel *BC* latitudine planorum prisma constituentium, amovetur. Quod si amoveatur adhuc longiùs absimilium radiorum commistio perfectior fortasse evadet, sed colores purpurei et caerulei ad *P* ac flavi rubeique ad *T* latiores fient et interjectum spatium album minuetur, donec totum destruatur ab istis coloribus occupatum.

to their separations or a little greater. Next attach that paper to any of the prism's plane sides. Let that side covered with the paper be *ACED* (Fig. II, 12) and the slits cut in the paper be designated by the letter λ; let their position be parallel to *EC*, the intersection of the prism's refracting sides, or its vertex. The paper, however, must cover that entire plane *ADEC*, so that light transmitted anywhere other than through those slits does not disturb the experiment. Then place the prism in the sun's light so that its rays either enter it through those slits or leave through them after having been refracted, and fix it in that position. After this is done, take another paper *PT* and hold it at a distance of two or three inches from the back side of the prism so that the light is terminated on it, and you will see just as many lines of colors as there are oblong slits λ. To each of these lines there correspond as many colors as usually appear by the power of prisms; namely, each slit assumes the function of one of the prisms used in the previous experiment and generates its characteristic colors, blue, red, and the others, just as if there were as many prisms as there are slits. Furthermore, if the paper *PT* is moved farther away from the prism, you will see those colored lines expand somewhat[7] and the intervening spaces diminish until they are consumed by the colors when they finally become contiguous. If the paper is moved still farther away, the colors produced by the various slits will begin to be more and more mixed, first reds with blues, then other colors with still others. In this way they will gradually dilute one another until, when the mixture is sufficiently perfect, they are converted into white, except at their ends *P* and *T* where there is almost no mingling and mixing. This happens when the paper *PT* is removed to a distance about ten or twelve times greater than *AC* or *BC*, the length of the planes forming the prism. If now it is removed still farther, the mixture of unlike rays will perhaps become more perfect; but the colors purple and blue at *P*, and yellow and red at *T* will become wider, and the white space in between will decrease until the entirety is destroyed when it is filled by those colors.

---

(7) I: *paulatim* (gradually).

[31 = 44. In illum notae]

In hisce autem experiendis cavendum est, ut oblonga foramina λ sint accuratè aequalia, et aequalibus distantijs ab invicem dissita ne luce magis copiosâ per aliquod ingressâ quàm per caetera colores exinde generati praevaleant caeteris et misturam perfectam conturbent: et sic vice albedinis colores apparebunt hinc illinc more furtuito sparsi. Illa verò distantia rimarum λ ut et earundem latitudo non malè statuitur esse pars digiti circiter duodecima aut ea major fortè si prisma satis amplum adhibeas. Quinetiam si cupias ut experimentum sit omnibus numeris absolutum, vice prismatum vitreorum vulgò venalium (quae sunt nimis gracilia) debes amplioribus uti, qualia possis efficere ex laminis vitreis utrinque perpolitis et conjunctis in morem vasculi prismiformis, quod vasculum impleatur aquâ clarissimâ et undique cemento obturetur. Non multùm refert quaenam sit / hujus longitudo,[8] sufficit ut sit ₂21/ trium digitorum. Sed refringentia latera debent esse quatuor vel sex digitos lata aut ampliùs, ut rimae praefatae λ cum distantijs earum fiant majores et plures et magis accuratae. Sin utaris angustioribus, qualia vulgo venduntur; colores externi juxta *P* ac *T* dilatando priùs destruent interjectam albedinem quàm perficiatur per remotionem papyri *PT*. Et illa praeterea quae in totum constant ex vitro, colore aliquo ut viridi flavo plerunque tinguntur, et radios ita tingunt in transitu ut albedinem perfectam exhibere nequeant.

[32 = 45. Objectio quòd albor ex destructione non misturâ colorum generatur.]

Jam verò audire videor objectionem ex receptis Philosophorum opinionibus depromptam: Dicat enim aliquis quòd colores revera et propriè loquendo non miscentur sed destruuntur potiùs; idáue eâ de causâ quòd umbrae vicinia quae necessaria est ad productionem colorum, tollitur cum radij per diversas rimas trajecti commisceri incipiunt; et praeterea quod radijs sic mixtis quorum motus inter se dissentiunt, necesse est ut isti motus destruant alterutros, quibus cessantibus color omnis perit et in albedinem convertitur. Sic Cartesianus aliquis contendat forte, quod cum globuli miscentur quorum rotationes contrariantur sibi, necesse est ut impediant sese et alternos motus destruant: Et sic alij objiciant alia.

[33 = 46. Responsio multiplex: Primò quòd illi colores non destruuntur ex umbrae confinio sublato.]

Sed responsio multiplex in promptu est: et imprimis inquam quòd cùm umbrae coloribus interjectae primùm evanescunt removendo papyrum *PT*, colores tamen non ideò pereunt neque minimùm immutantur donec incipiant misceri per remotiorem distantiam papyri; et albedo non producitur donec per distantiam adhuc remotiorem mistura radiorum omnis generis evadat perfecta. Unde confinium umbrae non est necessarium ad colores producendos, neque albedo generatur ex isto sublato.

[34 = 47. Secundò, Neque probabilitèr ex motuum contrarietate.]

Secundò colores qui primò omnium miscentur, nimirùm purpureus sive violaceus et rubeus, videntur maximè esse omnium dissimiles,[9] propterea quod adversas colorum extremitates occupent. Quamobrem itaque motus eorum contrarij non / destruunt sese neque color albus generatur antequam ₂22/ caeteri etiam colores omnes misceantur?

---

(8) *Editio princeps*: latitudo.
(9) Originally (as I): absimiles.

In doing these experiments, however, one must ensure that the oblong holes λ are exactly equal and placed at equal distances from each other, lest a greater quantity of light enters through one hole than through the others, and the colors generated from it overwhelm the others and disturb the perfect mixture, and thus instead of whiteness colors will appear randomly scattered about. In fact, fix both the distance of the slits λ and their width, not inappropriately, at about a twelfth of an inch, or perhaps greater than that if you use a sufficiently wide prism. Moreover, if you want the experiment to be perfect in all respects, instead of the glass prisms commonly sold (which are too slender) you must use broader ones, such as those you can make from glass plates highly polished on both sides and joined together in the form of a small prism-shaped vessel; the vessel should be filled with very clear water and sealed all around with cement. Its length[8] does not matter much—three inches is sufficient—but its refracting sides must be four or six inches broad, or more, so that those slits λ together with their intervening spaces become larger, more numerous, and more accurate. But if you use narrower ones, as are commonly sold, the outer colors at *P* and *T* by their expansion will destroy the white in between before it can be perfected by moving back the paper *PT*. Those prisms, moreover, that are made wholly of glass are often tinged with some color, such as green or yellow, and they so tinge the rays in their passage that they are unable to display a perfect white.

[31 = 44. Notes on that experiment.]

But now I seem to hear an objection drawn from the received opinions of philosophers. Some indeed would assert that truly and properly speaking the colors are not mixed but rather are destroyed, because the adjoining shadow necessary for the production of colors is destroyed when the rays transmitted through the various slits begin to be mixed together. Besides, when the rays whose motions are opposed to one another are thus mixed, it necessarily follows that those motions destroy one another; and with the motions ceasing, all color disappears and is transformed into whiteness. Thus some Cartesian might perhaps contend that when the little balls are mixed whose rotations are opposed to one another, it necessarily follows that they impede one another and mutually destroy their motions. Similarly others would object to other things.

[32 = 45. An objection that white is generated by the destruction and not by the mixture of colors.]

But a multiple response is at hand: First of all, I say that when the shadows that lie between the colors first vanish by moving back the paper *PT*, nevertheless the colors do not consequently disappear, nor are they changed in the least until they begin to be mixed on account of the more remote distance of the paper. White is not produced until, on account of a still more remote distance, the mixture of rays of every kind proves to be perfect. Consequently, the border of shadow is not necessary to produce the colors nor is the whiteness generated by its elimination.

[33 = 46. A multiple response: First, those colors are not destroyed by eliminating the border of shadow.]

Second, the colors that are mixed first of all, namely, purple or violet and red, appear to be the most dissimilar[9] of all, because they occupy opposite ends of the colors. Why therefore do their contrary motions not destroy each other and is the color white not generated before all the other colors are also mixed?

[34 = 47. Second, nor probably by the contrariety of the motions.]

[35 = 48. Tertiò,
Quòd radij per idem
medium confusè
transientes non
agunt in se invicem]

Tertio cuique licet observare idque nullo negotio quòd colores non omninò mutantur trajiciendo radios per medium quantumvis luminosum. Sic colores prismatum sunt ijdem sive trajiciantur per spatium illuminatum, sive tenebris involutum. Et res omnes eodem modo coloratae cernuntur sive conspiciantur cùm lumen solis trajicitur per intermedium spatium sive cùm excluditur. Id quod secus esset si lux in lucem per idem medium transeuntem posset agere. Quinimo si radij duobus prismatibus refracti sese decussent, postquam ab invicem discreti sunt, eosdem colores efficient quos alias efficerent si non omninò miscerentur. Quod non posset evenire si radij diversis coloribus tincti sibi mutuò per eadem spatia transeuntibus mutationem aliquam inducerent.

[36 = 49. Quartò
Quòd albor
praefatus perit si
quilibet color e
misturâ tollatur.]

Quartò, cùm in illâ distantiâ papyrum *PT* fixeris ubi colores albedinem optimè componunt: statuatur alia papyrus *YZ* ad distantiam duorum vel trium digitorum a prismate et in eâ notentur lineae coloratae: tum exscindantur istae partes papyri in quas dictae lineae cecidere, factis eo pacto rimis oblongis *ν–* parallelis et aequalibus, ut et aequè latis ac distantibus. Deinde papyrus ista *YZ* in locum suum restituatur tres digitos circiter a prismate distantem, ut per rimas ejus lux colorata trajiciatur ad alteram papyrum *PT* longinquiorem. Quo facto possis observare quòd si parùm deprimas papyrum *YZ* ut purpureos colores et caeruleos superioribus labris rimarum ejus impingentes intercipiat, et transmittat caeteros: albedo ad papyrum *PT* convertitur in rubeum colorem aut citrium vel flavum; sin attollas eam ut rubei et flavi labris inferioribus intercipiantur, caeterique perlabantur; albedo ista convertetur in purpureum indicum et caeruleum. Perinde ut fieri oporteret in mixtura colorum: Nam unis e mixturâ sublatis alteri debent ad propriam speciem et formam restitui.

[37 = 50. Quintò,
quòd colores, cùm
decussando
segregantur iterum,
ad propriam
speciem redeunt.]

Quintò, papyro *YZ* sublatâ, et reliquis stantibus: papyrum alteram *PT* in meditullio albedinis acu perfora ut lucis ejus albae portiuncula trajiciatur quam deinceps excipe in aliam papyrum isti *PT* ad distantiam quatuor vel sex digitorum / postpositam: et vice albedinis colores iterum apparebunt. At ₂23/ quomodo colores illi de novo generari potuissent si destruerentur in productione potiùs quàm miscerentur, non video. Concedendum est itaque quòd tantùm miscentur: et quòd radij varijs coloribus tincti, et promanantes a diversis rimis λ λ decussant sese in dicto foramine acu effecto, et postea divergentes ab invicem gradatim segregantur et segregati proprios iterum colores depingunt: quemadmodum posthac fusiùs explicabitur.[10] Ad eundem praeterea modum si speculum aliquod planum et exiguum *K* statuas in medio albedinis ad *PT* papyrum effectae, ita quidem ut aliquos ex albificantibus radijs aliorsum, veluti ad *ππ* reflectat: lux alba sic reflexa degenerabit in colores, quos videre est ad *ππ*, papyrum objiciendo. Etenim radij tincti cum diversis coloribus et in albedinem ad speculum *K* commisti inclinantur ad se invicem propterea quod adveniunt a diversis fissuris λ, λ, λ, λ, λ, λ. Atqui tantum divergunt a speculo postquam reflectuntur quantum antea convergebant. Divergentes itaque paulatim dissocientur ac dissociati proprios colores

---

(10) In §II, 53 = 62.

Third, anyone can observe, and with no difficulty, that colors are not at all changed by transmitting rays through a medium however luminous. Thus the prismatic colors are the same whether they pass through an illuminated space or one enveloped in darkness. All colored things are perceived in the same way whether they are viewed when the sun's light is transmitted through the intermediate space or when it is excluded. This would not happen if light could act on light while passing through the same medium. Indeed, if rays refracted by two prisms cross each other, after they are separated from one another they will produce the same colors that they would otherwise produce if they were not at all mixed. This could not occur if rays imbued with different colors induced some change on each other while passing through the same space.

[35 = 48. Third, rays passing confusedly through the same medium do not act upon one another.]

Fourth, when you have fixed the paper *PT* at that distance where the colors best compose white, place another paper *YZ* two or three inches away from the prism and observe the colored lines on it. Next cut out those parts of the paper on which those lines fell, in this way making oblong slits $v \ldots v$, parallel, equal, and as wide as their separation. Then replace that paper *YZ* to its proper position about three inches from the prism so that the colored light may be transmitted through its slits to the other, more distant paper *PT*. Having done this, you can observe that if you lower the paper *YZ* a bit so that it intercepts the purple and blue colors striking the upper edges of its slits and it transmits the others, the whiteness at the paper *PT* is converted into a red, orange, or yellow color. But if you raise it so that the reds and yellows are intercepted by the lower edges, while the others pass through, that whiteness will be converted into purple, indigo, and blue. Just as ought to happen in a mixture of colors, for when one has been removed from the mixture, the others must be restored to their own species and nature.

[36 = 49. Fourth, that white vanishes if any color is removed from the mixture.]

Fifth, the paper *YZ* having been removed and everything else remaining the same, with a needle pierce the other paper *PT* in the middle of the white so that it may transmit a small portion of its white light; then receive that on another paper placed four or six inches behind *PT*, and instead of whiteness colors will appear again. But how those colors could have been produced de novo if they were destroyed in the production [of white] rather than mixed, I do not understand. It must therefore be conceded that they are only mixed, and that the rays imbued with various colors and flowing from the different slits λ, λ cross each other in that hole made with the needle and diverging afterward are gradually separated from one another, and once separated they again depict their own colors, as will be explained more fully later.[10] Similarly, if you put some small plane mirror *K* in the middle of the whiteness produced at the paper *PT* so that it of course reflects some of the white-making rays elsewhere, such as to $\pi\tau$, the white light thus reflected will degenerate into colors, which can be seen at $\pi\tau$ by inserting a paper. For the rays imbued with different colors and mixed together into white at the mirror *K* are inclined to one another because they arrive from different openings λ, λ, λ, λ, λ, λ; yet they diverge just as much after they are reflected from the mirror as when they converged before. By diverging, therefore, they are gradually separated, and once separated they will

[37 = 50. Fifth, when the colors are separated again by crossing, they return to their own species.]

non secus exhibebunt quàm si nunquam fuissent commisti. Liquet ergo quod in misturâ, radiorum diversicolorum dispositiones ad efficiendos varios colores non destruuntur; ut ut albedinem exhibeant dum commisceantur sibi.

Lect 5
[38]
Adhaec lamina *K,* si valdè obliquetur ad radios in ipsam incidentes, non ampliùs alba apparebit, sed vel cum rubeo vel caeruleo colore imbuta, prout vel versus verticalem angulum, vel versus basin Prismatis inclinatur. Id quod nullo modo accideret, si alba lux quâcum illuminatur, homogenea esset, quandoquidem alba et specularia corpora reflectendo lucem non mutant colorem ejus. Sed hoc ex eo evenire fatendum est quòd in speculum, quando incidentibus radijs admodum obliquatur, pauciores ex obliquioribus radijs in illud incidant inque reflexâ luce major sit copia radiorum minùs obliquorum qui perinde[1] praedominantur, et proprium colorem ostendunt, quem non possent exerere, si ad albedinem lucis incidentis producendam, non tantum cum alijs coloribus miscerentur, sed revera / transmutarentur in uniformem ₂24/ albedinem. Caeterum nota quod in isthoc experimento faciendo praestat laminam non perpolitam, sed superficie nonnihil aspera (qualis est nummi argentei vel chartae &c) praeditam adhibere.[2]

[39 = 51. Sextò, res illustratur per misturam diversicolorum pulverum. Et quòd ex pulveribus omnium colorum debitè mistis fuscus producitur.]
Praeterea[3] vulgò notum est quod ex pulveribus diversicoloribus inter se commistis color novus emergit, tamen si pulveres isti inspiciantur Microscopijs, omnes videntur tincti proprijs coloribus. Adeo ut ex mixturâ pulverum colores proprij non destruantur, sed permiscendo tantùm color novus eliciatur. Verùm ijdem planè colores ex mixtura colorum prismatum ac pulverum producuntur: Sic pulvis caeruleus cum flavo mixtus producit viriditatem, et eadem viriditas etiam producitur ex mixtura radiorum tinctorum cum caeruleo et flavo. Et proinde non dubium est, quin colores novi ex coalescentibus prismatum coloribus, non facta assimilatione, sed mistura tantum, simil[it]er oriantur. Caeterum ut nullum dubitandi locum relinquerem; effeci ut pulveres colorum principalium quos prismata generant rubei, flavi, viridis, caerulei et purpurei in proportione certâ miscerentur: et licet albedo perfecta non prodibat, tamen isti colores ad sensum periêre, et quoddam genus albedinis fuscum et obscurum, sive mediocre inter albedinem perfectam et nigredinem producebatur. Quod nostro proposito non minus inservit, quàm si albedo perfecta prodijsset, quandoquidem fuscus ille ab albo perfecto tantùm differt quantitate lucis non autem specie coloris, ut exinde pateat, quòd producitur ex albo cum nigredine contemperato. Neque expectandum est, ut mihi videtur alium quam fuscum colorem e tali pulverum misturâ generari: Nam cum pulveres colorati intromittant maximam partem lucis istam fere solam reflectentes, quae apta est ad exhibendos / proprios colores, ut ostendetur postea:[4] eorum ₂25/ mixtura maximam quoque partem lucis intromittet: Unde pro albedine perfectâ talis color generandus est, qualis efficitur ex albedine et nigredine

---

(1) Read: *proinde* (consequently).

(2) In the *Opticks* Newton frequently used an oblique mirror to observe the colors compounding white; see Bk. I, Pt. II, Expts. II, III, X, XII.

(3) I: *Denique* (Finally).

exhibit their own colors no differently than if they had never been mixed. It is therefore evident that in a mixture of diversely colored rays the dispositions to produce different colors are not destroyed, howsoever they may exhibit whiteness when they are mixed together.

In addition, if the plate *K* is very oblique to the rays incident upon it, it will no longer appear white, but colored either with the color red or blue according as it is inclined either toward the prism's vertex angle or toward its base. This would in no way occur if the white light illuminating it was homogeneous, since white specular bodies do not change the color of light by reflecting it. It must be acknowledged, however, that this happens, because when the mirror is very oblique to the incident rays, fewer of the more oblique rays fall on it; and in the reflected light there is a greater quantity of less oblique rays that consequently[1] predominate and display their own colors, which they could not reveal, if in producing the whiteness of the incident light, they were not only mixed with the other colors but were in fact transmuted into uniform whiteness. Furthermore, note that in doing this experiment it is better to use a plate that is not highly polished but one possessing a somewhat rough surface, such as a silver coin, a sheet of paper, and so forth.[2]

Moreover,[3] it has been commonly observed that when diversely colored powders are mixed together a new color emerges; yet if those powders are examined with microscopes, they all are seen to be colored with their own colors. Consequently, their own colors are not destroyed by a mixture of the powders, but rather, by mixing, only a new color is brought forth. Clearly the same colors are produced from a mixture of the colors of prisms as well as those of powders. Thus a blue powder mixed with a yellow one produces green, and the same green is also produced from a mixture of rays imbued with blue and yellow. Consequently, it cannot be doubted that new colors similarly arise from a coalescence of prismatic colors and are not made by assimilation but only by mixture. Moreover, to leave no room for doubt I caused powders of the principal colors generated by prisms, red, yellow, green, blue, and purple, to be mixed in a definite proportion. Although a perfect white did not appear, still to the senses those colors had vanished, and a certain kind of white, gray and dark, or a mean between perfect whiteness and blackness, was produced. This serves our purpose no less than if perfect whiteness had appeared; since that gray differs from a perfect white only in the quantity of light but not in the species of color, it is clear from this that it is produced from white tempered with black. Nor, it seems to me, must one expect anything other than a gray color to be generated from such a mixture of powders. Since colored powders admit the greatest part of light and reflect almost solely that part disposed to exhibit their proper color, as will be shown later,[4] a mixture of them also admits the greatest part of the light. Consequently, instead of perfect whiteness, a color ought to be generated such as is made by mixing whiteness and blackness, that is, gray. You may

Lecture 5
[38]

[39 = 51. Sixth, the matter is illustrated by a mixture of diversely colored powders: When powders of all colors are duly mixed together gray is produced.]

(4) In §§II, 71–2.

mixtis, id est, fuscus. Attamen non eo inficias quin tales forte pulveres inveniantur, praesertim inter mineralia, qui tantum lucis reflectant, ut mixti exhibeant albedinem perfectiorem quam hactenus vidi e mixturis effectam. Caeterùm quòd pulveres coloribus tantùm quinque praecipuis tinctos miscebam non ideò cogitandum est albedinem ex quinque solis productam fuisse, sed ex omnigenis. Nam in omnium corporum coloribus alij latent principalibus commixti licèt minùs fortes ut a principali[5] superati non cernantur. Sic in caeruleo pulvere latent cyaneus, et indicus alijque gradus omnes usque ad viridem aut flavum fortassis ex una parte, et ad intensum purpureum ex alterâ: Ut ut caeruleus eò solus appareat quòd sit caeteris longè copiosior.

Experientijs hisce admonitus in mentem praeterea revocabam, quòd corpuscula quae conspiciuntur in radijs solaribus huc illuc volitantia varios colores exhibent, modò quisquam ea diligenter observet in cubiculo quaquaversum luci occluso praeter unicum foramen per quod illuminantur. Et tamen cum isti pulvisculi in acervum congregantur nullus omninò color apparet praeterquam fuscus.

[40]   Non minùs apposita est observatio quòd cùm aqua sapone in ea soluto paululum inspissata et in spumam agitando conversa fuerit, postquam paululum constitit spuma, in singulis bullulis ex quibus conglomeratis efformata est, innumeri omnis generis colores acutiùs inspicienti apparuere, et tamen spuma ad tantam distantiam spectata ubi colores in singulis bullulis ab invicem distingui nequibant apparuit perfectè candida.[6]

[41 ≈ 52. Tertius modus miscendi colores prismatis in albedinem.]   / Patet itaque colores prismatum revera non destrui ad albedinem producendam sed commisceri[7] tantùm; quandoquidem emergant immutati cùm radij coeuntes decussavere et per subsequentem divergentiam iterum dissociantur; ‖ et proprios etiam colores exhibent cùm aliquae copiosiùs quàm caeterae reflectuntur; atque subalbus color e mistura pulverum omnigenis coloribus praeditorum, ut et albedo perfectior e diversicoloribus bullulis sine aliqua congredientium colorum mutatione similiter emergat. ‖ Ad haec cùm rei dignitas postulare videatur, ut nullus non moveatur lapis, praeter modos praecedentes componendi albedinem lubet adhibere tertium, et quartum deinde quo praedicta faciliùs experiri possis, et magis fortè cum evidentiâ. ₂₆/

Posito quòd sol illuceat obscurato cubiculo per unicum tantùm foramen *F* (fig II, 13) cui Prisma *ABC* affigitur ingressam lucem refringens ad *PT*: juxta

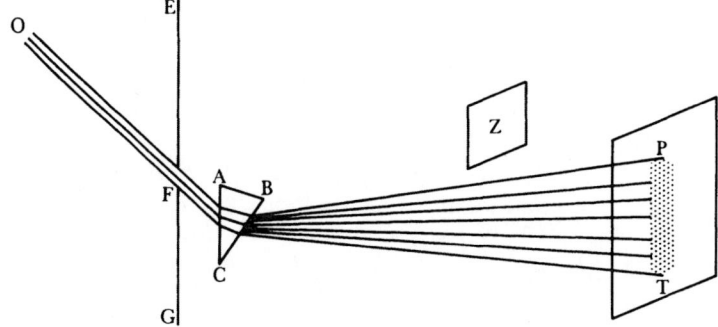

Figure II, 13

even perhaps succeed in finding such powders, especially among minerals, that reflect so much light that when mixed they exhibit a more perfect white than I have hitherto seen produced from mixtures. Moreover, because I mixed powders colored with only the five principal colors, it must not therefore be imagined that white had been produced from only five, but from all kinds. For latent in the colors of all bodies are other, though less strong, colors mixed with the principal ones, so that being overwhelmed by the principal one[5] they are not perceived. Thus in a blue powder, cyan and indigo and all the other gradations up to green or perhaps yellow are latent on the one side, and up to an intense purple on the other side. However, blue alone is seen there, because it is far more abundant than the others.

In doing these experiments, moreover, I recalled to mind the suggestion that the particles seen flitting about in the sun's rays exhibit various colors, provided that one observes them carefully in a room closed off to light everywhere except for a single hole through which they are illuminated. Yet when those fine powders are gathered in a heap, no color at all appears besides gray.

It is no less appropriate to observe that when water thickened somewhat [40] by soap dissolved in it has been converted into a froth by shaking it, after the froth stands a while, in the individual bubbles that together form the froth, innumerable colors of every kind appeared to someone rather astutely examining them. Yet when the foam was viewed at such a large distance that the colors in the individual bubbles could not be distinguished, it appeared perfectly white.[6]

It is therefore evident that prismatic colors are in fact not destroyed in producing whiteness, rather they are only mixed,[7] since they emerge unchanged when the rays after coming together have crossed, and by subsequently diverging they are again separated; ‖ that they also exhibit their own colors when some are more copiously reflected than others; and that a whitish color emerges from a mixture of powders possessing every kind of color, as well as a more perfect white from diversely colored bubbles without any change in the colors gathered together. ‖ Moreover, since the merit of the subject seems to require that no stone be left unturned, besides the preceding ways of compounding whiteness it is satisfying to employ a third and then a fourth way whereby you can undertake that more easily and perhaps with greater clarity.

[41 ≈ 52. A third way of mixing prismatic colors into whiteness.]

Assume the sun to shine into a darkened room through only a single hole *F* (Fig. II, 13) to which there is affixed the prism *ABC* that refracts the entering

---

(5) I: *principali colore* (the principal color).

(6) In his *Touching Colors,* Boyle had noted both the whiteness of the froth of shaken water (Pt. I, Ch. III, pp. 29–30) and the colors of soap bubbles (Pt. III, Expt. XIX, pp. 243–4). Newton also set forth this example in his reply to Hooke, 11 June 1672 (*Correspondence,* 1:186 = *Phil. Trans.,* 7 (1672):5102), and in the *Opticks* (Bk. I, Pt. II, Prop. V, Expt. XIV).

(7) *Patet itaque colores prismatum . . . destrui . . . commisceri*] I: *Imò quòd colores prismatum . . . destruantur . . . commisceantur* (Indeed, prismatic colors are . . .).

colores in papyrum *PT* sic projectos teneatur alia papyrus *Z* ut illuminetur a coloratâ luce quam altera papyrus *PT* reflectit. Quo facto papyrus *Z* sic illuminata radijs omnium colorum a *PT* confusè reflexis, apparebit alba. De hoc autem specimine maximè luculento et facili juvabit observare sequentia.

[42 ≈ 53. In eundem notae.] Primò quòd auferendo papyrum *PT* nè lucem ampliùs ad *Z* reflectat: e consequente defectu lucis in *Z* cognoscas eam illuminari per solam lucem coloratam a *PT* reflexam.

Secundò si papyrum *Z* ipsi *PT* valdè vicinam teneas, ut una pars ejus magis illuminetur ab uno colore, et alia ab alio: ipsa *Z* non apparebit alba, sed ejus partes coloribus[8] istis tingentur quibus sunt vicinissimae. Sin ipsa *Z* ad majorem a *PT* distantiam transferatur ut omnes ejus partes aequaliter ferè ab omnibus coloribus illuminentur: ex illa colorum mixtura generabitur albedo. Pari ratione si quemlibet e coloribus ad papyrum *PT* tendentibus intercipias, ne reflectatur ad *Z*, illud *Z* non amplius albescet, sed evadet coloratum pro mixturâ quam caeteri colores in ipsam *PT* prolapsi componunt.

/ Denique quòd albedo illa *Z* non destruendo colores sed tantùm miscendo ₂27/ generatur exinde pateat quòd colores *PT* cernuntur beneficio radiorum non secus oculo mixtim incidentium, quàm papyro *Z*. Itaque si colores destruerentur potiùs quàm miscerentur ad *Z*, etiam destruerentur ad corneam tunicam oculi, vel pupillam: ubi tamen certissimum est, quòd misceantur tantùm, ut decussantes, postea divergant ad varias partes Retinae et sic excitent phantasmata propria. Quinimò si radij tincti cum diversis coloribus dum per eadem spatia confusè transeunt possent in se invicem agere et dispositiones mutare quas quilibet habent ad expingendos proprios colores; omnes omnium rerum colores conturbarentur, ac se mutuò transmutarent dum per aera transmittuntur; ubique scilicet radijs aliorum corporum omnigenis coloribus tinctorum occurrentes. Et sic in coloribus visibilium nulla esset certitudo, constantia nulla.

[43 = 54. Quartus ejusdem rei peragendae modus, caeteris illustrior.] Quartum praeterea modum descripturus quo colores in albedinem misceri possent, pono *ABC* (fig II, 14) esse prisma foras ante foramen *F* dispositum,

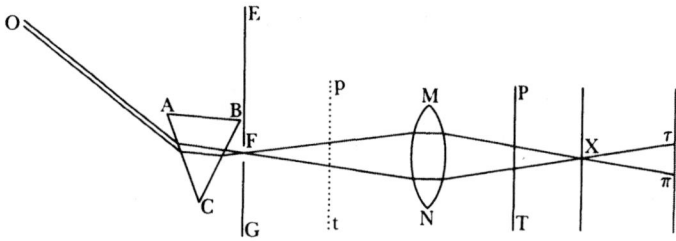

Figure II, 14

quod refractam lucem in obtenebratum cubiculum transmittat versus *MN*. Tum lentem *MN* convexam sume, cujus focus est ad distantiam semipedis vel pedis unius duorumve (quale est objectivum vitrum Perspicilli bipedalis;) et eam statue paulo plus distantem a foramine *F* quàm focus distat a se; ita scilicet ut lux colorata per eam deinceps trajiciatur, sicut videre est in sche-

light to *PT*; near the colors thus projected upon the paper *PT* hold a second paper *Z* so that it is illuminated by the colored light that the first paper *PT* reflects. When this has been done, the paper *Z*, illuminated in this way by rays of every color reflected confusedly from *PT*, will appear white. It will be useful, however, to note the following concerning this particularly excellent and easy example.

First, by removing the paper *PT* so that it no longer reflects light to *Z*, you may, as a consequence of the absence of light at *Z*, perceive that it is illuminated solely by the colored light reflected from *PT*. [42 ≈ 53. Notes on the same.]

Second, if you hold the paper *Z* extemely close to *PT* so that one part of it is illuminated more by one color and another part by another color, then *Z* will not appear white, but its parts will be colored with those colors[8] to which they are nearest. If now *Z* is shifted farther away from *PT* so that all its parts are illuminated almost equally by every color, white will be generated from that mixture of colors. ‖ For the same reason, if you intercept any of the colors tending to the paper *PT* so that it is not reflected to *Z*, that *Z* will no longer be white but will become colored according to the mixture that the other colors proceeding to *PT* compound.

Finally, it is evident that the white at *Z* is generated not by destroying but only by mixing the colors, and accordingly that the colors at *PT* are perceived by means of rays falling mixedly upon the eye just as upon the paper *Z*. Therefore, if the colors were destroyed rather than mixed at *Z*, they would also be destroyed at the cornea of the eye, or rather the pupil, where nonetheless it is most certain that they are only mixed, inasmuch as upon crossing they afterward diverge to different parts of the retina and thus excite their own sensations. Indeed, if rays imbued with different colors could act upon one another while passing confusedly through the same space and could change the dispositions they each have to depict their own color, then all colors of all things would be thrown into confusion, and they would mutually transmute each other while they passed through the air, that is, everywhere running into rays of other bodies imbued with every sort of color. Thus there would be neither certainty nor stability in the colors of visible things.

A fourth way, moreover, will be described whereby colors can be mixed into white. I assume that (Fig. II, 14) *ABC* is a prism placed outside in front of the hole *F*, which transmits the refracted light into a darkened room toward *MN*. Then take a convex lens *MN*, whose focus is distant by half a foot, or one or two feet (such as the object glass of a two-foot telescope), and place it slightly farther from the hole *F* than the focus is distant from the lens, specifically, so that the colored light is then cast through it, as is seen in the [43 = 54. A fourth way of accomplishing the same thing more clearly than the others.]

---

(8) *Editio princeps*: ab omnibus coloribus.

mate. Sit autem ejus latitudo, sive apertura tanta ut omnes radios transmittat. Deinde cùm lentem in dicto situ stabilitam feceris, ponè statuatur papyrus *PT* in quam radij hi[9] refracti terminentur. Eamque primo colloca proximè ad lentem, deinde ad majorem distantiam continuato motu transfer, et videbis colores purpureum *P* / rubeumque *T* contrahi et eousque minui, dum omnes ₂28/ convertantur in albedinem, puta ad *X* quatuor, vel sex pedes aut longius forte distantem a lente, pro convexitate ejus vel positione. Deinde si papyrum adhuc longius transferas, colores iterum emergent sed in situ contrario, rubeo ad *τ* conspecto et purpureo ad *π*. Neque ulla inter eos ad *PT* et *ππ* differentia intercedit praeterquam quòd situs sit contrarius. Scilicet a lente *MN* effectum est ut omnes radij venientes ab aliquot punctis foraminis *F* in totidem iterum punctis congregentur ad papyrum *X*: Et sic omnes omnium specierum tum purpuram ad *P*, tum rubedinem ad *T*, tum alios alibi colores efficientium convergunt ad *X*, et ibi confuse miscentur ad albedinem generandam: De qua imagine albâ et orbiculari monebam supra.[10] Postea verò cùm sese decussavere in *X*, radij *PX* tendunt ad *π*, et *TX* ad *τ*, adeò ut ijdem colores expingantur ad *P* et *π* per eosdem radios *Pπ*, et ijdem ad *T* et *τ* per eosdem *Tτ*, et sic de alijs. Unde liquet iterum quòd dispositiones radiorum dissimilium,[11] ad diversos colores producendos non destruantur per eorum mixturam, quandoquidem eosdem expingunt cùm segregantur quos ante mixturam expingebant.

[44 = 55. In
eundem nota.]

Porro si radios cujusvis coloris intercipias, interponendo corpus aliquod opacum prope lentem *MN*, et caeteros facias missos: videbis non modò colores interceptos e papyris *PT* ac *ππ* tolli, sed et albedinem *X* destrui, et ejus vice colorem aliquem, qualis efficitur per mixturam radiorum praeterlabentium, generari. Sic si radios intercipias ostendentes rubeum ad *N*: rubedo *T* ac *τ* tolletur et albedo *X* convertetur in caeruleum. Vel si sistas tum rubeum ad *N* tum purpureum / ad *M*, et intermedios flavum viridem et caeruleum ₂29/ praeterlapsos mittas: ex eorum mistura viriditas producetur ad *X*. Et sic praetermittendo quos velis et sistendo alios, pro arbitratu possis experiri mixturas quaslibet et explorare qui color inde generabitur; modò pretium laboris experientiam illam judicaveris.

Lect 6
[45 = 56. Quo more
radij diversicolores
in albentem lentis
focum convergunt.]

Verum cum experimenti hujus dignitas videatur exigere ut summâ cum diligentiâ retegatur et penitius explicetur, dum plura de coloribus simul complectitur et exhibet quam in unico tantùm experimento solent latere: non gravabor modum copiosiùs ostendere quo[1] radij miscentur ad *X*, et nonnulla postmodum scitu non indigna patefacere. Itaque concipiantur tales refractiones in Prismate fieri, ut radij incidant in varios circulos ad Lentem *MN* qui varios gradus refractionis patiuntur; prout explicui in praecedentibus.[2]

---

(9) I: *bis* (the twice).
(10) A marginal note in I refers to §§16, 18 [= I, 16, 19].
(11) Originally (as I): *absimilium*.
(1) *modum . . . quo*] Originally (as I): *copiosiùs ostendere quo pacto*.

figure. However, let its width or aperture be large enough to transmit all the rays. Then, when you have fixed the lens in the designated position, behind it place the paper *PT* on which these[9] refracted rays are outlined. First place it very close to the lens, and then shift it with a continuous motion to a greater distance; you will see the colors purple *P* and red *T* contract and continuously diminish until they are all converted into white, suppose at *X*, at a distance of four or six feet, or perhaps more, from the lens according to its convexity or position. If you then shift the paper still farther away, the colors will emerge again but in a reversed situation, red visible at *τ* and purple at *π*. Nor is there any difference between those colors at *PT* and *ππ* except that their situation is reversed, namely, the lens *MN* causes all rays coming from the many points of the hole *F* to gather again into just as many points at the paper *X*. Thus all rays of every sort, making purple at *P* as well as red at *T* and the other colors elsewhere, converge to *X* and are there confusedly mixed to generate white; I pointed out this white and circular image above.[10] But after they have crossed one another at *X*, the rays *PX* tend to *π* and *TX* to *τ*, so that the same colors are depicted at *P* and *π* by the same rays, *Pπ*, and the same colors at *T* and *τ* by the same rays, *Tτ*, and similarly for the others. Consequently, it is again evident that the dispositions of dissimilar[11] rays to produce diverse colors are not destroyed by their mixture, since they depict the same colors when they are separated as those they depicted before the mixture.

If, moreover, you intercept the rays of any color by inserting some opaque body near the lens *MN*, while allowing the others to pass by, then you will see not only the intercepted colors vanish from the papers *PT* and *ππ* but also the white at *X* destroyed and in its place some color generated such as is made by the mixture of the rays passing by. For instance, if you intercept the rays displaying red at *N*, the red at *T* and *τ* will vanish, and the white at *X* will be converted into blue; or if you stop both the red at *N* and the purple at *M*, and let the intermediate ones, yellow, green, and blue, pass by, from their mixture green will be produced at *X*. And thus by allowing those you choose to pass, while stopping the others, you can test at will any mixtures whatsoever and examine the color that will thereby be generated, provided that you considered that experiment worth the effort.

[44 = 55. A note on this.]

Since the value of this experiment truly seems to require that it be disclosed with the utmost diligence and explicated rather thoroughly, while it also includes and shows more about colors than usually is concealed in only a single experiment, I will not be reluctant to show more fully how[1] the rays are mixed at *X* and then to reveal several things worth knowing. Accordingly, imagine such refractions to occur in the prism that the rays fall on the lens *MN* in different circles that experience different degrees of refraction, just as I explained in the preceding.[2] Let *PQRST* (Fig. II, 15) be the oblong

Lecture 6
[45 = 56. How the diversely colored rays converge into the white focus of the lens.]

---

(2) A marginal note in I refers to §12 [= I, 12].

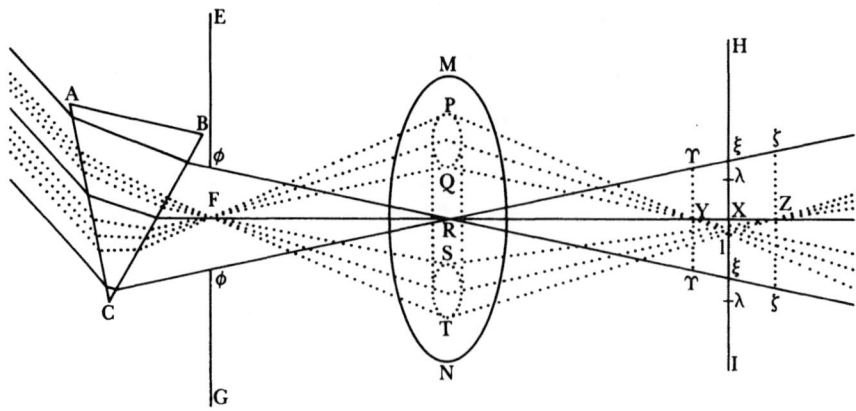

Figure II, 15

Sitque *PQRST* (fig II, 15) oblonga imago composita ex praecedentibus[3] circulis et in Lentem projecta, quorum circulorum extremi duo sunto *PQ* purpureus et *ST* rubeus. Porro sit *φFφ* diameter foraminis per quod lux in lentem trajicitur cujus foraminis punctum aliquod ut *F* primò consideremus a quo venientes radij dictos circulos *PQ*, *ST* totamque imaginem *PT* efformant. Et praeterea cum radij eundem[4] quemlibet circulum efformantes sint homogenei:[5] ponatur quòd lens sit tali figura praedita ut eos omnes ad eundem illum circulum, (puta rubeum *ST*) pertinentes versus[6] punctum quoddam *Z* exactè refringat. Quod fieri posse per lentem co[n]vexis Hyperbolis terminatam, ut et per lentes aliter formatas Cartesius in Dioptrica et Geometriâ edocuit.[7] Est itaque *Z* focus radiorum *FS FT* et caeterorum uniformiter rubeorum, et recta *FZ* ducta erit axis lentis; Praetereà cùm radij *FP*, *FQ* caeterique conficientes alterum extremum circulum *PQ* colorem purpureum ostendant, et propterea magis refringantur quàm alteri tendentes ad *ST*; illi ideò convergent[8] ad punctum quoddam aliquantò propinquius / quàm *Z*, ₂30/ veluti ad *Y*; ut ij facile percipient qui norunt focos lentium esse tanto propinquiores sibi, quanto major est earum vis refractiva. Liquet itaque radios in coloribus et refractionibus absimiles ad diversos focos convergere. Sed cùm eadem lens pluribus focis haud queat adaptari, et ideò cùm *Z* supponatur focus in quem omnes radij ad circulum rubeum *ST* pertinentes exactè conveniant: radij pertinentes ad alterum circulum *PQ* purpureum, omnes in ejus focum *Y* exactè convenire nequeunt. Attamen eorum concursus juxta *Y* in axe tam proximè accuratus erit ut quoad sensum et experientiam omnem habeatur pro accurato. Quinetiam si Lens *MN* ponatur sphaericè convexa ut neuter focorum *Y* vel *Z* strictè loquendo possit esse accuratus, tamen quantum ad praesentia spectat pro accuratis habere liceat.[9] Itaque concipiendo quod radij manantes [ab *ST* convergant ad *Z* et quòd alteri manantes][10] a *PQ* convergant ad *Y*, et ibi decussantes divergant itidem: patebit quod hi duo radiorum penicilli concurrent et miscebuntur in spatio focis *Y* et *Z* intermedio veluti ad *l*, modo Lentis centrum *R* ponatur intermedium circulis *PQ* et *ST*. Ad eundem modum radij caeterorum generum convergent in alios focos ipsis

image composed of the preceding[3] circles and projected onto the lens, and let its two outermost circles be *PQ*, purple, and *ST*, red. Furthermore, let $\phi F\phi$ be the diameter of the hole through which the light is cast onto the lens, and first let us consider any point of this hole, such as *F*, from which the rays come and form those circles *PQ* and *ST* and the whole image *PT*. Since then the rays forming any one[4] circle are homogeneous,[5] let it be assumed that the lens possesses a shape such that it refracts all those rays belonging to that same circle (for example, the red, *ST*) exactly toward[6] some point Z. This can happen by means of a lens bounded by convex hyperbolas as well as by lenses shaped otherwise, as Descartes has taught in the *Dioptrics* and *Geometry*.[7] Consequently, Z is the focus of the rays *FS*, *FT*, and of the others uniformly red; and when the line *FZ* is drawn, it will be the axis of the lens. Moreover, since the rays *FP*, *FQ*, and the others composing the other outermost circle *PQ* display the color purple and are thus more refracted than the others tending to *ST*, they therefore converge[8] at some point slightly closer than Z, such as at *Y*, as will be readily comprehended by those who know that the foci of lenses are closer to themselves the greater their refractive force is. It is therefore evident that rays dissimilar in colors and refractions converge to different foci. But since the same lens is not at all able to be adapted to several foci, and since Z is assumed to be the focus at which all rays belonging to the red circle, *ST*, meet exactly, then rays belonging to the other, purple circle, *PQ*, cannot all meet exactly at its focus *Y*. Nonetheless, their intersection near *Y* on the axis will be so nearly accurate that according to sense and all experience it may be considered as accurate. Besides, if the lens *MN* is assumed to be spherically convex, so that neither of the foci *Y* or Z strictly speaking can be accurate, still, insofar as present matters are concerned, it is proper to consider[9] them as accurate. Consequently, by conceiving that the rays flowing [from *ST* converge to Z, and that the others flowing][10] from *PQ* converge to *Y* and after crossing there diverge again, it will be clear that these two pencils of rays will intersect and be mixed in the space between the foci *Y* and Z, such as at *l*, provided that the lens's center *R* is assumed to be between the circles *PQ* and *ST*. In the same way the other kinds of rays will converge to other foci between *Y* and Z, and

(3) I: istis (those).

(4) Added.

(5) Originally (as I): conformes sibi (similar to themselves).

(6) eos . . . versus] Originally (as I): eos omnes cujusdam e circulis, (puta rubei *ST*) versus (all those rays of some one circle . . .).

(7) See Lect. 5, note (10).

(8) *Editio princeps*: emergent.

(9) habere liceat] Originally (as I): habeantur (they may be considered).

(10) The words in brackets from I were omitted in transcription and consequently from the *editio princeps*.

Y et *Z* intermedios ac tanto propinquiores ipsi *Y* quanto major est eorum passio refractiva. Sic focus viridiformium radiorum cadet in medio spatio veluti ad *X*; radijque caeruliformes[11] convenient citiùs inter *X* et *Y* et flaviformes longinquius inter *X* et *Z* ac caeteri colores intermedij in spatijs intermedijs: Eorumque penicilli sese decussabunt ultra citraque locum *l*; ita tamen ut istae decussationes sint eo densiores quanto sint ipsi *l* viciniores et ut spatium *Xl* sit minimum per quod omnes radij transeunt manantes ab eodem puncto *F*. Non dissimili modo radij venientes ab alio quovis puncto foraminis, ut *φ*, si sint rubriformes convergent ad *ζ*; sin purpuriformes ad *Υ*; et ad intermedium aliquod punctum si sint intermedij generis et eorum concursus densissimus erit in loco medio, veluti ad *ξλ*. / Atque adeò ex radijs ab integro ₂31/ foramine *φF[φ]* manantibus foci maxime refrangibilium jacebunt in superficie quadam *ΥΥΥ* ad lentem proximâ, foci minimè[12] refrangibilium jacebunt in alia superficie *ζZζ* a lente remotissimâ, focique mediocriter refrangibilium jacebunt in alijs intermedijs superficiebus. Et sic omnes omnium radiorum foci totum spatium *ΥζζΥ* a superficiebus istis integratum occupabunt, et in eo praecipuè penicilli decussabunt et commiscebuntur.

[46 = 57. De coloribus in extremitate foci illius, propter exilitatem vix conspicuis.]

Jam ex hac descriptione venit observandum quòd cùm papyrus *HI* teneatur in medio dicti spatij *ΥζζΥ* ut in eam radij terminentur ubi est densissimus eorum concursus et mixtura ad albedinem generandam perfectissima: radij viridiformes tendentes ad focos in papyro sitos in eam incident intra literas *ξξ*, sed rubriformes venientes ab *ST*, ac tendentes ad focos in superficie *ζZζ* sitos ut dictum est, incident in papyrum intra literas *λλ* paulo vicinius ad *I*. Et pari modo purpuriformes incident in eundem locum *λλ* dum tendunt a *PQ* ad focos sitos in superficie *ΥΥΥ*. Caeteri autem radij cadent in alia spatia inter *ξξ* & *λλ* mediocria ipsisque *ξξ* tanto viciniora quanto foci eorum minùs absint a papyro. Liquet itaque quod totum spatium *ξXlλ* non debet albescere, sed pars ejus tantum media inter literas *ξ* et *λ* interiores sita ubi scilicet colores omnes commiscentur. Etenim in extremitate *ξ* versus *H* radij viridiformes cadunt soli; qui proinde tingent extremitatem istam cum viriditate. Ad alteram autem extremitatem versus *I* nulla miscetur viriditas sed purpura tantùm cum rubore. Qui dicta perpendet, etiam facilè concipiet,[13] quòd cùm papyrus paululum transferatur ultra citraque, colores alij praeter viriditatem apparebunt ad extremitatem imaginis versus *H*, scilicet inter *P* et *Υ* purpureus apparebit extimus, inter *Υ* et *ξ* caeruleus, et viridis ad *ξ*, deinde flavus inter *ξ* et *ζ*, ac rubeus denique ad *ζ* et postea perpetuò. Ad alteram autem imaginis extremitatem versus *I* sitam rubeus erit extimus a *T* usque ad *λ* ubi commiscetur purpurae; Quae quidem mixtura dat pallidum quendam colorem nunc ad rubeum nunc ad / caeruleum nonnihil vergentem pro variâ proportione ₂32/ mistorum. At ultra *λ* purpura semper conspicietur. Caeterùm cùm distantia inter *Υ* et *ζ* valdè parva sit et multò magis distantia inter *X* et *l* sive *ξ* et *λ*, hoc est latitudo limbi colorati: propter summam ejus exilitatem conspectui vix patebit, sed totum spatium *ξXlλ* nisi acriùs observanti apparebit album.

---

(11) In the *Lectiones opticae* Newton had here explained his new terminology, "caeruliformes" (blue-making), and so on, but he deleted it when he decided to introduce it earlier in §II, 19.

(12) maxime . . . minimè] *Editio princeps*: minime . . . maxime.

(13) I: percipiet (perceive).

they will converge closer to *Y* the greater their refractive susceptibility is. Thus, the focus of the green-making rays will fall in the middle of the space, such as at *X*; the blue-making[11] rays will meet sooner, between *X* and *Y*; the yellow-making rays farther off, between *X* and *Z*; and the other intermediate colors in the intermediate spaces. Their pencils will cross one another on both sides of the place *l*, yet in such a way that these crossings are denser the closer they are to *l*, and the space *Xl* is the smallest through which all rays flowing from the same point *F* pass. In a not dissimilar way, the rays coming from any other point of the hole, such as ϕ, if they are red-making, will converge to ζ, but if they are purple-making, to Υ, and to some intermediate point if they are of an intermediate kind; and their most dense intersection will be in the middle region, such as at $\xi\lambda$. Hence, of the rays flowing from the entire hole $\phi F\phi$, the foci of the most refrangible rays will be in some surface $\Upsilon Y\Upsilon$ nearest to the lens, the foci of the least[12] refrangible ones will lie in another surface $\zeta Z\zeta$ most remote from the lens, and the foci of the mean refrangible ones will lie in other intermediate surfaces. In this way all the foci of all the rays will occupy the entire space $\Upsilon \zeta\zeta \Upsilon$ made up by those surfaces, and the pencils will cross and be mixed together primarily in that space.

Now from this description it can be observed that when the paper *HI* is held in the middle of that space $\Upsilon\zeta\zeta\Upsilon$ so that the rays are terminated on it where their intersection is densest and the mixture for generating whiteness is most perfect, the green-making rays tending to the foci situated on the paper will fall upon it between the letters $\xi\xi$, but the red-making ones coming from *ST* and tending to the foci situated on the surface $\zeta Z\zeta$, as has been described, will fall upon the paper between the letters $\lambda\lambda$ a little closer to *I*. In a like way, the purple-making rays will fall upon the same place $\lambda\lambda$ while they tend from *PQ* to the foci situated on the surface $\Upsilon Y\Upsilon$. The other rays, however, will fall in the other intermediate spaces between $\xi\xi$ and $\lambda\lambda$ and be nearer to $\xi\xi$ as their foci are less distant from the paper. It is evident, therefore, that the entire space $\xi Xl\lambda$ ought not to become white, but only its middle part located between the inner letters ξ and λ, namely, where all the colors are mixed. For at the end ξ toward *H*, there fall only green-making rays that consequently color that end green; but at the other end toward *I* no green is mixed, but only purple together with red. Whoever carefully considers what has been described will also easily understand[13] that when the paper is shifted back and forth somewhat, other colors besides green will appear at the end of the image toward *H*: specifically, purple will appear outermost between *P* and Υ, blue between Υ and ξ, green at ξ, next yellow between ξ and ζ, and finally red at ζ and ever after. At the other end of the image located toward *I*, however, red will be outermost from *T* to λ where it is mixed with purple; this mixture in fact yields a certain pale color inclining somewhat now to red and now to blue according to the different proportion of the mixtures. But beyond λ purple will always be observed. Since, however, the distance between Υ and ζ is very small and the distance between *X* and *l* or ξ and λ (that is, the width of the colored band) is even smaller, on account of its extreme thinness the band will scarcely be evident to sight, whereas the entire space $\xi Xl\lambda$ will appear white except to a rather sharp observer.

[46 = 57. Concerning the colors at the extremity of the focus that are scarcely visible because of their thinness.]

[47 = 58. Dictorum colorum observatio.]

Cùm haec advertissem, experiebar deinde an responderent praeconceptis: et licèt malè successisset primò dum utebar angustâ lente: posteà tamen cùm adhibui lentem eâ de causâ latiorem ut angulus *XYl* sive $\xi T\lambda$ et inde $\xi\lambda$, sive *Xl* hoc est latitudo dicti limbi colorati fieret major, quod optabam evenit. Adhibeatur itaque lens cujus latitudo, sive apertura sit trium digitorum aut major eo, foci autem longinquitas pro lubitu pedum trium vel quatuor; tum ea collocetur ad distantiam sex vel octo pedum a foramine $\phi F\phi$, ut colores *PQRST* in eam prolapsi usque ad extremitates ejus extendantur nullis tamen praeterlabentibus. Deinde papyrus *HI* ponè collocata transferatur ultra citraque, et ad extremitatem imaginis versus *H* videbis omnes prismatum colores a purpura ad rubedinem usque gradatim successivos; sed ad alteras imaginis partes versus *I*, inter purpuram ad ζ et rubedinem ad T conspicuam, neque viriditas neque alius quispiam ex intermedijs coloribus apparebit nisi forte qui fiunt ex rubeo et purpureo mixtis; Quemadmodum ex eo cognoscas quòd cùm intercipis extremitatem purpurae ope corporis opaci juxta lentem ad *P* interpositi, ille limbus imaginis versus *I* fiet rubeus; sin extremitas rubedinis ad *T* intercipiatur, limbus idem fiet purpureus. Et hinc est quòd transitus a purpura ad rubedinem ex hac parte imaginis fit multo celerior quàm ex alterâ versus *H* ubi colores omnes interveniunt. Caeterùm cùm dictorum colorum latitudo tam exigua sit (videlicet haud major centesimâ parte digiti) ut nisi vitra sint bene polita et a venis libera et insuper experientis diligentia et curiositas solito major, forte excidet proposito. Quamobrem in majorem evidentiam rei et experiendi copiam addo, / quòd si microscopium sumas atque ita disponas ut papyrum aliquam affixam laminae super quam objecta collocantur contemplanda, distincte ampliet; Dein ita statuas ut imago lucida $\xi Xl\lambda$ incidat in istam payrum: colores in ejus limbo sic ampliatos videbis sat[14] manifestos. ₂33/

[48 = 59. Quintus modus albedinem componendi quarto ferè similis.]

Verùm cùm mistura radiorum quoad colores dissimilium[15] non sit adeò perfecta in hoc specimine quin ut e coloribus aliqui in extremitate albedinis appareant (licèt tam exigui ut incautus forte non advertat;) placet insuper observare quòd si vice lentis refractoriae speculum concavum accuratè formatum et perpolitum adhibeas, dicta mistura fiet omnibus numeris perfecta. Etenim irregularitas illa qua refractiones ita perturbantur, in reflexionibus nulla est, sed radij quoscunque colores depingentes, et utcunque refrangibiles ad eosdem tamen angulos reflectuntur in quibus incidunt. Quamobrem si *MN* (fig. II, 16) sit speculum Ellipticum cujus foci sint *F* et *X*. radij omnes a puncto *F* manantes, cujuscunque sint generis sive purpuram ad *P*, sive rubedinem ad *T*, sive alios alibi quoscunque colores ad speculum exhibentes, omnes[16] accuratè convenient in eodem puncto *X*. Quinimo licet speculum *MN* non sit ex ellipticâ figurâ segmentum, sed e sphaericâ; modò semidiameter sphaerae, hoc est distantia ejus a focis praedictis *F* et *X*, satis magna sit, puta trium pluriumve pedum, et distantia focorum valdè parva, puta non plusquam unius digiti: si haec inquam ponantur, radij ab *F* manantes adeò propemodum convenient in *X* ut istud *X* quoad sensum pro exacto foco

---

(14) Added.   (15) Originally (as I): absimilium.   (16) I: tamen omnes (all still).

When I had recognized these things, I then tested whether they would agree with my preconceptions. Although the test turned out poorly at first when I employed a narrow lens, yet afterward, when I consequently used a wider lens so that the angle XYl or $\xi T\lambda$ and hence $\xi\lambda$ or Xl (that is, the width of that colored band) would become greater, it turned out as I hoped. Therefore, use a lens whose width or aperture is three inches or greater but whose focal length (as you choose) is three or four feet; then place it six or eight feet away from the hole $\phi F\phi$, so that the colors $PQRST$ flowing into it extend up to its edges, yet none flow beyond. Next shift the paper HI placed behind it back and forth, and at the end of the image toward H you will see all the successive prismatic colors continuously from purple to red; but at the other parts of the image toward I, between the purple visible at ζ and the red at T, neither green nor any of the other intermediate colors will appear, except perhaps those that arise from a mixture of red and purple. On this basis you may understand that when you intercept the extremity of the purple with the aid of an opaque body inserted next to the lens at P, that band of the image toward I will become red; but if the extremity of the red is intercepted at T, the same band will become purple. Hence it happens that the transition from purple to red on this side of the image occurs much more quickly than on the other side toward H where all the colors intervene. Since, however, the width of those colors is so small (specifically, not greater than one hundredth of an inch), unless the glasses are well polished and free from veins, and moreover unless the experimenter's diligence and care are unusually great, he will perhaps not succeed. Accordingly, for the greater clarity of the matter and ability to do the experiment, I add that if you take a microscope and set it so that it distinctly enlarges some paper affixed to the plate on which objects to be studied are placed, and then you arrange it so that the bright image $\xi Xl\lambda$ falls upon that paper, you will see the colors in its band enlarged in this way clearly enough.[14]

[47 = 58. Observation of those colors.]

But since the mixture of rays dissimilar[15] with respect to their colors is not perfect in this example, so that some of the colors on the extremity do not appear white (even if they are so slight that a careless person might perhaps not observe them), it is appropriate to note in addition that if instead of a refractive lens you were to use an accurately shaped and highly polished concave mirror, that mixture will become perfect in all respects. For that irregularity that so disturbs refractions does not exist in reflections; on the contrary, rays depicting any color whatsoever and however refrangible are notwithstanding reflected at the same angle at which they are incident. Accordingly, if MN (Fig. II, 16) is an elliptical mirror whose foci are F and X, all rays flowing from the point F—whatever kind they may be, whether exhibiting purple at P, red at T, or any other colors at all elsewhere on the mirror—will all[16] meet accurately at the same point X. Indeed, even if the mirror MN is not a segment of an elliptical figure but of a spherical one, provided that the radius of the sphere (that is, its distance from the specified foci F and X) is sufficiently great, say, three or more feet, and the distance of the foci is very small, say, not more than one inch; if, I say, these things are assumed, the rays flowing from F will meet so nearly at X that that X may be sensibly considered as the exact focus. In the

[48 = 59. A fifth way to compound whiteness nearly similar to the fourth.]

habeatur. Et eodem modo radij manantes ab alijs punctis ut ϕ ipsi *F* vicinis, in alijs ut ξ ipsi *X* vicinis quam proxime convenient. Et sic omnes omninò colores reflectentur a speculo *PT* in unumquodque punctum imaginis $\xi X \xi$ totamque exhibebunt albam.

[49]   Sunt et alij modi componendi albedinem, quemadmodum si vice Lentis speculive duo Prismata *ILM* et *KMN* (fig II, 17) in situ ad consimile Prisma

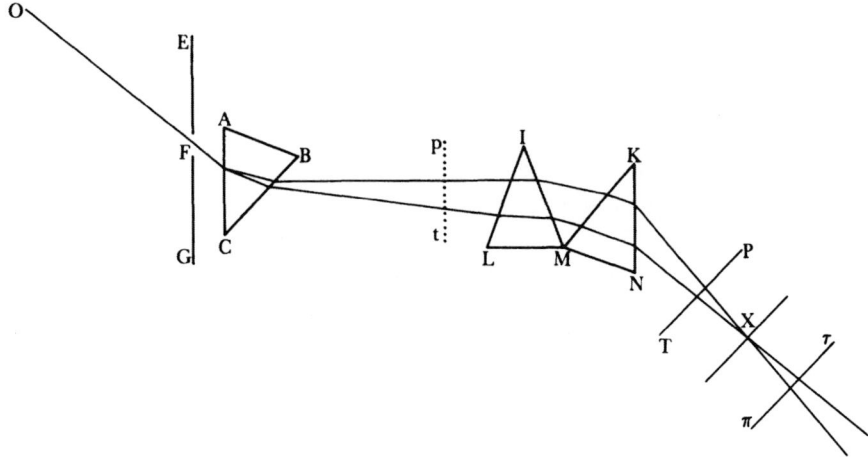

Figure II, 17

ABC parallelo ad distantiam aliquot pedum juxta posita adhibeantur, quae radios in contrarias partes refringant, faciantque versus *X* convergere quos Prisma *ABC* divergentes effecerat: colores ad *X* con/gregati component al- ₂34/ bedinem ac post decussationem sub proprijs (ut antea) formis ad $\pi\tau$ denuò apparebunt.[17]

[50]   Opportuna hic alia subit assertionis demonstratio, quod colores in concursu non destruuntur ad albedinem eliciendam, sed commiscentur tantum. Utpote Rotam dentibus undique in Perimetro cons[t]i[tu]tam ita collocabis juxta duo prismata *ILM* et *KMN* vel juxta Lentem *MN* in praecedenti experimento, ut e coloribus aliqui in dentem aliquem impingant, dum caeteri per intervallum inter illum et proximum dentem praeterlabantur, et in chartam ad praefatum colorum concursum *X* excipiantur; Tum Rotam imprimis lentè circumvolve; et videbis singulos colores in chartam sine aliquâ albedinis apparitione successivè procidere. Postea vero si Rotam tam celeri motu circumagi facias ut succenturiantes colores propter velocitatem consecutionis ab invicem distingui nequeant, transmigrabunt in albedinem, eamque quoad sensum homogeneam sine aliquâ colorum apparitione ex quibus celerrimè se mutuò consequentibus albedo illa efficitur. Et hanc albedinem e coloribus illis successivè commistis componi per se manifestum est.[18]

---

(17) Newton also suggested this way of compounding white in the *Opticks,* Bk. I, Pt. II, Prop. V, Expt. X, p. 105 = Add. 3975, f. 105ʳ.

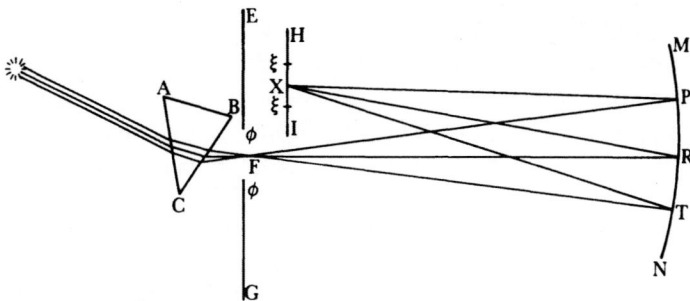

Figure II, 16

same way, rays proceeding from other points near *F*, such as *φ*, will meet exceedingly closely at other points near *X*, such as *ξ*. Thus absolutely all colors will be reflected from the mirror *PT* to every single point of the image *ξXξ*, and the entire image will exhibit white.

There are also other ways of compounding white. For instance, if instead of a lens or mirror one were to use two prisms, *ILM* and *KMN* (Fig. II, 17), placed next to one another several feet away from and in a position parallel to the wholly similar prism *ABC*, and were to let them refract the rays in the opposite direction and make those that the prism *ABC* caused to diverge to converge to *X*, then the colors gathered at *X* will compound white, and after crossing they will again appear at *ππ* according to their own nature, as before.[17] [49]

Another, apt demonstration comes to mind here of our assertion that colors are not destroyed in their concurrence to produce whiteness, but are only mixed. To wit, place a wheel constructed with teeth all around its perimeter next to the two prisms *ILM* and *KMN*, or next to the lens *MN* in the preceding experiment, so that some of the colors strike some tooth, while the rest flow by through the interval between that tooth and the next one and are received on a sheet of paper at the concourse of the colors *X*. Then turn the wheel slowly at first, and you will see the individual colors successively fall upon the sheet without any appearance of whiteness. But afterward if you make the wheel turn around with such a rapid motion that the colors taking each others' place cannot be distinguished from one another because of the speed of their succession, then they will pass into whiteness—and it will be sensibly homogeneous without any appearance of the colors from which that whiteness is produced when they very rapidly follow one another. And that this whiteness is compounded from these successively mixed colors is evident in itself.[18] [50]

---

(18) Neither here nor in his reply to Hooke, on 11 June 1672, where he also sets forth this experiment, does Newton explain why white is perceived even though all colors are not simultaneously present; *Correspondence,* 1:182–3 = *Phil. Trans.,* 7 (1672):509[7–8]. In the *Opticks* (Bk. I, Pt. II, Prop. V, Expt. X), where the toothed wheel is replaced by an oversized comb, he explains that the color mixing occurs in the "sensorium." When a light ray of a particular color strikes the eye, it makes, he supposes, its characteristic impression or vibration on the retina, whence it is propagated through the optic nerve to the brain, where it persists for about one second. Therefore, when the colors follow one another so quickly that their impressions do not die away, a confused mingling of the impressions of all the colors, that is, whiteness, results.

Quinetiam albedo non tantum ad locum concursûs *X* e commistis coloribus componitur, sed etiam ad foramen *φFφ* ubi lux modo transijt Prisma et colores nondum apparuere; ‖ quandoquidem omnes radij quibuscunque coloribus affecti qui ad punctum quodvis imaginis *ξXξ* convergunt, ab alio quodam puncto foraminis *φFφ* manarunt; et sic ijdem radij ad utrumque spatium *φFφ* et *ξXξ* miscentur et utriusque albedinis eadem est compositio.

Atque haec clariora fient observando primò quod rei alicujus utcunque figuratae et applicatae ad foramen *φFφ* umbra distinctè projicitur in papyrum radios excipientem ad *X.* Quinimò bullularum aëris in Prismate latentium (sicut vitris omnibus contingere solet) umbras videre licet[1] ad instar macularum / in dictam papyrum projectas. Id quod nullo pacto contingere ₂35/ potuisset nisi radij manantes ab aliquot punctis ipsius *φFφ* in totidem punctis rursus convenirent ad *ξXξ.* Et licet non exactè conveniant in ijsdem punctis manantes ab ijsdem cum Lens refractaria vice speculi adhibetur, ut in figuris II, 14 & 15, et proinde colores nonnullos generent in confinio lucis et umbrae sicut fusè explicui; tamen spatium in quod conveniunt, tantillum est ut pro puncto sensibili ferme habeatur.

Secundo, si Lentem in fig. II, 14 ita statuas ut aequidistet a focis ejus *F* et *X* in medio posita, ac deinde colores excipias in papyrum *PT* tum ultra lentem versus *X* tum citra versus *F* alternis temporibus admotam: possis observare quod colores eodem plane modo apparent, diminuuntur, et in albedinem paulatim convertuntur dum dicta papyrus motu lento et continuo transfertur ad *F,* atque dum transfertur ad *X.* Adeò ut divergentia colorum ab *F* et convergentia ad *X* omninò similis sit. Pari ratione si papyrus *ππ* lentè moveatur ad *X* juxta et *pt*[2] moveatur ad *F,* ijdem colores conspicientur in utrâque et eodem modo desinent in albedinem hoc tantum excepto quod eorum situs contrariatur propter decussationem radiorum in *X.* Atque adeò divergentia colorum ab utrisque *F* et *X* similis[3] est. Quid itaque concludendum est[4] exindè quàm quod eodem modo commiscentur et ad *F* antequam divaricarunt ab invicem et ad *X* ubi rursus congregantur in albedinem? Sed ut comparatio modò facta evadat illustrior,[5] venit observandum porrò quod cùm papyrus statuitur ipsi *F* contigua et amovetur deinde versus *pt* et postea statuitur ad *X* et amovetur versus *ππ*; Quod, inquam, albedo ad *F* et *X* in utroque casu primò degenerabit in colores secundum extremitates ejus dum in meditullio manet alba. Cujus rei ratio non est alia quàm quòd radij ‘divergentes perinde[6] segregantur in confinio lucis et umbrae. / Sic posito quod radij divergant a spatio *Fφ* (fig: II, 18) alij quidem paralleli

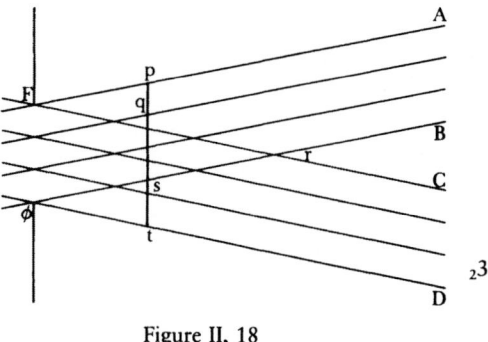

Figure II, 18

---

(1) videre licet] I: videbis (you will see).

(2) I: papyrus [*pt*] (the paper [*pt*]).

Moreover, white is compounded from a mixture of colors not only at their place of concurrence X, but also at the hole $\phi F\phi$ when the light has just passed through the prism and colors have not yet appeared. ‖ Since all rays—endowed with any color whatever—that converge to some point of the image $\xi X\xi$ flow from some other point of the hole $\phi F\phi$, the same rays are thus mixed at each space, $\phi F\phi$ and $\xi X\xi$, and the composition of each white is the same.

Furthermore, these things will become clearer by observing first that the shadow of any object, however it is shaped and placed at the hole $\phi F\phi$, is distinctly projected onto the paper receiving the rays at X. Indeed, it is possible to see[1] the shadows of the little air bubbles that are concealed in the prism (as usually occurs in all glass) projected onto that paper in the shape of spots. This could in no way occur unless the rays flowing from several points of $\phi F\phi$ came together again in just as many points at $\xi X\xi$. Although the rays flowing from the same points do not come together exactly in the same points when a refractive lens is used instead of a mirror (as in Figs. II, 14 and 15), and they consequently generate some colors in the boundary of light and shadow, as I have explained in detail, nevertheless, the space in which they meet is so small that it may be considered as nearly a sensible point.

Second, if you set the lens in Fig. II, 14 so that being placed in the middle it is equidistant from its foci F and X, and then you receive the colors on the paper PT, alternately moved both on the far side of the lens toward X and on the near side toward F, then you can observe that the colors clearly appear in the same way: They are diminished and gradually converted into white, both when that paper is shifted with a slow and continuous motion to F and when it is shifted to X. Hence the colors' divergence from F and convergence to X are entirely similar. In a like manner, if the paper $\pi\tau$ is moved slowly toward X, just as pt[2] is moved toward F, the same colors will be perceived in each, and they will similarly end in white, with this exception only, that their position is reversed because of the crossing of the rays at X. Hence the colors' divergence from both F and X is similar.[3] What therefore ought to be concluded[4] from this, other than that they are mixed in the same way, both at F before they have spread out from one another and at X where they are again gathered together into white? But for the comparison just made to become clearer,[5] it appears that in addition it should be observed that when the paper is placed contiguous to F and then moved away toward pt, and when it is afterward placed at X and moved away toward $\pi\tau$, then, I say, the white at F and X will in each case initially degenerate into colors along its edges while remaining white in the middle. The reason for this is just that the diverging rays are initially[6] separated at the boundary of light and shadow. Thus, assuming that the rays diverge from the space $F\phi$ (Fig. II, 18)—for

Lecture 7
[51 ≈ 60. Light leaving a prism is compounded from colors, even if they are not yet apparent, no differently than afterward when the colors have been gathered into the same space.]

[52 = 61. It is proved from this that in the fourth and fifth ways of compounding whiteness only those rays converge to the same space that diverged from the same space.]

[53 = 62. And that the divergence of the colors from the prism is very similar to their divergence from the white focus of a lens.]

---

(3) I: omninò similis (entirely similar).
(4) concludendum est] I: concludatur (is to be concluded).
(5) Originally continued (as I): ab instantiâ aliquâ (by some example).
(6) Read (as I): primò (initially).

tendentes ad *AB* atque alij ad priores inclinati sed inter se paralleli tendentes ad *CD*; Prima segregatio fiet in extremitatibus juxta lineas *FA* et *φD*, ultimaque in medio veluti ad *r*. Nam, lineâ *pt* inter *Fφ* et *r* ductâ, videre est quòd parallelae juxta extremitates *pq* et *st* ab invicem segregantur, sed mixtim transeunt per intermedium spatium *qs*.

Tertiò sicut lens *MN* in fig: II, 14 refringendo radios divergentes ab *F* facit ut convergant ad *X* et ibi conficiant albedinem; eodem modo si isti radij postquam decussavêre divergentes ab *X* iterum trajiciantur per aliam lentem *μν* (fig II, 19) priori similem et similiter positam inter focos ejus *X* et *ξ*, (id

[54 = 63. Atque etiam quòd divergentes a dicto foco non secus in alium focum congregari possunt, quàm divergentes a [prismate].]

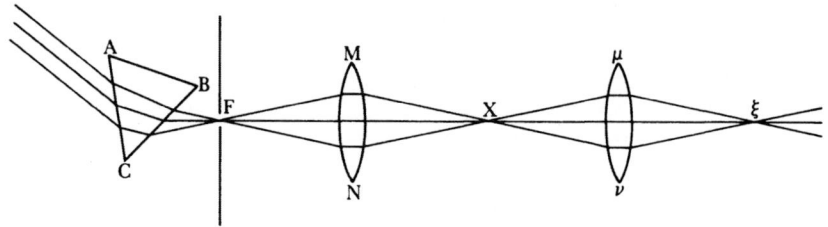

Figure II, 19

est, in aequali ab utrisque distantiâ:) colores sic ad *ξ* secundâ vice congregati albedinem rursus component sicut ante composuerant ad *X*, hoc tantùm interposito discrimine quòd apparebunt in limbo albedinis ad *ξ* duplo latiores quàm (e mox ostensis) apparent ad *X*; atque insuper in situ contrario. At speculis ut dictum est adhibitis quae lucem aliquoties repercutiant, isti colores erunt nulli; atque adeo penicilli *FX* et *Xξ* evadent omnino similes et similis fiet decussatio et commistura radiorum ad *F, X,* et *ξ*. Concludendum est itaque quòd lux, cùm modo trajicitur per prisma, licèt albedinem exhibeat, tamen constat ex radijs heterogeneis confusè mixtis et ab invicem per divergentiam mox discessuris, qui postquam ita segregantur proprijs apparent formis, sin iterum congregantur albedinem rursus componunt, et sic praeterea in infinitum.

[55 ≈ 64. Imò lucem e coloribus ante omnem refractionem componi]

Imò verò lux non solum componitur ex omnium colorum radijs ut egreditur prismate et nondum discernitur in colores istos, sed etiam cùm nondum attigit prisma et antecedenter ad omnem refractionem. Et inde non mirum est quòd cùm segregatur in colores virtute prismatis radios inaequaliter refringentis, et colores iterum commiscentur ope lentis aut alio quovis modo praemonstrato, quòd, inquam, rursus componant albedinem. ‖ Neque hoc solùm exinde confirmatur quòd lux e coloribus composita primigenae luci persimilis sit, sed etiam ex eo quod radij penitus[7] differant refrangibilitate. Et conceptus non / est durior quòd differunt coloribus; imò cùm eidem refrangibilitatis ₂37/ gradui color idem perpetuo competat (ut purpureus maximè refrangibilibus, rubeus minimè refrangibilibus, et sic porrò) quid aliud ab ista cognatione innuitur quàm quòd sint congenita et fortassè quòd a communi quadam causâ dependeant. Sed in hujus rei majorem evidentiam ostendam praeterea quod radiorum solis aequaliter incidentium quaedam genera reflecti possunt

example, some parallel and tending to *AB*, and others inclined to these but parallel to one another and tending to *CD*—then the initial separation will occur at the edges near the lines *FA* and *φD* and the final one at the middle, just as at *r*. For when the line *pt* has been drawn between *Fφ* and *r*, it is possible to see that the parallels near the edges *pq* and *st* are separated from one another, but that they pass through the intermediate space *qs* mixed together.

Third, just as the lens *MN* (Fig. II, 14) by refracting the rays diverging from *F* causes them to converge to *X* and produce white there, in the same way, if those rays, after they have crossed and diverged from *X*, are again cast through another lens *μν* (Fig. II, 19)—similar to the first one and similarly placed between its foci *X* and *ξ*, that is, equidistant from both— then the colors gathered in this way at *ξ* for a second time again compose white, just as they had previously composed it at *X*, with only this difference being introduced: They will appear twice as wide in the band of whiteness at *ξ* as they appear at *X* (from what was then shown), and moreover in a reversed position. But as has been described, when mirrors that reflect the light several times are used, there will be none of those colors; hence the pencils *FX* and *Xξ* will prove to be completely similar, and also the crossing and mixture of the rays at *F*, *X*, and *ξ* will become similar. It must be concluded, therefore, that when light has just passed through a prism, although it exhibits whiteness, it nonetheless consists of confusedly mixed heterogeneous rays soon about to separate from one another by diverging, which after they are thus separated, appear according to their own nature; but if they are again gathered together, they once more compound whiteness, and so on to infinity.

[54 = 63. And also that diverging from that focus they can be gathered at another focus no differently than those diverging from a [prism].]

In fact light is compounded of rays of all colors not only as it leaves the prism and is not yet separated into these colors, but also when it has not yet reached the prism and prior to any refraction. Consequently, it is not surprising that when light is separated into colors by the power of a prism to refract rays unequally, and the colors are again mixed by the aid of a lens or by any other way previously shown, that, I say, they again compound white. ‖ This may be confirmed not only from the fact that light compounded from colors is very similar to original light, but also because the rays differ completely[7] in refrangibility. And the concept is not harder that they differ in colors; indeed, since the same color perpetually corresponds to the same degree of refrangibility (for instance, purple to the most refrangible rays, red to the least refrangible, and so on), what else is meant by that relation other than that they are created together, and perhaps that they may depend on some common cause. But to make this more evident, I shall show hereafter that at equal incidence some kinds of the sun's rays can be reflected, while others are

[55 ≈ 64. Light in fact is compounded of colors before any refraction.]

---

(7) "Penitus" may with equal sense be translated here as "internally."

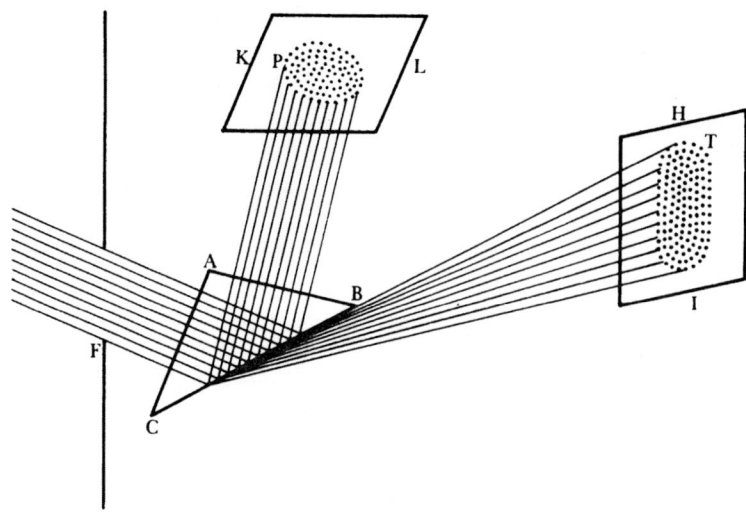

Figure II, 20

[56 = 65. Ut ex eo
pateat quòd aliqui
colores reflecti
possunt dum alij per
prisma trajiciantur.
Hujusque rei
experiendae modus
adducitur.]

‖dum alia per reflectentem superficiem trajiciuntur, adeoque diversos colores diversis radijs ante omnem refractionem inesse:[8] ‖ sit *ABC* (fig. II, 20) prisma quod excipit radios in obscurum cubiculum per foramen *F* uno digito latum trajectos, eosque refringit ad papyrum vel parietem *HI* ijs obsistentem apud *T*. Porrò autem cùm superficies prismatis *BC* non omnes refringat radios versus *T* sed et plurimos reflectat, eos apud *P* siste etiam cum aliâ papyro *KL* in morem albae imaginis foramini *F* persimili terminante. Deinde converte prisma circa axem ejus secundum ordinem literarum *ABCA* et videbis tum amplitudinem colorum ad *T*, tum quantitatem lucis ad *P* augeri perpetuò donec tandem cùm refractio ad planum *BC* fit maximè obliqua, colores ad *T* incipiant evanescere et reflecti ad *P*, purpureus primò deinde caeruleus viridis et flavus, ac denique ruber. Cujus quidem lucis accessu imago *P* fiet multò lucidior quàm antea. Interea verò dum colores a *T* gradatim evanescunt videbis albedinem *P* paululum mutari, et nonnihil vergere ad caeruleum, per accessum nempe purpurei et caerulei qui primò reflectuntur.[9] Id quod nullo modo accidisse potuisset nisi radijs, prout a sole veniunt, discrimen interesse concedatur. Scilicet quòd ex ijs quidam ad efficiendos rubeum et flavum dispositi pertinaciùs et cum minori refractione penetrent superficiem *BC* et versus *T* prolabantur, dum alij ad exhibendum purpureum et caeruleum parati superficiem dictam aut penetrent languidiùs majores refractiones patientes, aut si nequeant penetrare propter nimiam eorum obliquitatem / tum faciliùs et citiùs reflectuntur ad *P*. Ijs primò omnium reflexis ₂38/ quorum potentia ad istam superficiem penetrandam sit minima, id est, purpuriformibus, et caeteris deinde suo ordine prout incidentia fit magis obliqua donec rubriformes ultimò reflectantur obliquitate tantâ debilitati ut non sint ampliùs potentes dictae superficiei resistentiam superare. Atque haec facilè constabunt ijs qui norunt quòd quo major est vis refractiva superficiei cujusvis, eò citius et ad minorem obliquitatem radij reflectentur; et quò minor eo magis obliqui penetrabunt.

transmitted through the reflecting surface, and hence that different colors are ‖
in different rays before any refraction.[8] ‖ Let *ABC* (Fig. II, 20) be a prism
that receives the rays cast into a dark room through a one-inch-wide hole *F*
and refracts them to a paper or a wall *HI* that stops them at *T*. Moreover,
since the prism's surface *BC* does not refract all the rays toward *T* but in
addition reflects very many, also stop those at *P* with another paper *KL* that
terminates them in the form of a white image very similar to the hole *F*. Next
rotate the prism about its axis in the order of the letters *ABCA*, and you will
see both the extent of the colors at *T* and the quantity of light at *P* continually increase until finally, when the refraction at the plane *BC* becomes very
oblique, the colors at *T* begin to vanish and to be reflected to *P*: first purple,
then blue, green, and yellow, and last red. Naturally, with the entrance of this
light, the image *P* will become much brighter than before. But at the same
time as the colors gradually vanish from *T*, you will see the white at *P* change
slightly and incline somewhat to blue, namely, by the entrance of purple and
blue, which are reflected first.[9] This in no way could have occurred unless it
be conceded that the difference is present in the rays just as they come from
the sun: specifically, that some of those—disposed to produce red and yellow—more vigorously penetrate the surface *BC* with a smaller refraction and
go on toward *T*, while others—fit to exhibit purple and blue—either weakly
penetrate that surface and experience greater refractions, or, if they are
unable to penetrate it because of their excessive obliquity, are reflected to *P*
more easily and quickly. Those are reflected first of all whose power to
penetrate that surface is the least, that is, the purple-making rays, and then
the others in their own order according as their incidence becomes more
oblique, until the red-making rays, having been weakened by such a great
obliquity that they are no longer able to overcome the resistance of that
surface, are finally reflected. These things will readily be evident to those who
have learned that the greater the refractive force of any surface, the more
quickly and at a smaller obliquity will the rays be reflected; and the smaller
the refractive force, the more the oblique rays will penetrate.

[56 = 65. As is evident because some colors can be reflected while others are transmitted through a prism. And the method of testing this is set forth.]

---

(8) adeoque . . . inesse] Originally (as the beginning of §65): Verum ut hoc discrimen inesse
radijs antecedenter ad refractiones ostendam (But that I may demonstrate that this difference is
in the rays prior to refraction).

(9) I continues: at postquam caeteri etiam colores viridis flavus et ruber reflectuntur a *T*,
albedo ad *P* redintegrabitur (but after the other colors, green, yellow, and red, are also reflected
from *T*, the white at *P* will be restored).

[57 = 66. Notanda quaedam] De hoc autem experimento convenit observare:[10] Primò quòd cùm praedicta variatio albedinis ad *P* sit admodum parva propter exuberantiam lucis albae collatae ad reflexum caeruleum; itaque cavendum est ne prismate utaris quod ex vitro conflatur, tincto cum colore aliquo, ne lucem ad *P* reflexam ita tingat ut difficile sit dictam variationem observare. Praestat adhibere prisma ex laminis vitreis tenuibus et perpolitis confectum et aquâ lympidissimâ repletum.

Secundò licet mutatio dicta sit parva, tamen satis est ad ostendendum quòd radij retinent eosdem colores cùm reflectuntur quos exhibent cùm trajiciuntur per superficiem *BC*: Siquidem tingunt albedinem [*P*] colore suo quantùm liceat tam paucis tingere. Colores itaque suos habuêre priùs, et eosdem retinent sive refringantur, sive reflectantur: licèt in mixturis plerunque celati lateant donec eruantur (non autem fiunt) virtute Prismatum.

Tertiò ex luce ad priorem speciem albedinis per reflexionem omnium colorum a *T* restitutâ, quid aliud denotatur quàm albedinem istam per misturam colorum omnium reproduci. Scilicet cùm rubor ultimò reflexus admiscetur caeteris coloribus antea reflexis, reflexorum colorum mistura tunc perfecta est ad albedinem componendam quae superadditur albedini priùs existenti in [*P*].

Quartò nequa oboriatur suspicio quòd refractiones in superficiebus *AC* et *AB* ad ingressum radiorum in Prisma / et egressum factae, possint aliquid ₂39/ conducere ad effectus hosce producendos; observare licet[11] quòd effectus ijdem producuntur, cujuscunque licet magnitudinis statuatur angulus *ACB*; hoc est quaecunque sit refractio superficiei *AC*: modò angulus *ABC* ponatur ejusdem magnitudinis atque angulus *ACB*: alias enim pro imagine alba ad [*P*] generabuntur colores. Experimentum itaque nullatenus dependet a refractionibus superficierum *AC* et *AB*; imò possis efficere quod cùm colores partim reflectuntur ad [*P*] et partim trajiciuntur ad *T*, radij perpendiculariter incidant in *AC* emergantque ex *AB*, et sic neutrâ superficie refringantur; modò statuas angulùm *ACB* ut et *ABC* esse 40^{grad}, 40′[12] circiter; et ijdem tamen effectus producentur.

[58 = 67. In majorem rei evidentiam ostenditur quosdam colores alijs facilius reflecti.] Caeterùm in majorem evidentiam et explicationem modi quo praedicta fiunt, liceat experiri per lucem in colores discretam, quòd purpureus primò et caeteri deinde (quisque suo ordine) reflectuntur. Etenim (in fig II, 21) sint *ABC* et *αβκ* duo prismata parallela quorum alterum *ABC* projicit colores in alterum *αβκ* ad distantiam duodecim vel plurium pedum. Tum Prismate *αβκ*

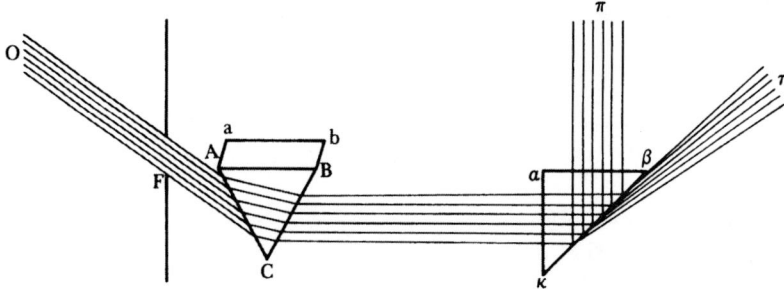

Figure II, 21

About this experiment, however, it is fitting to observe:[10] First, since the aforementioned variation of white at *P* is very small, because of the abundance of white light compared to the reflected blue, you must therefore be careful that you do not use a prism that is melted from glass having some color, lest it so color the light reflected to *P* that it is difficult to observe that variation. It is better to use a prism made from thin, highly polished glass plates and filled with very clear water. [57 = 66. Some things to be noted.]

Second, although that change is small, it is nonetheless sufficient to show that the rays preserve the same colors when they are reflected as they exhibit when they are transmitted through the surface *BC*, since they color the white at [*P*] with their own color, insofar as it is possible to color with so few. Therefore, they possessed their own colors previously, and they preserve the same colors whether they are refracted or reflected, although generally the colors are latent, hidden in the mixture until they are drawn out—but not made—by the power of prisms.

Third, as a consequence of the light having been restored to the former kind of white by the reflection of all colors from *T*, what else is implied other than that that white is reproduced from a mixture of all colors. Namely, when the red reflected last is mixed with the other colors already reflected, the mixture of reflected colors is then completed to compose white, which is added over and above the white previously existing at [*P*].

Fourth, lest the suspicion arise that the refractions that are made in the surfaces *AC* and *AB* by the rays entering and departing from the prism can somehow combine to produce these effects, one may observe[11] that the same effects are produced at whatever magnitude the angle *ACB* is set, that is, whatever the refraction of the surface *AC* is, provided that the angle *ABC* is assumed to be the same magnitude as the angle *ACB*, for, otherwise, instead of a white image at [*P*], colors will be generated. The experiment, therefore, in no way depends on the refractions of the surfaces *AC* and *AB*. In fact, you may arrange it so that when the colors are partly reflected to [*P*] and partly transmitted to *T*, the rays fall upon *AC* and emerge from *AB* perpendicularly and are thus refracted by neither surface, provided that you set the angle *ACB* as well as *ABC* to be approximately 40°40′;[12] and still the same effects will be produced.

However, for greater clarity and a fuller explanation of the way in which the preceding occurs, one may prove with light separated into colors that purple is reflected first and then the other colors, each in its own order. For (in Fig. II, 21) let *ABC* and *αβκ* be two parallel prisms, one of which, *ABC*, projects the colors onto the other, *αβκ*, twelve or more feet away. Then [58 = 67. For greater clarity of the matter, it is shown that some colors are more easily reflected than others.]

---

(10) convenit observare] Originally (as I): juvabit observare sequentia (it will help to observe the following).

(11) observare licet] Originally (as I): juvabit observare (it will help to observe).

(12) 40^{grad}, 40′] *Editio princeps*: grad. 40.

circa axem ejus secundum ordinem literarum α, β, κ, α, converso donec tanta sit obliquitas radiorum in superficiem βκ incidentium ut incipiant ad π reflecti non ampliùs potentes penetrare ad τ; videbis omnes purpuriformes primò omnium[13] reflecti, caeterosque deinde suo ordine.

[59 = 68. Idem aliter ostenditur, circumstantiâ tantùm variatâ] Veruntamen quia purpuriformes radij paulo magis refringuntur in primo prismate *ABC*, et ideo magis inclinantur ad superficiem βκ secundi Prismatis αβκ quàm caeteri; poterit objici quod eâ de causâ primo omnium reflectuntur. Quamobrem (in fig: II, 22.) duo prismata statuantur non parallela sibi

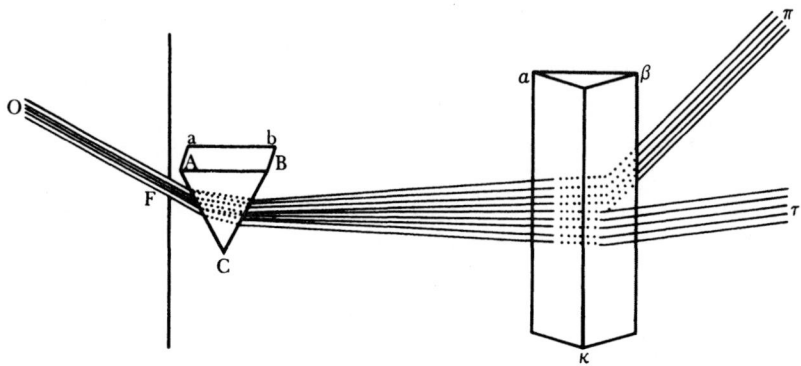

Figure II, 22

invicem, sed in transverso situ, ut omnicolores radij quasi ad eosdem angulos incidant in praefatam superficiem βκ. Quo posito possis observare convertendo prisma[14] αβκ circa axem ejus secundum ordinem literarum α, β, / κ, ₂40/ α,[15] quòd radij purpuriformes primò omnium reflectuntur, et ultimo rubriformes; coloribus ad π continuò translatis prout a τ dispareant.

[60 = 69. Idem adhuc aliter.] Sunt et alij praeterea modi quibus experiri liceat quòd ex radijs similiter incidentibus quaedam genera penitus reflecti possunt dum alia partim transmittuntur. Quemadmodum si *EFG* (fig II, 23) sit operculum fenestrae ad *F* terebratum, et foràs statuatur Prisma *ABC* quod lucem solis foramen *F* ingressuram intercipiat et refringat versus φ, ad illud φ pedibus ab *F* duodecim aut longiùs postpositum statuatur opacum corpus εφγ quod lucem sistat, dempto parvo foramine φ per quod aliqua pars lucis nempe violacea longiùs trajiciatur ad *Y*. Istud autem φ non sit semisse digiti latius. Deinde prae manibus sumatur aliud Prisma αβκ et ad radios transversè positum statuatur a posticâ parte foraminis φ, circaque axem ejus convertatur donec videas lucem violaceam, postquam ab ejus basi βκ obliquissimè refracta fuerit versus τ, totam a τ disparuisse modò, et ad π reflecti. Luce violaceâ tam obliquè ad π reflexâ ut ad τ statim pervasura esset, modò ex angulari motu Prismatis secundum ordinem literarum α, β, κ, α,[15] facto, angulus κYφ vel minimùm augeretur; Prisma istud αβκ in eo situ figatur. Tum alterum Prisma *ABC* motu circa axem ejus nunc hac nunc illac parùm convertatur, ut colores quos

---

(13) Added.   (14) I: prisma posterius (the second prism).   (15) Read: α, κ, β, α.

turning the prism $\alpha\beta\kappa$ about its axis in the order of the letters $\alpha\beta\kappa\alpha$ until the obliquity of the rays falling upon the surface $\beta\kappa$ is so great that they begin to be reflected to π, they being no longer able to penetrate to τ, you will see all the purple-making rays reflected first of all,[13] and then the others in their own order.

Nevertheless, because the purple-making rays are refracted a little more in the first prism ABC and are therefore more inclined than the others to the surface $\beta\kappa$ of the second prism $\alpha\beta\kappa$, it could be objected that for that reason they are reflected first of all. Consequently (in Fig. II, 22), set the two prisms not parallel to each other but in a transverse position so that rays of all colors fall at almost the same angles upon that surface $\beta\kappa$. After this has been arranged, by turning the prism[14] $\alpha\beta\kappa$ around its axis in the order of the letters $\alpha\beta\kappa\alpha$[15] you can observe that the purple-making rays are reflected first of all and the red-making ones last, with the colors continuously shifting to π as they disappear from τ. [59 = 68. The same thing is shown differently, with only one feature changed.]

There are in addition other ways whereby one may prove that at similar incidence some kinds of rays can be totally reflected while others are partly transmitted. For instance (Fig. II, 23), if EFG is a window shutter pierced at [60 = 69. The same thing still differently.]

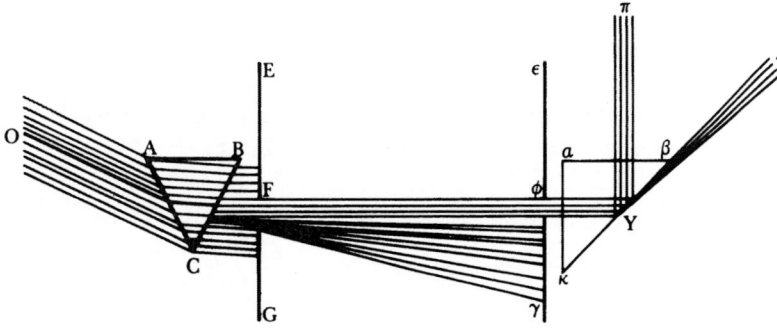

Figure II, 23

F, and outside a prism ABC is placed that intercepts the sun's light about to enter the hole F and refracts it toward ϕ, then at ϕ (located twelve or more feet behind F) place an opaque body $\epsilon\phi\gamma$ that stops the light except for the small hole ϕ through which some part of the light, namely, the violet, is transmitted farther to Y. Do not, however, let that hole ϕ be wider than half an inch. Then take another, readily accessible prism $\alpha\beta\kappa$ placed transverse to the rays and set it at the rear of the hole ϕ; turn it around its axis until you see that all the violet light, after it has been most obliquely refracted by its base $\beta\kappa$ toward τ, has just then vanished from τ and is reflected to π. When the violet light has been so obliquely reflected to π that it would immediately penetrate to τ, if as a result of the prism's angular motion made in the order of the letters $\alpha\beta\kappa\alpha$,[15] the angle $\kappa Y\phi$ increased even the least bit, fix that prism $\alpha\beta\kappa$ in that position. Then rotate the other prism ABC with a slight to and fro motion around its axis so that the colors it projects onto the obstacle

projicit in obstaculum εγ paululum attollantur, eoque pacto omnes successivè transmittantur per foramen φ in posterius Prisma αβκ. Et videbis quòd cùm flavedo transmittitur ad Y, illi radij non omnes ad π reflectentur, sed plurimi perrumpent superficiem βκ et ad τ pertingent. Et cùm rubor ad Y transmitti-tur, illi radij fortiùs adhuc perrumpent ut ex copiâ perrumpentis lucis et minori ejus refractione constet. Neque mirum videatur quòd purpuriformes radij sint minùs potentes penetrare superficiem βκ quàm rubriformes quan-doquidem prismatibus eodem modo dispositis, antehac[16] / ostendi quod ₂41/ majorem refractionem patientur; posito scilicet angulo κYφ tanto ut omni-geni radij possint superficiem βκ penetrare.

[61 = 70. Et proinde cùm e radijs solaribus alij alijs, pro specie colorum quos postmodum exhibent, faciliùs reflectuntur; constat lucem solis ex illis coloribus componi.]

Jam cùm radij qui citiùs et faciliùs reflectuntur in experimento ad fig II, 20 tradito (nempe purpuriformes) etiam citius et faciliùs reflectantur in experi-mentis duobus novissimè recitatis, cùm eadem ijsdem radijs semper eveniant; liquet quòd hoc non fit ex contingentiâ sed ex praedispositione radiorum, et quod antecedenter ad omnem reflexionem aut refractionem quidam ad exhi-bendos quosdam colores sunt apti et faciliùs reflexibiles, alij vero alijs colori-bus et progrediendi viribus afficiuntur. Neque aliud experimentis jam recitatis discrimen interesse videtur, quàm quòd in primo radij omnium formarum, prout a sole adveniunt confusè mixti, incidant in Prisma quod rubriformes transmittit et reflectit caeruliformes; in reliquis autem duobus experimentis dissimiles radij priùs discernuntur ab invicem quàm incidunt in dictum Prisma.

Lect 8
[62 = 71. Alius modus quo lux solis partim reflecti potest et partim refringi.]

Ad haec lubet alium adducere modum quo dissimilitudo radiorum in luce solis mixtorum innotescat, non multò dissimilem ei ad fig. II, 20 ostenso, sed conspectui jucundiorem et aequè scientificum. In fig. II, 24 sunto AαBβC et

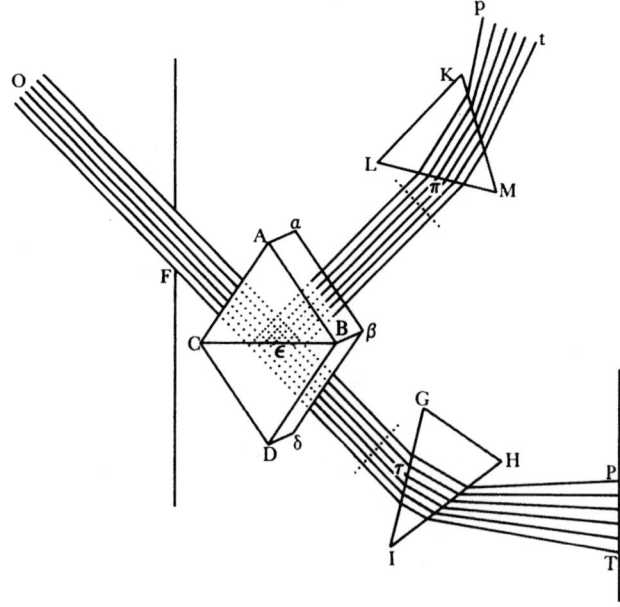

Figure II, 24

εγ are raised a little, and in this way all the colors may be successively transmitted through the hole φ onto the rear prism αβκ. You will see that when the yellow is transmitted to Y not all those rays will be reflected to π, but very many of them will break through the surface βκ and reach τ. When the red is transmitted to Y those rays will break through still more vigorously, as is evident from the quantity of light breaking through and its smaller refraction. Nor should it seem surprising that the purple-making rays are less able to penetrate the surface βκ than the red-making ones, since when the prisms were arranged in the same way, I showed earlier[16] that they will undergo a greater refraction, to be sure, with the angle κYφ having been set so large that rays of every sort could penetrate the surface βκ.

Now since the rays that are more quickly and easily reflected in the experiment presented at Fig. II, 20—namely, the purple-making ones—are also more quickly and easily reflected in the two experiments most recently recited, since the same things always occur with the same rays, it is evident that this does not occur from a contingency but from a predisposition of the rays, and that prior to any reflection or refraction some are disposed to exhibit certain colors and are more easily reflexible, whereas others are endowed with other colors and forces of proceeding. Nor in the experiments just now recited does there appear to be any difference present other than that in the first one rays of every nature, just as they arrive confusedly mixed from the sun, fall upon the prism, which transmits the red-making rays and reflects the blue-making ones; whereas in the other two experiments the dissimilar rays are separated from one another before they fall upon that prism.

[61 = 70. Consequently, since some of the solar rays are more easily reflected than others according to the species of color they exhibit afterward, it is manifest that the sun's light is compounded of those colors.]

Moreover, it is desirable to bring forward another way whereby the dissimilarity of the rays mixed in the sun's light may become known; it does not differ much from that shown at Fig. II, 20, but it is visually more pleasing while being equally scientific. In Fig. II, 24 let AαBβC and BβDδC be two

Lecture 8 [62 = 71. Another way whereby the sun's light can be partly reflected and partly refracted.]

---

(16) In I there is a marginal note here to §37 [= II, 15].

BβDδC duo prismata ita juxta se posita et colligata, ut duo ex eorum planis *CBβ* conveniant sibi et coincidant, excepto tantùm quòd nonnihil aeris in morem tenuissimae laminae intercedat ijs: Id quod eveniet ultrò, siquidem haud queas Prismata tam arctè constringere quin tantum intercedet aeris quantum proposito sufficiet. Porrò in majorem rei evidentiam convenit ut anguli *ACB* et *CBD* sint aequales[1] proximè, eò ut plana *AαC* et *BβδD* fiant parallela, licèt hoc non sit omninò necessarium. His praemissis statuantur dicta prismata juxta foramen *F*, ut lux ingressa per ea trajiciatur versus *τ*, primo permeans superficiem *AαC*, deinde intermediam superficiem *BβC*, et inde per *BβδD* prolapsa in papyrum ad *τ* collocatam, / quam albedine tingit ₂42/ tanquam si non omninò transiret Prismata, sed vitrum parallelis planis *AαC* et *BβδD* terminatum. Praeterea cùm intermedia superficies *BβC* lucem ei incidentem non omnem transmittat ad *τ* sed multam reflectat, quae aliquò exibit e prismate *ABC* per superficiem ejus *AαβB*, puta versus *π*: ad illud *π* statuatur alia papyrus quae lucem hanc similiter albicantem terminet. Quod ubi feceris converte prisma quadrangulare (ex duobus triangularibus colligatis confectum) motu lento circa axem ejus secundum ordinem literarum *ABDCA*; tandemque videbis quod albedo ad *π* ac *τ* degenerabit in colores; flavedine primò, deinde rubedine ad *τ* conspectâ, caeruleo autem colore ad *π*; donec post intensissimam rubedinem ad *τ* color et lux omnis evanescat inde, et caeruleus ad *π* iterum transformetur in albedinem aliquantò lucidiorem quàm antea. Utpote dum prismata circa communem axem, ut dictum est, convertuntur radiorum in mediam superficiem *BβC* (hoc est in laminam aeris prismatibus interjectam) prolapsorum incidentia continuò fit obliquior, donec tanta sit eorum obliquitas ut nequeant ampliùs penetrare dictam laminam progredique ad *τ*, sed abinde reflectantur ad *π*. Quod accidet cum angulus *FεC* (obliquitas incidentium) sit graduum ferè quinquaginta. Radij autem purpuriformes minimè omnium potentes penetrare dictam laminam aëream, reflectentur primò, et albedinem priùs reflexam ad *π* nonnihil tingent eorum colore dum ex radijs praeterlabentibus[2] ad *τ* flavedo imperfecta aut potiùs color inter flavum et viridem mediocris componitur. Postea caeruleus, et viridis deinde reflexus paulò magìs tinget lucem in *π* cum colore caeruleo (licèt admodum diluto propter exuberantiam a[l]bedinis commixtae,) manebitque rubor in *τ*, qui mox per flavedinis hactenus commixtae reflexionem fiet intensior, donec ipse etiam denuò reflexus albedinem in *π* redintegret.

[63 = 72. Penitiùs hic ostenditur quinam e radijs solaribus reflectuntur et quinam transmittuntur:

Caeterùm ut hoc specimen evadat illustratius sumatur aliud Prisma *GHI* quod a posticâ parte prismatum *ABCD* ita collocetur ut lucem *Oετ* per ea transmissam refringat versus / *PT* et in colores permutet; violaceo in *P*, rubeo ₂43/ in *T* caeterisque in intermedia loca projectis. Tum prismata colligata circa communem axem (ut prius) rotentur donec lux alba versus *τ* transmissa

---

(1) In I a marginal note refers to §90[= II, 91].
(2) I: perlabentibus (flowing through).

prisms juxtaposed and fastened together in such a way that two of their planes, *CBβ*, meet one another and coincide, except for some air that comes between them in the form of a very thin plate. This will occur spontaneously, provided that you do not have the power to compress the prisms too closely, so that not enough air comes between them as will suffice for the purpose. Moreover, for greater clarity of the experiment, it is appropriate that the angles *ACB* and *CBD* be almost equal[1] so that the planes *AαC* and *BβδD* may consequently become parallel, although this is not altogether necessary. With these matters having been premised, set those prisms near the hole *F* so that the light entering through it is transmitted toward *τ*, by first passing through the surface *AαC*, then the intermediate surface *BβC*, and from there proceeding through *BβδD* to the paper placed at *τ*, which it colors white, just as if it had not passed through the prisms at all but through a glass bounded by the parallel planes *AαC* and *BβδD*. Furthermore, since the intermediate surface *BβC* does not transmit to *τ* all the light falling on it, but reflects much of it, which will pass out of the prism *ABC* through its surface *AαB* to some place, suppose toward *π*; at *π* place another paper that similarly terminates this whitish light. When you have done this, turn the quadrangular prism (made from the two fastened triangular ones) with a slow motion about its axis in the order of the letters *ABDCA*. After a while you will see that the white at *π* and *τ* will degenerate into colors: First yellow and then red is visible at *τ*, but the color blue is visible at *π*, until finally, after the most intense red is visible at *τ*, all color and light vanishes from there [*τ*], and the blue at *π* is again transformed into a white somewhat brighter than before. That is to say, when the prisms are rotated about their common axis, as has been described, the incidence of the rays proceeding into the middle surface *BβC* (that is, into the plate of air lying between the prisms) becomes continually more oblique, until their obliquity is so large that they are no longer able to penetrate that plate and continue to *τ*, but are then reflected to *π*. This will occur when the angle *FεC* (the obliquity of the incident rays) is about fifty degrees. The purple-making rays, however, the least able of all to penetrate that plate of air, will be reflected first, and they will slightly color the whiteness already reflected to *π* with their color; while an imperfect yellow, or rather a mean color between yellow and green, is compounded from the rays flowing by[2] to *τ*. Afterward, the blue, and then the green reflected a little more, will color the light at *π* with the color blue (although quite diluted because of the overabundance of white mixed with it); while the red will remain at *τ* and will thereafter become more intense by the reflection of the yellow previously mixed with it, until after also being reflected the red itself will restore the whiteness at *π*.

But so that this example may become more illustrative, take another prism *GHI,* so placed behind the prisms *ABCD* that it refracts the light *Oετ* transmitted through it toward *PT* and changes it into colors: violet projected to *P,* red to *T,* and the other colors to the intermediate spaces. Then rotate the fastened prisms about their common axis (as before) until the white light transmitted toward *τ* begins to grow yellow, and you will see that the color

[63 = 72. It is more thoroughly shown here which of the solar rays are reflected and which are transmitted, and

atque adeò hoc
non casu sed
praedispositione
radiorum evenire]

incipiat flavescere; et videbis quod color purpureus in *P* simul evanescet. Id quod arguit purpuriformes radios non ampliùs ad Prisma *GHI* pertingere sed a superficie *CBβ* primò omnium ad *π* reflecti; et lucem *ετ* ideò flavescere quòd purpura e misturâ tollitur quâ priùs albedinem exhibuit. Ad eundem modum si prismata *ABCD* diutiùs rotentur, videbis reliquos colores a *π* ad *τ*[3] successivè disparere prout lux *ετ* plus plusque rubescit; et cùm fit ruberrima, tum solam rubedinem in *τ* manere. Quod manifestò convincit hanc lucem *ετ* non aliunde rubescere, quàm quòd a radijs aliorum colorum per superficiem *CBβ* reflexis secernitur.

Simili ratione si cum prismate quarto *KLM* refringas radios ad *π* reflexos, et colores eo pacto productos et in album parietem projectos duodecim pedes aut longiùs distantem animadvertas; videbis quòd cùm lux *ετ* incipit viridè flavescere, purpura in *p*, quam prisma hoc elicit e luce *επ*, plusquam caeteri colores augebitur, per accessum nempe purpurae quae tunc in *P* disparuit; caeterisque deinde coloribus in *pt* gradatim fiet accessus prout a *PT* disparent; donec cum omnis color a *PT* disparuit, colores ad *pt* non ampliùs augeantur. Hoc autem discrimen quo violaceus et caeruleus ad *pt* augmentum suum omne paulò citiùs obtinent quàm rubeus aut flavus, tam exile est, ut observator, nisi sit attentus, aegrè advertat.

[64 = 73. Tertius
modus quo lux solis
partim reflecti potest
et partim refringi.]

Ut istis denique finem imponamus lubet alium adducere modum quo quaedam genera radiorum luci solis intermista partim transmitti possint, dum alia reflectuntur. Nempe si duas laminas vitreas *CB* (fig II, 25) planè perpolitas et ad invicem applicatas secundum planitiem earum connectas, easque vasi *RQ* aquae pleno immergas, extremitate superficierum juxtapositarum undique cerâ vel pice priùs obturatâ, ut aqua / non interrepat et expellat aërem, qui more laminae tenuissimae ut dictum est, interjacebit vitris. Si haec, inquam, fiant, possis efficere dictorum vitrorum talem esse situm, ut (illucente sole) aer interjectus caeruliformes radios reflectat versus *π* et transmittat rubriformes versus *τ*; atque alias omnes apparentias modò recensitas exhibeat.

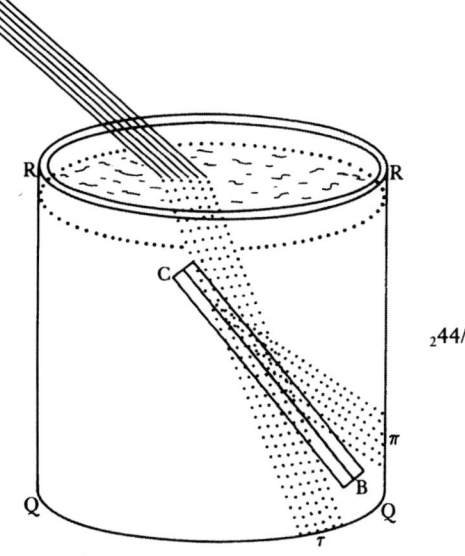

Figure II, 25

[65 ≈ 74, 75.
Notandum Quòd
colores hic fiunt
a parallelis
superficiebus, et
quod lux postquam
reflexa vel per

Caeterum de hisce experiendi modis notandum venit primò quod colores hic a parallelis superficiebus producuntur quarum aliquae recurvant radios quantum aliae incurvant atque adeò quae mutuos effectus destruerent, si quos in immutandis intrinsecis dispositionibus radiorum quoad eorum colores, ut opinantur Philosophi, producerent. Deinde[4] quòd lux postquam per istas parallelas[5] superficies trajicitur licet alba sit manifestò tamen constet ex

purple at *P* will simultaneously vanish. This makes it clear that the purple-making rays no longer reach the prism *GHI* but are reflected first of all from the surface *CBβ* to *π*; and the light *ετ* therefore becomes yellow because the purple, with which it previously exhibited white, is removed from the mixture. In the same way, if the prisms *ABCD* are rotated further, you will see the remaining colors disappear successively [from *P* to *T*]⁽³⁾ as the light *ετ* becomes more and more red; and when it becomes reddest, then the red alone remains at *τ*. This manifestly proves that this light *ετ* becomes red from nothing else other than that it is separated from the rays of the other colors that are reflected by the surface *CBβ*.

[thus that this does not occur by chance but by a predisposition of the rays.]

In a similar manner, if you refract the rays reflected to *π* with a fourth prism *KLM*, and you observe the colors produced in this way and projected onto a white wall twelve or more feet away, you will see that when the light *ετ* begins to become a greenish yellow, the purple at *p*, which this prism draws out from the light *επ*, will increase more than the other colors, namely, by the entrance of the purple that has then disappeared at *P*. And thereafter the entrance of the other colors at *pt* will occur gradually according as they disappear from *PT*, until when all color has disappeared from *PT*, the colors at *pt* no longer increase. But this difference whereby the violet and blue attain their whole increase at *pt* somewhat sooner than the red or yellow is so small that unless the observer is attentive, he will scarcely notice it.

As we finally set a limit to these things, it is appropriate to bring forth another way by which some kinds of rays intermingled in the sun's light can be partly transmitted while the others are reflected. Specifically (Fig. II, 25), if you join together two highly polished glass plates *CB* by connecting them to each other along their flat surfaces, and you plunge them into a vessel *RQ* that is full of water, after having first sealed the edges of the juxtaposed surfaces everywhere with wax or pitch, so that water does not creep in and drive out the air that will lie between the glasses in the form of a very thin plate (as has been described)—if, I say, these things are done—you can arrange it so that the position of those glasses is such that when illuminated by the sun, the interposed air reflects the blue-making rays toward *π* and transmits the red-making rays toward *τ*. And also, it may exhibit all the other appearances just now related.

[64 ≈ 73. A third way whereby the sun's light can be partly reflected and partly refracted.]

Moreover, concerning these ways of experimenting it can first be noted that colors are here produced by parallel surfaces, some of which bend the rays back as much as the others bend them in, which would consequently destroy their mutual effects, if—as philosophers believe—they produced any effects by changing the intrinsic dispositions of the rays with respect to their colors. Next,⁽⁴⁾ after the light is transmitted through those parallel⁽⁵⁾ surfaces,

[65 ≈ 74, 75. It ought to be noted that colors are made here by parallel surfaces, and that after light has been reflected from or

---

(3) Newton's "a *π* ad *τ* [from *π* to *τ*]" does not make sense; possible alternatives are: "a *P* ad *T*" (adopted in the translation), "a *τ* ad *π*," and "a *PT*."

(4) **I** (beginning §75): Secundò (Second).

(5) Omitted in *editio princeps*.

parallelas superficies
trajecta fuit, e
coloribus
componitur;][6]

heterogeneis[7] radijs, quandoquidem eorum aliqua genera penitùs reflecti possunt ad π dum alia ad τ partim trajiciuntur. Et eâdem ratione constat reflexam albedinem similiter compositam esse, siquidem (ut dixi) redintegrata est, cum rubor omnium ultimus reflectitur a τ. ‖ Et haec ex eo etiamnum summè confirmantur quod a solâ vitrorum obliquitate sine aliquâ refractionis vel reflexionis novâ modificatione efficiuntur.[8]

[66 → 75. Quam
tamen ejusdem
naturae cum
immediatâ solis luce
credimus.]

Lux itaque quamvis uniformis esset quae a sole immediatè profluit, postquam tamen unquam reflexa vel refracta fuit, constat ex heterogeneis radijs. Et ejusmodi est ea lux omnis quae per vitreas fenestras trajicitur vel quam planetae nubesve ad nos reflectunt; imò lux omnis a sole aut lucernis quibusvis derivata, siquidem aliqualem saltem refractionem ab Atmosphaerâ (ut dicunt Astronomi) patitur, ut taceam quae in objectis, denuoque in oculi tunicis ante visionis actionem impressam fiunt. Jam si nihil aliud ostenderam, fuisset aliquid prodijsse tenus, siquidem omnia visibilium phaenomena nobis per ejusmodi lucem exhibentur. Atqui cùm solis lux immediata albere censeatur / et ille color non sit ex primitivis, sed per misturam generari ostendatur, ₂45/ et cum nullum inter lucem originalem et illam quae a diversicoloribus radijs componitur, sensibile discrimen intercedat, haud dubitandum est quin

[67 → 76. Conclusio,
quòd reflectio vel
refractio non
mutat radiorum
dispositiones; neque
adeò discrepantiam
quoad colores
inducit.]

utraque sit ejusdem naturae.[9] Imò verò certissimum est, siquidem (in Prop 2) ostenditur, quod inhaerentes dispositiones vel formae radiorum quibus apti sunt ad proprios colores exhibendos nec destrui possunt, nec ullo modi vi secundariae refractionis mutari. Et par est ratio de refractione primariâ.[10] Concludendum est itaque, quod istae dispositiones sunt insitae radijs ab eorum origine, quamvis proprios colores antequam heterogenei ab invicem virtute refractionis secernantur exhibere nequeant.

[68]

Caeterùm de eo quòd dixi lucis colorem album esse, et tamen sol aliquantulum flavescere videtur, notandum est, quòd caeruliformes radij ab Atmosphaerâ prae caeteris conturbantur, (ut caeruleus ejus color innuit,)[11] et inde quod e directis solaribus radijs flaviformes praevalere solent et efficere ut sol flavescat qui secus fortasse appareret albus. Et ad hunc effectum Atmosphaera circa solem forte[12] conglobata potest etiam conducere. Utut non eo inficia[ri]s quin aliquod radiorum genus in originali luce s[a]epissimè redundet, quandoquidem flammarum et siderum diversi sunt colores.[13]

[69]

De lucis et albedinis compositione haec satis. Quod autem nigredo ex omnibus coloribus similiter composita sit, et in solo lucis defectu ab albedine differat, ex eo manifestum est quod nigrorum in radijs solaribus intra cubicu-

---

(6) Since in revision Newton altered the division into paragraphs in §§65, 66, we have accordingly shifted the headings for these articles.

(7) I: *dissimilibus* (dissimilar).

(8) In §II, 89 ≈ 92 Newton again invokes these experiments against modification theories.

(9) Newton extends this argument with a new instrument in §II, 78 = 94.

(10) Newton did not in Prop. 2 prove color immutability for the primary refraction; see Lect. II, 3, note (18).

(11) Although Newton here has left the mechanism for the greater disturbance of the blue rays unspecified, in Prop. 7 of his "Discourse of Observations," sent to the Royal Society on 7 December 1675, he conjectured that the blue or azure of the sky was the same as the blue of the first order in "Newton's rings": "For all vapors when they begin to condense & coalesce into

although it is white, it nevertheless manifestly consists of heterogeneous[7] rays, because some sorts of them can be totally reflected to π while others are partly transmitted to τ. By the same argument it is evident that the reflected white is similarly compounded, since (as I said) it is restored when the red is reflected from τ last of all. ‖ These things may also be highly confirmed, since they are produced solely by the obliquity of the glasses without some new modification of a refraction or a reflection.[8]

transmitted through parallel surfaces, it is compounded from colors.][6]

Therefore, even if light that flowed directly from the sun were uniform, nevertheless after it has ever been reflected or refracted, it consists of heterogeneous rays. Of this sort is all that light that is transmitted through glass windows or that the planets or clouds reflect to us, indeed, all light derived from the sun or any lamp, since it experiences at least some sort of refraction from the atmosphere (as astronomers assert); though I will not mention what occurs in objects and again in the membranes of the eye before the action of vision is impressed. Now, if I had shown nothing else, it would have been somewhat of an advance, since all phenomena of visible things are exhibited to us by light of this sort. Yet, since the sun's direct light is perceived to be white, and that color is not one of the primitives but may be shown to be generated by a mixture; and since there is no sensible difference between original light and that which is compounded from diversely colored rays, it must not be doubted that both are of the same nature.[9] Indeed this is most certain, since it is shown (in Prop. 2) that the inherent dispositions or forms of the rays whereby they are disposed to exhibit their own colors can neither be destroyed nor changed in any way by the force of a secondary refraction; and the reason is the same for the primary refraction.[10] It must therefore be concluded that those dispositions are innate to the rays from their origin, although they cannot exhibit their own colors before the heterogeneous rays are separated from one another by the power of refraction.

[66 → 75. And we nonetheless believe it is of the same nature as the sun's direct light.]

[67 → 76. Conclusion: Reflection or refraction does not change the rays' dispositions, and thus does not introduce a difference with respect to their colors.]
[68]

However, inasmuch as I have said that the color of light is white, and yet the sun appears to be somewhat yellow, it ought to be observed that the blue-making rays are disturbed by the atmosphere more than the others (as its blue color indicates);[11] and hence of the direct solar rays the yellow-making ones usually predominate and cause the sun, which would otherwise perhaps appear white, to become yellow. The sun's atmosphere, which perhaps[12] forms a sphere around it, can also contribute to this effect. You should not, however, on that account deny that some kind of ray in the original light is frequently in excess, since the colors of flames and stars are diverse.[13]

This is sufficient for the composition of light and whiteness. That blackness, however, is similarly compounded of all colors and differs from whiteness solely by the absence of light is evident, because the borders of black things, placed in the solar rays admitted into a room otherwise sealed off,

[69]

---

small parcells become first of that bignesse whereby such an Azure must be reflected before they can constitute clouds of other colours" (Add. 3970, f. 515ʳ = Birch, *History of the Royal Society,* 3:302; and incorporated in the *Opticks,* Bk. II, Pt. III, Prop. VII). See also Lect. II, 9, note (10).   (12) Added.   (13) See the "New theory," Prop. 8.

lum (alias obtenebratum) intromissis, positorum, termini omnigenis coloribus tincti apparent, si prismate juxta oculum interposito inspiciantur; quod singulos prismatis colores seorsim incidentes pari intentione reflectunt, idque longè debiliori quàm alba corpora; et quòd alba defectu lucis nigrescere videntur, ita ut corpus quod revera albius est, in debiliori luce nigrius apparere possit.

/ Denique de cinereis caeterisque non primitivis coloribus propositio manifesta est siquidem cinereos ex albo et nigro, caeterosque omnes ex rubro flavo et caeruleo componere norunt Pictores.[14]   ₂46/

PROP: 4. PRIMITIVI COLORES PER COMPOSITIONEM COLORUM SIBIMET UTRINQUE CONFINIUM EXHIBERI POSSUNT.

[70]   Hoc varijs modis (perinde ut in albedinis compositione[15] sistendo aliquos e coloribus antequam compositionem ingrediuntur) tentari potest; et ipse aliquos expertus sum quibus constitit luteum a croceo et subflavo, porraceum a subflavo et Thalassino (vel etiam minus perfectè à luteo et cyaneo,) et cyaneum a Thalassino et Indico, aliosque omnes colores a coloribus hinc et inde conterminis componi posse.[16] Quinetiam indicus cum rubei extremitate con-

---

(14) The painters' primaries, or colorants, red, yellow, and blue, which since the mid-eighteenth century have been nearly universally adopted, appeared rather suddenly at the beginning of the seventeenth century. The idea that all colors could be composed from three primaries, red, yellow, and blue (together with white and black), while these could not themselves be compounded, was independently proposed and demonstrated by de Boodt and Savot in 1609 and by d'Aguilon in 1613. Only d'Aguilon, however, who collaborated with the Flemish painter Peter Paul Rubens, set the number of chromatic primaries at three without qualification, and he further clarified these ideas with what may be the first color-mixing diagram; *Opticorum*, Bk. I, Prop. XXXIX, pp. 38–41. De Boodt compromised the triad by introducing two sorts of red, "ruber" and "miniatus" (*Gemmarum*, Bk. I, Ch. XV, p. 25), and Savot introduced only two colors, "rubrum, & *bleu,*" but then divided red into "one of thick parts and another of thin parts," or red and yellow (*jaune*) (*Nova-antiqua*, Chs. IV, VIII, ff. 6, 7ᵛ–8ᵛ). See the excellent series of papers by Charles Parkhurst, "Aguilonius' optics and Rubens' color," *Nederlands Kunsthistorisch Jaarboek*, 12 (1961):35–49; "A color theory from Prague: Anselm de Boodt, 1609," *Allen Memorial Art Museum Bulletin*, 29 (1971):2–10; "Louis Savot's 'Nova-antiqua' color theory, 1609," *Album amicorum J. G. van Gelder*, ed. J. Bruyn et al. (The Hague, 1973), pp. 242–7; and "Camillo Leonardi and the green-blue shift in sixteenth-century painting," *Intuition und Kunstwissenschaft: Festschrift für Hans Swarzenski*, ed. Peter Bloch et al. (Berlin, 1973), pp. 419–25.

The new triad of primary colors was gradually assimilated into the scientific tradition, especially after 1647, when Kircher in his widely read *Ars magna* (Bk. I, Pt. III, Ch. II, pp. 67–8) clearly set forth d'Aguilon's theory. Boyle gave a clear account of the primaries and of the mixing rules of "the Painters Art" in his *Touching Colours* (Pt. III, Expt. XII, pp. 219–21), and Newton took some notes, though none too accurately, on this experiment in his first essay "Of Colours" (§12, Add. 3996, f. 124ᵛ). William Petty also adopted these three primaries in his "An apparatus to the history of the common practices of dying," read on 7 May 1662 and included in Thomas Sprat's *History of the Royal Society* (London, 1667), pp. 284–306. In his notes on Sprat's *History* (p. 302) Newton succinctly entered Petty's first rule: "All yᵉ materialls (wᶜʰ of themselves doe coulor) are Red yellow & blew, from wᶜʰ (wᵗʰ fundamentall white) ariseth yᵗ greate variety wee see in dyed stuffs" (Add. 3958, f. 7ᵛ). For the composition of gray and Newton's assumption that the same mixing rules apply to pigments and lights, see the following article and §51 = II, 39. One must not confuse Newton's innumerable primary or simple colors with the three painters' primaries; see Lect. II, 1, note (10), this volume. Newton's simple colors are the irreducible elements of white light decomposed by refraction, whereas the painters' primaries are the elements out of which all colors may be compounded.

appear colored with every sort of color if they are viewed with a prism inserted next to the eye; because they reflect the individual prismatic colors incident separately with equal intensity, though very much weaker than white bodies; and because white bodies in the absence of light appear to become black, so that a body that is in fact rather white can appear rather black in weaker light.

Finally, concerning gray and other nonprimitive colors, the proposition is evident, since painters have known that grays are compounded from white and black and all the others from red, yellow, and blue.[14]

PROPOSITION 4. PRIMITIVE COLORS CAN BE EXHIBITED BY THE COMPOSITION OF THE NEIGHBORING COLORS ON EACH SIDE OF THEM.

This can be tested in various ways, exactly as in the composition[15] of [70] white, by stopping some of the colors before they enter into the composition. I myself have tried some whereby golden yellow was made from saffron and pale yellow, leek green from pale yellow and sea green (or even less perfectly from golden yellow and cyan), and cyan from sea green and indigo; and all the other colors can be compounded from the colors bordering on each side.[16] Moreover, indigo tempered by mixing with the extremity of the red

---

(15) Originally: *constitutione.*

(16) On the translation and sequence of the colors see Lect. II, 3, note (17), this volume. Newton's proposition is valid but requires certain restrictions on the location of the colors. This can be seen from his parenthetical remark that he could make leek green (*porraceum*), though only imperfectly, from golden yellow (*luteum*) and cyan, for it should yield white or at least a very pale or whitish green. Newton in fact probably did observe such a color. When Huygens suggested that not all colors of the spectrum are necessary to compound white, as Newton claimed, but that it could probably be composed of just two, yellow and blue (Oldenburg to Newton, 18 January 1672/3, *Correspondence,* 1:255–6 = *Phil. Trans.,* 8 (1673):6086), Newton could only counter that "I remember I once tried by graduall succession the mixture of all paires of uncompounded colours, & though some of them were paler & nearer to white then others yet none could be truly called white" (Newton to Oldenburg, 3 April 1673, *Correspondence,* 1:265 = *Phil. Trans.,* 8 (1673):6110). Newton, however, had no motivation to interpret this ambiguous result, a pale, nearly white green, to be white rather than green. To have done so would have violated the virtually universally accepted rule that mixtures of pigments and of lights follow identical rules; see Lect. 4, note (31). Moreover, it would have compromised his theory of color, for it would have granted the existence of two sorts of white light—one compounded of only two colors, and the sun's light compounded of innumerable colors—without his being able to explain the difference.

The proposition does, however, give quite good results when the colors to be mixed are near each other and on the same side of the green. The composition of golden-yellow (*luteum*), the more perfect leek green (*porraceum*), and cyan, were confirmed by Hermann von Helmholtz; *Helmholtz's Treatise on Physiological Optics,* trans. from 3rd German ed., ed. James P. C. Southall, 3 vols. ([Rochester, N.Y.], 1924–5), 2:128–9, which improves on his earlier "Ueber die Theorie der zusammengesetzten Farben," *Annalen der Physik und Chemie,* ser. 2, 87 (1852):45–66, esp. 58, 63–4 (the English translation, "On the theory of compound colours," in the *Philosophical Magazine,* ser. 4, 4 (1852):519–34, omits Helmholtz's footnote to the *Optical Lectures*). Newton reports observations elsewhere that show that he was aware of the range of applicability of his rule. In the initial statement of this proposition (§II, 5), he relates that orange and green make yellow, and in the version in the "New theory" (Prop. 6) that red and yellow makes orange. However, in §II, 97, and also in the "New theory," he explains that if the colors are too far apart (for example, red and green, orange and indigo, or green and violet), they will not satisfactorily compound the intermediate color (yellow, green, or blue). He incorporated this proposition in Prop. IV of the *Opticks,* Bk. I, Pt. II, but in Prop. VI of that same part he

temperatus purpurascebat, et minius cum extremâ purpurâ paululum conspersus coccineus evasit: tanquam si inter colorum extremitates intercederet affinitas qualis est in sonis inter octavae terminos.[17]

Ijdem colores a coloratis pulveribus componi possunt, sed minùs perfectè, ut opinor, propterea quòd ipsi componentes ex alijs coloribus (quorum aliqui sunt dissimiliores) componuntur.[18]

Caeterum ne nimius hic sim, breviter dicam quo pacto prismatici colores in hos effectus producendos optimè misceri possunt. Nempe prisma *GDE* (fig II, 26) ex pellucidissimis et perpolitis lamellis vitreis in vasculum aquae ple-

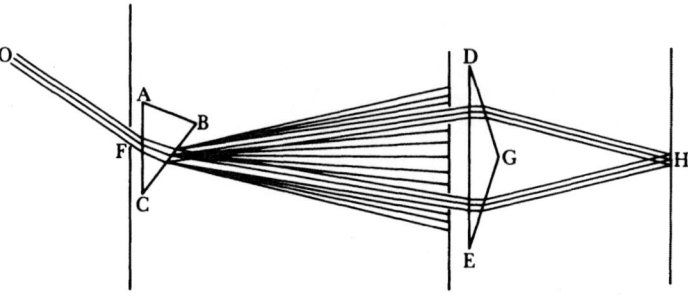

Figure II, 26

num coaptatis efficiatur quo radiorum in colores per divergentiam discretorum duo quaelibet genera juxta diversos angulos *D* et *E* sat acutos et aequales transmissa, ad invicem versus *H* cogantur.

Lect 9   PROP 5. CORPORUM NATURALIUM COLORES E GENERE RADIORUM DERI-
[71] VANTUR QUOS MAXIMÈ REFLECTUNT.[1]

Hoc e praemonstratis tantâ necessitate et evidentiâ consequitur ut supervacaneum esse videatur me aliquid de industriâ hic / in probationem ejus ₂47/ moliri; utpote cùm ostensum sit[2] quòd nullius generis uniformium radiorum color per reflexionem a corpore physico mutari possit, sed unumquodque colore radiorum tinctum appareret quibuscum illuminatur. Si corpus cujuscunque subdialis coloris a solis rubriformibus radijs in tenebroso cubiculo illuminetur, rubescit: Si flaviformibus illuminetur, flavescit; si viridiformibus, virescit; et sic praeterea.

[72]   Sed in hujus rei majorem evidentiam observandum est insuper quod unumquodque corpus proprium colorem prae caeteris seorsim incidentibus copiosè reflectit. Sic Cinnabaris in luce rubea maximè resplendet, in viridi minùs, et adhuc minùs in caerulea. Sic Indicum in violacea et caerulea luce maximè resplendet, et splendor ejus gradatim diminuitur prout in rubeam lucem per gradus intermedios continuò transfertur[.] Sic Porrus lucem viridem plusquam rubeam aut purpuream reflectere conspicitur. Et sic in alijs. Et quo corpora sub dio sunt intensiorum et magis specificorum colorum, eo minùs in alienâ luce resplendent.

became purple, and vermillion mingled with a touch of extreme purple turned out scarlet, just as if there was an affinity between the extremities of the colors as there is in the sounds between the termini of an octave.[17]

The same colors can be compounded from colored powders but less perfectly, I believe, because the components are themselves compounded of other colors, some of which are dissimilar.[18]

But lest I dwell on this too long, I will briefly relate how prismatic colors can be best mixed to produce these effects. Namely (Fig. II, 26), make a prism *GDE* from very clear and well-polished glass plates joined into a vessel filled with water, whereby any two kinds of rays separated into colors by diverging and transmitted toward the opposite angles *D* and *E*, sufficiently acute and equal, may be gathered toward one another at *H*.

PROPOSITION 5. THE COLORS OF NATURAL BODIES ARE DERIVED FROM THE SORT OF RAYS THAT THEY REFLECT MOST.[1] Lecture 9 [71]

This follows with such necessity and clarity from what has already been shown that it may appear superfluous at this point for me to attempt to do anything purposely with respect to proving it, since it was shown[2] that the color of no kind of uniform rays could be changed by reflection from a physical body, but that one and the same body appeared colored with the color of the rays that illuminate it. If a body of any color in daylight is illuminated by the sun's red-making rays in a dark room, it becomes red; if it is illuminated by yellow-making rays, it becomes yellow; if by green-making ones, it becomes green; and so on.

But for greater clarity of the matter it must in addition be observed that [72] one and the same body reflects its own color more copiously than other colors when they are incident separately. For instance, in red light cinnabar shines most brightly, in green light less, and in blue still less; indigo shines most brightly in violet and blue light, and its brightness is gradually diminished as it is continuously shifted into red light through the intermediate gradations; a leek is seen to reflect green light more than red or purple; and similarly for the others. And the more intense and specific the colors of bodies are in daylight, the less brilliantly they will shine in a different light.

proposed an improved and more general rule (though not without its flaws) for compounding colors using his color-mixing circle. With this new rule both the resultant hue and saturation of any number of colors from any parts of the spectrum—not just neighboring pairs—could be determined.

(17) Newton emended these results in the *Opticks*; see Lect. 1, note (10), this volume. In Lect. II, 11, he further pursues the musical analogy.

(18) See Lect. 4, note (31).

(1) This proposition was also included in the "New theory," Prop. 13, and the *Opticks*, Bk. I, Pt. II, Prop. X. On earlier views of the colors of natural bodies see the Introduction, §1.

(2) In §II, 26.

Quamobrem ut haec faeliciùs[3] et magìs cum evidentia pertentes, corpora seligere oportet intensis coloribus et quàm poteris maximè simplicibus praedita. Id quod cognosces si, Prismate adhibito, seligas quae ad extremitates nigredini conterminas distinctiora apparent et minùs variegata.

Praeterea colores Prismatum quos in haec corpora projicis debent esse ab invicem per plures refractiones optimè discreti. Nam si colores per unici tantùm Prismatis juxta lucis ingressum positi refractionem secernantur, non color lucis incidentis, sed alius quidam inter corporis in aprico conspecti et lucis hujus incidentis colorem intermedius generabitur. Quemadmodum si hujusmodi lux flava in caeruleum corpus incidat, corpus illud non flavescet. Sed virebit potiùs propterea quòd plures e viridiformibus radijs in hâc flavâ luce latitantibus, quàm e flaviformibus / reflectere aptum sit.[4] Et sic rubeum ₂48/ corpus in viridi luce flavescere potest, et in caerulea luce virescere, si modò lux illa ab alijs commistis coloribus non bene purgetur. Et ob hanc causam summè cavendum est in faciendis his experimentis ut cubiculum fiat obscurissimum ne lux erratica cum prismatico colore commisceatur.

Denique quo coloris cujusvis a corporibus sub dio diversè coloratis reflexi quantitas meliùs innotescat, corpora illa in eâdem lucis quâlibet specie juxtapositâ confer, et videbis unumquodque in luce proprij coloris prae caeteris resplendere. Sic Indicum in caeruleâ vel purpureâ luce plusquam Cinnabaris resplendet, et minùs in rubeâ. Aut si forte (propter alterutrius coloris imperfectionem et obscuritatem) ambo aequaliter in luce violaceâ resplendere contingat, tum in rubeâ luce Cinnabaris fiet longe illustrior. Aut contra longè debilior in luce violaceâ si aequaliter resplendeant in rubeâ. Cinnabaris itaque plures e rubriformibus quàm alijs quibuslibet radijs reflectit et proinde rubet.[5] Indicum verò plures e caeruliformibus et purpuriformibus reflectit et proinde fit intermedij coloris. Et ad eundem modum si in albis corporibus fiat experimentum, constabit quod omnigenos reflectant aequaliter. Et sic in alijs.

[73]   Antequam huic de coloribus physicorum corporum propositioni finem impono, placet annotare de quibusdam apparentijs quantâ necessitate consequuntur nostris principijs, quae aliàs mirae viderentur, et explicatu difficillimae. Et imprimis quia corpora evadunt colorata reflectendo quaedam genera radiorum et intromittendo caetera, si aliquatenus transpareant, concludendum esse videtur, quod colores maximè transmittantur qui minimè reflectuntur, et inde quod alius sit eorum color cum transpiciuntur atque alius cum cernantur luce reflexâ. Et hoc quàm benè convenit cum experientiâ videre est in libro Mᵣⁱ Boylei de coloribus conscripto. Scilicet infusio ligni Nephritici

---

(3) Read "feliciùs" (successfully); the *editio princeps* has "faciliùs."

(4) Despite its agreement with the then accepted rules of color mixing (yellow and blue make green), the reflected light, according to Newton, is green because of the predominance of monochromatic green rays and not from a mixture of yellow and blue rays. This phenomenon was commonly invoked as an instance of color mixing; for example, by Boyle, who projected prismatic yellow onto a very vivid blue body to get green (*Touching Colours*, Pt. III, Expt. XIV, p. 226).

Therefore, to test this more successfully[3] and with greater clarity you must choose bodies endowed with intense and, as much as you can, simple colors. You will identify these if, with the use of a prism, you choose those bodies that appear more distinct and less variegated at their edges that are bordered by blackness.

Moreover, the prismatic colors that you project onto these bodies must be thoroughly separated from one another by several refractions. For if the colors are separated by the refraction of only one prism placed next to the light's entrance, the color of the incident light will not be generated, but some other one that is intermediate between the color of the body viewed in sunlight and that of this incident light. For instance, if yellow light of this kind falls upon a blue body, that body will not become yellow but rather green, because it is disposed to reflect more of the green-making rays latent in this yellow light than the yellow-making ones.[4] In this way a red body in green light can become yellow, and in blue light green, provided that the light is not completely purged of the other intermixed colors. For this reason one must be extremely careful in doing these experiments that the room is made very dark so that stray light is not mixed with the prismatic color.

Finally, to know better the quantity of any color reflected from bodies that are differently colored in daylight, place those bodies alongside one another in the same light of whatever sort you wish, and you will see each one shine brightly in the light of its own color compared with light of other colors. Thus indigo shines more brightly than cinnabar in blue or purple light, and less in red light. If perhaps (because of the imperfection and obscurity of either color) it happens that both shine equally in violet light, then in red light cinnabar will become much brighter; or, on the contrary, if they shine equally in red light, in violet light the cinnabar will become much weaker. Cinnabar therefore reflects more of the red-making rays than any others and consequently is red,[5] but indigo reflects more of the blue- and purple-making rays and consequently becomes the intermediate color. In the same way, if the experiment is done with white bodies, it will be manifest that they reflect all kinds equally; and similarly for the others.

Before I set an end to this proposition about the colors of physical bodies, I [73] propose to comment on some appearances that follow from our principles with such necessity and that would otherwise seem surprising and very difficult to explain. In the first place, because bodies become colored by reflecting some kinds of rays and letting in others, if they are in some measure transparent the conclusion seems unavoidable that those colors are transmitted the most that are reflected the least, and therefore that they are one color when they are looked through and another when they are seen by reflected light. One can see how very well this agrees with the experiment in Mr. Boyle's book, *Touching Colours*. Namely, when an infusion of *lignum nephriticum* is

---

(5) The color of cinnabar is again briefly treated in a passage added to §II, 79 ≈ 95.

quando adversâ[6] luce transpicitur rubea vel flava apparet, et caerulea cum cernitur ad plagas lucis incidentis.[7] E contra / vero aurum foliatum apparet ₂49/ flavum et transparet caeruleum[.][8] Sic vitri fragmenta per totam profunditatem colorata, qualia in antiquis Templorum fenestris reperiuntur; varios plerunque colores pro positione spectatoris exhibent. Et crassiorum laminarum vitri pellucidissimi (quale ad fabricanda Telescopia adhibentur) cum obversas oras aspexi caeruleum vidi reflexum, et flavum transmissum cùm perspexi.[9] Caeruleus autem maximè apparuit cùm illustrabatur jubare in obscuratum cubiculum immisso et a lente concavâ distracto, ne nimiâ luce color perfunderetur. Neque ullus dubito quin plurima existant hujus rei exempla siquis operae pretium duxerit in varijs liquoribus alijsque corporibus transparenter coloratis examen instituere; interea cavendo ne lux e pluribus plagis simul incidat.[10]

[74]   Quod autem isthoc non semper eveniat (quemadmodum in eâdem infusione ligni Nephritici cùm caeruleus color salibus acidis destruitur,[11] et in alijs plerisque quae undique sunt ejusdem coloris) ratio est quod corporibus non solum insit potestas reflectendi vel transmittendi radios sed etiam suffocandi et in se terminandi.[12] Sic aliqua obstruunt et retinent omnigenos radios, eoque pacto fiunt undique nigra; alia reflectunt quosdam, caeterosque supprimunt ut opaca colorata; alia quosdam supprimunt, caeterosque partim reflec-

---

(6) *Editio princeps*: diversa.

(7) In *Touching Colours*, (Pt. III, Expt. X, pp. 199–212) Boyle described the properties of *lignum nephriticum*, the wood from a small tree or shrub (*Eysenhardtia polystachya*) found in Mexico and used for treating kidney ailments. In his notes, "Of Colours," Newton succinctly recorded the main elements of Boyle's description: "47 take Lignum Nephriticum . . . put a handfull of thin slices of it into 3 or 4 pound of pure spring water[,] after it hath infused there a night put yᵉ water into a cleare violl, & if you see yᵉ light through it it appeares of a golden colour . . . But if your eye is twixt yᵉ liquor & light it appeares ceruleous: &c Acid salts destroy yᵉ blew colour . . . " (Add. 3996, f. 134ᵛ). As George Gabriel Stokes discovered, in a fluorescent substance such as *lignum nephriticum* the rays more refrangible than blue are absorbed and reradiated (not reflected) at the surface as less refrangible, mostly blue rays; "On the change of refrangibility of light," *Phil. Trans.*, **142** (1852):463–561. See in addition W. E. Safford, "*Lignum nephriticum*—its history and an account of the remarkable fluorescence of its infusion," *Annual Report of the Smithsonian Institution*, 1915:271–98; J. R. Partington, "Lignum nephriticum," *Annals of Science*, **11** (1955):1–26; and E. Newton Harvey, *A History of Luminescence from the Earliest Times until 1900* (Philadelphia, 1957), Ch. XI.

(8) Gold, like most metals, acquires its color by selective reflection; it reflects predominantly yellow rays and transmits the complementary blue. Newton learned of the colors of gold leaf from Boyle's *Touching Colours* (Pt. III, Expt. IX, pp. 198–9) and described it in "Of Colours" (§39, Add. 3996, f. 134ʳ).

(9) The appearance of the colored glass is most likely due to scattering, and that of the clear glass to fluorescence.

(10) Newton has here set forth both the fundamental assumption of his theory of the color of natural bodies—namely, that some colors are reflected and the remainder, their complement, are transmitted or let into the body—and the archetypal phenomena upon which it is based—the colors of an infusion of *lignum nephriticum*, gold leaf, and colored glass. As early as his second essay "Of Colours," he had laid down the essential elements of his theory and invoked the same three phenomena, as he later would in the "New theory," Prop. 11, and the *Opticks*, Bk. I, Pt. II, Prop. X. After briefly describing the colors of gold leaf and *lignum nephriticum* in the first two

looked through facing[6] the light, it appears red or yellow, while it appears blue when it is viewed on the side of the incident light.[7] Conversely, a gold leaf looked upon is yellow, and looked through is blue.[8] Similarly, pieces of glass colored throughout their whole thickness, such as those found in the windows of old churches, frequently exhibit various colors according to the position of the observer. When I looked at the facing side of rather thick plates of very clear glass (such as those used to make telescopes), I saw blue reflected, and when I looked through, I saw yellow transmitted.[9] It appeared bluest, however, when it was illuminated with a beam admitted into a darkened room and spread out by a concave lens, so that the color would not be drowned out by too much light. I do not at all doubt that very many examples of this condition exist, if anyone considered it worthwhile to undertake an examination of various liquids and other transparently colored bodies, while being careful that no light is incident simultaneously from several directions.[10]

This, however, does not always happen (for instance, in the same infusion [74] of *lignum nephriticum* when the blue color is destroyed by acidic salts,[11] and in many others that are the same color on all sides), because there is in bodies not only a power to reflect or transmit rays, but also one to stifle and terminate them within themselves.[12] Thus, some bodies stop and retain all kinds of rays, and in that way they become completely black; others reflect some and suppress the rest, as opaque colored bodies; others sup-

---

entries in "Of Colours," he noted that: "3 The flat peices of some kinds of Glase will exhibit ye same Phaenomena wth Lignum Nephriticum. And these Phaenomena of Gold & Lignum Nephriticum are represented by y^e Prisme in y^e 37th experiment as also in y^e 22th & 24th Experiment" (Add. 3975, f. 2^r). Newton is referring here to his own experiments later in the essay (compare Hall, "Further experiments," p. 28): Experiments 22 and 24 involve total reflection from the base of a single prism (as in §65 = II, 56) and from the air film between two prisms forming a parallelepiped (as in §71 = II, 62), respectively; and the 37th experiment records the sequence of reflected and transmitted interference colors produced by that air film. In these experiments colors are manifestly produced by reflection, and whatever colors are not reflected are transmitted, just as seems to occur in the archetypal phenomena. Indeed, by the spring of 1672 Newton had adopted the colors of thin films as the physical basis for his theory of the color of natural bodies and assumed that their colors are produced by particles of the same size and density as that of a thin film producing those colors; see his letter to Oldenburg of 21 May 1672, *Correspondence*, 1:160.

(11) On the destruction of the blue, see note (7), this lecture; and also "Of Colours," §5, Add. 3975, f. 2^r.

(12) Newton has here introduced into optics the concept of absorption (*potestas suffocandi et in se terminandi*) as an independent physical process in the coloration of natural bodies. Selective absorption—and not, as Newton would have it, selective reflection—is the principal process involved in the color of most natural bodies. White light falling on a body becomes colored by actually penetrating the body, where most of the rays are absorbed, while the remaining rays, which determine the body's color, are then diffusely reflected outward from the body's flaws, impurities, and sides. By reflecting light from the surface of colored bodies, such as colored liquids, it is in fact readily established that they are generally not colored by reflection, since the reflected light retains its original color and does not acquire that of the body; Kepler described such an experiment in a note to Ch. I, Props. 24 and 25, appended at the end of his *Ad Vitellionem paralipomena*, *Gesammelte Werke*, 2:368.

tunt et partim transmittunt, ut transparenter colorata quae circumcirca ejusdem sunt coloris; et alia quosdam reflectunt caeterosque transmittunt ut in
exemplis jam allatis constitit.[13] Atque ita praeterea.

[75]   Porro quod liquoris colorati varia crassities aliquando speciem coloris vari-
are potest cum nostris principijs quam optimè consentit. Sic infusio ligni
Nephritici pro variâ ejus crassitie vel flavum vel rubeum
colorem referre potest. Cujus rei rationem ut intelligas concipe quod liquor ille sit aptissimus ad reflectendum purpuriformes et caeruliformes radios, ineptissimus ad reflectendum
rubiformes, et mediocriter aptus ad reflectendum mediocres.
Et (in fig II, 27) posito / *ABC*[14] vitro coniformi hujus infu-
sionis pleno, sit *FI* crassities ejus cum aureo colore splenden-
tis, *EH* major crassities ubi incipit rubescere, ac *DG* crassities ubi fit intensioris et subobscuri ruboris. Et cùm caeruliformes et purpuriformes radij citissimè reflectantur, ut ex eo
patet quòd unius tantùm guttulae crassities ad eos colores
reflectendos et spectantibus exhibendos sufficit: ex illis pau-
cissimi penetrabunt ad profunditatem *FI*. Sed plurimi viridiformes et adhuc
plures flaviformes unà cum rubriformibus trajicientur, ex quâ misturâ fiet ille
color aureus. At per profunditatem *EH* pauci e flaviformibus transibunt, et
pauciores e viridiformibus: Ac soli fere rubiformes ad usque profunditatem
DG pervadere valebunt. Quinimo ex illis etiam complures in itinere reflectentur, et inde rubor trajectus subobscurus evadet.[15]

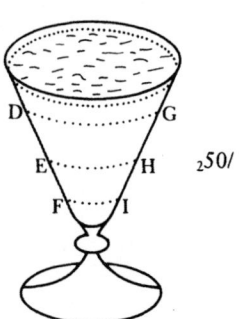

$_2$50/

Figure II, 27

[76]   Ad eundem ferè modum cum lux per plura corpora diversis coloribus
pellucidè tincta trajicitur, color ille ex adverso videbitur qui facillimè pertran-
sit omnia.[16] Quod si nullus potest omnia pertransire, utcunque seorsim pellu-
cida existant, conjunctim tamen evadent maximè opaca. Quemadmodum si
lamina *AB* transmittat solos rubiformes, et *CD* solos caeruliformes; cùm
juxta-ponuntur transmittent nullos. Cujus quidem rei exemplum habes in

---

(13) That is, by the phenomena described in the preceding article. Of these four categories of
colored bodies only transparent bodies, which reflect and transmit the same color, caused Newton difficulty, for they compromised his underlying scheme that some colors are reflected while
the rest are transmitted or absorbed. In his second essay "Of Colours," he had recognized and
attempted to resolve the conflict between the colors exhibited by his archetypal phenomena and
the obvious fact that most bodies are the same color all around: "4 But Generally bodys w^{ch}
appear of any colour to ye eye, appear of ye same colour in all positions; Nay Gold if it bee not
soe very thin as to bee transparent appeares onely yellow & perhaps ye yellow colour of Lignum
Nephriticum would vanish if ye tincture bee strong & ye liquor of a greate thicknesse. And
perhaps there are many coloured bodys wch if made so thin as to bee transparent would appeare
of one colour when looked upon & of another when looked through" (Add. 3975, f. 2^r).
Newton is suggesting that the reflected and transmitted colors are complementary in all bodies,
not just in those like *lignum nephriticum,* with the only difference being their thicknesses and
relative absorptive powers. Nonetheless, he has here abandoned that suggestion and implicitly
acknowledged that it is contradicted by simple observation with transparent colored bodies. In
the *Opticks,* however, he took the opposite tack, reasserted the suggestion in "Of Colours," and
to maintain the overall scheme that the reflected and transmitted colors are always complemen-

press some and partly reflect and partly transmit the rest, as transparent colored bodies that are the same color all around; and others reflect some and transmit the rest, as is established by the examples just related;[13] and so on.

Furthermore, the fact that a varying thickness of a colored liquid can [75] sometimes vary the species of color agrees quite well with our principles. Thus, an infusion of *lignum nephriticum* according to its varying thickness can return either a yellow or red color. To understand the reason for this, imagine that liquid to be most disposed to reflect purple- and blue-making rays, least disposed to reflect red-making ones, and moderately disposed to reflect the intermediate ones. Assuming (Fig. II, 27) *ABC*[14] to be a conical glass filled with this infusion, let *FI* be its thickness when shining with a gold color, *EH* the greater thickness where it begins to become red, and *DG* the thickness where it becomes a more intense, dark red. Since the blue- and purple-making rays are reflected soonest, because it is evident that a thickness of only one drop is sufficient to reflect and exhibit those colors to an observer, very few of those will penetrate to the depth *FI*. Most green-making rays, however, and even more yellow-making together with red-making ones will be transmitted, from which mixture that gold color will arise. Yet few yellow-making rays and fewer green-making ones will pass through the depth *EH*, and almost solely red-making ones will be able to reach as far as the depth *DG*; indeed very many of those too will be reflected en route, and hence the transmitted red will turn out dark.[15]

In nearly the same way, when light is transmitted through several trans- [76] parent bodies with different colors, that color that passes through most easily of all will be seen on the opposite side.[16] But if no color can pass through all the bodies, then however transparent they may be separately, together they will nevertheless turn out especially opaque. For instance, if the plate *AB* transmits only red-making rays and *CD* only blue-making, when they are juxtaposed they will transmit none. In fact, in Mr. Hooke's *Micrographia*

---

tary, he suggested an experiment (Bk. I, Pt. II, Prop. X, pp. 140–1) to show that transparent bodies are not truly the same color by reflection and transmission; this experiment, which Newton apparently never performed, is such a late addition to the *Opticks* that it is not in the main corpus of the manuscript sent to the printer (Add. 3970, f. 127ʳ). Compare Kepler's experiment described in the preceding note.

(14) The Letters *ABC* are lacking in Newton's figure. The colors are viewed with the eye on one side of the vessel, for example, *DEF*, facing the light on the other side, *GHI*.

(15) The varying colors of a red fluid described in the *Opticks* (Bk. I, Pt. II, Prop. X, p. 138 = Add. 3970, ff. 126–7) are not those of *lignum nephriticum*, but rather Newton seems to have substituted Hooke's description of a tincture of aloes, which is described in note (18), this lecture.

(16) Even if only briefly, Newton has explained here that this common method of mixing colors, that is, passing light through superposed colored glass plates, is not at all an instance of additive color mixing; compare Boyle, *Touching Colours*, Pt. III, Expt. XIII, pp. 221–4; della Porta, *De refractione*, Bk. IX, Prop. VII, p. 197; and Kircher, *Ars magna*, Bk. I, Pt. III, Ch. IV, p. 74.

Micrographiâ M^{ri} Hookij, de caeruleo et rubeo liquore qui seorsim trans-paruêre,[17] at conjunctim fuere opaci.[18]

Denique huc referri potest quòd cùm aliquis e coloribus Prismaticis per corpus transparentèr coloratum trajicitur intermedius color emergit. Sic vi-riditate v.g. in vitrum transparenter rubeum incidente, flaviformes radij qui in illa viriditate commisti latent, prae caeteris vitrum fortasse pervadent, effi-cientque ut lux emergens flavescat.[19]

[77 ≈ 93. Instrumentum describitur quocum omnia de coloribus hactenus tradita dilucidissimè probentur.]

Sed videor officij limites excessisse in campum physicum nimis expatiatus. Visum quidem fuerit haec attigisse ut universa rerum consensio pateret, sed sisto gradum, ac tandem coronidis loco instrumentum quoddam haud inele-gans describam quo praefata omnia summâ cum evidentiâ tentari possunt.

Fig. II, 28

/ Sit[20] *ABCac* prisma quod radios per foramen *F* in obscuratum cubiculum ₂51/ transmissos refringat versus lentem *MN* ut colores quos efficit in *p, q, r, s, t* per lentem deinde trajiciantur ad *X* et ibidem commisceantur in albedinem componendam[21] sicut in praecedentibus ostendi.[22] Deinde aliud prisma *DEGgd* priori parallelum ad locum *X*, ubi albedo redintegrata est, statuatur, quod lucem versus *Y* refringat. Hujus autem prismatis verticalis angulus *Gg* sit aequalis angulo verticali *Cc* prismatis anterioris, aut eo fortè minor, et similiter positus ut incidentes radios in parallelismum reducat quos prisma anterius dispersit. His positis observabis, an lux ad *Y* (pedes aliquot distans) trajecta aequè alba maneat ac fuerit in *X*, vel sensim abeat in colores. Si penitus appareat alba, tunc prismata cum lente debitè disposuisti: sin aliqui colores ad *Y* cernantur Prisma *DEG* circa suum axem eo modo parum con-verti debet, ut colores minuantur; et cum penitus evanuêre, et lux in totum albescit, siste prisma. Quod si nequeas hoc modo efficere quin lux inter transeundum ab *X* ad *Y* ex aliquâ sui parte transmigret in colores, lentem *MN* paulo longiùs a prismate *ABC* transfer, et loco *X* rursus invento ubi colores in albedinem accuratissimè convergunt, in eo statue prisma *DEF* ut priùs, et rursus experire an possis lucem sine coloribus ad *Y* projicere et cum eo usque mutaveris positiones prismatum et lentis dum effeceris lucem ad *Y* trajectam quàm minimè possis coloratam, prismata cum lente in eo situ figantur idque vel ope trabis, ut in schemate describitur, vel tubi aut instru-menti cujusvis in eum finem fabricati.

[78 = 94. Ejus usus describitur.]

Cùm habeas hanc machinam e prismatibus et lente ut dictum est, composi-tam; ope lucis per eam transmissae cuncta possis experiri quae hactenus

---

(17) *Editio princeps*: apparuêre.

(18) In his *Micrographia* (Observation X, pp. 73–4) Hooke described the varying colors of hollow prisms filled with a blue tincture of copper, and a red tincture of aloes, and Newton took notes on his observation (Add. 3958, f. 2^r = Newton, *Unpublished Papers*, p. 404). According to Hooke all colors are made from varying degrees of blue and yellow (which, he maintained, was a diluted red) and their mixtures. Consequently, when he laid his prisms over one another, he expected them to produce all varieties of color and not to become opaque. Newton chose to explain Hooke's unexpected and unexplained result in the "New theory" (Prop. 12) and later in the *Opticks* (Bk. I, Pt. II, Prop. X), where he freely admitted that he had never tried the experiment. He defended his interpretation in his replies to Pardies and Hooke on 13 April and 11 June 1672 (*Correspondence*, 1:142, 179–80 = *Phil. Trans.*, 7 (1672):4093, 5093–[4]), and then again for Lucas on 5 March 1677/8 (*Correspondence*, 2:256).

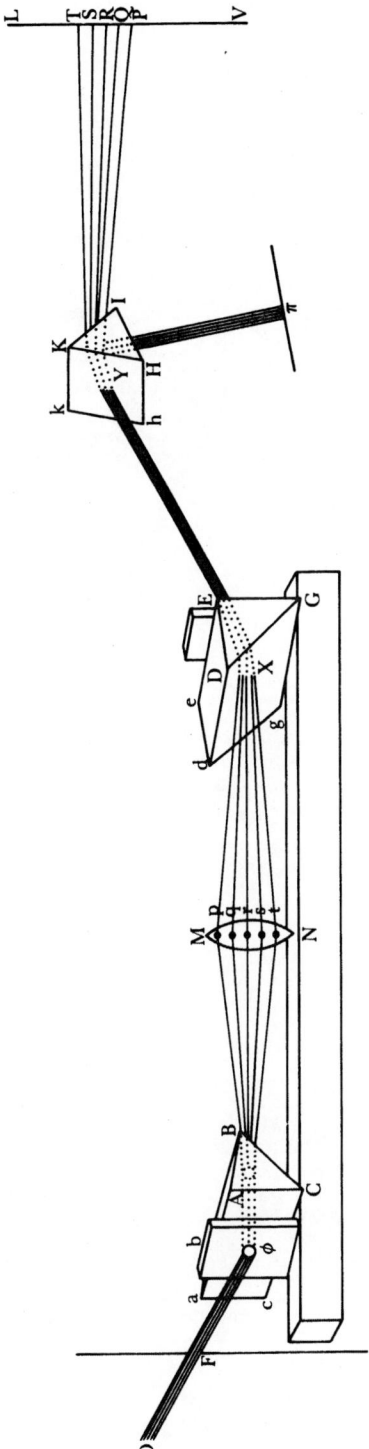

Figure II, 28

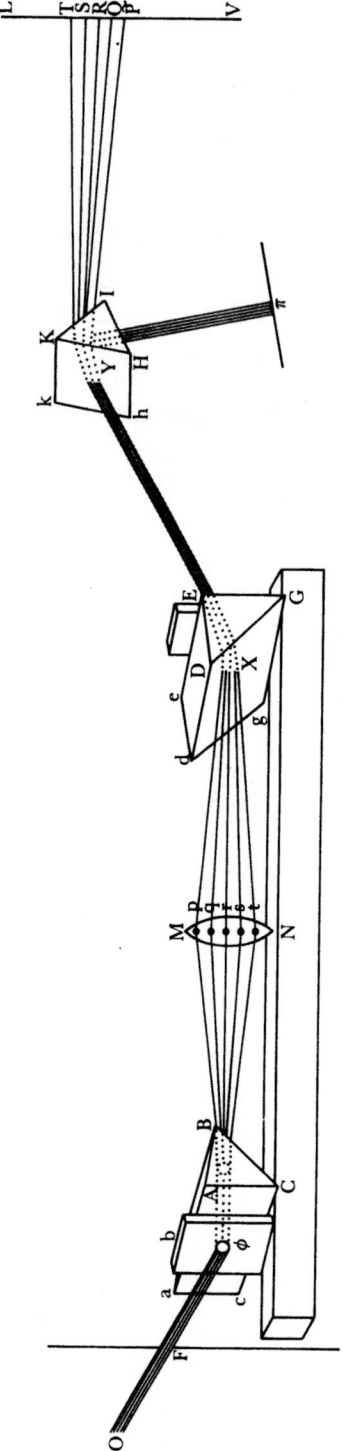

Figure II, 28

518

you have an example of this concerning a blue and a red liquid that separately were transparent[17] but together were opaque.[18]

Finally, it can be related here that when some prismatic colors are passed through a transparent colored body, the intermediate color arises. Thus, for example, when green falls on a transparent red glass, the yellow-making rays that are latent, being mixed in that green, probably pass through the glass more than the others and cause the emerging light to become yellow.[19]

But I seem to have exceeded the limits of my office, having digressed too much into the field of physics. It may have indeed seemed that I undertook these matters so that the universal agreement of things would be evident, but I halt, and, finally, instead of a conclusion I will describe a certain not inelegant instrument whereby all the preceding can be tested with the greatest clarity.

[77 ≈ 93. An instrument is described whereby everything hitherto related about colors may be proved very clearly.]

Let[20] *ABCac* be a prism that refracts the rays transmitted into a darkened room through the hole *F* toward the lens *MN*, so that the colors that it makes at *p, q, r, s,* and *t* may then be transmitted through the lens to *X* and there be mixed together to compound[21] white, just as I showed in the preceding.[22] Next, parallel to the first prism at the position *X*, where the white has been reconstituted, place another prism, *DEGgd*, that refracts the light toward *Y*. However, let this prism's vertex angle *Gg* be equal to or perhaps smaller than the first prism's vertex angle *Cc* and similarly placed so that it may restore to parallelism the incident rays dispersed by the first prism. After these things have been arranged, you will observe whether the light transmitted to *Y* (a few feet away) remains just as white as it was at *X* or is sensibly transformed into colors. If it appears completely white, then you have properly placed the prisms together with the lens. But if any colors are perceived at *Y*, the prism *DEG* must be turned slightly around its axis in such a way that the colors diminish. When they have completely vanished and the light becomes entirely white, stop the prism. If, however, in this way you cannot prevent the light from changing into colors in some part while passing from *X* to *Y*, shift the lens *MN* a little farther from the prism *ABC*. After again locating the place *X* where the colors converge most accurately into white, fix the prism *DEF* there as before, and once more test whether you can project the light to *Y* without colors. When you have altered the positions of the prisms and the lens to such a point that you have made the light transmitted to *Y* as little colored as you can, fix the prisms together with the lens in that position with the aid of a wooden beam (as is represented in the figure), a tube, or any other instrument constructed for that purpose.

Fig. II, 28

When you have constructed this device from a lens and prisms as I have described, then by means of the light transmitted through it you can test

[78 = 94. Its use is described.]

(19) Once more (see notes (4) and (16), this lecture) Newton shows that what had been considered to be an instance of true color mixing was really subtractive color mixing; compare Boyle, *Touching Colours*, Pt. III, Expt. XIV, p. 225.

(20) I: Quamobrem sit (Wherefore, let).

(21) Added.

(22) In I there is a marginal note to §54 [= II, 43].

fuerunt tradita; Haec enim lux *XY* jubari a sole directo persimilis est, et
easdem omnes apparentias / exhibet, ac si a foramine *F* rectà promanasset, ₂52/
nullam omnino refractionem passa; Adeoque ejusdem esse constitutionis facile credamus. Et tamen cùm in sua principia componentia, hoc est in radios
diversorum generum, apud lentem *MN* discreta fuerit, facilè erit modos examini subjicere quibus posthac in colores converti potest, idque tantùm sistendo hoc vel illud radiorum genus apud *MN*, ut constitutio lucis *XY* quoad
ejus conversionem in colores pateat.

[79 ≈ 95. Et
illustratur exemplis]
     Quemadmodum si desideretur ut sensui planissimè pateat quod prisma
convertit lucem in colores non transmutando proprietates ejus intrinsecas, sed
segregando tantum radios ad excitandum varia colorum phantasmata dispositos, ex quibus lux omnis albens constituitur; nihil aliud agendum est quam
ut prisma aliquod *HIK* ita statuatur, ut lucem *XY* excipiat, et refringendo
transmutet in colores *P, Q, R, S, T,* in papyrum aliquam *LV* procidentes.
Deinde si colorem quemlibet apud lentem *MN* interposito obstaculo sistas,
videbis eundem colorem a papyro *LV* deficere. Sic purpuram *p* obstruendo,
disparebit purpura *P,* caeteris coloribus non omninò mutatis (dempto fortè
caeruleo, quatenus aliquid purpurae commixtum habeat.) Sic viridem *r* intercipiendo, viridis *R* evanescet: et sic de alijs. Atque ita videre est quòd ijdem
colores apud papyrum *LV* et apud lentem *MN* pertinent ad eosdem radios,
ijsque non communicantur a refractione lentis[23] *HIK,* siquidem praeexiste-
bant segregati quidem ad lentem *MN,* et congregati in luce *XY.* Ad eundem
modum si cupias experimenta penitus rimari, quibus aliqua genera radiorum
omninò reflecti possint, dum alia (licet similiter incidentia) partim transmittantur: Prisma *HIK* circa axem ejus converte donec altera pars colorum
(violacea nempe et caerulea) postquam obliquissimè refracta fuerit versus *LV*
abinde penitus dispareat versus π reflexa:[24] parte tamen alterâ ad *LV* pervadente; Deinde si dimidium colorum rubedinem versus intercipias ad *MN,*
rubor et flavus disparebunt ab *LV,* et lux ad π reflexa fiet admodum caerulea. Sin alterum / dimidium purpuram versus intercipias, rubor apud *LV* non ₂53/
mutabitur. Sed lux in π (propter ablatum purpureum et caeruleum) flavescet
aut rubescet. Id quod indicat purpuriformia et caeruliformia radiorum genera
penitus ad π reflecti; dum caetera partim refringuntur ad *LV.* ‖ Praeterea si
corpus aliquod coloratum v.g. Cinnabaris hac luce *XY* illuminetur, sub proprio colore perinde apparebit quasi in luce subdiali constitutum aspiceres.
Quod si caeruliformes et viridiformes radios juxta lentem perlapsuros intercipias, rubor ejus intendetur: At cum rubiformes radios intercipis, Cinnabaris
non ampliùs rubebit: sed flavedinem aut viriditatem aliumve quemvis colorem, pro specie radiorum quos praetermittis, induet.[25] ‖ Nec secus alia colorum phaenomena, quae prismata ab immediatâ Solis luce eliciunt, ope lucis
hujus *XY* poteris experiri; et intercipiendo quodvis radiorum genus apud
MN, eorum causas intueri.

---

(23) Read (with the *editio princeps*): prismatis.
(24) *Editio princeps*: deflexa.
(25) This experiment was introduced in §II, 72.

everything that has been related thus far. This light *XY* is in fact very similar to a beam of direct light from the sun and exhibits all the same phenomena as if it had proceeded directly from the hole *F* without having experienced any refraction at all. Hence we may easily believe it to be of the same constitution. Yet when it has been separated at the lens *MN* into its principal components, that is, into rays of different kinds, it will be easy to put to a test the ways by which it can afterward be converted into colors merely by stopping either this or that kind of ray at *MN*, so that the constitution of the light *XY* with respect to its conversion into colors may be manifest.

For example, if it is desired to make it very clear to sense that a prism converts light into colors, not by transmuting its intrinsic properties but only by separating the rays that are disposed to excite various sensations of colors and from which all white light is constituted, nothing other needs to be done than to place some prism *HIK* so that it receives the light *XY* and by refraction transmutes it into the colors *P, Q, R, S,* and *T,* which fall onto some paper *LV.* Then if at the lens *MN* you stop any color by inserting an obstacle, you will see the same color depart from the paper *LV.* Thus by blocking off the purple *p,* the purple *P* will disappear with the other colors not being changed at all (except perhaps blue, insofar as it may have some purple intermingled); and by intercepting the green *r,* the green *R* will vanish; and similarly for the others. Thus it is possible to see that the same colors belong to the same rays at the paper *LV* and the lens *MN* and that they are not imparted to them by the refraction of the prism[23] *HIK,* since they preexisted, having in fact been separated at the lens *MN* and gathered together into the light *XY.* In the same way, if you want to investigate thoroughly experiments in which some kinds of rays can be totally reflected while others (although similarly incident) are partly transmitted, turn the prism *HIK* around its axis until one part of the colors (namely, violet and blue), after having been most obliquely refracted toward *LV,* completely disappears from there and is reflected[24] toward π, while the other part still reaches *LV.* If then at *MN* you intercept half of the colors toward red, red and yellow will disappear from *LV,* and the light reflected to π will become completely blue. But if you intercept the other half toward purple, the red at *LV* will not be changed, but the light at π will become yellow or red because the purple and blue have been removed. This shows that the purple- and blue-making kinds of rays are totally reflected to π, while the others are partly refracted to *LV.* ‖ Moreover, if some colored body, for instance, cinnabar, is illuminated with this light *XY,* it will appear with its own color, just as you would see it constituted in daylight. If now you intercept the blue- and green-making rays flowing by next to the lens, its red will be increased. Yet when you intercept the red-making rays, the cinnabar will no longer be red but will assume yellow or green or some other color, according to the sort of rays that you let pass.[25] In just the same way, with the aid of this light *XY* you will be able to test the other phenomena of colors that prisms produce from the sun's direct light and, by intercepting rays of any kind at *MN,* observe their causes.

[79 ≈ 95. And it is illustrated by examples.]

[80 ≈ 96. In
constructionem
praefati instrumenti
notae quaedam.]

Siquis autem velit instrumentum quale jam descripsimus ad experimenta
hujusmodi instituenda conficere, lentem adhibeat latam tres digitos, et
ampliùs, quae radios parallelos ad focum duos pedes circiter distantem congregat: atque ita prismata distabunt octo pedibus, et conficient instrumentum
satis magnum quò omnia strictiùs examini subjiciantur. Quod ad positionem
lentis attinet, si prismatum anguli verticales *ACB* et *DGE* sint aequales, puto
60 vel 70^{graduum}, ipsa aequaliter ab utrisque distabit; sin alter angulus sit major
altero lens illi prismati vicinior collocetur cujus angulus verticalis existit major. Et nota quòd jubar *XY* per spatium eo latius diffunditur, quo lens statuitur anteriori prismati *ABC* vicinior: atque adeò siquando opus sit amplo
jubare, debes tantùm efficere, ut lens sit aliquantò vicinior anteriori prismati,
quàm posteriori, et adhibere prisma posterius, cujus angulus verticalis sit
tanto ferè minor quàm angulus verticalis anterioris. / Denique si velis ut ₂54/
colores in lentem illam procidentes, sint magis discreti[26] et ab invicem distracti quàm more jam descripto contingat, eâ nempe de causâ ut singula
radiorum genera pro lubitu distinctiùs sive magis sejunctim intercipiantur; (Id
quod in experimentis nonnullis necessarium duco), nihil aliud agendum est
quàm ut lux per duo parva foramina *F* et *φ* ab invicem longè distantia priùs
trajiciatur quàm incidat in prismata:[27] Vel ut alia lens non procul ab anteriori prismate collocetur,[28] quae apta sit ut lucem a longinquo foramine *F*
divergentem, congreget ad alteram subsequentem lentem *MN*. ‖ Caeterùm
hoc instrumentum sic rectè disponere invenio molestissimum esse, ut et effectus ejus haud ita distinctos et sensui patentes ac in praecedentibus ubi per
pauciores refractiones et majora vitrorum intervalla ostendebantur.[29] Et eapropter Auditores imprimis illa simpliciora et faciliora experimenta examini
consultiùs subjicient.

SECT. 2da.
DE VARIJS COLORUM PHAENOMENIS.

1. DE PHAENOMENIS LUCIS PER PRISMA AD PARIETEM TRAJECTAE.

Lect 10
[81 → 77. Colorum
vulgare
phaenomenon
explicatur.]

Hucusque fundamenta struxi quibus colorum quocunque modo effectorum
phaenomena explicari possunt; Effectuum verò quos supra minùs attigi jam
causas particulares et immediatas, non Geometrarum (quibus, scio supervacaneum videbitur;) sed aliorum gratia sigillatim describam. Malo enim hic aliqua quae plerisque[1] superflua fortasse videbuntur interserere, quàm quicquam alicujus momenti omittere quod incautis et praejudicio laborantibus,
difficultatem subministrare possit.

---

(26) *Editio princeps*: directi.
(27) There is a note in I here to §17 [=I, 18].
(28) There is another note in I to §18 [=I, 19].
(29) This frank admission is unique in Newton's four descriptions of his instrument.
 (1) Originally continued: Auditoribus (in the audience).

If anyone wishes to construct an instrument such as we have just described for undertaking experiments of this kind, then he should use a lens three inches wide or larger, which gathers parallel rays to a focus about two feet away. The prisms will thus be eight feet apart and will make the instrument large enough so that everything may be put to a rather strict test. As to the position of the lens, if the prisms' vertex angles, *ACB* and *DGE,* are equal, suppose 60° or 70°, the lens will be equally distant from each; but if one angle is greater than the other, place the lens nearer to that prism whose vertex angle is greater. Also note that the beam *XY* is spread through a wider space the nearer the lens is placed to the first prism *ABC.* Therefore, if you ever require a wide beam, you need only to arrange for the lens to be somewhat nearer to the first prism than to the second and to use a second prism whose vertex angle is smaller than the vertex angle of the first one in about the same proportion. Finally, if you want the colors falling on that lens to be more parted[26] and separated from one another than occurs in the way just now described, namely, so that the individual kinds of rays may be intercepted at will more distinctly or separately (something I consider necessary in several experiments), nothing other needs to be done than to pass the light through the two small holes *F* and ϕ, far apart from one another, before it falls on the prisms;[27] or else, not far from the first prism, to place another lens[28] that is suitable for gathering the light diverging from the farther hole *F* at the other, subsequent lens *MN.* ‖ I find, however, that this instrument is very troublesome to set up properly, and also that its effects are not so distinct and evident to the senses as in the preceding experiments, where they were displayed with fewer refractions and a greater distance between the glasses.[29] On that account my audience will especially put those simpler and easier experiments to a more deliberate test.

[80 ≈ 96. Some notes on the construction of that instrument.]

SECTION 2

VARIOUS PHENOMENA OF COLORS

I. THE PHENOMENA OF LIGHT TRANSMITTED THROUGH A PRISM TO A WALL

Thus far I have erected the foundations whereby the phenomena of colors produced in any way can be explained, but now I will describe individually the particular and immediate causes of the effects that I have not previously treated, not for the sake of geometers (to whom, I know, it will appear unnecessary) but for others. In fact, I prefer here to add some things that to many[1] will perhaps appear superfluous, rather than to omit anything of some importance that could cause difficulty to the careless and those burdened by prejudice.

Lecture 10

[81 → 77. The common phenomenon of colors is explained.]

[82 ≈ 82. De varijs
phaenomeni
circumstantijs; Luce
juxta basem
prismatis terminatâ]

/ Et imprimis circa Prismatis vulgò notos effectus (quorum causam abunde ₂55/
satis retexi) circumstantiae nonnullae supersunt explicandae: utpote cur
primitivi colores non omnes eliciuntur cum lux (cujus radios ab origine heterogeneos Prisma per inaequales refractiones dispergit) non transit per angustum foramen sicut passim in praecedentibus supposui, sed ex unicâ tantum parte limitatur. ‖ Verbi gratia si corpus aliquod opacum *FG* (fig II, 29)
soli interponatur et Prismati juxta basem
ejus *AB,* quod umbram projiciat in *MP,*
colores efficiat in spatio *PT* et lucem per-
mittat in ipsum *NT* influere: In *PT* confinio lucis et umbrae nulli colores generabuntur praeter purpureum et caeruleum
cum varijs eorum gradibus. Et ratio est
quod ex radijs omnium formarum qui
transeunt per extremitatem dicti corporis
opaci *FG,* soli purpuriformes propter

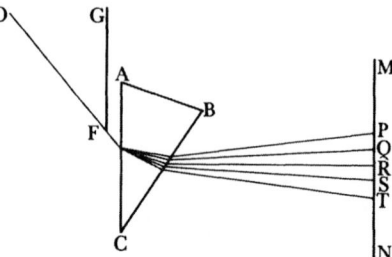

Figure II, 29

maximam eorum refractionem possunt ad *P* usque deflecti; unde color purpureus ibi conspicietur. Deinde caeruliformes cum paulò minùs refrangibiles
existant, incident in totum spatium *NQ,* non potentes ulteriùs versus *M*
deflecti quàm ad *Q.* Atque ita duae radiorum species eaeque solae incident in
Q et colorem ex purpureo et caeruleo compositum exhibebunt. Praeterea
viridiformes minùs adhuc refrangibiles, in spatio *NR* non ultra extendentur
quàm ad *R;* flaviformes autem terminabuntur in *S.* Quare tres tantùm species
colorum miscebuntur ad *R,* et color ex ijs omnibus (nempe ex purpureo
caeruleo et viridi) generabitur. At cùm purpureus et viridis commixti producant caeruleum, ut facilè est ex antedictis[2] experiri; liquet colorem ad *R* non
fore alium quàm caeruleum. Denique cùm radij rubriformes minime omnium
refringantur, ut in spatium *NT* incidentes non magis deflectantur versus *M*
quàm ad *T,* liquet quòd in dicto spatio / *NT* fiet mistura colorum omnium, et ₂56/
proinde albescet; sed in ipso *S* (ubi color omnis dempto rubeo miscetur)
caeruleus ad viriditatem nonnihil vergens apparebit, sed maximè dilutus
propterea quod solus rubor ex albedinis compositione desit.

[83 = 83. Vel juxta
verticem ejus]

Porrò si corpus opacum *φγ* soli interponatur et Prismati juxta verticem
ejus *C,* sicut videre est in schemate II, 30:
Inter obscuratum spatium *NT* et lucidum
PM cernes alios duos colores, rubeum in
T et flavum in *R;* idque propter jam dictas rationes. Quippe radij prout apti sunt
ad hos ordine colores (rubeum, flavum,
viridem, caeruleum et violaceum) generandos, intenduntur[3] per spatia *MT, MS,*
MR, MQ et *MP,* cùm soli rubriformes
extendantur usque ad *T,* caeteris propter
majorem refractionem citiùs terminatis;
necesse est ut iste color in *T* sit rubeus.

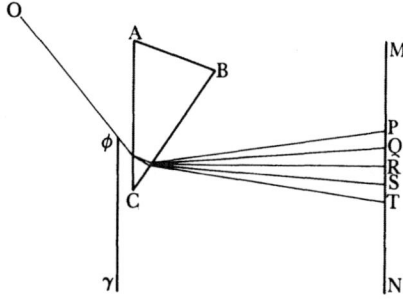

Figure II, 30

In the first place, some features remain to be explained concerning the commonly observed prismatic effects (the cause of which I have sufficiently amply revealed): namely, why not all primitive colors are elicited when light (whose originally heterogeneous rays the prism disperses by unequal refractions) passes not through a narrow hole, as I supposed at different places in the preceding, but is bounded on only one side. ‖ For instance (Fig. II, 29), if some opaque body *FG* is placed between the sun and the prism near its base *AB* and casts a shadow at *MP*, then it produces colors in the space *PT* and lets light flow into *NT*. In *PT*, the boundary of light and shadow, no colors will be generated besides purple and blue together with their various gradations. The reason is that of the rays of every nature that pass by the end of that opaque body *FG*, the purple-making ones alone, because of their greatest refraction, can be deflected as far as *P*, whence the color purple will be seen there. Next, the blue-making rays, since they are somewhat less refrangible and are unable to be deflected farther toward *M* than to *Q*, will fall in the whole space *NQ*. Thus two sorts of rays, and only those, will fall at *Q*, and they will exhibit a color compounded of purple and blue. Moreover, the still less refrangible green-making rays will extend into the space *NR* no farther than *R*, while the yellow-making ones will terminate at *S*. Consequently, only three sorts of colors will be mixed at *R*, and a color will be generated from all of them, that is, from purple, blue, and green. Since, however, purple and green mixed together produce blue, as is easily proved from what was said earlier,[2] it is evident that the color at *R* will be none other than blue. Finally, since the red-making rays are refracted least of all, so that falling in the space *NT* they are deflected no farther toward *M* than to *T*, it is clear that in that space *NT* a mixture of all colors will occur, and consequently it will become white. But at *S* (where every color except red is mixed) blue inclining somewhat to green will appear, but very diluted, since red alone is wanting from the composition of white.

Moreover, if an opaque body *ϕγ* is placed between the sun and the prism near its vertex *C*, as can be seen in Fig. II, 30, between the darkened space *NT* and the bright one *PM* you will see two other colors, red at *T* and yellow at *R*, for the reasons already related. Specifically, according as the rays are disposed to generate these colors in order (red, yellow, green, blue, and violet), they extend[3] through the spaces *MT*, *MS*, *MR*, *MQ*, and *MP*. Since only red-making rays extend as far as *T* (the others end sooner because of their greater refraction), it necessarily follows that that color at *T* is red.

‖[82 ≈ 82. On various features of the phenomenon: When the light is terminated near the prism's base.]

[83 = 83. Or near its vertex.]

---

(2) There is a marginal note in I here to Lects. 4, 5, and to an unnumbered lecture.
(3) I: *extenduntur*.

Item cùm tria radiorum genera in R incidant, color ex istis (nempe rubeo, flavo, et viridi) compositus ibidem cernetur: rubeus autem et viridis flavum constituunt,[4] atque adeò flavus apparebit in R. Praeterea cum omnium formarum radij misceantur in P, et postea perpetuò versus M; spatium istud PM apparebit album. Nec secus constat quòd citrius in S, et in Q flavus ad viriditatem vergens apparebit, sed adeò dilutus tamen, et caeruleo redundans ut nomen viriditatis non mereatur.

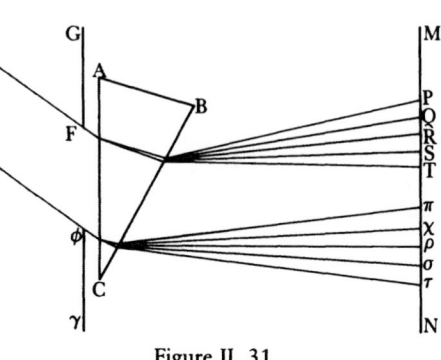

Figure II, 31

[84 ≈ 84. Vel utrinque (Ubi viriditatis productio bella describitur.)] Tertiò si opaca duo corpora GF et γφ (fig II, 31) Soli et Prismati interponantur ut radij inter utrunque quasi per oblongam rimam prismati parallelam transeant; atque distantia Fφ sit satis magna; pro utroque termino F et φ generabuntur colores, purpureus nempe ad P et caeruleus ad R per terminum [F]: atque flavus ad ρ et rubeus ad τ per terminum φ, sicut modo explicatum fuit: Eritque Tπ spatium album utrisque coloribus interjectum. Jam si obstacula GF et γφ ad se invicem paululùm admoveantur, ut intermedium spatium Fφ / evadat angustius, isto pacto spatium album ₂57/ quoque Tπ fiet angustius, donec tandem evanescat et colores utrinque coeant. Sin spatium Fφ magis adhuc coarctetur, viriditas in medio colorum emerget vice albedinis quae jam evanuit. Quae quidem viriditas antea non apparuit propter commisturam radiorum heterogeneorum, quibus involuta latuit: jam vero heterogeneis istis per obstacula duo sibi propius admota alternè interceptis; ea[5] paulatim detegitur, patet, et evadit perfectior; donec (cum dictum Fφ satis angustum est) ab omni ferè mistura liberatur, et eruitur, propriaque specie non minùs quàm caeteri colores elucet. Et hinc in transitu colligitur, quòd viriditas inter colores medietatem exactè obtinet, non magis ad rubeum vergens quàm violaceum, neque ad flavum quàm caeruleum; ‖ hoc est in specie coloris et respectu multitudinis radiorum ad colores utrinque pertinentium. Nam in gradu refrangibilitatis minus differt a parte rubeâ flavâque, et in aliâ quadam proprietate (cui jam explicandae non est locus)[6] minùs differt a parte purpureâ et caeruleâ.

Praeterea cum albedo[7] Tπ per angustiam pervij[8] spatij Fφ incipit evanescere, colores etiam paulatim contractiores apparebunt,[9] ita ut cùm istud[10] Fφ fit valde angustum, flavus ad rubeum, et caeruleus ad violaceum quasi duplo vicinior evadat quàm cùm amplitudo ejus[11] permisit albedinem in medio colorum produci.[12] Et ut quinque colores (viriditate jam internatâ)

---

(4) In I there is another marginal note to an unnumbered lecture.

(5) *Editio princeps*: viriditas ea.

(6) Newton is alluding to his discovery that the mean thickness of the air film that produces the colors of any order of "Newton's rings" occurs for the color between the green and yellow; see his "Of the colours of plated transparent substances" (Observation 15, Add. 3970, f. 521ʳ) from the spring of 1672; this was incorporated into Observation XIV of both the "Discourse of

Likewise, since three kinds of rays fall upon *R*, the color compounded from those (that is, from red, yellow, and green) will be seen at the same place; but red and green make yellow,[4] so that yellow will appear at *R*. Moreover, since rays of every nature are mixed at *P* and thereafter continuously toward *M*, that space *PM* will appear white. No differently it is clear that orange will appear at *S* and yellow inclining to green at *Q*, but yet so diluted and overflowing with blue that it does not deserve the name green.

Third (Fig. II, 31), if two opaque bodies *GF* and *γφ* are placed between the sun and the prism so that the rays pass between both, as it were, through an oblong slit parallel to the prism, and if the distance *Fφ* is sufficiently large, then colors will be generated by each of the edges *F* and *φ*; namely, purple at *P* and blue at *R* by the edge [*F*], and yellow at *ρ* and red at *τ* by the edge *φ*, as was just explained; and there will be a white space *Tπ* lying between both colors. If now the obstacles *GF* and *γφ* are moved slightly toward one another so that the intermediate space *Fφ* becomes narrower, in this way the white space *Tπ* will also become narrower, until it finally vanishes, and the colors on each side coalesce. But if the space *Fφ* is contracted still more, green will emerge in the middle of the colors in place of the white that just vanished. This green in fact did not appear earlier, because of the mixture of heterogeneous rays in which it was enveloped and hidden. But now when these heterogeneous rays are intercepted one after another by the two obstacles brought nearer to each other, the green[5] gradually is uncovered, stands visible, and becomes more perfect, until (when the specified *Fφ* is sufficiently narrow) it is nearly freed and extracted from the whole mixture and shines with its own species no less than the other colors. It is gathered in passing from this that green occupies exactly the middle among the colors, inclining no more to red than to violet, nor to yellow than to blue; ‖ that is, in species of color and with respect to the multitude of rays belonging to the colors on each side. For it differs less in degree of refrangibility from the red and yellow side, and in a certain other property (which it is not now opportune to explain)[6] it differs less from the purple and blue side.

Furthermore, when the white[7] *Tπ* begins to vanish because of the narrowness of the open[8] space *Fφ*, the colors will also gradually appear more contracted.[9] Thus when that[10] *Fφ* becomes very narrow, the yellow turns out nearly two times closer to the red, as does the blue to the violet, than when the size of it[11] allowed white to be produced in the middle of the colors.[12] Consequently, five colors (green now springing up in between) will

[84 ≈ 84. Or on both sides (where the beautiful production of green is described).]

Observations" from 1675, and the *Opticks*, Bk. II, Pt. I. In §II, 100, added in the *Optica*, he uses his musical division to determine the relative degree of refrangibility of the different colors, with the mean degree of refrangibility occurring at the boundary of green and blue.

(7) Praeterea cum albedo] I: Praeterea observandum est quòd cùm praefata albedo (Furthermore, it ought to be observed that when that white).   (8) Added.

(9) colores . . . apparebunt,] I: intermedij colores paulatim fiunt viciniores, flavus videlicet ad rubeum et caeruleus ad violaceum. (the intermediate colors gradually become closer, namely, the yellow to the red, and the blue to the violet.)

(10) I: spatium (the space).   (11) I: dicti *Fφ* (of the specified *Fφ*).

(12) in . . . produci.] I: in medio cerni. (to be seen in the middle.)

non occupent plus spatij quam eorum duo priùs occupavêre. Cujus rei ratio ex schematum inspectione patebit animadvertenti quod[13] flavus ad ρ et caeruleus ad *R* ex heterogeneis radijs compositus mutatur in ferè uniformem flavum ad loca *S* et σ incidentem et in fere uniformem caeruleum ad loca *Q* et χ similiter incidentem,[14] heterogeneis radijs[15] e misturâ per angustiam spatij *Fϕ* magnâ ex parte[16] sublatis.

[85 = 85. Vel juxta alterutrum triangularem limitem]   / Quartò si lux terminetur obstaculo *Gγ* cujus extremitas perpendiculariter $_2$58/ transversa est ad longitudinem Prismatis, colores omninò nulli virtute termini illius generabuntur. Etenim ponamus parallelos radios *OF Oϕ* caeterosque (in fig II, 32) juxta extremitatem dictam *Gγ* in prisma *ABC* prolapsos, ibi-

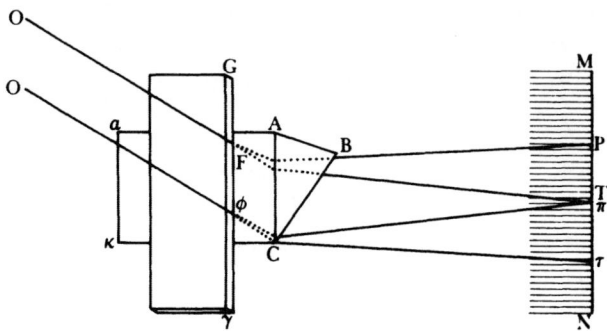

Figure II, 32

demque refractos esse ad *PT* et $\pi\tau$; atque *MN* esse umbram ipsius *Gγ*. Jam licèt purpuriformes *FP* et $\phi\pi$ magis refringantur quàm rubriformes *FT* et $\phi\tau$, tamen istâ refractione secundum terminum umbrae factâ, ita ut ex dictis radijs nulli[17] magis deflectant versus umbram quàm caeteri; palam est quod ubicunque purpuriformes incidunt, rubriformes etiam incident in eundem locum; et e contra. Quod idem de radijs intermedijs pari modo concipiatur. Et sic radijs omnium specierum ubique per extremitatem umbrae commixtis, umbra benè definietur sine aliquo colore (praeter album vel fuscum ex luce et umbra mixtis) conspecto. Sed cavendum est, ne colores per limites prismatis *Aα* vel *Cκ* generati habeantur pro generatis a limite *Gγ*. Quamobrem prismata[18] quae ex vitro in totum fiunt, ad examen hujus et proximè praecedentis commode instituendum sunt nimis exigua, propterea quòd colores per extremitatem verticis et basis producti interjectum spatium album haud relinquent satis amplum in quo generatio colorum praedictis modis probetur. Itaque ut prisma conficiatur ex vitris planis et bene politis, qualia ad specula conficienda[19] adhibentur, moneo; quibus in morem cunei connexis, et in vasculum dein prismiforme completis (ut supra dictum)[20] vasculum istud impleatur aquâ limpidissimâ et occludatur; et sic prismata ad arbitrium ampla conficias.

[86 = 86. Vel undique. (Ubi rursus de praefata viriditate.)]   Quintò ut omnia jam uno comprehendam specimine, sit *Gγ* (fig II, 33) corpus opacum orbiculari foramine *Fϕ* unum duosvè digitos lato pertusum, per quod lux in prisma trajiciatur ubi cum refracta fuerit, projicitur in papy-

occupy no more space than two of them had previously occupied. The reason for this will be clear to anyone who inspects the figure by noting that[13] the yellow at ρ and the blue at R, composed of heterogeneous rays, are changed into a nearly uniform yellow falling at the places S and σ, and into a nearly uniform blue similarly falling at the places Q and χ,[14] with the heterogeneous rays[15] having been for the most part[16] removed from the mixture by the narrowness of the space $F\phi$.

Fourth, if the light is terminated by an obstacle $G\gamma$ whose edge is transverse and perpendicular to the prism's length, no colors at all will be generated by virtue of that edge. For, let us assume (in Fig. II, 32) that the parallel rays OF, $O\phi$, and others flow alongside that edge $G\gamma$ into the prism ABC and are there refracted to PT and $\pi\tau$, and that MN is the shadow of $G\gamma$. Now, although the purple-making rays FP and $\phi\pi$ are refracted more than the red-making ones FT and $\phi\tau$, nevertheless, since that refraction is made along the boundary of the shadow so that none[17] of those rays are more deflected toward the shadow than the others, it is clear that wherever purple-making rays fall, red-making ones will also fall at the same place, and conversely. In the like way, let the same thing be understood for the intermediate rays. Thus with rays of every sort mixed together everywhere along the edge of the shadow, the shadow will be well defined without any visible color (except white or gray from a mixture of light and shadow). But you must be careful that the colors generated by the prism's edges $A\alpha$ and $C\kappa$ are not taken for those generated by the edge $G\gamma$. Consequently, prisms[18] made entirely from glass are too small for suitably carrying out an examination of this and the immediately preceding, because the colors produced by the edge of the vertex and of the base will not leave a sufficiently wide interposed white space in which the generation of colors may be tested in the specified ways. I therefore recommend that a prism be made from flat, well-polished glasses such as those used to make[19] mirrors. After joining them in the shape of a wedge and then finishing it into a small prismatic vessel (as described above),[20] fill that small vessel with very clear water and seal it. In this way you can make prisms as large as you wish.

Fifth, that I may now include everything in one example (Fig. II, 33), let $G\gamma$ be an opaque body perforated with a one- or two-inch wide circular hole $F\phi$ through which light is cast onto a prism, where after being refracted it is

[85 = 85. Or near either triangular end.]

[86 = 86. Or on all sides (and once again about that green).]

---

(13) ex . . . quod] I: patebit ex figuris tribus praecedentibus, contemplanti modum quo (will be clear from the three preceding figures to anyone who considers the way in which).

(14) in ferè uniformem flavum . . . incidentem] I: in flavum ad S vel σ & caeruleum ad Q vel χ constantem ex solis homogeneis (into yellow at S or σ and blue at Q or χ, consisting solely of homogeneous rays).

(15) heterogeneis radijs] I: caeteris (the others).

(16) magnâ ex parte] Added.

(17) Originally transcribed erroneously as "multi," the rendering adopted in the *editio princeps*.

(18) Originally (as I): de prismatibus monendum volo, quod (I wish to advise that prisms).

(19) *Editio princeps*: conspicienda.

(20) Namely, in §II, 31 = 44.

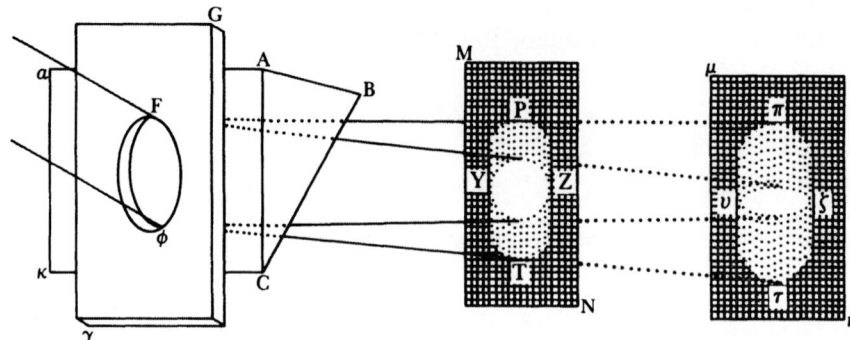

Figure II, 33

rum vel quodvis album corpus *MN* quasi semisse pedis a prismate postpositum et videbis illuminatum spatium *PYTZ* rotundum ad modum foraminis *Fϕ*, album in ejus medietate, et duobus / semilunulis colorum terminatum, ₂59/ purpureo et caeruleo ad *P* flavo autem et rubeo ad *T*: qui colores paulatim deficiunt versus *Y* et *Z* ubi nulli omnino conspiciuntur. Praeterea si papyrum ad majorem distantiam paulatim distuleris, velut ad *μν*; videbis colores distendi et augeri, et intermediam albedinem usque comminui dum prorsus evanescat totumque spatium coloribus rubeo flavo, caeruleo et purpureo tinctum appareat. Et papyrum longius differendo, viriditas e medio emerget et crescet tum amplitudine spatij tum perfectione speciei: Totumque spatium coloratum distrahetur in oblongam formam. Quorum omnium rationes ex supradictis depromantur.

[87 ≈ 87. Vel ad distantiam aliquam a prismate.]   Adhaec si lux obstaculo ad quamvis distantiam post prisma collocato terminetur consimilis erit colorum generatio. Sit v.g. in (fig II, 34) obstaculum *Gγ* perforatum in *F* et ad distantiam pedis unius aut amplius post prisma *ABC* collocatum Prisma autem satis amplum adhibeatur, (quale ex laminis vitreis ut supra possis efficere) ‖ ne lux omnis prius abeat in colores quam attingat[21] foramen *F*: et lux illa[22] postquam transijt per *F* non secus convertetur[23] in colores apud *P, Q, R, S, T,* quàm contigit[24] in praecedentibus. Scilicet inspicienti schema[25] patebit quomodò radij diversorum generum inaequaliter refracti convergunt a diversis partibus prismatis ad istud[26] *F,* ubi (ut et hinc inde versus *G* et *γ*) componunt albedinem; sed inibi decussantes divergunt postea, diversique colores in diversa spatia *P, Q, R, S, T,* tendunt. Et hinc[27] cum radij repagulo[28] quolibet *H* ex utravis parte prismatis intercipiuntur, e coloribus *P, Q, R, S, T* aliqui tollentur. Si radios nempe vertici *C* vicinos intercipias, tolles purpureum *P*; vel tolles rubeum *T*, si intercipias eos

---

(21) lux . . . attingat] I: lux priùs discernatur in colores quàm permeet (the light be separated into colors before it goes through).

(22) lux illa] I: ista lux alba (that white light).

(23) I: degenerabit (it will degenerate).

(24) I: factum erat.

(25) inspicienti schema] I: ex solâ schematis contemplatione (solely from studying the figure).

(26) I: foramen (the hole).

projected onto a paper or any white body *MN* placed about half a foot
behind the prism. You will see the illuminated space *PYTZ* to be round like
the hole *Fϕ*, white in its middle, and bounded by two crescents of colors,
with purple and blue at *P*, but yellow and red at *T*; and these colors gradually decay toward *Y* and *Z*, where none at all are visible. Moreover, if you
gradually move the paper away to a greater distance, for example, to *μν*, you
will see the colors expand and increase and the intermediate white diminish,
until it vanishes completely, and the entire space appears colored with the
colors, red, yellow, blue, and purple. By moving the paper farther away,
green will emerge from the middle and increase in both the size of its space
and the perfection of its species, and the entire colored space will be stretched
out into an oblong shape. The reasons for all this may be derived from what
has already been described.

Furthermore, if the light is terminated by an obstacle placed at any distance ‖ [87 ≈ 87. Or at any
behind the prism, the generation of colors will be very similar. For example ‖ distance from the
(Fig. II, 34), let there be an obstacle *Gγ* perforated at *F* and placed at a ‖ prism.]

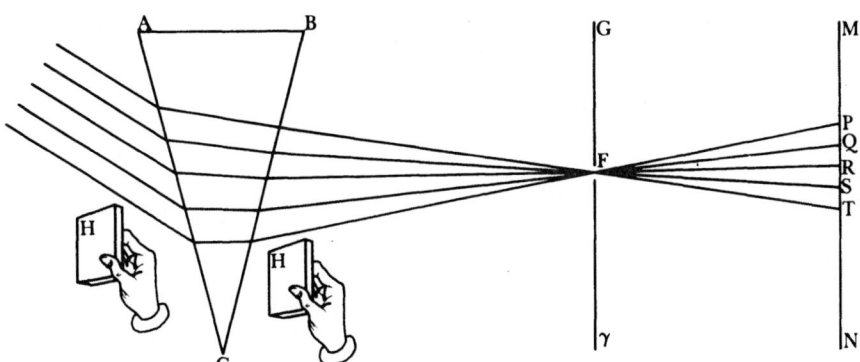

Figure II, 34

distance of one or more feet behind the prism *ABC*. However, use a suffi-‖
ciently large prism (such as you can make from glass plates, as above) ‖ lest
all the light pass into colors before it arrives at[21] the hole *F*; and after that
light[22] passes through *F*, it will be converted[23] into colors at *P*, *Q*, *R*, *S*, and
T just as occurs[24] in the preceding. Namely, to anyone examining the
figure[25] it will be clear how the unequally refracted rays of different kinds
converge from different parts of the prism to that[26] *F*, where (as well as on
both sides toward *G* and *γ*) they compound white; but upon crossing there,
they afterward diverge, and the different colors tend to different spaces *P*, *Q*,
R, *S*, and *T*. Hence[27] when the rays are intercepted by some barrier[28] *H* on
either side of the prism, some of the colors *P*, *Q*, *R*, *S*, and *T* will be removed.
Specifically, if you intercept the rays near the vertex *C*, you will remove the
purple *P*; or you will remove the red *T* if you intercept those rays near the

(27) Et hinc] **I:** Atque haec fortè clariora fient experienti quòd (These things will perhaps
become clearer to someone doing the experiment, because).   (28) **I:** obstaculo (obstacle).

basi *AB* vicinos. Et sic de alijs.[29] Ita ut quoslibet pro arbitrio possis tollere vel efficere[30] ut quilibet solus appareat.

[88 = 88. Vel alio quovis modo.]

Denique si lux ex unicâ tantum parte pone prisma limitetur, vel si duo statuantur limites, ijque vel ad easdem vel ad oppositas partes prismatis; vel quocunque alio more lux terminetur; modus quo colores exinde generantur ex antedictis / facile patebit; ut jacturam temporis fecero de hâc re plura ₂60/ verba facturus. Quinetiam si duo vel plura prismata quocunque modo inter se disponantur peritus optices facile explorabit causam.

[89 ≈ 92, 93. Adversus philosophorum Hypotheses notae.]

Caeterum de modo tollendi quoslibet colores in fig II, 34 per interpositionem obstaculi *H* hic obitèr notandum venit quantum[31] ista circumstantia adversatur Hypothesibus Philosophorum quae de coloribus huc usque fuerunt excogitatae. Ex illis enim positis refracta lux ad eas semper partes cum caeruleo et violaceo terminanda est versus quas fit refractio; quandoquidem gyrationes globulorum ex opinione Cartesij, vel partes anteriores pulsuum aetheris obliquè vibrantis ex hypothesi Mᵣⁱ Hookij[32] per viciniam quiescentis Medij ad eas semper partes impediuntur et hebescunt. Et tamen hic videre est, quod admoto obstaculo *H* ut radios vertici prismatis vicinos intercidat[33] possis violaceum et caeruleum tollere, et efficere ut viridis vel flavus, aut etiam ruber ad eas partes maneant extimus versus quas refractio peragitur. Nec Hypothesis eorum tutior est qui supponunt colores ex lucis et umbrae misturâ generari; nam eadem videtur esse in eorum confinio mistura sive aliqui e radijs ante refractionem limite *H* intercidantur, sive omnes per Prisma liberè transeant.

Hujusmodi etiam Hypotheses ex alijs experimentis sparsim occurrentibus everti possent, modo id instituto[34] meo necessarium ducerem:[35] Quemadmodum ex illis ubi lucem partim reflecti posse et partim transmitti docebam;[36] nam lux transmissa dabat flavum, rubeumvè; idque in meditullio ejus ubi a nullo[37] quiescente medio vel tenebris terminabatur. ‖ Sic etiam maximè valet quod ostendi colorem lucis ex uniformibus radijs constantis non posse per quaslibet refractiones mutari: ‖ Caeterùm non opus est ut Hypotheses[38] refutem quae ex inventâ tandem veritate suâ sponte corruent.

[90 ≈ 89. E praefatis modus deducitur albedinem e coloribus componendi]

/ Phaenomenis jam ante explicatis affinia sunt sequentia quae circa com- ₂61/ positionem albedinis versantur: Prismata duo *ABC* et *αβκ* (fig II, 35) quorum anguli verticales *ACB* et *ακβ* aequentur, ita parallelis axibus dispone ‖ ut alterius linea verticalis *C* cum *β* extremitate basis alterius conveniat, planis *BC* et *βκ* in directum jacentibus. Quo facto si sol transluceat ea in papyrum *MN*, octo vel duodecim digitos postpositam, colores quidem generabuntur ad

---

(29) I: reliquis (the rest).   (30) I: facere.

(31) Caeterum . . . quantum] I: Alterum quod notandum venit, est de modo tollendi quoslibet colores in fig 36 per interpositionem corporis *H*, quantùm nempe (There is another thing that can be noted about the way (in Fig. 36 [= Fig. II, 34]) of removing any colors by inserting a body *H*, namely, how much).

(32) Mᵣⁱ Hookij] I: aliâ quadam (a certain other).

(33) Et . . . intercidat] I: Attamen ostensum est ad fig 36 quòd, obstaculo *H* ex utrâvis parte prismatis interjecto ut radios ipsius vertici *C* vicinos intercipiat (Nevertheless, it has been shown at Fig. 36 [= Fig. II, 34] that when the obstacle *H* is inserted on either side of the prism so that it intercepts the rays near the prism's vertex *C*).

base *AB*; and similarly for the others.[29] So that you can, as you wish, remove any or cause[30] any one alone to appear.

    Finally, if the light is bounded on only one side behind the prism, or if two boundaries are erected either on the same or on opposite sides of the prism, or the light is terminated in any other way whatsoever, the way in which colors are consequently generated will readily be evident from the preceding, so that I would waste time in discussing this further. Moreover, if two or more prisms are arranged among themselves in any way whatsoever, the expert optician will easily investigate the cause.

    About the way (in Fig. II, 34) of removing any colors by inserting an obstacle *H*, it can, however, be noted here in passing how much[31] that circumstance is opposed to the hypotheses of the philosophers that have hitherto been devised about colors. For from those assumptions refracted light must always be terminated with blue and violet on those sides toward which the refraction occurs, since the rotations of the little balls (in Descartes's opinion) or the foremost parts of the pulses of the obliquely vibrating aether (in Mr. Hooke's[32] hypothesis) are always impeded and weakened by the neighboring quiescent medium on those sides. Nevertheless, it is possible to see here that when the obstacle *H* is applied so that it cuts off the rays near the prism's vertex,[33] you can remove the violet and blue and make green or yellow or even red remain outermost on those sides toward which the refraction is executed. ‖ Nor is the hypothesis of those who suppose colors to be generated from a mixture of light and shadow more secure, for the mixture appears to be the same on their boundary, whether some of the rays are cut off by the edge *H* before refraction, or all of them pass freely through the prism.

    Hypotheses of this kind could also be overthrown from other experiments encountered here and there, if only I considered[35] it necessary for my purpose:[34] for instance, from those where I taught that light could be partly reflected and partly transmitted,[36] for the transmitted light displayed yellow or red, especially in its middle, where[37] it was not bounded by any quiescent medium or darkness. ‖ Likewise it is also particularly impressive that I have shown that the color of light consisting of uniform rays cannot be changed by any refractions. ‖ But it is unnecessary for me to refute hypotheses[38] that will spontaneously collapse from the truth having at last been discovered.

    The following phenomena, involving the composition of whiteness, are related to those just previously explained. Place (Fig. II, 35) the two prisms *ABC* and $\alpha\beta\kappa$, whose vertex angles *ACB* and $\alpha\kappa\beta$ are equal, with their axes parallel, ‖ so that the vertex line *C* of the former meets the edge of the base β of the latter with their planes *BC* and $\beta\kappa$ lying in a straight line. After this is done, if the sun shines through them onto the paper *MN*, placed eight or twelve inches behind, colors will indeed be generated at *M* and *N* by the

Marginal notes:

[88 = 88. Or in any other way.]

[89 ≈ 92, 93. Notes opposed to the hypotheses of the philosophers.]

[90 ≈ 89. From the preceding a way to compound whiteness from colors is deduced.]

---

    (34) I: proposito.   (35) I: judicarem (I judged).

    (36) There is a marginal note here in I to §§71, 73 [= II, 62, 64].

    (37) idque . . . nullo] I: nec tamen ab ullo (and yet . . .).

    (38) I: Hypotheses ejusmodi (hypotheses of this sort). This final sentence began §93.

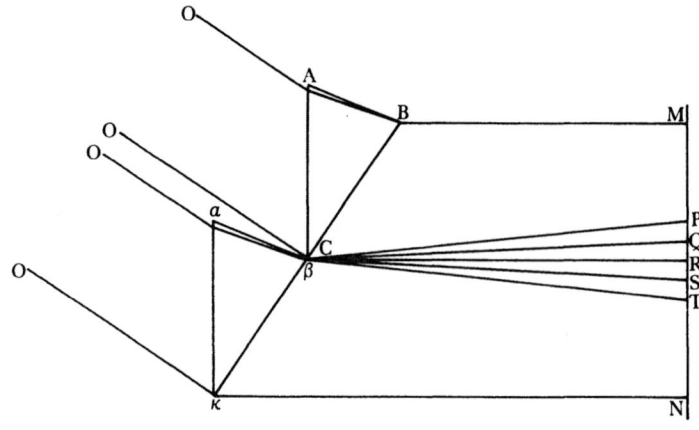

Figure II, 35

M et *N* per exteriores prismatum terminos *B* et *κ*, non autem per interiores *C* et *β* sed medium spatium *PT* totum apparebit album. Sin alterutrum prisma tollas alterius extremitas *C* vel *β* generabit colores ad *PT,* ac dein si restituas, albedo etiam restituetur. Scilicet albedo ista componitur e coloribus ab extremitate *C* et *β* prismatis utriusque prolapsis. Id quod facile constet ex praefatis. Nam radij purpuriformes ab utroque prismate refracti limitantur in eodem puncto *P*; ita ut ab uno prismate manantes incidant in *PM*, ab altero in *PN*, et ab utroque simul in totum *MN*, non secus quam si omnes ab unico prismate venissent. Eodem modo caeruliformes extenduntur per totum spatium *MN*: et eorum terminus communis est *Q* prout manant a diversis prismatibus. Et sic de caeteris. Quare omnigeni radij commiscentur in unaquâque parte spatij *PT,* et albedinem ideò component. Sin alterutrum prisma tollas, puta *ABC*, vel lucem ei potiùs occludas, tum radijs rubriformibus ab *MT*, flaviformibus a[b] *MS*, viridiformibus ab *MR*, caeruliformibus ab *MQ*, et purpuriformibus ab *MP* sublatis, manebunt rubriformes in *NT*, flaviformes in *NS*, viridiformes in *NR*, caeruliformes in *NQ*, et purpuriformes in *NP*. Adeoq́ue purpureus apparebit in *P*, et caeruleus in *R* ut ostendimus ante.[39] Et simili ratione si lux occludatur alteri prismati *αβκ* ne permeet, rubor apparebit in *T* et flavedo in *R*.

[91 = 90. Utrum anguli prismatum sint aequales cognoscere.] / In istis autem experiendis requiritur ut anguli *ACB* et *ακβ* sint aequales. ₂62/ Id quod tentabis si prismata secundum longitudinem eorum ita connectas ut duo ex planis dictos angulos comprehendentibus puta *BC* et *βκ* (fig II, 36) fiant contigua et reliqua duo *AC* et *ακ* sibi opposita. Quo facto si radij solis ingressi foramen *F* pergunt ad eundem locum *S* cum trajiciuntur per dicta prismata perpendiculariter ad eorum latera *AC* et *ακ*, atque cum liberè progrediuntur, nullo interjecto obstaculo; tum plana *AC* et *ακ* sunt parallela, et anguli *ACB* et *ακβ* aequales. Sin istud non eveniat, sunt inaequales: in quo casu notetur praeterea quod inclinando plana *BC* et *βκ* (in fig II, 35) vel ab invicem reclinando, possis albedinem in *PT* haud secus componere ac si dicti anguli fuissent aequales et plana *BC* et *βκ* in directum jacentia.

prisms' outer edges, *B* and *κ*, but not by their inner ones, *C* and *β*, while the entire middle space *PT* will appear white. But if you remove either prism, the other's edge, *C* or *β*, will generate colors at *PT*; and then if you restore it, the white will also be restored. That white is of course compounded from the colors proceeding from the edges *C* and *β* of each prism. This may be readily ascertained from the preceding: For the purple-making rays refracted by each prism terminate at the same point *P*, so that those flowing from the one prism fall on *PM*, those from the other on *PN*, and those from both together on all of *MN*, just as if they had all come from a single prism. In the same way the blue-making rays extend through the entire space *MN* and their common terminus is *Q*, according as they flow from the different prisms, and similarly for the others. Consequently, rays of every kind are mixed together in every single part of the space *PT*, and they will thus compound white. But if you remove either prism, say *ABC*, or rather if you block out its light, then after the red-making rays have been removed from *MT*, the yellow-making from *MS*, the green-making from *MR*, the blue-making from *MQ*, and the purple-making from *MP*, the red-making ones will remain in *NT*, the yellow-making in *NS*, the green-making in *NR*, the blue-making in *NQ*, and the purple-making in *NP*. Hence, purple will appear at *P* and blue at *R*, as we showed before.[39] By similar reasoning, if the light is blocked out at the other prism *αβκ* so that it does not pass through, red will appear at *T* and yellow at *R*.

[91 = 90. To know whether the prisms' angles are equal.]

In doing these experiments, however, it is required that angles *ACB* and *ακβ* be equal. You will test this if you join the prisms along their lengths so that two of the planes, suppose *BC* and *βκ* (Fig. II, 36), that contain those angles become contiguous, and the other two, *AC* and *ακ*, become opposite to one another. After this has been done, if the sun's rays entering the hole *F* continue to the same place *S* when they are transmitted through those prisms perpendicularly to their sides *AC* and *ακ*, as when they proceed freely with no interposed obstacle, then the planes *AC* and *ακ* are parallel, and the angles *ACB* and *ακβ* are equal. If that does not happen, they are unequal. In this case it is to be observed, moreover, that by tilting the planes *BC* and *βκ* (Fig. II, 35) toward or away from one another, you may compound whiteness at *PT* just as if the specified angles were equal, and the planes *BC* and *βκ* lay in a straight line.

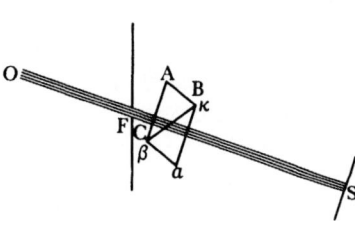

Figure II, 36

---

(39) In I there is a marginal note here to Figs. 31, 32 [= Figs. II, 29, 30].

[92 = 91. Alius
modus commiscendi
colores in
albedinem, priori
affinis.]
Figure II, 37

Quinetiam possis hoc idem cum unico tantum prismate perficere, dummodo satis magnum sit, puta cujus refringentia latera *AC* et *BC* sint sex vel octo digitos lata. Etenim sint *FG* et *ϕγ* duo corpora opaca, plana, rectangula, et ad prismatis planum *ACκα* secundum planitiem ejus sic applicata ut eorum

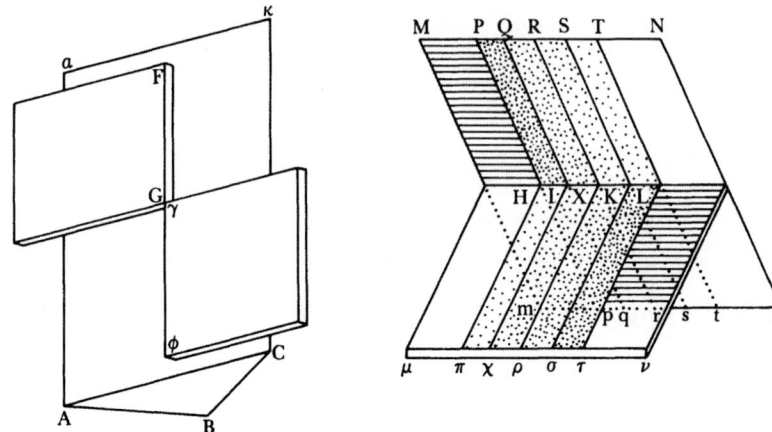

Figure II, 37

angularia puncta *G* et *γ* juxta plani istius centrum se mutuò contingant, et latera concurrentia (quorum *FG* et *ϕγ* sint ad axem Prismatis parallela) ex adverso jaceant in directum. Quo facto si lux refracta projiciatur in papyrum *MNX* pedes quasi duos distantem; obstaculum *FG* projiciet umbram in *MH*, purpuram efficiet in *PHIQ*, ac caeruleum colorem in *QILT* et permittet lucem in *LN*. [E contra verò obstaculum *γϕ* permittet lucem in *Hm*, rubedinem efficiet in *pHIq*, ac flavedinem in *qILt*, et projiciet umbram in *Ln*.][40] Dico jam si speculo aliquo *μνX* colores ex alterutrâ parte lineae *HL*, velut *HLpt*, ita reflectantur ut incidant in papyrum ad eundem locum cum coloribus *HLPT* ex alterâ parte; color omnis evanescet, totumque *HLPT* apparebit album. Nam purpuriformes radij a prismate ad *PHIQ* directè tendunt, / et ₂63/ caetera quatuor radiorum genera ad eundem locum reflectuntur a speculo, incidentes puta in *HIχπ*: Item purpuriformes et caeruliformes directe tendunt ad *QIXR* et caetera tria genera illuc reflectuntur ab *IXρχ*; et sic de reliquis. Adeo ut omnes omnium generum radij passim per spatium *PHLT* misceantur, ibidemque componant albedinem. Sed notandum est quòd cum lux reflectione semper debilitetur, radijs quamplurimis inter reflectendum amissis; exinde forsan eveniat quod lux directa nonnihil praevalebit reflexae, et color ejus dominabitur, nisi compensatio fiat ita papyrum inclinando ut directa lux paulò obliquiùs in eam incidat quàm reflexa de quâ re facilè judicium feras, ex perfectione albedinis emergentis.

Lect 11
[93]

Antequam ad aliud experimentorum genus transeo, e re[1] erit, ut formam imaginis coloratae, quam lux per arctum orbiculare foramen in tenebrosum cubiculum influens et per prisma deinde transmissa effingit, paulò magis articulatim inspiciamus, et singulorum ejus colorum dimensiones ac distantias

Moreover, you may also accomplish this with only a single prism, provided that it is sufficiently large, for example one whose refracting sides *AC* and *BC* are six or eight inches wide. For, let *FG* and *φγ* be two plane, rectangular, opaque bodies attached to the prism's plane *ACκα* along its surface so that their angular points *G* and *γ* touch each other near the center of that plane and the concurrent sides (of which *FG* and *φγ* are parallel to the prism's axis) lie opposite one another in a straight line. After this is done, if the refracted light is projected onto the paper *MNX* about two feet away, the obstacle *FG* will cast a shadow on *MH* and will make purple at *PHIQ* and a blue color at *QILT*, while it will let light pass onto *LN*. [Conversely, the obstacle *γφ* will let light pass onto *Hm* and will make red at *pHIq* and yellow at *qILt*, while it will cast a shadow on *Ln*.][40] I now say that if with some mirror *μνX* the colors from either side of the line *HL*, such as *HLpt*, are so reflected that they fall on the paper at the same place as the colors *HLPT* from the other side, then all color will vanish, and the whole of *HLPT* will appear white. For the purple-making rays travel directly from the prism to *PHIQ*, and the other four kinds of rays are reflected by the mirror to the same place, falling, for instance, on *HIχπ*. Likewise, the purple- and blue-making rays travel directly to *QIXR*, and the other three kinds are reflected there from *IXρχ*, and similarly for the others. Hence all rays of every kind are mixed everywhere throughout the space *PHLT* and compose whiteness there. It must be noted, however, that since light is always weakened by reflection, with a great number of rays lost in the course of reflection, it may perhaps then turn out that the direct light will be somewhat stronger than the reflected light and its color will dominate, unless one compensates by inclining the paper so that the direct light falls on it slightly more obliquely than the reflected light. This you may easily judge from the perfection of the emergent whiteness.

Before I pass over to another kind of experiment, it will be worthwhile[1] for us to investigate in somewhat more detail the shape of the colored image formed by light flowing through a narrow, round hole into a dark room and then passing through a prism, and to examine diligently the dimensions of

[92 = 91. Another way, related to the previous one, of mixing colors into whiteness.]

Fig. II, 37

Lecture 11

[93]

---

(40) This sentence from I was doubtless inadvertently omitted in transcription.

(1) e re] *Editio princeps*: necessarium.

ab invicem, nec non refrangibilitatis gradus singulis radiorum generibus competentes sedulò rimemur.

[94]   Ostendebatur sub initio[2] quod ubi Prisma (cujus angulus verticalis erat quasi 63^{grad}) imaginem ad distantiam 22 pedum projiciebat, longitudo ejus erat 13$\frac{1}{4}^{dig}$ et latitudo 2$\frac{5}{8}^{dig}$. Adeoque centra extimorum circulorum ex quibus in longum dispositis imago illa constitit, distabant 10$\frac{5}{8}^{digit}$. Jam ad hanc distantiam sive distractam longitudinem imaginis,[3] caeteras ejus dimensiones referre convenit propterea quod ad absolutam ejus longitudinem (quae a magnitudine comp[o]nentium circulorum dependet) non habent certam relationem. Quò autem dimensiones ejus majori ἀκριβείᾳ investigarem loca ubi colores in suo genere perfectissimi eorumque confinia in transversam papyrum / incidebant ₂64/ calamo scriptorio notabam, et observationibus hujusmodi saepiùs repetitis, et inter se collatis[4] has tandem conclusiones sigillatim perdidici.

1. Caeruleus et violaceus ex unâ parte et viridis, flavus[5] ac rubeus ex altera imaginem bipartiebantur; adeo ut viridis et caerulei confinium (quod Thalassinum appellare possim) meditullium ejus occuparet.

2. Locus ubi porracea sive floridissima viriditas apparuit divisit imaginis distractam longitudinem in ratione 3 ad 5, utpote longitudine illa in 8 partes divisâ, viriditas illa tribus partibus a rubeo termino distabat et quinque partibus a purpureo.[6]

3. Spatium per quod viriditas omnis ad usque caerulei et flavi confinium distendebatur, fuit quasi sexta pars totius distractae longitudinis.

4. Caerulei et purpurei confinium sive Indicus perfectissimus a confinio rubei flavique sive a perfectissimo citrino quasi $\frac{7}{12}$ partibus totius distractae longitudinis distabat.

5. Denique haec Indici et citrini distantia per confinium viridis et caerulei in ratione 2 ad 3 dividebatur; ita scilicet ut confinium istud sive meditullium imaginis ab Indico $\frac{14}{60}$ partibus totius distractae longitudinis distaret et $\frac{21}{60}$ partibus a citrino.

[95]   Cum isthaec quanta potui diligentia observassem, non proprio tantùm sensui confisus sed (propter summam difficultatem praecisè distinguendi confinia colorum et loca maximae perfectionis) aliorum judicijs fretus;[7] imaginis dimensiones juxta haec inventa delineavi quemadmodum videre est in fig II, 38.[8] Scilicet centris X et Y 10$\frac{1}{4}^{unc.}$[9] distantibus et semidiametris 1$\frac{5}{16}^{unc:}$ semicirculos duos APC et BTD e regione descripsi et rectis AB et CD tangentibus connexui. Deinde linea XY (quam supra denominavi distractam longitudinem

---

(2) In §I, 37 ≈ 110.   (3) On the diminished length see Lect. 10, note (19).

(4) In describing the division of the spectrum in "An Hypothesis explaining y^e properties of Light" in 1675, Newton specified that he repeated his observations at "divers times, both in y^e same & divers days" and on "several papers," but never with divers prisms; Add. 3970, f. 544^v = *Correspondence*, 1:377.

(5) Omitted in *editio princeps*.

(6) Recall that Newton treats "violaceus" and "purpureus" as equivalent; see Lect. 1, note (10).

(7) In his "Hypothesis" Newton explains that a "friend" drew the lines, "partly because my own eyes are not very critical in distinguishing colours, partly because another to whom I had

each of its colors, their distances from one another, and also the degrees of refrangibility corresponding to the individual kinds of rays.

It was shown at the beginning[2] that when a prism (whose vertex angle [94] was about 63°) projected an image to a distance of twenty-two feet, its length was $13\frac{1}{4}$ inches and its breadth $2\frac{5}{8}$ inches, and so the distance of the centers of the outermost of the circles, which disposed lengthwise compose that image, was $10\frac{5}{8}$ inches. Now it is proper to relate its other dimensions to this distance, or the image's diminished length,[3] because they have no definite relation to its absolute length, which depends on the size of the component circles. To investigate its dimensions with greater accuracy I marked with a pen the places where the most perfect colors of their kind and their boundaries fell on a transverse paper, and having frequently repeated such observations and compared them to one another,[4] I finally drew these conclusions one by one:

1. Blue and violet on the one side, and green, yellow,[5] and red on the other side bisected the image, so that the boundary of green and blue (which I may call sea green) occupied its middle.

2. The place where leek green, or the most florid green, appeared divided the image's diminished length in the ratio of 3 to 5, inasmuch as when that length was divided into eight parts, that green was distant by three parts from the red end and five parts from the purple.[6]

3. The space through which all the green was extended, as far as the boundary of blue and of yellow, was about $\frac{1}{6}$ part of the entire diminished length.

4. The boundary of blue and purple, or the most perfect indigo, was distant by about $\frac{7}{12}$ parts of the entire diminished length from the boundary of red and yellow, or the most perfect orange.

5. Finally, this distance between indigo and orange was divided by the boundary of green and blue in the ratio of 2 to 3; that is, so that that boundary, or the middle of the image, was distant by $\frac{14}{60}$ parts of the entire diminished length from indigo and $\frac{21}{60}$ parts from orange.

When I observed these with as much care as I could, trusting not only my [95] own senses but also relying on the judgment of others (because of the extreme difficulty in precisely distinguishing the colors' boundaries and places of greatest perfection),[7] I drew the dimensions of the image according to these determinations as can be seen in Fig. II, 38.[8] Specifically, I drew the two opposite semicircles *APC* and *BTD* with centers *X* and *Y* $10\frac{1}{4}$[9] inches apart and with radii of $1\frac{5}{16}$ inches, and I joined them with the tangent lines *AB* and *CD*. Then dividing the line *XY* (which above I called the image's

---

not communicated my thoughts about this matter, could have nothing but his eyes to determin his fancy in making those marks" (Add. 3970, f. 544ᵛ = *Correspondence*, 1:376). In the *Opticks* (Bk. I, Pt. II, Prop. III, Expt. VII), where Newton also set forth the musical division of the spectrum, his "friend" became "an Assistant."

(8) The red end of the spectrum in Fig. II, 38 should be labeled *T* rather than *Q*, and the dotted lines representing the most perfect colors should not be in the middle of each color.

(9) This should be $10\frac{5}{8}$ inches.

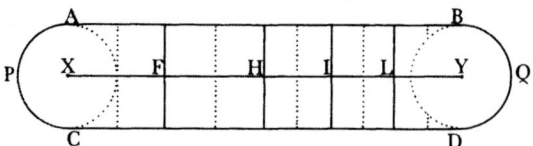

Figure II, 38

imaginis) in 60 partes aequales divisâ, sumpsi $LY = 9$, $IY = 20$, $HY = 30$, et $FY = 44$ partes ejusmodi. Et perpendiculis ad ista puncta erectis / Imaginem ₂65/ in quinque partes coloribus quinque insignioribus competentes distinxi; parte *PF* referente expansionem violacei, et *FH* expansionem caerulei et sic deinceps. Quo facto coloratam lucem in hanc figuram projeci, ut constaret denuò an color quilibet intra limites sic assignatos contineretur. Et cùm tota imago totam occupabat figuram singuli etiam colores cum singulis partibus quàm optimè conveniebant. Interea verò in spatijs istis loca observabam (qualia in hoc schemate punctim notantur) ubi singuli colores saturi et in suo genere illustrissimi apparuêre.

[96] Jam horum locorum et limitum colores disterminantium intervalla non alia fore manifestum est etiamsi circulos ex quibus imago conflatur, per methodos saepiùs recensitas,[10] centris non mutatis quantumvis minueres, eâ scilicèt de causâ ut heterogenei plus segregarentur et colores evaderent simpliciores. Quippe cùm in ipsissimis rectilineis terminis *AB* et *CD* colores sint absolutè simplices, et colores in mediâ imagine prope lineam *XY* cum istis quibus interjacent marginalibus congeneri appareant. Ratio etiam suadet quòd Heterogeneorum mistura non sensibiliter mutet locum alicujus coloris siquidem hinc et inde venientes se mutuò contemperant.[11] Sic radij viridiformes et purpuriformes per caeruleum sparsi aequipollent, et adeò non dimovent aut conturbant colorem illum, ut soli (quamvis nulli purpuriformes[12] intermiscerentur) ibidem componerent et exhiberent. Sed hic excipienda sunt spatia circulis terminalibus *AC* et *BD* comprehensa, ubi contemperamentum illud ex parte exteriori gradatim deficit. Et ideò saturi ruboris, qui solus e praefinitis in circulum terminalem se extendit, positionem in imagine e parte marginali ubi transibit circulum expedivi, ut indicat figura II, 38. In his autem si quid haesites possis experimenta de novo instituere contrahendo imaginis latitudinem ut circuli caeteris paribus minores evadant; et nullus dubito quin omnia quadrabunt.

[97] / Caeterùm quamvis colorum confinia in lineas ad *E, F, H* et *I*[13] erectas ₂66/ incidebant tamen loca ubi saturi et intensi apparuêre non omnia constitêre in medio interjecti spatij. Nam caeruleus qui in sua specie illustrissimus erat et nullatenus purpurascens propius ad *F* cadebat quam ad *E*.[14] Et plenissima[15] flavedo videbatur esse aliquantulum propior ad *H* quam ad *I*.[16] Atque ita rubedo et purpura propiùs ad centra *X* et *Y* quam ad alteros limites intensae apparuerunt. Solaque viriditas in medio limitum *F* et *H*[17] effloruit. Unde

---

(10) See §§I, 18–9 = 17–8, and also Lect. II, 3.

diminished length) into 60 equal parts, I took $LY = 9$, $IY = 20$, $HY = 30$, and $FY = 44$ parts. After erecting perpendiculars at those points, I distinguished the image into five parts corresponding to the five more prominent colors, with part PF representing the expanse of the violet, FH that of the blue, and so on. Having done this, I projected the colored light onto this figure to confirm once more whether every color would be confined within the limits assigned in this way. And when the entire image filled the entire figure, the individual colors also coincided as well as possible with the individual parts. Meanwhile, I observed in those spaces the places (indicated by dots in this figure) where the individual colors appeared full and the most brilliant of their kind.

It is evident, moreover, that the intervals of these places and of the boundaries [96] separating the colors will be no different, no matter how much (following the methods frequently considered)[10] you diminish the circles making up the image without changing their centers, namely, because the heterogeneous rays are more separated and the colors turn out simpler. For indeed, in the straight edges AB and CD themselves, the colors are absolutely simple, and in the middle of the image near the line XY the colors are seen combined with those between whose borders they lie. Reason even suggests that the mixture of heterogeneous rays does not sensibly change the place of any color, inasmuch as in coming from both sides they temper one another.[11] For instance, the green-making and purple-making rays scattered through the blue are in balance, and hence they do not displace or disturb that color, since those alone (even with no purple-making[12] ones intermixed) would compound and exhibit the same color. But an exception must be made here for the spaces contained within the terminal circles AC and BD where that tempering gradually dies out on the outer side. Consequently, for the full red, which alone of the prescribed ones extends into the terminal circle, I adjusted its position in the image on the marginal side where it will cross the circle, as Fig. II, 38 shows. But if you are at all undecided in these things, you can do the experiments anew by contracting the image's breadth so that, other things being equal, the circles will become smaller, and I do not doubt that everything will be in agreement.

Although the boundaries of the colors fell on the lines erected at [F, H, I, [97] and L],[13] nevertheless not all the places where the full and intense ones appeared stood in the middle of the intervening spaces. For the blue that was the most brilliant of its sort and in no way purplish fell nearer to [H] than to [F];[14] the fullest[15] yellow appeared to be a little nearer to [I] than to [L];[16] and red and purple appeared nearer to the centers X and Y than to the other, bright boundaries, while green alone sprang up in the middle of the boundaries [H] and [I].[17] Whence the reason is clear why although yellow and blue

---

(11) Namely, by Prop. 4, §II, 70, that any spectral color can be compounded from its neighboring colors.   (12) Read: *caeruliformes* (blue-making).

(13) $E \ldots I$] Read (as *editio princeps*): F, H, I et L.

(14) $F \ldots E$] Read "$H \ldots F$"; the *editio princeps* mistakenly has "$F \ldots X$".

(15) It is possible that this reads "*planissima*" (clearest).

(16) $H \ldots I$] Read: $I \ldots L$.   (17) F et H] Read: H et I.

constat ratio quod etsi flavus et caeruleus commistione viridem componant, rubeus tamen et viridis propter majus intervallum non bene componunt flavum nec viridis et purpureus caeruleum.[18] Cum itaque colores juxta medium constipatiores sint ita ut inter flavum ac rubeum juxta et inter caeruleum et purpureum quasi triente majus intersit intervallum quàm inter viridem et flavum vel caeruleum sibi hinc et inde conterminum; quo imago in partes elegantiùs inter se proportionatas distinguatur, in numerum quinque insigniorum colorum duos alios, citrium videlicet inter rubeum et flavum, ac indicum inter caeruleum et violaceum asciscere convenit; Idque potissimè quòd post quinque insigniores, illi duo eminere videntur; spatiaque, ubi interserantur pro speciei perfectione satis ampla obtinent. Et sic exteriorum colorum redundans expansio praescindetur, omnesque ad quantitatem viriditatis politiori symmetriâ proportionati evadent.[19]

[98]   His itaque intertextis coloribus, observationes denuò instituebam et (ut breviter dicam) omnia comparuêre juxta ac si partes imaginis quas colores occupant proportionales essent chordae sic divisae ut singulos gradus in Octavâ resonare faciat.[20] Quod cùm tandem deprehendi, figuram imaginis in partes perinde divisi; ut videre est in schem: II, 39. Atque iterum tentavi

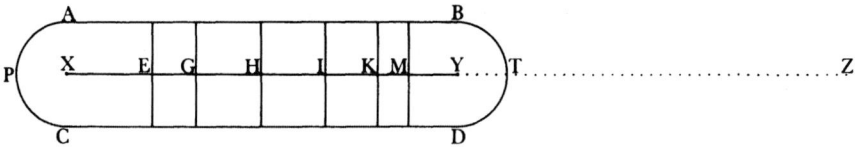

Figure II, 39

quàm benè cum his partibus colores convenirent. Scilicet imaginis distractâ longitudine *XY* productâ ad *Z*, ut *YZ* sit aequalis *XY* finge *XZ* chordam esse quam in *XY* ita dividere oportet quasi singula segmenta ad usque *Z* protensa singulos octavae gradus (sol, la, fa, ut, re, mi, fa, sol) edere deberent. Id / quod fiet bisecando *XY* in *H* et trisecando in *G* et *I*, rursusque trisecando *XI*   ₂67/

---

(18) While the interval between the most brilliant green and violet is much larger than that between yellow and blue, as Newton asserts, that between red and green is only slightly larger. On these color-mixing results see Lect. II, 8, note (16); and compare Lect. 7, notes (6), (7).

(19) Of Newton's three accounts of the division of the spectrum, only here is it that he explains his motivation for his sevenfold division and, in particular, his inclusion of indigo as a distinct spectral color; compare Robert Alexander Houstoun, "Newton and the colours of the spectrum," *Science Progress*, 12 (1917):250–64; *Light & Colour* (London, 1923), Ch. 1; Biernson, "Why did Newton see indigo in the spectrum?"; and H. L. Armstrong, "Comment on Newton's inclusion of indigo in the spectrum," *American Journal of Physics*, 40 (1972):1709.

(20) Until the publication of Newton's *Opticks*, Aristotle's *De sensu* remained the *locus classicus* for speculations on the analogy of musical and color harmonies. Aristotle held that all colors derive from various proportions of white and black, with some colors resulting from

in a mixture compound green, nevertheless red and green, because of their greater interval, do not compound yellow well, nor do green and purple compound blue well.[18] Since the colors are more crowded together near the middle so that between yellow and red, as well as between blue and purple, the interval is about one-third greater than that between green and yellow or blue, which border it on each side; therefore, in order to divide the image into parts more elegantly proportioned to one another, it is appropriate to admit to the five more prominent colors two others—namely, orange between red and yellow, and indigo between blue and violet—especially because after the five more prominent ones those two appear to be eminent, and the spaces where they are inserted remain sufficiently wide according to the perfection of their species. In this way the overflowing expanse of the exterior colors is cut off, and everything turns out proportionate to the quantity of green with a more refined symmetry.[19]

Consequently, after these colors had been interspersed, I once more made [98] observations, and (as I may briefly describe) everything appeared just as if the parts of the image occupied by the colors were proportional to a string divided so it would cause the individual degrees of the octave to sound.[20] When I finally discovered this, I divided the figure of the image into parts just as can be seen in Fig. II, 39, and I again tested how well the colors fit these parts. Specifically, extending the image's diminished length *XY* to *Z* so that *YZ* is equal to *XY*, imagine *XZ* to be a string that one must divide within *XY* as if the individual segments extending to *Z* are to emit the individual degrees of the octave (sol, la, fa, ut, re, mi, fa, sol). This will occur by bisecting *XY* at *H* and trisecting it at *G* and *I*, and, in turn, by trisecting *XI* at *E* and taking

---

simple ratios, such as 3:2 or 3:4, and others from incommensurable ratios: "and colours may, indeed, be analogous to harmonies. Thus, those compounded according to the simplest proportions, exactly as is the case in harmonies, will appear to be the most pleasant colours, e.g., purple, crimson, and a few similar species . . . Mixtures not in a calculable ratio will constitute the other colours" (*De sensu*, 439 b 33–440 a 3, p. 59). By the seventeenth century, ideas of harmony were ubiquitous, and it is consequently difficult to locate any specific sources for Newton's speculations. Although from the modern perspective Kepler seems a natural source to suspect, there is no evidence that at this time Newton was familiar with the *Harmonice mundi*, which, in any case, contains nothing on color harmonies. Nor is there any evidence to support the rumor related by Voltaire: "I have always heard it said that it was from Kircher that Newton had taken this discovery of the analogy of light and sound" (*Elémens de la philosophie de Neuton, mis à la portée de tout le monde* (Amsterdam, 1738), Ch. XIV, p. 178). Kircher had ordered the colors from white to black in two octaves with green in the middle and had assigned a different interval to each of thirteen colors; for example, white was a "semitonum," yellow a "semiditonus" (minor third), and black a "tonus major"; *Musurgia universalis sive ars magna consoni et dissoni in X libros digesta*, 2 vols. (Rome, 1650), 1:567–8. See Albert Wellek, "Farbenmusik," *Die Musik in Geschichte und Gegenwart: allgemeine Enzyklopädie der Musik*, ed. Friedrich Blume, 3 (Kassel/Basel, 1954), cols. 1811–22. Kircher's views were typical of musical analogies to color before Newton and were based on the presumed proportions of white and black in the colors subjectively ordered between those extremes. With his musical division of the spectrum Newton fundamentally altered the basis of the analogy by eliminating the roles of white and black and color-mixing concepts and by taking a natural ordering for the colors.

in *E* et capiendo *KY* quintam et *MY* octavam partem totius *XY*.[21] Et semitonia *EG* et *KM* Indicum et citrium referent, caeterique quinque toni *XE, FG, GH, HI* et *K*[22] caeteros quinque praecellentes colores. Quorum singuli, cum tota colorum congeries in totam figuram adaequatè incidit intra has singulas respectivè partes comprehensi fuerunt. Inque meditullijs harum partium circiter color quilibet in propria specie illustrissimus et intensissimus apparuit, etiam purpura et rubedo quamvis ultra versus *P* ac *T* marcescente luce exundarunt.

[99]   Caeterum haec non adeò precisè observare ac definire[23] potui quin ut fateri cogar ea posse paulò aliter fortasse constitui. Quemadmodum si inter *XZ* et *YZ* sumantur undecim mediae proportionales,[24] quarum *EZ* secunda sit, *FZ* tertia, *GZ* quinta, *HZ* septima, *IZ* nona, et *KZ*[25] decima; haec etiam imaginis distributio cum colorum expansionibus sat bene convenire videbitur. Nam differentiae adeò minutae quales inter hanc et superiorem distributionem intercedunt, acutissimo sensu judice vix comparituros errores efficere possunt.

Quantùm verò distributiones istae differunt ex adjunctis numeris patebit, quorum superiores ad chordam 720 partium ratione musica divisam respiciunt, et inferiores ad eandem chordam quàm proximè divisam ratione Geometricâ

360 . 320 . 300 . 270 . 240 . 216 . 202½ . 180. Chordâ Musicè divisâ.
360 . 321 . 303 . 270 . 240 . 214 . 202 . 180. Chordâ Geometricè divisâ.[26]

Superiorem verò distributionem potiùs adhibui, non tantùm quòd cum phaenomenis optimè convenit, sed quòd fortasse aliquid circa colorum harmonias (qualium Pictores non penitùs ignari sunt, sed ipse nondum satis

---

(21) Newton has divided the spectrum *XEGHIKMY* (or half the string) as follows, 1, $\frac{7}{9}$, $\frac{2}{3}$, $\frac{1}{2}$, $\frac{1}{3}$, $\frac{1}{5}$, $\frac{1}{8}$, 0. He presented a particularly clear summary of his musical division in the *Opticks* (Bk. II, Pt. II, Prop. III, Expt. VII, p. 92 = Add. 3970, f. 97ʳ), where he tells us to conceive the segments of the whole string "to be in proportion to one another as the numbers 1, $\frac{8}{9}$, $\frac{5}{6}$, $\frac{3}{4}$, $\frac{2}{3}$, $\frac{3}{5}$, $\frac{9}{16}$, $\frac{1}{2}$, & so to represent the Chords of the Key, & of a Tone, a third minor, a fourth, a fift, a sixt major, a seventh & an eighth above yᵉ Key . . . " This is a just diatonic scale (G, a, b♭, c, d, e, f, g) whose intervals are $\frac{9}{8}$, $\frac{16}{15}$, $\frac{10}{9}$, $\frac{9}{8}$, $\frac{10}{9}$, $\frac{16}{15}$, $\frac{9}{8}$; where $\frac{9}{8}$ is a major tone, $\frac{10}{9}$ a minor tone, and $\frac{16}{15}$ a semitone. By considering the divisions of the spectrum as musical intervals, or ratios of lengths rather than absolute lengths, Newton has more than satisfied his demand that the spectrum have "a more refined symmetry" with respect to the green—indeed, it is perfectly symmetrical. In his study of modes in his short tract, "Of Musick" (Add. 4000, ff. 138ʳ–43ʳ) from late 1665, he presented this scale as the second of "yᵉ 12 Modes in their order of Elegancy" and represented it in his own notation as *s, t, u, o, p, q, r, s* for the "key, 2ᵈ, 3ᵈ minor, 4ᵗʰ, 5ᵗ, 6ᵗ major, 7ᵗʰ, 8ᵗʰ" (§9, ibid., f. 139ᵛ). Newton later explained that "These degrees have of old beene expressed by yᵉ Six notes, ut, re, mi, fa, sol, la, the 7ᵗʰ note being omitted as being a discord to yᵉ key in yᵉ first mode. But of late yᵉ usuall notes are sol, la, mi, fa, sol, la, fa, hitherto expressed by yᵉ letters *o.p.q.r.s.t.u*" (§11, ibid., f. 142ʳ). The older representation gives us the present scale. Afterward, in the division of the spectrum in the "Hypothesis" of 1675 and in the *Opticks,*

KY to be the fifth and *MY* the eighth part of the whole *XY*.[21] The semitones, *EG* and *KM*, will represent indigo and orange, and the other five tones, *XE*, [*GH, HI, IK,* and *MY*],[22] the other five preeminent colors. Each of these, when the entire mass of colors fell alike on the whole figure, was respectively contained within these individual parts. Approximately in the middle of these parts each color appeared the most brilliant and intense of its own species, although purple and red still overflowed toward *P* and *T* with languishing light.

I could not, however, so precisely observe and define[23] this without being [99] compelled to admit that it could perhaps be constituted somewhat differently. For instance, if between *XZ* and *YZ* one takes eleven mean proportionals,[24] of which *EZ* is the second, [*GZ*] the third, [*HZ*] the fifth, [*IZ*] the seventh, [*KZ*] the ninth, and [*MZ*][25] the tenth, this distribution of the image will also seem to fit the colors' expanses sufficiently well. For such quite minute differences that occur between this and the above distribution can produce errors hardly visible to the keenest judge.

But how much these distributions differ will be clear from the appended numbers, of which the upper ones pertain to a string of 720 parts divided in a musical ratio, and the lower ones to the same string divided as exactly as possible in a geometrical ratio:

360.320.300.270.240.216.202½.180. A string divided musically
360.321.303.270.240.214.202 .180. A string divided geometrically[26]

I have, to be sure, preferred to use the upper distribution, not only because it agrees with the phenomena very well, but also because it perhaps involves something about the harmonies of colors (such as painters are not altogether

---

Newton adopted the more modern notation, *sol, la, fa, sol, la, mi, fa, sol,* by replacing the *ut, re* with *sol, la.* In the *Opticks* he uses the same scale for his color-mixing circle (Bk. I, Pt. II, Prop. VI); for the thicknesses of the air films producing the principal colors in "Newton's rings" (Bk. II, Pt. I, Observation XIV); for the construction at the beginning of Bk. II, Pt. II (p. ₂31 = Add. 3970, f. 155ʳ); and for the colors of thick plates (Bk. II, Pt. IV, Observation VIII). Since Newton defined the modes according to the position of the semitones, the slightly different division of the string in Bk. II, Pt. IV, Observation V, on thick plates, is the same mode, for it can readily be shown that the intervals are $\frac{9}{8}, \frac{16}{15}, \frac{10}{9}, \frac{9}{8}, \frac{9}{8}, \frac{16}{15}, \frac{10}{9}$. Therefore all musical divisions invoked in the *Opticks* are the same mode. Susi Jeans is preparing a monograph on Newton's music theory, and I am indebted to her for much of my understanding of this subject.

(22) *XE ... K*] Read: *XE, GH, HI, IK et MY.*

(23) *ac definire*] Omitted in *editio princeps.*

(24) That is, equal temperament, with $2^{1/12}$ (= 1.05946) being the equal-tempered half note.

(25) *EZ ... FZ ... GZ ... HZ ... IZ ... KZ*] Read: *EZ ... GZ ... HZ ... IZ ... KZ ... MZ.*

(26) In November 1665 Newton calculated the lengths of "A string (720) divid[e]d into 12 (Geometrically progresionall) parts, yᵗ it may sound yᵉ 12 exact ½ notes in an Eight" in order to compare them with the lengths when "yᵉ string 1 or 720 is to bee divided yᵗ it may sound all yᵉ Musicall notes & halfe notes in an eight" (Add. 4000, f. 105ᵛ). To get the two divisions tabulated here—which, in fact, are for a string divided into 360 parts—he merely had to halve and round off the values from his earlier table.

perspectas habeo)[27] sonorum concordantijs fortasse analogas, involvat:[28] Quemadmodum verisimiliùs videbitur animadvertenti affinitatem quae est inter extimam purpuram ac rubedinem, colorum extre/mitates, qualis inter ₂68/ Octavae terminos (qui pro unisonis quodammodò haberi possunt) reperitur.

[100]   Ex his demum proportiones sinuum refractionis cuique radiorum generi competentium (ratione Mechanicâ) determinantur. Utpote ad vitrum aeri contiguum cùm sinus radiorum hinc et inde extimorum sint ut 68 ad 69, divide intermediam unitatem in ratione partium hujus imaginis et orientur $68 . 68\frac{1}{8} . 68\frac{1}{5} . 68\frac{1}{3} . 68\frac{1}{2} . 68\frac{2}{3} . 68\frac{7}{9} . 69$ pro sinubus ad confinia terminosque singulorum septem colorum pertinentibus, respectu communis sinus incidentiae $44\frac{1}{4}$ cùm refractio fit e vitro.[29] Cùm verò fit in vitrum pro sinubus istis adhibe numeros $68 . 68\frac{2}{9} . 68\frac{1}{3} . 68\frac{1}{2} . 68\frac{2}{3} . 68\frac{4}{5} . 68\frac{7}{8} . 69$, existente communi sinu incidentiae 106.[30] Et pro sinubus ad radios ubi colores sunt in proprijs

---

(27) In his "Hypothesis" of 1675, Newton offered the examples of the harmony of "golden & blew" and the discord of "red & blew" (Add. 3970, f. 544ʳ = *Correspondence,* 1:376). His most comprehensive account of color harmony, however, was in a draft for a projected Prop. 12 for Bk. II of the *Opticks*: "For instance green agrees with neither blew nor yellow for it is distant from them but a note or tone above & below[.] Nor doth Orange for the same reason agree with yellow or red: but Orange agrees better wᵗʰ an Indigo·blew then wᵗʰ any other colour for they are fifts. And therefore painters to set off Gold do use to lay it upon such a blew. So Red agrees well wᵗʰ a sky coloured blew for they are fifts & yellow wᵗʰ Violet for they are also fifts. But this harmony & discord of colours is not so notable as that of sounds because in two concord sounds there is no mixture of discord ones, in two concord colours there is a great mixture, each colour being composed of many others" (Add. 3970, ff. 348–9). In the *Opticks* itself he only briefly refers to color harmony in Query 14.

(28) Newton's doubled reluctance—note the repeated "fortasse" (perhaps)—here to commit himself to the analogy between musical and color harmonies may well arise from its derivative nature, following from his explanation of the periodic colors of thin films by means of aethereal vibrations excited by incident light corpuscles. Newton similarly maintained that when light corpuscles strike the retina, they excite vibrations in the aether contained within it. These vibrations are propagated into the brain where they "affect the soule wᵗʰ a sensation of various colours according to their various proportions, something after the manner that various sounds are produced by various proportions of the vibrations of the Air. The harmony and discord also wᶜʰ the more skilfull Painters observe in colours may perhaps be effected & explicated by various proportions of the aethereall vibrations as those of sounds are by the aëreall. To which end I would suppose the vibrations causing the deepest scarlet to be to those causing the deepest violet as two to one; for so there would be all that variety in colours wᶜʰ wᵗʰin the compasse of an eight is found in sounds, & the reason why the extreames of colours Purple & scarlet resemble one another would be the same that causes Octaves (the extreames of sounds) to have in some measure the nature of unisons" (Add. 3970, f. 528ᵛ). These views, sketched in "An Hypothesis hinted at for explicating all the afforesaid properties of light," which he had intended to enclose with his reply to Hooke in the spring of 1672, present a physical model for color harmony that is altogether lacking in the musical division of the spectrum. Newton never offered any physical reason why the spectrum's length, which is proportional to the chromatic dispersion (see the next article), should also be divided musically. It seems that his intensive work in the early 1670s on "Newton's rings" stimulated him to develop the musical division of the spectrum, for no musical analogies appear in his optical work before 1672 when he had resumed his earlier investigations of interference colors. In fact, he introduced his musical division of the spectrum in the "Hypothesis" of 1675 as a consequence of his theory of aethereal vibrations.

Newton, it seems, intended to provide empirical support for the analogy between color and

unacquainted with, but which I myself have not yet sufficiently studied)[27] perhaps analogous to the concordances of sounds.[28] It will even appear more probable by noting the affinity existing between the outermost purple and red, the extremities of the colors, such as is found between the ends of the octave (which can in a way be considered as unisons).

Finally, from these the proportions of the sines of refraction corresponding [100] to each sort of ray are determined by a mechanical procedure. Specifically, for glass contiguous to air, when the sines of the outermost rays on each side are as 68 to 69, divide the intermediate unit in the ratio of the parts of this image; and there will result 68, $68\frac{1}{8}$, $68\frac{1}{5}$, $68\frac{1}{3}$, $68\frac{1}{2}$, $68\frac{2}{3}$, $68\frac{7}{9}$, 69 for the sines pertaining to the boundaries and ends of the seven individual colors with respect to a common sine of incidence $44\frac{1}{4}$, when the refraction is made from glass.[29] But when it is made into glass, instead of those sines use the numbers 68, $68\frac{2}{9}$, $68\frac{1}{3}$, $68\frac{1}{2}$, $68\frac{2}{3}$, $68\frac{4}{5}$, $68\frac{7}{8}$, 69, with the common sine of incidence being 106.[30] Moreover, for the sines pertaining to the rays where the colors are the

---

sound from his observations of the colors of thin films, which showed—or so he believed—that the ratio of the "vibrations" of the extreme red to the extreme violet are in the 2:1 ratio of the octave; see the quotation from the "Hypothesis" of 1672 in the preceding paragraph. Yet in the accompanying Observation 14, when he illuminated two lenses with the sequence of spectral colors, he found that for colored rings of the same order the ratio of the spaces between the lenses for the extreme red and violet (which is proportional to their vibrations) was "greater then 3 to 2 & lesse then 5 to 3. By the most of my observations it was as 9 to 14" (Add. 3970, f. 521ᵛ; read "14 to 9"). Although Newton eliminated his reference to the 2:1 ratio of the extreme vibrations in the revision of the "Hypothesis" in 1675, in numerous drafts for the *Opticks* from the early 1690s he diligently attempted to reintroduce the 2:1 ratio of the octave; see, for instance, the various drafts for the aborted "fourth book," which further develops the ideas of the "Hypothesis," such as Props. 10, 16, and 19 (Add. 3970, ff. 374ᵛ, 337ᵛ, 335ᵛ), and especially his most extensive draft on the analogy between color and sound, Prop. 12 for Bk. II (ff. 348–9). In other assessments, however, he more realistically concluded that the vibrations may be "in yᵉ proportion of 5 to 3 to make yᵉ 5 principal colours red yellow green blew violet answer to yᵉ tones in a sixt major, sol, la, mi, fa, sol, la" (Prop. 19, f. 336ʳ). In the *Opticks* itself Newton finally abandoned his attempts to establish specific ratios for the vibrations of the colors—indeed, he abandoned the vibrations and replaced them with "fits"—and besides the numerological use of musical scales to represent particular divisions or proportions of physical quantities (see note (21), this lecture), he confined himself to some qualitative speculations in Queries 12–14 and 23. On Newton's use of aethereal vibrations and "fits" see Richard S. Westfall, "Uneasily fitful reflections on fits of easy transmission," *Texas Quarterly*, 10, no. 3 (1967):86–102.

In the "Hypothesis" of 1675 Newton also justified the analogy between colors and sound by appealing to the principle that "yᵉ analogy of nature is to be observed" (Add. 3970, f. 544ʳ = Correspondence, 1:376). Some years later in a letter to John Harrington, an Oxford undergraduate, dated 30 May [1698] Newton elaborated this idea and explained that "I am inclined to believe some general laws of the Creator prevailed with respect to the agreeable or unpleasing affections of all our *senses;* at least the supposition does not derogate from the wisdom or power of God, and seems highly consonant to the macrocosm in general" (*Correspondence*, 4:275).

(29) The extreme values for the index of refraction are from §I, 38 = 111; for the fractional division see note (21), this lecture. In taking the angular dispersion, dD, to be proportional to the chromatic dispersion, dn, or the differences of the sines of refraction, dR, Newton is essentially using the approximation (see Lect. 18, note (7), this volume) that near minimum deviation $dD = 2\tan i \, (dn/\overline{n})$ or $dD = (2\overline{R}\tan i)dR$, where $n = R/I$ and $2\overline{R}\tan i$ are constants; see the *Opticks*, Bk. I, Pt. II, Prop. III, Expt. VII, pp. 92–3 = Add. 3970, f. 97ᵛ.

(30) See §I, 40 = 113.

speciebus perfectissimi pertinentibus, numeri inter hos numeros intermedij adhiberi possunt.

Sic ad aquam aeri conterminam ubi extremi sinus refractionis sunt ut 90 ad 91; sinus intermedios per consimilem intermediae unitatis dissectionem (statuendo scilicet esse 90 . $90\frac{1}{8}$. $90\frac{1}{5}$ &c. vel 90 . $90\frac{2}{9}$. $90\frac{2}{3}$ &c)[31] elicere possis. Ast hic memento determinationes hasce non esse praecisè Geometricas sed tam proximè tamen accuratas quàm exigunt hujusmodi res practicae, et quicquam ampliùs moliri, praeter computandi taedium, affectatam et inanem curiositatem argueret.

[101]   Sunt et aliae circa hos colores circumsta[n]tiae quas jam determinare potuissem, quemadmodum variae eorum formae et expansiones pro varijs positionibus Prismatis circa axem convolventis; vel pro variâ materiâ refractivâ ex quâ Prisma fabricatur, quâve circundatur; vel etiam pro varia magnitudine ejus anguli verticalis. Sed ea omnia ex ostensis in parte priori (conferendo cum jam explicatis) sat manifestantur; ut et effectus, quantùm scio, omnes quos vel unica tantùm refractione vel utcunque pluribus, et quâvis terminatione lucis elicere liceat.

/ 2. De phaenomenis Lucis per Prisma in oculum transmissae.   ₂69/

Lect 12   Post explicata colorum a parietibus alijsve objectis reflexorum phaeno-
[102] mena, ordo postulat ut ad affines objectorum trans Prismata prope oculum interposita conspicuorum apparentias explicandas jam animum adjiciam. Et cùm doctrinam in quinque propositionibus supra[1] traditam per prioris generis experimenta solummodo probaverim et hoc experimentorum genus, eò quod non sit adeo simplex, consulto reticuerim: explicationem ejus jam fusè tradere non pigebit.[2]

[103]   Hic ideò imprimis recordari oportet objectorum mediante refractione visorum, quòd non[3] in proprijs locis sed alijs quibusdam videntur, a quibus videlicet refracti radij rectà ad oculum tendunt.[4] Atque adeò si ita refringantur ut qui fluunt ab ijsdem partibus objecti a diversis locis directe ad oculum veniant; objectum illud in totidem locis apparebit. Sit e.g. (fig II, 40) X objectum, O oculus, et BC

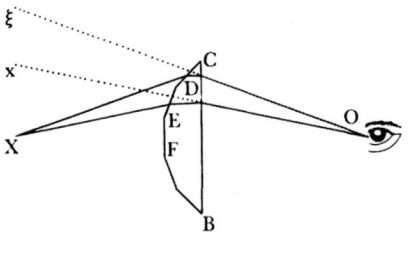

Figure II, 40

---

(31) For the index of refraction of water used here see §I, 44 = 117. Newton clearly intends his musical division of the spectrum to apply to all substances, or to describe a property of light and not matter. In physical terms his premise asserts (erroneously) that each color always occupies a fixed proportion—red, for instance, one eighth—of the spectral length, independent of that length and the refracting substance. The musical division is equivalent to a partial dispersion law, inasmuch as it specifies how the chromatic dispersion, dn or dR, varies through the entire spectrum, where R represents the sine of refraction. It is a linear law, since (for refraction to a rarer medium) $R = R_P - \lambda . \Delta R$, where $\Delta R = R_P - R_T$, P and T indicate extreme

most perfect of their own species, the intermediate numbers between these numbers can be used.

Thus for water bordering air, where the extreme sines of refraction are as 90 to 91, you may find out the intermediate sines by a quite similar division of the intermediate unit (namely, by setting them to be 90, $90\frac{1}{8}$, $90\frac{1}{5}$, and so on, or 90, $90\frac{2}{9}$, $90\frac{1}{3}$, and so on).[31] But indeed remember here that these determinations are not precisely geometric but still as nearly accurate as practical matters of this kind require, and to attempt anything further would manifest, beyond the tedium of calculating, an affected and vain curiosity.

There are also other features concerning these colors that I could now [101] determine, for instance, their various shapes and expanses according to the various positions of the prism when it is rotated about its axis; or according to the different refractive material from which the prism is made or with which it is surrounded; or also according to the different size of its vertex angle. But all this is sufficiently manifest from what has been demonstrated in the first part (combining it with what has just been explained), as are all the effects, insofar as I know, which one may produce with either only one or any number of refractions and with any termination of the light.

2. The Phenomena of Light Transmitted through a Prism into the Eye

After the phenomena of colors reflected from walls or other objects have **Lecture 12** been explained, proper order requires that I now turn my attention to ex- [102] plaining the related appearances of objects seen through prisms interposed next to the eye. Since I proved the doctrine presented above in five propositions[1] solely by experiments of the former kind, and I was deliberately silent about the latter kind of experiments, because it is not so simple, I will not now be reluctant to present its explanation in detail.[2]

In the first place, therefore, it must be recalled that objects seen with an [103] intervening refraction are seen not in their own places but in some others, namely, those from which the refracted rays tend to the eye in a straight line.[4] Hence, if the rays are refracted so that those that flow from the same parts of an object come to the eye in a straight line from different places, that object will appear in just as many places. For example (Fig. II, 40), let X be the object, O the eye, and BC an interposed lens that is bounded by several

blue and red, and λ the spectrum of colors as it varies continuously between 0 for extreme violet and 1 for extreme red, assuming, for example, the value 7/8 at the border of orange and red. It can, moreover, be shown that the musical division of the spectrum embodies his linear dispersion law; see Lect. 11, note (6), and Shapiro, "Newton's 'achromatic' dispersion law," pp. 110–13.

(1) quinque propositionibus supra] Originally: libro praecedenti (the preceding book).

(2) In the *Lectiones opticae*, §77 → II, 81, Newton indicated his intention to treat these phenomena, but he never did get around to it.

(3) oportet . . . quòd non] *Editio princeps*: oportet quòd . . . imagines non.

(4) See Lect. 12, note (19).

lens interposita quae pluribus planis superficiebus *CD, DE, EF*, &c terminetur sicut ad objecta multiplicia reddenda fabricari solet.[(5)] Dein suppone haec plana radios in sese incidentes ita refringere ut oculum petant quasi a loco ξ venientes qui incidunt in planum *DC*, vel a loco x venientes qui incidunt in planum *DE* et sic porrò; et manifestum est tum ratione tum experientiâ suadente, quòd idem objectum X in diversis locis ξ et x ad instar plurium videbitur.

[104]   Ad eundem modum, stantibus jam positis, nisi quod vice Poligoni *BC* Prisma *ABC* (fig II, 41) substituatur, cùm e praedemonstratis[(6)] constat quòd e radijs versus oculum refractis purpuriformes, propter maximam refrangibilitatem longissime a lineâ rectâ oculum et objectum interjacente divaricant,

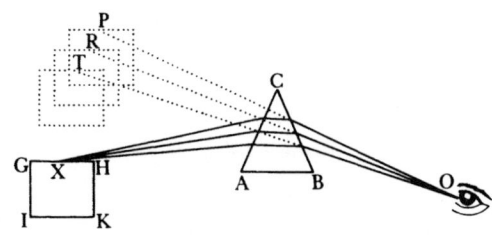

Figure II, 41

suppone quòd oculum petant quasi venientes a *P*, et quod rubiformes oculum petant quasi venientes a *T*, caeterique in[(7)] locis intermedijs, pro gradu refrangibilitatis fluant; et manifestum est quòd objecti, si ope purpuriformium radiorum / solummodò conspiceretur, imago foret ad *P*, idque caerulei coloris, sin radijs solummodò rubriformibus conspiceretur, imago ejus ad *T* existeret, idque rubei coloris; et ad *R* viridis appareret si modo solis[(8)] viridiformibus radijs conspiceretur: et sic praeterea. Quod si objectum duo tantùm radiorum genera simul emitteret, duplicem sortiretur imaginem. Sic emissis rubiformibus et purpuriformibus radijs, imago altera ad *T* rubea appareret, et altera ad *P* purpurea. Et sic denique si omne genus radiorum simul emitteret (ut solent corpora naturalia) tunc innumeras colorum gradatim differentium obtineret imagines per totum spatium *PT* ordine continuò dispositas; quae, cum in locis non penitus discretis formarentur, se mutuò obliterarent, efficerentque ut nil nisi confusa colorum series appareret.

[105]   Hoc pacto quidem colores omnigenos generari oportet cùm objecti lucidi nigredine vel tenebris terminati perexigua est apparens magnitudo; qualis est Solis vel Lunae aliorumvè siderum, aut foraminis in fenestra lucem a nubibus in obscurum cubiculum intromittentis. Quòd si expansius objectum intueamur quale ad [*X*] designatur, terminum ejus *GH* vertici prismatis propiorem imprimis animadvertamus, et manifestum est quod imaginum ejus ex varijs radiorum generibus formatarum purpurea longissimè omnium veluti ad *P* divaricante, color ille apparebit extimus. Imago autem viridis ad usque *R* translata cum parte aliqua purpureae imaginis ut et intermediae caeruleae ibidem coincidet et confundetur a qua mixtura caeruleum colorem generari oportet. Et rubea in *T* terminata cum partibus caeterarum omnium imaginum eousque extensis coincidet et colorem objecti ibi restituet, album puta, sed modò objectum[(9)] sit albi coloris.[(10)]

---

(5) Such a lens was described, for example, by della Porta, *Magiae naturalis*, Bk. XVII, Ch. X.

(6) *Editio princeps*: praemonstratis.

(7) Read: a (from).

(8) Omitted in *editio princeps*.   (9) Added.

plane surfaces *CD, DE, EF,* and so forth, just as is usually constructed to render objects multiple.[5] Next, suppose these planes refract the rays falling on them so that those falling on the plane *DC* aim for the eye as if they came from the place *ξ,* or those falling on the plane *DE* as if they came from *x,* and so on. It is evident by both the force of reason and of experience that the one object *X* will be seen as many in the different places *ξ* and *x.*

Similarly, with things remaining as already assumed except that in place of [104] the polygon *BC* the prism *ABC* is substituted (Fig. II, 41), since it is known from what has been previously demonstrated[6] that of the rays refracted toward the eye the purple-making ones because of their greatest refrangibility deviate farthest from the straight line between the eye and the object, suppose that they aim for the eye as if they come from *P,* that the red-making ones aim for the eye as if they come from *T,* and that the others flow from[7] the intermediate places according to their degree of refrangibility. It is evident that the image of the object, if it were viewed by means of purple-making rays alone, would be at *P* and of a blue color; but if it were viewed by red-making rays alone, its image would be at *T* and of a red color; and it would appear green at *R,* if it were viewed only[8] by green-making rays; and similarly for the others. But if the object simultaneously emitted only two kinds of rays, it would share a double image; for example, if red- and purple-making rays were emitted, one image at *T* would appear red and the other at *P* purple. Finally, if it simultaneously emitted every kind of ray (as natural bodies usually do), then it would possess innumerable images of gradually differing colors disposed in continuous order through the entire space *PT.* These would mutually obliterate one another, since they would not be formed in completely separate places, and they would cause nothing but a confused series of colors to appear.

In this way, in fact, colors of every sort must be generated when the [105] apparent magnitude of a bright object bounded by blackness or darkness is very small, as is the sun, moon, or other star, or a hole in a window admitting light from clouds into a dark room. But if we gaze upon a broader object, such as is represented at [*X*], we may in the first place consider its edge *GH* nearer to the prism's vertex. It is clear that since of all its images formed from different kinds of rays the purple deviates the farthest, as to *P,* that color will appear outermost. The green image, however, translated as far as *R,* will coincide and blend there with some part of the purple image as well as the intermediate blue, from which mixture the color blue must be generated. The red terminating at *T* will coincide with the parts of all the other images extended to there, and the color of the object will be restored there, suppose white, but only if the object[9] is the color white.[10]

---

(10) This explanation of boundary colors for images viewed through a prism corresponds to that for projected images in §§II, 82–3 ≈ 82–3. Directly viewed boundary colors was one of the first color phenomena that Newton was able to explain; see the Introduction, note (14). In the *Opticks* (Bk. I, Pt. II, Prop. VIII, p. 123 = Add. 3970, f. 118ʳ) he condensed this explanation into one terse paragraph that did not, in fact, explain the phenomenon; see also the "New theory," Prop. 9. He had, however, originally included, but then deleted, a simplified version of this explanation at the conclusion of Prop. 5, Expt. 18 of the "Fundamentum Opticae" (Add. 3970,

Et quemadmodum juxta limitem *GH* objectum purpureo et caeruleo fimbriatum apparebit sic in opposito limite *IK* per / consimile ratiocinium patebit alteros colores rubeum flavumque produci. ₂71/

Nec secus cum ejusdem objecti partes aliquae sunt alijs utcunque lucidiores, colores varij generari debent.

[106] Et quantitas anguli *POT* sub quo colores apparent erit maxima cum Prisma statuitur oculo vicinissimum eoque minor evadet continuò quò Prisma propiùs ad objectum collocatur. Quemadmodum si Prismatis ex vitro confecti angulus verticalis sit 60^{gr}, colores sub angulo $2^{gr}\ 2^{min.}$ circiter apparebunt cum proximè oculum disponitur, et $1^{gr}\ 1^{min}$ cùm in media inter oculum et objectum distantia statuitur, et quasi $30\frac{1}{2}^{min}$ cùm triplo plus distat ab oculo quàm objecto, et sic praeterea.[11] Hic autem suppono radios ad utramque superficiem ejus aequaliter refringi. Nam cùm positionem ad radios ex alterutra parte obliquiorem convertendo circa axem acquirit, ille angulus augebitur.[12] Suppono etiam Objectum satis lucidum esse ac tenebris densissimis terminatum, ut colores ad usque summas extremitates videri possint. Nam secus per latitudinem jam assignatâ minorem distendi videbuntur; ut ut de quantitate alijsque colorum circumstantijs in quibuslibet objectis sub dio conspectis apparentium; idque pro refractionibus utcunque factis, ex his facile est conjicere.

[107] Caeterum in allatae doctrinae illustrationem Phaenomena aliquot insigniora et minùs obvia ex abundanti jam brevitèr describere est animus.

Et imprimis accepto filo aliquo *PT* (fig II, 42) ejus alterum dimidium *PR* caeruleo colore tinxi atque alterum *RT* colore rubeo. Dein Prismate adhibito hoc filum intuebar, cujus a tergo, nisi locus erat tenebrosus, corpus aliquod nigerrimum statuebam, vidique praefata dimidia non in directum jacentia sed in duas lineas discreta quas in *πρ* et *στ* habes designatas. Scilicet caerulei

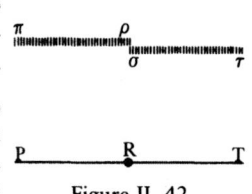

Figure II, 42

dimidij, propter majorem eorum radiorum refractionem imago paulò longiùs translata fuit. At linea tamen subobscura ipsius *PT* refracta apparuit, cujus partes in directum jacêre; et a quâ colores aliquantulum prostare visi sunt. Id quod / ex imperfectione et mistura in utrisque fili coloribus latente contigit: ₂72/ Nam quantò illustriores erant et simpliciores, ea linea tantò obscurior evasit et colores *πρ* et *στ* clariores et magis interrupti.[13]

---

f. 403ʳ), which corresponds to Prop. V, Expt. 14, of the *Opticks*, Bk. I, Pt. I. When Goethe in his *Beyträge zur Optik* (1791–2) used the phenomena of directly viewed boundary colors to oppose Newton's theory of color and to revive a modification theory in which colors arise from a mixture of light and shadow, Friedrich Albrecht Carl Gren in his critique of Goethe translated virtually the entirety of Newton's explanation (§§II, 103–6) save for a few sentences; "Einige Bemerkungen über des Herrn von Göthe Beyträge zur Optik," *Journal der Physik*, 7 (1793):3–21, esp. 10–13.

(11) Newton is here applying the rule he derived in §I, 115 = 190 for determining the apparent magnitude of an image viewed through a prism at minimum deviation: $(FV + VX)/FV = \widehat{NMO}/\widehat{GXH}$, where $\widehat{GXH}$ is the image's apparent magnitude, here $\widehat{POT}$; $\widehat{NMO}$ is the chromatic dispersion, assuming the object to be a point; *FV* is the distance between the prism and object, and *VX* the distance between the prism and eye, neglecting the prism's thickness. When the prism is next to the eye ($VX \approx 0$), then $\widehat{POT} = \widehat{NMO} = 2°2'$, which is a measure of the

Just as along the edge *GH* the object will appear fringed with purple and blue, so on the opposite edge *IK* it will be clear by quite similar reasoning that the other colors, red and yellow, are produced. Similarly, when some parts of the same object are in some way brighter than others, various colors must be generated.

The size of the angle *POT* under which the colors appear will be greatest [106] when the prism is placed closest to the eye and will become continually smaller the nearer the prism is placed to the object. For example, if the vertex angle of a prism made out of glass is 60°, the colors will appear under an angle of approximately 2°2′ when it is placed next to the eye, and 1°1′ when it is placed midway between the eye and the object, and approximately 30½′ when it is three times farther from the eye than from the object, and so on.[11] I suppose, here, however, that the rays are equally refracted at each of its surfaces, for when the prism acquires a position more oblique to the rays on each side by turning it about its axis, that angle will increase.[12] I also suppose the object to be sufficiently bright and bounded with the densest darkness so that colors can be seen as far as its outermost ends, for otherwise they will appear stretched out through a smaller extent than just designated. In any case, the size and other features of the colors appearing in any objects viewed in daylight—however the refractions are made—may easily be inferred from these things.

Furthermore, in illustration of this imparted doctrine I now intend to [107] describe briefly in addition several rather remarkable and less obvious phenomena.

In the first place, taking some thread *PT* (Fig. II, 42), I colored its one half *PR* the color blue and its other half *RT* the color red. Then using a prism I looked at this thread behind which, unless the place was dark, I fixed some extremely black body, and I saw those halves lying not in a straight line but separated into two lines, which you have represented in $\pi\rho$ and $\sigma\tau$. Specifically, the image of the blue half, because of the greater refraction of its rays, was translated slightly farther. But still there appeared a somewhat obscure refracted line of *PT*, the parts of which seemed to lie in a straight line, from which colors seemed to project a bit. This arose from the imperfection and from the mixture latent in both colors of the thread, for the more brilliant and simpler the colors were, the more obscure the line became, and the clearer and more separated the colors $\pi\rho$ and $\sigma\tau$ became.[13]

---

prism's chromatic dispersion. Consequently, when the prism is placed midway between the eye and object, or when $VX = FV$, then $\widehat{POT} = \frac{1}{2}\widehat{NMO} = 1°1′$; and when $VX = 3FV$, then $\widehat{POT} = \frac{1}{4}\widehat{NMO} = 30\frac{1}{2}′$.

(12) This claim is erroneous (see Lect. 18, note (14), this volume), for, as Marci had observed earlier, when the prism is rotated one way that angle decreases, and when rotated the opposite way it increases; *Thaumantias*, pp. 173–5.

(13) This is among the first of Newton's recorded optical experiments; see the Introduction, note (15). In his early writings he invoked it frequently, but because of the residual, obscure line that detracted from its demonstrative power, he came to prefer the experiment described in the next article; see "Of Colours," §6, Add. 3975, f. 2ᵛ = Hall, "Further experiments," p. 28; and also Expt. 6 in his draft of a reply to Hooke in the spring of 1672, Add. 3970, f. 441ʳ.

[108]   Caeterùm cùm lux quam filum tenue reflectit perexigua
sit, praestat adhibere corpus aliquod expansius quale per
PN (fig II, 43) designatur, quod v. g. concipe papyrum esse
ex parte *PRLK* caeruleam et ex alterâ parte rubeam. Tum
Prismate juxta oculum interposito, nigroque corpore, aut
loco tenebroso pone hoc objectum sito, videbis imaginem
caeruleae partis paulò longiùs translatam esse, terminis $\pi\tau$
et $\kappa\nu$ in confinio colorum $\rho\sigma$ et $\lambda\mu$ ut ante diffractis. Sed
hic summè cavendum est ut papyrus cum saturis[14] et in-
tensis coloribus illinetur.[15]

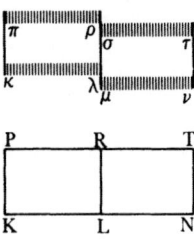

Figure II, 43

[109]   Huic affine est experimentum cùm statui duo Prismata ad foramina duo
quibus luci patuit aditus in tenebrosum cubiculum, ac in eo situ disposui ut
unius purpura et rubor alterius in eundem locum coirent. In quo loco fixi
papyri segmentum circulare, et non latius dimidio vel triente latitudinis colo-
ratarum imaginum, eapropter ut duobus illis solummodo coloribus illumi-
naretur. Quo facto papyrus[16] pallidi cujusdam coloris apparuit. Tum caeteris
utrinque coloribus objecto nigro terminatis, vel (quod satius erat) longiùs
projectis ut praefata papyrus nigredine vel tenebris circumcincta appareret,
tertium Prisma ad oculum applicavi, et ad distantias abinde pro arbitratu
varias me submovens, unicae illius subpallidae papyri geminam vidi imaginem, purpuream et rubeam. Et imago purpurea longiùs a papyro translata
erat, quàm rubea, prout major eorum radiorum refrangibilitas exegit. Rem
schemate II, 44to designatam habes ubi papyri πT imagines sunt *X* et *Y*.[17]

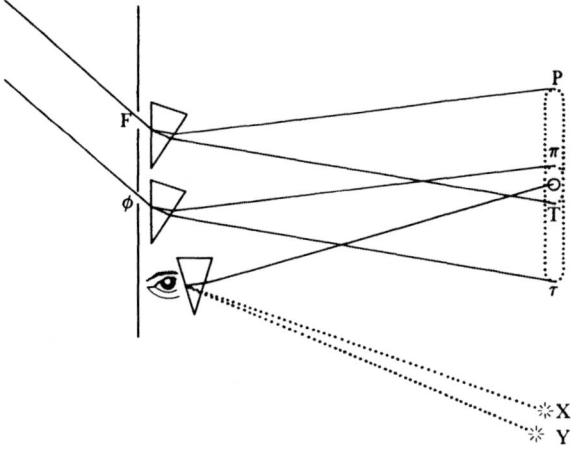

Figure II, 44

[110]   Ad eundem modum si duo pulverum genera, quorum alterum perfectè
rubrum est et alterum purpureum vel indicum, sine mistura Cyanei viridis aut
flavi parari possent; objectum aliquod perexiguum cum mistura pulverum
istorum crassè illitum geminam / imaginem exhiberet. Spectaculum forte cau-   ₂73/
sas ignorantibus mirandum. Sed vereor ut pulveres coloribus adeo simplicibus
praediti parari possint.

But since the light reflected by the thin thread is very meager, it is better to [108] use some more extensive body, such as that represented by *PN* (Fig. II, 43), imagining this, for example, to be a paper that is blue on the side *PRLK* and red on the other side. Then placing a prism next to your eye and setting a black body or a dark place behind this object, you will see the image of the blue part translated somewhat farther, with the edges $\pi\tau$ and $\kappa\nu$ broken as before at the boundary of the colors $\rho\sigma$ and $\lambda\mu$. But one must be exceedingly careful here that the paper is spread over with full[14] and intense colors.[15]

Related to this is an experiment where I placed two prisms at two holes, by [109] which an entry for the light was opened into a darkened room, and I arranged them in such a position that the purple of one and the red of the other came together in the same place. At this place I fixed a circular piece of paper no wider than a half or a third of the breadth of the colored image so that it would therefore be illuminated by only those two colors. When this was done, the paper[16] appeared to be some pallid color. Then with the other colors on both sides terminated by a black object or (which was better) projected farther so that the aforementioned paper appeared surrounded by blackness or darkness, I placed a third prism close to my eye, and removing myself arbitrarily to various distances, from there I saw a double image, purple and red, of that one somewhat pallid paper. Also the purple image had been translated farther from the paper than the red, as the greater refrangibility of its rays required. You have the matter represented in Fig. II, 44, where the images of the paper πT are *X* and *Y*.[17]

In the same way, if two kinds of powders could be obtained, one of which [110] is perfectly red and the other purple or indigo without a mixture of cyan, green, or yellow, then some very small object covered thickly with a mixture of these powders would exhibit a double image—a sight perhaps surprising to those ignorant of the causes. But I fear that powders endowed with such simple colors cannot be obtained.

(14) *Editio princeps*: crassis.

(15) Newton described this experiment in Expt. 6 of his spring 1672 draft letter for Hooke and in the *Opticks* (Bk. I, Pt. I, Prop. I, Expt. I), where he also performed this and the preceding experiment by illuminating the card and thread with red and blue lights, rather than by painting them; see Bk. I, Pt. I, Prop. II, Expt. VII, pp. 32–4 = Add. 3970, ff. 41–2.

(16) Added.

(17) Newton presented experiments related to this in his draft letter for Hooke (Expt. 5, Add. 3970, f. 441ʳ) and in the *Opticks* (Bk. I, Pt. I, Prop. II, Expt. VII, pp. 34–5 = Add. 3970, ff. 42–3).

[111]   His praeterea non multum absimile est cùm colores duorum Prismatum ita in parietem trajiciuntur, ut (unius rubore contingente purpuram alterius) in directum jaceant, quemadmodum videre est ad *Pτ* (fig II, 45,) et mediante Prismate parallel*ω*s interposito intuentur: Nam imagines non ampliùs in directum jacebunt, sed ab invicem apparebunt disjuncta. Sicut ad *MN* et *μν* designantur.[18]

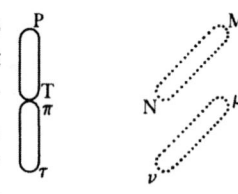

Figure II, 45

[112]   Atque ita si duo Prismata *A* ac *D* (fig II, 46) sic statuantur ut eorum colores ad locum *Pπ* adaequatè coincidant, in ordine tamen contrario,

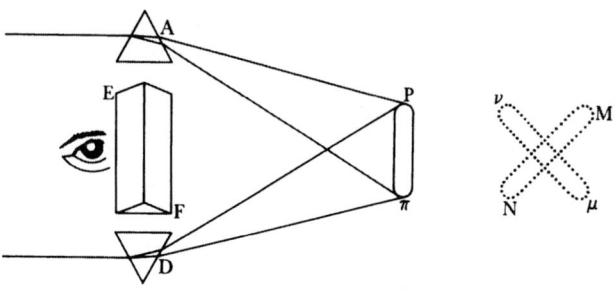

Figure II, 46

purpurâ alterius *A* cadente ad *π*, et rubore ad *P,* alterius autem *D* purpurâ cadente ad *P*, et rubore ad *π*: et per tertium Prisma *EF* imagini *Pπ* parallelum perspexeris; unicae *Pπ* duas decussantes imagines intueberis; alteram *MN* e coloribus Prismatis *D* productam, et alteram *μν* e coloribus prismatis *A*. Et quo longius ab objecto *Pπ* te conferas eo magis extremitates imaginum *M* a *ν* et *N* a *μ* distabunt.[19]

[113]   His etiam contraria sunt experimenta quod objecta duo (sive sunt circelli chartacei *X* et *Y* diversis coloribus illustrati sive diversorum prismatum paralleli vel decussantes colores ut *MN* et *μν*) ita possunt mediante alio Prismate conspici, ut in unum coalescere videantur.

[114]   Et praeter jam recensita perinsigne est hujusmodi experimentum quo objecta coloribus per interpositionem Prismatis denudantur, quibuscum nudo

Fig II, 47 oculo tincti apparent. Instantiam in solis imagine coloratâ [*PT*] accipe quae in parietem a Prismate *ABC* projecta, cum cernitur mediante alio parallelo Prismate *αβκ* manibus prehenso cujus vertex ad plagas versus rubeum colorem convertitur; si spectator se longiùs ab imagine gradatim amoveat, percipiet colores paulatim contrahi / et ad invicem eousque accedere donec tandem ₂74/

---

(18) This experiment is the direct equivalent of the one with projected images described in §II, 13 = 35. It was included in the *Opticks,* Bk. I, Pt. I, Prop. II, Expt. VII, p. 34 = Add. 3970, f. 41ᵛ.

(19) Newton invoked this experiment in his draft letter for Hooke to demonstrate that "the mixing of difform rays changeth not their degrees of refrangibility, nor their colorific dispositions" (Expt. 5, Add. 3970, f. 441ʳ) and again in the *Opticks* (Bk. I, Pt. I, Prop. II, Expt. VII, p.

Moreover, not very different from these is the case where the colors of two [111]
prisms are cast onto a wall so that (with the red of one touching the purple of
the other) they lie in a straight line, as can be seen at *Pτ* in Fig. II, 45, and
they are viewed by means of an interposed parallel prism: For the images will
no longer lie in a straight line but will appear separated from one another as
is represented at *MN* and *μν*.[18]

Furthermore, if two prisms *A* and *D* (Fig. II, 46) are placed so that their [112]
colors coincide evenly but in a contrary order at the place *Pπ*, with the purple
of the one prism *A* falling at *π* and the red at *P*, while the purple of the other
one *D* falls at *P* and the red at *π*, and if you look through a third prism *EF*
parallel to the image *Pπ*, you will observe two crossing images of the single
one *Pπ*: one, *MN*, produced from the colors of the prism *D* and the other,
μν, from the colors of the prism *A*. And the farther you move from the object
Pπ, the more distant will be the images' ends, *M* from *ν* and *N* from *μ*.[19]

There are also experiments opposite to these, because two objects (whether [113]
they are little paper circles *X* and *Y* illuminated with different colors or
parallel or crossing colors, and as *MN* and *μν* from different prisms) can be
viewed by means of another prism so that they seem to coalesce into one.

In addition to the experiment just related, there is a very remarkable one of [114]
this type whereby objects, by interposing a prism, are stripped of the colors
with which they are tinged to the naked eye. Receive the presence of the sun's
colored image [*PT*], which is projected onto a wall by the prism *ABC*. When Fig. II, 47

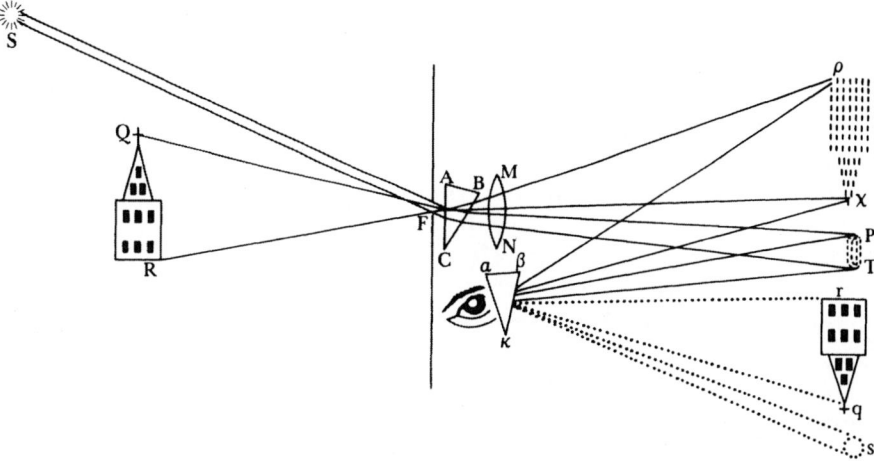

Figure II, 47

it is viewed by means of another, parallel prism *αβκ* held in the hands, and
its vertex is turned to the region toward the red color, if the observer gradually moves farther away from the image, he will perceive the colors to contract slightly and approach one another until finally, when they have been

34 = Add. 3970, f. 41ᵛ), where it was part of his demonstration that sunlight consists of
unequally refrangible rays.

uniti reficiant imaginem albam et circularem [s]. Id quod accidet cum spectatoris eadem est a coloribus distantia ac Prismatis *ABC,* si modo Prismatum anguli verticales aequentur. Et ratio ex eo manifesta est quod oblongam illam imaginem ex circulis sive circularibus imaginibus infinitè multis et in longum continuò dispositis efformatam esse docuerim, quarum quae sunt ad purpuream extremitatem longius per refractionem secundi Prismatis transferuntur, ut caeteras assequi possint et sic omnes coincidere.

Ad hunc modum cum objecta quaelibet ut *QR* (fig II, 47) foras posita confusas et coloratas eorum imagines ut χρ ad parietem per Prisma transmittunt, si mediante alio Prismate inspicias, possis imagines hasce coloribus denudare, et efficere praeterea ut distinctiores appareant quemadmodum ad *qr* videre est. Quoniam verò ad sufficientem copiam lucis requiritur ut foramen *F* sit amplum, per ejus autem amplitudinem transmissae imagines evadunt confusae; Lens aliqua convexa ut *MN* prope foramen istud statuenda est, quae radios a singulis punctis objecti foras positi venientes congreget in totidem alijs punctis ad parietem. Et insuper Prismata debent esse admodum transparentia, perpolita, et superficiebus accuratè planis terminata, inque situ quàm poteris exactè parallelo disposita.[20] Tanta quidem diligentia non requiritur ut imagines *q, r, s* sine coloribus appareant; sed ut inter tot ac tantas refractiones distinctae appareant, praeter accuratam fabricam vitrorum, requiritur experientis ingenium quo omnia rectè disponantur.

[115] Hic in cumulum praeterea adjici potest quod objecta quo simpliciori luce illuminantur, eo distinctiora per Prismata apparent: quippe cùm eorum per Prismata sub dio visorum confusio ex inaequali refrangibilitate illuminantium radiorum / oriatur.[21] Et hinc est quod solaris imaginis saepiùs commemoratae termini rectilinei (in quibus nullam esse heterogeneorum radiorum commisturam indicavi)[22] prae caeteris omnibus objectis distincti mediante Prismate appareant. Et sic muscae et similia animalia cum in rubeâ vel aliâ quavis luce simplici prismatibus elicitâ statuuntur, transvidentur solito dis-

₂75/

tinctiores. Quinetiam oculus Engyscopio armatus omnia hac simplici luce illustrata distinctiora cernit. Id quod insignem in contemplatione insectorum vel texturae aliarum rerum naturalium prae se usum ferre potest.[23]

[116] In tertia Propositione supra[24] de Phaenomenis quibusdam disserui ubi e radijs ad refringentem superficiem aequaliter inclinatis aliqua genera pervasêre, dum alia penitus reflectebantur; et illis affinia quaedam jam breviter attingere opportunum duco. Esto *S* spectatoris oculus (fig II, 48)

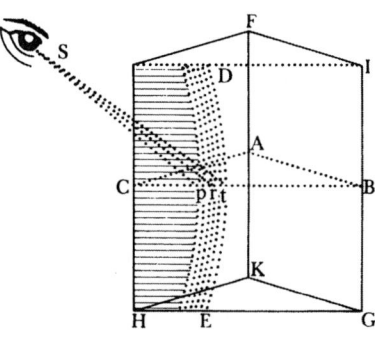

Figure II, 48

(20) In his draft reply to Hooke, Newton specified that the hole *F* be about one inch in diameter and the prisms' "verticall angles not too great, least by too great refractions they make a confusion even in the homogeneall rays" (Expt. 7, Add. 3970, f. 441ᵛ). In the *Opticks* (Bk. I,

united, they again make a white circular image [*s*]. This will occur when the distance of the observer and of the prism *ABC* from the colors is the same, provided that the prisms' vertex angles are equal. The reason is evident, because I have explained that that oblong image is formed from infinitely many circles, or rather circular images, disposed continuously lengthwise; and those circles that are at the purple end are translated farther by the refraction of the second prism so that they can catch up to the others and thus all coincide.

In this way when any objects, such as *QR* (Fig. II, 47), located outdoors transmit their confused and colored images, such as χρ, through a prism to a wall, if you view them by means of another prism, you can strip these images of their colors and moreover make them appear more distinct, as can be seen at *qr*. Since, however, for an adequate supply of light it is necessary that the hole *F* be wide, but because of its size the transmitted images turn out confused, one must set up behind that hole some convex lens, such as *MN*, that gathers the rays coming from the individual points of the object located outside into just as many other points at the wall. Moreover, the prisms must be completely transparent, highly polished, bounded by precisely plane surfaces, and placed as exactly as you can in a parallel position.[20] In fact, not too much care is required for the images *q, r,* and *s* to appear without colors; but for them to appear distinct amidst so many and such large refractions, besides the accurate fashioning of the glasses, a natural talent for doing experiments is required in order that everything may be correctly arranged.

It can, moreover, be added here that the simpler the light illuminating [115] objects, the more distinctly they appear through prisms, inasmuch as when viewed through prisms in daylight, their confusion arises from the unequal refrangibility of the illuminating rays.[21] Consequently, it happens that the frequently mentioned straight edges of the solar image (in which, I have indicated, there is no mixture of heterogeneous rays)[22] appear, by means of a prism, distinct beyond all other objects. Thus flies and similar living beings are seen through [a prism] more distinctly than usual when they are placed in red or any other simple light obtained from prisms. Indeed, an eye equipped with a microscope perceives everything illuminated by this simple light more distinctly. This can manifest an extraordinary benefit for the study of insects or of the structure of other natural bodies.[23]

In the third proposition above[24] I discussed certain phenomena where [116] some kinds of rays equally inclined to the refracting surface passed through while others were totally reflected, and I now consider it opportune to mention briefly some related to those. Let *S* (Fig. II, 48) be the observer's eye, in

Pt. II, Prop. V, Expt. XI) he invoked that part of this experiment pertaining to the observation of the sun's spectrum as a way to compound white from all spectral colors.

(21) This claim was made into a proposition in the *Opticks*, Bk. I, Pt. I, Prop. V.

(22) See §II, 23.

(23) Newton also made this suggestion in his letter for Hooke on 11 June 1672; *Correspondence*, 1:181 = *Phil. Trans.*, 7 (1672):509[6].

(24) tertia Propositione supra] Originally: libro 2ᵈᵒ (the second book). Specifically, in §§65–73 = II, 56–64.

quo lucem a nubibus sub dio ingressam planum *FG*, reflexam a plano *HI*, et plano *FH* egressam excipit, et cùm Prisma commodè statuitur ita ut radiorum e medietate basis *HI* versus oculum reflexorum angulus reflexionis sit quasi 50 graduum,[25] pars proximior basis remotiori aliquantulùm obscurior videbitur, et in utriusque partis confinio fimbria, qualis *DE*, subcaerulei coloris apparebit. Utpote cùm radij qui a remotiori parte basis ad oculum reflectuntur obliquiùs incidant quàm qui eo resiliunt a parte proximiori, talis potest assignari eorum circa medium basis obliquitas ut e proximioribus propter minorem obliquitatem aliqui perrumpere et refringi possint, dum remotiores propter majorem obliquitatem omnes ad oculum reflectantur. Sic ad vitrum cujus refractionem per rationem sinuum 42 ad 65 metimur, in plano *SABC* ad Prismatis longitudinem transverso, posito angulo *CtS* 49^{gr} 22^{min}; ang. *CrS* 49^{gr} 44^{min}; et ang. *CpS* 50^{gr} 5^{min};[26] *t* erit limes refractionis rubiformium radiorum ultra quem nulli superficiem *HI* penetrabunt qui propter debitam obliquitatem incidentiae ad oculum reflecti possunt, at a citeriori / parte *Ct* ₂76/ complures e radijs sic incidentibus propter minorem obliquitatem pervadere possunt et refringi, qui oculum peterent si modo reflecterentur. Et sic *r* erit limes viridiformium radiorum et *p* limes purpuriformium. Adeoque superficiei *IH* pars citima *Cp* propter complures radios omnis generis transmissos obscurior apparebit quàm pars ultima *tB* a quâ omnes qui oculum attingere possunt, eo reflectuntur. Et quia rubriformes a limite *t*, et viridiformes a limite *r* incipiunt ex parte pervadere, manifestum est quod ex illis pauciores a spatio *pt* ad oculum resilient quàm e purpuriformibus qui non priùs incipiunt pervadere quàm ad limitem *p*, ut et pauciores[27] quàm e caeruliformibus qui ad limitem inter *p* et *r* tantùm pervadere incipiunt. Et proinde in illo spatio purpura et caeruleus color aliquantulùm dominabitur. Deque totâ subcaeruleâ lineâ *DE* consimilis est discursus.

Haec autem linea non recta est, sed in morem arcus incurvata propterea quod puncta radios a basi prismatis ad oculum in angulo reflexionis dato resilientes reflectentia, ejusmodi curvam constituunt.[28]

Quod ad refractiones in superficiebus Prismatis *FG* et *FH* factas spectat, nihil refert in remotiori *FG* quaenam sint, dummodò radij e proximiori *FH*

---

(25) This 50° angle is not the angle of reflection (which is about 40°), but its complement; see Lect. 6, note (20).

(26) Since the index of refraction from glass to air is here 42/65, the critical angle is readily found to be 40°15' (see Lect. 6, note (5), this volume), and thus its complement, the angle *CrS*, will be 49°45', differing by only 1' from Newton's value. The value 42/65, which apparently comes from his table of refractions (see note (41) of the Introduction), differs slightly from the 53/82 adopted in §I, 30 ≈ 103 for the rays at the border of blue and green; but if the latter value is used, the 1' discrepancy vanishes. In Lect. 6, note (10), from Newton's measured indices of refraction we determined the critical angle for the extreme red and blue rays to be 40°37' and 39°55', and hence their complements $\widehat{CtS}$ and $\widehat{CpS}$ to be 49°23' and 50°5', respectively, in nearly perfect agreement with Newton's values. In the *Opticks* (Bk. I, Pt. II, Prop. VIII, Expt. XVI) these angles are slightly different, because different indices of refraction were used. However, the angles $\widehat{CpS}$ = 50°$\frac{1}{9}$ and $\widehat{CtS}$ = 49°$\frac{1}{28}$ found in all editions are typographical errors, their correct values being 50°9' and 49°28' as is found in the manuscript of the *Opticks* (Add. 3970, f. 119^r). Newton first described this blue band in his "Of Colours" (§23, Add. 3975, f. 4^v = Hall,

which he receives from the clouds in the sky light that has entered the plane *FG*, has been reflected from the plane *HI*, and has gone out by the plane *FH*. When the prism is suitably set so that the angle of reflection of the rays reflected from the middle of the base *HI* toward the eye is approximately 50°,[25] the closer part of the base will appear a bit darker than the farther part, and in the boundary of both parts a band of bluish color, such as *DE*, will appear. Namely, that since the rays that are reflected to the eye from the farther part of the base fall more obliquely than those that rebound from the closer part, one can assign such an obliquity to those near the middle of the base that some of the closer ones can break through and be refracted because of their smaller obliquity, while all the farther ones are reflected to the eye because of their greater obliquity. Thus, for glass, whose refraction we measure by the ratio of the sines 42 to 65, in the plane *SABC* transverse to the prism's length, set the angle *CtS* to be 49°22′, the angle *CrS* to be 49°44′, and the angle *CpS* to be 50°5′;[26] then *t* will be the limit of the refraction of the red-making rays beyond which none will penetrate the surface *HI*, and because of the appropriate obliquity of incidence, they can be reflected to the eye. Yet on the nearer side *Ct* a great number of the rays so incident, which would aim for the eye if they were reflected, are able to pass through and be refracted because of their smaller obliquity; and thus *r* will be the limit of the green-making rays, and *p* the limit of the purple-making ones. Hence the nearest part, *Cp*, of the surface *IH*, because of the great number of transmitted rays of all kinds, will appear darker than the farthest part, *tB*, from which all those rays that can reach the eye are reflected to it. Because the red-making rays begin in part to pass through from the limit *t* and the green-making ones from the limit *r*, it is clear that fewer of those rebound to the eye from the space *pt* than of the purple-making ones, which do not begin to pass through before the limit *p*, and also fewer[27] than of the blue-making ones, which only begin to pass through at the limit between *p* and *r*. Consequently, in that space a purple and blue color will dominate somewhat. And the explanation concerning the entire bluish line *DE* is quite similar.

This line, however, is not straight but curved inward in the shape of a bow, since the points reflecting the rays that rebound from the base of the prism to the eye at a given angle of reflection form a curve of this kind.[28]

As for the refractions made in the prism's surfaces *FG* and *FH*, it does not matter what they are in the farther one *FG*, provided that the rays emerge

"Further experiments," pp. 30–1), and again briefly in his draft reply to Hooke (Expt. 3, Add. 3970, f. 440ʳ).   (27) ut et pauciores] Originally: vel (or).

(28) The shape of this curve can be determined by adopting Newton's assumption that the reflected rays emerge from the prism normal to the face *FH*. To the base of the prism from the eye at *S*, drop a perpendicular that will intersect the red rays *St* at an angle equal to the critical angle *ι*. If the ray *St* is rotated about this perpendicular, generating a cone with vertex semi-angle *ι*, then all red rays within the circular base of this cone will be partially reflected, whereas all those outside of it will be totally reflected. Thus in the plane *HI* the boundary of the region of total reflection for red rays will be the (segment of a) circle with radius $h \tan \iota = h/\sqrt{[n_t^2 - 1]}$, where *h* is the height of the eye above the prism's base; this applies similarly for rays of other colors. This simple derivation is due to John Herschel, "Light," §555, pp. 446–7.

perpendiculariter emergant, angulo *KHG* existente quasi $40\frac{1}{4}^{\text{lgrad.}}$. Quod si angulus ille major existat colores in linea *DE*, adjuvante refractione paulò distinctiores evadent, et minùs distincti si sit minor. Major etiam oculi a Prismate distantia vel (quod perinde est) pupillae coarctatio colores nonnihil perficit.

[117]   Ad haec cùm duo Prismata parallelis axibus et basibus contiguis ad invicem applicantur et in eo situ colligantur, ijdem omnino effectus per radios ab aere intercluso reflexos / producentur, sed radij transmissi contrarios exhibebunt. Esto *ACDB* (fig II, 49) sectio utriusque Prismatis ad eorum longitudines perpendicularitèr transversa et *CB* contactus basium aut potiùs aer interclusus. Quippe Prismata vix queant tam arctè comprimi quin ut aer nonnullus in morem tenuissimae lamellaé maneat interclusus. His positis oculo *S* radios a *CB* lamella aeris interjecta

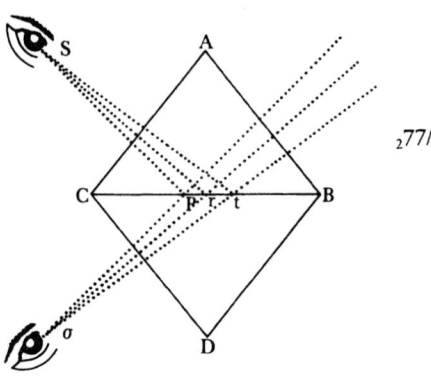

₂77/

Figure II, 49

reflexos excipienti omnia apparebunt ut ante, at oculo σ trajectos excipienti omnia cernentur contraria, spatio *tB* opaco et obscuro existente, et *Cp* t[r]anslucido, ac eorum confinio *pt* juxta *t* ruborem saturum juxtaque *r* citrium flavumque exhibente, qui color usque ad *p* gradatim diluitur, ubi in album *pC* desinit. Et hi colores longè intensiores et illustriores apparent quàm subcaeruleus color ex alterâ parte ad oculum *S* reflexus. Quorum quidem omnium rationes e supradictis patent, siquidem e radijs versus oculum σ tendentibus qui incidunt in superficiei partem *tB* omnes propter nimiam obliquitatem aliò reflectuntur solique rubriformes superficiem istam a *C* ad usque limitem *t*, viridiformes ad limitem *r*, et purpuriformes ad limitem *p* tantùm pervadere possunt.[29]

     Caeterùm hic cavendum est nequa lux in superficiem *CB* a parte *D* incidat, quae vel ad oculum σ reflexa vel transmissa ad oculum *S*, colores conturbet. Et insuper ne refractiones a superficiebus *AB* et *CD* factae ad effectus jam explicatos quicquam conducere videantur, praestat ut superficies istae statuantur parallelae, quo mutuos effectus (ex opinione receptâ) destruere possint.

/ 3. DE PHAENOMENIS LUCIS PER MEDIUM REFRACTIVUM PARALLELIS   ₂78/ PLANIS TERMINATUM TRANSMISSAE.

Lect 13   Transactis triangularium Prismatum phaenomenis: quae quadrangulis per
[118] parallela plana efficiuntur jam opportunè subveniunt enarranda. Id quod lubentius aggredior cùm Philosophi hactenus crediderint colores nullos hoc pacto generari, existimantes posteriorem superficiem effectus omnes per contrariam refractionem radijs auferre quos prior inducit,[1] et hoc pro experto habere rati quod in vitris fenestrarum aut alijs consimilibus nullos produci

perpendicularly from the closer one *FH* with the angle *KHG* being about $40\frac{1}{4}°$. But if that angle is larger, the colors in the line *DE* turn out a little more distinct with the help of the refraction, and if it is smaller, less distinct. A greater distance of the eye from the prism or, equivalently, a contraction of the pupil also perfects the colors somewhat.

Moreover, when two prisms are joined to one another with their axes [117] parallel and bases contiguous and are fastened in that position, entirely identical effects will be produced by the rays reflected from the enclosed air, but the transmitted rays will exhibit contrary ones. Let *ACDB* (Fig. II, 49) be a transverse section of both prisms perpendicular to their lengths, and *CB* the contact of their bases or rather the enclosed air. The prisms, to be sure, can hardly be so firmly compressed but that some air remains enclosed in the form of a very thin plate. These things having been set up, everything will appear to the eye *S* receiving rays reflected from the interposed plate of air *CB* as before, but to the eye *σ* receiving transmitted ones everything will be perceived reversed: the space *tB* being opaque and dark and *Cp* transparent, and their boundary *pt* exhibiting near *t* a full red, and near *r* orange and yellow, which color is gradually diluted as far as *p*, where it ends in white, *pC*. These colors appear far more intense and brighter than the bluish color reflected on the other side to the eye *S*. The reasons for all this are certainly clear from the preceding, since all the rays that tend toward the eye *σ* and fall on the side *tB* of the surface are reflected elsewhere because of their excessive obliquity; and only red-making ones can pass through that surface from *C* up to the limit *t*, green-making ones to the limit *r*, and purple-making ones to the limit *p*.[29]

But one must be careful here that no light falls onto the surface *CB* from the side *D*, which light when reflected to the eye *σ* or transmitted to the eye *S* would disturb the colors. Moreover, so that the refractions made by the surfaces *AB* and *CD* do not seem to contribute anything to the effects just explained, it is preferable that those surfaces be made parallel whereby (according to the received opinion) they can destroy their mutual effects.

3. The Phenomena of Light Transmitted through a Refractive Medium Bounded by Parallel Planes

Since the phenomena of triangular prisms have been completed, those that Lecture 13 are produced by the parallel planes of quadrangular prisms can now be [118] appropriately discussed. I undertake this rather gladly, since philosophers have hitherto believed that no colors are generated in this way, for they suppose that by a contrary refraction the second surface destroys all the effects in the rays that the first one induced.[1] Instead of testing this they consider it as certain, because in the glass of windows or in other quite

(29) Newton first described this experiment in "Of Colours," §25, Add. 3975, f. 5ʳ.

(1) This is clearly a reference to Descartes and Hooke; see Lect. 6, note (26).

videant. At in eo decepti sunt quod hujusmodi colorum quantitas et perfectio
dependet a distantia parallelarum superficierum. In laminis quidem vitreis
propter parvum superficierum intervallum colores sunt adeò tenues et exiles
et in spatio tam angusto comprehensi, ut effugiant sensus: at cum vitra magis
crassa adhibentur, aut potius vitrea vascula parallelepipeda aquae limpidissi-
mae plena, colores tunc liquidò generari cernuntur.[2]

[119]   Nam concipe *ABCD* (fig II, 50) esse vitreum vel aqueum parallelepipedum

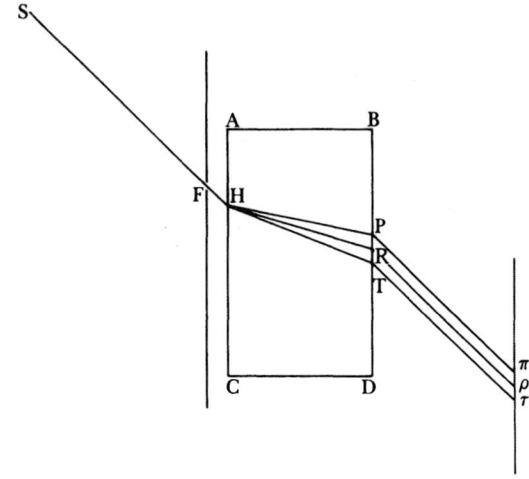

Figure II, 50

aere circumcinctum, cujus ex oppositis et parallelis planis duo lineis *AC* et
BD designentur. Et sol illud per exiguum foramen *F* obliquè transluceat.
Ejusque paralleli vel convenientes radij in anteriori superficie ad *H* ita debent
inaequaliter refringi ut ab invicem deinde divergant usque dum incidant in
posteriorem superficiem ad *PT* et ibidem colores omnigenos depingant[3] perinde ut supra sat fusè explicui. Jam cùm propter parallelismum refringentium
superficierum radij tantùm a posteriori recurventur quantum incurvantur a
priori, necesse est ut sibi ipsis ex aere secundum *SH* incidentibus emergant
paralleli, adeòque distantias ac positiones / acquisitas in infinitum servent, et   ₂79/
elicitos colores eousque sine aliquâ variatione promittant.[4] Quemadmodum
si *HP* e refractis ad *H* sit purpuriformis radius et *HT* rubriformis, eorum
denuò refracti *Pπ* ac *Tτ* incidentibus [*S*]*H*, adeoque sibimet ipsis paralleli
emergent, et proinde purpuram et rubedinem quàm ad *P* ac *T* exhibuere, ad
quamlibet distantiam *ππ* immutatam transferent, et sine quavis uspiam variatione conservabunt; purpureo a *P* in *π* translato, rubeo a *T* in *τ*, caeterisque a
locis intermedijs in loca correspondentia.

[120]   Hoc equidem praecisè deberet evenire si modo radij secundum eandem *SF*
vel parallelas lineas in hoc Prisma inciderent, siquidem tunc emergerent paralleli; at cùm inclinantur ad invicem uti de promanantibus a diversis partibus
solaris disci contingit, tunc etiam emergent inclinati et eapropter mutationes
quasdam in ulteriori translatione patientur. Utpote circuli a singulis radiorum

similar cases they do not see any produced. But they are deceived in this, because the quantity and perfection of colors of this kind depend on the distance of the parallel surfaces. In fact, because of the small interval of the surfaces in glass plates, the colors are so fine and subtle and contained within such a narrow space that they escape the senses; but when thicker glasses are used, or preferably little parallelepipedal glass vessels full of very clear water, then colors are clearly perceived to be produced.[2]

For instance, imagine *ABCD* (Fig. II, 50) to be a glass or water parallele- [119] piped surrounded by air; let its opposite and parallel planes be represented by the two lines *AC* and *BD*; and let the sun shine obliquely through that narrow hole *F*. Then its parallel or coincident rays must be unequally re- fracted in the first surface at *H*, so that they afterward diverge from one another until they fall on the second surface at *PT* and depict[3] every kind of color there, as I explained sufficiently amply above. Now, since the rays, because of the parallel refracting surfaces, are bent back by the second sur- face just as much as they are bent in by the first one, they must emerge parallel to themselves when they were incident from the air along *SH*; hence they preserve indefinitely their acquired distances and positions and reveal[4] the colors drawn out to that point without any change. For example, if *HP* is a purple-making ray from among those refracted at *H*, and *HT* is a red-mak- ing one, their rays, *Pπ* and *Tτ*, refracted once more, will thus emerge parallel to the incident ones [*S*]*H* and so to one another. Consequently, they will transfer with any unchanged distance, *ππ*, the purple and red that they exhib- ited at *P* and *T* and will preserve them everywhere without any variation, with the purple translated from *P* to *π*, the red from *T* to *τ*, and the others from the intermediate positions to corresponding positions.

Precisely this, of course, ought to occur, provided that the rays fell on this [120] prism along the same line *SF* or parallel ones, since they would then emerge parallel. But when they are inclined to one another, as happens on account of their flowing from different parts of the solar disk, then they will also emerge inclined and consequently experience some changes in the translation on the farther side. Namely, the circles that are made by the individual kinds of rays

---

(2) Given the force of Newton's rejection of Descartes's and Hooke's view in this lecture, it is somewhat disconcerting to find that he afterward in the *Opticks* essentially repudiated its sub- stance to side with them. He relates there that he "found" that whenever light emerges from parallel (or inclined) refracting surfaces "in lines parallel to those in which it was incident, it continues ever after to be white" (Bk. I, Pt. II, Prop. III, Expt. VIII, Add. 3970, f. 98ʳ = *Opticks*, p. 94).

(3) Originally: *expingant*.

(4) Read: *promant*.

generibus effecti, ex quibus in longum dispositis, colorata solis imago in superficiem *BD* procidens constituitur, propter divergentiam ṛadiorum in foramine *F* decussantium eo dilatiores evadunt quo radij longiùs post emergentiam fluunt; dum eorum centra quae radijs a centro Solis secundum eandem quampiam lineam ante refractionem affluentibus illuminantur, easdem post refractionem distantias et positiones inter se perpetuò conservant. Et hinc est quòd spatium πρτ solari luce in tenebrosum cubiculum immissa illuminatum, eò magis dilatetur et in orbicularem formam contrahatur, quo longiùs post Prisma terminatur; et viriditas in medio *R*, si qua sit, paulatim transmigret in albedĩnem; vel si nulla sit, sed propter angustiam Prismatis hujus aut amplitudinem foraminis lucem intromittentis albedo medietatem colorum occupet, eadem albedo sensim dilatetur. Sed colores tamen hinc inde non diluuntur, nec in spatium angustius contrahuntur, utut minùs luminosi propter dilatationem imaginis evadant.

[121] / Ad haec si mediante parallelepipedo(5) intueamur visibilia, coloribus non ₂80/
secus tingentur quàm si Prisma triangulare adhiberetur, praesertim si parallelepipedum ad pertransientes radios sat obliquetur, ut multùm refringat, et objecta sint admodum propinqua. Nam si objecta longinqua sunt, sive intervallum istud intercedat Parallelepipedum et objecta, sive Parallelepipedum et oculum, utcunque refractio per obliquitatem Parallelepipedi fiat magna, colores tamen non generabuntur. Sit [X] (fig II, 51) punctum lucidum radios per

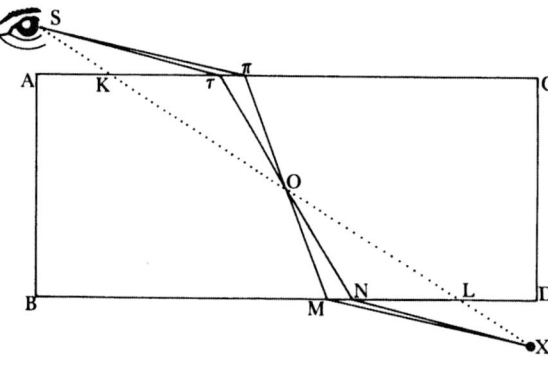

Figure II, 51

Parallelepipedi refringentia plana *AC* et *BD* ad oculum *S* emittens et manifestum est quod ducta *SτNX* quae rubiformem radium designet et *SπMX* quae designet purpuriformem, hi radij ad utramque superficiem aequaliter refringentur, adeoque triangula π*Sτ*, *MXN* similia conficient, purpuriformi radio propter majorem refrangibilitatem hinc et inde apud π et *M* plus vergente a directo tramite quàm rubiformi; Unde necesse est ut sese alicubi intra Prisma decussent quemadmodum videre est ad *O*, iterum conficientes triangula π*Oτ* et *MON* similia, sive trapezium *SπOτ* simile trapezio *XMON*. Adeoque oculum petent tanquam si primariò fluxissent ab eodem *O*, et refractionem ab unicâ tantùm superficie *AC* passi fuissent. Et hinc non tantùm sequitur co-

and are disposed lengthwise to form the sun's colored image falling upon the surface *BD* become wider the farther that the rays flow after emerging, because of the divergence of the rays crossing in the hole *F,* whereas their centers, which are illuminated by rays flowing from the sun's center along some one and the same line before refraction, perpetually preserve the same distances and positions among themselves after refraction. Consequently, the space $\pi\rho\tau$ illuminated by solar light admitted into a dark room becomes more enlarged and drawn into an orbicular shape the farther behind the prism it is terminated. Also, the green in the middle *R,* if there is any, passes somewhat into white; or if, because of the narrowness of this prism or the size of the hole admitting the light, there is no green but white fills the middle of the colors, then the same white is gradually expanded. Nevertheless, the colors on each side are neither diluted nor drawn into a narrower space, however less bright they may become because of the expansion of the image.

Moreover, if we look at visible objects by means of a parallelepiped, they [121] will appear tinged with colors just as if a triangular prism were used; particularly if the parallelepiped is sufficiently inclined to the rays passing through, so that it refracts them very much, and the objects are quite close. For if the objects are far away, whether that distance stands between the parallelepiped and the objects or between the parallelepiped and the eye, however great the refraction may become because of the inclination of the parallelepiped, colors will still not be generated. Let [*X*] (Fig. II, 51) be a luminous point sending out rays through the parallelepiped's refracting planes *AC* and *BD* toward the eye *S.* Then drawing *S*τ*NX* to represent a red-making ray and *S*π*MX* a purple-making one, it is clear that these rays will be equally refracted at each surface so that they will form the similar triangles π*S*τ and *MXN,* with the purple-making ray, because of its greater refrangibility, diverging on each side, at π and *M,* more from a straight path than the red-making one. Accordingly, they must cross one another somewhere within the prism, as can be seen at *O,* again making similar triangles, π*O*τ and *MON,* or rather trapezoid *S*π*O*τ similar to trapezoid *XMON.* Hence they will aim for the eye as if they had flowed originally from that same *O* and had undergone refraction only by the one surface *AC.* It consequently follows not only that colors

---

(5) Newton's scribe, Wickins, here and in sequel in fact wrote "parallelelipedo," a slip that Newton himself corrected in the next paragraph. In line with Newton's intentions, we have replaced the erroneous "l" with a "p" and otherwise corrected Wickins's spelling of this word.

lores generari, sed et angulum $\pi S\tau$, sive colorum apparentem latitudinem, aliasque circumstantias pro qualibet oculi positione determinari posse. Quemadmodum manifestius erit si conferas cum experimento quo objecta in aquam altè immersa obliquè inspicienti coloribus nonnihil tincta videntur propter refractionem stagnantis superficiei. Nam *AC* superficiem stagnantis aquae, et *O* objectum aliquod immersum quod spectator *S* intuetur referre potest. Quod quidem *O* facilè invenies ducendo rectam *SX* quae refringentes superficies secet in *K* et *L* ac dividendo in *O* ut sit *SK . LX :: SO . OX,* sive :: *KO . OL.*[6]

[122]   / Quinimo ad haec experienda pro Parallelepipedo vas optimè adhiberi ₂81/ potest quod in fundo transfoditur et vitri laminâ perpolitâ et Horizonti parallelâ resarcitur ut aquam cohibere possit. Nam cùm aqua ad altitudinem pedis aut ampliùs infunditur, lux per vitri et aquae istius parallelas superficies obliquè trajecta colores pro more explicato producet, possisque suc[c]essivè collocando objecta ad *X* et *O* phaenomena conferre.[7] Id quod etiam fieri potest disponendo duo vitrea Prismata triangularia ut *ACE, BDF* (fig II, 52)

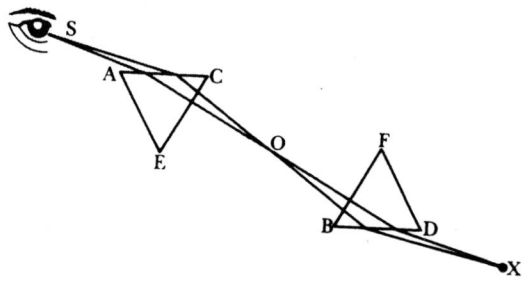

Figure II, 52

ad distantiam pedis aut amplius in eo situ ut eorum latera correspondentia *AC* ad *BD* et *CE* ad *BF* evadant parallela, et radij per interiores superficies perpendiculariter proxime trajiciantur. Tunc enim exteriores *AC* et *BD* refringentia plana Parallelepipedi referent. Et propter vitri majorem vim refractivam quàm aquae colores elicientur magis illustres.

[123]   Et haec de colorum a parallelis planis genesi monuisse sufficiat, nisi forte juvet annotare diversitatem effectuum qui ab hisce producuntur et a triangularibus Prismatibus. Cujusmodi sunt 1° Quod colores cum in papyrum projiciuntur, splendidiores evadunt per auctam papyri longinquitatem si modo prisma sit triangulare; sin Parallelepipedum, hebescunt. 2° Cum objecta per prismata triangularia transpiciuntur colores itidem splendidiores evadunt ex objectorum aucta longinquitate, at secus fit in Parallelepipedis. 3°. Cum Sol translucet Prisma triangulare, colores oriuntur terminando lucem ex utravis parte Prismatis; at cum translucet Parallelepipedum, colores non oriuntur terminando lucem a posteriori ejus parte. Cujus rei ratio est quod heterogenei radij a triangulari prismate divergentes fiunt, adeoque post emergentiam plus plusque segregantur; at in parallelismum restituuntur[8] emergentes e Paralle-

are generated but also that the angle $\pi S\tau$, or the colors' apparent breadth, as well as other features, can be determined for any position of the eye. This will be clearer if, for example, you compare it with the experiment whereby objects deeply immersed in water when viewed obliquely are seen somewhat tinged with colors because of the refraction of the still surface. For, AC can represent the surface of the still water and O some immersed object that the observer S views. In fact, you will easily find O by drawing the line SX, which cuts the refracting surfaces at K and L, and by dividing it at O such that $SK : LX = SO : OX$ (or) $= KO : OL.$[(6)]

To be sure, in doing these experiments instead of a parallelepiped you can [122] very well use a vessel that is pierced in its base and patched with a well-polished glass plate parallel to the horizon so that it can hold water. For when it is filled with water one or more feet deep, light cast obliquely through the parallel surfaces of this glass and water will produce colors in the way explained, and you can compare the phenomena by successively placing objects at X and O.[(7)] This can also be done by arranging two triangular glass prisms, such as ACE and BDF (Fig. II, 52), one or more feet apart in such a position that their corresponding sides prove to be parallel, AC to BD and CE to BF, and the rays may be transmitted through the interior surfaces nearly perpendicularly. Indeed, then the exterior surfaces AC and BD will represent the parallelepiped's refracting planes, and because of the greater refractive force of glass than of water, more brilliant colors will be elicited.

Let these things suffice for instruction about the genesis of colors by paral- [123] lel planes, except that it may perhaps help to note the difference of the effects that are produced by these and by triangular prisms. They are of the following kind: (1) When colors are projected onto a paper, they become brighter by an increase in the paper's distance only if the prism is triangular; but if it is a parallelepiped, they become fainter. (2) When objects are viewed through triangular prisms, the colors likewise become brighter by an increase in the objects' distance; but it turns out otherwise in parallelepipeds. (3) When the sun shines through a triangular prism, colors arise by terminating the light on either side of the prism; but when the sun shines through a parallelepiped, colors do not arise by terminating the light on its rear side. The reason for this is that heterogeneous rays are made divergent by a triangular prism, so that after emerging they are more and more separated, but emerging from a parallelepiped they are restored to parallelism[(8)] and no longer recede from

---

(6) Since the triangles $S\pi K$ and XML are similar, then $S\pi : MX = SK : LX$; and since the triangles $S\pi O$ and XMO are also similar, then $S\pi : MX = SO : OX$. Newton's result, $SK : LX = SO : OX = KO : OL$, follows by permuting the first proportion and then, "dividendo," subtracting the denominators from the numerators.

(7) Newton first described the colors cast by a water-filled parallelepiped in "Of Colours" (§15, Add. 3975, f. 3^v = Hall, "Further experiments," p. 30) and then again in the first experiment of his draft reply to Hooke (Add. 3970, f. 439).

(8) in parallelismum restituuntur] Originally: parallelismum redipiscuntur (they recover their parallelism).

lepipedo, et non amplius ab invicem / recedunt. [4°.] Denique notum est quod ₂82/
colorum in extremam partem oculi in solem vel lucernam per Prisma triangulare respicientis quilibet astans videbit ordinem ei contrarium quem videt ipse
spectator. At cum Parallelepipedum adhibetur, idem erit ordo colorum in
utroque casu, propter decussationem radiorum in Parallelepipedo ubi spectator transpicit: Quemadmodum inspicienti schemata manifestum est.[9]

[124]   Et ex hac effectuum diversitate Phaenomenon componitur quo colorum ad
diversas distantias diversi fiunt ordines. Utpote per vas aqueum *ABCD* (fig II,

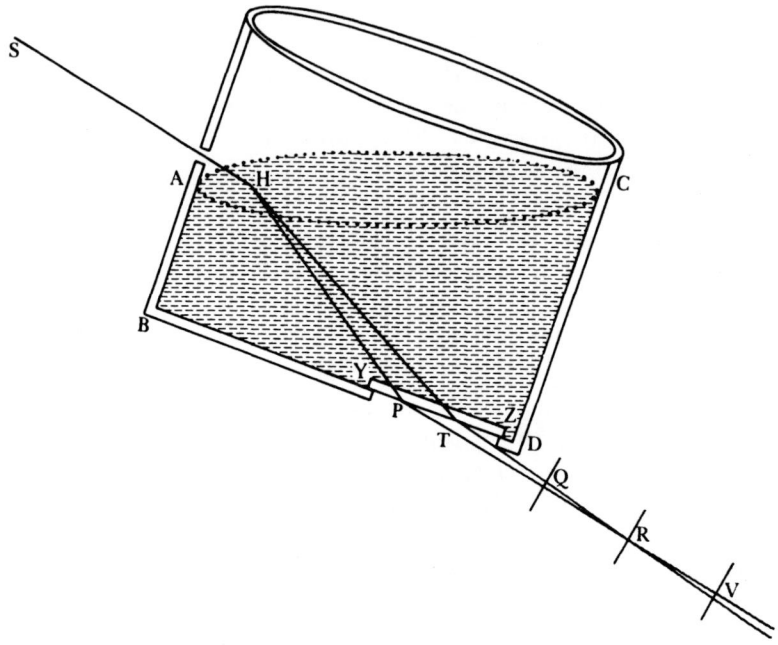

Figure II, 53

53) (in cujus fundo *YZ* refert laminam vitream quam in superioribus Horizonti parallelam esse supposui) jubare trajecto; si vas ad partes solem versus
allevetur ut fundum ejus magis obliquetur ad perlabentem lucem quam superior stagnans superficies; heterogenei radij propter majorem in egressu refractionem convergentes evadent, adeoque decussando mutabunt situm. Si lucem
chartâ proxime egressum excipias purpura cadet infra ruborem et chartam
longiùs differendo in loco decussationis per commisturam evanescent con-
versi in albedinem, ac postea de novo emergent in ordine contrario, quemadmodum videre est ad *Q, R,* et *V*.[10]

[125]   Ad aliud experimentum jam transeo his quodammodo affine quo colores
non a parallelis quidem superficiebus generantur sed a superficiebus ita in-
Fig II, 54 clinatis ut interposita reflexione parallelarum rationem habeant. Sit *SF* linea
coloribus omnigenis irradiata, quorum purpurei, dum ad *F* ingrediuntur
Prisma, refringuntur versus *H,* et rubei versus *G*; abinde verò reflectuntur ad
K et *I,* unde egredientes refringuntur denuò ad *M* et *L.* Dico jam si Prismatis

one another. [4] Finally, it is known that the colors in the outermost part of the eye when looking at the sun or a lamp through a triangular prism will be seen by anyone standing nearby in an order opposite to that which the observer himself sees. But when a parallelepiped is used, the order of the colors will be the same in each case because of the crossing of the rays in the parallelepiped when the observer looks through, as is evident to anyone who examines the figure.[9]

From this difference of effects a phenomenon is devised whereby the order [124] of the colors becomes different at different distances. Namely (Fig. II, 53), when a sunbeam passes through a vessel of water *ABCD* (in whose base *YZ* represents a glass plate, which in the preceding I assumed to be parallel to the horizon), if the vessel is raised on the side toward the sun so that its bottom is more oblique to the light flowing through than its upper, still surface, then the heterogeneous rays, because of the greater refraction upon exiting, will become convergent, and consequently by crossing they will change their situation. If you receive the exiting light on a nearby sheet of paper, the purple will fall below the red; and by moving the paper farther away into the crossing place, they will vanish by mixing together and being converted into white; and afterward they will emerge anew in the contrary order; as, for example, you can see at *Q*, *R*, and *V*.[10]

I now pass to another experiment related in [125] a way to these, whereby colors are in fact generated not by parallel surfaces, but by surfaces so inclined that by an intervening reflection they have the relation of parallel surfaces. Let *SF* be a line casting forth colors of every kind, Fig. II, 54 with the purple ones upon entering the prism at *F* being refracted toward *H* and the red toward *G*; but from there they are reflected to *K* and *I*, where upon leaving they are refracted once more to *M* and *L*. Now I assert that if the

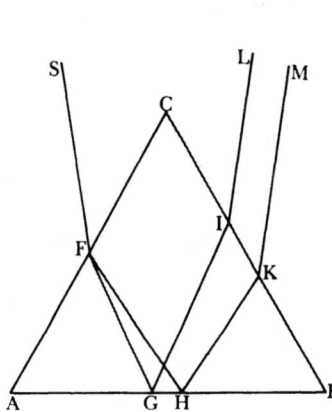

Figure II, 54

---

(9) Newton had investigated the differences between the colors produced by prisms and parallelepipeds in "Of Colours," §§14–21, Add. 3975, ff. 3ʳ–4ʳ. On the order of directly viewed prismatic colors also see the *Opticks*, Bk. I, Pt. II, Prop. IX, p. 133 = Add. 3970, f. 123ʳ; and *Correspondence*, 2:14, note 10.

(10) Newton's experimental arrangement here is virtually identical to one described by Hooke in the *Micrographia* (Observation IX, p. 58, and Plate VI, Fig. 2), except that Newton observed the colors after they were refracted a second time and left the vessel, rather than within it as Hooke did. With this difference, the phenomenon turns out directly contrary to that observed by Hooke, who saw the same color always on the same side of the beam, as is required by his theory.

anguli *ABC* et *CAB* aequentur, emergentes radij *IL* et *KM* paralleli erunt. Nam in triangulis *FGA, IGB*, cùm anguli *A* et *B* ex Hypothesi aequentur ut et anguli *FGA* et *IGB* propter aequalitatem incidentiae et reflexionis triangula erunt similia, angulique / *AFG, BIG* aequales. Atque adeò aequalis erit ₂83/ refractio in *F* et *I*, et inde anguli *CFS, CIL* aequales. Et eâdem ratione patebit angulum *CKM* angulo eidem *CFS* aequalem esse, adeóque radios *IL* et *KM* parallelos. Jam cùm radij *IL* et *KM* secundum eandem lineam *SF* successivè incidentes non secùs emergant paralleli quàm in praecedentibus ubi superficies refringentes erant parallelae, eadem omnia phaenomena quae ibi ostensa sunt huic competere certum est. Quemadmodum lucem solis coloribus tingi si Prisma satis amplum adhibeatur ut spatium *FGI* vel *FHK* sufficiat ad efficiendam sensibilem divergentiam radiorum antequam per iteratam refractionem in parallelismum reducantur, sed ejusmodi colores non perfectiores per longinquitatem obstaculi quo interciduntur evadere. Item istos colores, si oculo postposito immediatè excipiantur, eo magis manifestos fore quò objectum quod intuemur sit oculo propinquius, ut et eo magis quo anguli *CAB* et *CBA* majores existant; et eundem denique ordinem servare cùm in obversum oculum directè mittuntur, atque cùm cernuntur ad parietem aliudve obstaculum terminati. Haec inquam evenire debent si amplum Prisma adhibeatur (quale ex aqua vitro circundata fabricari possit,) et anguli *A* et *B* constituantur aequales. At in angustis prismatibus distantia radiorum *IL* et *KM* minor est, quàm ut colorum sensibilis possit esse latitudo; et cum anguli *C*[11] et *B* sunt inaequales perinde est ac si refringentes superficies in praecedentibus non sint parallelae, et similes sunt effectus.

[126]   Quod de coloribus dicitur cum unica tantum reflexio refractionibus intervenit facilè applicatur ad alios casus ubi plures interveniunt. Sed placet aliquod praeterea de reflexionibus exponere quibus geminantur[12] effectus, quos / solae refractiones exhibere possunt. Sit *SF* (Fig: II, 55) ut priùs linea diversis ₂84/ coloribus successivè irradiata qui versus *p, t*, aliaque intermedia loca pro gradibus refrangibilitatis [refracti][13] a Prismatis latere *BC* reflectantur ad *M*,

Figure II, 55

prism's angles *ABC* and *CAB* are equal, the emerging rays *IL* and *KM* will be parallel. For considering the triangles *FGA* and *IGB*, since the angles *A* and *B* are equal by hypothesis, as are the angles *FGA* and *IGB*, because of the equality of incidence and reflection, then the triangles will be similar, and the angles *AFG* and *BIG* will be equal. Hence the refraction at *F* and at *I* will be equal, and consequently the angles *CFS* and *CIL* will be equal. By the same reasoning it will be clear that the angle *CKM* is equal to the same angle *CFS*, and thus that the rays *IL* and *KM* are parallel. Now, since the rays *IL* and *KM* incident successively along the same line *SF* emerge parallel, just as in the preceding experiments where the refracting surfaces were parallel, it is certain that all the same phenomena that were shown there apply to this. For instance, the sun's light is tinged with colors if a wide enough prism is used so that the space *FGI* or *FHK* is sufficient to produce a sensible divergence of the rays before the renewed refraction returns them to parallelism; but colors of this kind do not become more perfect in proportion to the distance of the obstacle intercepting them. Likewise, if those colors are received by an eye placed directly behind, they will be more evident the nearer the object that we view is to the eye, and also the greater the angles *CAB* and *CBA* are. Finally, they keep the same order when they are sent directly into an eye facing them and when they are perceived terminated on a wall or other obstacle. These things, I say, must occur if a wide prism is used (such as can be made from water enclosed by glass), and the angles *A* and *B* are made equal. But in narrow prisms the distance of the rays *IL* and *KM* is too small for the colors' breadth to be sensible; and when the angles [*A*] and *B* are unequal, it is exactly as if the refracting surfaces in the preceding were not parallel, and the effects are similar.

What is said about colors when only one reflection occurs between the [126] refractions may easily be applied to other cases where several intervene. But it is desirable to relate something in addition about reflections and how they duplicate[12] the effects that refractions alone can exhibit. Let *SF* (Fig. II, 55) as before be a line successively casting forth different colors which [after being refracted][13] to *p*, *t*, and the other intermediate places, according to their degrees of refrangibility, are reflected by the side of the prism *BC* to *M*,

(11) Read: *A*.
(12) *Editio princeps*: generantur.
(13) We follow the *editio princeps* with this addition.

N, &c: ubi iterum impingentes in latus *AC* refringuntur denuò ad *PT*: et colores ad *PT* perinde apparebunt atque ad *ππ* apparerent si modò radij *Fp*, *Ft*, &c. per duplum Prisma *ABC* (i.e. per Prisma cujus angulus verticalis *AC*[β] sit duplo major hujus angulo verticali *ACB*) rectà fluxissent ad *μν* et inde versus *ππ* refringerentur. Nam pares sunt omnes utrobique anguli sive a plano *BC* per *AC* versus *PT* resiliant radij, sive longiùs per *BC* pergant ad *ππ*. Utpote ang *CtN* (= ang *Btf*) = ang *Ctv*, et inde ang *CNt* = ang *Cvt* adeoóque ang *CNT* = ang *Cvτ*. Atque idem in alijs radijs intellige. Cùm autem praecipuae colorum ad *ππ* circumstantiae in superioribus tradantur, crambem jam reponerem si quid campliùs de persimilibus phaenomenis ad *PT* instituerem dicere.

4. De Phaenomenis Lucis per Media sphaericè terminata transmissae. Deque Iride.

Lect 14.
Octob 1672
[127]

Hactenus colores refractionibus planarum superficierum generatos contemplati sumus, jam de sphaericis superficiebus agendum est, et imprimis de lentibus seu figuris a duabus diversarum sphaerarum portionibus comprehensis. Ejusmodi autem Lens esto *MN* in fig II, 56 per quam lux solaris juxta *Fφ* necnon undique terminata transmittitur, sitque *HK* focus ad quem postea convergat. Et cùm radij similiter incidentes non omnes similiter refringantur: concipe quod radiorum secundum *OF* incidentium purpuriformes refringantur ad *K* rubiformes ad *H* et viridiformes ad / punctum intermedium *r*, et pari ₂85/ ratione de radijs secundum *Oφ* incidentibus concipe quod purpuriformes tendunt ad *H*, rubiformes ad *K* ac viridiformes ad *r*. Atque idem de radijs undique (juxta lentis peripheriam) terminatis[1] concipe: Et patebit primò si radij a papyro *DL* priùs terminentur quàm ad locum[2] *HK* conveniant, q[uò]d

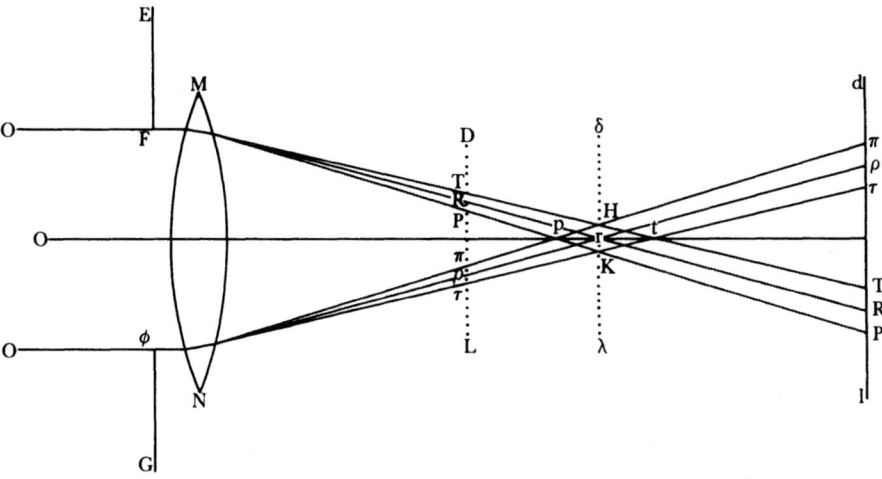

Figure II, 56

N, and so forth, where again striking against the side *AC* they are refracted once more to *PT*. The colors will appear at *PT* just as they would appear at *ππ*, if only the rays *Fp, Ft,* and so on, had proceeded directly through twice the prism *ABC* to *μν* (that is, through a prism whose vertex angle *AC[β]* was twice as large as this one's vertex angle *ACB*), and then they were refracted toward *ππ*. For all the angles on both sides are equal, whether the rays rebound from the plane *BC* through *AC* toward *PT* or they continue farther through *BC* to *ππ*; inasmuch as angle *CtN* (= angle *BtF*) = angle *Ctν*, and thus angle *CNt* = angle *Cνt*, and consequently angle *CNT* = angle *CνT*. The same thing must also be understood for the other rays. Since, however, the principal features of the colors at *ππ* were treated above, I would now repeat the same old story if I should undertake to say anything more about the very similar phenomena at *PT*.

4. THE PHENOMENA OF LIGHT TRANSMITTED THROUGH SPHERICALLY BOUNDED MEDIA. AND THE RAINBOW.

Hitherto we have considered the colors generated by the refractions of plane surfaces; now spherical surfaces must be treated, and first lenses or figures comprehended by two portions of different spheres. Now let *MN* (Fig. II, 56) be such a lens through which the sun's light, bounded at *Fφ* and also all around, is transmitted, and let *HK* be the focus to which it afterward converges. Since not all similarly incident rays are similarly refracted, imagine that of the rays incident along *OF*, the purple-making ones are refracted to *K*, the red-making to *H*, and the green-making to the intermediate point *r*; and for the same reason, imagine that of the rays incident along *Oφ*, the purple-making ones tend to *H*, the red-making to *K*, and the green-making to *r*; and also imagine the same thing for the rays terminated[1] all around along the periphery of the lens. First, it will be clear that if the rays are intercepted by the paper *DL* before they arrive at the focus[2] *HK*, the color red ought to be

Lecture 14
October 1672
[127]

---

(1) Omitted in *editio princeps.*
(2) Read (as *editio princeps*): focum (focus).

color rubeus in confinio lucis et umbrae deberet undique conspici. Utpote si lineae *FH, Fr* et *FK* ipsam *DL* in punctis *T, R,* et *P* secent; *FH* quidem in puncto *T, Fr* in puncto *R,* et *FK* in puncto *P*: posito similiter quòd φ*H,* φ*r,* et φ*K* eandem *DL* in punctis π, ρ, ac τ respectivè secent, et productis etiam *FH* et φ*K* donec sibi in *t* occurrant, ut et *FK* et φ*H* donec occurrant in *p*; constabit punctum *t* longiùs distare a lente quàm punctum *p,* quandoquidem cadit ultra locum[(2)] *HK, p* verò citra. Et proinde puncta *P* et π interjacent punctis *T* et τ. Constabit etiam purpuriformes radios per totum spatium *P*π solummodo dispergi, propterea quòd per integrum spatium *F*φ in lentem parallelè incidentes versus locum *p* refringantur. Et sic radij viridiformes spatium *R*ρ occupabunt, ut et rubiformes spatium *T*τ extra quod nulli omninò ex radijs parallelè incidentibus (nisi contingenter et nullâ certâ lege, propter bullulas quasdam aliaque vitia in vitro latentia, refracti) possint divaricare. Quare spatium *P*π a radijs omnium colorum illuminatum debet albescere. At cùm purpuriformes desint a spatijs *R* et ρ caeterorum mistura debet exhibere flavum. Atque ità cùm soli rubiformes extendantur ad *T* ac τ in locis *T* ac τ rubor apparebit et spatium illuminatum *P*π (quod orbiculare concipe) duobus colorum circulis rubeo flavoque tingetur. Haec equidem eveniunt cum charta *DL* inter Lentem et punctum *p* collocatur. Et colores tantò perfectiores evadunt, quò / charta sit puncto *p* propinquior et cum ₂86/ statuitur ad ipsum *p* albor e medio penitus evanescere deberet si modò radij a diversis partibus solaris disci ad Lentem manantes inciderent paralleli.[(3)] Quod si charta paulo longiùs amoveatur, uti ad *r* ubi viridiformes radij concurrunt, adversi colores ubique ad illam distantiam miscebuntur et se invicem ita delebunt ut vix alius quàm albor apparebit. Si charta deinceps adhuc longiùs transfertur puta ad *dl*: invertetur radiorum ordo, et puncta *T* ac τ interjacebunt punctis *P* et π, adeoóque spatium *T*τ ab omnibus coloribus illuminabitur, et proinde albescet, et in spatijs circa *R* et ρ ad quae rubor non extenditur caeruleus componetur, et violaceus apparebit in extremitate summâ *P* et π. Qui quidem colores non tantùm manifestiores sunt quàm rubor et flavus per interpositionem chartae inter lentem et focum ut prius emergentes, sed perpetuò manifestiores evadunt, quò charta adhuc longiùs amovetur.

[128]   Latitudo spatij sic tincti coloribus e praemonstratis petenda est,[(4)] vel etiam sic facilè determinari potest. Cùm differentia refractionis radiorum in refrangibilitate maximè discrepantium et similiter incidentium sit quasi septuagesima[(5)] pars totius refractionis ut ex ante ostensis patet; et cùm angu-

---

(3) Newton's diagram and description assume parallel incident rays, as if the sun were a point source, but taking the sun's finite size into account does not alter the described order of the colors, just their extent.

(4) Namely, the *Optica*, Pt. I, Prop. 37.

(5) Read: vicesimus quintus ($\frac{1}{25}$). Newton has mistakenly rounded off the value of the ratio $(P - T)/P = \Delta n/n$ of the greatest difference of refraction to the whole refraction, rather than the correct $(P - T)/(P - I) = \Delta n/(n - 1)$, where $P = 69\frac{1}{2}$, $T = 68\frac{1}{3}$, and $I = 44\frac{1}{3}$, as was determined in §I, 38 = 111. Although this slip is propagated through the rest of the calculation, he elsewhere uses the correct value. In repeating this simple derivation and calculation for Huygens, who

visible all around at the boundary of light and shadow. Seeing that if the lines *FH*, *Fr*, and *FK* intersect *DL* at the points *T*, *R*, and *P* (specifically, *FH* at the point *T*, *Fr* at the point *R*, and *FK* at the point *P*); and similarly, assuming that φ*H*, φ*r*, and φ*K* intersect the same *DL* in the points π, ρ, and τ, respectively; and furthermore, extending *FH* and φ*K* until they meet at *t*, and also *FK* and φ*H* until they meet in *p*; it will be evident that the point *t* is farther from the lens than the point *p*, since it falls beyond the focus[2] *HK*, while *p* falls before it. Consequently the points *P* and π lie between the points *T* and τ. It will also be evident that the purple-making rays are dispersed only through the whole space *P*π, because being incident parallel on the lens through the entire space *F*φ, they are refracted toward the place *p*. Similarly, the green-making rays will occupy the space *R*ρ, and the red-making ones the space *T*τ, beyond which none of the parallel incident rays can at all spread (unless they are refracted accidentally and by no definite law because of any bubbles and other flaws hidden in the glass). Therefore, the space *P*π, illuminated by rays of all colors, must be white. But since the purple-making rays cease at the spaces *R* and ρ, the mixture of the rest must exhibit yellow. Thus, since only red-making rays extend to *T* and τ, red will appear at *T* and τ, and the illuminated space *P*π (which is imagined to be circular) will be tinged with two circles of colors, red and yellow. This indeed occurs when the paper *DL* is placed between the lens and the point *p*. The colors turn out more perfect the nearer the paper is to the point *p*; and when it is placed at *p* itself, the white ought to vanish completely from the middle, provided that the rays flowing from different parts of the solar disk to the lens were incident parallel.[3] If now the paper is moved a little farther away, as to *r*, where the green-making rays meet, contrary colors will be mixed everywhere at that distance and will obliterate one another so that hardly anything other than white will appear. If the paper is in turn shifted still farther, suppose to *dl*, the order of the rays will be inverted, and the points *T* and τ will lie between the points *P* and π, so that the space *T*τ will be illuminated by all colors and will therefore be white. In the spaces around *R* and ρ, to which red does not extend, blue will be compounded, whereas violet will appear in the outermost ends *P* and π. These colors are in fact not only more evident than the red and yellow that arise by interposing the paper between the lens and focus, as earlier, but become continually more evident as the paper is moved still farther away.

The width of the space thus tinged with colors ought to be sought from [128] what has previously been shown,[4] or it can also be easily determined as follows. Since the difference of refraction of rays differing most in refrangibility and similarly incident is about $\frac{1}{70}$[5] part of the whole refraction, as is clear from what was shown before, and since the angle *HFK* represents the differ-

---

questioned the underived value of the lateral chromatic aberration given in the "New theory" (*Correspondence*, 1:95 = *Phil. Trans.*, 6 (1671/2):3079), Newton explained that "by my Principles the difference of refraction of the most difform rays is about the 24th or 25[t] part of their whole refraction" (Newton to Oldenburg, 8 July 1672, *Correspondence*, 1:212). In fact, the best value is 24$\frac{1}{2}$, since (where *R* = 69) the mean refraction *R* − *I*, rather than *P* − *I*, should be used.

lus *HFK* designet differentiam refractionis, angulusque *Frϕ* summam refractionum utrinque ad *F* et *ϕ* factarum, hoc est duplum refractionis juxta alterutrum *F* vel *ϕ*: angulus *HFK* erit quasi septuagesima[5] pars semissis anguli *Frϕ* sive $\frac{1}{140}$[6] pars totius *Frϕ*. et proinde subtensa *HK* quasi $\frac{1}{140}$[6] pars latitudinis *Fϕ* per quam luci patet aditus, aut ea fortasse paulo major.[7] Denique cùm sit *Fr . FR* :: *HK . TP*,[8] vel *τπ*; dabitur[9] intervallum *TP* vel *τπ* quod quaerebatur. Siquis autem cupit ut haec exactiùs determinentur, computatio non est adeo difficilis quin ut ipse adhibito calamo perficiat.[10] Quod ad lentes utrinque concavas attinet, e jam ostensis facilè constabit eas / lucem ₂87/ trajectam in ejus extremitate cum caeruleo tingere. Quae verò de lentibus utrinque convexis vel concavis dicuntur, de convexo-concavis aequipollentibus sunt etiam intelligenda.

[129]   Sunt et alia Phaenomena quas de lentibus explicare possem sed cùm oculi pars anterior (humor nempe Crystallinus ac tunica cornea)[11] speciem lentis radios ad retinam congregantis referat, de ipsâ maluissem nonnulla dicere. Eorum tamen quae de lente jam explicui nolo aliquid enixè repetere, cùm ad oculum facile applicentur, ut ut expertu satis difficilia sint propterea quòd aegrè possimus efficere ut oculi pars anterior et posterior ad invicem ita accedant, aut ab invicem recedant, sicut de lente et papyro Lucem terminante descripsi. Quapropter radij ut plurimùm eo modo in retinam procident quo posui terminatos esse in payrum *δλ*,[12] atque adeò propter misturam dissimilium quae ab oppositis partibus pupillae adveniunt colores mutuò delebuntur et convertentur in album si objectum quod intuemur sit album, aut in illum quemlibet colorem quocum objectum tingitur, siquidem ille tunc caeteris debet praevalere.

[130]   Caeterùm ex hisce detegitur modus quo omnia quae nudis oculis intuemur, possint ita tingi coloribus ac si Prisma interponeretur, licèt multò minùs manifestò: Idque si radij per alteram partem pupillae transituri ab interpositione digiti vel cujuslibet obstaculi propè oculum intercipiantur dum radij

---

(6) Read: $\frac{1}{50}$.

(7) This simple physical derivation, when the slip is corrected, agrees with the more rigorous derivation in Pt. I, Prop. 37; indeed, it is virtually identical with the numerical example at the conclusion of §I, 153. Newton assumes here that the incident rays make a small angle with the axis of the lens, so that $\widehat{HFK} \approx HK/Fr$ and $\widehat{Fr\phi} \approx F\phi/Fr$, or $\widehat{HFK}/\widehat{Fr\phi} \approx HK/F\phi$. He then in effect argues that $\widehat{HFK}/\frac{1}{2}\widehat{Fr\phi} = \Delta r/(r - i) \approx \Delta n/(n - 1)$, since $\widehat{HFK}$, or the angular dispersion Δr, is proportional to the difference of refraction Δn, and $\frac{1}{2}\widehat{Fr\phi}$, or the deviation $r - i$ for a single ray, is proportional to the whole refraction $n - 1$. Thus there results $\widehat{HFK}/\widehat{Fr\phi} \approx HK/F\phi \approx 2\Delta n/(n - 1)$, and the lateral chromatic aberration *HK* is given in proportion to the aperture of the lens *Fϕ*. In his letter for Huygens, Newton concluded that the chord *HK*, or the aberration, will be "about a 49th part of . . . the diameter of the lens; or in round numbers about a fifti[e]th part" (*Correspondence*, 1:213).

(8) This follows by Euclid's *Elements*, Bk. VI, Prop. 2.   (9) Originally: liquet (is evident).

(10) Originally continued: ubi focos lentium praefinivi faciendo (where I prescribed the foci of the lenses by making).

(11) The "crystallinus humor," or simply "crystallinus," is the lens of the eye. In general,

ence of refraction and the angle $Fr\phi$ the sum of the refractions made on both sides at F and ϕ (that is, double the refraction at either F or ϕ), the angle HFK will be about $\frac{1}{70}$[5] part of half of angle $Fr\phi$, or $\frac{1}{140}$[6] part of the whole of angle $Fr\phi$; and consequently the chord HK will be about $\frac{1}{140}$,[6] or perhaps a little more, of the width $F\phi$ through which an entry for the light lies open.[7] Finally, since $Fr : FR = HK : TP$[8] (or $\tau\pi$), the interval TP or $\tau\pi$ that is sought will be given.[9] If anyone, however, wishes to determine these more exactly, the computation is not too difficult to perform with a pen.[10] As for lenses that are concave on both sides, it will readily be evident from what has already been shown that those tinge the transmitted light blue on its edge. But what has been said about lenses that are convex or concave on both sides must also be understood about equivalent convexo-concave ones.

There are also other phenomena concerning lenses that I could have ex- [129] plained, but since the anterior part of the eye (namely, the crystalline humor and the cornea)[11] represents a type of lens that gathers rays at the retina, I preferred to say something about that. Still I do not wish to repeat laboriously any of those things I have already explained about lenses, since they may easily be applied to the eye however difficult they may be to test satisfactorily, because we can hardly make the anterior and posterior part of the eye approach or recede from one another, as I have described for the lens and the paper terminating the light. Accordingly, the rays for the most part will fall upon the retina in the way in which I have assumed them to be terminated on the paper $\delta\lambda$.[12] Indeed, because of the mixture of different rays arriving from opposite sides of the pupil, the colors will be mutually obliterated and converted into white (if the object we view is white) or into whatever color the object is colored, since that one must then predominate over the others.

From these things, however, a way is revealed whereby everything we view [130] with our naked eyes can be as tinged with colors as if a prism were interposed, although they are much less evident. This occurs if the rays about to pass through one side of the pupil are intercepted by inserting a finger or any other obstacle near the eye, while the rays about to enter the other side are

---

humors are fluid or liquid parts of the body, and tunics, such as the cornea, are membranes or coatings.

(12) Despite the eye being a compound lens and his unelaborated qualification, "for the most part [ut plurimùm]," Newton clearly recognizes, and is about to demonstrate, that the eye suffers from chromatic aberration just as any simple lens. Nonetheless, in the first half of the eighteenth century it was frequently argued that the eye must be achromatic, because in everyday experience we see what is about us without any color distortion. This misconception in part inspired the quest to develop achromatic lenses. Because they believed that the eye, as a lens, is achromatic, David Gregory, Chester Moor Hall, and Leonhard Euler all affirmed that compound achromatic lenses could in principle be constructed, and Hall in about 1732 even succeeded in constructing one. See Gregory, *Catoptricae et dioptricae*, Prop. XXIV, Scholium, pp. 98–9; Euler, "Sur la perfection des verres objectifs des lunettes," *Mémoires de l'académie des sciences de Berlin* (1747) [1749]:274–96 = *Opera omnia*, ser. iii, 6:1–21, esp. §11, p. 5; and also Jesse Ramsden's account of Hall's discovery, based on conversations with Hall, "Some observations on the invention of achromatic telescopes," a paper read to the Royal Society in 1789 and published by N. V. E. Nordenmark and Johan Nordström, "Om uppfinningen av den akromatiska och aplanatiska linsen," *Lychnos*, (1938):1–52; (1939):313–84, esp. pp. 368–9.

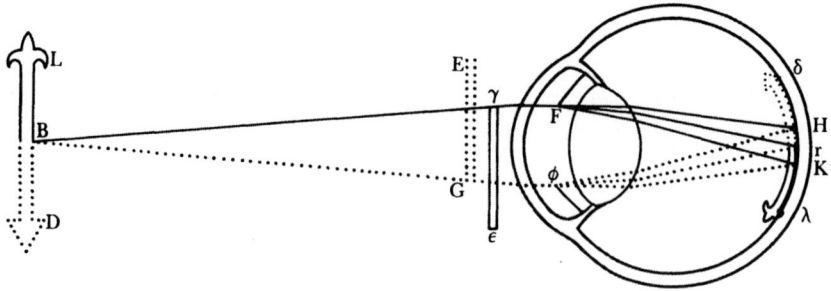

Figure II, 57

ingressuri alteram partem liberè transire permittantur. Hujusce vero rei duos casus non pigebit explicare; alterum cùm radios intercipimus ad partes versus objectum lucidius, posito nempe quòd objecta duo, album et nigrum, juxtaposita intueamur: et alterum cum radios intercipimus ad partes versus nigrius. Sit ergo in fig II, 57 *LB* objectum lucidum, et *BD* obscurum / quorum ₂88/ terminus communis sit *B* a quo radij in oculum δλ juxta oppositas partes pupillae *Fφ* promanantes sint *BF* et *Bφ*. Radij autem secundum lineam *BF* in oculum pergentes pro gradu refrangibilitatis refringantur versus *H, r,* et *K*. E contra verò qui pergunt in linea *Bφ*, refringantur versus *K, r,* et *H* caeteraque gradatim intermedia loca: prout de lente modo explicui. Ponamus jam quod εγ sit obstaculum quo omnes radij propè φ lapsuri intercipiantur, praetermissis *Bγ* et ejusmodi alijs per *F* solummodo tendentibus, et constabit primò quòd ex radijs a diversis partibus objecti *LB* manantibus, qui veniunt a partibus versus *L,* in retinam incidant propiùs ad λ quàm qui veniunt a parte *B,* siquidem in pupillâ decussant: et sic *BD* deberet radios versus *Hδ* emittere. Sed cùm illud *BD* propter nigredinem nullos penè radios in oculum jaculetur, retina λδ non ultra versus δ illuminabitur quàm ad *H*. Quinimò non ad *H* usque illuminabitur nisi a radijs rubiformibus. Viridiformes enim terminabuntur in *r,* et purpuriformes in *K*; spatio λ*K* a purpuriformibus, λ*r* a viridiformibus, et λ*H* a rubiformibus illuminato. Quamobrem spatium λ*K* propter omnium radiorum misturam albescet ad instar objecti *BL,* sed in exiguo spatio *HK* quod termino *B* respondet colores generabuntur, rubeus quidem ad *H* propter solos rubiformes radios illuc tendentes, et flavus ad *r* propter misturam viriditatis flavedinis ac rubedinis. Jam cùm omnia videantur pro more imaginum in oculum receptarum, constat objectum *LB* juxta extremitatem ejus *B* non distinctè cerni, sed coloribus rubeo & flavo tingi.

[131]   Ad eundem modum si transferatur obstaculum εγ, et caeteris stantibus objecto interponatur et oculo secundum adversas partes pupillae, prout videre est ad *EG,* eò ut radij juxta *F* intercipiantur radijque *Bφ* in oculum praetergrediantur:[13] constabit e contra quòd ex radijs a toto *BL* / prosilienti- ₂89/ bus purpuriformes occupabunt spatium *Hλ,* viridiformes spatium *rλ,* et rubiformes spatium *Kλ,* quare, spatio *Kλ* ut priùs albescente, color violaceus jam debet apparere in *H,* et caeruleus in *r*: et eapropter objecti *LB* extremitas *B* jam alijs tingetur coloribus, violaceo et caeruleo.[14]

(13) *Editio princeps*: praeter *BG* ingrediantur.

allowed to pass freely through. It will certainly not be burdensome to explain the two cases of this: the one, when we intercept the rays on the side of the brighter object, assuming, of course, that we view two juxtaposed objects, one white and one black; and the other, when we intercept the rays on the side of the blacker one. Therefore, in Fig. II, 57 let *LB* be a bright and *BD* a dark object and *B* their common boundary; and from there let *BF* and *Bφ* be rays flowing into the eye *δλ* along opposite sides of the pupil *Fφ*. Moreover, let the rays proceeding into the eye along the line *BF* be refracted toward *H*, *r*, and *K* according to their degree of refrangibility; but conversely, let those proceeding along *Bφ* be refracted toward *K*, *r*, and *H*, and by degrees to the other intermediate places, as I have just explained for a lens. Let us now assume that *εγ* is an obstacle by which all rays about to slip by near *φ* are intercepted, while letting by only *Bγ* and the others of its kind tending alongside *F*. It will be clear, in the first place, that of the rays flowing from various parts of the object *LB*, those that come from the parts toward *L* fall on the retina nearer to *λ* than those that come from part *B*, since they cross in the pupil, and thus *BD* must send out rays toward *Hδ*. But since *BD* on account of its blackness casts hardly any rays into the eye, the retina *λδ* will be illuminated in the direction of *δ* no farther than to *H*. In fact, it will be illuminated as far as *H* only by red-making rays; for the green-making rays will end at *r* and the purple-making at *K*, with the space *λK* illuminated by purple-making rays, *λr* by green-making ones, and *λH* by red-making ones. Consequently, the space *λK* will be white like the object *BL* because of the mixture of all rays, but in the small space *HK*, corresponding to the end *B*, colors will be generated: indeed, red at *H*, because red-making rays alone tend there, and yellow at *r*, because of the mixture of green, yellow, and red. Now, since all things are seen by the nature of images received in the eye, it is evident that the object *LB* will not be distinctly perceived near its end *B*, but will be tinged with the colors red and yellow.

Similarly, with everything else remaining unchanged, if the obstacle *εγ* is [131] shifted and placed between the object and the eye on the opposite side of the pupil, as can be seen at *EG*, so that the rays near *F* are intercepted while the rays *Bφ* go past[13] into the eye, it will be evident, on the contrary, that of the rays springing forth from the whole of *BL*, purple-making ones will occupy the space *Hλ*, green-making the space *rλ*, and red-making the space *Kλ*. Therefore, with the space *Kλ* being white as before, the color violet now ought to appear at *H* and blue at *r*. Consequently, the end *B* of the object *LB* will now be tinged with the other colors, violet and blue.[14]

---

(14) In a very late addition to the *Opticks* (Bk. I, Pt. II, Prop. VIII, p. 124), which is not in the manuscript (Add. 3970, f. 118ʳ), he again briefly and elliptically referred to this experiment. To convince Hooke of the value of his mathematical science of color, Newton claimed in his draft reply to him in 1672 that the colors produced in this experiment could in fact be "computed"; Expt. 9, Add. 3970, f. 433ᵛ. In 1808 it was still widely believed that the eye was achromatic. To prove the contrary, Carl Mollweide performed an experiment similar to Newton's before he came upon Newton's full account of it in the *Lectures,* which he then included in his "Ueber die Farbenzerstreuung in menschlichen Auge," *Annalen der Physik,* 30 (1808):220–34. Some twenty years earlier, Venturi also cited this experiment to demonstrate the chromatic aberration of the eye ("Considerazioni ottiche," p. 273), and he devised a number of new ones.

[132]   Et Ad eundem modum si duo quaelibet objecta vel ejusdem objecti diversae partes juxta-positae gradu lucis differant: etsi alterum non sit omniño nigrum tamen colores apparebunt in eorum communi termino rubeus quidem et flavus cum obstaculum ad partes versus objectum obscurius, violaceus autem et caeruleus cum ad partes versus objectum lucidius interponitur. Et ut paucis rationem denuò comprehendam; Necesse est ut radij ex unâ quavis parte pupillae colores producant cùm radij ex adversâ parte sistuntur a quorum omnium misturâ oritur temperamentum albedinis. An isthaec verò Phaenomena vulgò observantur haud scio. Sane non sunt inventu nec expertu tam difficilia, nec ab ijs quae Cartesius sub fine capitis Undecimi de Meteoris[15] edocuit tam aliena quin cuiquam potuissent occurrere; nisi forte quòd colores illi propter tenuitatem vix sint sensibiles. Experimentum itaque fiat per objecta longinqua quorum alterum sit nigerrimum et alterum satis candidum ad feriendum sensum, sed non tanta luce resplendens ut sensum obtundat vel pupillam constringat. Nam hujusmodi effectus sunt eò magis manifesti quo pupilla sit laxior et majori aperturâ radijs ingredientibus pateat.

[133]   Sunt et alij insigniores effectus, irides nempe vel coronae quales D Cartesius circa candelam quondam / observabat et in meteoris[16] explicuit. Et cum   ₂90/ illae solent apparere quando oculi figura ex aliquâ vi extrinsecus illatâ vitiatur. Necesse est ut a curvaturâ aliquâ vel plicâ in tunicis ejus de novo formatâ oriantur. Crystallino autem vis non imprimitur nisi mediantibus humoribus quibus undique cingitur.[17] Et cùm fluida facillimè cedant pressuris, humores illi vim quamlibet illatam ita per totam molem diffundent, ut Crystallinum vix possint inaequaliter premere, neque ideo figuram ejus vitiare.[18] Id enim experti sunt qui aquis altè submerguntur. Nam etsi tota aquarum moles incumbat illis, pressuram haud sentiunt, quae tamen foret maximè sensibilis si corporum submersorum partes ita premerent inaequaliter ut figuras eorum violare conarentur.[19] Restat ergo ut hujusmodi coronarum sive Iridum generatio vitiosis configurationibus tunicae corneae illatis tribuantur, idque eo magìs quòd radij maximam refractionem in exteriori ejus superficie patiantur, et proinde per leviora ejus vitia a recto tramite detorqueri possint. Utut non pernegem quin ijs qui laborant oculis, rugae aliquae (propter humorum defectum aut excessum) in Crystallini superficiebus non minùs quàm in tunicâ corneâ possint efformari. Nec non aliae etiam colorum causae possunt evenire, sed cum earum infinita sit varietas, et illae sint eminentiores quae a

---

(15) *Les Meteores* (*Meteora*) has only ten "discours" ("capita") and it is possible that Newton is referring to the colors seen about candles that are described at the end of Discourse IX and that he considers in the next article.

(16) *Meteora*, Ch. IX, § 7, *Oeuvres*, 7:713–4, 351–4.

(17) That is, by the aqueous humor on the anterior side and the vitreous humor on the posterior side.

(18) Newton's argument is a sound application of his newly wrought concept of hydrostatic pressure, which he developed just a few years earlier in an untitled manuscript known by its *incipit* "De gravitatione et aequipondio fluidorum" (Add. 4003 = Newton, *Unpublished Papers*, pp. 89–156); see Shapiro, "Light, pressure, and rectilinear propagation: Descartes' celestial optics and Newton's hydrostatics," *Studies in History and Philosophy of Science*, 5 (1974):239–96.

In the same way, if any two objects or different parts of the same object [132] that are juxtaposed differ in their degree of light—even if one is not completely black—colors will still appear on their common boundary, in fact, red and yellow when the obstacle is inserted on the side toward the darker object, but violet and blue when it is on the side of the brighter object. To express the explanation briefly again: The rays on any side of the pupil necessarily produce colors, when on the opposite side one stops the rays from the mixture of all of which the due proportion of whiteness arises. But whether these phenomena are generally observed, I do not know. They are certainly not so difficult to discover or to test, nor so unlike those that Descartes taught at the end of Chapter 11 of the *Meteorology*[15] that they could not have occurred to anyone, unless perhaps those colors are scarcely perceptible because of their fineness. The experiment, therefore, should be done with distant objects, one of which is very black and the other bright enough to strike the senses, though not shining with so much light that it stuns the senses or constricts the pupil. For effects of this kind are more evident, the wider the pupil and the greater its aperture open to the entering rays.

There are also other more prominent effects, namely, the rainbows or [133] coronas such as Descartes sometimes observed around a candle and explained in the *Meteorology*.[16] Since they usually appear when the eye's shape is deformed by some externally applied force, they necessarily arise from some newly formed curvature or crease in its membranes. However, a force is not impressed on the crystalline [humor] except by the mediation of the humors that surround it on all sides;[17] and since fluids yield very easily to pressures, those humors diffuse any applied force through the entire mass, so that they can scarcely press the crystalline [humor] unequally and so cannot deform its shape.[18] Indeed, those who have been deeply submerged in water have proved it. For although the entire mass of water may press them, they do not feel the pressure, which nevertheless would be most perceptible if the parts of their submerged bodies were pressed so unequally that they endeavored to deform their shapes.[19] It remains, therefore, that the generation of coronas or rainbows of this kind is to be attributed to defective configurations inflicted upon the cornea—and all the more so, because the rays undergo the greatest refraction in its exterior surface and consequently can be deflected from a straight path by its slightest defect. I do not, however, completely deny that in those who have trouble with their eyes some wrinkles (because of an excess or defect of the humors) can be formed in the surfaces of the crystalline [humor] no less than in the cornea. Certainly other causes of the colors can also arise, but since their variety is infinite and those that are sought in the defective shapes of the cornea are more prominent, I will not be

---

(19) Newton probably drew this example from Boyle's *Hydrostatical Paradoxes* (Oxford, 1666). In his notes in "Out of Philosophicall Transactions" on Oldenburg's review of Boyle's book (*Phil. Trans.*, 1 (1665/6):173–6) he jotted down, "Why divers feele not ye pressure of ye incumbent water; Descartes answer to this unsatisfactory" (Add. 3958, f. 11^r).

vitiosis tunicae corneae figuris petuntur non gravabor earum aliquod specimen exhibere unde caeterarum causae facilè patebunt.

Lect. 15
[134]
Notissimum est quod mollium partes non solum pressioni cedunt in quas vis immediatè imprimitur sed et aliae etiam partes semotae prout vim partium immediatè pressarum sustinent. Et ipse nonnunquam observavi in / laminis convexo-concavis et ex materiâ mediocriter rigidâ confectis (quales ₂91/ ex corijs bubulcis[1] in morem segmenti superficiei sphaericae contundendo formari possunt) quòd cùm in meditullio seu vertice premuntur, non solùm ibi cedunt tactui, sed et undique ad instar vallis annularem collem depressae vertici circumductum comprehendentis intus flectuntur, idque citiùs et magis manifestò, si sint paulò rigidiores juxta verticem, quàm propè periferiam. V. g. in fig II, 58: Sit *knκ* lamina sphaericè convexo-concava quae circulari

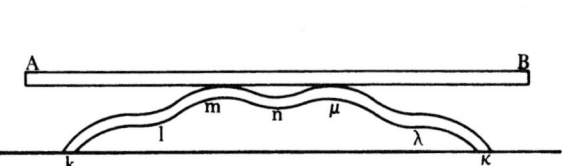

Figure II, 58

ejus extremitati tanquam basi incumbens mole aliquâ planâ et ad basin ejus parallelâ *AB* prematur; et manifestum erit quòd haec lamina maximè cedet pressioni in vertice *n,* ubi ab incumbente mole primò contingitur. Sed in alijs etiam locis ut in *λ* et *l* possit etiam intus recedere dum in locis intermedijs ut *m* et *μ* partes assurgunt. Atque hâc ratione configurationem acquiret haud dissimilem aquae undulanti, puteolo *n* referente centrum undarum, et ripa *mμ* referente undarum primam valle *lλ* circundatam. Et ad eundem modum possibile est ut tres vel plures valles premendo descendant quorum culmina internata sint pluribus undis se invicem subsequentibus consimilia. Et hujusmodi configurationes cessante pressione possunt aliquamdiu conservari, gradatim tamen evanescentes. Nam utprimùm pressio cessat cavitas *n* cessabit fortè et partes ibi in convexitatem assurgent et gradatim fient plus plusque convexae donec redeat figura quam ante pressionem habuêre; et sic caeterarum partium figurae ad pristinum statum gradatim redibunt. Jam / cum tunica cornea ad modum praefatum convexo-concava sit et mediocriter ₂92/ rigida et circa medietatem ejus paulò crassior, et proinde rigidior quàm juxta periferiam, et siquando figura ejus ab externâ pressione vitietur, probabile sit illam pressionem circa meditullium ejus maximâ ex parte contingere; itaque potest aliquando forsan accidere quod cùm premitur, non solum in apice cedat pressioni, sed quod in pluribus etiam circulis apici concentricis, parùm ascendat et alternis vicibus descendat,[2] et hujusmodi rugae concentricae possunt etiam ex defectu humorum quo tunicae flaccescunt, nec non ex alijs fortè causis accidere, et quantumvis exiguae sint possunt tamen radios ad alias atque alias partes retinae refringere, et sic efficere ut alij atque alij colorum circuli appareant. Sed ut videamus quo pacto ex hujusmodi rugis

reluctant to present some example of them, whence the causes of the others will readily be evident.

It is very well known that in soft bodies not only do the parts in which a force is immediately impressed yield to pressure, but also the other remote parts, insofar as they sustain the force of the immediately pressed parts. I myself have sometimes observed in thin convexo-concave sheets made out of moderately rigid material (such as can be formed from cowhides[1] by pounding them into the shape of a segment of a spherical surface) that when they are pressed in the middle or vertex not only do they yield to touch there but they are also bent in all around like a valley enclosing an annular hill that surrounds the depressed vertex, and all the more quickly and manifestly if they are a little more rigid close to the vertex than near the periphery. For example, let *knκ* (Fig. II, 58) be a thin, spherically convexo-concave sheet that lies on its circular edge as a base and is pressed upon by some flat mass *AB* parallel to its base, and it will be clear that this sheet will yield most to the pressure at its vertex *n*, where it is first touched by the incumbent mass. But at other places, such as at λ and *l*, it can also recede inward, while at intermediate places, such as at *m* and μ, its parts rise up. For this reason it will acquire a configuration not unlike undulating water, with the little pit *n* resembling the center of the waves and the wall *mμ* resembling the first of the waves surrounding the valley *lλ*. In the same way, it is possible that, by pressing, three or more valleys may descend with their peaks arising in between, similar to several waves following one another. Configurations of this kind can be maintained for some time after the pressure ceases, although they gradually vanish. For as soon as the pressure ceases, the cavity *n* will, as it happens, cease, and the parts there will rise up into the convexity and gradually become more and more convex, until the shape they had prior to the pressure returns; and in this way the shapes of the other parts will gradually return to their original state. Now, since the cornea is like that convexo-concave sheet, moderately rigid and near its middle a little thicker and consequently more rigid than along its periphery, and if its shape is ever deformed by external pressure, it is probable that that pressure occurs mostly around its middle. Therefore, it can perhaps sometimes happen that when it is pressed, not only does it yield to the pressure at its apex, but it also rises slightly and successively falls[2] in several circles concentric to the apex. Concentric wrinkles of this sort can also occur from a defect of humors whereby the membranes become flaccid, and perhaps from other causes too; and however small they may be, they can still refract the rays to different parts of the retina and thus make different circles of colors appear. But so that we

---

(1) Read: bubulis.
(2) The initial "assurgat" (rises up) was a slip.

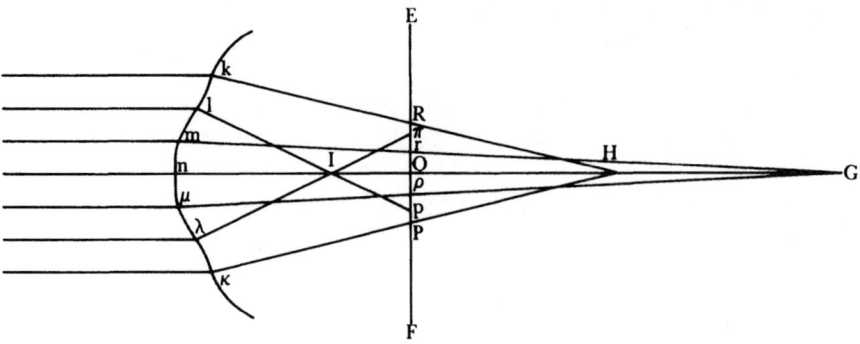

Figure II, 59

colores generari debent, ponamus radios e longinquo manantes sive parallelos in superficiem *k*κ (fig II, 59) ita ut dictum est intortam incidere,[3] et in ea refractos sisti deinde ab aliâ opacâ superficie *EF,* et cùm hujus superficiei partes depressiores radios ad puncta remotiora[4] congregent quàm partes ascendentes sive magis acclives, ponamus quòd radij circa meditullium ejus *mn*μ ubi maximè deprimitur congregantur ad *G.* Et quòd à partibus *l* et λ maximâ acclivitate surgentibus congregantur ad *I,* et sic quod a partibus *k* et κ ubi rursus deprimitur congregantur ad *H,* et quod ab intermedijs partibus còngregantur ad intermedia puncta. Ductis ergo *mG* & μ*G, lI* et λ*I, kH* et κ*H* occurrentibus superficiei seu obstaculo *EF* in punctis *r* et ρ, *p* et π, *R* et *P,* nec non axe *GHI* / occurrente eidem *EF* in puncto *O,* ut et refringenti ₂93/ superficiei *k*κ in puncto *n,* et posito quod illa *EF* interjaceat punctis *H* et *I;* manifestum erit perpendenti refractiones hujus *k*κ in singulis ejus punctis a centro *n* successivè ad extremitates *k* vel κ, quod radij prout longiùs ab *n* versus *m* per refringentem superficiem trajiciantur, incidant in obstaculum *EF* longiùs ab *O* versus *r* ad usque certum terminum puta dum ad radium *mr* deventum sit, deinde quòd facto regressu incidant propiùs ad *O* et postea ad alteras ejus partes pergant donec iterum fiat elongatio maxima[5] velut in *p* cùm deventum est ad radium *lp.* Tum denuò revertuntur radiorum occursus idque continuò, prout ab *l* versus *k* procedit refractio donec tertiò terminentur quemadmodum in *R* occursu radij *kR.* Ad eundem modum, lux inter *n* et κ refracta terminabitur in punctis ρ, π et *P.* Atque etiam si plures essent rugae, plures forent lucis terminationes. Caeterùm de luce per spatium *r*ρ diffusâ, cùm causa quòd extravagatur punctum *O* usque ad terminos *r* et ρ sit ejus parva refractio prope *m* et μ, sequitur quòd radij minùs refrangibiles hoc est rubiformes debent magis extravagari, et proinde terminus lucis *r* vel ρ debet rubedine tingi. Et sic de luce per spatium π[*p*] diffusâ, cùm causa quòd extravagatur punctum *O* usque ad terminos *p* et π sit ejus nimia refractio prope *l* et λ, sequitur quòd radij magis refrangibiles hoc est purpureum et caeruleum pingentes debent longiùs deviare, et colores eorum in exteriori parte termini *p* et π depingere, unde in interiori ejusdem termini parte rubiformes radij ad suos etiam colores depingendos debent praevalere. / Et simili ₂94/

may comprehend how wrinkles of this kind must generate colors, let us assume that rays flowing from far off, or parallel, fall[3] on that crooked surface *k*κ (Fig. II, 59), as has been described, and after having been refracted in it are then stopped by another, opaque surface *EF*. Since the more depressed parts of this surface gather the rays at more distant[4] points than the raised or more ascendent parts, let us assume that the rays near its middle *mn*μ where it is most depressed are gathered at *G*; that those from the parts *l* and λ rising with the greatest ascent are gathered at *I*; likewise that those from the parts *k* and κ, where it is again depressed are gathered at *H*; and that those from the intermediate parts are gathered at the intermediate points. Then draw *mG* and μ*G*, *lI* and λ*I*, and *kH* and κ*H*, which meet the surface or obstacle *EF* in the points *r* and ρ, *p* and π, and *R* and *P*; also draw the axis *GHI*, which meets the same *EF* in the point *O*, as well as the refracting surface *k*κ in the point *n*; and assume that *EF* lies between the points *H* and *I*. It will be evident to anyone who considers the refractions of *k*κ in each of its points successively from its center *n* to its ends *k* or κ that insofar as the rays pass through the refracting surface farther from *n* toward *m*, they fall on the obstacle *EF* farther from *O* toward *r* up to a certain bound, suppose until the ray *mr* is reached. Then they make a return and fall closer to *O*, and afterward they proceed to its opposite side until the greatest[5] elongation again occurs, as at *p*, when the ray *lp* is reached. Thereupon the intersections of the rays turn back once more—of course, continuously—according as the refraction proceeds from *l* toward *k*, until they are bounded for the third time, such as at *R*, the intersection of the ray *kR*. In the same way, the light refracted between *n* and κ will be bounded at the points ρ, π, and *P*. Moreover, if there were more wrinkles there would also be more bounds of the light. But concerning the light expanded through the space *r*ρ, because it spreads beyond the point *O* up to the bounds *r* and ρ, and its refraction is small near *m* and μ, it follows that the less refrangible rays, that is, the red-making ones, must spread farther, and consequently the bound of the light *r* or ρ must be colored red. Likewise, concerning the light expanded through the space π[*p*], because it spreads beyond the point *O* to the bounds *p* and π, and its refraction is very large near *l* and λ, it follows that the more refrangible rays, that is, those depicting purple and blue, must deviate farther and depict their colors on the outer side of the bound *p* and π; whence on the inner side of the same bound the red-making rays must predominate to depict

---

(3) Omitted in *editio princeps*.

(4) The original "propinquiora" (nearer) was, once more, just a slip. Since the focal distance of a spherical refracting surface is directly proportional to the radius, the greater the radius is, that is, the flatter or more depressed the surface, the more distant the focal point. But if the surface is so depressed that it becomes concave, as it is about the point *n* in the leather sheet in Fig. II, 58, then the rays diverge; and Newton was therefore careful to draw the corresponding portion of the lens in Fig. II, 59 as flat.

(5) Added.

ratione radij circa *k* et *κ* refracti si sint rubiformes tendent ad exteriorem partem termini *R* et *P,* et ad interiorem si caeruliformes. Et sic tres habebuntur irides *RP* extra rubea et intra caerulea, *pπ* extra caerulea et intra rubea, *rp* extra rubea quae etiam debet esse intra caerulea nisi fortè quòd color ille a rubeo propter parvitatem refractionis in *m* et *μ* haud satis secernitur ut fiat sensibilis, et praeterea[6] quòd multum obscuratur a copiâ lucis undique per *rOp* locum imaginis lucidae quam cingunt irides, sparsae. Harum verò iridum formae et relationes inter se possunt varijs modis mutari, idque non tantùm e varijs formis quas superficies *kκ* possit induere, sed etiam è varijs distantijs inter hanc *kκ* et obstaculum *EF.* Ut si statuatur paulo magis distantes quàm designavi, circuli *RP* et *πp* possint coincidere et mutuos colores delere coeuntes in albicantem circulum. Sin magis adhuc distent, iris *πp* cadet extra iridem *RP.* Quod si *EF* statuatur ad locum *I,* haec iris *πp* evanescet, et potest etiam coincidere cum Iride *rp* si *EF* paulo ultra vel citra locum *I* statuatur. Jam verò horum omnium ad oculum facilis est applicatio; posito quod obstaculum *EF* fundum ejus referat et *kκ* tunicam corneam ab externâ vi aut interno aliquo vitio perperam curvatam. Quinetiam ex his non modo generalis causa harum iridum declaratur sed pro quibuslibet ejusmodi particularibus apparentijs causae etiam particulares assignari posse videntur. Quemadmodum sicui fax appareat unicâ tantum iride cincta cujus pars exterior rubet, interior verò vel alba vel fortè nonnihil caerulea appareat: exinde concludi posse videtur quòd tunica cornea circa medietatem ejus sit paulò depressior / quàm solet esse sine aliquâ rugâ qualem ad *lλ* descripsi. Efficit enim illa ₂95/ depressio ut radij ab eodem puncto objecti venientes ad puncta longè post retinam convergant et qui proinde in Retinâ spatium aliquod (quale est *rOp*) occupabunt cujus periferia (ut modo ostendi) rubeo colore ad exteriorem ejus partem tingetur et albo vel dilutè caeruleo ad interiorem. Et quò major hujusmodi iris appareat eò magis ad interiorem ejus partem debet caeruleo tingi. Potest etiam hujusmodi iris propter annularem rugam accidere modo tunicae corneae figura in meditullio non simul vitietur.

[135] Quod si duae irides appareant illud ex utrâque causâ conjunctè petendum est, cornea nempe tum in medio tum juxta periferiam pupillae depressa. In hujus rei illustrationem adhibeamus casum quem Cartesius de seipso in Meteoris Cap. 9 ad hunc modum describit. Cùm noctu, inquit, navigarem, et totâ illâ vesperâ caput cubito innisus, manu oculum dextrum clausissem, altero interim versus coelum respiciens, candela ubi eram allata est, et tunc aperto utroque oculo, duos circulos flammam coronantes aspexi, colore tam acri et florido quàm unquam in arcu coelesti me vidisse memini. *AB* (fig II, 60) est maximus, qui ruber erat in *A*; et caeruleus in *B*: *CD* minimus, qui etiam ruber in *C,* sed albus versus *D,* ubi ad flammam usque extendebatur. Oculo dextro postea iterum clauso notavi has coronas evanescere; et contra

---

(6) *Editio princeps*: propterea.

their own color too. For a similar reason, the rays refracted near k and κ, if they are red-making, will tend to the outer side of the limit R and P, and to the inner side if they are blue-making. Thus three bows will be produced: RP, red outside and blue inside; $p\pi$, blue outside and red inside; and rp, red outside, and blue must also be inside, unless indeed that color—because of the smallness of the refraction at m and μ—is not sufficiently separated from the red to become sensible, and moreover[6] unless it is greatly obscured by the quantity of light extended throughout $rO\rho$, the location of the bright image that the bows encircle. But the shapes of the bows and the relations between them can be changed in various ways, and not only by the different shapes that the surface $k\kappa$ can assume but also by the different distances between the surface $k\kappa$ and the obstacle EF. For instance, if it is set a little more distant than I have represented it, the circles RP and πp can coincide and obliterate their mutual colors, combining into a whitish circle. If, however, they are still more distant, the bow πp will fall outside the bow RP. But if EF is set at I, this bow πp will vanish; and it can also coincide with the bow rp, if EF is set slightly beyond or before the place I. Now the application of all this to the eye is indeed easy, assuming that the obstacle EF represents its base and $k\kappa$ its cornea defectively curved by some external force or some internal flaw. Moreover, from these things not only is the general cause of these bows made clear, but also for any particular appearances of this sort particular causes can apparently be assigned. For example, if to someone a torch appears surrounded by only one bow whose exterior part is red but whose interior part appears either white or perhaps somewhat blue, from this it seems possible to conclude that his cornea is somewhat more depressed about its middle than it usually is, without any wrinkle such as I have drawn at $l\lambda$. For that depression causes the rays coming from the same point of the object to converge at points far beyond the retina, and these consequently will occupy in the retina some space (such as $rO\rho$) whose periphery (as I just showed) will be colored with the color red on its exterior side and white or diluted blue on its interior. The larger a bow of this kind appears, the more it must be colored blue on its interior side. Bows of this sort can also occur on account of annular wrinkles, provided that the shape of the cornea in the middle is not also deformed.

If, however, two bows appear, this must be sought from both causes to- [135] gether, namely, the cornea being depressed both in the middle and near the periphery of the pupil. To illustrate this matter let us consult the case that Descartes described about himself in Chapter 9 of the *Meteorology* in this way:

> When I was sailing at night and that entire evening I was leaning my head on my arm and had covered my right eye with my hand, while looking toward the sky with the other, a candle was brought to where I was. Then opening both eyes, I saw surrounding the flame two circles with colors as vivid and lively as I ever remember seeing in the rainbow, AB [Fig. II, 60] is the largest, which was red at A and blue at B, and CD the smallest, which was also red at C but white toward D, where it extended up to the flame. Afterward, having again closed my right eye, I observed these

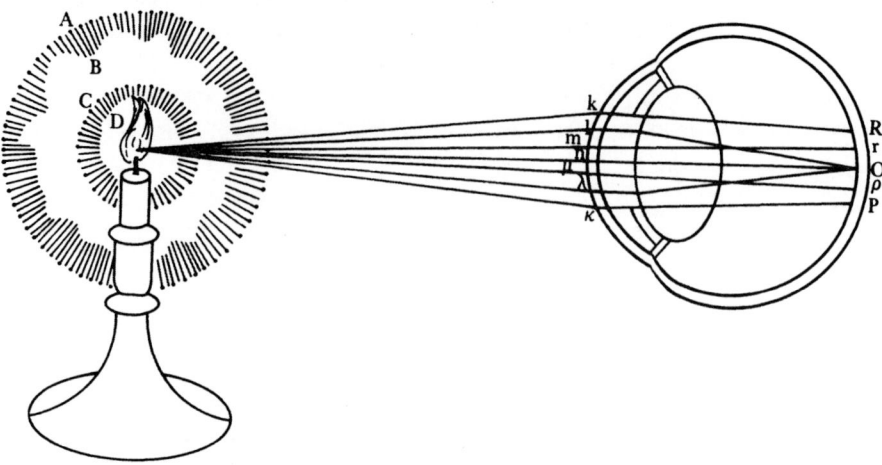

Figure II, 60

illo aperto et sinistro clauso permanere: Unde certò cognovi illas non aliunde oriri, quàm ex nova conformatione vel qualitate, quam dexter oculus acquisiverat, dum ipsum ita clausum tenueram, et propter quam non modo maxima pars radiorum quos ex flamma admittebat ipsius imaginem in *O* ubi congregabantur, pingebant, sed etiam nonnulli ex ijs ita detorquebantur, ut per totum spatium *rρ* spargerentur ubi pingebant coronam *CD* et nonnulli alij per totum spatium / *RP* ubi coronam *AB* etiam pingebant.[7] Cùm itaque   ₂96/ Cartesius haec viderit postquam per totam vesperam cubito innixus erat, rugae quales explicui potuerunt imprimi, unde necesse erat ejusmodi coronas apparere, et quòd tres coronae non apparebant, illâ scilicet non apparente cujus partem exteriorem caeruleam esse descripsi, et partem interiorem rubeam, id ex eo venire debuit quòd radij in *l* et *λ* refracti ex quibus hanc coronam generari oporteret, haud citiùs quàm ad retinam convergebant, aut potiùs non tam cito; non enim probabile videtur quod tunicae corneae pars aliqua ab externâ pressione possit fieri solito convexior, et nisi hoc eveniat radij illi non possunt citiùs quàm ad retinam convenire. Illa verò tertia corona non potest apparere nisi citiùs (ut ad *I*) conveniant. Si longè ultra convergant, coronam tunc quidem deberent efficere, sed cujus pars exterior rubesceret et tunc tres coronae in exteriori earum parte rubeae conspicerentur. Sed in hisce videar nimius, praesertim cùm tanta causarum varietas non solum a tunicâ corneâ sed humore Christallino et aliunde etiam peti possit ut haud sit difficilè plures assignare quae eosdem quoslibet effectus diversis temporibus producant.[8] Nescio tamen an operae pretium sit annotare causam radiorum a lucidis corporibus hinc inde ad instar trabium in longum

---

(7) *Meteora*, Ch. IX, §7, *Oeuvres*, 6:713–4; the Latin differs at many points from the original French, ibid., pp. 351–2. Newton has slightly altered Descartes's figure (and the text referring to it) to make it correspond with his own figure and explanation.

(8) Descartes too did not insist on any particular explanation of these coronas and recognized that a variety of causes could produce them, such as circular wrinkles or some opacity in the

coronas vanish, and, on the contrary, after it was opened and the left one closed, they remained. From this I recognized for certain that they arise from no other source than some new configuration or quality that the right eye had acquired when I had thus kept it closed. In consequence, not only the greatest part of the rays that it admits from the flame depicts its image at O, where they are gathered together, but also some of those are deflected so that they are spread through the entire space *rp*, where they depicted the corona *CD*; and some others are spread through the entire space *RP*, where they also depicted the corona *AB*.[7]

Since, therefore, Descartes saw these things after he had leaned on his arm all evening, such wrinkles as I have explained were able to be impressed whereby coronas of this kind would necessarily appear. That three coronas did not appear—namely, that one not appearing whose exterior side I described as blue and interior side as red—ought to follow from the fact that the rays refracted at *l* and λ, from which this corona must be generated, did not converge more quickly, or rather not as quickly, at the retina. For it does not seem probable that part of the cornea can become more convex than usual by some external pressure, and unless this happens those rays cannot meet more quickly at the retina; but that third corona cannot appear unless they do meet more quickly (as at *I*). If they converged farther on the other side, then they must certainly produce a corona, but one whose exterior side is red; and then three coronas that are red on their exterior side would be perceived. But I may seem excessive in these things, especially since such a great variety of causes can be sought not only from the cornea but also from the crystalline humor and elsewhere that it is not difficult to ascribe several causes that may produce the same effects whatsoever at various times.[8] Yet I do not know whether it is worthwhile to comment on the cause of the rays extending lengthwise from luminous bodies all around, like beams, when we look at

---

cornea or crystalline lens, or a change in temper or shape of the humors or membranes; *Meteora*, Ch. IX, §7, *Oeuvres*, 6:714, 352–3. The physician William Briggs, who established a friendship with Newton while at Cambridge, very succinctly recounted Newton's explanation of the coronas in his *Ophthalmo-graphia, sive oculi ejusque partium descriptio anatomica* (Cambridge, 1676), Ch. III, §4, pp. 15–17. The *Ophthalmo-graphia* was reissued in 1685—and frequently thereafter—in a combined edition with his *Nova visionis theoria* which bore a dedicatory letter from Newton; *Ophthalmo-graphia, sive oculi ejusque partium descriptio anatomica. Cui accessit 'Nova visionis theoria,' regiae societati Londin. proposita*, 2nd ed. (London, 1685), pp. ₂[iii–vi] = *Correspondence*, 2:417–9. The *Nova visionis theoria* was a Latin translation of two papers that Briggs had read to the Royal Society in 1682 and then published in the *Philosophical Collections* and *Philosophical Transactions*; see *Correspondence*, 2:377–8, 381–5. The translation was undertaken with Newton's encouragement ("hortatu doctissimi D. *Newtoni*, Matheseos in Academ. *Cantab.* Professoris dignissimi" according to Briggs's "Praeloquium ad lectorem," *Ophthalmo-graphia* . . . 'Nova visionis theoria' (1685), p. ₂[viii]). Philippe de la Hire rejected the idea that these coronas could be caused by wrinkles and attributed them to irregularities in the shape of the cornea; "Dissertation sur les differens accidens de la vue," Pt. I, §§XXIII–XXIV, in his *Mémoires de mathématique et de physique* (Paris, 1694), pp. 233–302, esp. 247 = *Mémoires de l'académie royale des sciences. Depuis 1666 jusqu'à 1699*, 9 [Paris, 1730]: 350–422, esp. 364–5. In fact, these colors most probably arise from the swelling and consequent "cloudiness" of the cornea caused by the extended pressing on the eye.

protensarum cum oculis pene clausis aspicimus. Nempe humiditas quae inter cilia et tunicam corneam versatur secundum extremitates ciliorum parùm assurgit. Sicut aqua vasi imposita altiùs assurgit ubi a vase terminatur quam alibi; quo pacto fit ut aliqui radij ab hac humiditate priùs refringantur quàm attingunt tunicam corneam et sursum detorqueantur in confinio superioris cilij ac deorsum in confinio inferioris.[9]

Lect 16   Superest jam mirum illud coelestis arcûs spectaculum ad cujus explicatio-
[136]   nem Cartesius viam stravit, Huic enim / debetur quod in guttis aquae pluvia- ,97/
lis decidentibus efformari cognoscimus. Quemadmodum ex eo constat quòd nunquam videtur nisi coelo pluente, quod sole pluviam decidentem illustrante in vicis nonnunquam apparuit quasi non in coelo collocatus sed in aere vicino super oppositarum domuum parietibus effixus vel potius interjectus, quod aqua per artificium aliquod in altum sparsim ejaculata iridem ostendit et quod gramen rore matutino quasi guttulis minutissimis conspersum colores etiam iridis exhibet. Huic etiam debetur ingeniosissima de refractionibus gut-tae earumque limitibus inventio.[1] Sed causam physicam minus foeliciter ag-gressus est. Hanc itaque ut intelligatis concipite radium *AN* (fig II, 61) in

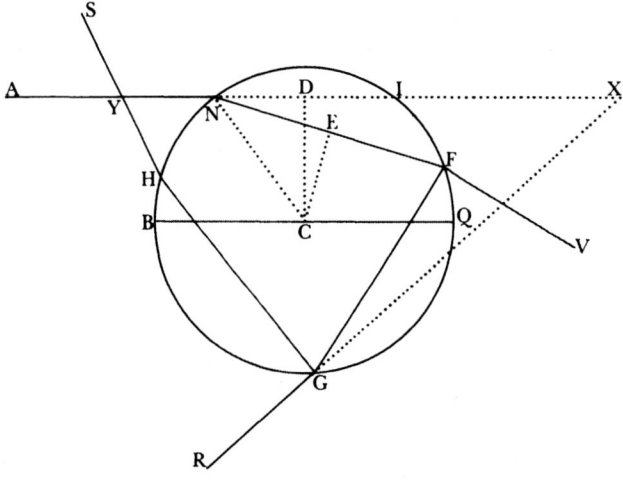

Figure II, 61

(9) Descartes very briefly explained that the large rays (*magnos quosdam radios*) visible around a flame are caused by straight wrinkles crossing at the center of the eye, though he did not specify that squinting is necessary for their appearance; *Meteora*, Ch. IX, §7, *Oeuvres*, 6:714, 353. Briggs, in his *Ophthalmo-graphia*, Ch. III, §4 (1676, p. 17), so closely paraphrased Newton's explanation here (with attribution to "amicus doctiss. D. *Newton*") that he no doubt had access to a manuscript copy of the *Optica*. Jacques Rohault proposed that this phenomenon was caused by reflection—not refraction, as Newton supposed—of light rays from tear drops; *Rohault's System of Natural Philosophy, Illustrated with Dr. Samuel Clarke's Notes Taken Mostly out of Sir Isaac Newton's Philosophy,* trans. John Clarke, 2 vols. (London, 1723), Pt. I, Ch. XXXV, 1:280–1; an augmented translation of Rohault's *Traité de physique* [1671]. La Hire rejected both Descartes's and Rohault's solution and independently presented one similar to Newton's. He mentioned that he had intended to publish his explanation separately before he

them with nearly closed eyes. Namely, the fluid that is between the eyelids and the cornea rises up slightly along the ends of the eyelids, just as water placed in a vessel rises higher where it is bordered by the vessel than elsewhere. In this way it turns out that some rays are refracted by this fluid before they reach the cornea and are deflected upward into the confines of the upper eyelid and downward into the confines of the lower one.[9]

There now remains that remarkable spectacle of the rainbow for whose **Lecture 16** explanation Descartes prepared the way, for it is due to him that we know [136] that it is formed in falling drops of rain water. For instance, it is manifest because it is never seen except in a rainy sky; when the sun illuminates falling rain, it sometimes seemed in towns as if it were not located in the sky but in the air near us and affixed, or rather interposed, above the walls of facing buildings; water shot upward in a spray by some device displays a rainbow; and a meadow sprinkled with morning dew, as it were, with extremely tiny drops, also exhibits the colors of the rainbow. A most ingenious discovery about the refractions of the drops and their limits is also due to him,[1] but he treated their physical cause less happily. That you may therefore understand this, conceive of the ray *AN* (Fig. II, 61) to fall on the globe *NFG* at *N* and

came across Briggs's *Ophthalmo-graphia*; "Dissertation," Pt. I, §§LVI–LX, *Mémoires de mathématique*, pp. 273–9 = *Mémoires de l'académie royale des sciences. Depuis 1666 jusqu'à 1699*, 9:392–7. Robert Smith translated portions of La Hire's account in his *Compleat System* ("General Remarks," §§46–9, 2:8), but Newton is not mentioned.

(1) In a change made in the manuscript of the *Opticks*, where he sets forth a full account of the rainbow, Newton was no longer so fair, or generous, to Descartes. After observing "yᵗ this Bow is made by refraction of yᵉ sun's light in drops of falling rain," he added: "This was understood by some of the Ancients & of late more fully discovered & explained by the famous Antonius de Dominis Archbishop of Spalato in his Book De Radijs Visus et Lucis, published by his friend Bartolus at Venice in the year 1611 & written above twenty years before. For he teaches there how the [interior] Bow is made in round drops of rain by two refractions of yᵉ suns light [& one reflexion between them & the exterior by two refractions] & two sorts of reflexions between them in each drop of water, & proves his explications by Experiments . . . The same explications Des-Cartes hath pursued in his Meteors [& mended that of yᵉ exterior bow]" (Bk. I, Pt. II, Prop. IX, Add. 3970, ff. 119ᵛ, 120ʳ = *Opticks*, pp. 126–7; square brackets indicate Newton's insertions). In an earlier interlineation replaced by the preceding, Newton's history was far less objectionable and simply stated, "as the Ancients believed & of late Des-Cartes & before him the famous Antonius de Dominis . . . have more fully discovered & explained" (Add. 3970, f. 120ʳ). Although Newton has vastly exaggerated, indeed misrepresented, the contributions of de Dominis, who ignored both the second refraction upon leaving the drop and the second reflection in the secondary bow, he was careful to remain silent on Descartes's most significant contribution, namely, that the bows' radii are determined by extrema of the deviation when more rays emerge in the same direction than at any other angle; see Lect. I, 14 note (19), and Lect. I, 15 note (1). Newton at his death possessed a copy of de Dominis's *De radiis visus et lucis in vitris perspectivis et iride tractatus* (Venice, 1611) (Harrison, *Library of Newton*, no. 535). He had arrived at his judgment of de Dominis's contribution no later than early 1698; see David Gregory's memoranda of 20 February 1698, in *Correspondence*, 4:266. For de Dominis's explanation of the rainbow, see Boyer, *Rainbow*, pp. 188–92, 252; R. E. Ockenden, "Marco Antonio de Dominis and his explanation of the rainbow," *Isis*, 26 (1936):40–9; and August Ziggelaar, "Die Erklärung des Regenbogens durch Marcantonio de Dominis, 1611. Zum Optikunterricht am Ende des 16. Jahrhunderts," *Centaurus*, 23 (1979):21–50.

globum *NFG* ad *N* incidere et inde versus *F* refringi ubi rursus vel refringitur versus *V* vel forte reflectitur ad *G*, et si posterius eveniat tunc iterum in *G* vel refringitur ad *R* vel reflectitur ad *H*, et sic deinceps, ita ut radijs globum ingredientibus aliqui, ut *NFV*, statim egrediantur nullam reflexionem passi, alij, ut *FGR*, post unam reflexionem, et alij, ut *GHS*, post duas alijque post tres vel etiam plures. Jam verò cùm guttae pluviales respectu distantiae ab oculo spectatoris sint admodum exiguae ut physice pro punctis haberi possint, non opus est ut earum magnitudines omnino consideremus, sed angulos tantum quos incidentes cum emergentibus radijs comprehendunt. Nam ubi anguli illi maximi sunt vel minimi, emergentes radij sunt solito confertiores et quia diversis radiorum generibus diversi competunt anguli maximi vel minimi, singula ad diversas plagas confertissimè tendentia in ijsdem praevalebunt ad colores proprios exhibendos. Anguli itaque maximi vel minimi quos singulorum generum / emergentes radij cum incidentibus possunt constituere determinandi sunt ut horum Phaenomen$\omega\nu$ rationes rectè percipiamus. $_2$98/

[137]   Scilicet in coroll: 1 et 2 prop 35 [Pars I] ostensum est emergentem radium *GR* ad incidentem *AN* minime[2] inclinari cum sit 3*RR* . *II* − *RR* :: *CN*^{q} . *ND*^{q}. Et *I* . 2*R* :: *ND* . *NE*. posito nempe *I* ad *R* ut sinus incidentiae ad sinum refractionis. Et ex hinc inventis *ND* et *NE* dabitur positione *RG*.

Sit e.g. pro radijs maximè refrangibilibus sinus incidentiae ad sinum refractionis, sive *I* ad *R*, ut 185 ad 138,[3] prout in aqua pluviali proximè comperi, et erit 57132 . 15181 (:: 3*RR* . *II* − *RR*) :: *CN*^{q} . *ND*^{q}. Adeoque *DN* = $\sqrt{\dfrac{15181}{57132}}$*CN*^{q} seu = $\dfrac{5155}{10000}$*CN* unde per tabulam sinuum datur arcus *NI* 62^{gr} 4^{min}.[4] Praeterea cùm sit *I* . 2*R* :: *ND* . *NE*, hoc est

$$185 . 276 :: \frac{5155}{10000}CN . NE;$$

erit *NE* = $\dfrac{7691}{10000}$*CN*. Et inde etiam per tabulam sinuum datur arcus *NF* 100^{gr}. 32^{min}.[5] Subduc jam duplum arcus *NF* ex aggregato arcus *NI* et arcus 180^{gr} sive sem[ic]irculi, et restabit 41^{gr} 0^{min} pro inclinatione radij *RG* ad radium *AN*, sive pro angulo *AXR*;[6] productis nempe *AN* et *RG* donec in *X* conveniant. Et hic angulus est sub quo intimus sive caeruleus limbus Iridis hujus apparere debet.[7]

Ad eundem modum pro radijs minimè refrangibilibus posito sinu incidentiae ad sinum refractionis ut 183 ad 138,[8] uti dimensus sum; invenietur *ND* = $\dfrac{5028}{10000}$*CN*, et *NE* = $\dfrac{7583}{10000}$*CN*; indéque per Tabulam sinuum arcus *NI* erit 60^{gr} 22^{min}, & arcus *NF* 98^{gr} 38^{min}, adeoque angulus *AXR* 43^{gr} 6^{min}, sub quo

---

(2) Read (as *editio princeps*): maximè (most).

(3) This value and that in the next article for the greatest and least indices of refraction (where $(n − 1) / \Delta n = 23$) differ slightly from those calculated by the dispersion model in §I, 44 = 117, and have a mean value of precisely 4/3.

(4) *ND/CN* = sin $\widehat{NCD}$, where $\widehat{CND}$ = *i*, and thus $\widehat{NI}$ = 2$\widehat{NCD}$ = π − 2*i*.

(5) *NE/CN* = sin $\widehat{NCE}$, where $\widehat{CNE}$ = *r*, and thus $\widehat{NF}$ = 2$\widehat{NCE}$ = π − 2*r*.

then to be refracted toward *F* where it is either refracted anew to *V* or, as it happens, reflected to *G*; and if the latter occurs, then it is either refracted once again at *G* to *R* or reflected to *H*; and so on. Consequently, some of the rays entering the globe, such as *NFV*, immediately leave without undergoing a reflection, while others, such as *FGR*, leave after one reflection, and others, such as *GHS*, leave after two, and still others after three or even more. Since now raindrops are in fact extremely small with respect to their distance from an observer's eye, so that physically they can be considered as points, we need not at all consider their size, but only the angles that the incident rays make with the emergent ones. For when those angles are greatest or least, the emergent rays are unusually concentrated, and because to the different sorts of rays there correspond different greatest or least angles, each of them, tending most concentratedly to different places, will predominate at that place and exhibit its own color. The greatest or least angles that the emergent rays of each kind can make with the incident ones must therefore be determined so that we may properly comprehend the rules for these phenomena.

Specifically, it was shown in Corollaries 1 and 2, Prop. 35 [Part I] that the [137] emergent ray *GR* is least[2] inclined to the incident one *AN*, when

$$3R^2 : (I^2 - R^2) = CN^2 : ND^2, \text{ and } I : 2R = ND : NE,$$

assuming, namely, *I* is to *R* as the sine of incidence to the sine of refraction. Having thus found *ND* and *NE*, *RG* will be given in position.

For example, for most refrangible rays let the sine of incidence to the sine of refraction, or *I* to *R*, be as 185 to 138[3] (as I found approximately in rain water), and then

$$57132 : 15181 (= 3R^2 : [I^2 - R^2]) = CN^2 : ND^2.$$

Hence $DN = \sqrt{[(15181/57132) \, CN^2]} = (5155/10000) \, CN$, which by the table of sines gives the arc *NI* as 62°4′.[4] Moreover, since $I : 2R = ND : NE$, that is, $185 : 276 = (5155/10000) \, CN : NE$, then $NE = (7691/10000) \, CN$; and from this, also by the table of sines, the arc *NF* is given as 100°32′.[5] Now subtract twice the arc *NF* from the sum of the arcs *NI* and 180° (or a semicircle), and there will remain 41°0′ for the inclination of the ray *RG* to the ray *AN*, or the angle *AXR*,[6] namely, extending *AN* and *RG* until they meet in *X*. This then is the angle at which the innermost or blue band of this rainbow should appear.[7]

In the same way, for least refrangible rays, assuming the sine of incidence to the sine of refraction to be 183 to 138[8] (as I measured it), it will be found that $ND = (5028/10000) \, CN$ and $NE = (7583/10000) \, CN$, and then by the table of sines the arc *NI* will be 60°22′ and the arc *NF* 98°38′, so that the

---

(6) Thus the primary bow's radius $A\widehat{X}R = \widehat{NI} + \pi - 2\widehat{NF} = 2(2r - i)$. By so closely following Descartes's calculation of the radius (*Meteora*, Ch. VIII, §10, *Oeuvres*, 6:706, 338), Newton unnecessarily complicated his own calculation, for he could have directly determined *i* and *r*, since $ND/CN = \cos i$ and $NE/CN = \cos r$.

(7) Et . . . debet.] Added, while the *editio princeps* continues: sive minima ejus semidiameter.

(8) See note (3), this lecture.

extimus sive rubeus hujus Iridis limbus apparebit. Itaque maxima ejus semi-diameter est 43^{gr} 6^{min}, et minima 41^{gr} 0^{min}, et Orbitae latitudo sive crassities[9] 2^{gr} 6^{min} circiter, vel potius 2^{gr} 37^{min}, addita diametro Solis.[10] Sed cùm colores in extremitatibus ad utrumque limbum debiliores sint quàm quae propter nubium conterminarum splendorem videri possint, sensibilis / ejus crassities ₂99/ duos gradus vix excedet.[11]

[138]   Haud secus determinantur exterioris Iridis dimensiones. Nam ostensum est in Coroll: 1 et 2 Prop 36, [Pars I] emergentem radium *HS* ad incidentem *AN* maximè[12] inclinari cum sit $8RR \cdot II - RR :: NC^q \cdot ND^q$. Et $I \cdot 3R :: ND \cdot NE$. Quamobrem pro radiorum maximè refrangibilium sinubus *I* et *R*, substitutis numeris 185 et 138, ut supra; obtinebuntur $ND = \frac{3157}{10000}CN$, et $NE = \frac{7064}{10000}CN$. Et inde per Tabulam sinuum arcus *NI* 36^{gr}. 48^{min}, et arcus *NF* 89^{grad} 53^{min}. Atque adeò angulus $AYS = 52^{grad}$ 51^{min} qui erit minima[13] semidiameter Iridis hujus. Et similiter pro radiorum minimè refrangibilium sinubus *I* et *R* substituendo numeros supra positos 183 & 138, emergent $ND = \frac{3079}{10000}CN$, et $NE = \frac{6965}{10000}CN$. Unde per Tabulam sinuum eliciuntur arcus *NI* 35^{gr} 52^{min}, et arcus *NF* 88^{gr} 18^{min}. Adeoque angulus *AYS* erit 49^{gr} 2^{min}, Iridis nempe minima semidiameter. Quamobrem si a maximâ semidiametro 52^{gr} 51^{min} auferatur minima 49^{gr} 2^{min}, et residuo addatur semidiameter[14] Solis 31^{min} emerget hujus Iridis crassities 4^{gr} 20^{min}.[15] Sed propter majorem hujus quàm interioris Iridis obscuritatem colores vix ultra crassitiem trium graduum vel trium et semissis videri posse conjicio.

[139]   Jam denique ut harum Iridum rationes conspectui distinctè exhibeam,[16]

---

(9) et minima . . . crassities] *Editio princeps*: A quâ si auferatur minima semidiameter 41 grad. 0 min. emergit iridis crassities.

(10) In the *Opticks* (Bk. I, Pt. II, Prop. IX, pp. 129–31 = Add. 3970, ff. 121^r, 122^r), Newton took for the least and greatest indices of refraction 108 and 109 to 81 (with $\nu = (n - 1)/\Delta n = 27\frac{1}{2}$), which he now calculated according to the linear dispersion law, and he found 42°2′ for the radius of the red band and 40°17′ for the blue one, thus making the bow's breadth 2°15′ including the sun's angular diameter. Before adopting this value for the indices of refraction, the manuscript shows that he calculated the radii using 98 and 99 to 73$\frac{1}{2}$ ($\nu = 25$), and its breadth turned out to be 2°26′.

(11) When Newton added the newly calculated breadth of the bow, 2°15′, in the manuscript of the *Opticks* (see the preceding note), he also added two observations in nearly perfect agreement with that value, one of 2°15′ and one of 2°10′ (*Opticks*, Bk. I, Pt. II, Prop. IX, p. 132 = Add. 3970, ff. 121^v, 122^r). Before that, when the calculated breadth was 2°26′, he had an observed breadth of "about 2^{degr}," which he then altered to a more conservative "about 1^{degr} & 30′ or something more," and he finally settled upon "about 1^{degr} & 30′ or 40′ " (ibid., f. 122^r).

(12) Read (as *editio princeps*): minimè (least).

(13) Read (as *editio princeps*): maxima (greatest). Newton calculates the secondary bow's radius, $A\widehat{YS}$, by again following Descartes, where

$$A\widehat{YS} = 3\widehat{NF} - \widehat{NI} - \pi = 3(\pi - 2r) - (\pi - 2i) - \pi = \pi - 2(3r - i);$$

see notes (4)–(6), this lecture.

(14) Read (as *editio princeps*): diameter.

angle AXR will be 43°6', at which angle the outermost or red band of this rainbow will appear. Its greatest radius is therefore 43°6', its least 41°0', and the breadth or thickness of the ring[9] approximately 2°6', or rather 2°37' when the sun's diameter is added.[10] But since the colors at the extremities of both bands are too weak to be seen, on account of the brightness of the bordering clouds, its sensible thickness will scarcely exceed 2°.[11]

The dimensions of the exterior rainbow are determined no differently. For [138] it was shown in Corollaries 1 and 2, Prop. 36 [Part I] that the emergent ray HS is most[12] inclined to the incident one AN when

$$8R^2 : (I^2 - R^2) = NC^2 : ND^2,$$

and $I : 3R = ND : NE$. Therefore, upon substituting for the sines I and R of the most refrangible rays the numbers 185 and 138, as above, there will result $ND = (3157/10000) \, CN$ and $NE = (7064/10000) \, CN$, and then by the table of sines the arc NI is 36°48' and the arc NF is 89°53'. Consequently, the angle AYS will be 52°51', which will be the least[13] radius of this rainbow. Likewise, upon substituting for the sines I and R of the least refrangible rays the numbers 183 and 138 assumed above, there will result $ND = (3079/10000) \, CN$ and $NE = (6965/10000) \, CN$; whence by the table of sines the arc NI is found to be 35°52' and the arc NF to be 88°18'. Hence the angle AYS will be 49°2', namely, the rainbow's least radius. Consequently, if the least radius, 49°2', is subtracted from the greatest radius, 52°51', and the remainder is added to the sun's radius,[14] 31', the thickness of this rainbow will come out as 4°20'.[15] But because of the greater obscurity of this than the interior rainbow, I guess that colors can hardly be seen beyond a thickness of 3° or $3\frac{1}{2}$°.

Finally, so that I may now clearly present the explanation of the appear- [139] ance of these rainbows,[16] let E, F, and G be drops arbitrarily scattered Fig. II, 62

(15) With the different values for the indices of refraction adopted in the *Opticks* and related in note (10), this lecture, Newton found 54°7' for the radius of the blue band and 50°57' for the red, thus giving a breadth of 3°40' including the sun's angular diameter. With the value initially adopted ($\nu = 25$), he had found 54°26' for the blue band and 50°57' for the red, giving a total breadth of 4°.

(16) As an undergraduate at Cambridge Newton was introduced to the Aristotelian theory of the rainbow and its formation through Magirus's commentary. According to this theory, the colors of the rainbow are to be attributed both to a mixture of light and shadow and to a weakening of sunlight reflected from different parts of the clouds; see Magirus, *Physiologiae*, Bk. IV, Ch. V, §§17, 22–5, pp. 161–2; and also the Introduction, note (3). De Dominis, in his *De radiis* (Ch. XIII, p. 56), adopted a similar explanation. The explanations of greatest concern to Newton at this time were those of Descartes, who attributed the colors to rotations acquired by the light corpuscles in their refractions in the raindrops, and of Hooke, who attributed them to an obliquity acquired by the pulses of light; see the Introduction, §1.

The only surviving draft of any portion of the text of the *Lectures* is of these two concluding paragraphs. This draft, now in the Turner Collection at Keele University, England and reproduced in Plate III, is on the upper half of one side of a sheet that Newton subsequently used for notes on ancient chronology, both on the lower half of the same side and the reverse. (This latter fact may well account for its survival.) In the following notes we indicate a number of significant variants and changes in this draft, which we will call the "Turner MS," and we include its somewhat different figure in the English translation.

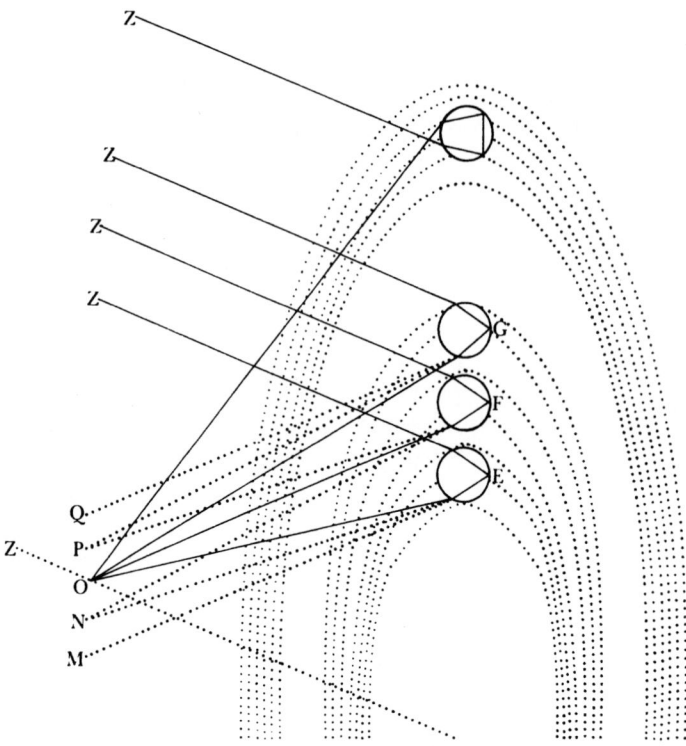

Figure II, 62

Fig II, 62 sunto *E, F,* & *G* guttae per aëra utcunque sparsae;[17] *ZE, ZF,* & *ZG* radij
solares parallelè[18] incidentes in guttas; *EM, EN,* & *EO* radij diversè refrangibiles e gutta *E* post unam reflexionem emergentes; atque *FN, FO, FP,* & *GO,
GP, GQ* consimiles radij emergentes e guttis *F* ac *G*: nempe *EO, FP, GQ*
maximè refrangibiles, et *EM, FN, GO* minimè refrangibiles &c.[19] Jam si
spectantis oculus ad *O* consistat, ex Hypothesi[20] manifestum est quod e
radijs quos gutta *E* post unam reflexionem emittit soli maximè refrangibiles
seu rubiformes,[21] (qualis / *EO*) impingent in oculum, reliquis ut *EN* et *EM*   ₂100/
propter minorem refractionem praeterlabentibus. Et proinde rubor[22] ad *E*[23]
conspicietur. E radijs autem quos gutta *G* post unam reflexionem[24] emittit
maximè refrangibiles qualis *GQ* praeteribunt oculum propterea quod radio
EO paralleli sunt,[25] & alterius generis radij puta minimè refrangibiles seu
caeruliformes[21] (qualis [*G*]*O*) in eum impingent; unde caeruleus color[22]
apparebit in *G*. Et simili discursu[26] gutta *F* in medio inter *E* ac *G* posita
radios mediocriter refrangibiles, ut *FO*, in oculum immittent reliquis, ut *FN,
FP* utrinque praeterlabentibus: indéque viriditas cernetur ad *F*. Eademque est
ratio guttarum omnium ad easdem cum his guttis apparentes distantias ab
axe *OZ* qui per Solem & oculum transit, positarum: et proinde ad distantias
illas[27] colores undique apparebunt, hoc est arcus variegatus[28] cujus interior

(17) sunto ... sparsae] Turner MS originally: Esto *AB* Iris exterior *CD* interior, *F* et *I* guttae
in medijs Iridum; *E, G, H,* & *K* aliae guttae hinc & inde in extremitatibus atque *O* spectantis

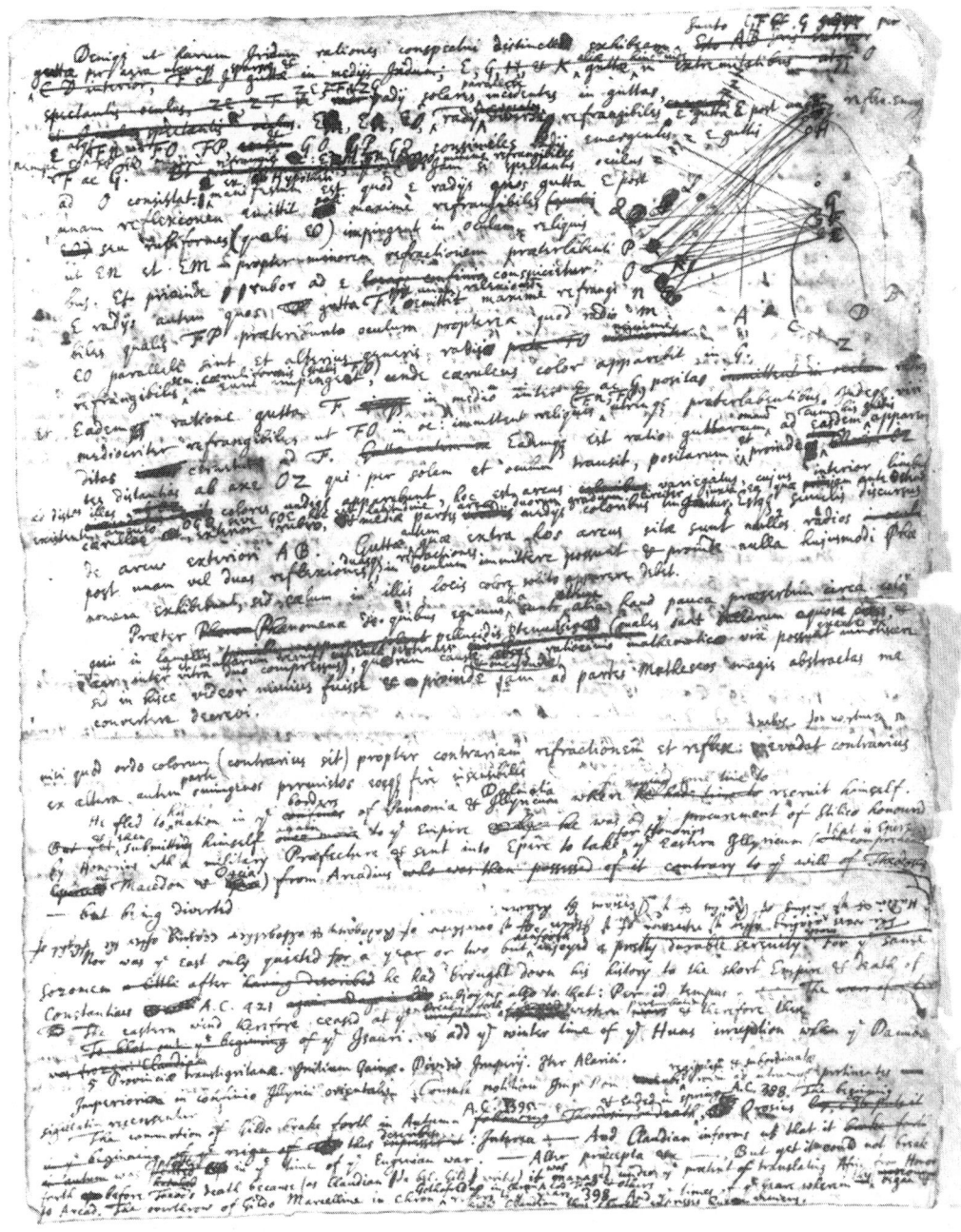

Plate III. A draft of the final two paragraphs of the *Optica* (§§II, 139, 140) together with notes on chronology. (By permission of the University Library, Keele, England.)

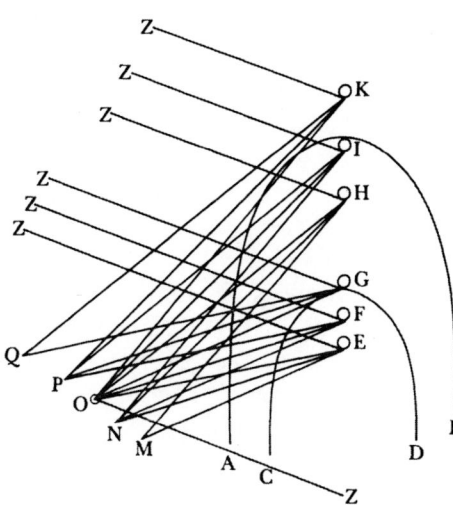

through the air;[17] *ZE, ZF,* and *ZG* parallel[18] solar rays incident on the drops; *EM, EN,* and *EO* differently refrangible rays emerging from the drop *E* after one reflection; and *FN, FO,* and *FP* and *GO, GP,* and *GQ* completely similar rays emerging from the drops *F* and *G,* specifically, *EO, FP,* and *GQ* being most refrangible rays, and *EM, FN,* and *GO* least refrangible ones, and so forth.[19] Now, if the observer's eye is situated at *O,* by hypothesis[20] it is evident that of the rays that the drop *E* emits after one reflection only the most refrangible, or red-making,[21] such as *EO,* will strike the eye, with the others, such as *EN* and *EM,* slipping by on account of their smaller refraction. Consequently, red[22] will be seen at *E.*[23] Of the rays, however, which the drop *G* emits after one reflection,[24] the most refrangible ones, such as *GQ,* will go by the eye since they are parallel to the ray *EO;*[25] whereas rays of a different kind, say the least refrangible or blue-making ones,[21] such as [*G*]*O,* will strike it. Accordingly, the color blue[22] will be visible at *G.* By a similar argument,[26] the drop *F* located midway between *E* and *G* will send mean refrangible rays, such as *FO,* into the eye, with the others on each side, such as *FN* and *FP,* slipping by. Thus green will be perceived at *F.* The reasoning is the same for all drops placed at the same apparent distances as these drops from the axis *OZ* that passes through the sun and the eye. Consequently, colors will appear at those distances in all directions, that is, a variegated[28] arc, whose interior band is colored blue,

---

oculus (Let *AB* be the exterior rainbow, *CD* the interior one, *F* and *I* drops in the middle of the rainbows, *E, G, H,* and *K* other drops at the extremities on both sides, and *O* the observer's eye).

(18) Added in Turner MS.

(19) nempe . . . refrangibiles [&c].] Added in Turner MS.

(20) ex Hypothesi] Added in Turner MS.

(21) These should be interchanged to read (as *editio princeps*) "caeruliformes" (blue-making) and "rubiformes" (red-making), respectively.

(22) In accordance with the preceding note, interchange these to read (as *editio princeps*) "caeruleus color" (the color blue) and "rubor" (red), respectively.

(23) Turner MS originally continued: locaque confinia (and adjacent places).

(24) post unam reflexionem] Added in Turner MS.

(25) In the draft figure Newton drew rays of the same color from different drops, such as *GQ* and *EO,* to be parallel, but in redrawing the figure for the *Optica* he failed to preserve this fundamental feature.

(26) Et simili discursu] Turner MS: Eademque ratione (By the same reasoning).

(27) ad distantias illas] Turner MS originally: circa axem OZ circundatas erit.

(28) Turner MS originally: coloribus variegatus.

limbus caeruleo, exterior rubro, & mediae partes medijs coloribus tingantur, existente angulo *OGQ*, sive *GOE*, hoc est latitudine arcûs, duorum circiter graduum juxta ea quae jam ante ostendi.[29] Estque similis discursus de arcu exteriori nisi quod ordo colorum propter contrariam inflexionem[30] radiorum contrarius evadat.[31] Guttae autem quae extra hos arcus ex una parte[32] sitae sunt radios omninò nullos post unam vel duas reflexiones duasque refractiones[33] in oculum immittent, ex altera autem parte omnigenos permistos eosque ferè insensibiles,[34] et proinde nulla hujusmodi phaenomena exhibere possunt, sed coelum in illis locis colore solito apparebit.[35]

[140]   Praeter phaenomena colorum de quibus egimus sunt adhuc alia haud pauca, (praesertim circa colores pertenuium lamellarum pellucidarum,[36] quales sunt bullarum aquosi / orbes, & aer intra vitra duo compressus, mul- ₂101/ tarumque rerum cuticulae pertinues,)[37] quorum causae & mensurae absque ratiocinijs mathematicis vix possunt accuratè determinari:[38] sed in hisce videor nimius fuisse, & proinde jam ad partes Matheseos magis abstractas me convertere decrevi.[39]

---

(29) existente . . . ostendi] Added in Turner MS. De Dominis (*De radiis,* Ch. XIII, p. 57), as had Theodoric of Freiberg about three centuries earlier, appreciated that the different colors of the rainbow cannot be seen by the eye to issue from the same drop; on Theodoric see Boyer, *Rainbow,* pp. 116–17.

(30) Turner MS: refractionem et reflex[ionem]. Newton is here adopting Barrow's usage of "inflectio" meaning both refraction and reflection (*Lectiones XVIII,* Lect. I, §XI, pp. 21–2). This should not be confused with either Hooke's use of the same term to denote curved light rays, or his own later application of it to diffraction.

(31) nisi . . . evadat] Added in Turner MS.

(32) ex una parte] Added. That is, in Alexander's dark band between the bows.

(33) duasque refractiones] Added in Turner MS.

(34) ex . . . insensibiles] Added in Turner MS.

(35) Newton gave a very terse summary of this explanation of the rainbow's colors in the "New theory," Prop. 10, which Halley cited and utilized in his "De iride," p. 720. See the remarkable color photograph taken by Roy L. Bishop on 7 October 1979 of both the primary

exterior band red, and middle parts intermediate colors, with the angle *OGQ* or *GOE,* that is, the breadth of the arc, being about two degrees, according to what I showed just before.[29] The argument is similar for the exterior arc, except that because of the reversed inflection[30] the order of the colors becomes reversed.[31] The drops, however, that are situated outside this arc on the one side[32] send no rays at all to the eye after one or two reflections and two refractions,[33] whereas those drops on the other side send all kinds of rays mixed together, and those are almost imperceptible[34] and consequently can exhibit no phenomena of this kind, and the sky will appear with its usual color in those places.[35]

Besides the phenomena of colors that we have treated, there are still not a [140] few others (especially the colors of very thin transparent plates,[36] such as the aqueous globes of bubbles, air compressed between two plates, and very thin skins of many bodies)[37] whose causes and measures can scarcely be accurately determined[38] without mathematical reasoning. But I seem to have lingered too long on these matters, and consequently I have now decided to turn to the more abstract parts of mathematics.[39]

---

and secondary rainbows over Newton's birthplace in "Rainbow over Woolsthorpe manor," *Notes and Records of the Royal Society of London,* 36, no. 1 (1981):3–11, and frontispiece.

(36) pertenuium lamellarum pellucidarum] Originally: qui in pertenuibus lamellis pellucidis vid[eri solent] (which are usually seen in very thin transparent plates). This replaced in the Turner MS: qui in lamellis apparere solent pellucidis pertenuibus.

(37) multarumque . . . pertinues] Added in Turner MS and originally continued: imò omnium communium (indeed of all common ones). Newton resumed his earlier investigations of the colors of thin films in the first years of the 1670s, that is, while (ostensibly) delivering his optical lectures.

(38) causae & mensurae . . . vix possunt accuratè determinari] Turner MS originally: causae . . . vix possunt innotescere (causes can scarcely become known . . .). Newton then added "& mensurae" and changed "innotescere" to "exacte de[terminari]."

(39) Namely, to his lectures on algebra that he began to deliver in October 1673; see *Mathematical Papers,* 5.

Bibliography

All works cited in the notes are included here except for Newton's optical correspondence published in the *Philosophical Transactions*.

Aguilon, François d'. *Opticorum libri sex philosophis juxta ac mathematicis utiles.* Antwerp, 1613.

Alberti, Leon Battista. *On Painting and On Sculpture. The Latin Texts of 'De pictura' and 'De statua.'* Edited and translated by Cecil Grayson. London, 1972.

Alembert, Jean le Rond d'. *Opuscules mathématiques, ou mémoires sur différens sujets de géométrie, de méchanique, d'optique, d'astronomie, &c.* 8 vols. Paris, 1761–80.

Alhazen [Ibn al-Haytham]. *Opticae thesaurus.* In *Opticae thesaurus. Alhazeni arabis libri septem, nunc primùm editi . . . Item Vitellonis Thuringopoloni libri X.* Edited by Friedrich Risner. Basel, 1572; rpt. New York, 1972.

Andrade, E. N. da Costa. "Newton's early notebook." *Nature* 135 (1935):360.

Anon. Review of Barrow's *Lectiones opticae & geometricae.* 2nd ed. (London, 1674). *Journal des sçavans* (18 November, 1675):268–71.

Aristotle. *De sensu and De memoria.* Text, translation, and commentary by G. R. T. Ross. Cambridge, 1906.

Armstrong, H. L. "Comment on Newton's inclusion of indigo in the spectrum." *American Journal of Physics* 40 (1972):1709.

Auzout, Adrien. "Considerations of Monsieur Auzout upon Mr. Hook's new instrument for grinding of optick-glasses." *Philosophical Transactions* 1 (1665):57–63.

Barrow, Isaac. *Lectiones XVIII, Cantabrigiae in scholis publicis habitae; in quibus opticorum phaenomenωn genuinae rationes investigantur, ac exponuntur.* London, 1669. In *The Mathematical Works of Isaac Barrow* 2:1–153.

—*The Mathematical Works of Isaac Barrow, D. D.* Edited by William Whewell. 2 vols. Cambridge, 1860.

—*See also* Anon., Review of Barrow's *Lectiones opticae & geometricae.*

Bate, John. *The Mysteryes of Nature, and Art: Conteined in Foure Severall Tretises, the First of Water Workes, the Second of Fyer Workes, the Third of Drawing, Colouring, Painting, and Engraving, the Fourth of Divers Experiments, as wel Serviceable as Delightful: Partly Collected, and Partly of the Authors Peculiar Practice, and Invention.* London, 1634.

Bechler, Zev. "Newton's search for a mechanistic model of colour dispersion: a suggested interpretation." *Archive for History of Exact Sciences* 11 (1973):1–37.

—"Newton's 1672 optical controversies: a study in the grammar of scientific dissent." *The Interactions Between Philosophy and Science.* Edited by Yehuda Elkana. Atlantic Highlands, N. J., 1974, pp. 115–42.

—"A less agreeable matter: the disagreeable case of Newton and achromatic dispersion." *British Journal for the History of Science* 8 (1975):101–26.

Bennett, A. G. "Some unfamiliar British contributions to geometrical optics." *Transactions of the International Ophthalmic Optical Congress, 1961.* British Optical Association. London, [1962], pp. 274–91.

Bernoulli, Jakob. "Curvae dia-causticae, earum relatio ad evolutas, aliaque nova his affinia." *Acta eruditorum* (1693):244–9.

Biernson, George. "Why did Newton see indigo in the spectrum?" *American Journal of Physics* **40** (1972):526–33.

Birch, Thomas. *The History of the Royal Society of London, for Improving of Natural Knowledge, from Its First Rise. In Which the Most Considerable of Those Papers Communicated to the Society, Which Have Hitherto Not Been Published, Are Inserted in Their Proper Order, as a Supplement to the Philosophical Transactions.* 4 vols. London, 1756–7; rpt. New York/London, 1968.

Bishop, Roy L. "Rainbow over Woolsthorpe Manor." *Notes and Records of the Royal Society of London* 36, no. 1 (1981):3–11, and frontispiece.

Boegehold, H. "Zur Vor- und Frühgeschichte der achromatischen Fernrohrobjektive." *Forschungen zur Geschichte der Optik* 3 (1943):81–114.

Boodt, Anselm de. *Gemmarum et lapidum historia . . .* Hanau, 1609.

Bošković, Rudjer Josip. *Dissertationes quinque ad dioptricam pertinentes.* Vienna, 1767.

Bouma, P. J. *Physical Aspects of Colour.* 2nd English ed. New York, 1971.

Boyer, Carl B. *The Rainbow: From Myth to Mathematics.* New York/London, 1959.

Boyle, Robert. *Experiments and Considerations Touching Colours. First Occasionally Written, Among Some Other Essays, to a Friend; and Now Suffer'd to Come Abroad as the Beginning of an Experimental History of Colours.* London, 1664; rpt. New York/London, 1964.

Briggs, William. *Ophthalmo-graphia, sive oculi ejusque partium descriptio anatomica.* Cambridge, 1676.

—*Ophthalmo-graphia, sive oculi ejusque partium descriptio anatomica. Cui accessit 'Nova visionis theoria,' regiae societati Londin. proposita.* 2nd ed. London, 1685.

Charleton, Walter. *Physiologia Epicuro-Gassendo-Charltoniana: Or a Fabrick of Science Natural, upon the Hypothesis of Atoms, Founded by Epicurus, Repaired by Petrus Gassendus, Augmented by Walter Charleton.* London, 1654; rpt. New York/London, 1966.

—*Exercitationes de differentiis & nominibus animalium.* 2nd ed. Oxford, 1677.

Clairaut, Alexis Claude. "Mémoire sur les moyens de perfectionner les lunettes d'approche, par l'usage d'objectifs composés de plusieurs matières différemment réfringentes." *Mémoires de l'académie royale des sciences* 1756 [Paris, 1762]: 380–437.

Coddington, Henry. *A Treatise on the Reflexion and Refraction of Light, Being Part I of a System of Optics.* Cambridge, 1829.

Cohen, I. Bernard. *Franklin and Newton. An Inquiry into Speculative Newtonian Experimental Science and Franklin's Work in Electricity as an Example Thereof.* Memoirs of the American Philosophical Society, 43. Philadelphia, 1956.

—"Versions of Isaac Newton's first published paper." *Archives internationales d'histoire des sciences* 11 (1958):357–75.

—*Introduction to Newton's 'Principia.'* Cambridge, Mass., 1971.

Cohen, Morris R. and I. E. Drabkin. *A Source Book in Greek Science.* Cambridge, Mass., 1958.

Descartes, René. *Oeuvres de Descartes.* Edited by Charles Adam and Paul Tannery. 13 vols. Paris, 1897–1913.

Dominis, Marc Antonio de. *De radiis visus et lucis in vitris perspectivis et iride tractatus.* Venice, 1611.

Euclid. *Euclidis elementorum libri XV. breviter demonstrati, operâ Is. Barrow.* Cambridge, 1655.

—*The Thirteen Books of Euclid's Elements.* Translated from the text of Heiberg with introduction and commentary by Thomas L. Heath. 2nd ed. 3 vols. Cambridge, 1925; rpt. New York, 1956.

Euler, Leonhard. *Leonhardi Euleri opera omnia.* Sub auspiciis societatis scientiarum naturalium helveticae. Edited by Ferdinand Rudio et al. 73 vols. to date. Leipzig/Berlin/Zurich, 1911–.

Gregory, David. *Catoptricae et dioptricae sphaericae elementa*. Oxford, 1695.

Gregory, James. *Optica promota, seu abdita radiorum reflexorum & refractorum mysteria, geometrice enucleata* . . . London, 1663.

—*See also* Turnbull.

Gren, Friedrich Albrecht Carl. "Einige Bemerkungen über des Herrn von Göthe Beyträge zur Optik." *Journal der Physik* 7 (1793):3–21.

Grimaldi, Francesco Maria. *Physico-mathesis de lumine, coloribus, et iride, alijsque adnexis libri duo*. Bologna, 1665; rpt. Bologna, 1963.

Hall, A. Rupert. "Sir Isaac Newton's note-book, 1661–65." *Cambridge Historical Journal* 9 (1948):239–50.

—"Further optical experiments of Isaac Newton." *Annals of Science* 11 (1955):27–43.

—"Newton's first book (I)." *Archives internationales d'histoire des sciences* 13 (1960):39–54.

Halley, Edmond. "De iride, sive de arcu coelesti, dissertatio geometrica, qua methodo directâ iridis utriusque diameter, data ratione refractionis, obtinetur: Cum solutione inversi problematis, sive inventione rationis istius ex data arcus diametro." *Philosophical Transactions* 22 (1700–1)[1702]:714–25.

Harrison, John. *The Library of Isaac Newton*. Cambridge, 1978.

Harvey, E. Newton. *A History of Luminescence from the Earliest Times until 1900*. Memoirs of the American Philosophical Society, vol. 44. Philadelphia, 1957.

Heath, Thomas L. *A History of Greek Mathematics*. 2 vols. Oxford, 1921.

Helden, Albert van. "The telescope in the seventeenth century." *Isis* 65 (1974):38–58.

Helmholtz, Hermann von. "Ueber die Theorie der zusammengesetzten Farben." *Annalen der Physik und Chemie*, ser. 2, 87 (1852):45–66.

—"On the theory of compound colors." *Philosophical Magazine*, ser. 4, 4 (1852):519–34. (A translation of the preceding.)

—*Helmholtz's Treatise on Physiological Optics*. Translated from the third German edition. Edited by James P. C. Southall. 3 vols. [Rochester, N. Y.], 1924–5; rpt. New York, 1962.

Hendry, John. "Newton's theory of colour." *Centaurus* 23 (1980):230–51.

Herman, R. A. *A Treatise on Geometrical Optics*. Cambridge, 1900.

Hermann, Jakob. "Méthode géométrique & générale de déterminer le diamétre de l'arc-en-ciel, quelque hypothèse de la refraction qu'on suppose dans l'eau, ou dans toute autre liqueur transparente. Et le diamétre de l'arc-en-ciel étant donné par observation, en trouver la raison de la refraction." *Nouvelles de la republique des lettres* 32 (1704):658–71.

Herschel, John Frederick William. "Light." *The Encyclopaedia of Mechanical Philosophy*. London, 1848, pp. 341–586. Reprinted from *The Encyclopaedia Metropolitana*, 1830.

Hooke, Robert. *Micrographia: Or Some Physiological Descriptions of Minute Bodies Made by Magnifying Glasses. With Observations and Inquiries Thereupon*. London, 1665; rpt. New York, 1961.

Houstoun, Robert Alexander. "Newton and the colours of the spectrum." *Science Progress* 12 (1917):250–64.

—*Light & Colour*. London, 1923.

Huxley, G. L. "Two Newtonian studies. I. Newton's boyhood interests." *Harvard Library Bulletin* 13 (1959):348–54.

Huygens, Christiaan. *Oeuvres complètes de Christiaan Huygens*. Publiées par la société hollandaise des sciences. 22 vols. The Hague, 1888–1950.

Kargon, Robert. "Newton, Barrow, and the hypothetical physics." *Centaurus* 11 (1965):46–56.

Kepler, Johannes. *Dioptrice seu demonstratio eorum quae visui & visibilibus propter conspicilla non ita pridem inventa accidunt*. Augsburg, 1611; rpt. Cambridge, 1962.

—*Gesammelte Werke.* Edited by Walther von Dyck and Max Caspar. 17 vols. to date. Munich, 1937–.

Keynes, Geoffrey. *A Bibliography of Dr. Robert Hooke.* Oxford, 1960.

Kircher, Athanasius. *Ars magna lucis et umbrae, in decem libros digesta.* Rome, 1646.

—*Musurgia universalis sive ars magna consoni et dissoni in X libros digesta.* 2 vols. Rome, 1650.

Kundt, August. "Ueber anomale Dispersion." *Annalen der Physik und Chemie,* ser. 2, **144** (1871):128–37.

La Hire, Philippe de. *Mémoires de mathématique et de physique.* Paris, 1694.

—"Dissertation sur les differens accidens de la vue." *Mémoires de l'académie royale des sciences. Depuis 1666 jusqu'à 1699* 9 [Paris, 1730]:350–422.

Larmor, J. "On the absolute minimum of optical deviation by a prism." *Proceedings of the Cambridge Philosophical Society* 9 (1896):108–10.

Laymon, Ronald. "Newton's advertised precision and his refutation of the received laws of refraction." *Studies in Perception.* Edited by Peter K. Machamer and Robert G. Turnbull. Columbus, Ohio, 1978, pp. 231–58.

Lejeune, Albert. *Recherches sur la catoptrique grecque d'après les sources antiques et médiévales.* Académie royale de Belgique. Classe des lettres et des sciences morales et politiques. Mémoires, vol. 52, fasc. 2. Brussels, 1957.

Levene, John R. "Sir George Biddell Airy, F.R.S. (1801–1892) and the discovery and correction of astigmatism." *Notes and Records of the Royal Society of London* **21** (1966):180–99.

—*Clinical Refraction and Visual Science.* London/Boston, 1977.

L'Hospital, Guillaume François Antoine de. *Analyse des infiniment petits, pour l'intelligence des lignes courbes.* 2nd ed. Paris, 1715.

Lohne, Johannes A. "Thomas Harriott (1560–1621), the Tycho Brahe of optics. Preliminary notice." *Centaurus* 6 (1959):113–21.

—"Newton's 'proof' of the sine law and his mathematical principles of colors." *Archive for History of Exact Sciences* 1 (1961):389–405.

—"Zur Geschichte des Brechungsgesetzes." *Sudhoffs Archiv für Geschichte der Medizin und der Naturwissenschaften* 47 (1963):152–72.

—"Isaac Newton: the rise of a scientist 1661–1671." *Notes and Records of the Royal Society of London* 20 (1965):125–39.

—"Regenbogen und Brechzahl." *Sudhoffs Archiv für Geschichte der Medizin und der Naturwissenschaften* 49 (1965):401–15.

—"Thomas Harriot als Mathematiker." *Centaurus* 11 (1965):19–45.

—"Dokumente zur Revalidierung von Thomas Harriot als Algebraiker." *Archive for History of Exact Sciences* 3 (1966):185–205.

—"Fermat, Newton, Leibniz und das anaklastische Problem." *Nordisk Matematisk Tidskrift* 14 (1966):5–25.

—"The increasing corruption of Newton's diagrams." *History of Science* 6 (1967): 69–89.

—"Experimentum crucis." *Notes and Records of the Royal Society of London* 23 (1968):169–99.

—"Newton's table of refractive powers: origins, accuracy, and influence." *Sudhoffs Archiv für Geschichte der Medizin und Naturwissenschaften* 61 (1977):229–47.

Lohne, Johannes A. and Bernhard Sticker. *Newtons Theorie der Prismenfarben. Mit Übersetzung und Erläuterung der Abhandlung von 1672.* Neue Münchner Beiträge zur Geschichte der Medizin und Naturwissenschaften, 1. Munich, 1969.

MacLean, J. "Geschiedenis van de Kleurentheorie in de zestiende Eeuw." *Scientiarum Historia* 9 (1967):23–39.

—"Kleurentheorie in de Periode 1600–1635." *Scientiarum Historia* 9 (1967): 126–47.

—"De Kleurentheorie van de Aristotelianen en de Opvattingen van de la Chambre, Duhamel en Vossius in de Periode 1640–1670." *Scientiarum Historia* **10** (1968):208–25.

—"De Kleurenleer van de Aanhangers der Corpusculairtheorie." *Scientiarum Historia* **12** (1970):1–22.

Magirus, Johannes. *Physiologiae peripateticae libri sex cum commentariis: Additis insuper notis quibusdam marginalibus, in posterioribus editionibus omissis: unà cum definitionibus, divisionibus, axiomatis, Graecè, ex Aristotele petitis . . . Omnia haec infinitis mendis repurgata, pluribus locis restituta, & jam denuo summa cum cura diligentiáque excusa.* Cambridge, 1642.

Maignan, Emmanuel. *Perspectiva horaria sive de horographia gnomonica tum theoretica, tum practica libri quatuor.* Rome, 1648.

Mamiani, Maurizio. *Isaac Newton filosofo della natura. Le lezioni giovanili di ottica e la genesi del metodo newtoniano.* Università degli studi di Parma. Pubblicazioni della facoltà di magistero, 2. Florence, 1976.

Manuel, Frank E. *A Portrait of Isaac Newton.* Cambridge, Mass., 1968.

Marci, Marcus. *Thaumantias: Liber de arcu coelesti deque colorum apparentium natura, ortu, et causis . . .* Prague, 1648; rpt. Prague, 1968.

Mariotte, Edme. *Oeuvres de M. Mariotte . . . comprenant tous les traitez de cet auteur, tant ceux qui avoient déja paru séparément, que ceux qui n'avoient pas encore été publiés . . .* 2 vols. New ed. The Hague, 1740.

Maurolico, Francesco. *Theoremata de lumine, et umbra, ad perspectivam, & radiorum incidentiam facientia. Diaphanorum partes, seu libri tres . . . Problemata ad perspectivam & iridem pertinentia.* 2nd ed. Lyons, 1613.

Mills, A. A. and P. J. Turvey. "Newton's telescope: an examination of the reflecting telescope attributed to Sir Isaac Newton in the possession of the Royal Society." *Notes and Records of the Royal Society of London* **33** (1979):133–55.

Mollweide, Carl. "Ueber die Farbenzerstreuung in menschlichen Auge." *Annalen der Physik* **30** (1808):220–34.

Newton, Isaac. *Opticks: Or, a Treatise of the Reflexions, Refractions, Inflexions and Colours of Light. Also Two Treatises of the Species and Magnitude of Curvilinear Figures.* London, 1704; rpt. Brussels, 1966.

—*Optice: sive de reflexionibus, refractionibus, inflexionibus & coloribus libri tres.* Translated by Samuel Clarke. London, 1706.

—*Opticks: Or a Treatise of the Reflections, Refractions, Inflections and Colours of Light.* 3rd ed., corrected. London, 1721.

—*Optical Lectures Read in the Publick Schools of the University of Cambridge, Anno Domini, 1669. By the late Sir Isaac Newton, then Lucasian Professor of the Mathematicks. Never before Printed. Translated into English out of the Original Latin.* London, 1728.

—*Isaaci Newtoni, eq. aur. in academiâ Cantabrigiensi matheseos olim professoris Lucasiani lectiones opticae, annis MDCLXIX, MDCLXX & MDCLXXI. In scholis publicis habitae: et nunc primum ex MSS. in lucem editae.* London, 1729.

—*Isaaci Newtoni, equitis aurati, opuscula mathematica, philosophica et philologica.* Edited by Giovanni di Castiglione. 3 vols. Lausanne/Geneva, 1744.

—*Isaaci Newtoni optices libri tres: accedunt ejusdem lectiones opticae, et opuscula omnia ad lucem & colores pertinentia sumpta ex transactionibus philosophicis.* Padua, 1749.

—*Isaaci Newtoni optices libri tres: accedunt ejusdem lectiones opticae, et opuscula omnia ad lucem & colores pertinentia sumpta ex transactionibus philosophicis.* Padua, 1773.

—*Isaaci Newtoni opera quae exstant omnia.* Edited by Samuel Horsley. 5 vols. London, 1779–85.

—*Sir Isaac Newton's Mathematical Principles of Natural Philosophy and His System of the World.* Translated into English by Andrew Motte in 1729. The translations revised, and supplied with an historical and explanatory appendix, by Florian Cajori. Berkeley, 1934.

—*Lekcii po Optike.* Translated by S. I. Vavilov. Moscow/Leningrad, 1946.

—*Isaac Newton's Papers & Letters on Natural Philosophy and Related Documents.* Edited by I. Bernard Cohen. Cambridge, Mass., 1958.

—*The Correspondence of Isaac Newton.* Edited by H. W. Turnbull, J. F. Scott, A. Rupert Hall, and Laura Tilling. 7 vols. Cambridge, 1959–77.

—*Unpublished Scientific Papers of Isaac Newton. A Selection from the Portsmouth Collection in the University Library, Cambridge.* Edited and Translated by A. Rupert Hall and Marie Boas Hall. Cambridge, 1962.

—*The Mathematical Papers of Isaac Newton.* Edited by D. T. Whiteside, with the assistance in publication of A. Prag. 8 vols. Cambridge, 1967–81.

—*Isaac Newton's 'Philosophiae naturalis principia mathematica.' The Third Edition (1726) with Variant Readings.* Edited by Alexandre Koyré and I. Bernard Cohen, with the assistance of Anne Whitman. 2 vols. Cambridge, Mass., 1972.

—*The Unpublished First Version of Isaac Newton's Cambridge Lectures on Optics 1670–1672. A Facsimile of the Autograph, now Cambridge University Library MS. Add. 4002.* Introduction by D. T. Whiteside. Cambridge, 1973.

Nordenmark, N. V. E. and Johan Nordström. "Om uppfinningen av den akromatiska och aplanatiska linsen." *Lychnos* (1938):1–52; (1939):313–84.

Ockenden, R. E. "Marco Antonio de Dominis and his explanation of the rainbow." *Isis* **26** (1936):40–9.

[Oldenburg, Henry]. "Of Monsieur Hevelius's promise of imparting to the world his invention of making optick glasses; and of the hopes given by Monsieur Hugens of Zulichem, to perform something of the like nature; as also of the expectations, conceived of some ingenious persons in England, to improve telescopes." *Philosophical Transactions* **1** (1665):98–9.

[—] "Of Monsieur de Sons progress in working parabolar glasses." *Philosophical Transactions* **1** (1665):119–20.

[—] Review of Robert Boyle's *Hydrostatical Paradoxes* (Oxford, 1666). *Philosophical Transactions* **1** (1665/6):173–6.

—*The Correspondence of Henry Oldenburg.* Edited by A. Rupert Hall and Marie Boas Hall. 11 vols. to date. Madison/London, 1965–.

Parkhurst, Charles. "Aguilonius' optics and Rubens' color." *Nederlands Kunsthistorisch Jaarboek* **12** (1961):35–49.

—"A color theory from Prague: Anselm de Boodt, 1609." *Allen Memorial Art Museum Bulletin* **29** (1971):2–10.

—"Louis Savot's 'Nova-antiqua' color theory, 1609." *Album amicorum J. G. van Gelder.* Edited by J. Bruyn et al. The Hague, 1973, pp. 242–7.

—"Camillo Leonardi and the green-blue shift in sixteenth-century painting." *Intuition und Kunstwissenschaft: Festschrift für Hans Swarzenski.* Edited by Peter Bloch et al. Berlin, 1973, pp. 419–25.

—"Alberti's color scheme and some antecedents." *A Conference on Color and Technique in Renaissance Painting, Italy and the North.* Temple University, September 22 and 23, 1980 (forthcoming).

Partington, J. R. "Lignum nephriticum." *Annals of Science* **11** (1955):1–26.

Porta, Giovanni Battista della. *Magiae naturalis libri viginti.* Leyden, 1651.

—*De refractione optices parte libri novem.* Naples, 1593.

[Portsmouth Collection]. *A Catalogue of the Portsmouth Collection of Books and Papers Written by or Belonging to Sir Isaac Newton, the Scientific Portion of Which Has Been Presented by the Earl of Portsmouth to the University of Cambridge.* Prepared by H. R. Luard, G. G. Stokes, J. C. Adams, and G. D. Liveing. Cambridge, 1888.

Ptolemy, Claudius. *L'Optique de Claude Ptolémée dans la version latine d'après l'arabe de l'émir Eugène de Sicile.* Edited by Albert Lejeune. Université de Louvain, Recueil de travaux d'histoire et de philologie, ser. 4, fasc. 8. Louvain, 1956.

Rayleigh (Baron). "Optical topics in part connected with Charles Parsons." *Nature* **152** (1943):676–82.

Rohault, Jacques. *Rohault's System of Natural Philosophy, Illustrated with Dr. Samuel Clarke's Notes Taken Mostly out of Sir Isaac Newton's Philosophy.* Translated by John Clarke. 2 vols. London, 1723.

Rohr, Moritz von, ed. *Geometrical Investigation of the Formation of Images in Optical Instruments.* Translated by R. Kanthack. London, 1920.

Sabra, A. I. *Theories of Light from Descartes to Newton.* London, 1967.

Safford, W. E. "*Lignum nephriticum*—its history and an account of the remarkable fluorescence of its infusion." *Annual Report of the Smithsonian Institution,* 1915:271–98.

Savot, Louis. *Nova, seu verius nova-antiqua de causis colorum sententia.* Paris, 1609.

Scaliger, Julius Caesar. *Exotericarum exercitationum liber XV. De subtilitate, ad Hieronymum Cardanum.* Lyons, 1615.

Scheiner, Christoph. *Oculus hoc est: fundamentum opticum* . . . London, 1652.

Schuster, John Andrew. "Descartes and the scientific revolution, 1618–1634: an interpretation." Ph.D. dissertation, Princeton University, 1977.

Scott, Joseph F. *The Scientific Work of René Descartes (1596–1650).* London, [1952].

Shapiro, Alan E. "Light, pressure, and rectilinear propagation: Descartes' celestial optics and Newton's hydrostatics." *Studies in History and Philosophy of Science* 5 (1974):239–96.

—"Newton's definition of a light ray and the diffusion theories of chromatic dispersion." *Isis* 66 (1975): 194–210.

—"Newton's 'achromatic' dispersion law: theoretical background and experimental evidence." *Archive for History of Exact Sciences* 21 (1979):91–128.

—"The evolving structure of Newton's theory of white light and color." *Isis,* 71 (1980):211–35.

[Smethwick, Francis]. "Of the invention of grinding optick and burning-glasses, of a figure not-spherical, produced before the Royal Society." *Philosophical Transactions* 3 (1667/8):631–2.

Smith, David Eugene. "Two unpublished documents of Sir Isaac Newton." *Isaac Newton, 1642–1727. A Memorial Volume Edited for the Mathematical Assocation* by W. J. Greenstreet. London, 1927, pp. 16–34.

Smith, Robert. *A Compleat System of Opticks in Four Books, viz. A Popular, a Mathematical, a Mechanical, and a Philosophical Treatise. To Which Are Added Remarks upon the Whole.* 2 vols. Cambridge, 1738.

Sprat, Thomas. *The History of the Royal-Society of London, for the Improving of Natural Knowledge* [London, 1667]. Edited with critical apparatus by Jackson I. Cope and Harold Whitmore Jones. St. Louis/London, 1959.

Stiegler, Karl. "Das Problem der sphärischen Aberration und seine Lösung durch Isaac Newton. Ein Beitrag zur Geschichte der Theorie der optischen Instrumente." *Technikgeschichte* 44 (1977):121–52.

Stokes, George Gabriel. "On the change of refrangibility of light." *Philosophical Transactions* 142 (1852):463–561.

Turbayne, Colin M. "Grosseteste and an ancient optical principle." *Isis* 50 (1959): 467–72.

Turnbull, Herbert Westren, ed. *James Gregory Tercentenary Memorial Volume. Containing his Correspondence with John Collins and his Hitherto Unpublished Mathematical Manuscripts, together with Addresses and Essays Communicated to the Royal Society of Edinburgh, July 4, 1938.* London, 1939.

Venturi, Giambattista. "Considerazioni ottiche." *Memorie di matematica e di fisica della società italiana* 3 (1786):268–77.

—"Indagine fisica sui colori." *Memorie di matematica e di fisica della società italiana* 8, ii (1799):699–754.

Voltaire, François Marie Arouet de. *Elemens de la philosophie de Neuton, mis à la portée de tout le monde.* Amsterdam, 1738.

Vossius, Isaac. *De lucis natura et proprietate.* Amsterdam, 1662.

Wallis, Peter and Ruth Wallis. *Newton and Newtoniana 1672–1975. A Bibliography.* Folkestone, Kent, 1977.

Wellek, Albert. "Farbenmusik." *Die Musik in Geschichte und Gegenwart: allgemeine Enzyklopädie der Musik.* Edited by Friedrich Blume. vol. 3. Kassel/Basel, 1954. cols. 1811–22.

Westfall, Richard S. "The development of Newton's theory of color." *Isis* 53 (1962):339–58.

—"The foundations of Newton's philosophy of nature." *The British Journal for the History of Science* 1 (1962):171–82.

—"Newton's reply to Hooke and the theory of colors." *Isis* 54 (1963):82–96.

—"Isaac Newton's coloured circles twixt two contiguous glasses." *Archive for History of Exact Sciences* 2 (1965):181–96.

—"Newton defends his first publication: the Newton-Lucas correspondence." *Isis* 57 (1966):299–314.

—"Uneasily fitful reflections on fits of easy transmission." *Texas Quarterly* 10, no. 3 (1967):86–102.

—*Never at Rest: A Biography of Isaac Newton.* Cambridge, 1980.

Whiston, William. *Praelectiones physico-mathematicae Cantabrigiae in scholis publicis habitae. Quibus philosophia illustrissimi Newtoni mathematica explicatius traditur, & facilius demonstratur.* Cambridge, 1710.

—*Sir Isaac Newton's Mathematick Philosophy More Easily Demonstrated . . . Being Forty Lectures Read in the Publick Schools at Cambridge.* London, 1716; rpt. New York/London, 1972.

Witelo. *Perspectiva. See* Alhazen.

Young, Thomas. "On the mechanism of the eye." *Philosophical Transactions* 91 (1801):23–88.

—*A Course of Lectures on Natural Philosophy and the Mechanical Arts.* 2 vols. London, 1807.

—*Oeuvres ophtalmologiques de Thomas Young.* Translated by Marius H. E. Tscherning. Copenhagen, 1894.

Ziggelaar, August. "Die Erklärung des Regenbogens durch Marcantonio de Dominis, 1611. Zum Optikunterricht am Ende des 16. Jahrhunderts." *Centaurus* 23 (1979):21–50.

Index

References pertaining to the text and translation of the *Optical Lectures* are given to the page numbers only of the English translation.

aberrations:
  chromatic, 11, 13–14, 49, 81, 281–3,
    433; absent in reflection, 12, 125,
    129, 485, 491; comparison with
    spherical aberration, 12, 41, 49,
    281–3, 427–9; correcting for, 13,
    178n, 201n, 579n; of eye, 579–83;
    of plano-convex lens, 35, 425–9; of
    thin lens, 575–9
  spherical, 11, 24, 47–9, 81, 281–3,
    433; circle of least confusion, 409–
    11; comparison with chromatic ab-
    erration, 12, 41, 49, 281–3, 427–9;
    of concave mirror, 407n; correcting
    for, 429; of plano-convex lens, 13,
    40–1, 405–11, 427–9
absorption, 33, 513–15
Adam, Charles, 7n
aether, 130n
  vibrations of, 9, 85, 161, 437, 533,
    546n
Aguilon, François d', *Opticorum,* 51n,
    506n
air:
  color of, 83, 435, 505
  thin film of, 137, 141, 501, 503,
    513n, 563
  see also atmosphere
Alberti, Leon Battista, *De pictura,* 82n
Alembert, Jean le Rond d', *Opuscules
    mathématique,* 179n, 195n
Alhazen (Ibn al-Haytham), 9, 106n,
    168n
  Opticae thesaurus, 107n, 175n, 217n
anaclastic problem:
  explained, 38–9
  Barrow's solution, 351
  Newton's solution, 227–9
Andrade, E. N., 2n
aplanatic surfaces:

explained, 41
  see also Cartesian ovals
Apollonius, *Conics,* 405n
Aristotle:
  Aristotelian color theory, 3–4, 83n;
    Newton's attack on, 28, 81–5,
    433–5, 437; opposition to, by Al-
    berti, 82n: opposition to, by me-
    chanical philosophers, 2, 4–5
  on color, 2–4, 83n, 460n, 542n; defi-
    nition of, 3, 83, 435; rainbow, 3
  De sensu, 2–3, 83, 542n
  Meteorologica, 3
Armstrong, H. L., 542n
ashen (*cinereus*):
  compound of white and black, 507;
    of white and blue according to
    Kircher, 463n
  equated with gray (*fuscus*), 82n, 437,
    463
  see also gray (*fuscus*)
astigmatism, 38, 216n, 414n
astrolabe, 173n
atmosphere:
  blue rays disturbed more than yellow,
    505
  refraction in, 505
  variation of density and pressure, 189,
    327
  see also air
Auzout, Adrien, 78n

Barrow, Isaac, 14–15, 18, 20, 26, 39,
    46n, 89n
  on anaclastic problem, 229, 351
  on color, 15, 142n
  Euclidis Elementorum, 57n, 207n,
    214n, 239n, 256n
  Lectiones XVIII, xix, 14–15, 38, 56n,
    125n, 142n, 183n, 197n, 204n,